Handbook of Experimental Pharmacology

Volume 82

Radioimmunoassay in Basic and Clinical Pharmacology

Contributors

L. Bartalena · K. Brune · F. Celada · I. Christensson-Nylander
G. Ciabattoni · D. Cocchi · M.G. Currie · J. Dawes · F. Dray
E. Ezan · D.M. Geller · J. Grassi · V. Guardabasso · E.C. Hayes
E. Knoll · G.A. Limjuco · J. Maclouf · S. Mamas · S. Mariotti
M.L. Michener · P.J. Munson · M.G. Murphy · P. Needleman
G. Nilsson · F. Nyberg · C.W. Parker · C. Patrono · M. Pazzagli
D.S. Pepper · B.A. Peskar · B.M. Peskar · A. Pinchera
P. Pradelles · M. Reinke · D. Rodbard · C. Rougeot · D. Schwartz
M. Serio · Th. Simmet · S. Spector · L. Terenius · L.E. Underwood
J.H. Walsh · H. Wisser · H.C. Wong · R.S. Yalow · H.J. Zweerink

Editors

C. Patrono and B.A. Peskar

Springer-Verlag
Berlin Heidelberg New York
London Paris Tokyo

Carlo Patrono, Professor of Molecular Pharmacology

Institute of Pharmacology
Catholic University of the Sacred Heart
Largo Francesco Vito, 1
I-00168 Rome

Bernhard A. Peskar, Professor of Pharmacology and Toxicology

Institute of Pharmacology and Toxicology
Ruhr-University Bochum
Im Lottental, P.O. Box 102148
D-4630 Bochum 1

With 129 Figures

Library of Congress Cataloging-in-Publication Data. Radioimmunoassay in basic and clinical pharmacology. (Handbook of experimental pharmacology; v. 82) Includes index. 1. Radioimmunoassay. 2. Biomolecules – Analysis. 3. Drugs – Analysis. I. Bartalena, L. II. Patrono, Carlo, 1944– . III. Peskar, B.A. (Bernhard A.) IV. Series. [DNLM: 1. Hormones – analysis. 2. Peptides – analysis. 3. Radioimmunoassay. W1 HA51L v. 82/QW 570 R1294] QP905.H3 vol. 82 615'.1 s 87-9659 [QP519.9.R3] [615'.1901]

ISBN-13: 978-3-642-71811-3 e-ISBN-13: 978-3-642-71809-0
DOI: 10.1007/978-3-642-71809-0

Softcover reprint of the hardcover 1st edition 1987

2122/3130-543210

List of Contributors

L. Bartalena, Istituto di Metodologia Clinica e Medicina del Lavoro, Cattedra di Endocrinologia e Medicina Costituzionale, Università degli Studi di Pisa, Viale del Tirreno, 64, I-56018 Tirrenia (Pisa)

K. Brune, Institut für Pharmakologie und Toxikologie, Universität Erlangen-Nürnberg, Universitätsstr. 22, D-8520 Erlangen

F. Celada, Cattedra di Immunologia dell'Università degli Studi di Genova, Viale Benedetto XV, 10, I-16132 Genova

I. Christensson-Nylander, Department of Pharmacology, University of Uppsala, Box 591, S-75124 Uppsala

G. Ciabattoni, Istituto di Farmacologia, Università Cattolica del Sacro Cuore, Facoltà di Medicina e Chirurgia "Agostino Gemelli", Largo Francesco Vito, 1, I-00168 Roma

D. Cocchi, Università degli Studi di Milano, Dipartimento di Farmacologia, Chemioterapia e Tossicologia Medica, Via Vanvitelli, 32, I-20129 Milano

M. G. Currie, Medical University of South Carolina, Department of Pharmacology, 171 Ashley Ave., Charleston, SC 29425, USA

J. Dawes, MRC/Scottish National Blood Transfusion Service, Blood Components Assay Group, 2 Forrest Road, Edinburgh EH1 2QW, Great Britain

F. Dray, U 207, INSERM, Institut Pasteur, Unité de Radioimmunologie Analytique, 28, rue du Docteur Roux, F-75724 Paris Cedex 15

E. Ezan, U 207, INSERM, Institut Pasteur, Unité de Radioimmunologie Analytique, 28, rue du Docteur Roux, F-75724 Paris Cedex 15

D. M. Geller, Washington University School of Medicine, The Edward Mallinckrodt Department of Pharmacology, Box 8103, 660 South Euclid Ave., St. Louis, MO 63110, USA

J. Grassi, Section de Pharmacologie et d'Immunologie, Laboratoire d' Etudes Radioimmunologiques, CEA, Departement de Biologie CEN-Saclay, F-91191 Gif sur Yvette Cedex

V. Guardabasso, Istituto di Ricerche Farmacologiche "Mario Negri", Via Eritrea, 62, I-20157 Milano

E. C. Hayes, Immunology Research, Merck Sharp & Dohme Research Laboratories, P.O. Box 2000, Rahway, NJ 07065-0900, USA

E. Knoll, Abteilung für Klinische Chemie, Robert-Bosch-Krankenhaus, Auerbachstr. 110, D-7000 Stuttgart 50

G. A. Limjuco, Immunology Research, Merck Sharp & Dohme Research Laboratories, P.O. Box 2000, Rahway, NJ 07065-0900, USA

J. Maclouf, U150, INSERM - LA334 CNRS, Hôpital Lariboisière, 6 rue Guy Patin, F-75475 Paris Cedex 10

S. Mamas, U207, INSERM, Institut Pasteur, Unité de Radioimmunologie Analytique, 28, rue du Docteur Roux, F-75724 Paris Cedex 15

S. Mariotti, Istituto di Metodologia Clinica e Medicina del Lavoro, Cattedra di Endocrinologia e Medicina Costituzionale, Università degli Studi di Pisa, Viale del Tirreno, 64, I-56018 Tirrenia (Pisa)

M. L. Michener, Washington University School of Medicine, The Edward Malinckrodt Department of Pharmacology, Box 8103, 660 South Euclid Ave., St. Louis, MO 63110, USA

P. J. Munson, Laboratory of Theoretical and Physical Biology, National Institute of Child Health and Human Development, National Institutes of Health, Building 10, Room 8C312, Bethesda, MD 20892, USA

M. G. Murphy, Department of Pediatrics, Division of Endocrinology, The University of North Carolina at Chapel Hill, Clinical Sciences Building 229H, Chapel Hill, NC 27514, USA

P. Needleman, Washington University School of Medicine, The Edward Malinckrodt Department of Pharmacology, Box 8103, 660 South Euclid Ave., St. Louis, MO 63110, USA

G. Nilsson, Department of Physiology, Faculty of Veterinary Medicine, The Swedish University of Agricultural Sciences, P.O. Box 7045, S-75007 Uppsala 7

F. Nyberg, Department of Pharmacology, University of Uppsala, Box 591, S-75124 Uppsala

C. W. Parker, Department of Internal Medicine, Division of Allergy and Immunology, Washington University School of Medicine, Box 8122, 660 South Euclid Ave., St. Louis, MO 63110, USA

C. Patrono, Istituto di Farmacologia, Università Cattolica del Sacro Cuore, Facoltà di Medicina e Chirurgia "Agostino Gemelli", Largo Francesco Vito, 1, I-00168 Roma

M. Pazzagli, Dipartimento di Fisiopatologia Clinica, Unità di Endocrinologia, Università degli Studi di Firenze, Viale Morgagni, 85, I-50135 Firenze

D. S. Pepper, Scottish National Blood Transfusion Service, Headquarters Unit Laboratory, 2 Forrest Road, Edinburgh EH1 2QN, Scotland, Great Britain

B. A. PESKAR, Institut für Pharmakologie und Toxikologie, Ruhr-Universität Bochum, Im Lottental, D-4630 Bochum 1

B. M. PESKAR, Abteilung für Experimentelle Klinische Medizin, Ruhr-Universität Bochum, Im Lottental, D-4630 Bochum 1

A. PINCHERA, Istituto di Metodologia Clinica e Medicina del Lavoro, Cattedra di Endocrinologia e Medicina Costituzionale, Università degli Studi di Pisa, Viale del Tirreno, 64, I-56018 Tirrenia (Pisa)

P. PRADELLES, Section de Pharmacologie et d'Immunologie, Laboratoire d'Etudes Radioimmunologiques CEA, Departement de Biologie CEN-Saclay, F-91191 Gif sur Yvette Cedex

M. REINKE, Institut für Pharmakologie und Toxikologie, Universität Erlangen-Nürnberg, Universitätsstr. 22, D-8520 Erlangen

D. RODBARD, Laboratory of Theoretical and Physical Biology, National Institute of Child Health and Human Development, National Institutes of Health, Building 10, Room 8C312, Bethesda, MD 20892, USA

C. ROUGEOT, U207, INSERM, Institut Pasteur, Unité de Radioimmunologie Analytique, 28, rue du Docteur Roux, F-75724 Paris Cedex 15

D. SCHWARTZ, Washington University School of Medicine, The Edward Malinckrodt Department of Pharmacology, Box 8103, 660 South Euclid Ave., St. Louis, MO 63110, USA

M. SERIO, Dipartimento di Fisiopatologia Clinica, Unita di Endocrinologia, Università degli Studi di Firenze, Viale Morgagni, 85, I-50135 Firenze

TH. SIMMET, Institut für Pharmakologie und Toxikologie, Ruhr-Universität Bochum, Im Lottental, D-4630 Bochum 1

S. SPECTOR, Department of Neuroscience, Roche Institute of Molecular Biology, Roche Research Center, Building 102, Nutley, NJ 07110, USA

L. TERENIUS, Department of Pharmacology, University of Uppsala, Box 591, S-75124 Uppsala

L. E. UNDERWOOD, Department of Pediatrics, School of Medicine, Division of Endocrinology, The University of North Carolina at Chapel Hill, Clinical Sciences Building 229H, Chapel Hill, NC 27514, USA

J. H. WALSH, Center for Ulcer Research and Education, Veterans Administration Wadsworth, Wilshire and Sawtelle Bvds., Building 115, Room 115, Los Angeles, CA 90073, USA

H. WISSER, Abteilung für Klinische Chemie, Robert-Bosch-Krankenhaus, Auerbachstr. 110, D-7000 Stuttgart 50

H. C. Wong, Department of Medicine, University of California, Los Angeles, CA 90024, USA

R. S. Yalow, Solomon A. Berson Research Laboratory, Veterans Administration Medical Center, 130, West Kingsbridge Road, Bronx, NY 10468, USA

H. J. Zweerink, Immunology Research, Merck Sharp & Dohme Research Laboratories, P.O. Box 2000, Rahway, NJ 07065-0900, USA

Preface

Thirty years have elapsed since the first description by S. A. BERSON and R. S. YALOW of the basic principles of radioimmunoassay (RIA). During this period of time, RIA methodology has been instrumental to the growth of many areas of biomedical research, including endocrinology, oncology, hematology, and pharmacology. It has done so by providing a relatively simple universal tool allowing, for the first time, the detection of endogenous mediators that are present in body fluids at concentrations as low as 10^{-12}–10^{-10} *M*. The fundamental nature of this discovery and the wide-ranging fall-out of basic and clinical knowledge derived from its application have been acknowledged by the many honors tributed to its pioneers, including the Nobel Prize awarded to Dr. YALOW 10 years ago.

Although several excellent books have been published during the past decades covering various aspects of RIA methodology, we felt the need, as pharmacologists, for a comprehensive discussion of the methodological and conceptual issues related to the main classes of mediators of drug action and to drugs themselves. Thus, we gladly accepted the challenge provided by the invitation to edit a volume of the *Handbook of Experimental Pharmacology* on *Radioimmunoassay in Basic and Clinical Pharmacology*. We tried to balance the emphasis placed on more general aspects of the RIA methodology and that on specific mediators. A potentially endless list of substances of pharmacological interest was necessarily limited by considerations of space and general interest, although we did make some last-minute adjustments in order to include very recent and exciting developments.

A number of introductory chapters provide the reader unfamiliar with RIA with all the basic information concerning the production and characterization of antibodies, labeling techniques, statistical aspects, and validation criteria. Moreover, each of these chapters provides the experienced reader with further insight into problems related to the development and validation of RIA for newly discovered mediator(s). In the following chapters, the emphasis is placed on the technical details relevant to each class of compounds and on specific aspects of their application to basic and/or clinical pharmacological studies. New developments in this area such as monoclonal antibodies and nonradioactive labeling techniques are also given adequate coverage.

We hope this book will represent a valuable working instrument in the hands of those investigators interested in measuring changes in the synthesis and metabolism of a variety of endogenous mediators, as well as in the kinetics of drugs, as related to experimental or clinical models of disease and to pharmacological intervention.

We are indebted to Mrs. P. TIERNEY, Mrs. A. ZAMPINI, and Ms. K. BUSCHEY for invaluable help in the handling and editing of the manuscripts, to Mrs. D. M. WALKER for providing all the necessary support and advice from the publisher, and to all the authors for generously sharing with us the effort and time that this project required.

Rome C. PATRONO
Bochum B. A. PESKAR

Contents

CHAPTER 3

Production of Antisera by Conventional Techniques

CHAPTER 4

Production of Monoclonal Antibodies for Radioimmunoassays

CHAPTER 5

Radioiodination and Other Labeling Techniques

CHAPTER 19

Radioimmunoassay of Leukotrienes and Other Lipoxygenase Products of Arachidonate

CHAPTER 20

Radioimmunoassay of Cyclic Nucleotides

CHAPTER 23

Radioimmunoassay of the Somatomedins/Insulin-like Growth Factors

CHAPTER 24

Radioimmunoassay of Drugs and Neurotransmitters

CHAPTER 1

Radioimmunoassay: Historical Aspects and General Considerations

R. S. YALOW

A. Historical Aspects

Radioimmunoassay (RIA) could be considered a serendipitous discovery in that it arose as fallout from investigations into what would appear to be an unrelated study. To test the hypothesis (MIRSKY 1952) that maturity-onset diabetes might be a consequence of abnormally rapid degradation of insulin by an enzyme, insulinase, which was shown to be widely distributed in the body, we administered radioiodine-labeled insulin as a tracer to study the distribution and turnover of insulin in diabetic and nondiabetic subjects (BERSON et al. 1956). We observed that the labeled insulin disappeared more slowly from the plasma of subjects with a history of insulin treatment than from the plasma of diabetic or nondiabetic subjects who had never received animal insulin. We soon demonstrated that the slower rate of removal was a consequence of the binding of insulin to an acquired antibody. Almost immediately we appreciated that the methodology used to study the kinetics of reaction of insulin with insulin-binding antibody and to determine the binding capacity of that antibody could be applied reciprocally to determine the concentration of insulin in body fluids. Although we first used the word immunoassay to describe the general methodology in 1957 (BERSON and YALOW 1957), it was not until several years later that our assay had sufficient sensitivity to measure the concentrations of insulin in the circulation of humans (YALOW and BERSON 1959).

During the first decade after its discovery RIA was used primarily to measure the concentration of peptide hormones. Since these substances are present in plasma in the unstimulated state at concentrations as low as $10^{-12}-10^{-10}$ M, the sensitivity and specificity of RIA were required to study the dynamic interactions between the peptide hormones and the substrates which they regulate and by which they are in turn regulated. By the 1970s RIA methodology had spread from endocrinology, its first home, into many other areas of medicine, including pharmacology and toxicology (OLIVER et al. 1968; MAHON et al. 1973), infectious diseases (WALSH et al. 1970), oncology (THOMSON et al. 1969), and hematology (NOSSEL et al. 1974).

B. Principle, Practices, and Pitfalls

RIA is simple in principle, but there has been an increasing appreciation that there are many problems and pitfalls that one must guard against in applying the

Labeled antigen		*Specific antibody*		*Labeled antigen-antibody complex*
Ag^*	+	Ab	$\rightleftharpoons$	$\overline{Ag^* - Ab}$
(F)		+		(B)
			Unlabeled antigen	
		Ag	in known standard solutions or unknown samples	
		$\upharpoonleft\downharpoonright$		
		$\overline{Ag - Ab}$		
		Unlabeled antigen-antibody complex		

Fig. 1. Competing reactions that form the basis of radioimmunoassay (RIA)

methodology to the hundreds of substances now measured by RIA (YALOW 1973).

I. Principle

Measurements by RIA are made by comparing the inhibition of binding of radio-labeled antigen to specific antibody by unknown samples with the inhibition by known standard solutions of unlabeled antigen (Fig. 1). RIA differs from traditional bioassay in that it is an *immunochemical* method in which the measurement depends only on the interaction of chemical reagents in accordance with the law of mass action. There is no requirement for the labeled and unlabeled antigen to be identical chemically or biologically. Furthermore, immunochemical activity may or may not be identical with, or even reflect, biologic activity. A necessary, but not sufficient condition for the validation of an RIA procedure is that the apparent concentration of the unknown be independent of the dilution at which it is assayed. Under some circumstances, however, assays may be clinically useful even if they cannot be validated because of immunochemical differences between standards and unknowns.

II. Practices

The essential requirements for RIA include suitable reactants, labeled antigen and specific antibody, and some technique for separating the antibody-bound from free labeled antigen since under the usual conditions of assay the antigen–antibody complexes do not spontaneously precipitate.

1. Labeled Antigen

It is obvious that it is inadvisable to employ in the assay an amount of labeled antigen whose immunochemical concentration is large compared with the concentration of unlabeled antigen in the unknown. Thus, the use of a tracer of $10 \times 10^{-12}\ M$ to measure a hormone concentration of $10^{-12}\ M$ means that a random 5% error in the tracer would produce an error of 50% in the concentration of the unknown. In particular, measurement of peptide hormones requires high specific activity tracers and for this purpose ^{125}I has become the radioisotope of choice. Its longer half-life ($t_{1/2}=60$ days) than that of ^{131}I ($t_{1/2}=8$ days) permits the preparation of labeled antigens with useful shelf lives of up to several months.

For nonpeptidal hormones and drugs that are generally present in much higher concentrations than the peptide hormones, ^{3}H-labeled tracers may be employed. However ^{3}H-labeled tracers do have the limitation that liquid scintillation counters are required for their detection, resulting in the cost of acquiring and disposing of the organic solvent containing the scintillator. For these reasons, even when there is no requirement for high specific activity, the method of choice is usually to couple the substance to a moiety containing a residue that can be iodinated even if the substance itself does not contain a tyrosyl or histidyl residue.

2. Specific Antibody

Most peptide hormones are satisfactorily immunogenic in a variety of experimental animals when the peptide is administered as an emulsion in Freund's adjuvant. We have generally used commercial or low purity preparations to take advantage of their possible slight denaturation which might render them more "foreign," thereby enhancing their antigenicity. Small peptides or nonpeptidal substances, which are not of themselves antigenic, may be rendered so by covalently linking them to a higher molecular weight carrier substance such as a protein, polypeptide, or polysaccharide.

Since the presence of other immunologic reactions does not interfere with the reaction between labeled antigen and its specific antibody, immunization with several unrelated antigens can be performed simultaneously. The antibody concentration and the sensitivity and specificity of antibodies directed toward the various antigens appear to be unrelated. Immunization with several antigens has the advantage of reducing the number of animals to be immunized and bled by a factor equal to the number of antigens used simultaneously.

Monoclonal antibodies have now come into use in RIA procedures. The development of such antibodies provides a virtually unlimited amount of homogeneous antibodies directed against a specific antigenic site. However, the energy of interaction of these antibodies with antigen usually reflects the mean energy of the heterogeneous antibodies produced in the immunized animal. In RIA employing a heterogeneous antiserum it is traditional to dilute the antiserum sufficiently so that only the small fraction of the total antibodies with the highest energy of interaction binds the antigen. Thus, heterogeneous antibodies are more likely to provide highly sensitive assays than are those employing monoclonal antibodies. However, the latter are more likely to be highly specific than heterogeneous antibodies. Thus, the choice of the source of the antibodies may depend on whether sensitivity or specificity is the limiting factor. For research laboratories requiring only limited amounts of antisera, the simplicity of antibody production in animals compared with the increased effort required for production of monoclonal antibodies must be considered.

3. Separation Methods

Since RIA is used for measurements of hundreds of substances, often with very different chemical properties, it is not surprising that there are a very large number of separation techniques that have proven to be quite useful. For the most part these techniques are some variation of two basic methods: precipitation of antigen–antibody complexes; adsorption of either free antigen or antibody to solid-phase material. Techniques for precipitation of the complexes include use of a second antibody or an organic solvent such as polyethylene glycol. Materials used to absorb free antigen include cellulose, charcoal, silicates, or ion exchange resins. Antibody has been absorbed to plastic tubes or complexed to dextran or glass.

III. Pitfalls

Unlike many analytical chemistry techniques RIA is not a procedure for the direct determination of a substance. Therefore, when using RIA one must be ever alert to distinguish fact from artifact and to appreciate that not every reduction in the binding of labeled antigen to antibody is due to the presence of the specific substance of interest. Some of the problems relate to nonspecific interference in the immune reaction; others relate to the heterogeneity of the molecular forms of the antigen or the presence of related antigens.

1. Nonspecific Interference in the Immune Reaction

The chemical reaction of antigen with antibody can be inhibited by a variety of factors and there may be substances in plasma which alter or destroy the reactants. Antigen–antibody reactions are generally pH dependent and dissociate at extremes of acidity or alkalinity. Most are pH independent in the range 6.5–8.5. However, it has been demonstrated that in some assay systems, usually those with very basic antigens, the binding of antigen to antibody is maximal at pH 5 and the immune reaction is markedly inhibited at an alkaline pH (Kajubi et al. 1981). To maximize the sensitivity of the assay, the pH range should be chosen to optimize the binding of antigen to antibody and both standard and unknown incubation mixtures should be within that range.

Formation of antigen–antibody complexes can also be affected by the presence of a variety of substances, including proteins, salts used for buffering, anticoagulants such as heparin, bacteriostatic agents such as thimerosal, and enzyme inhibitors such as aprotinin (Yalow 1973). Not every substance interferes in all immune reactions to the same extent. Thus, the possible effect of each substance must be tested or, alternatively, it must be assured that the milieu of the unknown sample is identical with that of the known standards.

2. Presence of Heterogeneous Molecular Forms

The assay of peptide hormones, whether by RIA or bioassay, has been complicated by the demonstration that many, if not most, of these substances are found in more than one form in plasma and tissue extracts. Since the different molecular forms may differ in biologic activity as well as in secretion and degradation rates, measurement of immunologic activity per se may not suffice to characterize the unknown peptide. Complications can also be introduced by the presence of immunologically related peptides with different biologic activities. For example, gastrin, cholecystokinin, and cerulein share the same carboxy terminal pentapeptide and may react identically or completely differently, depending on whether the specific antiserum is directed toward the amino or carboxy terminal portion of the different molecules.

The problem of several related forms of the same antigen is relevant to the application of RIA in pharmacology as well as in endocrinology. Structurally related compounds or metabolites of the drug to be assayed may be immunoreactive with some antisera, but not with others and may or may not constitute a problem, depending on the purpose of the assay. For instance, if the clinical problem relates to the efficacy or toxicity of a particular drug, then the question as to whether or not the assay measures only the biologically active form is relevant. If the question relates simply to whether or not a drug has been taken surreptitiously, then the reactivity of metabolites or the variation in immunoreactivity with the exact form of the drug may be irrelevant.

C. Conclusions

RIA has proven to be a powerful tool in many diverse areas of biomedical investigation and clinical medicine. It is probably no exaggeration now to state that if there is a need to measure an organic substance of biologic interest and there is no other easy method to do so, some perspicacious investigator will develop the needed immunoassay.

Acknowledgment. This work was supported by the Medical Research Program of the Veterans Administration.

References

Berson SA, Yalow RS (1957) Kinetics of reaction between insulin and insulin-binding antibody. J Clin Invest 36:873

Berson SA, Yalow RS, Bauman A, Rothschild MA, Newerly K (1956) Insulin-I^{131} metabolism in human subjects: demonstration of insulin binding globulin in the circulation of insulin-treated subjects. J Clin Invest 35:170–190

Kajubi SK, Yang R-K, Li HR, Yalow RS (1981) Differential effects of non-specific factors in several radioimmunoassay systems. Ligand Q 4:63–66

Mahon WA, Ezer J, Wilson TW (1973) Radioimmunoassay for measurement of gentamicin in blood. Antimicrob Agents Chemother 3:585–589

Mirsky IA (1952) The etiology of diabetes mellitus in man. Recent Prog Horm Res 7:432

Nossel HL, Yudelman I, Canfield RE, Butler VP Jr, Spanondis K, Wilner GD, Qureshi GD (1974) Measurement of fibrinopeptide A in human blood. J Clin Invest 54:43–53

Oliver GC Jr, Parker BM, Brasfield DL, Parker CW (1968) The measurement of digitoxin in human serum by radioimmunoassay. J Clin Invest 47:1035–1042

Thomson DMP, Krupey J, Freedman SO, Gold P (1969) The radioimmunoassay of circulating carcinoembryonic antigen of the human digestive system. Proc Natl Acad Sci USA 64:161–167

Walsh JH, Yalow RS, Berson SA (1970) Detection of Australia antigen and antibody by means of radioimmunoassay techniques. J Infect Dis 121:550–554

Yalow RS (1973) Radioimmunoassay. Practices and pitfalls. Circ Res [Suppl 1] 32:I/116–I/128

Yalow RS, Berson SA (1959) Assay of plasma insulin in human subjects by immunological methods. Nature 184:1648–1649

CHAPTER 2

Basic Principles of Antigen-Antibody Interaction

F. CELADA

A. Introduction to the Immune System

The immune system may be viewed as the locus of a multitude of molecular interactions involving a limited number of genetically determined structures. Most of these structures act as membrane receptors (i.e., communication organs of cells) or as secreted factors. These are called antibodies (Ab), T cell receptors, or class I and II molecules coded for in the major histocompatibility complex (MHC).

They often bind to each other and/or to a large, virtually infinite, number of molecules "external" to the system, the antigens (Ag). All immune interactions are reversible, based on "weak" forces, and governed by the law of mass action. This confers a decisive role on the specificity, a property by which a given structure binds to one antigen with 10^3–10^5 times higher affinity than to any other molecule of similar size or composition. I shall list here seven types of "immune interactions" occurring at various levels of the response. They are all essential to the functioning of the machinery and it may be informative to compare them, although in many cases our knowledge is incomplete.

1. Free antibody–antigen (to be described later; Fig. 1 a).

2. Ig membrane receptor on B lymphocyte–antigen (Fig. 1 b). The actual binding interaction is identical to 1. If the antigen bears many determinants and thus engages many receptors, modifications of the geography of the membrane occur (capping). In all cases receptor–Ag complexes are internalized with subsequent "digestion" of antigen (processing) and reexpression of limited fragments of the antigen on the B cell surface; at this stage the processed antigen is (with high probability) bound to class II MHC proteins (Fig. 1 c).

3. Class II MHC protein–processed antigen. The existence of this interaction is based on strong, but indirect evidence. "Processing" of the antigen can be substituted by proteolysis (Fig. 1 d).

4. Triangular binding: T helper cell receptor–processed antigen–class II syngeneic MHC molecule. This is the only way a T–B interaction occurs between cells from the same individual. The result is: (a) triggering of the T helper (Th) cell [increased production of interleukin-2 (IL-2), increased production of IL-2 membrane receptors, binding of IL-2 to IL-2 receptors, proliferation of the Th cell into a blast clone]; and (b) triggering of the B cell (by IL-2 and/or B cell growth factors) and consequent maturation, clonal proliferation, and Ab secretion (Fig. 1 e).

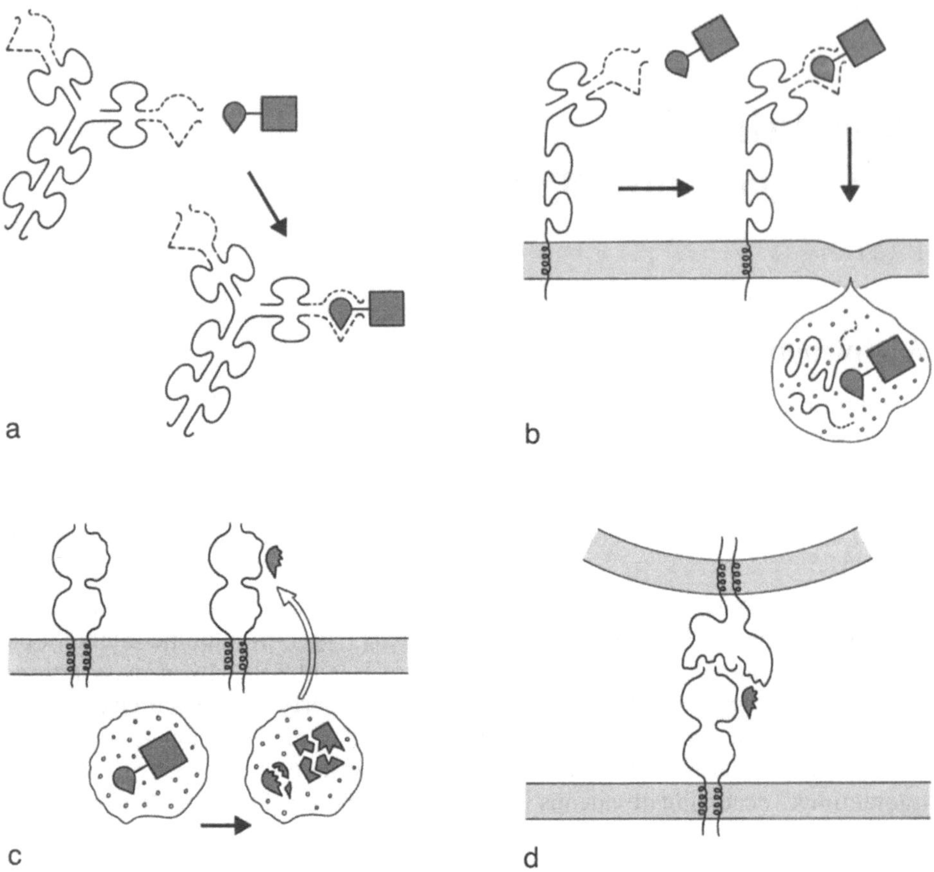

Fig. 1. a An antibody meets an antigen and binds an epitope. The variable parts of the heavy and light chain are indicated by *broken lines*. **b** A receptor on a B-lymphocyte binds an antigen and the complex is interiorized in a cytoplasm vesicle. **c** The antigen, separated from the Ig receptor, is digested and fragments are reexposed on the outside in close connection with a class II MHC molecule. **d** A T cell, by means of its receptor, makes contact with the complex syngeneic class II molecule + antigen fragment on the surface of a B cell (or other antigen-presenting cell).

5. Bilateral Th cell receptor–class II allogeneic MHC molecule. The triggering of the Th cell is identical to that in 4 (Fig. 1 g).

6. Triangular binding T cytolytic (Tc) cell receptor–foreign antigen (e.g., viral capsid)–class I syngeneic MHC molecule. The result in this example is killing of the viral antigen-bearing syngeneic cell (Fig. 1 f).

7. Bilateral binding Tc cell receptor–class I allogeneic MHC molecule. The result is the killing of transplanted cells.

The functioning of the immune system as it has evolved in the higher animals is based on the existence of at least three families of lymphocytes, endowed with clonotypic receptors (i.e., different on cells of a given population, but identical in the progeny of a single member of that population). The three families are T helper, T effector or cytolytic, and B. Less well defined is a fourth family, the T

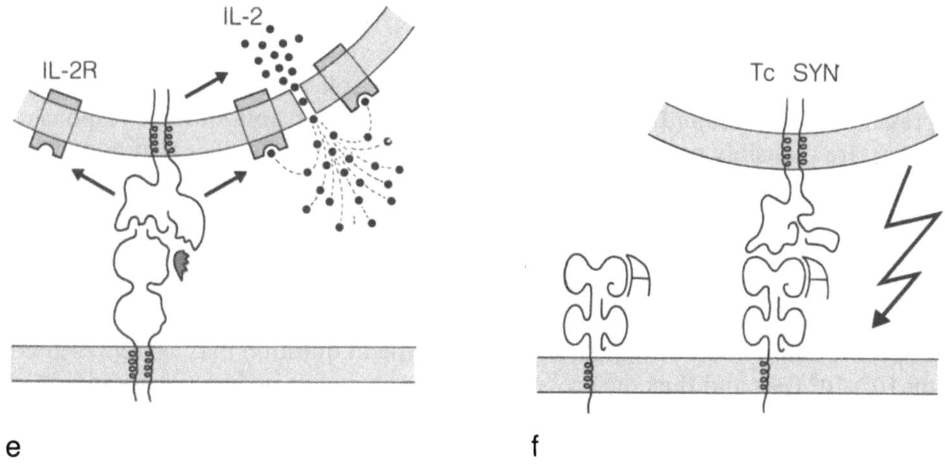

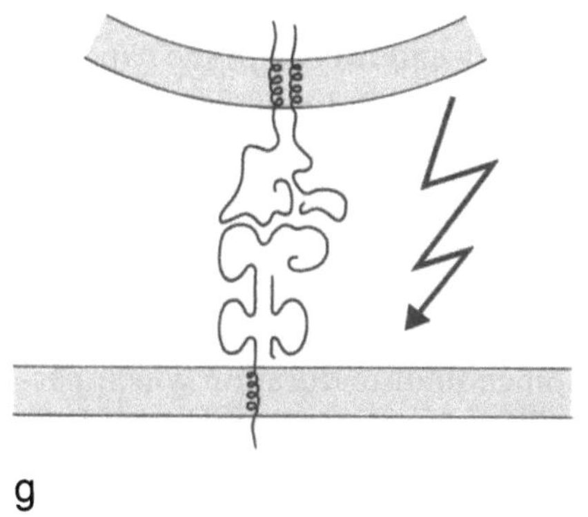

Fig. 1. e Following the contact in **d**, the T cell is activated. It exposes IL-2 receptors on the surface and secretes IL-2 (part of which is captured by its own receptors). **f** A T cytolytic cell makes contact with a complex class I MHC molecule + nominal antigen on the surface of a syngeneic cell (e.g. a cell infected by a virus). As a consequence,the target cell is killed. **g** A T cytolytic cell makes contact with a foreign cell, displacing a class I MHC molecule on the surface. The target cell is killed

suppressor cell. Each of these cell types has: (a) a large library of donotypic specificities – in the case of B-lymphocytes this is larger than 10^8, allowing the recognition of any possible external or internal structure – in the case of T-lymphocytes the size of the library is not yet known; and (b) the property – through appropriate binding of their surface receptors – to become stimulated, i.e., enter a state of intense nucleic acid and protein synthesis, preceding both proliferation and synthesis of more copies of their clonotypic molecules (for membrane display or for secretion) or of a series of hormone-like factors, as in the case of the T helper cells.

Hosting cells in fast proliferation is a potential danger for complex organisms; thus, rituals have evolved which may limit stimulation or block the succession of proliferation and differentiation of the triggered cells. One of these rituals is a sort of double-double-check, which can be sketched as follows. Cell 1 is triggered if: (a) its clonotypic receptor is bound and, (b) at a very short distance cell 2 also happens to be triggered. Cell 2 in turn will be triggered in a *restricted* fashion if its clonotypic receptor is bound. This only occurs in the presence of both the appropriate Ag and a syngeneic isotypic molecule coded by the MHC. This ritual imposes stringent requirements on the immunogen and limits the probability of accidental stimulation. Also, when successful triggering, following appropriate stimulation, yields a "useful" immune response, safeguards have evolved – interpreted today as means of avoiding uncontrolled proliferation (too much of a good thing is harmful). These safeguards are: (a) the still incompletely known suppressor cell actions; and (b) the well-studied B and/or T anti-idiotype responses. These are plausibly performed by the same T and B cell families that cope with the external antigens and are a by-product of the phenomenon of clonal expansion, e.g., a single B cell, bearing as a clonotypic antigen a certain amino acid combination on the variable parts of its immunoglobulin (Ig) receptor, has virtually no probability of meeting lymphocytes of the same organism "specific" for that determinant (called antibody idiotype). If, however, a clone is formed following a specific stimulation of that B cell, the idiotype in question may be represented on 10^6–10^8 cells and thus the probability of being recognized by the immune system will be 10^6–10^8 times higher. The effect of an anti-idiotype response will be a block in the growth of the clone, either by competition with antigen binding or by cytotoxicity by T effector cells.

In a provisional summary, the immune system can be regarded as representing 2%–5% of the weight of an organism, evolutionarily endowed with the capacity of reacting to any changes – endogenous or exogenous – in the molecular landscape. This is achieved by a highly specific recognition capacity, a highly selective capacity (Ag_1 vs Ag_2 or vs Ag_n; hetero- vs autoantigen) obtained through the mandatory use of cell cooperation and the combined reading of several filter repertoires, and a number of severe regulatory mechanisms. The result is high sensitivity to environmental changes with particular provisions to cope with repeated or cyclical events. This is favored by the phenomenon of maturation which permits, through repeated selection, the "best possible" response to be mounted.

If the evolutionary goal of the immune system is to maintain the primitive status quo, then this goal is never achieved, since at the end of the succession of anti-antigen and anti-idiotype responses, the equilibrium reached is at a new level of information (memory) and prepared to cope differently with the next environmental change.

This introduction has hopefully conveyed the impression that all parts, and all interactions, of the immune system are necessary and important. It is only because, partly for historical reasons, incomparably more is known about the Ag–Ab reaction that this chapter will deal solely with the principles of this particular encounter.

B. Antigens

Antigens are defined by their property of being bound by antibodies through their antigen-binding fragments (Fab), irrespective of their capacity to induce a response when introduced in an organism.

I. Chemical Nature

Natural antigens are soluble or cell-bound molecules of protein, glycoprotein, lipoprotein, or polysaccharide nature (LANDSTEINER 1936; SELA 1969; KABAT

1968). Macromolecular antigens (>2000 dalton) are as a rule also immunogens; that is, they are capable of stimulating the production of the corresponding antibodies when introduced into an immunocompetent organism.

II. Antigenic Determinants (Epitopes)

The binding by the specific antibody takes place between a well-defined portion of the antibody, the domain resulting from the co-folding of the variable parts of the light and the heavy chain (VL and VH) and a portion of the antigen, called antigenic determinant. The antibody-binding site is also called the paratope, while epitope is a synonym of antigenic determinant (JERNE 1960).

III. Haptens and Carriers

Small molecules (~500 dalton) can be artificially linked in a covalent fashion to protein carriers and function as epitopes. They are called haptens, are antigenic, nonimmunogenic; their experimental use has led to a better understanding of immune interactions (see Sect. D.IV) and the processes involved in the initiation of a response.

It has been shown that a B cell, in order to be triggered by a hapten, has to cooperate with a T cell recognizing a determinant of the carrier molecule (MITCHISON 1971 a, b). The carrier–hapten model, although useful, remains an artifact; in natural antigens it seems probable that cooperating T and B cells see geographically distinct epitopes, not necessarily of different nature.Natural soluble macromolecular antigens bear many antigenic determinants. As a rule, the determinants of the same protein antigen are different from each other, except when the protein is a polymer of identical protomers. Polysaccharide and sometimes glycoprotein antigens have repetitive determinants. When whole cells are considered as antigens, epitopes are also repetitive, but their density on the membrane is variable.

IV. Size of Determinants

The size of epitopes is determined by the dimensions that can be accommodated within the VH–VL domain. It corresponds to 3–6 saccharide residues in the case of polysaccharides (KABAT 1966) and 3–4 amino acids in the case of protein antigen (SCHECHTER et al. 1970). Haptens may be smaller or larger than this. In the first case, the antibody will also bind to a part of the carrier; in the second case, each antibody may recognize only a part of the hapten.

V. Sequential and Conformational Epitopes

The clusters of atoms that constitute an epitope may be amino acids ordered in sequence on the polypeptide chain, or happen to be near each other in the three-dimensional arrangements of the molecule although belonging to distant segments of the polypeptide chain – or even to different chains, in the case of polymeric antigens. The latter type of determinants, called conformation-dependent or conformational, are related to the general shape of the antigen molecule and are affected (and often destroyed) by mild treatments that cause conformational changes (Sela et al. 1967). Such events do not in the least disturb sequential determinants. In protein macromolecules there is a prevalence of conformational determinants, many of which are also dominant as immunogens. The existence of conformation-dependent recognition allows high selectivity, which has an evolutionary advantage.

C. Antibodies

Antibodies, or immunoglobulins, are specialized complex proteins synthesized by B-lymphocytes and by their final maturation product, plasma cells. The basic structure of any antibody is a heterodimer H – L, made up of a heavy H and a light L chain, 50–60 and 20 Kilodalton, respectively. The two chains are synthesized by the same cell, but are coded by genes on different chromosomes. An organism produces immunoglobulins which are heterogeneous in heavy chain class or isotype (μ, δ, γ with subclasses ε and α) and in light chain type (K or λ). Each of the five heavy chains can combine with either a K or a λ type L chain to form the basic heterodimer (Natvig and Kunkel 1973). Both H and L polypeptide chains have a distinctive portion at the amino terminus, where the sequence of the first 110 amino acids varies drastically among antibodies of the same individual, but produced by different cells. On the other hand, the remaining 100 amino acids of the L chain and the remaining 330 amino acids of the H chain are remarkably similar among all antibodies of the same isotype in the species. The variable parts of the heavy and of the light chain (VH and VL) intermingle to form the first domain of the heterodimers. A cleft in this domain constitutes the paratope (see Sect. C.I). Each competent B cell, once a laborious gene rearrangement is achieved for both H and L chains, sets up to produce a single model of H–L. H and L assemble in the cytoplasm and the resulting heterodimers are mounted on the cell surface, the carboxy terminus of H actually reaching inside, a hydrophobic near-terminal section being rooted in the membrane, with the amino terminal section of H and the entire L outside. They are the immunoglobulin receptors, and from this moment the B cell is specific for whatever epitope the receptors can bind. As a rule, the first heterodimer to be synthesized by any B cell is μK or $\mu\lambda$. Only a handful out of millions of B cells will ever meet an antigen that can be recognized by their receptors. The binding event, together with T–B cooperation, will represent a positive selection in the Darwinian sense. The selected cell will eventually proliferate to become a clone. With cell stimulation and clonal selection rounds, in connection with successive sensitizing events, the B cell will undergo a series of changes in the genes concerned with H and L that will mark the clone's life history:

1. A substitution through rearrangements and RNA splicing of the hydrophobic (membrane anchoring) carboxy terminus with a hydrophilic peptide (secretory switch) (Rogers et al. 1980), allowing the Ig to be secreted as free antibody; the quaternary conformation varies depending on the isotype, but in all cases it is a double heterodimer or a multiple thereof (5 × 2HL plus one joining

peptide J in the case of IgM and 2 × 2HL plus a "secretory fragment," added when passing through a mucosal epithelium, in the case of IgA). In all cases, the two HL units of the double heterodimer are identical. This symmetry allows Ab polyvalency and prevents one antibody molecule binding to two different epitopes.

2. The entire μ constant gene is deleted by rearrangement and substituted with the constant gene coding for one of the other isotypes. Thus, during the secondary response, the secreted antibodies will have the same variable, but different constant parts as compared with those secreted by the same clone during the primary response (isotypic switch) (Davis et al. 1980).

3. During stage 2, or during the rapid proliferation following antigen challenge, a number of point mutations may affect the VH and VL genes. Therefore, antibodies produced during the secondary response may be different in fine specificity from those produced by the cells from the same clone during the primary response, and subclones may be generated (Gerhart et al. 1981). This allows for a second round of selection by antigen for the most effective (i.e., affine) subclone, and explains the higher affinity and specificity reached after repeated antigenic stimulations.

I. The Ab Combining Site (Paratope)

The paratope is a three-dimensional space within the VH–VL domain of the antibody heterodimer (Givol 1973; Poljak 1975).

1. VH and VL Structures

The VH and VL fragments have marked similarities in size and structure. They both contain 110–115 amino acids, and they both have cysteine residues in positions 40 and 80 which allows an intrachain S–S bond. By comparing V regions from different antibodies, it has been possible to identify within the primary structure of both VH and VL segments of low variability and high variability (hypervariable regions). The segments of low variability flank the two cysteines that are always present. There are three hypervariable regions in both VH and VL (Wu and Kabat 1970). It has been demonstrated by affinity labeling experiments that it is amino acids within the hypervariable regions which have the closest contact with the antigen epitope during the binding (referred to as CDR, contact determining residues) (Wu and Kabat 1970).

2. Genetic Control of V Regions

Hypervariable regions I and II of heavy and light chains are encoded in the VH and VK or V segments, respectively. Region III on both H and L is encoded in the J segment; variability to III is increased significantly by recombinational inaccuracy at the V–D, D–J, and J–C junctions in H and V–J and J–C junctions in L.

3. Paratope Diversity

The theoretical diversity of antibody paratopes (and thus the number of specificities attainable by the immune system) is extremely large ($>10^9$). It results from the factorial multiplication of many variables: the number of V gene segments present in the germ line; the number of D segments; the number of JH segments, the number of VL and JL segments in the germ line; the probability of recombination inaccuracy; the frequency of somatic point mutation – especially important at the secondary response; the assortment of heavy and light chains (e.g., combination of 10^3 H and 10^3 L yields 10^6 VH (Leder 1982). Not all theoretical specificities are actually realized. Nevertheless, since cross-reactivity must be taken into account (the property of a given paratope to be able to bind with graded affinity to two or more nonidentical epitopes), the theoretical expectation agrees with the observed efficiency of the immune system.

4. Idiotypes

The VH–VL combination that constitutes a paratope is a structure unique to one (or perhaps a few) clones in the whole organism. Parts of this clonotypic feature may be recognized as new (thus, foreign) by the rest of the immune system. These particular epitopes are called idiotypes, and the collection of idiotypes found on a single immunoglobulin molecule are its idiotype. Idiotypes are extremely interesting: they have served to map variable region genes and to characterize different cellular receptors. They are the focus of Jerne's network theory, proposing that the system employs idiotypes as targets for recognition by anti-idiotypic regulatory cells and molecules (Jerne 1974).

D. The Immune Interaction

Epitope and paratope interact through a noncovalent bond. The resulting link is determined by a number of attractions between the two entities (intrinsic). The binding of the antibody molecule to a macromolecular antigen is determined by the number of immune interactions that may be realized between the two. Since Ab is always polyvalent, the possibility of multiple binding depends on the antigen bearing more identical epitopes. This (see Sect. A, interaction type 2) often happens with polysaccharides and is less common with proteins, where it is limited to the cases of multimeric macromolecules, exhibiting the same epitope on each protomer. Double binding in the case of IgG, IgA, and IgE, and up to quintuple binding in the case of IgM (a maximum of five of the possible ten bonds have been verified) increase the strength of the link exponentially, with important functional consequences.

I. The Forces Involved

The immune interaction, like any noncovalent protein bond, consists of multiple finite attractions which take place between atomic clusters of the paratope and the epitope. These forces are classified as:

1. Coulombic: electrostatic attraction between positive and negative charges
2. van der Waals: attractions between electron clouds
3. Hydrogen bonds: attraction of the same hydrogen atom by two different atoms
4. Hydrophobic forces: "cooperation" of hydrophobic groups (e.g., CH_2) in excluding water molecules; this is possibly the single most significant force in immune interactions

A common feature of these attractions is that they vary as inverse powers of distance. This means that their effects virtually disappear with minimal increments of distance between the atomic clusters involved.

II. Paratope–Epitope Fit

The consequences of the facts listed in Sect. D.I are the following. Strong intrinsic attractions are only possible if many weak attractions coincide in the binding (of course, the cooperative gain is exponential). For this to come about a topologically close fit between the Ab combining site and the epitope is required. The closer the fit (usually shown as a key–lock relation) the higher is the probability of many weak bonds occurring. The resulting attraction is called intrinsic affinity. Figure 2 shows the complementary shapes of paratope and epitope as revealed by high resolution X-ray diffraction analysis.

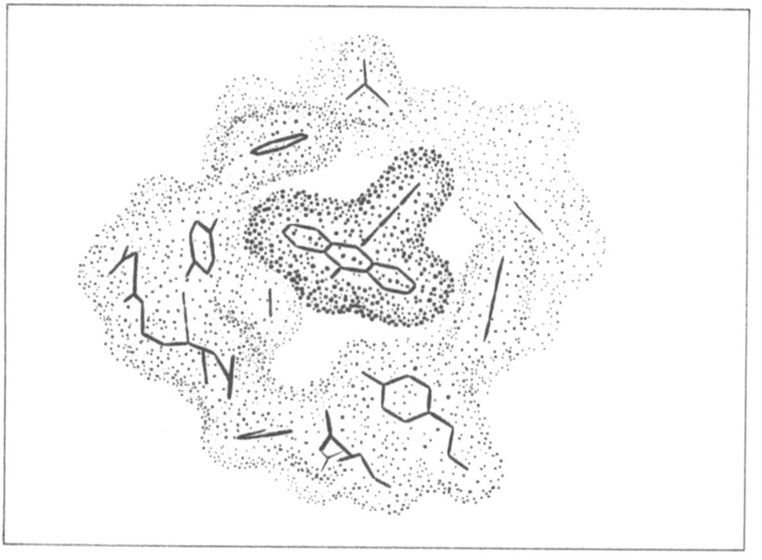

Fig. 2. Complementary shapes of paratope and epitope by X-ray diffraction analysis and computerized representation of water-accessible surfaces

III. Specificity

A given combining site can accommodate with different precision epitopes of similar structure. This discriminating capacity is called the specificity of the antibody. Although based on affinity, it is not related to the absolute affinity of the Ab. Thus, specificity is a relative concept, not measurable in quantitative terms. As a rule, the antigen against which the antibody has been raised will be bound with the highest affinity. An exception to this rule is the rare occurrence of heteroclytic antibodies which are able to bind with highest affinity epitopes different from the one against which they were raised. The same concepts of cross-reactivity and of specificity apply when a given epitope is confronted with several, similar antibody-combining sites.

IV. Affinity

As the fully reversible Ab–Ag interaction reaches an equilibrium, the amount of bound Ag depends on the affinity of the bond and is determined by the molar concentrations of free antigen and antibody, following the law of mass action

$$\mathrm{Ab} + \mathrm{Ag} \underset{k_2}{\overset{k_1}{\rightleftharpoons}} \mathrm{AbAg}$$

hence

$$\frac{[\mathrm{Ab\,Ag}]}{[\mathrm{Ab}]\,[\mathrm{Ag}]} = K$$

where $K = \frac{k_1}{k_2}$, represents the binding constant.

There is no substantial difference in terms of forces involved, range, specificity between the binding of Ag and Ab, and that of enzyme and substrate. Thus, the same basic methods, already developed in enzymology, can be applied to measure affinity constants. In the case of Ab–Ag interaction (where no product will be formed) the binding constant K is defined as the molar concentration of free epitope necessary to reach, at equilibrium, 50% saturation of the antibody-combining sites.

1. Measurement of Ab-Hapten Affinity

The actual measurement of affinity (WEIR 1976) takes advantage of the small size of the hapten (H). Ab is separated from H by a dialysis membrane which allows free circulation of H, but not of Ab. To attain high sensitivity, H is radiolabeled. The same concentration of antibodies is confronted with increasing concentrations of H in a series of parallel experiments. At equilibrium, on one side of the membrane H will be found in two forms, AbH and free H, while on the other side only free H, at the same concentration, remains. For each experiment, bound H is calculated simply by subtracting free H from total H, while free H will be the sum of free H on both sides of the membrane. Once free and bound H have been obtained from each experiment, there are two ways of plotting the data.

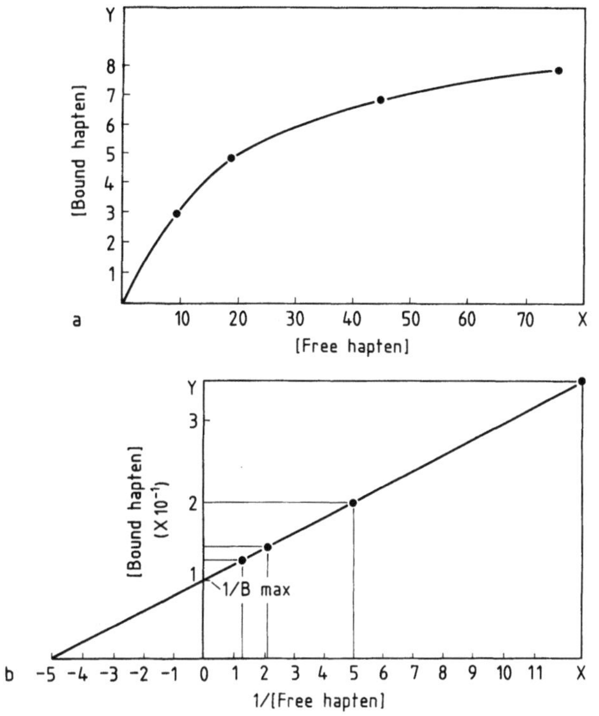

Fig. 3. a Saturation of a fixed amount of antibody by increasing concentrations of hapten. **b** Lineweaver–Burk transformation of same curve

a) The Lineweaver–Burk Plot

On the abscissa 1/[H]; on the ordinate 1/[AbH], both [H] is and [AbH] plotted being expressed in moles. By extrapolating to $1/[H]\rightarrow 0$, the intersection with the ordinate indicates B_{max} (maximum H bound by Ab). The value of 1/[H] that corresponds to 2 B_{max} is $1/K_m$, (where K_m is the Michaelis constant), and corresponds to the reciprocal of [H] at which the antibody is half-saturated (Fig. 3).

b) The Scatchard Plot

For this plot also, the molarity of the antibody is required in addition to free and bound hapten: this means that a purified Ab preparation must be used. In this plot, $y = R/[H]$ and $x = R$; where $R = [AbH]/[Ab_t]$. The resulting slope is $-K$, the y intercept is nK and the x intercept is n, where n is the number of H molecules bound per Ab molecule, i.e., the antibody valency. The affinity constant K is readily calculated by determining the slope (Fig. 4).

2. Heterogeneity

An unexpected by-product of measuring antibody affinity is the demonstration of the heterogeneity of combining sites among antibodies produced by an organism against a homogeneous hapten or determinant. This results in a curvature in the Lineweaver–Burk double-reciprocal plot of the Ab–hapten saturation curves, which can only be explained by the presence in the serum of combining sites, endowed with different affinity toward the same hapten.

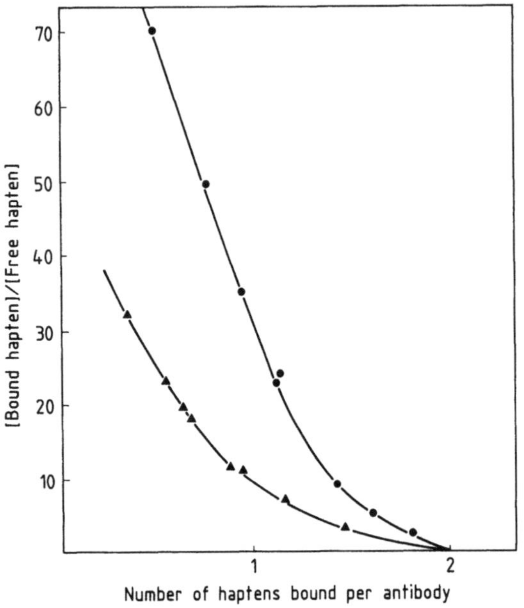

Fig. 4. Scatchard plot of two polyclonal antibodies

In the past, efforts have been applied to quantify the extent of heterogeneity, e.g., according to SIPS (1949) the values on the abscissa are raised to a power between 1 and 0.1 until the curvature of the graph disappears. The corresponding power is called the heterogeneity index and taken as an indication of the number of different antibodies present in the Ag-purified antiserum and of the spread of their affinity (WEIR 1976). The affinity constant calculated by this procedure has been considered an "average" affinity. All these conclusions are only tentative since there are too many unknowns (number of clonal antibodies, distribution of affinities, relative amount of antibody with a given affinity) to reach any meaningful conclusion.

3. Monoclonal Antibodies

There is no heterogeneity of affinity in a monoclonal antibody and no bending of the double-reciprocal functions (VELICK et al. 1960), as in the determination of affinity of enzymes.

4. Alternative Ways to Measure Affinity

Physical separation of bound and free Ag is not necessary if a functional way exists to discriminate between the two forms. Two examples among many are: (a) the method of fluorescence quenching; and (b) Ab-dependent enzyme activation.

a) Fluorescence Quenching

Immunoglobulin has a measurable natural fluorescence, which is lost when it binds the hapten dinitrophenol (DNP), which absorbs the emitted wavelength.

From the decrease in fluorescence it is thus possible to quantitate bound DNP (VELICK et al. 1960).

b) Enzyme Activation

Antibodies may influence the conformation of antigens (see Sect. F). In the notable case of β-galactosidase, the binding of a single Ab paratope activates a molecule of defective enzyme. Thus, it has been possible to quantitate the bound Ag by determining enzyme activity, and for the first time to measure intrinsic affinity of Ab to the natural epitope of a macromolecular antigen (CELADA et al. 1973).

E. Effects of Ab–Ag Interaction

At variance with enzyme–substrate binding, the immune interaction is not directly linked with a catalytic step. Instead, it causes the formation of a rather stable Ag–Ab complex and prepares the disposal of the antigen through a series of antigen-nonspecific mechanisms. Examples are given for the situation of membrane-bound antibody and free Ab.

I. Free Ag Binds Ig Receptor

In the case of antigen being bound by a receptor on the B-lymphocyte, the complex is interiorized (with a rearrangement of the membrane architecture that becomes macroscopic when the bindings are multiple: the capping phenomenon) and the antigen is degraded by proteolytic enzymes in the lysosomes. This process is actually the basis for the reexpression on the membrane of fragments of the antigen, in close connection with class II molecules, coded by the major histocompatibility complex, where they may be recognized by a specific T cell, thus initiating the T–B cooperation.

II. Free Ag Binds Free Ab

In the case of free Ab binding free Ag, complexes of varying size may be formed, depending on the valency of Ab and the number of epitopes available on Ag. If the encounter takes place in body fluids – as in most natural situations – the precipitation does not occur, even in the so-called equivalence zone of the Ag/Ab ratio. This is because the complement is present in large excess (except in extreme pathologic cases) and it contains factors, within the alternative pathway chain, that inhibit physical precipitation by binding to the crystallizable fragment (Fc) moieties and interdigitating between them, with the result of preventing large aggregates.

The principal action of the immune complex is to signal its existence to phagocytic and scavenger cells. This is done by binding (at the antibody hinge region) Clq molecules and thus triggering the complement cascade. The immediate and medium-term effects are the initiation of the phenomenon of inflammation (through the liberation of C3a and C5a), the attraction by opsonization of phagocytic cells, and the display on the complex of C3b, which favors the binding by any cell, e.g., granulocytes and monocytes, that bear C3 receptors. The next step is phagocytosis or pinocytosis, with consequent digestion and disposal of the complex.

III. Free Ab Binds Cell-Associated Ag

A third possibility is the binding of free Ab to cell-bound Ag. This will happen in responses to bacteria, protozoa, tumors, viruses, and self antigens. With the exception of certain antiviral antibodies that may neutralize the virus by simply binding to or near sites critical for phenomena such as penetration into the target cells, all other cases require signalling to, and further action taken by, macrophages or more specialized cells. The activation of complement (as in Sect. E.II), ending with perforation of the target membrane, will cause direct killing of bacteria and of a certain fraction of tumor cells; killed cells will be disposed of by phagocytes opsonized by complement products. Most nuclear cells (e.g., tumor cells, or transplanted cells) are not killed by Ab + complement (C′). Instead, killer lymphocytes (belonging to the family of LGL, large granular lymphocytes) will be attracted and will make a lethal contact with the target through their Fc receptors, which have affinity toward the carboxy terminal domain of bound antibody.

F. Immune Interaction as a Signal: Role of Conformational Changes

I. Antibody

From the examples discussed in Sect. E, it appears that central questions in molecular immunology such as how the binding of antigen to membrane antibody induces the immune response; how the phagocyte knows which molecule should be disposed of; and how a nonspecific killer cell can zero in on the appropriate target, would eventually produce the same answer, if we were able to decode the language of the antibody molecule and to understand the transmission of a message from the Fab end, where antigen binds, to the Fc end of Ab, where the information is read.

In order to explore this question at the molecular level, three models have been devised for the generation of the immunologic signal between the Fab and Fc regions of antibodies: (a) the allosteric model; (b) the distortive model; and (c) the associative model. For IgG, using complement activation as a measure of immune response, most of the evidence indicates that association is a necessary step for signal generation (METZGER 1978). What is not clear, however, is whether association of IgG is a sufficient step for complement activation or whether a conformational change in IgG structure induced by antigen must also accompany the association. For IgM, on the other hand, a conformational change appears to be essential for complement activation, since a single molecule of IgM acquires the ability to bind and activate C1 when it becomes attached to a cell surface through two or more combining sites. IgM is already a pentameric molecule, and this "built-in" association also appears to be a necessary feature for efficient complement activation. Therefore, signal generation in IgM has been described as a mixture of models, associative, but also requiring a conformational change. There is direct evidence that at least small conformational changes occur when hapten is bound to antibody; however, these changes in structure may be restricted to the readjustment of the variable domain to a better binding conformation, and their biologic significance is not clear (CELADA et al. 1983).

II. Antigen

Since the two patners of a protein–protein interaction are subject to the same forces and thermodynamic conditions, the hypothesis that conformational transitions are part of the "antibody language" has been considerably strengthened by finding that, in a number of cases, antigen is changed in conformation after Ab binding (CELADA and STROM 1972). When two or more different kinetically accessible conformations are compatible with the primary structure of a protein, they can be envisaged to be in dynamic equilibrium, no matter how much this equilibrium is displaced in favor of one of them. The transition from one to another conformation may be marked by the disappearance, the new appearance, and/or the modification of conformational antigenic determinants. In each of these cases, the presence of specific antibody is expected not to be neutral, but rather to influence the conformation by increasing the stability of one form or inducing/selecting the change to the other form.

Therefore, the requirement for an antibody-induced "change in conformation" is for the antibody molecule to have a higher binding affinity directed against the less frequently presented form. To this type of antibody-mediated effect belong the classical examples by CRUMPTON (1966) where interaction with anti-apomyoglobin causes metmyoglobin to dissociate and release the heme, and by SELA et al. (1967) where antibodies directed toward synthetic amino acid sequences in a helical conformation cause amorphous fragments to assume a helical shape.

Possibly the most dramatic examples in terms of functional effects are those involving antibodies elicited against a structurally "normal" species, which are then confronted with genetically defective molecules. Here, the outcome may be the "cure" of the genetic damage at the product level, and this may be perceived as the new appearance of phenomena such as enzyme activity, as in the well-studied case of *Escherichia coli* β-galactosidase (ROTMAN and CELADA 1968).

References

Celada F, Strom R (1972) Antibody-induced conformational changes in proteins. Q Rev Biophys 5:395–425

Celada F, Macario AJL, Conway De Macario E (1973) Enzyme activation by antibodies: a method to determine the binding constant of the activating antibody towards one determinant of E. Coli β-D-galactosidase. Immunochemistry 10:797–804

Celada F, Schumaker VN, Sercarz EE (eds) (1983) Protein conformation as an immunological signal. Plenum, New York

Crumpton MJ (1966) Conformational changes in sperm whale metmyoglobin due to combination with antibodies to myoglobin. Biochem J 100:227–234
Davis MM, Kim SK, Hood L (1980) Immunoglobulin class switching: developmentally regulated DNA rearrangements during differentiation. Cell 22:1–2
Eisen HN (1980) Immunology, 2nd edn., Harper and Row, Hagerstown
Gerhart P, Johnson ND, Douglas R, Hood L (1981) IgG antibodies to phosphorylcholine exhibit more diversity than their IgM counterparts. Nature 291:29–34
Givol D (1973) Structural analysis of the antibody combining site. Contemp Top Mol Immunol 2:27–45
Jerne NK (1960) Immunological speculations. Annu Rev Microbiol 14:341–358
Jerne NK (1974) Towards a network theory of the immune system. Ann Immunol 125c:373
Kabat EA (1966) The nature of an antigenic determinant. J Immunol 97:1–11
Kabat EA (1968) Structural concepts in immunology and immunochemistry, 2nd edn. Holt, Rinehart, and Winston, New York
Landsteiner K (1936) The specificity of serological reactions, 2nd edn. Harvard University Press, Cambridge
Leder P (1982) The genetics of antibody diversity. Sci Am 246:102–113
Metzger H (1978) The effect of antigen on antibodies: recent studies. Contemp Top Mol Immunol 8:119–152
Mitchison NA (1971 a) The carrier effect in the secondary response to hapten-protein conjugates. I. Measurement of the effect with transferred cells and objections to the local environment hypothesis. Eur J Immunol 1:10–17
Mitchison NA (1971 b) The carrier effect in the secondary response to hapten-protein conjugates. II. Cellular cooperation. Eur J Immunol 1:18–25
Natvig JB, Kunkel HG (1973) Human immunoglobulins: classes, subclasses, genetic variants and idiotypes. Adv Immunol 16:1–59
Poljak RJ (1975) X-ray diffraction studies of immunoglobulins. Adv Immunol 21:1–32
Rogers J, Early P, Carter C, Calame K, Bond M (1980) Two mRNAs with different 3′ ends encode membrane-bound and secreted form of immunoglobulin μ chain. Cell 20:303–312
Rotman B, Celada F (1968) Antibody-mediated activation of a defective β-D-galactosidase extracted from an Escherichia Coli mutant. Proc Natl Acad Sci USA 60:660–667
Schechter B, Schechter I, Sela M (1970) Antibody combining site to a series of peptide determinants of increasing size and defined structure. J Biol Chem 245:1438–1447
Schechter B, Conway Jabobs A, Sela M (1971) Conformational changes in a synthetic antigen induced by specific antibodies. Eur J Biochem 20:321–324
Sela M (1969) Antigenicity: some molecular aspects. Science 166:1365–1374
Sela M, Schechter B, Schechter I, Borek M (1967) Antibodies to sequential and conformational determinants. Cold Spring Harb Symp Quant Biol 32:537–539
Sips R (1949) On the structure of a catalyst's surface. J Chem Phys 16:490
Velick SF, Parker CW, Eisen HN (1960) Excitation energy transfer and the quantitative study of the antibody hapten reaction. Proc Natl Acad Sci USA 46:1470–1482
Weir DM (1976) Handbook of experimental immunology, 3rd edn. Blackwell Scientific, London
Wu TTE, Kabat EA (1970) An analysis of the sequences of the variable regions of Bence Jones proteins and myeloma light chain and their implications for antibody complementarity. J Exp Med 132:211–250

CHAPTER 3

Production of Antisera by Conventional Techniques

G. CIABATTONI

A. Introduction

An immunogen may be considered as any chemical structure capable of inducing a specific immune response after injection. Immunogens are not characterized by a specific chemical sequence, but are molecules which share the common property of evoking the synthesis of specific antibodies and reacting with them. To possess immunogenic properties, a molecule has to satisfy the following requirements: (a) to be a heterogeneous substance for the animal species in which it is injected; (b) to undergo slow metabolic clearance or excretion after injection; (c) to be of adequate size and complexity to be recognized by the immunologic system; and (d) to present active surface groups. Although isoantibodies or autoantibodies may be produced in several animal species, including humans, heterogeneity represents an essential feature for the immunogenic activity of a substance. Generally, the more different the species, the greater will be the specific immune response. To be immunogenic a heterogeneous substance has to diffuse slowly from the point of injection, and should not be excreted or metabolized quickly. Slow excretion as well as low metabolic clearance does not mean low solubility. To be immunogenic and to stimulate immunoglobulin production by B-lymphocytes, an antigenic structure has to be solubilized by macrophages.

Size represents an essential feature in determining the immunogenicity of a molecule. However, size itself is not enough to make a substance immunogenic. Synthetic polymers, such as nylon or polystyrene, are unable to induce antibody production. Probably this is due to the lack of molecular complexity of these compounds. Despite their high molecular weight, synthetic polymers are built up of very simple units, i.e., one or two very simple repeated monomers. Polymers of amino acids, which have a more complex chemical structure, have been shown to induce antibody production (SELA 1969, 1971). Natural antigens, on the other hand, are usually very complex molecules. Even the simplest proteins are made up of at least 15 different amino acids, which may be arranged in a number of sequences and steric configurations. Generally speaking, the more complex the antigen is in terms of size and chemical structure, the more suitable it is in inducing a strong immune response, although no general rule exists about the minimal size required for immunogenicity, and no close connection has been found between molecular weight and antigenic strength. Very small natural proteins like glucagon (molecular weight 2600) are immunogenic. However, the bigger the antigen, the higher is the probability of chemical complexity of the molecule and, therefore, the presence of antigenic determinants in the molecule, which are responsible for the immunogenic strength.

Substances with a low molecular weight are poor immunogens or are completely unable to induce antibody generation. It is necessary to conjugate a small antigenic determinant to a big molecule, preferably with high chemical complexity, to evoke an immune response. This type of reaction allows production of a very large number of antigens, which are not naturally occurring. A substance of low molecular weight coupled to a large molecule represents a hapten. This technique was originally due to LANDSTEINER (1945). Conjugating a substance of low molecular weight to a protein carrier and injecting the conjugate into an experimental animal, he observed an increase in the heterogeneous population of antibodies. These reacted with both the small coupled molecule and the protein carrier. Landsteiner and co-workers conjugated a number of aromatic amines to the tyrosine, histidine, and lysine residues of proteins. Immunization led to the production of specific antibodies directed against the hapten, the protein, or the hapten–carrier junction.

The carrier provides the antigen with the immunologically necessary size and reactive groups on its surface. Although these reactive groups are not required in determining specificity toward the hapten, they participate in inducing the immune reaction by activating the B-lymphocytes, thereby leading to the generation of a heterogeneous population of immunoglobulins. Haptens may thus be considered single antigenic determinants. Natural antigens, like proteins, have a large number of different antigenic determinants and could be regarded as molecules with a lot of natural haptens. In the years following the Second World War it was possible to prepare antigens by attachment to synthetic polypeptides or naturally occurring proteins of small haptens such as dinitrophenyl, *p*-azobenzenarsonate, or penicilloyl groups. In the 1960s, antibodies were raised with specificity directed toward sugars, nucleosides, peptides, coenzymes, vitamins, and several other naturally occurring small molecules. In the following years, it has been possible to synthesize immunogens leading to antibodies of nearly any specificity desired.

B. Substances Which Are Able to Evoke Immunogenic Responses Per Se

Proteins or large peptides, such as most of the peptide hormones, are satisfactorily immunogenic in a variety of experimental animals and can be directly injected as an emulsion in Freund's adjuvant. The immune response of the host will lead to the production of antisera, composed of a mixture of site-directed antibodies, each of which is directed against a different antigenic site in the peptide structure. Since the binding site of an immunoglobulin can accommodate only a few amino acids, no single antibody molecule can make close contact with the entire surface of a peptide larger than this. As a consequence, antisera produced in this way specifically recognize only small portions of the larger peptide as major antigenic determinants, the other areas being unable (or less able) to elicit antibody formation. Intermediate-sized peptides (10–30 amino acids) may contain only one or two highly antigenic regions, and partial sequences containing this region may interact fully with antisera raised against the whole molecule. Generally, molecules

Table 1. Immunogens and haptens

Immunogenic substances (conjugation to carriers not necessary)	Poor immunogenic substances (conjugation to carriers improves immunogenicity)	Haptens (conjugation to carriers necessary to produce antisera)
Insulin	Atrial peptides	Eicosanoids
Proinsulin	Glucagon	Steroids
FSH-LH	Insulin C peptide	Bile acids
hCG	Gastric inhibitory polypeptide (GIP)	Serotonin
Human chorionic somatomammotropin (hCS)	Vasoactive intestinal polypeptide (VIP)	T_3 and T_4
Prolactin	Cholecystokinin	α-, β-MSH
hGH	Pancreozymin	cAMP and cGMP
Parathyroid hormone (PTH)	Gastrin, pentagastrin	Bradykinin
TSH	Motilin	Bombesin and bombesin-like peptides
Calcitonin	Neurotensin and related substances	Melatonin
Placental lactogen (PL)	Secretin	Angiotensin II
Erythropoietin	Oxytocin	Substance P and related neuropeptides
Human pancreatic polypeptide (HPP)	Vasopressin	Enkephalins
Bovine pancreatic polypeptide (BPP)	Somatostatin	Atropine
Nerve growth factor (NGF)	Urogastrone	Folic acid
Relaxin	GnRH	25-Hydroxy and 1,25-dyhydroxy Vitamin D_3
Myelin basic protein	TRH	Drugs
Carcinoembryonic antigen (CEA)	ACTH	
α-Fetoprotein (αFP)	Calmodulin	
Trypsin, chymotrypsin	S 100 protein	
Pancreatic amylase and lipase		
Glial fibrillary acidic protein (GFA)		
Microtubule-associated proteins (MAPs)		
Neurofilament proteins		
Australia antigen (Au)		
Rheumatoid factor		

of this size are not very immunogenic and production of antibodies may be difficult, unless administered in a bound form, i.e., covalently attached or tightly linked with a carrier substance. Table 1 lists some substances which have been found to be immunogenic per se, substances which are poor immunogens by themselves, and substances acting as haptens.

Compounds of the first series are usually capable of raising antisera when injected emulsified in complete or incomplete Freund's adjuvant (see Sect. E.II). Chemical attachment to a larger carrier or polymerization are not required to improve immunogenicity. Pituitary gonadotropins (FSH and LH), thyrotropin (TSH), parathyroid hormone (PTH), growth hormone (GH), prolactin, ACTH, placental lactogen (PL), human chorionic gonadotropin (hCG), calcitonin, and many other substances of high molecular weight and high molecular complexity are good antigens, capable of eliciting soluble antibody production in several animal species, including nonhuman primates.

In the early days of radioimmunoassay (RIA), Berson and Yalow observed that, for immunization with most peptide hormones, it is possible to use relatively impure antigens (for review see BERSON and YALOW 1973). Commercial or low purity hormonal preparations may present slight denaturations which render the hormones more heterogeneous for the animal species which receives them and, thereby, enhance their antigenicity. For example, the same authors reported that immunization of guinea pigs with crude bovine parathyroid hormone (about 5% purity), or with two preparations of semipurified porcine gastrin, gastrin A (0.5% gastrin) and stage II gastrin (10% gastrin), resulted in the production of antisera suitable for measurement of human parathyroid hormone (molecular weight about 9000) or human gastrin (molecular weight about 2600), respectively (BERSON et al. 1963; YALOW and BERSON 1970a). Sera containing anti-gastrin antibodies with high avidity were noted in approximately 40% of the treated animals.

Antisera strongly binding human insulin were reported for the first time by BERSON and YALOW (1959a). This observation led in the following year to the immunoassay of endogenous plasma insulin in humans (YALOW and BERSON 1960). Since then, a number of investigators have successfully raised anti-insulin sera specific for the human hormone as well as for other animal species. Crystalline, lente, NPH, protamine, or regular pork insulin available for therapeutic purposes, and other preparations have been successfully used. However, the same authors (YALOW and BERSON 1971a) observed that crude preparations are not necessarily poor immunogens and on occasion may be superior to the purified antigen. Subsequently, studies performed by SCHLICHTKRULL et al. (1972) and ROOT et al. (1972) have shown that impurities of high molecular weight, which cocrystallize with insulin, may be largely responsible for antibody stimulation in rabbits as well as in diabetic patients, while highly purified insulin preparations are less immunogenic. This suggests that the inherent immunogenicity of this peptide may be increased if adsorbed on larger molecules, likely acting as a stimulant of the immune apparatus.

Immunogenicity of relatively small molecules may be enhanced by physical adsorption on carbon or latex microparticles or polyvinylpyrrolidone, or by conjugation to a larger carrier protein. Immunogenicity of human gastrin appears to be increased by conjugation to bovine serum albumin (BSA), as judged by the percentage of treated animals yielding high affinity antibodies suitable for RIA (REHFELD et al. 1972). Similarly, the immunogenicity and the resultant antibody-binding capacity and affinity for glucagon are increased by chemical binding or physical adsorption on larger molecules. In 1965 ASSAN et al. first demonstrated that the adsorption of glucagon to polyvinylpyrrolidone prior to emulsification with adjuvant greatly increases antibody production in rabbits. In 1970 GREY et al. reported the production of high affinity antibodies by coupling glucagon to hemocyanin and CUATRECASAS and ILLIANO (1971) obtained high affinity antisera using a glucagon–poly (L-lysine) conjugate. HEDING (1972) obtained good antisera production by coupling glucagon to glucagon, thus increasing the molecular weight and offering a repeated antigenic sequence to the B-lymphocytes of immunized animals. Nevertheless, YALOW and BERSON (1970b) have obtained antibodies to unconjugated commercial glucagon in Freund's adjuvant, which were at least as satisfactory as those obtained with crystalline glucagon conjugated to BSA or adsorbed to a silicate in suspension. Elsewhere, these authors described an antiserum useful at a dilution of $1:10^6$ from a guinea pig immunized with commercial glucagon (BERSON and YALOW 1973).

A similar case exists for arginine vasopressin (AVP). Like oxytocin, this hormone is considered not to be sufficiently immunogenic per se and conjugation with a larger carrier protein is thought to be necessary to evoke antibody production. In 1966 ROTH et al. obtained antibodies to unconjugated AVP in rabbits, even thought this substance circulates in that species. A possible explanation has been suggested by R. S. Yalow. The protein impurities of the crude extracts which were used in early RIAs could also serve as the larger protein carrier to which the small peptide can be coupled (STRAUS and YALOW 1977). Moreover, some denaturated hormone present in the standard injected into animals may improve immunogenicity by increasing heterogeneity to the immunized animal species.

Adrenocorticotropin (ACTH) is a peptide hormone of 39 amino acid residues. A 24 amino acid peptide (sequence 1–24) retains the activity of the parent hormone and the eicosapeptide (sequence 1–20) is also fully active. The 1–39 sequence is sufficiently immunogenic by itself to stimulate antibody production. In 1968 BERSON and YALOW reported production of ACTH antibodies by subcutaneous injection into guinea pigs with porcine ACTH emulsified in complete Freund's adjuvant. ACTH antibodies were detectable in almost all guinea-pigs after two or three injections. Other authors followed the same procedure and succeeded in producing anti-ACTH antibodies suitable for RIA of plasma ACTH (DEMURA et al. 1966; LANDON and GREENWOOD 1968; MATSUKURA et al. 1971). However, 1–24 ACTH or other smaller fragments derived from the parent hormone may be coupled to albumin, thyroglobulin, or other protein carriers and the conjugate successfully used for immunization (ORTH 1979). The aminoacid sequence of ACTH contains several antigenic determinants; thus, a heterogeneous population of antibodies directed toward single portions of the 1–24 sequence or other fragments of the molecule may be raised. By using single fragments of ACTH molecule for immunization, it is theoretically possible to induce antibody formation with specificity directed against a preselected sequence of the peptide chain.

Despite chemical structures similar to that of ACTH, α- and β-endorphin (amino acid sequences with opioid activity derived from β-LPH) are generally considered weak immunogenic substances and conjugation with protein carriers is required to obtain good antisera production. Antibodies specific for α- and β-endorphins have been described by several authors (GUILLEMIN et al. 1977; WEBER et al. 1982).

Calmodulin is a single polypeptide consisting of 148 amino acids, with a molecular weight of 16700. Although satisfactory immunogenicity could be expected on the basis of the relatively long amino acid sequence (compared with other polypeptide hormones successfully employed for antisera production), only monospecific antibodies suitable for immunofluorescence have been produced by immunizing animals with purified preparations emulsified in Freund's adjuvant

(ANDERSEN et al. 1978; DEDMAN et al. 1978), while repeated efforts failed to obtain antisera with titers high enough to enable development of RIA, even when calmodulin was injected in a coupled form or polymerized. It should be noted that calmodulin has been detected in most eukaryotic cells examined and the amino acid sequence is highly conserved throughout vertebrate and invertebrate species. In other words, it lacks tissue and species specificity, an attribute which is required by a strong immunogen. By contrast, good results have been obtained with dinitrophenylated (WALLACE and CHEUNG 1979) or performic acid oxidized calmodulin (VAN ELDIK and WATTERSON 1981) as antigen. The incorporation of several dinitrophenyl groups or performic acid oxidation renders calmodulin highly immunogenic in experimental animals by slightly modifying its chemical structure. For this polypeptide, the major immunoreactive site is contained in a unique region of the molecule. VAN ELDIK and WATTERSON (1981) identified this site within an 18-residue region in the COOH terminal domain of calmodulin. It is likely that the major antigenic determinants of this region are not substantially modified by the complexing procedure. Immunogenic strength might be increased by modifications induced in other areas possibly required to activate B-lymphocytes.

C. Special Problems with Small Peptides

Very small peptide hormones are generally poor immunogens per se. They consist of small molecules, often immunologically indistinguishable among different animal species, and when injected undergo a rapid enzymatic destruction both in blood and in tissue. Antisera may be easily produced in experimental animals injected with a conjugate of the small peptide with larger protein carriers. Successful antibody production to bradykinin has been reported only when the nonapeptide was injected as a complex with a protein carrier or in the larger kininogen form (WEBSTER et al. 1970). Similarly, antisera to angiotensins were produced in rabbits by angiotensin conjugated by carbodiimide condensation to albumin (GOODFRIEND et al. 1964) or succinylated polylysine (STASON et al. 1967). RIA measurement of bombesin-like peptides was made possible by immunization of animals with bombesin B-14 or B-9 polypeptide conjugated to heterologous serum albumin (WALSH and HOLMQUIST 1976). Good results were obtained by EBEID et al. (1976) and FAHRENKRUG et al. (1974) with purified porcine VIP coupled to BSA with carbodiimide. On the other hand, antisera to gastric inhibitory polypeptide (GIP) have been raised in guinea pigs by multiple injections with both the conjugated and unconjugated polypeptide (KUIZO et al. 1974). Indeed, immunogenicity of heterologous GIP seems to be adequate to evoke antibody generation by itself.

In spite of their low molecular weights, vasopressin and oxytocin are antigenic in many species (rabbits, guinea pigs, chickens, pigs, and humans). Antibodies have been successfully produced by immunization of rabbits with commercial vasopressin, or natural or synthetic AVP and lysine vasopressin nonapeptides emulsified in Freund's adjuvant (ROTH et al. 1966; WU and ROCKEY 1969; BEARDWELL 1971), as well as with unconjugated oxytocin (GILLILAND and PROUT 1965; GLICK et al. 1968). However, the percentage of animals producing antisera and the titers of the antisera produced were relatively low. Better results seem to be obtained using the natural or synthetic hormones coupled to bovine or human serum albumin (BSA or HSA) or equine γ-globulin with carbodiimide or glutaraldehyde as the coupling agents (PERMUTT et al. 1966; MILLER and MOSES 1969; WU and ROCKEY 1969; OYAMA et al. 1971; BEARDWELL 1971), or adsorbed to carbon microparticles injected directly into lymph nodes (CHARD et al. 1970).

Recently, RIA systems have been developed for atrial natriuretic factor (ANF) and related polypeptides. ANFs are a group of peptides possessing a potent diuretic and natriuretic activity and producing vasorelaxation. A major bioactive 28 amino acid sequence has been identified in α human and rat atrial natriuretic polypeptide (αhANP and αrANP). A 25 amino acid sequence (rat ANF-IV) has been coupled to thyroglobulin by TANAKA et al. (1984) and used as antigen to raise antibody with specificity directed toward rat ANF. A smaller common COOH terminal fragment of αhANP and αrANP, i.e., $\alpha ANP_{(17-28)}$, has been conjugated to bovine thyroglobulin by NAKAO et al. (1984) to produce an antiserum which recognizes αhANP and αrANP equally. Finally, a synthetic 26 amino acid fragment of ANF has been employed by GUTKOWSKA et al. (1984) to develop a direct RIA of ANF. In any case, because αANP or ANF and related peptides are small molecules, it was necessary to conjugate the peptide covalently to a larger protein for immunization. The same procedure had been followed a few years before for production of antibodies directed toward enkephalins. As these compounds are opiate-like pentapeptides, they should be considered true haptens rather than low immunogenic substances and, therefore, injected in a coupled form.

In summary, small peptides with 9–20 amino acid residues in the chain may be immunogenic per se in heterologous and, to a lesser extent, in homologous species. Their ability to elicit specific antibody formation is increased by coupling to a larger carrier protein. As the specificity of antibodies directed against them is limited to a small amino acid sequence, fragments, rather than the whole hormone, may be successfully employed to increase the specificity of immune response in experimental animals. Single fragments containing less than ten amino acids in sequence may be coupled by the reactions described in the following section.

D. Haptens Covalently Coupled to Protein Carriers

I. Preparation of Derivatives Containing Reactive Groups

When preparing a hapten–protein conjugate for immunization, it is necessary to select the site of coupling: a reactive group (i.e., carboxyl, amino, or other activated group) is required in this position to form a strong covalent linkage with carrier proteins. Peptide bonds are among the most stable chemical bonds, and various approaches have been reported to form this type of attachment between substances of low molecular weight and proteins or polypeptides.

Substances which offer a suitable coupling site, represented by a carboxyl or an amino group, may be directly coupled to protein carriers by forming peptide bonds between these reactive groups and amino or carboxyl groups of protein molecules. Prostaglandins and thromboxane fulfill this requirement. They are fatty acids containing a carboxyl group in a not particularly interesting part of the molecule. This group can be easily coupled to a side amino group on the protein molecule, leaving other regions of the carboxyl side chain exposed. Keto and hydroxyl groups are present in the ring and in the C-15 position (also C-6 for 6-keto compounds): these parts of the molecule are more important for recognition by the antibodies, whereas the structures close to and at the coupling site are of minor importance. Therefore, when coupled to protein carriers, prostaglandins are attached at the COOH terminus and the ring structure is exposed on the surface of the hapten. This is a very favorable condition: ring structures have been proved to be strong antigenic determinants and the coupling site should be as far as possible from these structures.

Hydroxyl and keto groups of small molecules can be used as anchoring points to prepare derivatives containing reactive groups. Steroid derivatives provided with free carboxyl groups suitable for conjugation with the amino groups of proteins have been prepared since 1957 (ERLANGER et al. 1957). Table 2 reports some steroid derivatives employed to form peptide bonds with several protein carriers.

Table 2. Some steroid derivatives suitable for coupling to protein carriers

Antigen types	Reference	Protein carrier[a]
Cortisone 21-hemisuccinate	ERLANGER et al. (1957)	BSA
	LIEBERMAN et al. (1959)	BSA
	BEISER et al. (1959)	BSA
	SRIVASTAVA et al. (1973)	
	ABRAHAM (1974a)	HSA
11-Desoxycorticosterone		
21-Hemisuccinate	LIEBERMAN et al. (1959)	BSA
	ERLANGER et al. (1959)	BSA
	BEISER et al. (1959)	BSA
	ABRAHAM (1974a)	HSA
3-(*O*-Carboxymethyl)oxime	ARNOLD and JAMES (1971)	BSA
Cortisone 21-hemisuccinate	UNDERWOOD and WILLIAMS (1972)	BSA
Cortisol		
21-Hemisuccinate	RUDER et al. (1972)	BSA
3-(*O*-Carboxymethyl)oxime	DASH et al. (1975)	BSA
	FAHMY et al. (1975)	BSA
3-(*O*-Carboxymethyl)oxime 21-acetate	COOK et al. (1973)	
6α/6β-Hemisuccinoxy	NISHINA et al. (1974)	BSA
Aldosterone		
18,21-Dihemisuccinate	HANING et al. (1972)	BSA
3-(*O*-Carboxymethyl)oxime	MAYES et al. (1970)	BSA
	ITO et al. (1972)	BSA
	VETTER et al. (1974)	RSA-BGG
3-(*O*-Carboxymethyl)oxime 18,21-diacetate	BAYARD et al. (1970)	RSA
	MIGEON et al. (1972)	BSA
3-(*O*-Carboxymethyl)oxime γ-lactone	FARMER et al. (1972)	BSA
Pregnenolone 20-(*O*-carboxymethyl)oxime	ERLANGER et al. (1959)	BSA

Table 2 (Coninued)

Antigen types	Reference	Protein carrier[a]
Progesterone		
20-(*O*-Carboxymethyl)oxime	ERLANGER et al. (1959)	BSA
3-(*O*-Carboxymethyl)oxime	FURUYAMA and NUGENT (1971)	BSA
11α-Hydroxyhemisuccinate	ANDERSON et al. (1964)	BSA
	LINDNER et al. (1972)	BSA
	ABRAHAM et al. (1975)	HSA
6-(Carboxymethylene)thioether	LINDNER et al. (1972)	BSA
17α-Hydroxyprogesterone		
3-(*O*-Carboxymethyl)oxime	YOUSSEFNEJADIAN et al. (1972)	BSA
	ABRAHAM et al. (1975)	HSA
20α-Hydroxyprogesterone		
3-(*O*-Carboxymethyl)oxime	ABRAHAM et al. (1975)	HSA
17,21-Dihydroxyprogesterone		
21-Hemisuccinate	ABRAHAM et al. (1971)	HSA
Dehydroepiandrosterone		
17-(*O*-Carboxymethyl)oxime	NIESCHLAG et al. (1972)	BSA
	SEKIHARA et al. (1972)	BSA
17β-Glucuronide	KELLIE et al. (1972)	BSA
3β-Succinate	ABRAHAM et al. (1974a)	HSA
Testosterone		
17-Chlorocarbonate	LIEBERMAN et al. (1959)	BSA
17-Hemisuccinate	BEISER et al. (1959)	BSA
	ABRAHAM et al. (1974)	HSA
3-(*O*-Carboxymethyl)oxime	ERLANGER et al. (1957)	BSA
	LIEBERMAN et al. (1959)	BSA
	BEISER et al. (1959)	BSA
	FURUYAMA et al. (1970)	BSA
	CHEN et al. (1971)	BSA
	ABRAHAM (1974a)	HSA
7α-Carboxymethylthioether	WEINSTEIN et al. (1972)	BSA
7α-Carboxyethylthioether	WEINSTEIN et al. (1972)	BSA
17β-Glucuronide	KELLIE et al. (1972)	BSA
Estrone		
3-Hemisuccinate	ABRAHAM (1974a)	BSA
3-(*O*-Carboxymethyl)ether	RAO and MOORE (1977)	BSA
Estradiol		
17β-Hemisuccinate	FERIN et al. (1968)	BSA
	ABRAHAM (1974a)	BSA
3-Succinate	ABRAHAM (1974a)	BSA
6-(*O*-Carboxymethyl)oxime	DEAN et al. (1971)	BSA
	LINDNER et al. (1972)	BSA
	JURJENS et al. (1975)	BSA
3-(*O*-Carboxymethyl)ether	RAO and MOORE (1977)	BSA
17β-Glucuronide	KELLIE et al. (1972)	BSA
Estriol		
3-Succinate	ABRAHAM (1974a)	HSA
3-(*O*-Carboxymethyl)ether	RAO and MOORE (1977)	BSA
3-(*O*-Carboxymethyl)oxime	LINDNER et al. (1972)	BSA

[a] BSA, bovine serum albumin; HSA, human serum albumin; RSA, rabbit serum albumin; BGG, bovine gamma-globulin.

The chemical reaction required to obtain the steroid derivative depends upon whether hydroxyl or keto groups of the steroid are used as anchoring points. Selective esterification of a hydroxyl group with succinic anhydride is among the most commonly used methods. The degree of reactivity of hydroxyl groups depends on the nature of the alcohol involved, being maximal for phenolic hydroxyl and progressively decreasing for primary, secondary, and tertiary alcohols. If the steroid hormone contains only one hydroxyl group, succination may be achieved by incubating the steroid with succinic anhydride in pyridine. However, tertiary alcohols are not esterified under these conditions: an acid-catalyzed reaction using *p*-toluenesulfonic acid instead of pyridine is required (ABRAHAM and GROVER 1971). (*O*-Carboxymethyl)oxime derivatives may be prepared with steroids possessing keto groups. The reaction of a keto group with (*O*-carboxymethyl)hydroxylamine yields the corresponding oxime. This reaction is usually base catalyzed and was reported for preparation of steroid conjugates with BSA by ERLANGER et al. (1957). These authors described testosterone-3-(*O*-carboxymethyl)oxime preparation by carrying out the reaction in ethanol under alkaline conditions. In the following years, other investigators prepared the same derivatives of a large number of steroids (Table 2). This chemical reaction may be carried out by refluxing the steroid with *O*-(carboxymethyl)hydroxylamine in pyridine or alkaline ethanol. Alkaline conditions are required to activate the amino group of aminoxyacetic acid. Finally, a third procedure has been used, i.e., the conversion of a hydroxyl group to a chlorocarbonate by reacting the steroid with phosgene. All these procedures have been described in detail by several authors (for review see ABRAHAM and GROVER 1971). Generally, these reactions require chemical expertise. Presently, oxime as well as hemisuccinate derivatives suitable for conjugation are commercially available from several chemical companies for nearly all the steroids.

Since then, antisera against 25-hydroxy- and 1,25-dihydroxycholecalciferol have been raised by immunizing animals with a conjugate of BSA and the 3-hemisuccinate derivatives of the parent compounds (BOUILLON et al. 1980, 1984; GRAY et al. 1981). 25-Hydroxyvitamin-D_3-3-hemisuccinate was prepared by 14 days incubation with succinic anhydride in pyridine (BOUILLON et al. 1984); 1-α,25-dihydroxyvitamin-D_3-3-hemisuccinate was prepared from 1-α,25-dihydroxy-7-dehydrocholesterol, by incubating the latter compound with 2,2,2-trichloroethylsuccinyl chloride in pyridine and subsequent passage through the 3-2′,2′,2′-trichloroethylsuccinate derivatives of 1-α,25-dihydroxyprecholecalciferol and 1-α,25-dihydroxycholecalciferol (BOUILLON et al. 1980). In any case, the anchoring point of the vitamin D_3 molecule was represented by the 3 position.

A similar approach was followed for cyclic nucleotides. In order to increase the possibility of obtaining specific antibody production, neither the purine rings nor the diester 3′,5′-bonds of cAMP or cGMP should be altered when conjugating with protein carriers. Consequently, the cyclic nucleotides have been succinylated at the 2′-*O*-position to yield 2′-*O*-succinyl cAMP (FALBRIARD et al. 1967) and 2′-*O*-succinyl cGMP (STEINER et al. 1972), respectively, and the free carboxyl groups of these derivatives have been conjugated to protein. Presently, the method for preparing 2′-*O*-succinyl nucleotides consists in reacting cAMP or cGMP with succinic anhydride in aqueous solution in the presence of triethylamine. Details of the reaction have been described by CAILLA et al. (1973, 1976) and HARPER and BROOKER (1975).

Raising antibodies suitable for RIA measurement of pharmacologic levels of different drugs in human sera has been achieved on several occasions by introducing reactive groups into the molecule. Synthesis of 3-*O*-succinyldigitoxigenin was described by YAMADA (1960) and used to develop a specific RIA of digitoxin by OLIVER et al. (1968). The method employs reaction with succinic anhydride in pyridine and allows the coupling of digitalis to a protein carrier for immunization. In the following years, very sensitive RIAs have been developed for a number of drugs. Antisera directed against the major classes of pharmacologic substances have been obtained by immunizing experimental animals with drugs or drug derivatives coupled to protein carriers (for review see BUTLER 1977). Drugs which possess reactive groups can be covalently linked to proteins by the usual methods (see Sect. D.II). Alternatively, acidic or amino derivatives can be employed. For example, antisera which react with methadone and with urine from users of this drug have been prepared by immunizing animals with protein conjugates of 4-dimethylamino-2,2-diphenylvaleric acid (chemically very similar to methadone, but containing a carboxyl group; LIU and ADLER 1973), or D- and L-methadolhemisuccinate (BARTOS et al. 1978). To yield the hemisuccinate derivative of methadone, reaction with succinic anhydride was carried out in dioxane (BARTOS et al. 1977). A similar approach led to RIA measurement of different groups of anabolics, i.e., nortestosterone and related steroids. The 17-carboxymethyloximes and 3-hemisuccinates of nortestosterone (HAMPL et al. 1979) as well as 19-norandrosterone and 19-noretiocholanolone (HAMPL et al. 1982) have been employed for conjugation with protein carriers.

RAHMANI et al. (1982) reported antisera produced against 5′-noranhydrovinblastine (a hemisynthetic derivative of vinblastine) by converting the parent compound into the monohydrazide derivative and, subsequently, into a reactive acid azide, which, in turn, was used for preparation of immunogen by reaction with BSA and glycyltyrosine.

Similar approaches have been that tempted for many other compounds. Production of antibodies directed against atropine was made possible by introducing a carboxyl group into the atropine (D,L-hyoscyamine) or L-hyoscyamine molecule and conjugating this product to BSA or HSA (WURZBURGER et al. 1977; VIRTANEN et al. 1980). Introduction of a free carboxyl group in this case was achieved by coupling these substances to diazotized *p*-aminobenzoic acid.

Oligopeptides having neither a tyrosine, nor histidine, nor free primary amine may present some difficulties to raise antisera. First, these substances are often difficult to conjugate by themselves; moreover, coupling through the COOH or NH_2 terminus directly to proteins generally results in attaching the peptide close to the carrier molecule so that little of the molecule remains exposed for immunologic recognition. In many instances, small oligopeptides possess only a single amino acid residue which can be easily employed for conjugation to a larger carrier. But even in this case, it may not be convenient to use this residue for conjugation if its location within the peptide chain may lead to an excessively tightened molecular configuration of the immunogen. The attachment of a prosthetic group to peptides is a technique which has been widely used to circumvent these difficulties. Activated aromatic amino acids, such as tyrosine or histidine, or short chain fatty acids which bear an activated phenyl group, such as *p*-hydroxyphenylacetic acid and *p*-hydroxyphenylpropionic acid, may be coupled to the NH_2 terminal amine of oligopeptides, thus introducing an activated arylamine, which, in turn, can easily be coupled to a protein carrier (YOUNGBLOOD and KIZER 1983). This procedure has been used for producing antisera directed against neuropeptides, such as the tripeptide prolylleucylglycinamide (MIF-I), which is difficult to couple through its NH_2 terminus because of the relative lack of reactivity of the secondary amine of proline (MANBERG et al. 1982). A similar approach to prepare NH_2 terminus extended peptides is represented by coupling a carboxybenzyloxy N-blocked ω-amino fatty acid to the C-blocked peptide, remove the N-blocking group by hydrogenation, and couple the amino fatty acid NH_2 terminus to a larger protein. For protein amides there is no need to block the COOH terminus before coupling. Both *N*-carbobenzyloxy-ω-amino-*n*-caproic and *N*-carbobenzyloxy-α-aminobutyric acids have been successfully used (YOUNGBLOOD and KIZER 1983). The result is represented by a more immunogenic conjugate, with the oligopeptide not closely bound to the protein carrier, but separated by a bridge of variable length.

II. Covalent Coupling to Protein Carriers

Several coupling methods are suitable for forming peptide bonds between carboxyl groups of haptens and amino groups of protein carriers, and vice versa.

1. Carbodiimide Method

In 1964 GOODFRIEND et al. first reported the use of water-soluble carbodiimides in immunology. These highly reactive compounds can couple substances containing several functional groups, including carboxylic acids, amines, alcohols, and thiols, with the formation of amides, esters, and so forth. Haptens containing a reactive carboxyl group can form peptide bonds with the NH_2 terminus or the side amino groups of protein carriers; but also the COOH terminus or the side carboxyl groups of proteins can react with amino groups of the hapten, if it possesses any. Bovine serum albumin, for instance, has 59 lysine residues, which provide sites sufficient in number to produce conjugates with carboxylic haptens, with highly specific antigenic properties. In any case, the presence of at least one reactive carboxyl or amino group in the hapten is required. Figure 1 illustrates such a procedure. The method is generally simple and rapid and can be performed at room temperature. As regards the choice of carbodiimide, many different compounds have been successfully employed by several investigators, and a very large number of recipes has been described which give excellent coupling reactions. Differences between recipes concern buffers, incubation volume, solution, purification of reactive compounds, stirring of solutions, nature and dimension of tube or vial for reaction, and so on. Moreover, differences among the various coupling procedures may be due to instability of the hapten, or to its peculiar characteristics (solubility, salt generation, denaturation, etc.), or to possible formation of other coupled products, either owing to low standard purity or chemical conver-

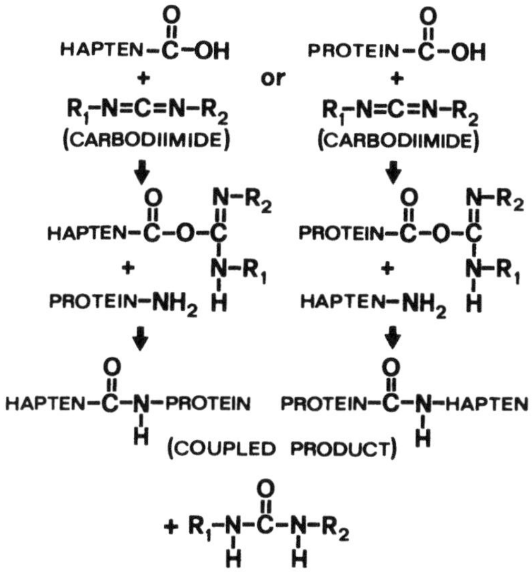

Fig. 1. Chemical reactions involved in the carbodiimide conjugation method

sion of hapten. Presently, it is impossible to include in one recipe all the detailed techniques that have been published. Investigators should check the nature of carbodiimide and incubation conditions against previously reported successful procedures. However, a general procedure may be summarized as follows. The total amount of protein carrier (such as bovine, human, or rat serum albumin, or other heterologous proteins) should be solubilized in water. Hapten, if soluble in water, may be dissolved in the same vial containing the protein solution. Otherwise, a separate solution in an appropriate solvent may be prepared. Relative amounts of protein and hapten depend on the desired molar ratio in the conjugate (see Sect. E.IV) and on the final yield of the coupling reaction.

Thus, if we consider a final conjugation of 10% added hapten, and if we wish a 10:1 molar ratio in the conjugate (hapten:protein), then a 100:1 initial molar ratio should be realized in the reacting mixture. Usually, a protein solution is first prepared. The vial or tube for chemical reaction should be carefully cleaned to avoid extraneous material which could interfere with the carbodiimide reaction. Polypropylene tubes are suitable for conjugation. pH adjustment is generally assumed to be important; a weak acid pH (5.0–5.5) should be obtained if the hapten is represented by an organic acid, while neutral pH (6.8–7.0) has usually been specified for peptide substances. With constant stirring, carbodiimide is added to the solution. Its amount is generally in molar excess. The reaction mixture, if compounds are completely dissolved, remains clear throughout the conjugation procedure. However, formation of visible precipitates or colloidal suspensions may occur. This may be caused by changes in the solubility of protein when substituents are added, or may be due to macroaggregate generation. Continuous stirring helps to prevent the latter event. However, these phenomena do not always follow the formation of conjugates. Reaction should be permitted to proceed at room temperature until a satisfactory yield is obtained. A longer time might cause chemical changes of the hapten molecule, or damage the protein. Usually, 24 h incubation is required with low carbodiimide concentrations and chemically stable haptens containing a single carboxyl group. Shorter incubations are usually employed for peptide hormones or substances which offer more than one possible reactive group. Removal of excess coupling reagent and estimation of yield are accomplished by gel filtration on Sephadex G-50 fine resin or by dialysis against water for 24 h. The contents of the dialysis bag or the column eluates are employed for later injection into animals (OLIVER et al. 1968; LEVINE and VAN VUNAKIS 1970; JAFFE et al. 1971; LEVINE and GUTIERREZ-CERNOSEK 1972). Estimation of yield may be performed by adding a known amount of radioactively labeled hapten to the reaction mixture, and counting a small fraction of the purified conjugate. This also helps to identify the presence of conjugate in column eluates. Otherwise, estimation of yield is based on the assumption that the content of a particular amino acid residue in the protein will increase in proportion to the amount of peptide hormone coupled (ORTH et al. 1979), or the hapten: carrier molar ratio may be calculated by UV analysis (see Sect. E.IV).

2. Mixed Anhydride Method

Since the original report of ERLANGER et al. (1957) this represents the most commonly used method of conjugating steroid derivatives to proteins. The oxime or hemisuccinate of the steroid is reacted in dioxane and tri-*n*-butylamine-containing medium with isobutylchlorocarbonate (also termed isobutylchloroformate) to form the mixed anhydride of the steroid; the mixed anhydride will then react with free side amino groups present on the protein molecule (LIEBERMAN et al. 1959). Figure 2 illustrates this chemical reaction. The same method has been successfully employed for several other products containing single carboxyl groups, including prostanoids, drugs, vitamin D, and other substances (OLIVER et al. 1968; JAFFE et al. 1971; KIRTON et al. 1972; BOUILLON et al. 1980, 1984).

Prostaglandins are 20-carbon, unsaturated fatty acids derived from arachidonic acid. Coupling with heterologous proteins may be achieved by dissolving

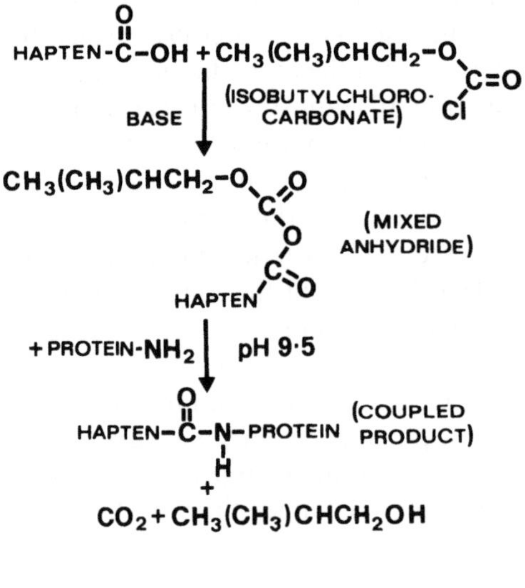

Fig. 2. Chemical reactions involved in the mixed anhydride coupling procedure

them in dioxane containing tri-*n*-butylamine and reacting with isobutylchlorocarbonate for 20 min at 4 °C; thereafter the mixture is added to water–dioxane solution containing the carrier protein. The mixture is stirred for 4 h at 4 °C, then dialyzed against water or PBS. A similar procedure has been reported for conjugating leukotriene B_4 (SALMON et al. 1982) and (glycine)monomethylester of *N*-trifluoroacetyl leukotriene D_4 (LEVINE et al. 1981) to BSA. Note in the latter example the conversion of the dimethylester of *N*-trifluoroacetyl leukotriene D_4 in the corresponding C-1 monoacid (eicosanoid carboxyl-free) to mask the glycine carboxyl group (remaining as methylester); this procedure is required to select the carboxyl group reacting with protein (LEVINE et al. 1981). The use of the eicosanoid carboxyl group for coupling to protein carrier permits the exposure of most of the carbon chain of the molecule and, particularly, of the sulfidopeptide group (glutathione for leukotriene C_4).

A molar excess of hapten to protein of 40–80 to 1 is generally employed in this coupling reaction to obtain a satisfactory level of incorporation in the conjugate. The molar ratio after coupling is in the range 7:1–10:1 hapten to protein; the final yield should be evaluated by adding a known amount of labeled hapten to the reaction mixture and calculating the bound radioactivity, or through analysis of the coupled product by UV absorbance after correction of protein absorbance at 280 nm (if BSA is employed).

3. Glutaraldehyde Method

This consists of an intermolecular cross-link, represented by five carbon atoms in a chain between two nitrogen atoms (RICHARDS and KNOWLES 1968). The coupling procedure may be realized by adding a glutaraldehyde solution dropwise to a medium containing hapten and protein in solution. Reaction with α,ω-dialdehydes, especially glutaraldehyde, has been widely employed for the intermolecular cross-linking of crystalline enzymes or to prepare protein polymers. A prerequisite for this particular conjugation procedure is the presence of a reactive amino group in the hapten molecule: it is required to allow formation of a bridge between its nitrogen atom and the one present in a homologous side amino group on the protein molecule. This reaction has been successfully employed to conjugate peptide hormones (REICHLIN et al. 1968) as well as substances containing a single reactive amino group, i.e., sulfidopeptide leukotrienes (AEHRINGHAUS et al. 1982). The coupling procedure in the latter example involves a reaction on the free amino group of the glutamyl residue of leukotriene C_4, thus leaving the most important parts of the hapten molecule (including the eicosanoid carboxyl group) unchanged.

Possible polymerization of the protein carrier in this coupling procedure should be considered. Reaction of antisera produced in this way with BSA polymers has been described (REICHLIN et al. 1968).

4. Schotten–Baumann Method

This has been commonly employed to conjugate a steroid to a protein when the chlorocarbonate derivative has been formed (LIEBERMANN et al. 1959). The Schotten–Baumann reaction yields amides when amines or amino acids are treated with an acid chloride. Alkaline conditions (pH 9.5) are required to permit the chlorocarbonate derivative to react with a side amino group of the protein: the product is a carbamate, in which the steroid is joined to the protein via an amide bond. However, this coupling procedure has not been extensively employed, because of its limited application to a few compounds (see ERLANGER et al. 1957 for further-details).

5. Periodate Oxidation Method

This is based on a technique originally described by ERLANGER and BEISER (1964) for conjugation of ribonucleosides and ribonucleotides to the ε-amino groups of lysine in a protein carrier. In this reaction, the ribonucleoside or ribonucleotide is first oxidized with sodium periodate at room temperature. In this way the ribose ring is opened, thus leaving two groups which can react with amino groups supplied by the lysine residues. After destruction of the excess periodate by the addition of ethylene glycol, the reaction product is coupled to protein at alkaline pH. Subsequent reduction with $NaBH_4$ stabilizes the linkage between the haptenic group and protein. This reaction is not restricted to ribonucleosides or ribonucleotides, but is applicable to the synthesis of protein conjugates of any glycoside possessing neighboring hydroxyl groups. This method has been also employed to conjugate digoxin to BSA (BUTLER and CHEN 1967; SMITH et al. 1970). In a first stage, digoxin, which consists of a steroidal aglycone linked to three digitoxose residues, is oxidized by sodium periodate. The digitoxose ring is opened and in a second stage reacted with NH_2 supplied by lysine and NH_2 terminal residues of the protein. Reaction details are described by BUTLER and CHEN (1967).

6. Diisocyanate Method

This method was originally described by SCHICK and SINGER (1961) to form covalent linkages between two protein molecules. SPRAGG et al. (1966) reported the synthesis of bradykinin coupled to poly(L-lysine) or rabbit serum albumin via NH_2 terminal arginine. The reaction employs a bifunctional reagent of low molecular weight, i.e., toluene-2,4-diisocyanate, the functional groups of which react at alkaline pH primarily with amino groups to give stable covalent linkages. The coupling reaction is carried out in two stages: hapten is first reacted with toluene-2,4-diisocyanate at 0 °C under continuous stirring for 45 min. During this step, a stable ureido linkage is formed between the NH_2 terminal group of the hapten and the isocyanate group in position 4 (SPRAGG et al. 1966). Thereafter, the excess unreacted diisocyanate is removed; this is followed by the addition of protein carrier to the diisocyanate-treated hapten to yield a second stable ureido linkage between an amino group of the protein and the isocyanate group in position 2. The final coupled product is represented by a double ureido linkage of hapten and protein, the toluene ring acting as a spacer. For details of the chemical reaction and procedure see SCHICK and SINGER (1961) and SPRAGG et al. (1966).

7. N,N′-Carbonyldiimidazole Method

In this procedure *N,N′*-carbonyldiimidazole is employed as the coupling reagent. The final product is represented by a peptide bond between the carboxyl group present in the hapten molecule and a side amino group of the protein. The carboxyl group of eicosanoids represents a suitable reactive site for this conjugation procedure. For this purpose, the compound to be coupled should be dissolved in dry dimethylformamide and *N,N′*-carbonyldiimidazole added to the mixture. Stirring under nitrogen is continued for 20 min and the resulting solution is then added to an aqueous solution of protein. The reaction is carried out at room temperature for 5 h and the coupled product separated by extensive dialysis, first against a water: dimethylformamide mixture (3:2), then against running water (AXÉN 1974). The intermediate step is represented by a reactive imidazolide derivative. The reaction of this compound with amines is sufficiently fast to allow coupling to protein with water as cosolvent.

8. Conjugation with Diazonium Compounds

Diazonium functional groups react readily with the phenolic (tyrosyl), hystidyl, amino, and carboxyl groups of proteins. A variety of reactions employing the bifunctional bis-diazotized benzidine (BDB) or bis-diazotized 3,3′-dianisidine (BDD) have been reported for the preparation of various protein–erythrocyte conjugates (LIKHITE and SEHON 1967). BDB is formed by reaction of an aromatic amine with nitrous acid to give a diazonium salt. This can react in alkaline solution with the appropriate group in the protein molecule (i.e., α- or ε-amino groups, imidazole, phenol, or carboxyl groups), thus allowing the formation of a benzidene bridge interposed between two proteins. This method has been used for coupling small peptides, such as thyrotropin-releasing hormone (TRH) to BSA (BASSIRI and UTIGER 1971), gonadotropin-releasing hormone (GnRH), to keyhole limpet hemocyanin (KLH) (JENNES and STUMPF 1983; NETT et al. (1973), and luteotropin-releasing hormone (LRH), and TRH to thyroglobulin (BRYCE 1974).

9. Conjugations with Other Coupling Agents

1,5-Difluoro-2,4-dinitrobenzene (DFDNB) reacts quite specifically with amino groups, allowing replacement of the two fluorine atoms (positions 1 and 5 of the benzene ring) with two amino groups. The most likely candidates for the amino acid side chain of the protein which can react with DFDNB are the ε-amino group of lysine, the terminal amino group, the phenolic hydroxyl group of tyrosine, the sulfhydryl group of cysteine and, possibly, the imidazole group of hystidine (RIVA et al. 1977). On the other hand, haptens which possess a side amino group can react with DFDNB to form an intermediate monoamino substitute, which, in turn, may react with a side amino group of the protein, the second fluorine atom of DFDNB being replaced at a slower rate. Moreover, the characteristic UV absorption of the reagent and its mono- and diamino-substituted derivatives allows one to follow the course of the coupling procedure and to quantitate the final adducts by UV spectroscopy (YOUNG et al. 1982). The final result of this technique is represented by hapten–protein conjugate with a spacer group (dinitrobenzene) interposed (see YOUNG et al. 1982 for details of this chemical procedure).

Enzyme-coupled immunoassay of insulin using *N*-Maleimidobenzoyl-*N*-hydroxysuccinimide ester as coupling reagent has been reported (KITAGAWA and AIKAWA 1976). This reagent reacts through acylation of amino groups by the active ester and formation of thioether bonds by addition of a thiol group to the double bond of the maleimide moiety. It results in a sufficiently stable complex with protein.

6-*N*-Maleimidohexanoic acid chloride, prepared from 6-aminohexanoic acid, has been employed for conjugating leukotriene C_4 to protein through the free α-amino group of the glutamyl residue (YOUNG et al. 1982). Also, leukotriene B_4 converted into the hydrazide has been successfully coupled to KLH by this method (HAYES et al. 1983). 6-*N*-Maleimidohexanoic acid chloride reacts with the amino group of the hapten to form an amide, which, in turn, may be coupled with a thiolated protein. The final product is represented by a hapten–protein complex with a molar ratio of 7–10:1. Hapten molecules are bound to the thiol groups of the protein, 6-*N*-Maleimidohexanoic acid acting as a spacer between the amino and thiol groups (YOUNG et al. 1982). The thiolated protein may be prepared by reaction with *S*-acetylmercaptosuccinic anhydride (KLOTZ and HEINEY 1962).

Finally, conjugation of an oxy group present in the hapten with side amino groups of the protein may be realized by reaction with *p*-nitrophenylchloroformate. This chemical reaction has been described for the conjugation of leukotriene B_4 to BSA via the 12-oxy group of the lipid (LEWIS et al. 1982). The resulting immunogen was a urethane derivative in which linkage had been established between a side amino group of BSA and the 12-oxy group of leukotriene B_4 through a carboxyl group, as represented by the formula: leukotriene B_4-*O-CO-NH*-albumin (LEWIS et al. 1982). For this purpose, the eicosanoid carboxyl and the 5-oxy groups were protected by conversion of leukotriene B_4 to the 5-benzoate of leukotriene B_4 methylester. This compound was reacted with *p*-nitrophenylchloroformate to form the 12-*p*-nitrophenoxycarbonyl derivative, which was subsequently bound to the lysine amino group of the protein. The methylester and benzoate groups were selectively saponified after conjugation. A molar ratio of approximately 11:1 (hapten:protein) was reported as the final yield of conjugate.

E. Immune Response to Hapten–Protein Conjugates

I. Effect of Method and Duration of Immunization

Rabbits and guinea pigs are the most frequently used animals for the immunization process, although sheep and goats have been successfully used, especially for producing large amounts of antisera. Relatively exotic species have occasionally been used to raise antisera, for example, leghorn chickens (YOUNG et al. 1969). Neither cows nor horses are mentioned in the literature for RIA purposes. Among rabbits, New Zealand white rabbits seem to produce the highest antibody amounts. In the early work of Yalow and Berson, rabbit antisera generally reacted much less energetically with peptide hormones than did guinea pig antisera and were therefore less satisfactory for use in RIA of plasma hormones

(Yalow and Berson 1964, 1970b; Berson and Yalow 1967). For this reason, guinea pigs were extensively employed in the early days of RIA. Moreover, the total amount of immunogen required to immunize a single animal is lower for the guinea pig, thus allowing a larger number of animals to be immunized with a given amount of immunogen. In this way, the probability of obtaining a satisfactory immune response by at least one animal is increased. However, the use of guinea pigs for RIA purposes suffers from some limitations. First of all, bleeding of the guinea pig is performed by cardiac puncture; if this procedure is not performed with great experience and ability, animals may die while the titer of antisera is still increasing. Moreover, the small size of guinea pigs requires subcutaneous injection of antigen–adjuvant mixture in the thigh. This offers the advantage of ileal lymphnodes being involved in immune response, but precludes intradermal injection of very small doses at a large number of different sites, a technique which appears to be very effective when the amount of immunogen is very limited and, in any case, helps to stimulate the immune apparatus maximally (Vaitukaitis et al. 1971). In addition, the amount of serum which may be obtained by a single bleeding is not very high compared with rabbits and sheep, and, finally, guinea pigs frequently suffer from enteric diseases or lung parenchymal infections, which in a short time can kill several animals in neighboring cages.

Guinea pigs are immunized at intervals of 2–4 weeks, to a total of 3–6 doses (Yalow and Berson 1964). About 10–15 days after the second, third, and subsequent immunizations, approximately 10 ml whole blood is taken by cardiac puncture into a plastic syringe while the animal is under light ether or chloroform anesthesia. Heparin or sodium EDTA may be used to prevent blood clotting. The plasma is separated by centrifugation and stored frozen. Rabbits are immunized from 3 to 16 months, sheep up to 30 months. Booster injections are repeated every 2–4 weeks and bleedings may be performed 15 days after the third and subsequent immunization. Sheep and goats are very easy to bleed by puncturing the jugular vein and collecting blood through the Vacutainer system. In rabbits, 10–30 ml blood can easily be collected from the central artery of the ear. For this purpose vasodilation is obtained by rubbing the ear with cotton soaked in xylol. After a few minutes, arteries are easily visible and can be cannulated. Extensive washing after bleeding prevents damage of the ear. Animals producing satisfactory antisera may subsequently be given booster immunizations every few months and bled 10–15 days afterward. Some characteristics of the antiserum can improve by the prolongation of the immunization, such as titer or specificity, but within limits. After some months of immunization, a stabilization of titer may be observed, while specificity of antibodies may decrease with prolonged immunization. In this case, the optimal blood taking period is generally 10–20 days after the last booster injection. An example of immunization is shown in Fig. 3. Two New Zealand rabbits received three injections of thromboxane B_2–HSA conjugate emulsified in complete Freund's adjuvant at 3-week intervals. About 10 μg coupled throboxane B_2 was given to each animal by multiple intradermal injections into the back. The first immunization was performed when animals were still very young (about 2 months); 15 days after the third injection animals were bled twice and reimmunized with about one-quarter of the previous dose of immunogen. A third bleeding was performed at the same time as the booster injec-

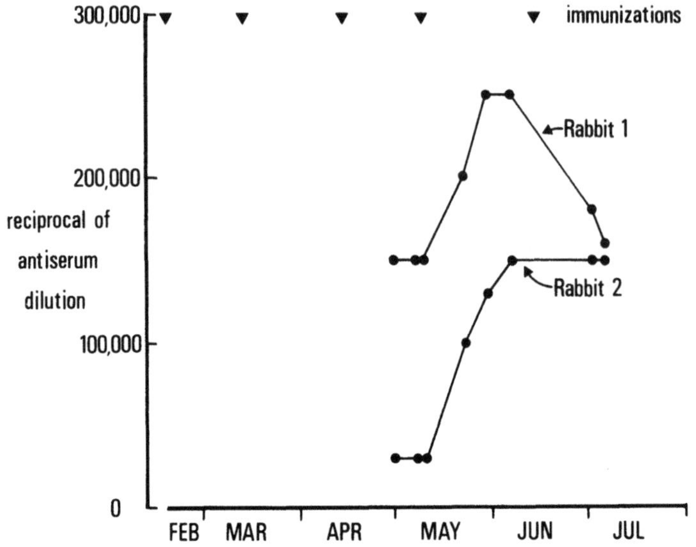

Fig. 3. Antibody response in two rabbits immunized with a thromboxane B_2–human serum albumin conjugate

tion. A sensible rise of titer was observed in both animals. After extensive bleeding (about 30–40 ml on two different occasions), animals were again immunized with a low dose of immunogen. No change of titer was observed in animal 2, while in animal 1 the titer significantly decreased.

What may happen after multiple injections of the same immunogen into an experimental animal and how may titer and specificity be influenced by the method and duration of immunization? In the induction of antibodies against hapten–carrier molecules, cooperation between B and T cells is required. After the first injection of a hapten–carrier conjugate, specific precursor lymphocytes, B cells as well as T cells, are activated. After the second injection of immunogen, antibody production by B cells requires the presence of carrier-specific activated T helper cells. These activated T helper lymphocytes are necessary in order to help B cells in the formation of an antibody, which is specific for the hapten–carrier complex. A secondary hapten response is absent if the second injection of antigen is performed with the hapten coupled to another carrier. For this reason, preferably only one coupling reaction should be performed (according to the chemical stability of the immunogen) with enough antigen to immunize the desired number of animals repeatedly. In this way, the structure of the hapten–carrier conjugate remains absolutely identical throughout the immunization. As for the type of antibody produced, the first injection of an antigen into an experimental animal leads to a stimulation of IgM-producing cells. Then, a large number of IgG-producing cells rise and a switch to IgG production may be observed. This step is not necessarily followed by an increase in the specificity of the antibodies, but only represents a shift of this specificity to a different immunoglobulin class (NOSSAL 1975). When hapten–carrier conjugates are used as antigens, specific T helper cells are required to switch antibody production from IgM to IgG.

At the start of the immune response, a wide variety of antibodies are present. However, during the course of immunization, the antibody production turns to those immunoglobulins with higher affinity. This is due to a selective proliferation of B-lymphocytes with a high affinity for the antigen when antigen concentration decreases. Memory cells are responsible for a quicker response of the IgG type after the second and following injections of antigen.

During immunization with hapten–carrier conjugates, the titer of antisera usually increases and reaches a plateau after a variable period ranging from 3–4 months to 6–8 months of immunization. A concomitant increase in affinity generally accompanies the titer increase, but affinity often does not reach a plateau an may still be rising after a longer period of immunization. As for the specificity, in most cases there is an increase during the first few months of immunization, followed by decreased specificity afterwards (see Sect. E.IV).

II. Role of Adjuvants in Antibody Production

A wide variety of organic as well as inorganic compounds are capable of potentiating immune responses by acting as adjuvants. These can be: (a) water and oil emulsions; (b) endotoxins; (c) lymphokines and monokines; (d) pharmacologic substances and hormones; and (e) synthetic polynucleotides (WEBB and WINKELSTEIN 1984). The most commonly used substance is represented by Freund's adjuvant, which consists of heat-killed cultures of *Mycobacterium tuberculosis,* dried and suspended in mineral oil and lanolin (Freund's complete adjuvant). Also, an incomplete form exists, which differs from the former in being free of bacteria. However, this Freund's incomplete adjuvant is a poor substitute for the complete form. When injected into experimental animals, antigens are usually emulsified with complete Freund's adjuvant. Its potentiating capacity appears to be due, at least in part, to a delayed antigen absorption, which, in turn, provides a prolonged antigenic stimulus of the immune system. However, this is not the only mode of action, since deprivation of bacterial components strongly reduces immunopotentiation of Freund's adjuvant. It is likely that some components of the bacterium itself may be potent adjuvants per se. For practical use, the emulsion with Freund's adjuvant should be the water in oil type, the water phase preferably one-third to one-fifth of the emulsion (v/v). The emulsion is prepared by repeated aspiration of the mixture into a syringe or by joining one syringe containing the aqueous phase to another syringe containing the adjuvant by means of a double-hub connector and then rapidly passing the contents to and from many times. Equipment specially designed for the preparation of emulsions also exists.

A variety of inorganic compounds, such as aluminum salts and calcium phosphate, have also been used as adjuvants. Presumably, their potentiating capacity is related to a slower release of antigen at the site of injection and to an inflammatory effect which intensifies the host reaction to antigen (WEBB and WINKELSTEIN 1984).

Finally, cellular as well as humoral responses to a wide variety of antigens may be enhanced by simultaneous injection of bacterial endotoxins. *Bordetella pertussis* vaccine injected in multiple intradermal sites in a different lymphatic drainage area from the primary immunization is among the most commonly used substances. Also, dried *Mycobacterium butyricum* may be employed to enrich Freund's adjuvant at a concentration of 5–10 mg/ml. Bacterial endotoxins strongly affect B cells and macrophages, but may also activate some T cells. A significant augmentation of antibody production to hapten–carrier conjugates with adjuvants has been clearly demonstrated (HAMAOKA and KATZ 1973).

III. Effect of Carrier Protein on Characteristics of Antisera

The assertion that the larger and more immunogenic the carrier substance, the higher is the success rate and the antibody titer is widely accepted, but requires some critical consideration. Hapten–carrier conjugates stimulate antibody production against the hapten, the carrier, and the site of junction. When conjugates are prepared with synthetic polymers, such as polyglutamic acid or polylysine, the lack of molecular complexity requires a further coupling, for example, to KLH or thyroglobulin prior to injection into animals. This technique has been successfully used to produce antisera against prostaglandin–polylysine conjugates (LEVINE et al. 1971; ATTALLAH and LEE 1973). In some cases, to improve immunogenicity the primary conjugate was absorbed on a second carrier, such as pneumococcus strain cells (STYLOS and RIVETZ 1972; RAZ and STYLOS 1973). Moreover, when comparing the ability of various immunogens to elicit formation of antibodies toward intermediate-sized peptides, such as neurotensin, it clearly appears that hemocyanin and thyroglobulin complexes give a higher success rate than the neurotensin–polyglutamic acid or neurotensin–polyglutamic acid–lysine complexes (CARRAWAY and LEEMAN 1976). The higher molecular weight of hemocyanin (about 2×10^6) and its higher complexity compared with synthetic polymers can easily explain the stronger immunogenicity of its conjugates. The problem, however, appears to be more complex if we consider that: (a) proteins of lower molecular weight and perhaps of lower complexity than hemocyanin and thyroglobulin have been successfully used as carriers; and (b) homologous instead of heterologous proteins may be used for conjugation. As for point (a), bovine serum albumin and human serum albumin are among the most commonly used carriers for small molecules and excellent antisera have been obtained from conjugates with these proteins by a number of investigators in all the fields explored by RIA. As for point (b), a hapten–carrier conjugate with a homologous carrier is capable of inducing antibodies to the hapten or to new antigenic determinants unmasked by the coupling reaction, but not against the carrier, because it is not immunogenic in the same species. However, in the homologous system, the temporal course of the antibody production is delayed compared with a heterologous carrier, and antibody titer reaches a plateau in a short time (RUBIN 1972; RUBIN et al. 1973). In any case, antibodies with a specificity to hapten will be raised. It is well documented that prior immunity against the carrier enhances the anti-hapten response to a subsequent challenge with the hapten–protein complex. This may be explained by activation of T helper cells, which co-operate in raising specific antibodies to the hapten. A reduced or very limited number of T helper cells are activated in the case of a homologous carrier, and this may explain the delayed temporal course and the weaker antibody production against a homologous hapten–carrier molecule.

Thus, the main role of the carrier molecule appears to consist in graduating the intensity of immune response, rather than the antibody specificity. Although different results have been reported by several authors with a wide number of carrier molecules, no convincing evidence has been so far provided to demonstrate a selective effect of different carrier proteins on anti-haptenic antibody specificity. The frequently reported differences in the titer, affinity, and specificity of antisera produced for RIA purposes may be explained on the basis of the different site of hapten attachment to the carrier, the hapten: carrier molar ratio, the portion of hapten molecule which is presented to recognition cells, and the mode of its presentation, rather than the nature of carrier protein employed. As a general rule, we may summarize the following points: (a) a heterologous carrier is generally preferable to a homologous one, because of the enhanced antibody production to hapten; (b) a carrier with satisfactory chemical complexity should be preferred, however, it is not necessary to employ highly sophisticated molecules of high molecular weight; (c) conjugates with more than one carrier should be used if no experience has yet been obtained with a particular substance; and (d) in the author's experience, serum albumin of heterologous species represents a suitable carrier, inexpensive and easily coupled, for many substances of molecular weight below 1000.

IV. Effect of Hapten: Protein Molar Ratio on Characteristics of Antisera

Estimation of the hapten: carrier molar ratio may be calculated by including a small, known amount of the labeled antigen when preparing the conjugate, or by UV analysis of the conjugate product. In the latter procedure, direct UV analysis of the purified immunogen reveals an absorption spectrum with one or more peaks characteristic for each substance. The light absorption of a hapten is generally little modified by attachment to a protein carrier. Assuming a given molar extinction coefficient ε corresponding to the bound hapten and subtracting the protein absorption, it is possible to calculate the number of moles of hapten per mole of protein carrier from the formula $A = \varepsilon C$, where A = absorbance, and C = concentration.

The molar ratio of a carboxylic hapten to a protein such as BSA has a theoretical maximum of 59, there being 59 free amino groups in the BSA molecule. This ratio, however, is never reached. Molar ratios generally ranging from 1 to 25 have been reported by a number of investigators using different coupling procedures, although higher molar ratios have occasionally been described.

The molar ratio, or the degree of substitution on the protein carrier appears to be important for the production of antisera directed against some substances, such as steroids. An important effect of hapten density on carrier proteins has been described for the production of antisera specific to aldosterone (VETTER et al. 1974) or some other steroids (NISWENDER and MIDGLEY 1970). With steroid–protein conjugates, a molar ratio of 18 was recommended (MIDGLEY et al. 1971). High titer antisera were obtained with albumin conjugates carrying more than 20 steroid molecules (KUSS et al. 1973). When the steroid: protein molar ratio was less than 10, the antibody titer was low (ABRAHAM 1974b). Also, synthetic antigenic determinants, such as dinitrophenyl groups, show an optimal immunogenicity at a given degree of substitution on the carrier molecule (QUIJADA et al. 1974).

On the other hand, the molar ratio in the conjugate does not seem to be as important for the production of prostaglandin antisera as for steroids (GRANSTÖM and KINDAHL 1978). All conjugates which have been reported with various degrees of substitution in the conjugate have given rise to good antisera. In this field, a hapten:protein molar ratio of approximately 10:1 is considered satisfactory by most investigators. Occasionally, very high degrees of substitution have been reported (DRAY et al. 1972).

It is not a general rule that the larger the degree of incorporation of the hapten into the carrier, the more pronounced is the antibody response. A maximal incorporation may be disadvantageous when conjugating small peptides or complex molecules with an appropriate number of reactive groups. Overcrowding the surface of the carrier may result in the formation of antigenic determinants made from parts of several hapten molecules. Moreover, antigenic groups may be sterically masked or hindered. The possibility of raising antisera with low specificity should be considered. In such a case, inserting a suitable spacer between hapten and carrier and allowing a high, but less than maximal degree of substitution to occur may help in increasing specificity of antisera.

Finally, it should be considered that highly substituted conjugates which possess a high number of antigenic determinants may stimulate B-lymphocytes with low affinity receptors and raise low affinity antibodies. Therefore, it seems reasonable to consider the size of the hapten and the surface area available on the carrier protein and not exceed a substitution of 20% of carrier weight (CARRAWAY 1979). On some occasions, this means a low hapten:protein molar ratio, depending on the size of the protein carrier.

V. Effect of Site of Hapten Linkage to Protein on Characteristics of Antisera

There is general agreement among investigators that antibodies directed against haptens are more specific for the position of the molecule protruding out of the carrier protein and less specific for the position of the hapten used for linkage to protein. In general, the more distant a certain structure is from the coupling site, the better it will be recognized by the antibodies. The hydroxyl and keto groups of biologically active compounds, such as steroids, eicosanoids, drugs, cyclic nucleotides, vitamins, and peptides confer biologic specificity to these molecules, but are also easily recognized by the antibodies. A few examples will explain this concept.

With regard to steroids, attachment to the protein through ring *A, B, C,* or *D* seems to confer a different specificity to the antibodies elicited, with substantial differences from steroid to steroid. Antibodies elicited by immunizing with a steroid hormone attached to a protein through one of its preexisting functional groups usually lack specificity toward the group used for the coupling reaction and the region of the hapten in its immediate vicinity. For example, if carbon C-3 of ring *A* of a Δ^4-3-keto C_{19} steroid is used as a site of attachment to the carrier protein, antibodies elicited against this conjugate will be most specific for ring *D* and least specific for ring *A* (ABRAHAM 1974b). When 17-β-estradiol, estriol, and progesterone molecules are attached to the carrier protein at carbon C-6 of ring *B* and C-11 of ring *C*, the specificity of antibodies produced against the 11-conjugate is greater for some portions of the molecule, such as ring *A* and the C-17 side chain (LINDNER et al. 1972). A comparison between different conjugates of Δ^4-3-oxosteroids bound to protein through several positions of the steroid molecule reveals a greater specificity of antibodies directed against C-7 conjugates (WEINSTEIN et al. 1972). Therefore, the specificity of antibodies appears to be markedly influenced by the position of the steroid molecule used for coupling to the peptide carrier, with a particular "proximity effect." This means that antibodies fail in discriminating between molecules differing but slightly in the vicinity of the point of attachment of the steroid to the carrier protein.

Similar problems arise when working with prostaglandins, thromboxanes, leukotrienes, and polyunsaturated hydroxy acids. The carboxyl group of these compounds is generally employed for conjugation to proteins, thus leaving the whole carbon chain protruding out of the carrier. Side hydroxyl and keto groups in the cyclopentane ring are the groups best detected by the antibodies. Prostaglandins of the α series are usually well distinguished from the homologous compounds of the β series, although the difference is represented by the steric configuration of the hydroxyl group in position 9. More problems are generated by dioic acids, which are urinary metabolites of prostaglandins in humans and in several animal species, and by leukotrienes. The former compounds possess two carboxyl groups. If the coupling procedure is performed in the usual way, the conjugation will occur randomly at either carboxyl group and the resulting antisera will presumably contain several clones of antibodies directed toward different portions of the molecule. Therefore, the specificity may be expected to be very low. A selective coupling of BSA to the ω-carboxyl group of 5α,7α,-dihydroxy-11-ketotetranorprosta-1,16-dioic acid (a urinary metabolite of prostaglandin $F_{2\alpha}$ in humans) has been described (GRANSTRÖM and KINDAHL 1976). The α-carboxyl group was protected by δ-lactone formation with the 5α-hydroxyl group. The resulting antiserum did not recognize the ω end of the molecule at all, but was specific for the tetranor side chain.

In the last few years, sulfidopeptide leukotrienes and leukotriene B_4 have represented a serious problem for raising antisera suitable for RIA. Immunization with a conjugate of leukotriene D_4 and BSA via the eicosanoid carboxyl group produced antibodies with comparable affinities for leukotrienes C_4, D_4, and E_4 and their 11-*trans* stereoisomers (LEVINE et al. 1981). The use of the carboxyl group as an anchoring point, thus leaving the sulfidopeptide subunit exposed for recognition by the immune apparatus, does not seem to be very effective in improving antibody specificity, although leukotrienes C_4, D_4, and E_4 differ in subunit composition. Better results have been reported by using coupling procedures which involve reactions on the free amino group of the glutamyl residue of leukotriene C_4 (AEHRINGHAUS et al. 1982; HAYES et al. 1983). These antibodies recognized both the glutathione and the fatty acid moiety as immunodominant parts of the leukotriene C_4 molecule. Thus, coupling the hapten via the single free amino group may be adavantageous, leaving the most characteristic parts of the hapten unchanged.

Two different ways have been followed to raise antibodies to leukotriene B_4 – coupling leukotriene B_4 to BSA via the 12-oxy group of the lipid (LEWIS et al. 1982), or conjugating it by the mixed anhydride method through the COOH terminus (SALMON et al. 1982). A satisfactory specificity was obtained in both cases. Studies of cross-reactivity with other prostanoids indicate that the C-5 and C-12 hydroxyl groups, with their relative positions determined by the 6-*cis*-8, 10-*trans*-triene structure, are important immunodeterminants (LEWIS et al. 1982; SALMON et al. 1982).

F. Characterization of Antisera

I. Titer

Antisera are first screened for titer. The optimum dilution of antiserum for immunoassay has been defined by YALOW and BERSON (1964) as the dilution at which ~50% of tracer is bound, i.e., "trace bound/free (B/F) ratio" ~1. According to most authors, the titer is the dilution of the antiserum at which 50% of the radioactivity present in the RIA system is bound. When characterizing a new antiserum, a preliminary testing is performed at low dilutions (below 1:1000) and with incubation periods ranging from a few hours up to 5–7 days (BERSON and YALOW 1968). Under some circumstances, equilibrium is not complete (i.e., maximum B/F ratio is not achieved in tubes containing only the labeled tracer with antibody) before 4–5 days (BERSON and YALOW 1968). If a negligible amount of labeled antigen is bound at 1:100–1:150 dilution, the antiserum will not be very useful. A further increase of titer is required before it could be employed for RIA purposes. If more than 50% of the radioactivity is bound, further dilutions must be tested. On the basis of these tentative results, antisera are examined over a narrower range of dilutions to find the final dilution in the assay yielding a trace B/F ratio of about 1.

It must be emphasized that the titer of antiserum or concentration of antibody cannot be used to define the quality of an antiserum. Except when a choice is to be made among different antisera exhibiting identical characteristics in terms of sensitivity and specificity, the titer is not made the basis for selection. The only requirement of antisera in terms of titer is that it should be great enough to allow a large enough number of assays to be performed over a reasonable period of time with a small volume. Except for this reason, a very high titer is not required. The measure of usefulness of an antiserum for assay of low concentrations of a given substance is determined by the slope of its standard curve. This slope is related to the equilibrium constant K, not to the titer of antiserum. Paradoxically, a comparison of various antisera for sensitivity for measurement of gastrin (BERSON and YALOW 1972), revealed that the antiserum with the lowest titer (1:2500) showed maximal initial slope of the standard curve. It should also be considered that titer is dependent on antibody-independent technical factors, i.e., specific activity of labeled tracer, radioactive tracer concentration in the assay system, or conditions of incubation (time, temperature, buffer, plasma, or albumin added to the mixture, etc.). Therefore, it is reasonable to expect differences in titers of the same antiserum from laboratory to laboratory.

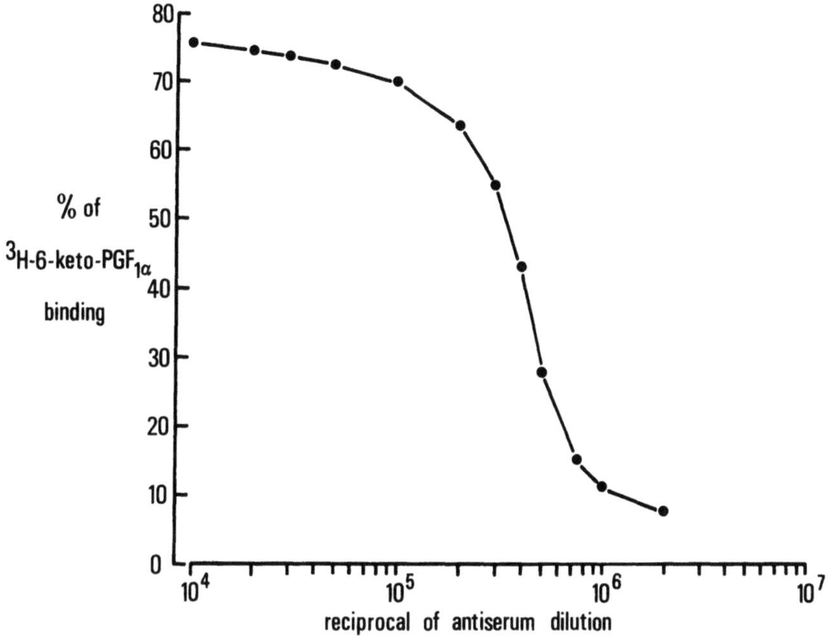

Fig. 4. Binding of the labeled tracer by an anti-6-keto-prostaglandin $F_{1\alpha}$ serum at different dilutions in the RIA system

Figure 4 depicts the percentage binding of labeled tracer by an anti-6-keto-prostaglandin $F_{1\alpha}$ serum at different final dilutions in the assay. The concentration of the labeled tracer (^{3}H-labeled 6-keto-prostaglandin $F_{1\alpha}$) was maintained constant all through the dilutions and was the lowest detectable by the counting equipment. Specific activity of the tracer was 160 Ci/mmol. At dilutions lower than 1 : 50 000 there is only a minimal increase of the binding. This means that the RIA system has excess antibody. In a range of dilutions from 1 : 100 000 to 1 : 750 000, the binding is nearly a linear function of the reciprocal of the final dilution of antibody in the assay. At dilutions exceeding 1 : 1 000 000, a residual minimal binding is still present, but this is not useful for RIA purposes.

Immunization of animals is generally monitored by titer. When it increases, booster injections are suspended and bleedings at regular intervals are made. However, although titer may continue to increase significantly during the course of immunization, the equilibrium constant *K* usually reaches a plateau and does not seem to increase significantly on continued immunization. A single bleeding when affinity has reached a plateau may produce enough antiserum to serve for millions of assays, or even more. In these cases, it is convenient to bleed animals frequently and to stock as much antiserum as possible for later use. Undiluted rabbit or guinea pig antisera may be stored in the frozen state for a decade or longer without evident loss of potency (Berson and Yalow 1973). Lyophilized aliquots may be stored indefinitely. Dilutions of 1 : 100–1 : 1000 in normal saline, containing 1% human or animal plasma and merthiolate (1 : 5000), are stable in a refrigerator (+4 °C) for more than 1 year. The working titer of a given antiserum sometimes seems to change. Generally, this is only an apparent change; the reason is almost always found in one of the other reagents. If the maximum binding with a relative excess of antiserum decreases, degradation of the labeled ligand is the most likely explanation. If the standard curve shifts to the right, this is usually a consequence of degradation of standard. If the curve shifts to the left, the error may reside in the standard dilutions. When the maximum binding and the initial slope of the standard curve decrease, contamination of the RIA system with specific or nonspecific interfering factors may be the reason. In this case, one should check separately all the reagents of the RIA system. However, sometimes no obvious explanation can be found for a change in the working titer. If a loss of binding capacity can be ascertained, the working dilution must simply be adjusted to give the normal binding. No change in affinity constant nor in specificity follows the titer decrease in such cases. Sometimes an apparent rise in titer may be observed. If it appears to be real, two explanations may be possible. One is progressive deterioration of the endogenous antigen (hormones or other endogenous substances) which is present in the antiserum or antiplasma, thus yielding gradually more binding sites available for exogenous tracer. In this way, the final dilution of antibody required in the RIA system to bind the same amount of tracer increases. A second possibility is represented by a higher specific activity of the labeled tracer. When a new labeled tracer preparation is to be used, its real specific activity should be properly considered and its immunoreactivity should be checked at different dilutions of antiserum.

II. Affinity

Sensitivity of RIA is influenced by several factors, but first it depends on the affinity constant of the antibody. The terms "avidity" and "affinity" are generally used to describe the energy of binding of a given antigen–antibody reaction. Avidity refers to the properties of the antibody and affinity to the ability of antigen to be bound at the specific antibody site. However, in both cases avidity and affinity may be expressed in terms of the association or equilibrium constant of the antigen–antibody reaction. To consider the theoretical basis of the immunochemical reaction we will refer to the model system developed by Solomon A. Berson and Rosalyn S. Yalow (BERSON and YALOW 1964, 1973). The simplified model originally proposed by these two authors consists of a univalent hormonal antigen Ag reacting reversibly with a single order of antibody-combining sites Ab to form an antigen–antibody complex

$$\mathrm{Ag} + \mathrm{Ab} \rightleftharpoons \mathrm{Ag\,Ab}\,, \tag{1}$$

$$K = \frac{k_1}{k_2} = \frac{[\mathrm{Ag\,Ab}]}{[\mathrm{Ag}][\mathrm{Ab}]} \tag{2}$$

were K is the equilibrium constant and k_1 and k_2 are the forward and reverse rate constants, respectively. The square brackets indicate molar concentrations. If we assume F (free) = [Ag] and B (bound) = [Ag Ab], then F and B will represent the concentrations of free antigen and antibody-bound antigen, respectively. If $[\mathrm{Ab}^0]$ is the total antibody concentration and [Ab] is the concentration of antibody still available at equilibrium, then $[\mathrm{Ab}^0] = [\mathrm{Ab}] + [\mathrm{Ag\,Ab}] = [\mathrm{Ab}] + B$. Thus Eq. 2

may be written

$$K = \frac{B}{F([\mathrm{Ab}]-B)}$$

or

$$\frac{B}{F} = K[\mathrm{Ab}^0]-B\,. \tag{3}$$

The association constant of a given antiserum can be calculated by the Scatchard plot or the Michaelis-Menten hyperbola. The equation

$$y = mx + b$$

expressing a linear relationship between an independent and a dependent variable (Scatchard plot) may be applied to Eq. 3, B being the independent and B/F the dependent variable. RODBARD et al. (1971) have developed a mathematical theory of RIA, showing the applicability of the Scatchard plot to calculation of the association constant of antisera. In Chap. 8, Rodbard et al. discuss these mathematical aspects in detail.

From a practical point of view, this type of plot allows a graphical determination of the affinity constant. The Scatchard plot may be constructed for labeled antigen alone or for labeled antigen in the presence of various amounts of unlabeled antigen. The latter method is widely employed. Molar antigen concentrations bound to antiserum at equilibrium are plotted against B/F (Fig. 5). In the present example, a standard curve of prostaglandin E_2, covering the whole work-

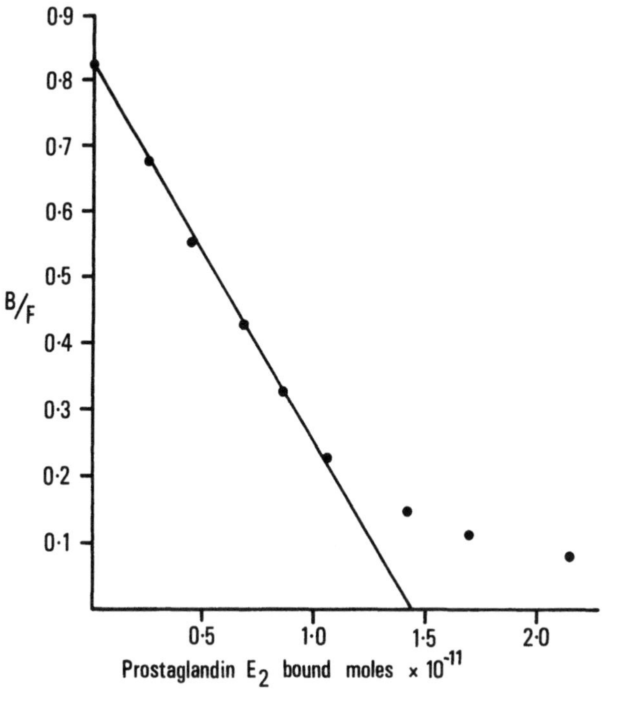

Fig. 5. Scatchard plot of amount of antigen (prostaglandin E_2) bound vs bound to free ratio (B/F)

ing range of the assay, was employed. Anti-prostaglandin E_2 serum was produced by immunizing a guinea pig with a prostaglandin E_2–HSA conjugate. The molar amounts of antigen bound to antibody were calculated on the basis of the added concentrations of standard, assuming identical immunologic behavior of labeled and unlabeled prostaglandin E_2. Complete equilibrium was reached after 24 h incubation at +4 °C. The slope of this line is equal to $-K$ (in the example of Fig. 5 it is approximately 1.64×10^{11} L/M). The x intercept is equal to the total number of binding sites, and corresponds to [Ab^0] in Eq. 3. The y intercept is equal to K [Ab^0]. If linearity is satisfactory in this type of graphical representation, only a single order of binding sites may be supposed in a given antiserum. This also means that satisfactory linearity is present throughout the working range of the assay. In practice, however, we generally do not deal with a single homogeneous order of antibody-combining sites, since conventionally produced antisera are usually quite heterogeneous in respect to affinities of the various classes of antibodies which they contain.

Figure 5 is an example of this heterogeneity: the relative positions of the single points on the left and right sides of the curve suggest at least two main components of the graphically determined line. The second component (not represented in Fig. 5) shows a significantly lower slope, and, therefore, a lower affinity constant K. More than one population of immunoglobulins may be supposed to be present in antisera or antiplasma obtained from animals injected with a given immunogen. This leads to a nonlinear relationship between B/F and the molar concentration of bound antigen, since the assay system is represented by a mixture of antibodies with different affinity constants. Equation 3 refers to a simplified model system. Actually, it may be considered satisfactory only for monoclonal antibodies. Nevertheless, the theoretical considerations developed by Berson and Yalow are applicable to more complicated real systems as well. Indeed, when antisera show a high affinity constant and are diluted extensively, it is frequently the case that the reaction involves only (or principally) a single class of antibody–combining sites remaining at significant concentration. This is why in Fig. 5 at optimal dilution of antiserum binding, about 35%–40% of labeled antigen, a single clone of immunoglobulins emerges and the affinity constant of antiserum is determined by these high affinity antibodies. A clearer example is given in Fig. 6, which represents a standard curve for thromboxane B_2 obtained with an antiserum containing at least two different orders of antibody-binding sites. The shape of the standard curve when represented in the conventional way (Fig. 6 b) is asymptotic, with a steepest segment in the low dose range, where the slope reaches a maximum value. The same curve is represented in a logit-log transformation in Fig. 6 a. It is evident that it consists of at least two different segments of different slope, the left one being the steeper. In the low concentration range, the linearity of the standard dose–response curve is mainly due to the activity of immunoglobulins with high affinity for the substrate. However, the amount of antigen they can bind is limited, or, in other words, they have a high affinity for the antigen, but a low or limited binding capacity. The other orders of antibody-binding sites which are present in the antiserum of Fig. 6 can account for the binding of higher amounts of antigen, but with a lower affinity. Thus, on increasing antigen concentration, the high affinity binding sites will be saturated and in-

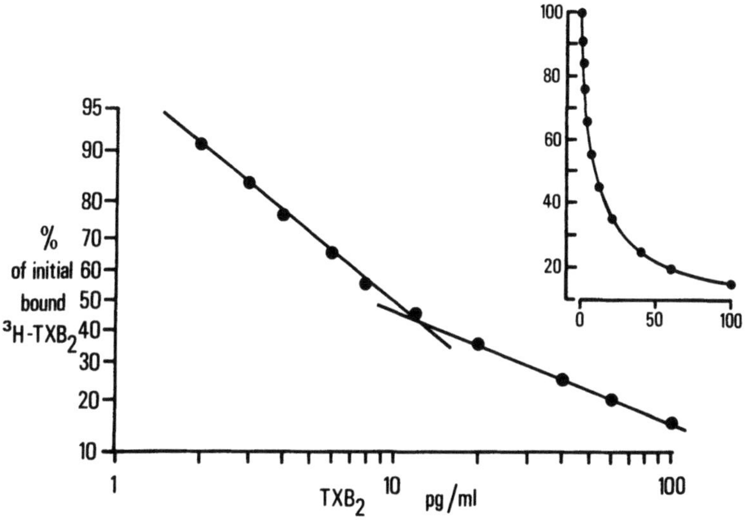

Fig. 6 a, b. Standard curve of an anti-thromboxane B_2 serum represented on an arithmetic scale (**b**) and in a logit-log transformation (**a**). In both **a** and **b** ordinate is percentage of initial bound labeled thromboxane B_2; abscissa is standard thromboxane B_2 concentration

creasing amounts of antigen will occupy sites with low affinity, but high binding capacity. It is obvious that to gain the maximum specificity of this RIA system the dilution of the samples must be adjusted to fall within the linear dose range.

III. Sensitivity

Sensitivity has been defined by different investigators either as the minimal detectable concentration in the RIA system or the slope of the standard dose–response curve. However, the former definition should be properly referred to as the detection limit of the assay, i.e., the smallest amount of unlabeled antigen (hormone or any other substance) that can be significantly distinguished from the complete absence of the measured compound. The detection limit of an assay corresponds to the lowest point of the standard curve which differs by two standard deviations from zero, or which causes at least 10% displacement of the maximum binding in the absence of unlabeled antigen. It is obvious that the lowest detectable concentration of a given substance depends on the limit of the standard curve plus the final dilution of the measured sample in the assay. If a given biologic sample can be measured at 1:1 dilution in the RIA system, the assay limit will be represented by a concentration which is double that at the lowest point of the standard curve. This, however, is not the sensitivity with respect to antiserum. BERSON and YALOW (1973) originally defined sensitivity in terms of the slope of the dose–response curve. In Eq. 3, B/F is a linear function of B. Although such a simple linear relationship cannot be theoretically derived for heterogeneous systems containing more than one clone of antibodies, experimental data generally closely approximate this simplified model. It is quite apparent that for a given value of B/F, the slope of the dose–response curve decreases with increasing K

$[Ab^0]$ or, conversely, the slope increases for decreasing values of $K[Ab^0]$, with a limiting tendency of $K[Ab^0]$ toward zero. This means that the sensitivity of the assay under certain conditions, i.e., incubation volume, time, temperature, labeled antigen-specific activity, and counting equipment sensitivity, is essentially regulated by the dilution of antisera and the slope $-K$ of the dose–response standard curve of the Scatchard plot. Conditions which increase the sensitivity are discussed in Chap. 7. If an adequate dilution of antiserum is realized in the RIA system, the sensitivity is closely dependent on the affinity constant of antibody K. For example, at $B/F=1$, $K[Ab^0]=1$. Under these conditions, the lowest detectable concentration of unlabeled antigen is a function of the molar antibody concentration, which, in turn, is the reciprocal of the association constant K. Thus, we can assume that the first condition regulating the sensitivity of an RIA is the association constant of antisera. All other conditions can be modified to improve sensitivity, but the theoretical limit of sensitivity achievable with an antiserum is set by its affinity constant.

IV. Specificity

Although employing radioactivity and requiring expensive counting equipment or sophisticated automated data handling systems, RIA relies on a biologic reaction, i.e., an antigen–antibody reaction and, for this reason, can be regarded as a sophisticated form of bioassay, substituting the classical smooth muscle strip with a soluble antibody as reactant. Like all immunologic reactions, RIA offers the potential advantage of great specificity. This specificity depends upon the inherent ability of the immunoglobulins present in the antiserum to recognize a given molecular structure. However, cross-reactions with compounds closely related to the antigen structure are always observed. Moreover, like all chemical reactions, an antigen–antibody reaction may be influenced by a series of nonspecific factors which induce a nonimmunologic inhibition of Ag–Ab binding.

Specificity of RIA has been defined by MIDGLEY et al. (1969) as the extent of freedom from interference by substances other than the one intended to be measured. This interference may be due to an immunologic cross-reactivity or to the nature of the medium where the Ag–Ab reaction takes place. It should be remembered that all an RIA does is to measure the degree of inhibition of the maximum binding of labeled antigen to antiserum. When working with a given antiserum, it is assumed that the displacement of labeled tracer is caused by the presence of endogenous concentrations of the substance against which the antibody is directed. Therefore, it is evident that everything causing a displacement of antibody binding with labeled antigen may be seen as endogenous reacting antigen. If the system is not examined critically, this inhibition will be interpreted as being caused by true concentrations of the measured compound.

Given a particular antiserum, the specificity of the reaction will depend on a constant, i.e., the affinity and conformation of the binding sites available on the antibody, and a variable, the composition of the biologic fluid to be examined and its degree of purification. The criteria for validating RIA measurements are discussed in detail in Chap. 9. Here, we will summarize the basic considerations related to the specificity of antisera produced by conventional techniques.

1. Immunologic Cross-Reactivity

Nonspecificity of the antibody may be due to cross-reaction by substances related to antigen, i.e., substances with a chemical structure similar to antigen, antigen fragments and derivatives, and products of antigen metabolism. An antibody-binding site can combine with different antigens with different values of *K*. As a binding site can accommodate only a few amino acids, or molecules with low molecular weight, or single regions of larger chemical structure, it is obvious that immunologic cross-reactivity will be the rule, rather than the exception, if antibody reacts with compounds showing a similar chemical structure, or steric arrangement, or common amino acid sequences. Moreover, considering the heterogeneity of antibody-binding sites present in the "classical" polyclonal antisera, at least some of these binding sites in any one antiserum are likely to have at least a weak association constant for immunoreactive sites in some other molecule.

The spectrum of possible interfering substances is obviously related to the type of antigen against which the antibody was produced. However, the ability of antigen-related compounds to react with a given antibody cannot, in general, be predicted and is usually different from antiserum to antiserum. In the case of antigens which possess several antigenic determinants, such as polypeptide hormones, antisera specificity may vary greatly, depending on the particular fragment which is best recognized by antibodies. Since the pioneering studies performed by Berson and Yalow, it appears that big molecules, such as insulin chains, may react with antisera in different ways: some antisera to insulin do not react with the separated A and B chains (BERSON and YALOW 1959b), or specifically recognize some portions of the molecule, such as the B chain lacking the COOH terminal 8 amino acid fragment (BERSON and YALOW 1963). The possibility that hormonal fragments, or polypeptides with similar amino acid sequences, are present in the biologic samples to be measured and contribute to the immunologic reaction cannot be excluded.

The problem is even more complex for neuropeptides, such as those from the tachykinin family (neurokinin A and B, substance P, eledoisin, kassinin, physalaemin, uperolein, phyllomedusin). The common amino acid sequence of a relevant portion of the molecule may represent a serious limit to antibody specificity and, therefore, to the precision of the assay. An important example is represented by the ACTH molecule. Cross-reactivity of anti-ACTH sera with α- and β-MSH and with γ- and β-LPH is usually present, because of the common peptide sequence of these hormones with ACTH. Characterization of an anti-ACTH serum should include definition of the portion of the ACTH molecule to which antibody is directed as well as cross-reactivity with different hormone fragments and synthetic analogs (ORTH 1979). The problem also exists for small molecules, which only occupy a single antibody-binding site. The exchange of a single atom of hydrogen and iodine in the structure of triiodothyroxine and thyroxine, the presence or absence of a single atom of hydrogen in cortisone and cortisol, and the steric configuration of the eicosanoid structure in sulfidopeptide leukotrienes and their 11-*trans* stereoisomers are examples of immunologic distinctions possible between structurally similar compounds. The cross-reactivity of an antiserum to structural analogs of the primary compound against which antibody specificity

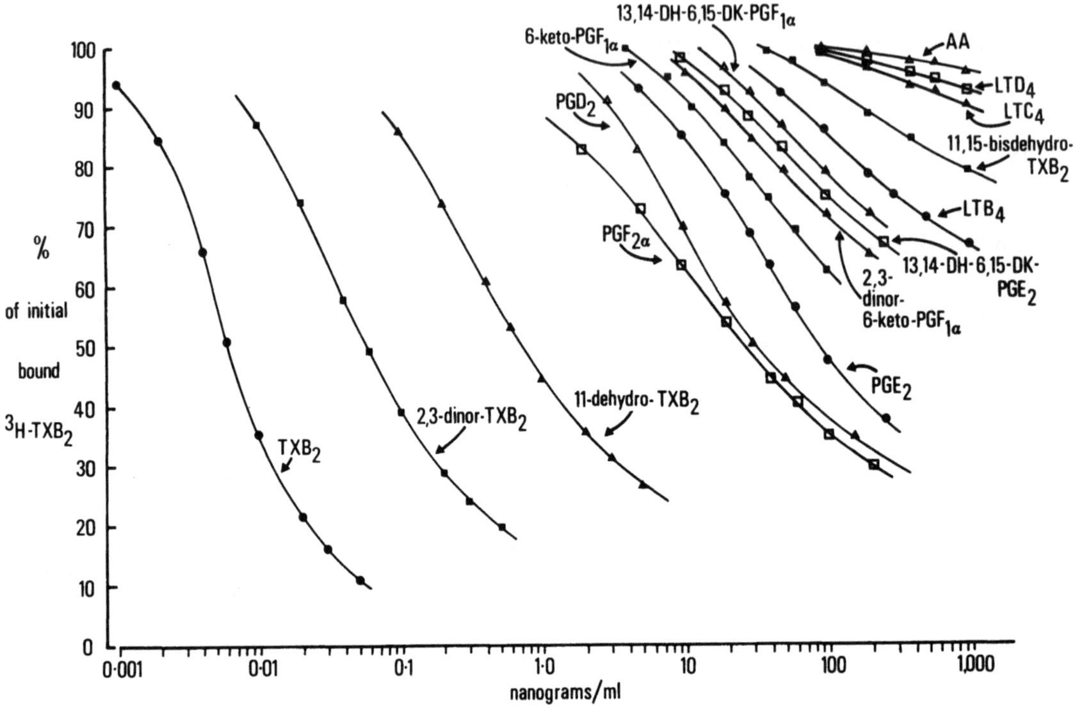

Fig. 7. Family of standard dose–response curves of an anti-thromboxane B_2 serum obtained with the primary antigen and with a series of cross-reacting substances. TXB_2 thromboxane B_2; PG prostaglandin; DH dihydro; DK diketo; LT leukotriene; AA arachidonic acid

is directed is generally illustrated by a set of standard curves. The curves are calculated by incubating the antibody with a constant amount of labeled tracer and at a constant final dilution, yielding the optimal maximum binding, in the presence of the unlabeled, related compounds at increasing concentrations. Figure 7 shows an example of this way of calculating the cross-reactivities of an anti-thromboxane B_2 serum. The standard curve with the highest slope should ideally be referred to the compound which has been used to elicit antibody production. Sometimes it may be that the specificity of an antiserum is primarily directed toward some related compound. This usually happens when the antigen undergoes enzymatic or nonenzymatic degradation after injection into an experimental animal. In the example of Fig. 7, the curve on the far left was obtained with standard thromboxane B_2. The other compounds tested are metabolites of thromboxane B_2, prostaglandins, and lipoxygenase products (i.e., leukotrienes and HETE). It is noteworthy that the highest cross-reactivity is shown with 2,3-dinorthromboxane B_2. The difference between this metabolite and thromboxane B_2 is represented by the loss of a C_2 fragment resulting from a single β-oxidation step, the remaining portion of the molecule being identical. For this reason, antibodies produced against thromboxane B_2 or 6-ketoprostaglandin $F_{1\alpha}$ coupled to proteins via the COOH terminus generally show a high degree of cross-reactivity with the related 2,3-dinor derivatives. Compounds with lower affinity to the antibody are required in larger amounts to displace the labeled tracer. These standard curves are to the right in the diagram. Figure 7 shows that the more structurally related a substance is to the antigen against which antibody specificity is directed, the higher is its cross-reaction with antiserum. If we calculate the concentration of each compound required to displace 50% of initially bound ^{3}H labeled thromboxane B_2 (IC_{50}), we will obtain a table of increasing concentrations corresponding to decreasing degrees of cross-reactivity (Table 3). If we assume the lowest IC_{50} (represented by the standard thromboxane concentration required to displace 50% of labeld Ag–Ab binding) as corresponding to 100% cross-reactivity, we can express the cross-reaction of all other tested compounds as a percentage of that shared by the primary antigen (Table 3). The higher IC_{50} the related tested compounds show with antibody, the higher is the specificity of antiserum.

This is the most commonly used way, but the percentage cross-reaction so defined represents only the cross-reaction at 50% displacement of antibody binding with radioactive tracer. A complete picture of the events over the total range of displacement of binding is not given by a simplified representation such as the one shown in Table 3. The percentage values which are generally reported by the vast majority of authors to define the specificity of antisera cannot be used for an accurate calculation of the amounts of cross-reacting substance, unless the displacement curves of the cross-reacting substance and of the primary compound are absolutely parallel. But even in this case, the percentage cross-reactivity of a given compound varies if its standard curves are determined in the presence of variable amounts of the primary antigen against which antibody has been elicited. The picture appears even more complex if a mixture of possibly cross-reacting substances is incubated with the antibody in the absence or in the presence of variable amounts of the primary compound. Moreover, the complex composition of biologic fluids and the impossibility of testing all the possible cross-reacting com-

Table 3. Immunologic specificity of an anti-thromboxane B_2 serum

Substance measured[a]	IC_{50}[b] (ng/ml)	Relative cross-reaction (%)
TXB_2	0.006	100.0
2,3-Dinor-TBX_2	0.057	10.5
11-Dehydro-TBX_2	0.741	0.81
$PGF_{2\alpha}$	026.0	0.023
PGD_2	30.0	0.02
PGE_2	85.7	0.007
2,3-Dinor-6-keto-$PGF_{1\alpha}$	>100.0	< 0.006
6-Keto-$PGF_{1\alpha}$	>100.0	< 0.006
13,14-Dihydro-6,15-diketo-$PGF_{1\alpha}$	>150.0	< 0.004
13,14-Dihydro-15-keto-PGE_2	>150.0	< 0.004
11,15-Bis-dehydro-TXB_2	>150.0	< 0.004
LTB_4	>150.0	< 0.004
LTC_4	>150.0	< 0.004
LTD_4	>150.0	< 0.004
Arachidonic acid	>150.0	< 0.004

[a] TX, thromboxane; PG, prostaglandin; LT, leukotriene.
[b] Concentration required to displace 50% of bound ^{3}H-labeled thromboxane B_2.

pounds, does not enable us to develop a mathematical model (although this is theoretically possible) by which we can subtract all the cross-reactivities from the measured immunoreactivity. Therefore, the validity of this approach is limited to giving us a relatively simplified picture of the ability of a given antiserum to discriminate among structural analogs of the compound which is to be measured. A more extensive discussion of criteria which must be satisfied to validate an RIA measurement, other than assessment of cross-reactivity with a limited number of compounds, can be found in Chap. 9.

2. Nonspecific Interfering Factors

a) Influence of pH and Ionic Environment

Although the antigen–antibody reaction is dependent on the pH of the buffer solution, it is evident that immune complexes are dissociated only by extremes of pH, while in the neutral and mild alkaline range the reaction is almost completely stable. On several occasions, Berson and Yalow reported that there is little evidence that peptide hormone systems exhibit a significant pH dependency in the range 7.0–8.5. The same authors have established that the buffer systems employed for RIA are of considerable importance, since high concentrations of salts as well as some buffers may inhibit the hormone–antibody reaction (BERSON and YALOW 1968). Thus, it is advisable to maintain a constant pH for both standards and unknowns and to use only well-known buffer solutions with moderate ionic strength. However, under some conditions both the pH and ionic environment of the RIA mixture may represent a serious problem. As for the influence of pH, it should be remembered that some nonpeptide substances may modify their chemical structure or steric configuration at different pH values.

The salt content of biologic samples like plasma, urine, gastric juice, synovial fluid, tears, and so on, may represent an important source of artifactual concentrations of the hormone or the substance which is intended to be measured. No general rule may be proposed in these situations. If the sensitivity of the assay allows a high dilution of the sample into the incubation mixture, the problem may be easily solved. If the content of the measured substance in the biologic fluid is too low to allow a convenient dilution rate, extraction and chromatographic separations are usually required. Sometimes, it is difficult to choose between the extraction and purification of the sample and the direct RIA measurement of the unextracted material. This problem usually exists for plasma and serum. Generally speaking, direct RIA of unextracted plasma represents the advantage of being free from interference introduced by solvent extraction and purification procedures, provided that antisera with high affinity constants are employed and that plasma is adequately diluted in the assay mixture. A dilution at least 1:10–1:15 of plasma samples is generally considered to be satisfactory to eliminate the nonspecific salt inhibition of the antigen–antibody reaction and the competitive binding represented by plasma proteins, in particular, serum albumin. Nevertheless, optimal conditions for specificity should be checked in any particular RIA system with any kind of biologic sample. The same biologic fluid, deprived of the substance which is to be measured, is usually employed to build up the dose–response curve with the standard substance. In any case, a comparison with an independent assay method should be considered to validate the results obtained with the RIA techniques.

b) Temperature Effects

The antigen–antibody reaction is temperature dependent. Even *K*, the equilibrium constant of the reaction (see Chap. 24) is influenced by temperature. The pioneering studies performed by Berson and Yalow more than 25 years ago show that higher *B/F* ratios are generally observed when the RIA system is equilibrated at 4 °C than at room temperature (18°–22 °C). Temperature may influence the rate of the antigen–antibody reaction, the ratio of products (i.e., antigen–antibody complexes) at equivalence, the adsorption of free antigen to the adsorbing agents, and the stability of the labeled and unlabeled antigens. Although the time required to attain equilibrium is sensibly longer, the highest specificity is generally obtained by incubating the mixtures at 4 °C. To shorten the incubation time, thus reducing the overall time required for an RIA to be performed, incubation is often carried out at room temperature within a few hours. However, this procedure should not be recommended either for research purposes or for clinical evaluation. The slight advantage of saving time is counterbalanced by lowering the sensitivity, precision, and accuracy of the assay. Moreover, incubating the RIA system for a short time at room temperature may result in disequilibrium or nonequilibrium at the time of separation of antibody-bound from free antigen, also reducing sensitivity and specificity. Finally, this procedure requires a higher concentration of antibody-binding sites in the assay mixture (i.e., a lower antiserum dilution) to yield an acceptable *B/F* ratio. Thus, binding sites of low affinity and high capacity may be involved in the immunochemical reaction.

c) Anticoagulants

It is well known that certain anticoagulants may interfere with the antigen–antibody reaction as well as the separation of antibody-bound and free antigen. HENDERSON (1970) first reported such interference from heparin, which was confirmed by the classical studies of YALOW and BERSON (1971 b). These authors explained the effect of heparin as being due to its polyanionic structure which exerts a salt effect, inhibiting the antigen–antibody reaction. From this point of view, very high concentrations of anticoagulants like sodium EDTA, citrate, citrate dextrose, and citrate phosphate dextrose, could also be expected to exert the same action on the immunologic reaction.

d) Labeled Tracer Damage

This problem mainly involves peptide hormones, since they are subject to proteolytic damage by enzymes in biologic fluids. But other nonpeptide substances may also undergo enzymatic as well as nonenzymatic degradation during incubation and/or separation of antibody-bound and free antigen. The incubation damage of labeled antigen may result both in a reduced binding to antibody and in errors in *B/F* determination, because the damaged tracer is improperly measured within the bound or the free fraction, depending on the separation method employed. Moreover, the extent of labeled tracer damage may vary greatly in the samples to be measured, as compared with the standard. This results in an apparent nonspecificity of the assay, since the concentrations of the measured substance in the unknowns will show a nonlinearity with standard upon serial dilution.

In other circumstances, the use of labeled tracers with low radiochemical purity (i.e., compounds which have undergone significant decomposition) leads to an apparently low binding with antiserum. If the purity is not checked, a large excess of antibody-binding sites (i.e., a low antiserum dilution) will be employed to increase the fraction of antibody-bound labeled tracer. The result will be an RIA system working within an apparently normal or low binding of labeled antigen, but, as matter of fact, working with a large excess of antiserum. This decreases the sensitivity of the assay (see Chap. 7), but also allows antibody-binding sites with low affinity constants to participate significantly in the immunochemical reaction, with a consequent loss of specificity. Therefore, to gain the maximum specificity of any RIA system with a given antiserum, the radiochemical purity of the labeled tracer should be checked. A simple way to assess the immunoreactivity of the tracer is to test it with an excess of antiserum. Nearly complete binding should be found under these conditions.

e) Influence of Extraction and Purification Procedures

Like all chemical reactions, the antigen–antibody reaction is influenced by the presence of residues or impurities of organic solvents, acids, salts, gases, and buffers employed for extraction and purification of samples. These methodological problems arise when the substance which is intended to be measured cannot be directly assayed in the sample, but must be extracted and subjected to chromatographic separation to yield a purified solution free from other substances that may react with antigen or antibody. When an RIA measurement includes some processing of the sample, a "blank" procedure should be inserted. This is the result obtained in the assay from a water or buffer sample, or better still a sample free from the endogenous substance subjected to exactly the same treatment as the biologic samples. The reason for such a procedure is that even the most rigorously performed extractions and purifications are likely to introduce some interfering material. The omission of this information makes it difficult to evaluate the measured concentrations of a given substance in the unknowns. If a significant blank value is measured by the RIA system, this inevitably implies a nonspecific effect, either on the binding of antigen to antibody or on the separation of antibody-bound and free antigen.

During the early days of RIA, peptide hormones were measured by direct RIA in plasma or serum. But, as soon as the clinical investigation was extended to more complex biologic fluids and nonpeptide substances were measured, or interference with plasma components were reported, most scientists extracted their samples and purified them extensively. Later, the problem arose as to whether a complex purification procedure is really necessary. This problem was particularly discussed for nonpeptide substances, like steroids, prostanoids, and other derivatives of fatty acid metabolism, which are present in biologic fluids at low concentration in a mixture with other metabolites and related compounds, and, on the other hand, are theoretically easily extractable by organic solvent systems. Although a general rule cannot be established, a purification procedure is generally required when the concentration of the substance is below the limit of detection of the assay, or the biologic samples contain undesirable factors that interfere with the assay, or the antiserum is not very specific and cross-reacts with other substances that occur in the sample and are chemically related to the measured compound. The degree of purification of the sample, i.e., the extraction and chromatographic steps which may be necessary, is strictly dependent on a series of variables, which are related to the composition of the biologic fluid, the chemical characteristics of the measured substance, the specificity of the antiserum employed, and the presence of cross-reacting compounds. The drawbacks of extracting and purifying the samples before the assay are substantially related to the introduction of nonspecific interfering factors by the procedure, the possibility of a reliable estimation of recovery for each sample, the prolongation of the time required to perform the assay, and a series of practical problems depending on the particular method employed. Thus, one may ask: when is it necessary to extract samples? An affirmative answer is obvious when measuring complex biologic fluids, like urine, gastric juice, or whole tissues that contain a very low concentration of the measured compound and too many interfering factors to be eliminated by sample dilution in the assay. However, in some cases, such as plasma, serum, or synovial fluid, extraction may introduce more problems than it solves. In most cases, albumin or protein binding of hormones or endogenous substances in blood appears to be the major problem. Nevertheless, this binding is generally weak and reversible, and in competition with a high specific antibody the affinity constant of albumin is two or three orders of magnitude lower. This means that endogenous protein interference in most cases will disappear upon dilution of the sample in the assay. In any case, the opportunity of measuring extracted or unextracted samples should be evaluated for any RIA system and the results validated according to the criteria discussed in Chap. 9.

References

Abraham GE (1974a) Radioimmunoassay of steroids in biologic material. Acta Endocrinol 75 [Suppl] 183:1–42

Abraham GE (1974b) Radioimmunoassay of steroids in biological material. In: Radioimmunoassay and related procedures in medicine. Proceedings of a Symposium, Istanbul, 10–14 September 1973. International Atomic Energy Agency, Vienna, Vol II, pp 3–29

Abraham GE, Grover PK (1971) Covalent linkage of steroid hormones to protein carriers for use in radioimmunoassay. In: Odell WD, Daughaday W (eds) Principles of competitive protein binding assay. Lippincott, Philadelphia, p 140

Abraham GE, Swerdloff R, Tulchinnsky D, Odell WD (1971) Radioimmunoassay of plasma progesterone. J Clin Endocrinol Metab 32:619–624

Abraham GE, Manlimos FS, Solis M, Garza R, Maroulis GB (1975) Combined radioimmunoassay of four steroids in one ml of plasma. I. Progestins. Clin Biochem 8:369–373

Aehringhaus U, Wölbling RH, König W, Patrono C, Peskar BM, Peskar BA (1982) Release of leukotriene C_4 from human polymorphonuclear leucocytes as determined by radioimmunoassay. FEBS Letters 146:111–114

Andersen B, Osborn M, Weber K (1978) Specific visualization of the calcium dependent regulatory protein of cyclic nucleotide phosphodiesterase (modulator protein) in tissue culture cells by immunofluorescence microscopy: mitosis and intercellular bridge. Eur J Cell Biol 17:354–364

Anderson GW, Zimmerman JE, Callahan FL (1964) The use of esters of N-hydroxy succinimide in peptide synthesis. J Am Chem Soc 86:1839–1842
Arnold ML, James VHT (1971) Determination of deoxycorticosterone in plasma: double isotope and immunoassay methods. Steroids 18:789–801
Assan R, Rosselin G, Drouet J, Dolais J, Tchobroutsky G (1965) Glucagon antibodies. Lancet II:590–591
Attallah AA, Lee JB (1973) Radioimmunoassay of prostaglandin A. Intrarenal PGA_2 as a factor mediating saline-induced natriuresis. Circ Res 33:696–703
Axén U (1974) *N,N'*-Carbonyldiimidazole as coupling reagent for the preparation of bovine serum albumin conjugates. Prostaglandins 5:45–47
Bartos F, Olsen GD, Leger RN, Bartos D (1977) Stereospecific antibodies to methadone. I. Radioimmunoassay of D,L-methadone in human serum. Res Commun Chem Pathol Pharmacol 16:131–143
Bartos F, Bartos D, Olsen GD, Anderson B, Daves GD Jr (1978) Stereospecific antibodies to methadone. II. Synthesis of D- and L-methadone antigens. Res Commun Chem Pathol Pharmacol 20:157–164
Bassiri RM, Utiger RD (1971) The preparation and specificity of antibody to thyrotropin releasing hormone. Endocrinol 90:722–727
Bayard F, Beitins IZ, Kowarski A, Migeon CJ (1970) Measurement of plasma aldosterone by radioimmunoassay. J Clin Endocrinol Metab 31:1–6
Beardwell CG (1971) Radioimmunoassay of arginine vasopressin in human plasma. J Clin Endocrinol Metab 33:254–260
Beiser SM, Erlanger BF, Agate FJ, Lieberman S (1959) Antigenicity of steroid-protein conjugates. Science 129:564–565
Berson SA, Yalow RS (1959a) Recent studies on insulin binding antibodies. Ann NY Acad Sci 82:338–344
Berson SA, Yalow RS (1959b) Species specificity of human anti-beef, pork insulin serum. J Clin Invest 38:2017–2025
Berson SA, Yalow RS (1963) Antigens in insulin determinants of specificity of porcine insulin in man. Science 139:844–845
Berson SA, Yalow RS (1964) Immunoassay of protein hormones. In: Pincus G, Thimann KV, Astwood EB (eds) The hormones, vol IV. Academic, New York, p 557
Berson SA, Yalow RS (1967) Principles of immunoassay of peptide hormones in plasma. In: Astwood EB, Cassidy CE (eds) Clinical endocrinology, vol II. Grune and Stratton, New York, p 699
Berson SA, Yalow RS (1968) Radioimmunoassay of ACTH in plasma. J Clin Invest 47:2725–2751
Berson SA, Yalow RS (1972) Radioimmunoassay in gastroenterology. Gastroenterology 62:1061–1084
Berson SA, Yalow RS (1973) Radioimmunoassay. General. In: Berson SA, Yalow RS (eds) Methods in investigative and diagnostic endocrinology, vol 2A. North-Holland, Amsterdam, pp 84–120
Berson SA, Yalow RS, Aurbach GD, Potts JT Jr (1963) Immunoassay of bovine and human parathyroid hormone. Proc Natl Acad Sci USA 49:613–617
Bouillon R, De Moor P, Baggiolini EG, Uskokovic MR (1980) Radioimmunoassay for 1,25-dihydroxycholecalciferol. Clin Chem 26:562–567
Bouillon R, Van Herck E, Jans I, Keng Tan B, Van Baelen H, De Baelen H, De Moor P (1984) Two direct (nonchromatographic) assays for 25-hydroxyvitamin D. Clin Chem 30:1731–1736
Bryce GF (1974) Development of a radioimmunoasay for luteotropin releasing hormone (LRH) and thyrotropin releasing hormone (TRH). Immunochemistry 11:507–511
Butler VP Jr (1977) The immunological assay of drugs. Pharmacol Rev 29:103–184
Butler VP, Chen JP (1967) Digoxin-specific antibodies. Proc Natl Acad Sci USA 57:71–78
Cailla HL, Racine-Weisbuch MS, Delaage MA (1973) Adenosine 3′,5′-cyclic monophosphate assay at 10^{-15} mole level. Anal Biochem 56:394–407

Cailla HL, Vannier CJ, Delaage MA (1976) Guanosine 3′,5′cyclic monophosphate assay at 10^{-15} mole level. Anal Biochem 70:195–202

Carraway R (1979) Neurotensin and related substances. In: Jaffe BM, Behrman HR (eds) Methods of hormone radioimmunoassay. Academic, New York, pp 139–169

Carraway R, Leeman SE (1976) Radioimmunoassay for neurotensin, a hypothalamic peptide. J Biol Chem 251:7035–7044

Chard T, Kitau MJ, Landon J (1970) The development of a radioimmunoassay for oxytocin: radioiodination, antibody production and separation techniques. J Endocrinol 46:269–278

Chen JC, Zorn EM, Halberg MC, Wieland RG (1971) Antibodies to testosterone-3-bovine serum albumin applied to assay of serum 17β-ol androgens. Clin Chem 17:581–584

Cook IF, Rowe PH, Dean PD (1973) Investigations into the immune response to steroid conjugates using corticoids as a model. I. A specific cortisol antibody. Steroids Lipids Res 4:302–309

Cuatrecasas P, Illiano G (1971) Production of anti-glucagon antibodies in poly-L-lysine "responder" guinea-pigs. Nature New Biol 230:60–61

Dash RJ, England BG, Midgley AR, Niswender GD (1975) A specific non-chromatographic radioimmunoassay for human plasma cortisol. Steroids 26:647–661

Dean PDG, Exley D, Johnson MW (1971) Preparation of 17β-oestradiol-6(*O*-carboxymethyl)-oxime-bovine serum albumin conjugate. Steroids 18:593–603

Dedman JR, Welsh MJ, Means AR (1978) Ca^{2+}-dependent regulator. Production and characterization of a monospecific antibody. J Biol Chem 253:7515–7521

Demura H, West CD, Nugent CA, Nakagawa K, Tyler FH (1966) A sensitive radioimmunoassay for plasma ACTH levels. J Clin Endocrinol Metab 26:1297–1302

Dray F, Maron E, Tillson S, Sela M (1972) Immunochemical detection of prostaglandins with prostaglandin-coated bacteriophage T4 and by radioimmunoassay. Anal Biochem 50:399–408

Ebeid AM, Murray P, Hirsch H, Wesdorp RIC, Fischer JE (1976) Radioimmunoassay of vasoactive intestinal peptide. J Surg Res 20:355–360

Erlanger BF, Beiser SM (1964) Antibodies specific for ribonucleosides and ribonucleotides and their reaction with DNA. Proc Natl Acad Sci USA 52:68–74

Erlanger BF, Borek F, Beiser SM, Lieberman S (1957) Steroid-protein conjugates. I. Preparation and characterization of conjugates of bovine serum albumin with testosterone and with cortisone. J Biol Chem 228:713–727

Fahmy D, Read G, Hillier SG (1975) Radioimmunoassay for cortisol: comparison of H^3-, Se^{75}-, and I^{125}-labeled ligands. J Endocrinol 65:45P–46P

Fahrenkrug J, Schaffalitzky De Muckadell OB (1977) Radioimmunoassay of vasoactive intestinal polypeptide in plasma. J Lab Clin Med 89:1379–1388

Falbriard JG, Posternak T, Sutherland EW (1967) Preparation of derivatives of adenosine 3′,5′-phosphate. Biochim Biophys Acta 148:99–105

Farmer RW, Roup WG Jr, Pellizzari ED, Fabre LF Jr (1972) A rapid aldosterone radioimmunoassay. J Clin Endocrinol Metab 34:18–22

Ferin M, Zimmering P, Lieberman S, Vande Wiele R (1968) Inactivation of the biologic effects of exogenous and endogenous estrogens by antibodies to 17β-estradiol. Endocrinology 83:565–571

Furuyama S, Nugent CA (1971) A radioimmunoassay for plasma progesterone. Steroids 17:663–674

Furuyama S, Mayes DM, Nugent CA (1970) A radioimmunoassay for plasma testosterone. Steroids 16:415–428

Gilliland PF, Prout TE (1965) Immunologic studies of octapeptides. II. Production and detection of antibodies to oxytocin. Metabolism 14:918–923

Glick SM, Wheeler M, Kagan A, Kumaresan P (1968) Radioimmunoassay of oxytocin. In: Back N, Martini L, Paoletti R (eds) Proceedings of international symposium on the pharmacology of hormonal polypeptides. Plenum, New York, pp 93–100

Goodfriend TL, Levine L, Fasman GD (1964) Antibodies to bradykinin and angiotensin: a use of carbodiimides in immunology. Science 144:1344–1346

Granström E, Kindahl H (1976) Radioimmunoassay for urinary metabolites of prostaglandin $F_{2\alpha}$. Prostaglandins 12:759–783
Granström E, Kindahl H (1978) Radioimmunoassay of prostaglandins and thromboxanes. In: Frölich JC (ed) Advances in prostaglandin and thromboxane research, vol 5. Raven, New York, pp 119–210
Gray TK, McAdoo T, Pool D, Lester GE, Williams ME, Jones G (1981) A modified radioimmunoassay for 1,25-dihydroxycholecalciferol. Clin Chem 27:458–463
Grey N, McGuigan JE, Kipnis DM (1970) Neutralization of endogenous glucagon by high titer glucagon antiserum. Endocrinology 86:1383–1388
Guillemin R, Ling N, Vargo T (1977) Radioimmunoassay for α-endorphin and β-endorphin. Biochem Biophys Res Comm 77:361–366
Gutkowska J, Thibault G, Januszewicz P, Cantin M, Genest J (1984) Direct radioimmunoassay of atrial natriuretic factor. Biochem Biophys Res Commun 122:593–601
Hamaoka T, Katz DH (1973) Cellular site of action of various adjuvants in antibody responses to hapten carrier conjugates. J Immunol 111:1554–1563
Hampl R, Picha J, Chundela B, Starka L (1979) Radioimmunoassay of nortestosterone and related steroids. J Clin Chem Clin Biochem 17:529–532
Hampl R, Stolba P, Putz Z, Protiva J, Kimlova I, Starka L (1982) Advances in immunoassay of anabolic steroids. In: Radioimmunoassay and related procedures in medicine. Proceedings of a symposium, Vienna, 21–25 June 1982. International Atomic Energy Agency, Vienna, pp 365–366
Haning R, McCracken J, St Cyr M, Underwood R, Williams G, Abraham G (1972) The evolution of titer and specificity of aldosterone binding antibodies in hyperimmunized sheep. Steroids 20:73–88
Harper JF, Brooker G (1975) Femtomole sensitive radioimmunoassay for cyclic AMP and cyclic GMP after 2′-*O*-acetylation by acetic anhydride in aqueous solution. J Cyclic Nucleotide Res 1:207–218
Hayes EC, Lombardo DL, Girard Y, Maycock AL, Rokach J, Rosenthal AS, Young RN, Zweerink HJ (1983) Leukotriene specific radioimmunoassay. J Immunol 131:429–433
Heding LG (1972) Immunologic properties of pancreatic glucagon: antigenicity and antibody characteristics. In: Lefebvre P, Unger RH (eds) Glucagon: molecular physiology, clinical and therapeutic implications. Pergamon, Oxford, pp 187–200
Henderson JR (1970) Serum insulin or plasma insulin? Lancet II:545–547
Ito T, Woo J, Haning R, Horton R (1972) A radioimmunoassay for aldosterone in human peripheral plasma including a comparison of alternate techniques. J Clin Endocrinol Metab 34:106–112
Jaffe BM, Smith JW, Newton WT, Parker CW (1971) Radioimmunoassay of prostaglandins. Science 171:494–496
Jennes L, Stumpf WE (1983) Preparation and use of specific antibodies for immunohistochemistry of neuropeptides. Methods Enzymol 103:448–459
Jurjens H, Pratt JJ, Woldring MG (1975) Radioimmunoassay of plasma estradiol without extraction and chromatography. J Clin Endocrinol Metab 40:19–25
Kellie AE, Samuel VK, Riley WJ, Robertson DM (1972) Steroid glucuronoside-BSA complexes as antigens: the radioimmunoassay of steroid conjugates. J Steroid Biochem 3:275–288
Kirton KT, Cornette JC, Barr KL (1972) Characterization of antibodies to prostaglandin $F_{2\alpha}$. Biochem Biophys Res Commun 47:903–909
Kitagawa T, Aikawa T (1976) Enzyme coupled immunoassay of insulin using a novel coupling reagent. J Biochem 79:233–236
Klotz IM, Heiney RE (1962) Introduction of sulfhydryl groups into proteins using acetyl mercaptosuccinic anhydride. Arch Biochem Biophys 96:605–612
Kuss E, Goebel R, Enderle H (1973) Steroids as immunochemical problems. I. Influence of oxo- and-or hydroxy-groups at C-16-C17 of estrogens on affinity to anti-estrone, anti-estradiol-17 alpha- and anti-estradiol-17 beta-antisera. Hoppe-Seylers Z Physiol Chem 354:347–364
Kuzio M, Dryburgh JR, Malloy KM, Brown JC (1974) Radioimmunoassay for gastric inhibitory polypeptide. Gastroenterology 66:357–364

Landon J, Greenwood FC (1968) Homologous radioimmunoassay for plasma levels of corticotrophin in man. Lancet I:273–276

Landsteiner K (1945) The specificity of serological reactions. Harvard University Press, Cambridge

Levine L, Gutierrez-Cernosek RM (1972) Preparation and specificity of antibodies to 15-keto-prostaglandin $F_{2\alpha}$. Prostaglandins 2:281–294

Levine L, Van Vunakis H (1970) Antigenic activity of prostaglandins. Biochem Biophys Res Commun 41:1171–1177

Levine L, Gutierrez-Cernosek RM, Van Vunakis H (1971) Specificities of prostaglandins B_1, $F_{1\alpha}$ and $F_{2\alpha}$ antigen-antibody reactions. J Biol Chem 246:6782–6785

Levine L, Morgan RA, Lewis RA, Austen KF, Clark DA, Marfat AM, Corey EJ (1981) Radioimmunoassay of the leukotrienes of slow reacting substance of anaphylaxis. Proc Natl Acad Sci USA 78:7692–7696

Lewis RA, Mencia-Huerta JM, Soberman RJ, Hoover D, Marfat A, Corey EJ, Austen KF (1982) Radioimmunoassay for leukotriene B_4. Proc Natl Acad Sci USA 79:7904–7908

Lieberman S, Erlanger BF, Beiser SM, Agate FJ Jr (1959) II. Aspects of steroid chemistry and metabolism. Steroid-protein conjugates: their chemical, immunochemical and endocrinological properties. Rec Prog Horm Res 15:165–200

Likhite V, Sehon A (1967) Protein-protein conjugation. In: Williams CA, Chase MW (eds) Methods in immunology and immunochemistry, vol 1. Academic, New York, pp 150–167

Lindner HR, Perel E, Friedlander A, Zeitlin A (1972) Specificity of antibodies to ovarian hormones in relation to the site of attachment of the steroid hapten to the peptide carrier. Steroids 19:357–375

Liu CT, Adler FL (1973) Immunologic studies on drug addiction. I. Antibodies reactive with methadone and their use for detection of the drug. J Immunol 111:472–477

Manberg PJ, Youngblood WW, Kizer JS (1982) Development of a radioimmunoassay for Pro-Leu-Gly-NH_2 (PLG or MIF-1): evidence that PLG is not present in rat brain. Brain Res 241:279–284

Matsukura S, West CD, Ichikawa Y, Jubiz W, Harada G, Tyler FH (1971) A new phenomenon of usefulness of radioimmunoassay of plasma adrenocorticotropic hormone. J Lab Clin Med 77:490–500

Mayes D, Furuyama S, Kem DC, Nugent CA (1970) A radioimmunoassay for plasma aldosterone. J Clin Endocrinol Metab 30:682–685

Midgley AR, Niswender GD, Rebar RW (1969) Principles for the assessment of the reliability of radioimmunoassay methods (precision, accuracy, sensitivity, specificity). Acta Endocrinol Suppl 142:163–184

Midgley AR, Niswender GD, Gay VL, Reichert LE (1971) Use of antibodies for characterization of gonadotropins and steroids. Recent Progress Hormone Res 27:235–301

Migeon CJ, Kowarski A, Beitins IZ, Bayard F (1972) Radioimmunoassay of plasma aldosterone. In: Heftmann E (ed) Modern methods of steroid analysis. Academic, New York, pp 471–491

Miller M, Moses AM (1969) Radioimmunoassay of vasopressin with comparison of immunological and biological activity in the rat posterior pituitary. Endocrinology 84:557–562

Nakao K, Sugawara A, Morii N, Sakamoto M, Suda M, Soneda J, Ban T, Kihara M, Yamori Y, Shimokura M, Kiso Y, Imura H (1984) Radioimmunoassay for α-human and rat atrial natriuretic polypeptide. Biochem Biophys Res Commun 124:815–821

Nett TM, Akbar AM, Niswender GD, Hedlund MT, White WF (1973) A radioimmunoassay for gonadotropin-releasing hormone in serum. J Clin Endocrinol Metab 36:880–885

Nieschlag E, Loriaux DL, Lipsett MB (1972) Radioligand assay for Δ^5-3β-hydroxysteroids. I. 3β-hydroxy-5 androstene-17-one and its 3-sulfate. Steroids 19:669–676

Nishina T, Tsuji A, Fukushima DK (1974) Site of conjugation of bovine serum albumin to corticosteroid hormones and specificity of antibodies. Steroids 24:861–874

Niswender GD, Midgley AR (1970) Hapten radioimmunoassay for steroid hormones. In: Peron FG, Caldwell BV (eds) Immunologic methods in steroid determination, chap 8. Appleton, New York

Nossal GJ (1975) Kinetics of antibody formation and regulatory aspects of immunity. Acta Endocrinol [Suppl] 78:96–116

Oliver GC Jr, Parker BM, Brasfield DL, Parker CW (1968) The measurement of digitoxin in human serum by radioimmunoassay. J Clin Invest 47:1035–1042

Orth DN (1979) Adrenocorticotropic hormone (ACTH). In: Jaffe BM, Behrman HR (eds) Methods of hormone radioimmunoassay. Academic, New York, pp 245–284

Oyama SN, Kagan A, Glick SM (1971) Radioimmunoassay of vasopressin: application to unextracted human urine. J Clin Endocrinol Metab 33:739–744

Permutt MA, Parker CW, Utiger RD (1966) Immunochemical studies with lysine vasopressin. Endocrinology 78:809–814

Quijada L, Kim YT, Siskind GW, Ovaru Z (1974) Immunogenicity of lowly and highly derivatized dinitrophenylated bovine γ-globulin in different strains of guinea-pigs. J Immunol 113:1296–1301

Rahmani R, Barbet J, Cano JP (1982) Development of a radioimmunoassay for 5′noranhydrovinblastine (navelbine), a new anti-tumor vinca alkaloid. In: Radioimmunoassay and related procedures in medicine. Proceedings of a symposium, Vienna, 21–25 June 1982. International Atomic Energy Agency, Vienna, pp 367–369

Rao PN, Moore PH (1977) Synthesis of new steroid haptens for radioimmunoassay. Part IV. 3-*O*-carboxymethyl ether derivatives of estrogens. Specific antisera for radioimmunoassay of estrone, estradiol-17β, and estriol. Steroids 29:461–469

Raz A, Stylos WA (1973) Specificity of prostaglandin A_1 antiserum against prostaglandin A_1 and B_1. FEBS Letters 30:21–24

Rehfeld JF, Stadil F, Rubin B (1972) Production and evaluation of antibodies for the radioimmunoassay of gastrin. Scand J Clin Lab Invest 30:221–232

Reichlin M, Schnure JJ, Vance VK (1968) Induction of antibodies to porcine ACTH in rabbits with non steroidogenic polymers of BSA and ACTH. Proc Soc Exp Biol Med 128:347–350

Richards FM, Knowles JR (1968) Glutaraldehyde as a protein cross-linking reagent. J Mol Biol 37:231–233

Riva F, Giartosio A, Turano C (1977) A pyridoxamine phosphate derivative. Methods Enzymol 46:441–447

Rodbard D, Ruder J, Vaitukaitis J, Jacobs HS (1971) Mathematical analysis of kinetics of radioligand assays: improved sensibility obtained by delayed addition of labeled ligand. J Clin Endocrinol Metab 33:343–355

Root MA, Chance RE, Galloway JA (1972) Immunogenicity of insulin. Diabetes 21 [Suppl] 2:657–660

Roth J, Glick SM, Klein LA, Peterson MJ (1966) Specific antibody to vasopressin in man. J Clin Endocrinol Metab 26:671–675

Rubin B (1972) Studies on the induction of antibody synthesis against sulfanylic acid in rabbits. I. Effect of the number of hapten molecules introduced in homologous protein on antibody synthesis against the hapten and the new antigenic determinants. Eur J Immunol 2:5–11

Rubin B, Schirrmacher V, Wigzell H (1973) Induction of antihapten antibody responses against haptens conjugated to autologous and heterologous proteins. Adv Exp Med Biol 29:133–139

Ruder HJ, Guy RL, Lipsett MB (1972) A radioimmunoassay for cortisol in plasma and urine. J Clin Endocrinol Metab 35:219–224

Salmon JA, Simmons PM, Palmer RMJ (1982) A radioimmunoassay for leukotriene B_4. Prostaglandins 24:225–235

Schick AF, Singer SJ (1961) On the formation of covalent linkages between two protein molecules. J Biol Chem 236:2477–2485

Schlichtkrull J, Branke J, Christiansen AH, Hallund O, Heding LG, Jørgensen KR (1972) Clinical aspects of insulin-antigenicity. Diabetes 21 [Suppl] 2:649–656

Sekihara H, Ohsawa N, Ibayashi H (1972) A radioimmunoassay for serum dehydroepiandrosterone sulfate. Steroids 20:813–824
Sela M (1969) Antigenicity: some molecular aspects. Science 166:1365–1374
Sela M (1971) Effect of antigenic structure on antibody biosynthesis. Ann NY Acad Sci 190:181–202
Smith TW, Butler VP Jr, Haber E (1970) Characterization of antibodies of high affinity and specificity for the digitalis glycoside digoxin. Biochem 9:331–337
Spragg J, Austen KF, Haber E (1966) Production of antibody against bradykinin: demonstration of specificity by complement fixation and radio-immunoassay. J Immunol 96:865–871
Srivastava LS, Werk EE, Thrasher K, Sholiton LJ, Kozera R, Nolten W, Knowles HC (1973) Plasma cortisone concentration as measured by radioimmunoassay. J Clin Endocrinol Metab 36:937–943
Stason WB, Vallotton M, Haber E (1967) Synthesis of an antigenic copolymer of angiotensin and succinylated poly-L-lysine. Biochim Biophys Acta 133:582–584
Steiner AL, Parker CW, Kipnis DM (1972) Radioimmunoassay for cyclic nucleotides. I. Preparation of antibodies and iodinated cyclic nucleotides. J Biol Chem 247:1106–1113
Straus E, Yalow RS (1977) Specific problems in the identification and quantitation of neuropeptides by radioimmunoassay. In: Galner H (ed) Peptides in neurobiology. Plenum, London, pp 39–60
Stylos WA, Rivetz B (1972) Preparation of specific antiserum to prostaglandin A. Prostaglandins 2:103–113
Tanaka I, Misono KS, Inagami T (1984) Atrial natriuretic factor in rat hypothalamus, atria and plasma: determination by specific radioimmunoassay. Biochem Biophys Res Commun 124:663–668
Underwood RH, Williams GH (1972) The simultaneous measurement of aldosterone, cortisol and corticosterone in human peripheral plasma by displacement analysis. J Lab Clin Med 79:848–862
Vaitukaitis J, Robbins JB, Nieschlag E, Ross GT (1971) A method for producing specific antisera with small doses of immunogen. J Clin Endocrinol Metab 33:988–991
Van Eldik LJ, Watterson DM (1981) Reproducible production of antiserum against vertebrate calmodulin and determination of the immunoreactive site. J Biol Chem 256:4205–4210
Vetter W, Armbruster H, Tschudi B, Vetter H (1974) Production of antisera specific to aldosterone: effect of hapten density and of carrier protein. Steroids 23:741–756
Virtanen R, Kanto J, Iisalo E (1980) Radioimmunoassay for atropine and L-hyoscyamine. Acta Pharmacol Toxicol 47:208–212
Wallace RW, Cheung WY (1979) Calmodulin. Production of an antibody in rabbit and development of a radioimmunoassay. J Biol Chem 254:6564–6571
Walsh JH, Holmquist AL (1976) Radioimmunoassay of bombesin peptides: identification of bombesin-like immunoreactivity in vertebrate gut extracts. Gastroenterology 70:948
Webb DR Jr, Winkelstein A (1984) Immunosuppression, immunopotentiation and anti-inflammatory drugs. In: Stites DP, Stobo JD, Fudenberg HH, Vells JV (eds) Basic and clinical immunology. Lange Medical, Los Altos, pp 271–287
Weber E, Evans CJ, Chang JK, Barchas JD (1982) Antibodies specific for α-N-acetyl-β-endorphins: radioimmunoassays and detection of acetylated β-endorphins in pituitary extracts. J Neurochem 38:436–447
Webster ME, Pierce JV, Sampaio MU (1970) Studies on antibody to bradykinin. In: Sicuteri F, Rocha e Silva M, Back N (eds) Bradykinin and related kinins: cardiovascular, biochemical and neural action. Plenum, New York, pp 57–64
Weinstein A, Lindner HR, Friedlander A, Bauminger S (1972) Antigenic complexes of steroid hormones formed by coupling to protein through position 7: preparation from Δ^4-3-oxosteroids and characterization of antibodies to testosterone and androstenedione. Steroids 20:789–812

Wu WH, Rockey JH (1969) Antivasopressin antibody. Characterization of high-affinity rabbit antibody with limited association constant heterogeneity. Biochemistry 8:2719–2728

Wurzberger RJ, Miller RL, Boxenbaum HG, Spector S (1977) Radioimmunoassay of atropine in plasma. J Pharmacol Exp Ther 203:435–441

Yalow RS, Berson SA (1960) Immunoassay of endogenous plasma insulin in man. J Clin Invest 39:1157–1175

Yalow RS, Berson SA (1964) Immunoassay of plasma insulin. In: Glick D (ed) Methods of biochemical analysis, vol 12. Wiley, New York, pp 69–96

Yalow RS, Berson SA (1970a) Radioimmunoassay of gastrin. Gastroenterology 58:1–14

Yalow RS, Berson SA (1970b) General aspects of radioimmunoassay procedures. In: Proceedings, symposium on "in vitro" procedures with radioisotopes in clinical medicine and research, Vienna, 8–12 September 1969: International Atomic Energy Agency, Vienna, SM-124/106, pp 455–472

Yalow RS, Berson SA (1971a) Introduction and general considerations. In: Odell WD, Daughaday WH (eds) Principles of competitive protein-binding assay. Lippincott, Philadelphia, pp 1–21

Yalow SA, Berson RS (1971b) Problems of validation of radioimmunoassays. In: Odell WD, Daughaday WH (eds) Principles of competitive protein-binding assay. Lippincott, Philadelphia, pp 374–400

Yamada A (1960) Synthesis of succinoyl and phthaloyl derivatives of digitoxigenin, gitoxigenin and oleandrigenin. Chem Abstr 54:6807

Young JD, Byrnes DJ, Chisholm DJ, Griffiths FB (1969) Radioimmunoassay of gastrin in human serum using antiserum against pentagastrin. J Nucl Med 10:746–748

Young RN, Kakushima M, Rokach J (1982) Studies on the preparation of conjugates of leukotriene C_4 with proteins for development of an immunoassay for SRS-A. Prostaglandins 23:603–613

Youngblood WW, Kizer JS (1983) Strategies for the preparation of haptens for conjugation and subtrates for iodination for use in radioimmunoassay of small oligopeptides. Methods Enzymol 103:448–459

Youssefnejadian E, Florensa E, Collins WP, Sommerville IF (1972) Radioimmunoassay of 17-hydroxyprogesterone. Steroids 20:773–781

CHAPTER 4

Production of Monoclonal Antibodies for Radioimmunoassays

K. Brune and M. Reinke

A. Introduction

I. Rationale for the Production of Monoclonal Antibodies

Specific antibodies (Abs) have proven most useful and versatile tools for the identification, quantification, and localization of minute amounts of small and large molecules in biologic materials, e.g., body fluids, specific cells, and other body components. So far the most widely used technique for the production of specific Abs consists in immunization of animals like rabbits, goats, or horses, monitoring of Ab formation in the serum, and selection of animals which produce serum containing Abs sufficiently specific for the use envisaged. Although this approach has yielded many valuable results, it has some deficiencies.

1. Disadvantages of Polyclonal Antisera

1. The serum produced by the sensitized animal contains the product of hundreds or even thousands of different Ab-secreting cell clones. These clones comprise a dynamic population in which the relative contribution of each cell clone to the antiserum is changing with time, implying a varying concentration of different Ab groups of different sensitivity and different Ab classes.

2. Within a vertebrate organism, an injected antigen is often metabolized rapidly, leading to metabolites which are immunogenic themselves. It has proven almost impossible to develop specific Abs against certain antigens because they are metabolized very quickly or these antigens are less immunogenic than their metabolites (cf. Chap. 18). Hence, all attempts to produce sufficiently specific antisera have been frustrated.

3. Certain types of experiment require very large quantities of Abs of relatively high specificity, but at the same time relatively low affinity toward the antigen in question (e.g., affinity chromatography). In animals, the production of such large quantities of high specificity, but only moderate affinity Abs has proven almost impossible.

4. Even the best Ab-producing animal will lose its Ab production with time and eventually die. Consequently, the production of a high quality antiserum is limited, and further supply will vanish after some time.

2. Advantages of Monoclonal Antibodies

These and many other problems have limited the use of conventionally generated antisera in biomedical research. Most of them, however, may be solved by the development of monoclonal Abs (mAbs) as first described by Köhler and Milstein (1975). Using this technique, it is possible to select cell clones which produce Abs of desired specificity and affinity. They may be stored or expanded to produce almost unlimited quantities of Abs for unlimited periods of time. In the beginning, only mouse myeloma cell lines were available. Fusion of these cells with mouse lymphocytes was the standard technique. The mouse, however, is a relatively poor antigenic responder, producing mostly Abs of lower affinity than rabbits and goats (Morgan 1984). Consequently, thousands of mouse/mouse hybridoma [1] clones have to be screened to select one producing sufficiently specific and sensitive Ab (for assay purposes). The development of stable rabbit hybridomas has so far not been achieved (Goding 1980). More successful was the development of rat/mouse hybridomas. Rats provide larger spleens with more lymphocytes for fusion and may lead to larger numbers of Ab-producing clones (Goding 1980; Hämmerling et al. 1981). Rat/rat hybridomas have also been produced. These hybridomas show a high Ab secretion rate which is lost only exceptionally (Clark et al. 1983). Lately, attempts to produce human/mouse and also human/human hybridomas have ben successful (Lane and Fauci 1983; Olsson et al. 1983). Stable clones producing satisfactory Abs will be of obvious importance in medicine.

II. Applications of Monoclonal Antibodies

The new options deriving from the possibility of producing mAbs are a rapidly expanding area. They range from serologic diagnostic procedures, monitoring of structural features of macromolecules, localization of specific cells (tumor cells) or molecules (cell marker, hormones, drugs in tissues, etc.) to highly sensitive purification techniques. mAbs will soon be applied in tumor and antimicrobial therapy (Goding 1980; Obrist 1983). These areas are still developing and are only mentioned as examples. In pharmacology, the possibility seems to be particularly attractive: to produce Abs against drugs, mediators, and hormones without any cross-reaction with their metabolites or structurally related compounds; or to produce Abs specific toward metabolites, but lacking cross-reactivity with their parent compounds. Specific mAbs appear to be particularly suited for this purpose.

III. Problems

Although the advantages of the hybridoma technique of producing mAbs are evident, many technical difficulties remain to be overcome. They concern all levels of the multistep technique, i.e., immunization and cell fusion as well as testing, selecting, and cloning of the hybridomas. In addition, care has to be taken in characterizing and purifying the resulting Abs. Altogether, it comprises a demanding and difficult technique.

[1] Monoclonal antibodies producing hybrid cells.

When spleen cells are fused with myeloma cells, the incidence of the formation of a heterokaryon[2] is 1%. The chance that such a heterokaryon develops into a stable synkaryon[3] is 1 in 1000. The chance that such a hybridoma produces an Ab of the desired quality is less than 1 in 1000. These probability data suggest that approximately 10^{-8} spleen cells have to be fused in order to yield one satisfactory stable, and suitable Ab-producing clone (HÄMMERLING et al. 1981). This rough estimate implies that many immunizations and fusions will end unsuccessfully. In addition, it means that large numbers of cell clones have to be assayed for Ab production and for the quality of the Abs produced. It should be added that unexpected events such as infections, technical problems, etc., may ruin an otherwise promising experiment. Moreover, about 50% of the hybridomas lose their ability to produce Abs or the clones themselves are lost by overgrowth of nonproducing but faster reproducing cells which may develop owing to chromosomal losses within 3–8 weeks after fusion. In order to avoid overgrowth by nonsecreting cells, frequent cloning of "good" hybridomas is necessary to maintain high quality clones, thus adding to the work involved (GODING 1980; OLSSON et al. 1983).

B. Techniques

I. Materials

1. Cells

Three different cell types are necessary for the production of hybridomas.

a) Myeloma Cells

It was the major contribution of KÖHLER and MILSTEIN (1975, 1976) to develop myeloma cell variants which could be fused with immune lymphocytes, bringing about the expression of immunoglobulins (Igs) of the immunoblast, even if they were not secreting Igs themselves. This success was based on experience with myeloma lines resistant to 8-azaguanine or 5-bromo-2′-deoxyuridine and lacking the enzymes hypoxanthineguanosinephosphoribosyl transferase or thymidine kinase (COTTON and MILSTEIN 1973; KÖHLER and MILSTEIN 1975). These cell lines do not grow in aminopterin-containing media, even if these media contain hypoxanthine and thymidine, because they cannot utilize the "rescue pathways" for DNA synthesis employing thymidine kinase and hypoxanthineguanosinephosphoribosyl transferase as "normal" cell types can (LITTLEFIELD 1964). These myeloma cell lines, therefore, will die in media containing hypoxanthine, aminopterin, and thymidine (HAT) in addition to the normal tissue culture medium constituents such as Dulbecco's modified Eagle's medium complemented with L-glutamine, sodium pyruvate, 2-mercaptoethanol, and fetal calf serum. Immunoblasts, lymphocytes, and most other spleen cells, on the other hand, do not proliferate under normal tissue culture conditions. Therefore, hybridomas resulting from the fusion of cells of the defective myeloma cell lines and normal spleen cells (carrying the enzymes for HAT resistance) will proliferate in HAT medium although both parent cell

[2] System of genetically different nuclei within one cell.
[3] Zygote nucleus generated by fusion of the heterokaryon nuclei.

Table 1. Cell lines used for hybridization

Cell line	Derivation	Drug resistance	Ig chains H	Ig chains L	References
Mouse					
X63-Ag8 (P3)	MOPC-21	8-Ag	γ_1	κ	KÖHLER and MILSTEIN (1975)
X63-Ag8.653	P3	8-Ag			KEARNEY et al. (1979)
SP2/0-Ag14 (SP2)	P3 × BALB/*c*	8-Ag			SHULMAN et al. (1978)
NSI/1-Ag4-1	P3	8-Ag		κ	KÖHLER and MILSTEIN (1976)
FO	SP2 × SP2	8-Ag			FAZEKAS DE ST. GROTH and SCHEIDEGGER (1980)
Rat					
210. RCY3. Ag1	Lou rat	8-Ag		κ	COTTON and MILSTEIN (1973)
Human					
U-266 AR_1	U-266	8-Ag	ε	λ	CROCE et al. (1980)
GM 1500 6TG-AL2	GM 1500	8-Ag	γ_2	κ	OLSSON and KAPLAN (1980)

lines will not. Hence, selection of clones resulting from successful cell fusions between myeloma cells and lymphocytes is easy. Table 1 gives information about some HAT-sensitive cell lines used so far successfully in hybridoma research.

b) B-Lymphocytes

For production of hybridomas, ideally B-lymphocytes are used because they are genetically programmed for the production of specific Abs to a certain antigen. These B cells should be in a state of proper activation and differentiation to ensure the production of hybrids, secreting high titers of a specific Ab. In mouse hybridoma production, B cells are recovered as spleen or lymphoid cells from adequately immunized female BALB/*c* mice used at the age of 4–8 weeks. This is because all myeloma cell lines used are of BALB/*c* origin, and sometimes it is necessary to implant the hybridomas into the peritoneal cavity of preferably female BALB/*c* mice (see Sect. B.II.4.a). Experiments with rat/mouse hybridomas reported so far have employed lymphocytes from, for example, Sprague-Dawley rats (TANAKA et al. 1985). Attempts to produce human hybridomas have sometimes made use of spleen cells from human spleen, but in most cases peripheral blood lymphocytes, separated from whole blood, were used. The use of lymphocytes from tonsils, lymph fluid, and bone marrow has, however, also been reported (for review see LANE and FAUCI 1983).

c) Feeder Cells

As stated before, hybridoma cells are at risk after fusion because they may lack growth-enhancing factors in their environment or they may succumb to minor microbial contaminations. For these reasons, many researchers employ feeder layers of cells which do not proliferate themselves, i.e., compete with the hybridomas for space, nutrients, etc., but secrete trophic factors and, in addition, may provide some antimicrobial defense. For mouse and rat hybridomas, macrophages have frequently been employed (HENGARTNER et al. 1978; FAZEKAS DE ST. GROTH and SCHEIDEGGER 1980). They appear to support cell proliferation by the production of monokines (READING 1982) if seeded in adequate density. Moreover, they clear cellular debris present in cell cultures after fusion owing to decaying myeloma cells in HAT medium and convey antimicrobial activity. Human/human hybridomas have been grown on monocyte or thymocyte layers. Mouse macrophages appear to be unsuitable for this purpose, they appear to phagocytose human hybridomas (OLSSON et al. 1983). The drawback of employing macrophages as feeder layers results from the possibility that these cells, being isolated from animal or human donors, may be contaminated with microorganisms, increasing the risk of microbial infection of the cultures and hybridoma loss. This risk may be overcome by adding endothelial cell growth supplement (ECGS) or human endothelial cell-conditioned supernatants (HECS) to the selection medium. Such supplemented medium has been successfully applied by ASTALDI et al. (1981, 1982) and PINTUS et al. (1983).

II. Methods

1. Immunization

a) In Vivo

Immunization as a prerequisite for the production of immunoblasts for hybridoma research follows the same principles as for the production of polyclonal Abs (for details see also Chap. 3 and 6). A few aspects ought to be remembered, particularly when it comes to immunization against small, soluble antigens as is usual in pharmacologic research. They are less immunogenic compared with, say all surface antigens (ROITT 1977). First, to enhance the rudimentary antigenicity of small, chemically well-defined molecules (haptens) it is necessary to bind these haptens covalently to larger protein molecules. It appears that the hapten-specific antigenicity is increased if carrier molecules alien to the animal species to be immunized are chosen. Most successful reports are based on the use of haptens conjugated to bovine serum albumin, keyhole limpet hemocyanin, or similar proteins. It appears that the antigenicity of the hapten is enhanced when larger numbers of hapten molecules are bound to the same carrier protein (ROITT 1977).

Second, the antigenicity may be enhanced further if metabolism of the antigen is difficult, so that it takes place slowly and preferentially in macrophages (ROITT 1977). Antigen incorporated in the oil droplets of Freund's adjuvant (FA) shows retarded metabolism and guarantees a reservoir of antigen. Adsorption to inorganic aluminum compounds shows a similar effect (ROITT 1977; COOPER 1981). In addition, complete FA contains heat-killed mycobacteria (*Mycobacterium tuberculosis*) which are assumed to initiate nonspecific stimulation of the reticuloendothelial system, thus enhancing the antigenicity of the antigen (COOPER 1981). Another factor apparently of proven value is the selection of the site of im-

Table 2. Immunization protocol. (Modified from Köhler and Milstein 1976)

Time	Dose of antigen (Ag) per mouse	Technical details
Zero time, 1st injection	5–100 μg; 0.5 ml	Ag solution + complete FA (1 + 1), s.c. and/or i.p.
3–21 days later, 2nd injection	5–20 μg; 0.5 ml	Ag solution + incomplete FA (1 + 1), s.c. and/or i.p.
Every 3–21 days, booster injections	5–20 μg; 0.5 ml	Ag solution in PBS, i.p.
3–1 days before fusion, last injection	50–100 μg; 0.5 ml	Ag solution in PBS, i.v.

Ag antigen; FA Freund's adjuvant; PBS phosphate-buffered saline

munization. In principle, administration at multiple subcutaneous sites guarantees a long-lasting reservoir while intramuscular injection guarantees rapid access to the lymphoid system (Cooper 1981). Intravenous application causes rapid distribution into blood-perfused elements of the lymphatic system, particularly the spleen. Consequently, immunization schedules such as those shown in Table 2 are recommended. Nevertheless, many alternative schedules have been proposed, depending on the personal experience and preference of the author (cf. Goding 1980; Stähli et al. 1980; Cianfriglia et al. 1983; Morgan 1984).

Third, it appears necessary to choose the right hapten dose for immunization. This decision is sometimes dictated by the small amounts of hapten available. If this is not the limiting factor, care should be taken to avoid both overdosage and underdosage. In both cases, immunization may fail owing to low or high dose tolerance (Cooper 1981). High doses may also induce the growth of lymphocytes expressing low affinity toward the antigen. They may lead to clones producing low affinity Abs. Finally, high doses injected for prolonged periods of time may activate additional lymphocyte clones reactive toward major metabolites or structurally related compounds which again will bring about the development of unwanted clones (Cooper 1981). Despite these principles, one may conclude that personal experience is most widely used as a guideline for selecting the right dosage of the immunogen.

b) In Vitro

Within the last few years, immunization in vitro has been tried successfully (Hengartner et al. 1978; Astaldi et al. 1982; Reading 1982; Olsson et al. 1983). The advantage of this method consists in the very small amounts of antigen required (Luben et al. 1982). In addition, the time of exposure to the immunogen can be kept short, and, for the production of human hybridomas in vitro, immunization is frequently the only acceptable method of immunization since in vivo immunization is precluded for ethical reasons. The principles of in vitro immunization are as follows.

Isolated blood lymphocytes are exposed for 5–7 days to the antigen under culture conditions (for details see Reading 1982). In order to increase the number

Table 3. Growth-stimulating substances for lymphocyte culture

Name	Abbreviation	References
B cell mitogens		
Reformalinized *Staphylococcus aureus,* Cowan I strain	STA	DOSCH et al. (1980), SCHUURMAN et al. (1980), HEITZMANN and COHN (1983)
Lipopolysaccharide from *Escherichia coli*	LPS	KEARNEY and LAWTON (1975a), KEARNEY and LAWTON (1975b), PASLAY and ROOZEN (1981), READING (1982)
T cell mitogens		
Phytohemagglutinin	PHA	KEARNEY and LAWTON (1975b), DOSCH et al. (1980)
Pokeweed mitogen	PWM	DOSCH et al. (1980), ASTALDI et al. (1982), READING (1982), OLSOON et al. (1983)
Lectin from *Canavalia ensiformis*	Con A	DOSCH et al. (1980), READING (1982)
Staphylococcal protein A	SpA	DOSCH et al. (1980)

of lymphoblasts, enhancing factors (Table 3) are added together with the antigen. Some authors also report the cocultivation of lymphocytes with human endothelial cells (ASTALDI et al. 1982) or the addition of thymocyte culture-conditioned medium (LUBEN et al. 1982; READING 1982) to improve lymphoblast proliferation. Lack of fetal calf serum in the culture media precludes the possible problem of competition with antigenic binding sites by serum components. This method allows for response to less immunogenic antigens which would be impossible in the presence of fetal calf serum (VAN NESS et al. 1984). Such media may require supplementation with SiO_2, 2-mercaptoethanol, and glutathione (BURGER 1982).

c) Secondary Immunization Procedures

In vivo and in vitro immunization may be combined to yield enhanced success rates. In other words, immunization is started in vivo as described, then lymphoid cells are transferred into in vitro cultures together with growth-enhancing factors or mitogens and small amounts of antigen (FOX et al. 1981; BUTLER et al. 1983). Alternatively, spleen cells from preimmunized animals may be inoculated into lethally irradiated syngeneic recipients. The transplanted lymphoid cells will proliferate actively, particularly cell clones, which find antigenic stimulation in the new environment. This is the case if one injects the antigen in question into the recipients. Approximately 4 days after adoptive cell transfer, the spleens of the recipients are removed and the cells fused as usual (FOX et al. 1981; BOHN and KÖNIG 1982).

2. Fusion

As soon as the serum of immunized animals shows a sufficiently high titer of specific binding of the immunogen, or after 5–7 days of lymphocyte culture with antigen and other factors (cf. Sect. B.II.1.b), fusion is attempted. In principle, the technique is still the same as originally described by KÖHLER and MILSTEIN (1975, 1976) (for fusion protocols see GODING 1980; PEARSON et al. 1980; HÄMMERLING et al. 1981; LØVBORG 1982; READING 1982; MORGAN 1984). Many modifications have, however, been introduced in order to increase the success rate. From our present knowledge, the following guidelines may be followed for optimal success.

a) Cells

For fusion, the myelomas should be in the logarithmic growth phase. They have to be separated from serum components which might interfere with the fusion (MORGAN 1984). The ratio of myeloma cells to lymphocytes should be around 1 : 2 according to BUTLER et al. (1983).

b) Fusion Medium

Since PONTECORVO (1976) and GALFRE et al. (1977) described the use of polyethylene glycol (PEG), all researchers initiate cell–cell fusion by this method. Formerly, Sendai virus was used (KÖHLER and MILSTEIN 1976). This agent works as well as PEG, but it is less easy to handle. The PEG used may be of different molecular weight. At present, most investigators use PEG of molecular weight 500–4000 or 6000 (DAVIDSON and GERALD 1976; GODING 1980). There is no evidence that either is better than the other. More important appears to be the purity of the PEG and its concentration during fusion. If the concentration is kept below 30%, the fusion frequency is low. If concentrations above 50% are reached, toxicity predominates. Consequently, most researchers use concentrations between 45% and 50% (DAVIDSON and GERALD 1976; GODING 1980; LØVBORG 1982). The fusion rate may also be influenced by the pH of the PEG solution. An optimum appears to be around pH 8 (SHARON et al. 1980). It should also be kept in mind that heat sterilization and storage in the light reduce effectiveness and stability of PEG (KADISH and WENC 1983).

c) Fusion Process

Cells should be incubated with the PEG solution at 21 °C (FAZEKAS DE ST. GROTH and SCHEIDEGGER 1980; KLEBE and MANCUSO 1981) for approximately 1 min (DAVIDSON and GERALD 1976). The addition of 5% dimethylsulfoxide to the PEG solution is believed to increase the fusion rate (LØVBORG 1982). Immediately after fusion, the PEG solution has to be diluted stepwise within 5–10 min (GODING 1980; MORGAN 1984). DAVIDSON and GERALD (1976), however, claim better success with rapid, one-step dilution. Apparently, it is more important to keep the cells for about 15 min in calcium-free medium which appears to increase the number of hybridomas formed (SCHNEIDERMAN et al. 1979). After cell fusion, transfer in either normal tissue culture medium or HAT medium is recommended, supplemented with HECS or ECGS, if feeder cells are not present (see Sect. B.I.1.c).

3. Selection of Hybridomas

a) Selection of Proliferating Clones

Immediately after fusion, cells may be transferred into HAT-containing medium (cf. Sect. B.I.1.a) as recommended by FAZEKAS DE ST. GROTH and SCHEIDEGGER (1980) and MORGAN (1984). Other authors suggest transfer into HAT medium 1–2 days after fusion according to KÖHLER and MILSTEIN (1976). It appears to be a matter of taste whether to use simple HAT medium or insulin-containing HAT medium, which has been found advantageous by BARTAL et al. (1984). Since only hybridomas formed between lymphocytes and myeloma cells can survive and actively proliferate in HAT medium, rapid cell decay occurs and the first supply of fresh medium is necessary only 1 week after fusion. In most cases, so-called demifeeding is practiced, i.e., half of the medium is carefully replaced by fresh medium in order to avoid removal of trophic factors. The cultures should, however, be monitored for color changes indicative of too low pH so that fresh medium can be supplied earlier, if necessary. After about 14 days in HAT medium, the cells are transferred for about 1 week in hypoxanthine-thymidine (HT) medium before culture in normal tissue culture medium has begun. During the whole period, all cultures have to be checked for microbial contamination. It appears advisable to start cloning early after transfer into normal tissue culture medium in order to avoid overgrowth of Ab-producing cells by nonproducers. As soon as culture supernatants show antigen-specific binding, a few cells should be cloned in soft agar medium according to the techniques described by SANDERS and BURFORD (1964), PLUZNIK and SACHS (1965), GODING (1980), and PEARSON et al. (1980), or by limiting dilution as described by HÄMMERLING et al. (1981) and MORGAN (1984). Both techniques imply the use of feeder layers or growth factors as described in Sect. B.I.1.c.

b) Selection of Antibody-Producing Hybridomas

The following aspects should be borne in mind. On the one hand, it is necessary to screen all hybridomas for Ab-producing clones. This can be achieved by virtually every assay capable of measuring nanogram quantities of antibodies, e.g., plaque-forming cell assay, indirect hemagglutination, immunoprecipitation, radioimmunoassay (RIA), indirect immunofluorescence, and enzyme-linked immunosorbent assay (ELISA) (for references see LANE and FAUCI 1983). Most widely used are the standard RIA and ELISA. Application of some of these test systems also allows for the determination of the specificity and cross-reactivity of the Abs produced. In addition, a definition of the immunoglobulin classes and subclasses is possible using the same methods. Normally, the Abs produced are IgGs, only in about 5%–7% of the clones does IgM predominate (CIANFRIGLIA et al. 1983). The subclasses are detected by standard methods such as immunodiffusion (Ouchterlony analysis) (for details see HÄMMERLING et al. 1981), RIA (e.g., STORCH and LOHMANN-MATTHES 1984), and ELISA (for details see HÄMMERLING et al. 1981; READING 1982).

4. Production of Monoclonal Antibodies

Two different methods are in principle applicable.

a) In Vivo

This technique is frequently applied in order to produce relatively large quantities of highly concentrated Abs. Mouse hybridomas derived from immunized BALB/*c* mice by fusion with BALB/*c*-derived myeloma cells are normally antigenically compatible with BALB/*c* mice. They can thus be transferred into the peritoneal cavity of female BALB/*c* mice in order to initiate production of an ascites fluid. In practice, the following procedure is widely used.

The animals are pretreated with about 0.5 ml pristane (2,6,10,14-tetramethylpentadecane) 3–30 days before injection of hybridomas. For best results (HOOGENRAAD et al. 1983), 2×10^6–2×10^7 hybridoma cells should be injected into the peritoneal cavity approximately 10 days after pristane injection. Tumor growth, leading to ascites production in the peritoneal cavity, should be visible about 2–4 weeks after cell injection. Mice may produce up to 10 ml fluid containing up to 25 mg Abs per milliliter. This concentration is about 100 fold higher than can be achieved in vitro. The method suffers from the fact that ascites fluid does not always develop. Then, irradiation of acceptor mice with 3–4 Gy (GODING 1980; HÄMMERLING et al. 1981) or inoculation of hybridomas into immunodeficient athymic nude mice or athymic nude rats, respectively (HÄMMERLING et al. 1981) may solve the problem. If rat/mouse hybridomas are used, Ab production in ascites fluid is only achieved in nude mice primed with pristane or irradiated, pristane-pretreated nude rats, since suppressor T cells have to be eliminated in the rats' spleen (NOEMAN et al. 1982). In order to achieve maximum yields, it is necessary to reconstitute the ascites fluid with saline after tapering and to harvest fresh ascites again after approximately 2 weeks. Obviously, ascites fluid contains all types of mouse or rat proteins in addition to Ab; the yield is variable, and animals are subjected to serious disease which is not always acceptable for ethical reasons.

b) In Vitro

In principle, mAbs may be produced in vitro under tissue culture conditions. On the other hand, hybridomas are grown under normal conditions, i.e., tissue culture medium, as described before, is used which of course contains all serum proteins, including Abs. If the Abs are to be used in a conventional RIA, dye-free culture medium should be used, otherwise, color quenching may be unavoidable. If pure Abs are to be produced, separation of the desired Ab from others and different serum components is necessary. To overcome these difficulties, a synthetic medium may be used which does not contain Abs. All known serum-free media are based on the investigations of ISCOVE and MELCHERS (1978). Most of these synthetic media contain a combination of normal serum components, especially albumin, transferrin, and some selenium, all dissolved in Dulbecco's modified Eagle's medium enriched with amino acids and vitamins (ISCOVE and MELCHERS 1978; BOTTENSTEIN and SATO 1979; MURAKAMI et al. 1982; KAWAMOTO et al. 1983; YSSEL et al. 1984). Published results (CHANG et al. 1980; MCHUGH et al. 1983) indicate that hybridomas grow in such media and produce similar amounts of Ab as in media containing serum. Growth, however, appears to be retarded and the cells appear particularly sensitive to "stressful" culture conditions, e.g., to very low and very high cell density in the culture (McHugh et al. 1983). Nevertheless, this technology opens a new field of Ab production by overcoming many difficulties of the separation procedures necessary to obtain highly purified antibodies for specific investigational purposes. This technique is rapidly expanding. It is applied in stationary cultures as well as in suspension cultures, especially in the production of mAbs in large quantity (e.g., Fazekas de St. Groth 1983; Marcipar et al. 1983). Both technologies, production of mAbs in vivo and in vitro, are now frequently applied.

In addition, biochemical purification methods are used to eliminate unwanted serum proteins from ascites or tissue culture supernatants. These methods are necessary even if serum-free media are used since cells produce not only antibodies, but also a large variety of metabolic products which have to be eliminated. The details of the purification procedure depend on the species and subclass of Ab. The most commonly used methods are affinity chromatography and ion exchange chromatography on DEAE columns (for references see Goding 1978, 1980; Bruck et al. 1982; Morgan 1984; Stanker et al. 1985).

C. Results and Outlook

Despite the obvious advantages of mAbs for pharmacologic purposes, this technique has not been widely applied so far. A list of mAbs developed against pharmacologically important molecules such as drugs, drug metabolites, mediators, and peptide or proteohormones is given in Table 4. Most investigations have concentrated on developing mAbs against large molecules, e.g., proteohormones, receptors, growth factors. Small molecules such as drugs, steroid hormones, or so-

Table 4. Substances of pharmacologic interest[a] against which monoclonal antibodies have been described

Substance	References
1. Drugs	
Alprenolol	Chamat et al. (1984), Sawutz et al. (1985)
Benzodiazepines	De Blas et al. (1985)
Cyclosporin	Quesniaux et al. (1985)
Digoxin	Margolies et al. (1981), Hunter et al. (1982), Zalchberg et al. (1983), Pincus et al. (1984), Buchman et al. (1985)
Digitoxin	Colignon et al. (1984)
Digitalin	Edelman et al. (1984)
Gentamicin	Place et al. (1984)
7-Hydroxychlorpromazine	Yeung et al. (1985)
Methotrexate	Kato et al. (1984)
Morphine	Glasel et al. (1983)
Nortriptyline	Marullo et al. (1985)

[a] Enzymes, receptor proteins, etc., are not included.

Table 4 (Continued)

Substance	References
2. Steroid hormones and thyroxine	
Dehydroepiandrosterone	FANTI and WANG (1984), KNYBA et al. (1985)
Deoxycorticosterone	AL-DUJAILI et al. (1984)
11-Hydroxyprogesterone	BROCHU et al. (1984)
Estradiol	FANTI and WANG (1984)
Progesterone	FANTI and WANG (1984)
Testosterone	KOHEN et al. (1982), FANTI and WANG (1983), WHITE et al. (1985)
Thyroxine	MOROZ et al. (1983), WILKE et al. (1985)
3. Peptide Hormones	
Angiotensin II	NUSSBERGER et al. (1984)
Endorphin	HERZ et al. (1982)
Enkephalins	JONES et al. (1983), PONTAROTTI et al. (1983), CUELLO et al. (1984), DEGUCHI and YOKOYAMA (1985)
Glucagon	GREGOR and RIECKEN (1985)
Insulin	BENDER et al. (1983), MADSEN et al. (1983), JORGENSEN et al. (1984), ZIEGLER et al. (1984), KUROCHKIN et al. (1985), MARKS et al. (1985), STORCH et al. (1985)
Somatomedin C	BAXTER et al. (1982)
Vasoactive intestinal polypeptide	GOZES et al. (1983)
Vasopressin	HOU-YU et al. (1982), ROBERT et al. (1985)
4. Proteohormones	
Adrenocorticotropin	WHITE et al. (1985)
Chorionic gonadotropin	FURUHASHI et al. (1983), STUART et al. (1983a), BERGER et al. (1984), TEH et al. (1984), CARAUX et al. (1985)
Growth hormone	RETEGUI et al. (1982), STUART et al. (1983b), JACOBELLO et al. (1984), RETEGUI et al. (1984), ASTON and INVANYI (1985)
Luteinizing hormone	SOOS and SIDDLE (1983), KNAPP and STERNBERGER (1984), CHOW et al. (1985)
Thyrotropin	RIDGWAY et al. (1982), JACOBELLO et al. (1984), BOETTGER et al. (1985)
Prolactin	STUART et al. (1982), ASTON et al. (1984), JACOBELLO et al. (1984), ASTON and IVANYI (1985), COOK et al. (1985), OOSTEROM and LAMBERTS (1985)
5. Mediators, messengers, transmitters	
Cyclic adenosine monophosphate	O'HARA et al. (1982)
Eicosanoid	
Prostaglandins	BRUNE et al. (1985), DAVID et al. (1985), TANAKA et al. (1985)
Leukotrienes	LEE et al. (1984)
Serotonin	MILSTEIN et al. (1983), LINDGREN et al. (1984)

Table 5. Characteristics of monoclonal antibodies against E- and F-type prostaglandins

Prostaglandins	Cross-reaction (%)			
	mAb-PGE_2 (E_2R_2)	mAb-PGE_2 (E_2R_1)	mAb-6-keto-$PGF_{1\alpha}$	mAb-$PGF_{2\alpha}$
PGE_2	100	100	1.83	1.32
PGE_1	5.54 (12)[a]	18.5	1.37	2.42
PGA_2	0.32 (5)	NT[b]	NT	NT
PGB_2	0.79	NT	NT	NT
PGD_2	0.12	< 1.3	<1	<1
$PGF_{2\alpha}$	0.17	21.6	0.67	100
$PGF_{1\alpha}$	0.33	NT	1.83	2.05
TXB_2	<0.05	<1.3	<1	<1
TXB_1	NT	NT	<1	<1
15-Keto-PGE_2	1.32 (2)	NT	NT	NT
15-Keto-13,14-dihydro-PGE_2	0.97	NT	NT	NT
11-Deoxy-15-keto-13,14-dihydro-11,16-cyclo-PGE_2	<0.05	NT	NT	NT
15-Keto-13,14-dihydro-$PGF_{2\alpha}$	<0.05	NT	NT	<1
15-Keto-$PGF_{2\alpha}$	<0.05	NT	NT	<1
6-Keto-$PGF_{1\alpha}$	0.07	100	100	NT
6,15-Diketo-$PGF_{1\alpha}$	0.07	NT	<1	NT
6,15-Diketo-13,14-dihydro-$PGF_{1\alpha}$	0.06	NT	<1	NT
Arachidonic acid	<0.05	NT	NT	NT
Sensitivity	0.03 (0.01)	0.22	1.1	2.0
Immunoglobulin subclass	IgG_1, κ-1 chains	IgG_1, κ-1 chains	IgG_1, κ-1 chains	IgG_1, κ-1 chains

[a] Values in parentheses give characteristics of a specific antiserum against PGE_2 (32–7) used for comparison.
[b] NT not tested.

called tissue hormones have not been used so frequently as antigens for the development of mAbs.

As an example, our results concerning the production of mAbs against prostaglandins (PGs) are listed in Table 5. The mAbs against $PGF_{2\alpha}$ and 6-keto-$PGF_{1\alpha}$ showed a high degree of specificity, displaying only about 1% cross-reactivity with the structurally related PGE_2, PGE_1, and $PGF_{1\alpha}$. These mAbs, however, allowed only for the measurement of fairly high concentrations of 6-keto-$PGF_{1\alpha}$ or $PGF_{2\alpha}$ (1–2 ng) when used in the conventional solution RIA employed by us routinely (Brune et al. 1981; cf. Chap. 19). In contrast, our first mAb against PGE_2 (E_2R_1) allowed for the detection of much smaller amounts of PGE_2 (>200 pg), but its cross-reaction with 6-keto-$PGF_{1\alpha}$, PGE_1, and $PGF_{2\alpha}$ was considerable. The second mAb against PGE_2 (E_2R_2) proved to be very specific, displaying some cross-reactivity only with PGE_1 (~5%), and 15-keto-PGE_2 (1%), but not with 6-keto-$PGF_{1\alpha}$ or a great variety of other PGs. It compared very well in that respect with a widely used antiserum (Table 5). In addition, this mAb, if employed in the conventional fluid phase RIA, allowed for the detection of small amounts of PGE_2 (30 pg). In further experiments (Fig. 1), it was shown that this

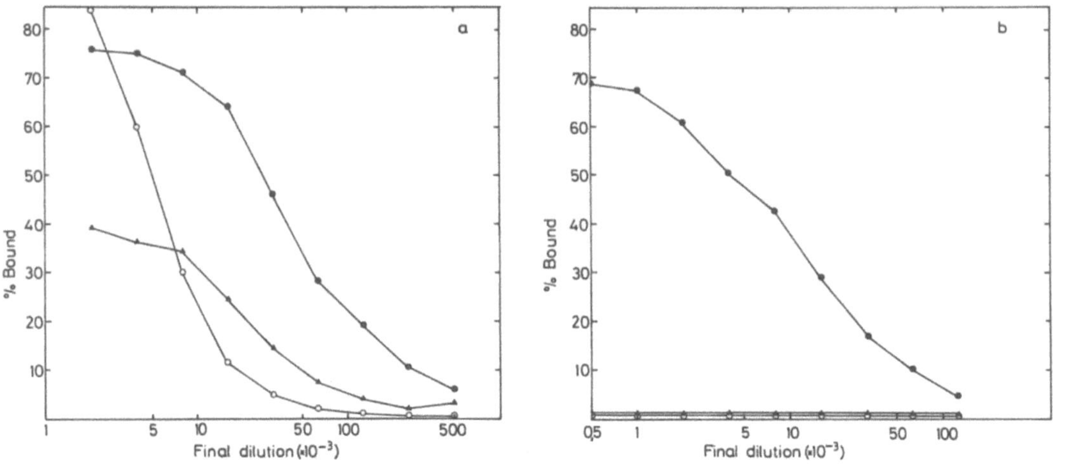

Fig. 1 a, b. Binding of ^{3}H PGE_2 (*full circles*); ^{3}H PGA_2 (*open circles*); and ^{3}H PGB_2 (*full triangles*) at equal specific activity by dilutions of **a** specific polyclonal antiserum (32–7) and **b** mA-PGE_2 (E_2R_2)

mAb was devoid of the ability to bind ^{3}H PGA_2 and ^{3}H PGB_2, which is usually displayed by antisera against PGE_2. As shown in Fig. 1, even at high concentrations, the mAb-PGE_2 does not bind PGA_2 or PGB_2, while the antiserum does.

Our results show that it is possible to raise hybridoma clones which produce mAbs against PGs in vitro. Some of these mAbs can be used in conventional RIAs, allowing for the direct and sensitive detection of biologically relevant PGs. For example, our mAb-PGE_2 (E_2R_2) is now routinely used in a standard RIA assessing PG release from macrophages, yielding almost identical results to a well-defined specific antiserum against PGE_2 (Table 5). The mAb-6-keto-$PGF_{1\alpha}$ and mAb-$PGF_{2\alpha}$ are not of comparable quality in conventional fluid phase assays owing to relatively high detection limits. They have, however, proven useful when employed in an immunosorbent assay on plastic surfaces. The mAb concentration in hybridoma supernatants is high enough for the direct use of these supernatants in the assay procedure. For example, the dilutions given in Fig. 1 are direct dilutions from pooled hybridoma supernatants harvested after approximately 3 days (cell content at that time $\sim 5 \times 10^6$ cells per milliliter). In other words, it is not necessary to initiate the production of mAbs in the peritoneal cavity of BALB/*c* mice for sufficient mAb concentrations. Furthermore, the data prove that it is possible to circumvent problems inherent in conventionally produced polyclonal antisera. The mAb-$PGF_{2\alpha}$, mAb-6-keto-$PGF_{1\alpha}$, and mAb-PGE_2 (E_2R_2) exert a high degree of specificity, i.e., these Abs can discriminate well between minor structural differences in the side chains of PGs and the ring structure. The mAb-PGE_2 (E_2R_2), for example, is able to detect minor changes in the side chain configuration in PGE_1 or in 15-keto-PGE_2, albeit with some cross-reactivity, but there is no detectable cross-reactivity with $PGF_{2\alpha}$, PGA_2, or PGB_2 which differ in the substituent configuration of the cyclopentane ring. The small (5%) cross-reactivity with PGE_1 is less than with most published anti-PGE_2

antisera. Furthermore, it is of little relevance under most experimental conditions owing to the low production of PGE_1 by most mammalian cells. Moreover, the inherent problem of PGE_2 antisera, namely, that they contain Abs against PGA_2 and PGB_2, is avoided by selecting a clone which can only produce one species of Abs, although one may argue that unlimited quantities are rarely necessary. By this procedure, one can obtain Abs which are much more specific than those produced by the elegant method of FITZPATRICK and BUNDY (1978) who employed the metabolically stable PGE_2 analog, 9-deoxy-9-methylene-$PGF_{2\alpha}$. How these mAbs compare with those reported elsewhere in the literature (cf. Table 4) will have to be investigated.

Finally, it should be mentioned that the examples given show that it is possible to produce specific Abs of a uniform standard in almost unlimited quantities. The advantage of selecting specific clones may be even more pronounced if monospecific Abs against macromolecules are to be produced. By feeding hybridomas with labeled amino acids, one can produce labeled Abs which should prove particularly valuable in assays based on labeled Abs instead of labeled antigens. This option, together with the relative ease of purifying mAbs from culture supernatants, may even initiate the displacement of conventional RIAs by assays based on purified Abs labeled with isotopes, fluorescence markers, or enzymes.

References

Al Dujaili EA, Hubbard AL, van Heyningen V, Edwards CR (1984) Production of high affinity monoclonal antibodies to deoxycorticosterone. J Steroid Biochem 20:849–852

Astaldi GCB, Janssen MC, Lansdorp PM, Zeijlemaker WP, Willems C (1981) Human endothelial culture supernatant (HECS): evidence for a growth-promoting factor binding to hybridoma and myeloma cells. J Immunol 126:1170–1173

Astaldi GCB, Wright EP, Willems Ch, Zeijlemaker WP, Janssen MC (1982) Increase of hybridoma formation by human lymphocytes after stimulation in vitro: effect of antigen, endothelial cells, and PWM. J Immunol 128:2539–2542

Aston R, Ivanyi J (1985) Monoclonal antibodies to growth hormone and prolactin. Pharmacol Ther 27:403–424

Aston R, Young K, van den Berg H, Ivanyi J (1984) Identification of Mr variants of prolactin with monoclonal antibodies. FEBS Lett 171:192–196

Bartal AH, Feit C, Hirshaut Y (1984) The addition of insulin to HAT medium (HIAT) enhances hybridoma formation. Dev Biol Stand 57:27–33

Baxter RC, Axiak S, Raison RL (1982) Monoclonal antibody against human somatomedin-C/insulinlike growth factor. I. J Clin Endocrinol Metab 54:474–476

Bender TP, Schroer J, Claflin JL (1983) Idiotypes on monoclonal antibodies to bovine insulin. I. Two public idiotypes on anti-bovine insulin hybridomas define idiotypically distinct families of hybridomas. J Immunol 131:2882–2889

Berger P, Kofler R, Wick G (1984) Monoclonal antibodies against human chorionic gonadotropin (hCG). II. Affinity and ability to neutralize the biological activity of hCG. Am J Reprod Immunol 5:157–160

Boettger I, Papst HW, Senekowitsch R, Kriegel H (1985) Performance of two new solid-phase ligand assays for TSH using monoclonal antibodies. Nucl Med Commun 6:195–207

Bohn A, König W (1982) Generation of monoclonal DNP-specific IgM and IgE murine antibodies on the efficacy of hybridization. Mol Immunol 19:193–199

Bottenstein JE, Sato GH (1979) Growth of a rat neuroblastoma cell line in serum-free supplemented medium. Proc Natl Acad Sci USA 76:514–517

Brochu M, Veilleux R, Lorrain A, Belanger A (1984) Monoclonal antibodies for use with 125iodine-labeled radioligands in progesterone radioimmunoassay. J Steroid Biochem 21:405–411

Bruck C, Portetelle D, Glineur C, Bollen A (1982) One-step purification of mouse monoclonal antibodies from ascitic fluid by DEAE Affi-gel blue chromatography. J Immunol Meth 53:313–319

Brune K, Rainsford KD, Wagner K, Peskar BA (1981) Inhibition by anti-inflammatory drugs of prostaglandin production in cultured macrophages. Naunyn Schmiedebergs Arch Pharmacol 315:269–276

Brune K, Reinke M, Lanz R, Peskar BA (1985) Monoclonal antibodies against B- and F-type prostaglandins. FEBS Lett 186:46–50

Buchman D, Miller D, Koch G (1985) Monoclonal antibodies to digoxin: comparison of in vitro and in vivo immunization. Hybridoma 4:173–177

Burger M (1982) Immunization of mouse spleen cell cultures in the absence of serum and its proteins using SiO_2 and 2-mercaptoethanol. Immunology 45:381–385

Butler JL, Lane HC, Fauci AS (1983) Delineation of optimal conditions for producing mouse-human heterohybridomas from human peripheral blood B cells of immunized subjects. J Immunol 130:165–168

Caraux J, Chichehian B, Gestin C, Longhi B, Lee AC (1985) Non-cross-reactive monoclonal antibodies to human chorionic gonadotropin generated after immunization with a synthetic peptide. J Immunol 134:835–840

Chamat S, Hoebeke J, Strosberg AD (1984) Monoclonal antibodies specific for beta-adrenergic ligands. J Immunol 133:1547–1552

Chang TH, Steplewski Z, Koprowski H (1980) Production of monoclonal antibodies in serum free medium. J Immunol Meth 39:369–375

Chow SN, Ho-Yuen B, Lee CY (1985) Application of monoclonal antibodies of human luteinizing hormone. J Appl Biochem 7:114–121

Cianfriglia M, Armellini D, Massone A, Mariani M (1983) Simple immunization protocol for high frequency production of soluble antigen-specific hybridomas. Hybridoma 2:451–457

Clark M, Cobbold S, Hale G, Waldmann H (1983) Advantages of rat monoclonal antibodies. Immunol Today 4:100–101

Colignon A, Edelman L, Scherrmann JM (1984) A comparative study between animals and monoclonal anti-digitoxin antibodies. Int J Nucl Med Biol 11:97–101

Cook KM, Aston R, Ivanyi J (1985) Topographic and functional assay of antigenic determinants of human prolactin with monoclonal antibodies. Mol Immunol 22:795–801

Cooper TG (1981) Biochemische Arbeitsmethoden. de Gruyter, Berlin, p 242

Cotton RGH, Milstein C (1973) Fusion of two immunoglobulin-producing myeloma cells. Nature 244:42–43

Croce CM, Linnenbach A, Hall W, Steplewski Z, Koprowski H (1980a) Production of human secreting antibodies to measles virus. Nature 288:488–489

Cuello AC, Milstein C, Couture R, Wright B, Priestley JV, Jarvis J (1984) Characterization and immunocytochemical application of monoclonal antibodies against enkephalins. J Histochem Cytochem 32:947–957

David F, Somme G, Provost-Wisner A, Roth C, Astoin M, Dray F, Theze J (1985) Characterization of monoclonal antibodies against prostaglandin E_2: fine specificity and neutralization of biological effects. Mol Immunol 22:339–346

Davidson RL, Gerald PS (1976) Improved techniques for the induction of mammalian cell hybridization by polyethylene glycol. Somatic Cell Genet 2:165–176

De Blas AL, Sangameswaran L, Haney SA, Park D, Abraham CJ Jr, Rayner CA (1985) Monoclonal antibodies to benzodiazepines. J Neurochem 45:1748–1753

Deguchi T, Yokoyama E (1985) Monoclonal antibody to enkephalins with binding characteristics similar to opiate receptor. Biochem Biophys Res Commun 126:389–396

Dosch H-M, Schuurman RKB, Gelfand EW (1980) Polyclonal activation of human lymphocytes in vitro. II. Reappraisal of T and B cell-specific mitogens. J Immunol 125:827–832

Edelman L, Colignon A, Scherrmann JM, Girre C, Dally S, Reviron J (1984) Preparation and experimentation of an antidigitalin monoclonal antibody: interest in human treatment. Dev Biol Stand 57:343–347
Fanti VE, Wang DY (1983) Characterization of monoclonal antibodies raised against testosterone. J Steroid Biochem 19:1605–1610
Fanti VE, Wang DY (1984) Simultaneous production of monoclonal antibodies to dehydroepiandrosterone, oestradiol, progesterone and testosterone. J Endocrinol 100:367–376
Fazekas de St. Groth S (1983) Automated production of monoclonal antibodies in a cytostat. J Immunol Meth 57:121–136
Fazekas de St. Groth S, Scheidegger D (1980) Production of monoclonal antibodies: strategy and tactics. J Immunol Meth 35:1–21
Fitzpatrick FA, Bundy GL (1978) A hapten mimic elicits antibodies recognizing prostaglandin E_2. Proc Natl Acad Sci USA 75:2689–2693
Fox PC, Berenstein EH, Siraganian RP (1981) Enhancing the frequency of antigen-specific hybridomas. Eur J Immunol 11:431–434
Furuhashi Y, Mano H, Hattori S, Goto S, Tomoda Y (1983) Monoclonal antibodies to human chorionic gonadotropin and its beta-subunit. Nippon Sanka Fujinka Gakkai Zasshi 35:2415–2420
Galfre G, Howe SC, Milstein C, Butcher GW, Howard JC (1977) Antibodies to major histocompatibility antigens produced by hybrid cell lines. Nature 266:550–552
Glasel JA, Bradbury WM, Venn RF (1983) Properties of murine anti-morphine antibodies. Mol Immunol 20:1419–1422
Goding JW (1978) Use of staphylococcal protein A as an immunological reagent. J Immunol Meth 20:241–253
Goding JW (1980) Antibody production by hybridomas. J Immunol Meth 39:285–308
Gozes I, Milnev RJ, Liu FT, Johnson E, Battenberg EL, Katz DH, Bloom FE (1983) Monoclonal antibodies against vasoactive intestinal polypeptide: studies of structure and related antigens. J Neurochem 41:549–555
Gregor M, Riecken EO (1985) Production and characterization of N-terminally and C-terminally directed monoclonal antibodies against pancreatic glucagon. Gastroenterology 89:571–580
Hämmerling GJ, Hämmerling U, Kearney JF (1981) Production of antibody producing hybridomas in rodent systems. In: Monoclonal antibodies and T cell hybridomas. Elsevier North-Holland, Amsterdam, chap 12, p 563
Heitzmann JG, Cohn M (1983) Production of stable human-human hybridomas at high frequency. In: Boss BD, Langman R, Trowbridge I, Dulbecco R (eds) Monoclonal antibodies in cancer. Proceedings of the 4th Armand Hammer cancer symposion. Academic, New York, p 157
Hengartner H, Luzzati AL, Schreier M (1978) Fusion of in vitro immunized lymphoid cells with X63Ag8. In: Melchers F, Potter M, Warner NL (eds) Lymphocyte hybridomas. Springer, Berlin Heidelberg New York, pp 92–99 (Current Topics Microbiologycal Immunology, vol 81)
Herz A, Gramsch C, Hoellt V, Meo T, Riethmueller G (1982) Characteristics of a monoclonal beta-endorphin antibody recognizing the *N*-terminus of opioid peptides. Life Sci 31:1721–1724
Hoogenraad N, Helman T, Hoogenraad J (1983) The effect of pre-injection of mice with pristane on ascites tumour formation and monoclonal antibody production. J Immunol Meth 61:317–320
Hou-Yu A, Ehrlich PH, Valiquette G, Engelhardt DL, Sawyer WH, Nilaver G, Zimmerman EA (1982) A monoclonal antibody to vasopressin: preparation, characterization, and application in immunocytochemistry. J Histochem Cytochem 30:1249–1260
Hunter MM, Margolies MN, Ju A, Haber E (1982) High-affinity monoclonal antibodies to the cardiac glycoside digoxin. J Immunol 129:1165–1172
Iscove NN, Melchers F (1978) Complete replacement of serum by albumin, transferrin, and soybean lipid in cultures of lipopolysaccharide-reactive B lymphocytes. J Expl Med 147:923–933

Jacobello C, Malvano R, Dotti C, Vecchi GF, Belloli S, Bonezzi M, Piroddi B, Albertini A (1984) Evaluation of IRMA kits using monoclonal antibodies for growth hormone, prolactin and thyrotropin determinations. J Nucl Med Allied Sci 28:271–276

Jones CA, Lane DP, Hughes J (1983) Monoclonal antibodies to leucine encephalin. Biochem Biophys Res Commun 113:757–764

Jorgensen PN, Wu CV, Pedersen PC, Patkar SA, Zeuthen J (1984) Specificity and cross-reactivity of monoclonal antibodies against bovine insulin. Dev Biol Stand 57:313–319

Kadish JL, Wenc KM (1983) Contamination of polyethylene glycol with aldehydes: implication for hybridoma fusion. Hybridoma 2:87–89

Kato Y, Paterson A, Langone JJ (1984) Monoclonal antibodies to the chemotherapeutic agent methotrexate: production, properties and comparison with polyclonal antibodies. J Immunol Meth 67:321–336

Kawamoto T, Sato JD, Le A, McClure DB, Sato GH (1983) Development of a serum-free medium for growth of NS-1 mouse myeloma cells and its application to the isolation of NS-1 hybridoma. Anal Biochem 130:445–453

Kearney JF, Lawton AR (1975a) B Lymphocyte differentiation induced by lipopolysaccharide. I. Generation of cells synthesizing four major immunoglobulin classes. J Immunol 115:671–676

Kearney JF, Lawton AR (1975b) B Lymphocyte differentiation induced by lipopolysaccharide. II. Response of fetal lymphocytes. J Immunol 115:677–681

Kearney JF, Radbruch A, Liesegang B, Rajewsky K (1979) A new mouse myeloma cell line that has lost immunoglobulin expression but permits the construction of antibody-secreting hybrid cell lines. J Immunol 123:1548–1550

Klebe RJ, Mancuso MG (1981) Chemicals which promote cell hybridization. Som Cell Genet 7:473–488

Knapp RJ, Sternberger LA (1984) High affinity monoclonal antibodies to luteinizing hormone – releasing hormone. Preparation and binding studies. J Neuroimmunol 6:361–371

Knyba RE, Wang DY, Fanti VE (1985) A solid-phase radioimmunoassay of serum dehydroepiandrosterone sulphate using a monoclonal antibody. J Steroid Biochem 22:427–429

Kohen F, Lichter S, Eshhar Z, Lindner HR (1982) Preparation of monoclonal antibodies able to discriminate between testosterone and 5 alpha-dihydrotestosterone. Steroids 39:453–459

Köhler G, Milstein C (1975) Continuous cultures of fused cells secreting antibody of predefined specificity. Nature 256:495–497

Köhler G, Milstein C (1976) Derivation of specific antibody-producing tissue culture and tumor lines by cell fusion. Eur J Immunol 6:511–519

Kurochkin SN, Egerov AM, Gavrilova EM, Rubtsova MYU, Cherednikova TV, Severin ES (1985) Study of the monoclonal antibody-insulin interaction. Adv Enzyme Regul 23:377–386

Lane HC, Fauci AS (1983) Establishment of human-human and human-mouse B cell hybrids ant their use in the study of B cell activation. In: Haynes BF, Eisenbarth GS (eds) Monoclonal antibodies: probes for study of autoimmunity and immunodeficiency. Academic, New York, chap 8, p 131

Lee JY, Chernov T, Goetzl EJ (1984) Characteristics of the epitope of leukotriene B4 recognized by a highly specific mouse monoclonal antibody. Biochem Biophys Res Comm 123:944–1950

Lindgren PG, Lundqvist M, Norheim J, Wilander E, Oberg K (1984) Silver stains and immunocytochemical analysis with monoclonal serotonin antibodies for liver metastases of endocrine tumors. A study on percutaneous biopsy specimens. Am J Surg 148:353–356

Littlefield JW (1964) Selection of hybrids from matings of fibroblasts in vitro and their presumed recombinants. Science 145:709–710

Løvborg U (1982) Monoclonal antibodies: production and maintenance. Heinemann, London

Luben RA, Brazeau P, Böhlen P, Guillemin R (1982) Monoclonal antibodies to hypothalamic growth hormone-releasing factor with picomoles of antigen. Science 218:887–889

Madsen OD, Cohen RM, Fitch FW, Rubenstein AH, Steiner DF (1983) The production and characterization of monoclonal antibodies specific for human proinsulin using a sensitive microdot assay procedure. Endocrinology 113:2135–2144

Marcipar A, Henno P, Lentwojt E, Roseto A, Broun G (1983) Ceramic-supported hybridomas for continuous production of monoclonal antibodies. Ann NY Acad Sci 413:416–420

Margolies MN, Mudgett-Hunter M, Smith TW, Novotny J, Haber E (1981) Monoclonal antibodies to the cardiac glycoside digoxin. In: Hämmerling GF, Hämmerling U, Kearney JF (eds) Research monographs in immunology. Elsevier/North-Holland, New York, vol 3, p 367

Marks A, Yip C, Wilson S (1985) Characterization of two epitopes on insulin using monoclonal antibodies. Mol Immunol 22:285–290

Marullo S, Hoebeke J, Guillet JG, Strosberg AD (1985) Structural analysis of the epitope recognized by a monoclonal antibody directed against tricyclic antidepressants. J Immunol 135:471–477

McHugh YE, Walthall BJ, Steimer KS (1983) Serum-free growth of murine and human lymphoid and hybridoma cell lines. Biotechniques 1:72–77

Milstein C, Wright B, Cuello AC (1983) The discrepancy between the cross-reactivity of a monoclonal antibody to serotonin and its immunohistochemical specificity. Mol Immunol 20:113–123

Morgan M (1984) Monoclonal antibody production. In: Spector S, Back N (eds) Modern methods in pharmacology. Liss, New York, p 29

Moroz LA, Meltzer SJ, Bastomsky CH (1983) Thyroid disease with monoclonal (immunoglobulin G lambda) antibody to triiodothyronine and thyroxine. J Clin Endocrinol Metab 56:1009–1015

Murakami H, Masui H, Sato GH, Sueoka N, Chow TP, Kano-Sueoka T (1982) Growth of hybridoma cells in serum-free medium: ethanolamine is an essential component. Proc Natl Acad Sci USA 79:1158–1162

Noeman SA, Misra DN, Yankes RJ, Kunz HW, Gill TJ III (1982) Growth of rat-mouse hybridomas in nude mice and nude rats. J Immunol Meth 55:319–326

Nussberger J, Mudgett-Hunter M, Matsueda G, Haber E (1984) A monoclonal antibody specific for the carboxy-terminus of angiotensin. II. Hybridoma 3:373–376

Obrist R (1983) Monoclonal antibodies as drug carrier in oncology. Trends Pharmacol Sci 4:375–379

Ohara J, Sugi M, Fujimoto M, Watanabe T (1982) Microinjection of macromolecules into normal murine lymphocytes by means of cell fusion. II. Enhancement and suppression of mitogenic responses by microinjection of monoclonal anti-cyclic AMP into B lymphocytes. J Immunol 129:1227–1232

Olsson L, Kaplan HS (1980) Human-human hybridomas producing monoclonal antibodies of predefined antigenic specificity. Proc Natl Acad Sci USA 77:5429–5431

Olsson L, Kronstrøm H, Cambon-De Mouzon A, Honsik C, Brodin T, Jakobsen B (1983) Antibody producing human-human hybridomas. I. Technical aspects. J Immunol Meth 61:17–32

Oosterom R, Lamberts SW (1985) Comparison of RIA kits with polyclonal antiserum and IRMA kits with monoclonal antibodies for the determination of plasma prolactin and somatotropin in patients with pituitary adenoma. J Clin Chem Clin Biochem 23:51–54

Paslay JW, Roozen KJ (1981) The effect of B-cell stimulation on hybridoma formation. In: Hämmerling GJ, Hämmerling U, Kearney JF (eds) Monoclonal antibodies and T cell hybridomas. Elsevier, Amsterdam, p 551

Pearson TW, Pinder M, Roelants GE, Kar SK, Lundin LB, Mayor-Withey KS, Hewett RS (1980) Methods for derivation and detection of anti-parasite monoclonal antibodies. J Immunol Meth 34:141–154

Pincus SH, Watson WA, Harris S, Ewing LP, Stocks CJ, Rollins DE (1984) Phenotypic and genotypic characterization of monoclonal anti-digoxin antibodies. Life Sci 35:433–440
Pintus C, Ransom JH, Evans CH (1983) Endothelial cell growth supplement: a cell cloning factor that promotes the growth of monoclonal antibody producing hybridoma cell. J Immunol Meth 6:195–200
Place JD, Thompson SG, Burg JF, Molinaro C, Ott RA, Jensen FC (1984) Production and characterization of monoclonal antibody to gentamicin. Hybridoma 3:187–193
Pluznik DH, Sachs L (1965) The cloning of normal "mast" cells in tissue culture. Cell Comp Physiol 66:319–324
Pontarotti PA, le Borgne De Kaouel C, Verrier M, Cupo AA (1983) Monoclonal antibodies against Met-enkephalin as probe in the central nervous system. J Neuroimmunol 4:47–59
Pontecorvo G (1976) Production of indefinitely multiplying mammalian somatic cell hybrids by polyethylene glycol (PEG) treatment. Somatic Cell Genet 1:397–400
Quesniaux V, Himmelspach K, van Regenmortel MH (1985) An enzyme immunoassay for the screening of monoclonal antibodies to cyclosporin. Immunol Lett 9:99–104
Reading CL (1982) Theory and methods for immunization in culture and monoclonal antibody production. J Immunol Meth 53:261–291
Retegui LA, Milne RW, Cambiaso CL, Masson PL (1982) The recognition by monoclonal antibodies of various portions of a major antigenic site of human growth hormone. Mol Immunol 19:865–875
Retegui LA, Keefer LM, Fryklund L, De Meyts P (1984) Monoclonal antibodies and specific cell surface receptors do not discriminate between human growth hormone prepared by DNA recombinants techniques and the native hormone. Acta Physiol Pharmacol Lat Am 34:193–197
Ridgway EC, Ardisson LJ, Meskell MJ, Mudgett-Hunter M (1982) Monoclonal antibody to human thyrotropin. J Clin Endocrinol Metab 55:44–48
Robert FR, Leon-Henri BP, Chapleur-Chateau MM, Girr MN, Burlet AJ (1985) Comparison of three immunoassays in the screening and characterization of monoclonal antibodies against arginine-vasopressin. J Neuroimmunol 9:205–220
Roitt I (1977) Leitfaden der Immunologie. Steinkopf, Darmstadt, p 158
Sanders FK, Burford BO (1964) Ascites tumours from BHK.21 cells transformed in vitro by polyoma virus. Nature 201:786–789
Sawutz DG, Sylvestre D, Homcy CJ (1985) Characterization of monoclonal antibodies to the beta-adrenergic antagonist alprenolol as models of the receptor binding site. J Immunol 135:2713–2718
Schneiderman S, Farber JL, Baserga R (1979) A simple method for decreasing the toxicity of polyethylene glycol in mammalian cell hybridization. Som Cell Genet 5:263–269
Schuurman RKB, Gelfand EW, Dosch H-M (1980) Polyclonal activation of human lymphocytes in vitro. I. Characterization of the lymphocyte response to a T cell-independent B cell mitogen. J Immunol 125:820–826
Sharon J, Morrison SB, Kabat EA (1980) Formation of hybridoma clones in soft agarose: effect of pH and of medium. Somat Cell Genet 6:435–441
Shulman MC, Wilde CD, Köhler G (1978) A better cell line for making hybridomas secreting specific antibodies. Nature 276:269–270
Soos M, Siddle K (1983) Characterization of monoclonal antibodies for human luteinizing hormone, and mapping of antigenic determinants on the hormone. Clin Chim Acta 133:263–274
Stähli C, Staehelin T, Miggiano V, Schmidt J, Häring P (1980) High frequencies of antigen-specific hybridomas: dependence on immunization parameters and prediction by spleen cell analysis. J Immunol Meth 32:297–304
Stanker LH, Vanderlaan M, Juarez-Salinas H (1985) One-step purification of mouse monoclonal antibodies from ascites fluid by hydroxylapatite chromatography. J Immunol Meth 76:157–169
Storch M-J, Lohmann-Matthes M-L (1984) A new and rapid method for immunoglobulin class and subclass determination of mouse monoclonal antibodies using a solid-phase immunoradiometric assay. J Immunol Meth 68:305–309

Storch MJ, Petersen KG, Licht T, Kerp L (1985) Recognition of human insulin and proinsulin by monoclonal antibodies. Diabetes 34:808–811
Stuart MC, Underwood PA, Boscato L (1982) A monoclonal antibody suitable for the radioimmunoassay of prolactin in human serum. J Clin Endocrinol Metab 54:881–884
Stuart MC, Underwood PA, Harman DF, Payne KL, Rathjen DA, Razziudin S, von Sturmer SR, Vines K (1983a) The production of monoclonal antibodies to human chorionic gonadotrophin and its subunits. J Endocrinol 98:323–330
Stuart MC, Walichnowski CM, Underwood PA, Hussain S, Harman DF, Rathjen DA, von Sturmer SR (1983b) The production of high affinity monoclonal antibodies to human growth hormone. J Immunol Meth 61:33–42
Tanaka T, Ito S, Hiroshima O, Hayashi H, Hayaishi O (1985) Rat monoclonal antibody specific for prostaglandin E structure. BBA 836:125–133
Teh CZ, Wong E, Lee CY (1984) Generation of monoclonal antibodies to human chorionic gonadotropin by a facile cloning procedure. J Appl Biochem 6:48–55
Van Ness J, Laemmli UK, Pettijohn DE (1984) Immunization in vitro and production of monoclonal antibodies specific to insoluble and weakly immunogenic proteins. Proc Natl Acad Sci USA 81:7897–7901
White A, Gray C, Corrie JE (1985) Monoclonal antibodies to testosterone: the effect of immunogen structure on specificity. J Steroid Biochem 22:169–175
White A, Gray C, Ratcliffe JG (1985) Characterization of monoclonal antibodies to adrenocorticotrophin. J Immunol Meth 79:185–194
Wilke TJ, Sheedy TJ, Hirning DA (1985) Laboratory and clinical experience with a monoclonal antibody – based radioimmunoassay for serum total thyroxine. J Clin Chem Clin Biochem 23:35–39
Yeung PK, McKay G, Ramshaw IA, Hubbard JW, Midha KK (1985) A comparison of two radioimmunoassays for 7-hydroxychlorpromazine: rabbit polyclonal antibodies vs. mouse monoclonal antibodies. J Pharmacol Exp Ther 233:816–822
Yssel H, de Vries JE, Koken M, van Blitterswijk W, Spits H (1984) Serum-free medium for generation and propagation of functional human cytotoxic and helper T cell clones. J Immunol Meth 72:219–227
Zalchberg JR, Healey K, Hurrell JGR, McKenzie IFC (1983) Monoclonal antibodies to drugs – digoxin. Int J Immunopharmac 5:397–402
Ziegler M, Dietz H, Keilacker H, Witt S, Ziegler B (1984) Monoclonal antibodies to human insulin and their antigen binding behaviour. Biomed Biochem Acta 43:695–701

CHAPTER 5

Radioiodination and Other Labeling Techniques

J. GRASSI, J. MACLOUF, and P. PRADELLES

A. Labeling for Immunologic Assay: Introduction

Immunologic assays are based on the use of two reagents: an antibody and a labeled molecule, whose properties directly limit the quality of the measurement. The antibody constitutes the principal element of the assay and defines the specificity of the method through the particular characteristics of its binding to the antigen.

The antibody–antigen interaction is characterized by a high bonding energy, which ensures the formation of the antibody–antigen complex, even when the two constituents are present at very low concentrations. This strong affinity of the antibody for the antigen is exploited in immunologic assays in order to obtain high sensitivity. In order to measure the very low concentrations of these complexes, it is necessary to introduce into the assay a labeled molecule, which provides a detectable signal at these concentration levels. The labeling can be of the antibody or of the antigen, depending on the type of assay used. The properties of this labeled molecule (i.e., tracer), greatly influence the characteristics of the assay, and in particular its sensitivity. First, we shall briefly define the characteristics that this tracer must possess in order that full advantage may be taken of the antibody properties.

The preparation of a tracer consists in using any physicochemical method to couple an antigen, or an antibody, with an atom or molecule capable of emitting a measurable signal. Many different probes have been employed to this effect. The characteristics and respective merits of each of these signals will be discussed later in this chapter. In many cases, the introduction of the tracer moiety leads to a chemical modification of the antigen or antibody. Any effects of such modifications of the immunologic properties of the molecule should be strictly minimized. In particular, any resultant decrease in the affinity of the antibody for the antigen will lead to a desensitization of the method. This problem is especially acute in the case of low molecular weight haptens. Here, the structure of the tracer must be based on that of the immunogen which served to prepare the antibodies, so as to reproduce as closely as possible the structural design implicated in recognition by the antibody sites.

The other essential characteristic of a tracer is its specific activity, that is to say, the amplitude of the signal it provides for a given mass of labeled molecule. Indeed, if the strong binding between antibody and antigen is to be exploited fully, it is necessary to be able to measure very low concentrations of tracer, and this calls for the highest possible specific activity. This specific activity depends, of course, to a large extent on the type of signal used. From this point of view, the best results have been obtained with radioactive, enzymatic, fluorescent, or luminescent tracers. The specific activity will also depend on the number of signal moieties bound per molecule. In fact, this number is often limited to a few units as experience has shown that the introduction of a large number of signal moieties per molecule is detrimental to the immunoreactivity of the tracer, to its stability, and to its binding to elements other than the antibodies (nonspecific binding). All these considerations should be borne in mind in order to ensure the quality of the assay. The preparation of a good tracer is often, therefore, the result of a compromise between different, and sometimes contradictory, requirements: maximal immunoreactivity, maximal specific activity, maximal stability, and minimal nonspecific binding.

In the pages that follow, we shall first describe radioactive labeling, which formed the historical basis of the development of analytic immunology and which continues to play a considerable role. We shall then consider the different nonisotopic labeling methods which have been developed over the last 15 years or so. The initial value of these methods was their ability to overcome certain practical problems associated with the use of radioactive tracers, and some of them have since proved to be of greater sensitivity than radioimmunologic methods.

B. Radioiodination

I. General Considerations

Radioimmunologic techniques were initiated by S. A. Berson and R. S. Yalow through the use of the radioactive isotope of iodine ^{131}I in the labeling of insulin, which was employed in a study of the metabolism of this hormone in diabetic patients (BERSON et al. 1956). The use by these authors of a radionuclide of high specific radioactivity was invaluable in the demonstration of anti-insulin antibodies. Theoretical studies as well as experimental data (EKINS 1978; YALOW 1980) have shown that in the case of conventional radioimmunologic assays (competition assay), the range of concentrations of the molecule to be assayed depends closely on that of the tracer. Indeed, to a first approximation, it can be considered that the lowest measurable concentration of antigen for a given assay will be of the same order of magnitude as the concentration of tracer used. Thus, in endocrinology, for example, it is often necessary to measure hormone concentrations between 10^{-10} and 10^{-12} *M* (YALOW and BERSON 1968), and this calls for radioactive tracers that are easily measured at such concentrations. In practice, this requirement means that the tracer has a specific radioactivity of 20–2000 Ci/mmol, a fact which limits the number of usable radioisotopes.

Many radioisotopes are used in biology (BAILY et al. 1976) and in Table 1 we present those most frequently employed in biochemistry, together with their nuclear characteristics. With the exception of ^{14}C, whose specific radioactivity is too low, it appears that each of these radioisotopes is a good candidate for tracer synthesis in radioimmunology. However, for various reasons, the choice is almost exclusively limited to tritium and iodine (e.g., lack of precursor; labeling technique tedious or harmful to the molecule; short half-life of radionuclide; radioactive

Table 1. Nuclear characteristics of the principal radioisotopes used in biochemistry

Radioisotope	Half-life	Specific activity (Ci/matom)[a]	Type of emission	Energy of emission (MeV)
^{14}C	5730 years	0.062	β	0.156
^{3}H (tritium)	12.2 years	29	β	0.018
^{35}S	87 days	1500	β	0.167
^{32}P	14 days	9200	β	1.710
^{125}I	60 days	2200	γ X	0.035 0.027–0.032
^{131}I	8 days	16000	β γ	0.247–0.806 0.080–0.723

[a] Specific radioactivity reached for 100% isotopic abundance. The SI unit for radioactivity is the becquerel (Bq), which is equal to 1 disintegration per second, 1 Ci = 37×10^9 Bq.

safety measures too costly or handling of radionuclide illegal in nonspecialist laboratories; detection of radioactive emission difficult, polluting, or expensive).

Before considering iodine labeling itself, it is worth emphasizing the theoretical value of tritium. As it is possible to replace one or more hydrogen atoms of a molecule by isotopic exchange, a tracer which is structurally identical to the antigen or hapten can be synthesized, thereby retaining the full affinity of the antibody for the molecule. Considerable progress has been made in this field since the pioneering studies of WILZBACH (1957) on tritiation by exchange in the presence of tritium gas, and it is now possible to produce tritiated molecules of high specific radioactivity (~100 Ci/mmol) without modifying the primary structure, in particular, in the areas of medicines (BUTLER 1973), steroids (ABRAHAM 1974), proteins (TACK and WILDER 1981), and peptides (FROMAGEOT et al. 1978).

Tritiated tracers initially enjoyed considerable success, but over the last 10 years, an increasingly marked preference has developed for iodinated tracers (^{125}I). Illustrative of this is the fact that of the 104 radioimmunologic kits produced by manufacturers and indexed in the United States of America in 1983 (CLINICAL CHEMISTRY 1983), 80 use ^{125}I, 15 use tritium, and 14 involve a choice of the two isotopes. The essential reasons for this derive from the fact that: (a) the higher specific activity of ^{125}I provides greater sensitivity; (b) radioactive ^{125}I is more easily measured (solid scintillation) than tritium (liquid scintillation); (c) it is difficult to produce tritiated tracers of high specific radioactivity ($\geqq$100 Ci/mmol) outside specialist centers of the nuclear industry; and (d) much progress has been made in iodine labeling methods, notably in the labeling of peptides and proteins.

1. Principles

Two situations may arise in the labeling of a molecule with iodine:

1. The molecule to be labeled possesses one or more atoms of stable iodine, ^{127}I. This is a rather unusual case among molecules of biologic interest and the best-known example is that of the iodothyronines. Labeling is performed either through isotopic exchange, or with a noniodinated precursor.

2. The molecule possesses no iodine atoms. The introduction of iodine can, therefore, either be effected directly (covalent linkage to a carbon atom of the molecule), or indirectly, by incorporating into the molecule another molecule which has been iodinated previously or which is capable of being iodinated (conjugation labeling). These labeling methods have been described in detail in several reviews (BOLTON 1977; ARGENTINI 1982; DEWANJEE and RAO 1983; REGOECZI 1984).

2. Choice of Radioactive Iodine

Iodine radioisotopes are much used in nuclear medicine. There are approximately 23 radioisotopes of iodine that can be produced in the nuclear industry (HORROCKS 1981). The more important are listed in Table 2. Among these, only ^{131}I and ^{125}I have been employed for the preparation of tracers in radioimmunology, because of their relatively long half-lives.

a) ^{131}I

^{131}I has a half-life of 8.06 days and a complex decay behavior involving the emission of β and γ radiations, the most important being β^- 0.606 MeV (89%) and γ 0.364 MeV (80%). Its theoretical specific activity is ~16000 Ci/matom. It is prepared by irradiation of Te metal or TeO_2. In fact, the commercially available preparations of reductant-free ^{131}I contain 80%–85% ^{127}I, the reason being that the tellurium targets contain three stable isotopes, ^{126}Te, ^{128}Te, and ^{130}Te, which produce respectively ^{127}I, ^{129}I, and ^{131}I. The first of these iodine isotopes is stable, and the second is weakly radioactive. They interfere by reducing the specific radioactivity of ^{131}I. $Na^{131}I$ is available commercially at a specific radioactivity of ~20 Ci per milligram iodine.

b) ^{125}I

^{125}I has a half-life of 60.2 days and a decay pattern as complex as that of ^{131}I (Fig. 1). However, unlike ^{131}I, it can be obtained at isotopic concentrations close to 100%. ^{125}I is produced by irradiation of a target of ^{124}Xe

$$^{124}_{54}Xe(n,\gamma) \rightarrow {}^{125}_{54}Xe\,(\text{radioactive}) \xrightarrow{\text{Electron capture}} {}^{125}_{53}I$$

Commercially available preparations of $Na^{125}I$ have a specific radioactivity of ~17 Ci/mg.

Nowadays, ^{125}I is preferred to ^{131}I in the synthesis of radioligands for analytic immunology because of its radiochemical purity, its long half-life, the efficiency of detection of its radioactive emission (FREEDLENDER 1968) by scintillation counters (NaI crystal; 80%), the absence of β radiation and consequent limit-

Table 2. Half-lives of the more important radioisotopes of iodine

Isotope	^{123}I	^{124}I	^{125}I	^{131}I	^{132}I
Half-life	13 h	4 h	60 days	8.09 days	2.3 h

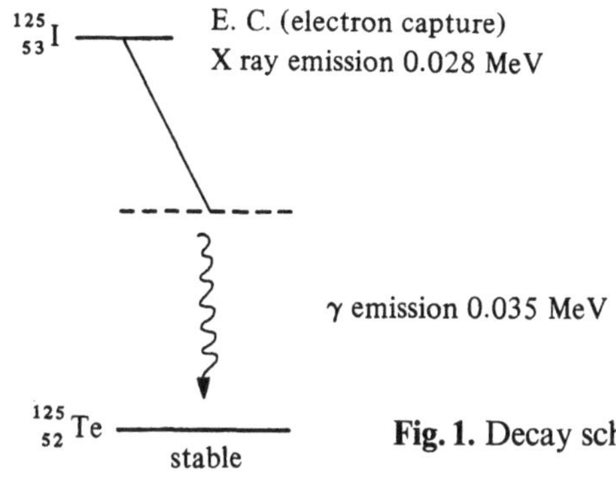

Fig. 1. Decay scheme of ^{125}I

ing of damage due to radiolysis, and because of the low energy of the γ radiation, which allows handling of the isotope with a minimum of precautions.

3. Purification of Iodinated Compounds

Whatever the radioiodination technique, the chemical reactions called into play lead to the formation of different compounds in the reaction mixture:

1. Molecules labeled with different proportions of iodine
2. Labeled molecules that have undergone structural modifications other than those due to the incorporation of iodine (polymers, oxidation and reduction products); these molecules may have lost their ability to bind to the antibodies
3. Unlabeled molecules that have not reacted with the iodinating agent
4. Mineral iodine in different chemical forms or reactive iodine not conjugated to the molecule
5. Iodination reagents
6. Neutralization reagents

It is therefore necessary to purify the tracer in order to obtain a well-defined molecular entity of high specific radioactivity and maximal immunoreactivity. The methods employed should give a high yield and good resolution, and should be simple to apply. Separation can be effected on the basis of different physicochemical properties of the labeled molecule:

1. Molecular weight: molecular sieve chromatography, dialysis
2. Charge: electrophoresis, ion exchange chromatography
3. Polarity: chromatography by adsorption on silica or cellulose, thin layer chromatography, high pressure liquid chromatography

a) Molecular Sieve Chromatography

Molecular sieving on a gel (Sephadex, Bio-Rad, Trisacryl) is one of the most commonly employed methods, particularly in the case of peptides, polypeptides, or proteins of molecular weight $\geqq 1000$ (Porath and Flodin 1959). This approach allows the removal of mineral iodine and molecules of low molecular weight by using a gel such as Sephadex G-10 or G-25 and a small chromatography column (Fig. 2). Combined use of a microcolumn and centrifugation results in a very

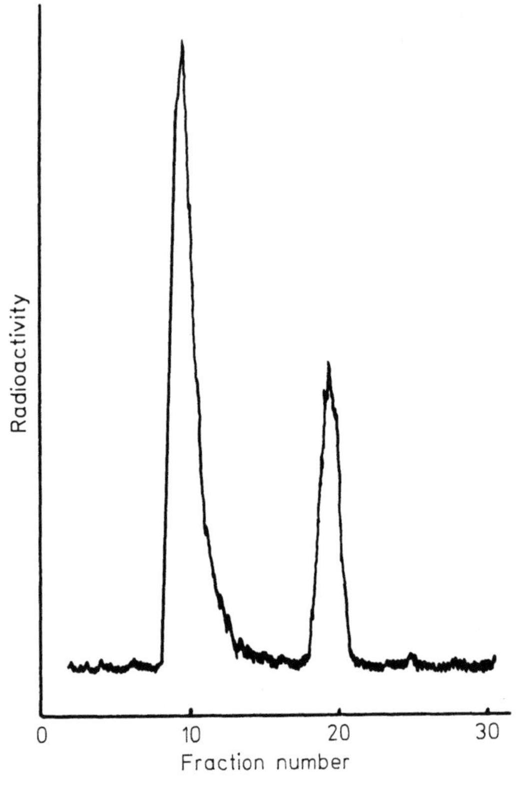

Fig. 2. Radiochromatography on Sephadex G-25 (coarse) of the reaction mixture for iodination of collagen. The column (30 × 1 cm) (Econo-column; Bio-Rad) was equilibrated and eluted with a phosphate buffer (0.1 *M*, pH 7.4) containing 0.4 *M* NaCl and 0.1% bovine serum albumin. The volume of the fractions collected was approximately 1 ml. The collagen was labeled by the chloramine-T method. The first peak corresponds to the void volume of the column and the second to mineral iodine. Detection of radioactivity in the elute was effected with a Geiger–Müller counter

rapid separation (TUSKYMSKI et al. 1980). Higher resolution gels can separate the different labeled molecular species from by-products of the reaction, such as polymers in the case of proteins (Fig. 3; AUBERT 1971). Dextran gels (Sephadex) are also known as weak cation exchangers (NEDDERMEYER and ROGERS 1968) and can form hydrophobic bonds with certain chemical groups (JANSON 1967). These special features can be taken advantage of to increase the performance of the filtration. Aromatic groups, such as tyrosine moieties of peptides and proteins, are especially implicated in these types of interactions. Hence, iodothyronines can be purified in a single step with an elution order inversely related to molecular weight (BISMUTH et al. 1971; KJELD et al. 1975), as can the mono- and diiodo compounds of peptides (GANDOLFI et al. 1971). Apart from some special cases, elution solvents are commonly of high ionic strength (0.1–1 *M* NaCl) and contain albumin (0.1%) in order to avoid adsorption phenomena and the retention of the molecule on the gel and column walls.

b) Dialysis

The use of a dialysis membrane of appropriate porosity allows high molecular weight molecules (>10000) to be separated from small molecules and free iodine. However, this method is not very commonly employed because of certain drawbacks, such as the time required (>2 h), quite frequent and significant adsorption

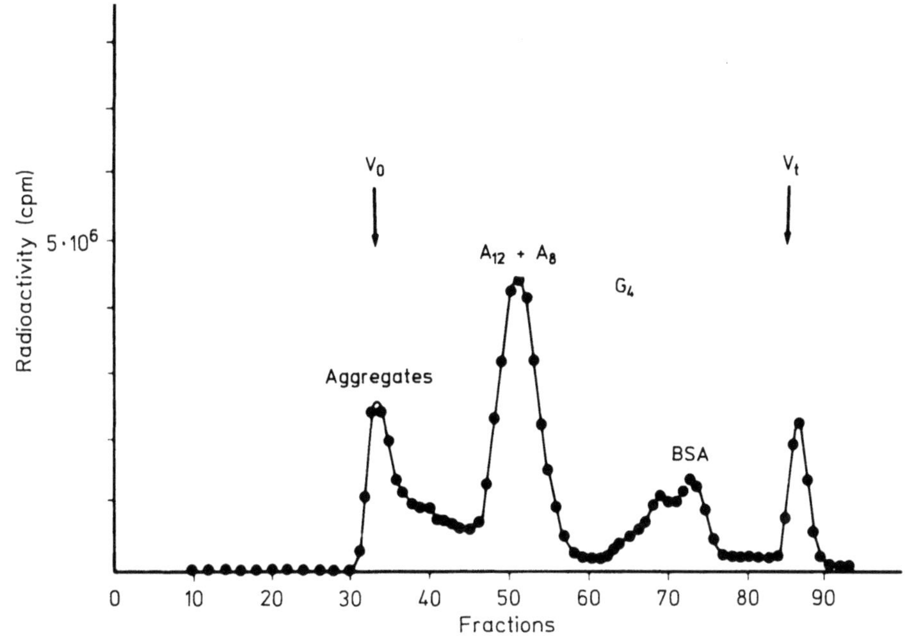

Fig. 3. Radiochromatography of the products of iodination of acetylcholinesterase (from *Electrophorus electricus,* EC 3.1.1.7) after elimination of mineral iodine (iodination by the chloramine-T method). The column (70 × 1.8 cm) was filled with Bio-Gel A 15 m, 200–400 mesh (Bio-Rad) and was equilibrated and eluted with a Tris-HCl buffer, pH 7.4, containing 1 *M* NaCl and 10^{-2} *M* $MgCl_2$. The fractions were 2 ml in volume. The first two peaks correspond to different labeled molecular forms of the enzyme (SIKORAV et al. 1984). Labeled bovine serum albumin and low molecular weight by-products were eluted last. The labeled albumin derives from the fact that, after labeling, the free iodine was eliminated by gel filtration with a phosphate buffer containing albumin as eluant (see Fig. 2). It is known that this protein binds free iodine noncovalently

of the product onto the membrane, and the volumes of radioactive effluent involved.

c) Electrophoresis

Electrophoresis on paper or gel, and more recently isoelectric focusing, which are common analytic methods in biochemistry, are only rarely used for quantitative purification of iodinated tracers (RAE and SCHIMMER 1974), and tend to be reserved rather for testing tracer purity (AUBERT 1971).

d) Ion Exchange Chromatography

The use of pH gradients in anionic, cationic, or mixed resins is a means of separating different iodinated species of peptides and proteins. This is because the introduction of one or two iodine atoms into the phenol or imidazole rings of tyrosine or histidine residues results in a modification of the pK_a of these residues (Table 3; SORIMACHI and UI 1975; LING et al. 1976). Elimination of mineral iodine is effectively achieved with Dowex X8 resin (Bio-Rad) (GANDOLFI et al. 1971).

Table 3. pK_a Values for iodinated derivatives of tyrosine and histidine

Compounds	pK_a	References
Tyrosine	10.1	
Monoiodotyrosine	8.2	(EDELHOCH 1962)
Diiodotyrosine	6.4	
Histidine	6	
Monoiodohistidine	4.18	(BRUNINGS 1947)
Diiodohistidine	2.72	

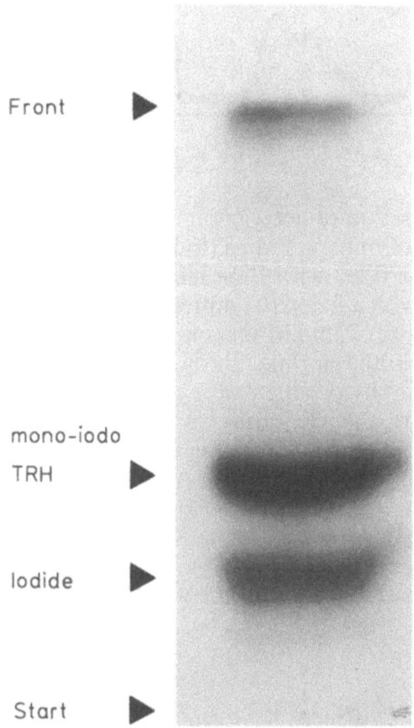

Fig. 4. Autoradiochromatogram of the reaction mixture from iodination of thyroliberin (thyrotropin-releasing hormone, TRH, PGlu-His-ProNH_2) analyzed by thin layer chromatography (Merck cellulose matrix; solvent butanol/acetic acid/water 75/10/25). The film used was Kodirex ready pack (Kodak). The mono- and diodo-derivatives of the peptide had been previously identified by means of iodine monochloride as iodinating reagent (PRADELLES 1977). Thyroliberin was labeled by the chloramine-T method to give maximal specific radioactivity (GROUSELLE et al. 1982)

e) Adsorption Chromatography

Thin layer adsorption chromatography on cellulose or silica is a preferred technique for the separation of low molecular weight molecules (peptides, steroids, drugs) as well as inorganic molecules such as the different forms of iodine (REGOESCZI 1984). It lends itself particularly well to the purification of radioactive tracers for the following reasons:

1. It is simple to implement and is well adapted for small quantitites of product and small sample volumes.

2. It is possible to separate the different types of molecule (mono- and diiodo compounds) as well as the reaction by-products (iodines, conjugation reagents, etc.) in a single step.

3. The positions of the different radioactive products can be established simply by autoradiography (Fig. 4).

4. The radioactive product can be recovered simply by scraping off the silica or cellulose powder at the appropriate spot, and can be rapidly eluted in a minimum of solvent.

High pressure liquid chromatography (HPLC) is a technique whose use is increasingly widespread, not only in the testing of tracer purity, but also in quantitative purification (SEIDAH et al. 1980).

4. Specific Radioactivity

Specific radioactivity is defined as the amount of radioactivity per unit mass or per molecule, and is generally expressed in μCi/μg or Ci/mmol. It is desirable to measure this value for each tracer preparation since it can be used to calculate the mass of tracer introduced in the assay, and to standardize it (BOLTON 1977). Theoretically, the introduction of an atom of ^{125}I per molecule from commercial solutions of $Na^{125}I$ leads to specific radioactivities of the order of 2000 Ci/mmol. In practice, this value depends on the possibility of separating labeled and unlabeled molecules (frequently the case for protein hormones). Determination of the specific radioactivity will vary in difficulty depending on the type of molecule (antigen, hapten) and the methods for purification and identification of the molecular species.

In the general case of macromolecules, the degree of incorporation (fraction of bound iodine versus fraction of free iodine) is measured on a sample aliquot by chromatography electrophoresis (CARO et al. 1975) or protein precipitation by trichloroacetic acid). In the case of small molecules (haptens), it is generally possible to purify the different molecular species and identify them (PRADELLES et al. 1978). The most favorable instance in this case is when the molecule possesses a tyrosine group. So, far a larger quantity of product, stable iodine, ^{127}I, is incorporated by the ICl technique for example (MORGAT et al. 1970), and after purification the mono- and diiodo species are identified by spectrophotometric analysis as function of pH (GEMMIL 1955), and the radioactive species are identified by simple comparison of the chromatographic profiles with those of nonradioactive iodinated species. In this situation, the determination of specific radioactivity involves only the measurement of the radioactivity of the identified product. Another approach for the measurement of tracer mass involves the use of radioimmunoassay calibration curves (WALKER 1977).

5. Iodination Damage

Chemical alteration of an antigen after iodination can result in a partial or complete loss of immunoreactivity. Such damage can have various causes: reaction with iodine, reaction with the iodinating reagents, reaction with impurities present in the radionuclide solutions, effects of internal radiation, and the decay catastrophe.

a) Reaction with Iodine

As will be described later in this account, the major organic groups capable of forming stable covalent bonds are the phenol and imidazole groups, which are present for example in tyrosine and histidine residues in peptides and proteins. Nonetheless, iodine can react with other organic groups and other amino acid residues by forming, most commonly, unstable derivatives which result in a chemical modification of the molecule (for review see RAMACHANDRAN 1956; KOSHLAND et al. 1963; REGOECZI 1984).

α) Effect of Iodine on Cysteine (R–SH). Cysteine reacts with I_2 through a series of complex reaction pathways. One of these pathways leads to the formation of disulfide bridges

$$\text{R–SH} + I_2 \rightarrow \text{RSI (sulfenyl iodide)} + H^+ + I^-$$

$$\text{R–SI} + \text{R–SH} \rightleftharpoons \text{R–S–S–R} + \text{HI}$$

The S–I bond is less stable than the C–I bond (NELANDER 1969).

β) Effect of Iodine on Cystine (R–S–S–R). The oxidation of disulfide bridges in proteins during iodination is considered as one of the factors involved in the loss of immunoreactivity of proteins (HUNG 1973).

$$\text{R–S–S–R} + 2I_2 + 3H_2O \rightarrow 2\text{R–}SO_3H + 4\text{HI}$$

γ) Effect of Iodine on Methionine (R–S–CH_3). The hydrolysis of an unstable iodine compound with methionine leads to its oxidation (ROSENBERG and MURRAY 1979)

$$\text{R–S–}CH_3 + I_2 \longrightarrow \left[\text{R–}\overset{\overset{\text{I}}{|}}{\underset{+}{\text{S}}}\text{–}CH_3\right] I^-$$

iodosulfonium

$$\downarrow \quad +H_2O$$

$$\text{R–}\overset{\overset{\text{O}}{\|}}{\text{S}}\text{–}CH_3 + 2\text{HI}$$

through an unstable intermediate (iodosulfonium ion).

δ) Effect of Iodine on Tryptophan. It has been clearly demonstrated (ALEXANDER 1974) that the effect of iodine is to oxidize the indole heterocycle to oxindole, and that peptide bonds may be broken.

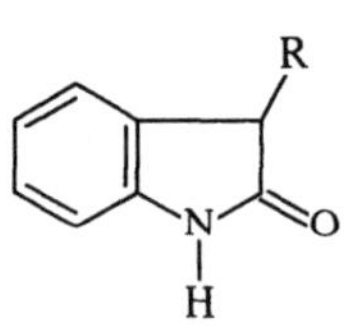

Oxindole

b) Reaction with Chemical Reagents

Most iodination methods employ oxidizing and reducing agents (see Sect. B.II), the former producing a reactive species through oxidation of iodine, and the latter stopping the reaction. The oxidizing agents often act directly on the molecule to be iodinated by modifying its immunologic properties. The effect of chloramine-T is well known (SHIMA et al. 1975), particularly on the methionine residues of peptides and proteins (STAGG et al.; SHECHTER 1975). In certain cases, the sensitivity of this residue to these agents calls for an exploration of other labeling methods (HEMMINGS and REDSHAW 1975).

c) Reaction with Impurities Present in Radionuclide Solutions

It has been shown that preparations of radioactive NaI without carrier and reductant for the labeling can contain different oxidizing molecular species of iodine, some of which have not been identified (CVORIC 1969). Iodates, in particular, seem to be the cause of failures often encountered with commercial batches of radioactive NaI. Solutions of iodine are kept at high pH and preferably at low temperature in order to minimize these undesirable effects.

d) Effects of Internal Radiation and the Decay Catastrophe

During labeling and storage, the tagged molecule is subject to radiation emitted by the radionuclide. In general, all types of radiation produce ionization and excitation of the molecule. ^{131}I and ^{125}I emit β and γ radiations, which result in chemical modifications of the tracer through complex mechanisms. In the case of tyrosine, for example, the C–I bond can be transformed into a C–OH bond (JIANG et al. 1975; BERRIDGE et al. 1979). In addition, the disintegration of ^{125}I and ^{131}I atoms leads to the formation of ^{125}Te and ^{131}Xe, respectively, whose positive charges differ and which can react with the electrons of neighboring molecules (CARLSON and WHITE 1963).

6. Storage of Label

The structural alterations of a molecule subjected to ionizing radiation are particularly important during storage. Radioimmunologists are very familiar with the time-dependent decrease in immunoreactivity. These drawbacks are usually mitigated by storing the tracer diluted in the presence of proteins (albumin, for example), which act as scavengers of free radicals (very reactive chemical species formed by irradiation). Free radicals combine with oxygen dissolved in the solvent to form peroxide radicals. Other types of molecule can also act as scavengers (e.g., alcohols, thiols) (COHEN 1976).

Irradiation-induced damage is particularly serious when the tracer is concentrated and of high specific radioactivity (KJELD et al. 1975). Breaking of the C–I bond leads to the formation of I_2 in the medium, which is very volatile and is a cause of atmospheric contamination (BOGDANOVE and STRASH 1975). In general, iodine molecules are also sensitive to light (photolysis), pH, and the presence of metal ions (BROWN and REITH 1967). Appropriate storage conditions should be found for tracers, depending on their chemical characteristics (for review see BAYLY and EVANS 1968).

7. Safety

Most radioiodinations are performed with a solution of radioactive NaI in sodium hydroxide, without reducer and with a total activity of 1–5 mCi. Apart from the dangers due to ionizing radiation emitted by the source, atmospheric pollution is one of the principal causes of contamination of personnel. Formation of volatile I_2 in the radioactive iodine source during storage, or in the reaction mixture after labeling, can result in internal contamination through inhalation. Contamination through direct contact may also occur, since iodine diffuses through the skin very rapidly and is absorbed by the thyroid gland. The radiotoxicity of iodine is now well established (TAYLOR 1981). The International Commission for Radiological Protection (ICRP) and the International Atomic Energy Agency (IAEA) publish regular reports describing the evaluation of contamination risks and giving recommendations for radiation protection. By way of illustration we cite, briefly, the risks involved in the handling of ^{125}I, as they are evaluated by the French atomic energy agency (Commissariat à l'Energie Atomique, CEN Saclay, France).

Risk of External Irradiation

- At a distance: irradiation of whole body
  Annual dose-equivalent limit: 50 mSv/year (5 rem/year)
  1 mCi gives approximately: 10 μSv/h (1 mrem/h at 30 cm)
  Upon contact with recipient: irradiation of hands
  Annual dose-equivalent limit: 500 mSv/year (50 rem/year)
  1 mCi gives approximately: 15 mSv/h (1.5 rem/h)

Risk of Internal Contamination Through Inhalation

- Annual limits on intake for workers (ALI): 2×10^6 Bq (5.4×10^{-5} Ci)
- Derived air concentration (DAC): 10^3 Bq/m^3 (2.7×10^{-8} Ci/m^3)

Conditions of Work are Divided Into Two Classes

- Working condition A: annual exposure might exceed three-tenths of the dose-equivalent limits
- Working condition B: annual exposure does not exceed three-tenths of the dose-equivalent limits (generally the case for most biologic research laboratories)

In terms of specific equipment we recommend

1. Handling of iodine, labeling, and purification should be performed under a well-ventilated hood [0.5 m/s air flow at the level of the glove holes (diameter 12 cm)] equipped with an activated charcoal filter in the air extraction system
2. Use of disposable gloves (surgical latex gloves)
3. Radioactive waste disposed regularly in sealed plastic bags
4. Use of a sensitive device for the monitoring of surface contamination
5. Monitoring of atmospheric contamination

II. Chemistry

1. Iodinating Reagent

The iodinating reagent is defined as the molecular form of iodine which reacts directly to incorporate iodine into a molecule. In organic synthesis, iodinating reagents are varied, and certain of them can be used for the labeling of organic molecules (ARGENTINI 1982). In the particular case of the preparation of iodinated radioligands intended for radioimmunology, the need to prepare a product with a high specific radioactivity and to effect its synthesis in a nonspecialist laboratory limits the number of these reagents. The nuclear industry offers only a restricted number of these high specific radioactivity reagents (MANI 1983). The major ones are as follows.

a) ^{125}ICl

Iodine monochloride is the most effective iodinating agent. Unfortunately, this molecule is not available from the nuclear industry at a sufficiently high specific activity (5–50 mCi/mmol) for the preparation of tracers in radioimmunology. However, it is possible to obtain this agent at specific radioactivities of at least 1000 Ci/mmol by isotopic exchange.

$$^{127}ICl + Na^{125}I \rightleftharpoons {}^{125}ICl + Na^{127}I.$$

b) $^{125}I_2$

Molecular iodine is an iodinating agent in addition reactions at double bonds, during the substitution of hydrogen at aromatic groups electrophilic substitution at unsaturated carbon atoms: tyrosine and histidine residues of peptides and proteins), and during halogen substitutions through the reaction of radicals. The production of iodine occurs in situ by oxidation of iodide

$$2^{125}I^- \xrightarrow{[O]} I_2 + 2e^-$$

In fact, these oxidation processes constitute the major methods for radioiodination which will be discussed further in this chapter. Iodine is an extremely reactive molecular species which generates numerous chemical reactions (HUGUES 1957). In dilute solution, these reactions are still poorly understood, and many oxidation states are involved (Table 4).

It is generally admitted that the H_2OI^+ cation is in fact the principal reactive species in electrophilic substitution reactions at unsaturated carbon atoms, which

Table 4. Oxidation states and chemical species of iodine

Principal oxidation state	−1	0	+1	+5	+7
Predominant chemical species	Iodide ion I^-	Free radical (atom) I	Iodonium ion I^+	Iodate ion IO_3^-	periodate ion IO_4^-

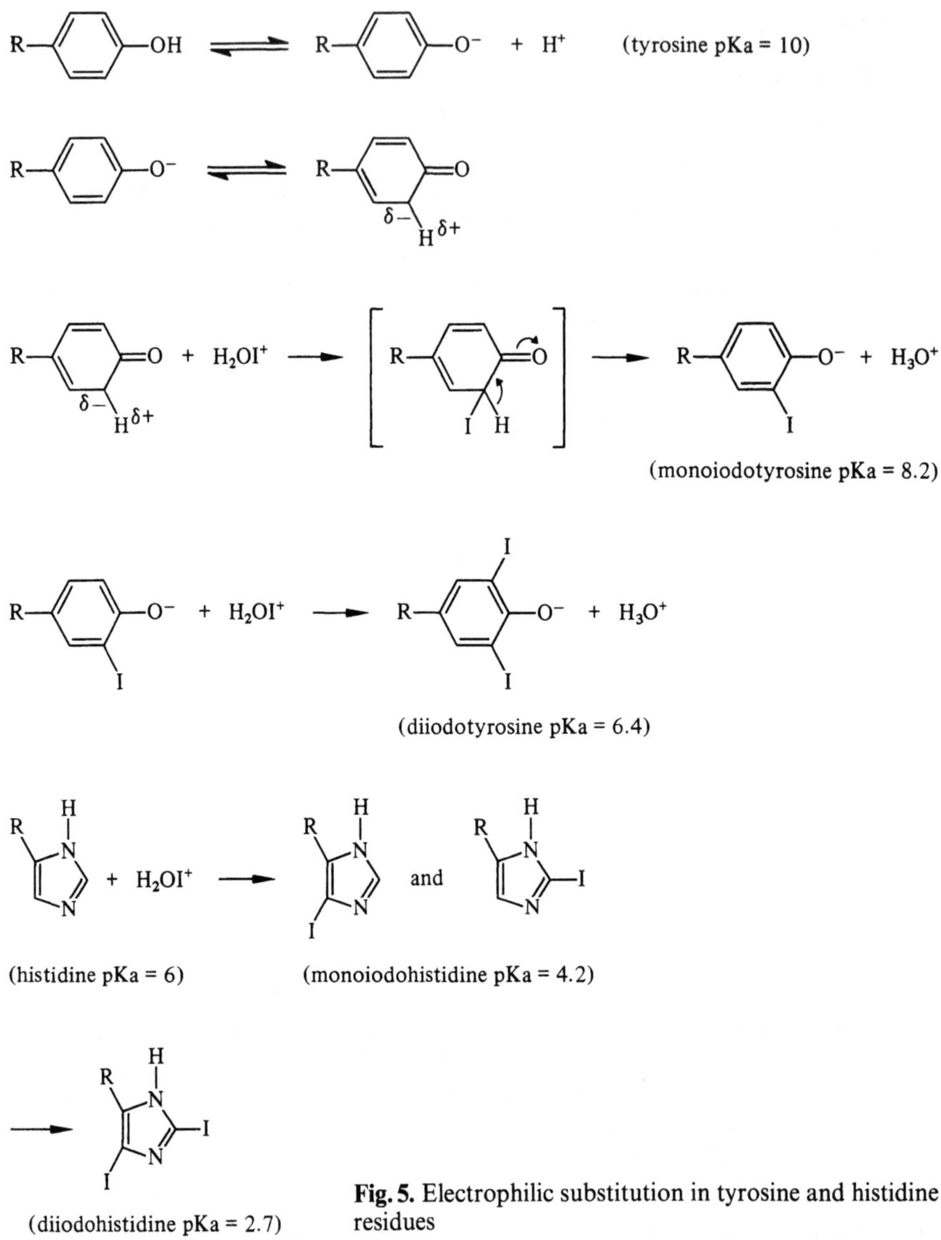

Fig. 5. Electrophilic substitution in tyrosine and histidine residues

constitute the majority of reaction mechanisms involved in methods for the preparation of iodinated radioligands

$$I_2 + H_2O \rightleftharpoons H_2OI^+ + I^-.$$

The H_2OI^+ cation dissociates to give hypoiodous acid HOI, which is remarkably stable in dilute solution (JIROUSEK 1981)

$$H_2OI^+ \rightleftharpoons HOI + H^+.$$

Figure 5 summarizes the electrophilic subsitution reaction mechanisms occurring at phenol and imidazole groups (tyrosine and histidine residues of peptides and proteins). The different mono- and dihalogen derivatives are stable and their physicochemical properties have been described (EDELHOCH 1962; BRUNINGS 1947).

c) $Na^{125}I$

This is the simplest iodine compound and the most frequently used. It is involved either in isotopic exchange with the stable iodine of iodine monochloride ICl, or in the substitution of diazonium salts (Sandmeyer's reaction). It is available in the nuclear industry in alkaline solution (~ 0.1 *M* NaOH) without carrier and reductant, and with a specific radioactivity of the order of 17 Ci/mg, an activity per unit volume of $\sim$100–600 mCi/ml and a radiochemical purity of greater than 99% ($<1\%$ ^{126}I).

2. Methods

a) Introduction

There are many methods for the insertion of radioactive iodine into a molecule, depending on its type (ARGENTINI 1982). They include nucleophilic substitution at aliphatic carbon atoms, electrophilic addition at double bonds, electrophilic substitution at aromatic groups, nucleophilic substitution at aromatic groups, use of radical reactions, etc. Curiously, nowadays the synthesis of iodinated tracers in radioimmunology, whatever the nature of the molecule (protein, peptide, steroid, lipid, nucleotide, sugar, drug, etc.), is essentially carried out by a single method, which can be schematized as follows:

Precursor: $Na^{125}I$
Method: oxidation of ^{125}I by a mild oxidizing agent
Mechanism: electrophilic substitution at an aromatic group.

Various reasons account for this impoverishment of methods. First of all, there are historical reasons. The discovery of the radioimmunologic technique arose through studies with proteins. Now since this type of molecule is particularly fragile, it was necessary to choose a method that was: (a) mild, so as not to damage the epitopes; (b) highly efficient, thus giving an appropriate specific radioactivity, and (c) easily implemented in the laboratory.

Thus, the classic studies of Berson and Yalow on insulin were performed following the method of PRESSMAN and EISSEN (1950) with mild oxidation of iodine by sodium nitrite. The iodine-binding sites on the protein were tyrosine and histidine residues. Many methods were subsequently proposed, varying essentially in terms of the type of oxidizing agent or iodinating agent used, but a few rare exceptions apart, the principle of the insertion of iodine at an aromatic group was established. Thus, as we shall see later in detail, labeling techniques can be summarized as follows:

α) First Case. The molecule has an iodine acceptor group such as: a tyrosine or histidine residue (peptides or proteins), a double bond (phospholipids), or another aromatic group (cytidine in nucleic acids, and all other natural or synthetic molecules). In this case, direct labeling is performed as described in the following section.

β) Second Case. The molecule does not possess an iodine acceptor group. In this case two approaches are possible. One consists in introducing into a specific site on the molecule an iodine acceptor group (phenol or imidazole ring in most cases) (CORRIC and HUNTER 1981), thus reducing the problem to that of the first case. The other approach involves the conjugation of the molecule with a small, previously radioiodinated molecule.

b) Iodine Monochloride Method

Since its development for the labeling of proteins (MCFARLANE 1958), this method has been subject to different modifications (HELMKAMP et al. 1960; SAMOLS and WILLIAMS 1961; MCFARLANE 1968). The iodine monochloride is rendered radioactive by isotopic exchange with radioactive NaI, a concomitant drawback being that the specific radioactivity of the iodine is reduced. However, it is possible, through oxidation of radioactive NaI, to retain in the ICl about 70% of the initial specific radioactivity of the iodine (HELMKAMP et al. 1967). This technique has the advantage of being quick, and of reducing the oxidation state of the iodine and limiting undesirable secondary reactions. In addition, the reagent is particularly stable in aqueous medium, thus allowing tight control of the electrophilic substitution reaction. Despite recent progress in improving the specific radioactivities of molecules labeled by this method (DORAN and SPAR 1980), it is a technique little used by radioimmunologists.

c) Methods Based on the Oxidation of Iodide

α) Oxidation by Chlorine Compounds. It is well known that chlorine (Cl) is more electronegative than iodine, and that it can remove two electrons from it, thus changing its charge from -1 to $+1$ and rendering it reactive in electrophilic substitution reactions. Thus, molecular chlorine (Cl_2) and sodium hypochlorite (NaClO) have been used as oxidation agents of radioactive NaI in the iodination of proteins (BUTT 1972; REDSHAW and LYNCH 1974).

It was Hunter and Greenwood who first used a chlorine compound (chloramine-T) as a mild oxidizing agent of iodine for high specific radioactivity labeling of proteins (HUNTER and GREENWOOD 1962; GREENWOODet al. 1963).

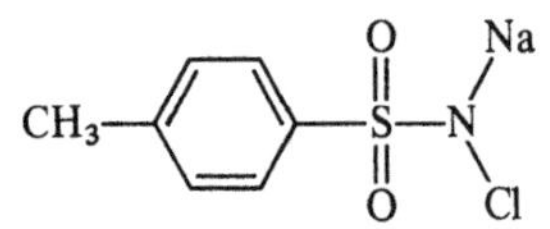

Chloramine-T (molecular weight 227.67) is nowadays the most commonly employed oxidizing agent in iodine labeling. This sulfonamide is considered as a mild oxidant, which in neutral aqueous solution releases hypochlorous acid, HOCl, which oxidizes iodide to I_2, and to ICl in acid medium. However, the reaction mechanisms are still subject to controversy, as witnessed by comparisons of the iodine incorporation yields in proteins, depending on whether chloramine-T or

I_2 is used (JIROUSEK 1981). The protocol of this method will vary depending on the type of molecule to be labeled and the level of specific radioactivity desired. A general basic outline can be specified:

1. The molecule to be labeled should be dissolved in a very small volume (5–100 µl) of a buffered medium (phosphate buffer 0.5 *M*, pH 7.4).
2. The chosen quantity of radioactive NaI (2–10 µl, 0.2–1 mCi) is added.
3. Chloramine-T (freshly dissolved in water or buffer) is added.
4. The reaction is stopped after 1–2 min by the addition of a reducing agent in water or buffer ($Na_2S_2O_5$, sodium metabisulfite).

This method is much used for the labeling of peptides and proteins (MCCONAHEY and DIXON 1980).

One of the major drawbacks of this technique is that the molecule to be labeled is in direct contact with the oxidizing agent, and in the case of proteins in particular this leads to secondary reactions (polymerizations, oxidation of methionines). These effects can be minimized by using the lowest possible ratio of chloramine-T to protein (ARGENTINI 1982). This method is employed for the iodination of groups other than phenol and imidazole residues, such as, for example, nucleic acids (SHAPOSHNIKOV et al. 1976), alkaloids from rye ergots (COLLIGNON and PRADELLES 1984), phospholipids (ANTONOV et al. 1985), and some steroids (JEFFCOATE 1982).

FRAKER and SPECK (1978) have proposed the use of another chlorine compound, 1,3,4,6-tetrachloro-3*a*,6*a*-diphenylglycoluril (molecular weight 431.91).

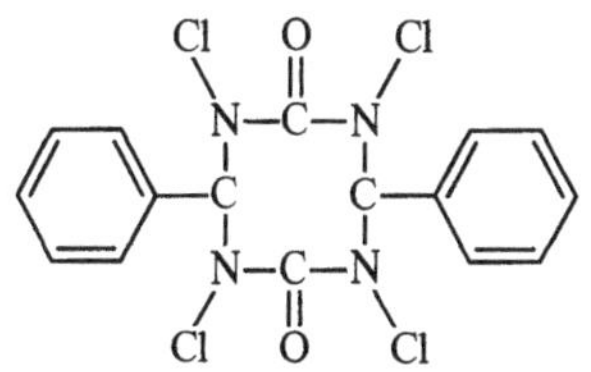

This product is available from the Pierce Chemical Company, Rodford, Illinois, United States, under the trade name of Iodo-gen. It is insoluble in water, but soluble in chloroform or dichloromethane. The principle of the method consists in coating the walls of the reaction tube with this reagent by evaporating the solvent, and then the molecule to be labeled and the radioactive iodine are introduced. Because the reagent is insoluble in aqueous media, its contact with the molecule to be labeled is weak, and secondary reactions are therefore minimized. According to some authors, higher yields and greater stabilities are obtained in the labeling of peptides and proteins with this method, as compared with the chloramine-T technique (SALACINSKI et al. 1981). Another approach involves the covalent immobilization of *N*-chlorobenzylsulfonamide on a polystyrene matrix (MARKWELL 1982).

β) Electrolytic Oxidation. Two decades ago, Rosa introduced the electrolytic conversion of iodide to iodine (ROSA et al. 1964). The principle of the method is simple and it can be applied to micro-quantities (DONABEDIAN et al. 1972). In an electrolysis cell, a platinum cathode is enclosed by a dialysis membrane, and at the anode, under the effect of an electric field, the reaction of radioactive sodium iodide produces reactive iodine

$$2I^- - 2e^- \rightarrow 2I \rightarrow I_2$$

The oxidized iodine formed reacts with the substrate and the solvent.

The advantage of this technique is the absence of chemical oxidation and reduction agents. Furthermore, the speed of formation of I_2 is controlled by the current strength and it is possible to work with dilute solutions of iodide. One drawback is that the duration of electrolysis is quite long (30–60 min) and the molecule to be labeled is in very dilute solution, conditions which will tend to favor protein denaturation.

γ) Enzymatic Oxidation. This method grew out of the in vivo synthesis of thyroid hormones. The labeling of proteins with radioactive iodine, as originally performed, used lactoperoxidase, which catalyzes the incorporation of iodine in the presence of hydrogen peroxide (H_2O_2) and sodium iodide (MARCHALONIS 1969). Following this principle, it has proved possible to label peptides and proteins with high specific radioactivities (THORELL and JOHANSSON 1971; MIYACHI et al. 1972). In general terms, the molecule to be labeled is dissolved in an appropriate buffer in the presence of radioactive NaI. The enzyme and hydrogen peroxide are then added successively. The reaction is stopped after a few minutes by the addition of a reducing agent (mercaptoethanol or cysteine), or simply by dilution with buffer. The concentrations of enzyme and hydrogen peroxide in the reaction medium determine the labeling yield. The pH of the medium can influence the speed of the reaction (REGOECZI 1984). This method also has the advantage that secondary reactions leading to protein denaturation are minimized. One hypothesis advanced is that the tyrosine side chains involved are those that are exposed at the surface of the molecule, and which form the complex with the enzyme (MORISON 1980). One of the disadvantages of this method is that the enzyme, being a protein, iodinates itself, and if the molecule to be labeled has physicochemical properties close to those of the enzyme (molecular weight, charge, etc.), an impure tracer will be contaminated by iodinated enzyme. It is advisable to use either a small quantity of enzyme, or enzyme coupled to an insoluble matrix, which can thus be readily removed from the reaction mixture (DAVID and REISFELD 1974). Imidazole side chains of peptides and proteins (KROHN and WELCH 1974), as well as other organic structures such as lipids and steroids (MERSEL et al. 1976; MATKOVICS et al. 1971) can be labeled by this technique. The effectiveness of this method is particularly shown for the labeling of the surface proteins (SEFTON et al. 1973). Other enzymes have been used in this type of labeling, such as myeloperoxidase, chloroperoxidase, and the glucose–glucose oxidase system (for review see REGOECZI 1984).

δ) Oxidation by Thallium Chloride. COMMERFORD (1971) has the first to label nucleic acids radioactively using iodine. The basic principle rests on the fact that thallium chloride $TlCl_3$ oxidizes the iodide and allows substitution of the hydrogen at C-5 of the cytidine by an iodine atom. This technique has also been proposed for labeling lipids in organic solvents (ANTONOV 1985).

ε) Sandmeyer's Reaction. This method is applied to molecules that possess a primary aromatic amino group. In the first step, the amino group is diazotized and is then replaced by iodine.

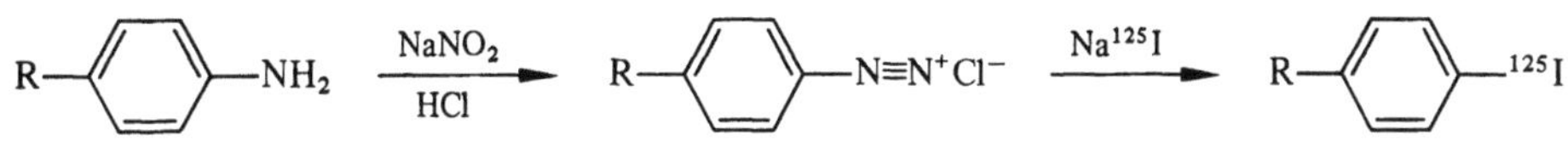

This technique has been known for many years (BLOCH and RAY 1946) and has been used in special cases such as, for example, radioimmunologic assay of drugs (CARDOSO and PRADELLES 1982).

3. Conjugation Labeling

The direct incorporation of iodine into a molecule for the synthesis of a radioimmunologically active iodinated tracer is doomed to failure if:

1. The molecule does not possess groups that can be iodinated (or if such groups are present they are inaccessible to iodinating agents).
2. The presence of iodine atoms in the molecule decreases its affinity for the antibodies, either because they are incorporated directly at the epitopes, or because they modify the tertiary structure of the molecule. Remember in this regard that the steric bulk of an iodine atom is close to that of benzene.
3. The molecule is sensitive to the reagents employed (oxidant, reductant).

For these reasons, the conjugation of the molecule with a group that is susceptible to iodination, or with an already iodinated group, presents an alternative to direct labeling.

Two strategies can be adopted, depending on the nature of the problem: either the molecule to be labeled X is reacted with an iodinated molecule of high specific radioactivity Y, which has been activated and will form a covalent bond with X, (conjugation labeling method), or, X is reacted with Y, which is activated, but not iodinated, and the product X–Y is then iodinated. We shall not elaborate further on this latter approach, which is only original in terms of the pathways of organic synthesis called into play according to the type of molecule. It is commonly used for the preparation of iodinated haptens in radioimmunology and interested readers can refer to reviews describing the different substrates that can be iodinated and the organic reactions used (CORRIE and HUNTER 1981; REGOECZI 1984).

a) Iodination with the Bolton–Hunter Reagent

The extensive use of ^{125}I-3-(4-hydroxphenyl)propionic acid *N*-hydroxysuccinimide ester for the labeling of peptides and proteins is due to BOLTON and HUNTER (1973). This derivative had originally been proposed by RUDINGER and RUEGG (1973). The principle of the method is simple, and three steps are involved.

α) First Step. Synthesis of the active ester of *N*-hydroxysuccinimide (NHS)

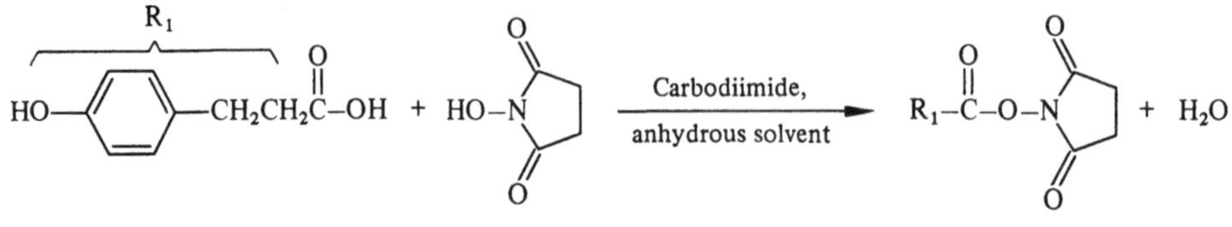

NHS

The esters of NHS are quickly hydrolyzed by amines to give an amide and have been commonly employed for peptide synthesis (ANDERSON et al. 1964).

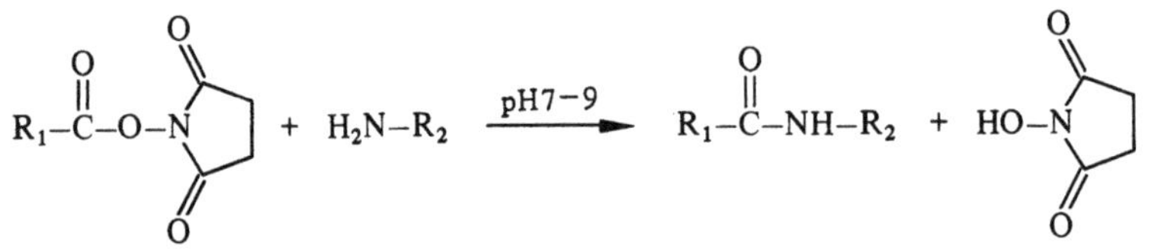

β) Second Step. Labeling with ^{125}I by the chloramine-T method. The labeled product (mono- or diiodo derivative) is rapidly extracted from the aqueous iodination reaction mixture by benzene, and is kept in this solvent. The reagent is commercially available in a ready-to-use form of high specific radioactivity (2000–4000 Ci/mmol) (Amersham International plc, New England, Nuclear).

γ) Third Step. A molecule X, which possesses a free amine (e.g., the lysine residue of proteins, is placed in the presence of reagent, after removal of solvent, for a short period.

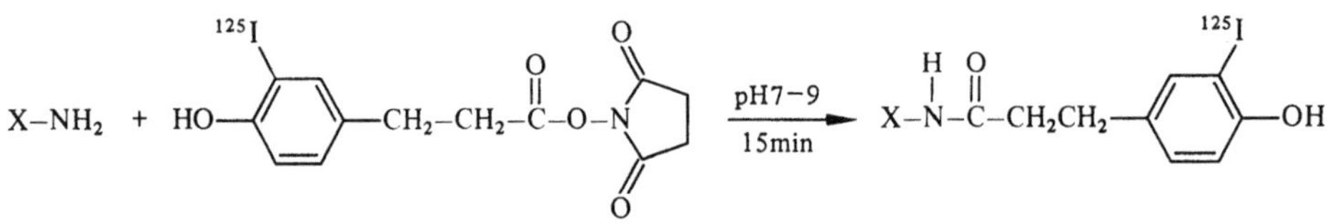

The reaction is stopped by the addition of excess lysine. This method has been most successful and can be used for the labeling of a great variety of molecules (LANGONE 1981).

b) Iodination with Other Conjugating Reagents

We shall cite some other compounds used in the labeling of proteins and other organic substances, but which, curiously, have not been used for the synthesis of tracers in radioimmunology.

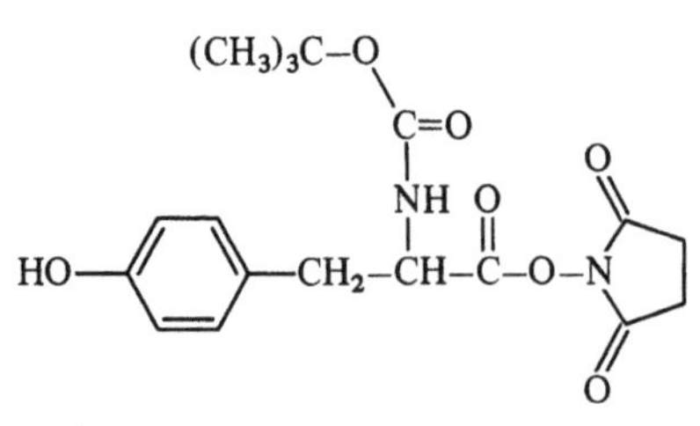

α) t-Butyloxycarbonyliodotyrosine N-Hydroxysuccinimide Ester ^{125}I.

This reagent, which is similar to the previous one, has the advantage of restoring the charge on the amino group of the molecule after having eliminated the *t*-butyl group (ASSOIAN et al. 1980).

β) Wood's Reagent

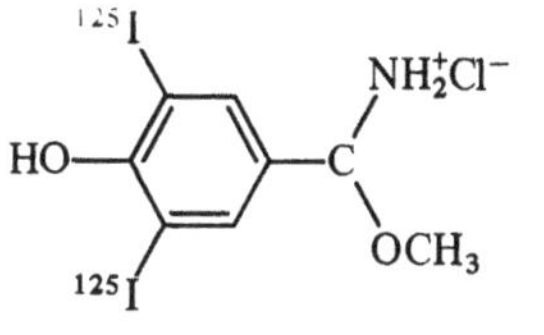

This imidoester reagent (3,5-diiodo-*p*-hydroxybenzimidate; WOOD et al. 1975) reacts with the ε-amino group of the lysine residues of proteins and, like the preceding reagent, does not alter the overall charge of the native protein.

γ) Diazotized Iodosulfanilic Acid

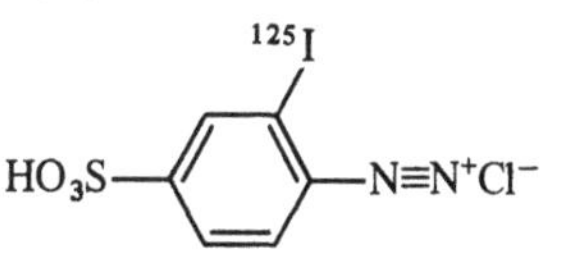

Diazotized sulfanilic acid has long been used as a specific reagent for the tyrosine and histidine residues of peptides and proteins through the formation of a colored azoic derivative (Pauly's reagent). Iodination and diazotization of sulfanilic acid give a reagent which is used in the labeling of cell membranes (HELMKAMP and SEARS 1970). It is available in kit form from New England Nuclear at a specific radioactivity of ~1000 Ci/mmol.

δ) Iodinated Aniline. Iodinated aniline has also been used according to the same reaction scheme for labeling proteins which are sensitive to oxidizing agents (HAYES and GOLDSTEIN 1975).

C. Nonisotopic Labeling

I. General Considerations

Despite the success of radioimmunoassays in many areas of biology, there has in recent years been an increasingly marked tendency to use methods that employ nonradioactive tracers. This trend is partly explained by the drawbacks involved in handling radioactive isotopes, but also because improvements in techniques have led to the appearance of substitution labels which can improve the performance of radioimmunoassays.

One of the major drawbacks of radioactive tracers is tied to the tightening up of legislation covering the use of radioactive products and the disposal of the corresponding waste. This radiation safety problem is further aggravated by the increased use of "excess reagent methods" (MILES and HALES 1973; EKINS 1973), such methods involving much larger quantities of tracer. In addition, radioactive tracers have a half-life limited by that of the radioelement involved. Shelf-life is reduced even further when high specific radioactivity radioisotopes are used, e.g., ^{125}I or ^{131}I, which provide the most sensitive assays. This limited lifetime of radioactive tracers poses additional problems of management and quality control.

In order to be performed under appropriate conditions, radioactivity measurements call for specialized and costly apparatus. For example, it is not possible to design radioimmunoassays that can be carried out in a limited environment (developing country, patient's home, doctor's surgery), whereas it is possible with other, nonisotopic markers (enzymes, erythrocytes, etc.). Lastly, it is not possible to design "homogeneous" radioimmunoassays that do not require a separation step, since the radioactive disintegration process is completely independent of the physicochemical environment. These methods, though, lend themselves to high

Table 5. The most commonly used abbreviations in the field of immunoassay

A. Main categories

Abbreviation	Full name
RIA	Radioimmunoassay
IRMA	Immunoradiometric assay
EIA	Enzyme immunoassay
FIA	Fluoroimmunoassay
LIA	Luminescence immunoassay
SIA	Spin immunoassay
MIA	Metalloimmunoassay
PIA	Particle immunoassay

B. Subcategories

Abbreviation	Full name	Corresponding main category
ELISA	Enzyme-linked immunosorbent assay	Heterogeneous EIA
EMIT[a]	Enzyme-multiplied immunoassay technique	Homogeneous EIA
EASEIA	Enzyme-associated substance enzyme immunoassay	Homogeneous EIA, sometimes classified as FIA (SLFIA)
SHEIA	Steric hindrance enzyme immunoassay	Heterogeneous EIA
SLFIA	Substrate-labeled fluoroimmunoassay	Homogeneous EIA (EASEIA) or FIA
USERIA	Ultrasensitive enzymatic radio-immunoassay	Heterogeneous EIA
TRFIA	Time-resolved fluoroimmunoassay	Heterogeneous FIA or MIA
DELFIA[a]	Dissociation-enhanced lanthanide fluorescence immunoassay	FIA, associated with TRFIA
FPIA	Fluorescence polarization immunoassay	Homogeneous FIA
FETIA	Fluorescence excitation transfer immunoassay	Homogeneous FIA
CLIA	Chemiluminescent immunoassay	LIA
BLIA	Bioluminescent immunoassay	LIA
PACIA	Particle counting immunoassay	PIA (latex)

[a] Registered trademark.

performance automation and enjoy a certain success in clinical analysis (EMIT, FPIA).

Very many labels have been proposed as alternatives to radioisotopes. Among them are enzymes, fluorescent or luminescent compounds, metals, bacteriophages, spin or particle labels. For each of these labels, several types of homogeneous or heterogeneous assays have been described, which involve labeling of antigens, haptens, or antibodies. As a consequence, a considerable amount of new methods, accompanied by a variety of abbreviations, have emerged, leading to a quite confusing situation. The most commonly used abbreviations are given in Table 5 together with the category of immunoassay to which they correspond. We have first restricted our attention to an examination of assays calling for enzymatic, fluorescent, or luminescent labels which are the most developed and seem to be of the greatest interest. With few exceptions, all these methods have the common property of involving a tracer which consists of a conjugate between an immunoreactant (antigen, hapten, or antibody) and a probe (enzyme, fluorescent compound, or luminescent compound). In a later section, we shall examine briefly some other labels, which have not yet undergone extensive development, either because they do not give sufficiently sensitive assays, or because they are still too arduous to employ.

II. Labeling with Enzymes

The first enzyme immunoassays (EIA) appeared in 1971, following the studies of VAN VEEMEN and SCHUURS (1971) and of ENGVALL and PERLMAN (1971). These assays were heterogeneous and employed covalent conjugates of an antigen and an enzyme. The first homogeneous assays appeared a year later (RUBENSTEIN et al. 1972). EIA has since experiment a considerable upward trend and has become the most widespread of nonisotopic immunoassay methods (SCHUURS and VAN VEEMEN 1977, 1980; O'SULLIVAN 1984; OELLERICH 1984).

Note first of all that enzymes occupy a separate place in the arsenal of nonisotopic probes. Indeed, in themselves they carry no directly measurable signal, and are only detected through the reactions they catalyze, by monitoring the transformation of their substrate, which itself bears some signal. So there are EIA associated with colorimetric, fluorescent, luminescent, radioactive, or turbidimetric measurements. In any case, some of these assays have been improperly categorized in terms of the type of signal measured, such as fluoroimmunoassay (SLFIA) or luminescent immunoassay (BLIA). In EIA, the enzyme acts as a powerful amplifier which can transform its substrate up to tens of thousands of times per second. By virtue of this amplification, it is possible to obtain tracers with a specific activity greater than that given by ^{125}I.

When the enzymatic assay calls for colorimetric detection, it is possible to envisage a qualitative or semiquantitative measurement based on a simple visual assessment. In this case, EIA allows sensitive detection of antigens or antibodies without the need for measuring equipment. This privileged position of EIA suggests that it will be a method of choice in developing countries (VOLLER 1982; DAMLE et al. 1982). It could also prove to be a powerful aid to general practitioners, who could perform reliable assays at their surgeries. The major drawback of EIA is linked to a certain fragility of the enzyme, whose activity and its measurement may be influenced by some components present in biologic samples. This limitation, which is encountered above all in homogeneous assays, has nonetheless been overcome in a great many cases.

1. Heterogeneous Assays

The principle of these assays does not call for a great deal of comment since they are closely based on the corresponding radioimmunoassays. When a solid phase is involved in the separation process, these assays are often categorized as enzyme-linked immunosorbent assay (ELISA). They have been applied successfully to the assay of antigens, haptens, and antibodies, either by using competition or excess reagent methods (O'BEIRNE and COOPER 1979; JARVIS 1979; SCHUURS and VAN VEEMEN 1977; SCHALL and TENOSO 1981).

An original approach was developed by CASTRO and MONJI (1981) who used an insoluble pseudosubstrate inhibitor of β-galactosidase (agarose-aminocaptroyl-β-D-galactactosylamine) as separation reagent. Here, the presence of specific antibody bound on the enzymatic tracer prevents its fixation on the affinity gel. This assay, called steric hindrance enzyme immunoassay (SHEIA), is applicable to both haptens and antigens.

The performance of a heterogeneous enzyme immunoassay is critically dependent on the choice of enzyme. The following criteria are applied in making this choice:

1. For the synthesis of the enzymatic conjugate, it is necessary to have at one's disposal large amounts (milligrams) of pure enzyme, which should also be available at a reasonable price.

2. The enzyme should lend itself well to the different chemical reactions involved in the preparation of conjugates, without suffering a loss of enzymatic activity. The conjugates obtained should also be of good stability.

3. It should be possible to detect the enzymes with the greatest possible sensitivity. Naturally, this sensitivity will depend on the turnover rate of the enzyme, but also on the sensitivity with which the enzymatic reaction products can be detected. The substrate should also be stable in the absence of enzyme, since too great an autolysis would limit the sensitivity of the assay.

4. The enzymatic assay should be as simple as possible to perform and should lend itself well to automation.

5. The enzyme should retain its activity in complex biologic media in which different substances may tend to inactivate it (inhibitors, proteases, etc.). Despite a commonly held belief, the presence of endogenous enzyme in biologic samples is not necessarily troublesome in the case of heterogeneous assays. The problem can in fact be resolved by the use of an appropriate separation method allowing a low nonspecific binding.

6. In order to ensure a rapid immunoreaction, the tracer and as a consequence the enzyme should be small in size since the rate of reaction is dependent on the diffusion coefficient of the reactants which is related to the molecular size.

Table 6 gives a list of different enzymes used in heterogeneous assays. Each enzyme more or less satisfied the criteria enunciated above. It is not possible at

Table 6. Enzymes used in heterogeneous EIA

Enzyme	Source	Enzyme Commission number
Peroxidase	Horseradish	1.11.1.7
Acetylcholinesterase	*Electrophorus electricus*	3.1.1.7
Alkaline phosphatase	Calf intestinal mucosa	3.1.3.1
β-Galactosidase	*Escherichia coli*	3.2.1.23
Glucose oxidase	Fungal	1.1.3.4
Glucose amylase	*Rhizopus niveus*	3.2.1.3
β-Lactamase	Bacterial	3.5.2.1

the present time to make out a case for the superiority of one of these enzymes over the others, since, with few exceptions (PORSTMANN et al. 1985), no serious systematic comparison has yet been carried out. Nevertheless, it is possible, on the basis of certain experimental data, to make a few comments, in particular with regard to the detection sensitivity, which conditions the sensitivity of the assays performed. Some of these data, those corresponding to the most commonly used enzymes, are summarized in Table 7.

It can be noted, for example, that if we confine ourselves to colorimetric enzymatic assays, acetylcholinesterase from *Electrophorus electricus* seems to be the best candidate since it is possible to detect this enzyme at levels close to 1 amol (10^{-18} mol) under standard measurement conditions (1 h incubation, 200 μl reaction mixture). Slightly lower sensitivity can be obtained with horseradish peroxidase, while other enzymes such as alkaline phosphatase or β-galactosidase give significantly inferior performance.

Even greater sensitivity can be achieved with fluorescent substrates. In this case, β-galactosidase appears to be the best candidate since, by using 4-methylumbelliferyl-β-galactoside, it is possible to measure less than 10^{-20} mol enzyme (ISHIKAWA et al. 1983). The sensitivity of detection in these enzymatic systems is remarkable, and clearly superior to that obtainable with ^{125}I (detection limit 1.4×10^{-17} mol ^{125}I, corresponding to a counting rate of 50 cpm for a specific activity of 2000 Ci/mmol, assuming a counting efficiency of 80%). This sensitivity also compares very favorably with that obtained with other nonisotopic systems, such as time-resolved fluoroimmunoassay (TRFIA) and luminescence immunoassay (LIA) (see following paragraphs). It should also be noted that even more sensitive detection can be achieved, either by increasing the incubation time for enzymatic detection, or by using analysis methods more sensitive than colorimetry or conventional fluorescence.

This approach has been developed in the ultrasensitive enzymatic radioimmunoassay (USERIA) (VAN DER WAART et al. 1976; HARRIS et al. 1979; HSU et al. 1981) which employs radioactive detection of the enzymatic reaction products. Attention should also be drawn to the recent development of enzyme amplification methods (STANLEY et al. 1985a, b; JOHANNSSON et al. 1985). The use of enzymatic tracers of high specific activity has allowed the development over the last few years of assays with sensitivities greater than those of the best corresponding radioimmunoassays (IMAGAWA et al. 1981, 1982, 1983; STANLEY et al. 1985b;

Table 7. Comparison of colorimetric assays for the most commonly used enzymes in heterogeneous EIA

Enzyme	Molecular weight (Number of catalytic subunits)	Turnover number (h^{-1} mol^{-1})	Substrates	Chromophore	Detection limit (mol)[a]
Acetylcholinesterase (*Electrophorus electricus*) EC 3.1.1.7	330000 (4)	1.8×10^8 (VIGNY et al. (1978)	Acetylthiocholine + 5,5'-dithiobis (2-nitrobenzoic acid)	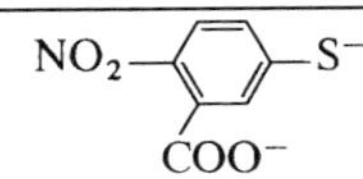 $\varepsilon_M = 1.36 \times 10^4$ (412 nm)	1.6×10^{-18}
Alkaline phosphatase (calf intestine) EC 3.1.3.1	100000 (2)	1.45×10^7 (MORTON 1957)	*p*-Nitrophenyl phosphate	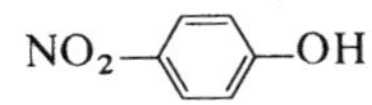 $\varepsilon_M = 1.85 \times 10^4$ (405 nm)	1.5×10^{-17} (2.6×10^{-16})
β-Galactosidase (*E. coli*) EC 3.2.1.23	540000 (4)	2.5×10^7 (ROTMAN 1961)	*O*-Nitrophenyl-β-D-galactopyranoside	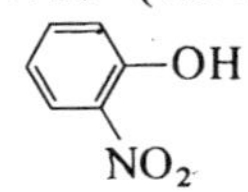 $\varepsilon_M = 4.7 \times 10^3$ (410 nm)	3.4×10^{-17} (2.6×10^{-17})
Peroxidase (horseradish) EC 1.11.1.7	40000 (1)		*O*-Phenylene diamine + H_2O_2	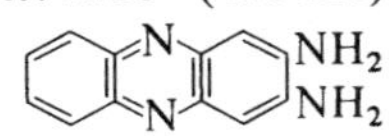 $\varepsilon_M =$? (492 nm)	3.6×10^{-18} [b] (5.9×10^{-18})

[a] Detection limit for each enzyme was calculated using the corresponding values for turnover number and ε_M. It is defined as the amount of enzyme producing an absorbance increase of 0.01 in 1 h, 0.2 ml volume, 0.5 cm pathlength. These conditions correspond to those used when EIA is performed in microtitration plates (see PRADELLES et al. 1985). Values in parentheses are those given by ISHIKAWA et al. (1983). Original values from these authors have been corrected for time, volume, and pathlength conditions in order to be consistent with our own calculations.

[b] Determined experimentally for pure crystalline enzyme (RZ = 3, Sigma, St. Louis, United States) using *O*-phenylene diamine (9.3×10^{-3} *M*) and H_2O_2 (10^{-2} *M*) as substrates.

HARRIS et al. 1979; SHALEV et al. 1980; CLARK and PRICE 1986; INOUE et al. 1986).

2. Homogeneous Assays

Homogeneous enzyme immunoassays depend on the effect exerted by the immunologic reaction on the activity of an enzyme. In these assays, the formation of the antigen–antibody complex will, in one way or another, result in a drop or an increase of the activity of the enzymatic tracer. Hence, it is possible to monitor directly the immunologic reaction by measuring the enzymatic activity of the reaction mixture. Unlike heterogeneous methods, these assays do not require a separation step.

Several types of EIA, all based on competition processes, but involving palpably different principles, have been described. Detailed description and performance of these different methods can be found in review articles (CROWL et al. 1980; SCHALL and TENOSO 1981; SCHUURS and VAN VEEMEN 1977).

Lysozyme (EC 3.2.1.17) was used as the enzymatic marker in the first homogeneous EIA published. In these assays, the binding of the antibody to the hapten–enzyme conjugate results in an inhibition of the enzyme by preventing the macromolecular substrate (bacterial cell wall peptidoglycans) from reaching the enzymatic active site (RUBENSTEIN et al. 1972; SCHNEIDER et al. 1973; BASTIANI et al. 1973).

Other methods have been described which use enzymes (glucose-6-phosphate dehydrogenase, malate dehydrogenase) that act on small molecule substrates. In this case, it seems that the inhibition of enzymatic activity observed during the immune reaction is due to conformational effects (ROWLEY et al. 1975). In some rare cases, the binding of antibody induces an increase of enzymatic activity (ULLMAN et al. 1979; JAKLITSCH et al. 1976). All these methods, known as EMIT (enzyme-multiplied immunoassay techniques), have been applied successfully to the assay of numerous haptens (hormones, medicines, drugs of abuse).

The development of homogeneous EIA applicable to macromolecular antigens is more difficult for a variety of reasons (for further details see CROWL et al. 1980). However, some results have been obtained by using antigen–β-galactosidase conjugates and a flexible macromolecular substrate (GIBBONS et al. 1980). Here too, the binding of antibody to the antigen–enzyme conjugate inhibits enzymatic activity by sterically hindering the access of substrate to the active site.

Other types of homogeneous assays have been described over the last few years. These assays, which have been called enzyme-associated substance enzyme immunoassays (EASEIA) (SCHALL and TENOSO 1981), use tracers which are conjugates between a hapten and different substances able to influence the activity of an enzyme, including cofactors (CARRICO et al. 1976; SCHROEDER et al. 1976a), substrates (BURD et al. 1977; LI et al. 1981; WORAH et al. 1981), or inhibitors (FINLEY et al. 1980; NGO and LENOFF 1980; BACQUET and TWUMASI 1984; DONÀ 1985). Here, the binding of the antibody to the conjugate modulates the activity of the enzyme by inhibiting the effect of the substance. As with EMIT, these methods apply essentially to the assay of low molecular weight haptens. They differ in that the enzymes are not involved in the preparation of the conjugate and as a con-

Table 8. Enzymes used in homogeneous EIA

Enzyme	Source	Enzyme Commission number
Lysozyme	Egg white	3.2.1.17
Glucose-6-phosphate dehydrogenase	*Leuconostoc mesenteroides*	1.1.1.49
Malate dehydrogenase	Pig heart mitochondria	1.1.1.37
β-Galactosidase	*Escherichia coli*	3.2.1.23

sequence should not be classified as enzyme immunoassays. We have included them in this category because the use of an enzyme as a detection system is a common property of all EASEIA. Some of these assays, those using conjugates between a hapten and a fluorescent substrate, have been categorized as substrate-labeled fluoroimmunoassays (SLFIA).

Table 8 gives a list of the principal enzymes used in homogeneous EIA (enzymes used in EASEIA are not included). With the exception of β-galactosidase, this list differs markedly from that given for heterogeneous assays. Indeed, although the criteria of choice defined previously remain valid for homogeneous assays, they are no longer paramount. The choice of enzyme will be dictated rather by the possibilities it offers for the modulation of its activity as a function of the immunologic reaction. Note that, for homogeneous EIA, it is very important that the enzyme used as tracer is absent from the biologic media involved, since the final enzymatic measurement will not be preceded by any separation method.

The rapidity, simplicity, and automation of homogeneous EIA methods have led to a considerable expansion of their use in clinical analysis. Most of the assays described until now relate to low molecular weight haptens (hormones, medicines, drugs of abuse) and are characterized by a rather low sensitivity (10 ng/ml–1 μg/ml).

This lack of sensitivity can be explained in a number of ways:

1. A high background in the enzymatic measurement tied to the fact that this is carried out in complex biologic media.

2. The enzymes most commonly used in EMIT (lysozyme, glucose-6-phosphate dehydrogenase, malate dehydrogenase) are not those with the best detection sensitivity.

3. In the case of EASEIA methods, the modulation of the enzymatic activity is not directly due to the immunologic reaction, but depends on other interactions (cofactor–enzyme, inhibitor–enzyme, substrate–enzyme), which are characterized by binding affinities well below those of antigen–antibody interactions. As a result, these systems only respond at higher hapten concentrations, and lose in terms of sensitivity. For assays involving hapten–substrate conjugates, the situation is even more unfavorable since, in this case, the amplification provided by the enzyme is lost.

Apart from the first point cited, all these limitations are not, however, inherent to homogeneous methods and it is to be hoped that the future will bring forth assays with sensitivities equivalent to those of heterogeneous methods. Some promising results have already been obtained in the case of thyroxine (FINLEY et al. 1980).

3. Preparation of Enzymatic Conjugates

There are different ways of conjugating an enzyme to a hapten, an antigen, or an antibody. Whichever method is used, it must preserve entirely the activity of the enzyme and the immunoreactivity of the immunologic partner. The conjugate must be stable and possess the lowest possible degree of nonspecific binding. Whatever the coupling procedure, it is often necessary to purify the enzymatic conjugate from the reaction mixture in order to eliminate unreacted products, inactive conjugate, or undesirable polymers. Dialysis, molecular sieve chromatography (ISHIKAWA et al. 1983; KENNEDY et al. 1976; SCHUURS and VAN VEEMEN 1977; PRADELLES et al. 1985; NAKANE and KAWOI 1974), and affinity chromatography (IMAGAWA et al. 1981, 1981; ISHIKAWA et al. 1983) can be used to achieve this goal.

a) Enzyme–Hapten Conjugates

The choice of a method for coupling the enzyme and hapten will depend, of course, on the nature of the hapten and above all on the chemical groups it possesses. In general, for the synthesis of the enzymatic conjugate, it is better to use a reaction scheme identical to that employed in the synthesis of the immunogen. In this way, one usually obtains a conjugate of maximal immunoreactivity, and hence a sensitive assay.

As suggested by some authors (O'SULLIVAN 1984; VAN VEEMEN and SCHUURS 1975), in certain cases, it can happen that the presence of the same cross-link on the immunogen and the conjugate leads to an insensitive assay of the native hapten, the cross-link being involved in antibody recognition. This problem has already been encountered in radioimmunology, and may be resolved by using different cross-links for the synthesis of the immunogen and the enzymatic conjugate (HUNTER 1982; VAN VEEMEN and SCHUURS 1975). Another solution is to subject the native hapten, before the assay, to a modification identical to the one it underwent during the preparation of the immunogen. This approach has been applied successfully to radioimmunoassays of cyclic AMP (CAILLA et al. 1973) and thyrotropin-releasing hormone (GROUSELLE et al. 1982).

It is not possible here to present an exhaustive review of the hapten–enzyme coupling methods described in the literature. Interested readers can refer to the reviews devoted to the subject (ERLANGER 1973; MARKS et al. 1980; RIAD-FAHMY et al. 1981; SCHUURS and VAN VEEMEN 1977) as well as to the original articles. It can be noted, however, that the coupling reaction very often consists in forming a peptide bond between a carboxyl group or a primary amine group of the hapten and the complementary side chains of the enzyme. Methods using mixed anhydrides (VAN VEEMEN and SCHUURS 1972, 1975; AL-BASSAM et al. 1979; RUBENSTEIN et al. 1972; SCHNEIDER et al. 1973), carbodiimides (DRAY et al. 1975), or activated esters of *N*-hydroxysuccinimide (PRADELLES et al. 1985) have been employed to this effect.

The stoichiometry of the hapten–enzyme complex obtained should also be taken into account, as it can influence the characteristics of the assay. In theory, the optimal hapten : enzyme ratio in the conjugate is 1 : 1, since this corresponds to a maximal specific activity for the tracer. Some authors have been able to verify this theoretical notion experimentally (PRADELLES et al. 1985; AL-BASSAM et al. 1979).

In order to obtain a sensitive assay, it is often preferable to perform the hapten–enzyme coupling in conditions such that a nonnegligible fraction of the enzyme population is not coupled. The excess uncoupled enzyme can be eliminated if necessary later, through a purification step. This operation however, is not always necessary, notably in the case of heterogeneous methods. Indeed, in this case, it is not problematical if part of the enzymatic activity involved is immunologically inactive, since an adequate separation method (e.g., solid phase) will ensure its elimination before the enzymatic measurement is performed.

b) Enzyme–Antigen or Enzyme–Antibody Conjugates

Whether it is an antigen or an antibody that is to be coupled to an enzyme, the same problem is involved: that of obtaining a protein–protein conjugate. Very many coupling methods exist for achieving this aim (for review see KENNEDY et al. 1976). Only a few of them have been applied successfully to EIA (AVRAMEAS et al. 1978; O'SULLIVAN and MARKS 1981).

In this type of reaction, the principal difficulty is to avoid autopolymerization reactions (enzyme–enzyme, antigen–antigen, antibody–antibody), and the formation of enzyme–antigen or enzyme–antibody macropolymers. This difficulty is bound up with the fact that each participant in the coupling reaction possesses many of the same reactive groups. The formation of polymerized conjugates has to be avoided, even if it may appear advantageous in terms of specific activity. The reason for this is that a polymerized tracer would induce an increase of nonspecific binding which would limit the sensitivity of the assay, especially in the case of excess reagent methods (EKINS 1973; EKINS and JACKSON 1984).

The glutaraldehyde method has certainly proved to be the most successful since its introduction by AVRAMEAS in 1969. It has the advantages of being reproducible, easy to perform, and of giving stable conjugates. Using a two-step coupling procedure (AVRAMEAS and TERNINCK 1971), it has been possible to obtain monovalent enzyme–antibody conjugates with peroxidase (stoichiometry 1 : 1).

However, this method has tended to be abandoned progressively because it suffers from some drawbacks. The conjugates are often obtained in low yield, with a significant loss of immunologic activity and marked polymerization. As a result they are characterized by a "useful signal : nonspecific binding" ratio that is rather low.

Use of the periodate method (NAKANE and KAWOI 1974; TIJSSEN and KURSTAK 1984) has given higher yields, at least with peroxidase. However, this method does not seem to be applicable to all enzymes and it also leads to polymers and marked losses of enzymatic activity (BOORSMA and STREEFKERK 1979; WILSON and NAKANE 1978; TSANG et al. 1984).

Over the last few years, it has become clear that the most valuable conjugates are those prepared either by methods that use the maleimide group through homobifunctional (dimaleimide) (KATO et al. 1975) or heterobifunctional reagents (*N*-succinimidyl-*m*-maleimidobenzoate derivatives) (KITAGAWA and AIKAWA 1976; YOSHITAKE et al. 1979), or by the pyridyldisulfide method (CARLSSON et al. 1978; KING et al. 1979). All these methods give a high yield and involve mild reactions which are not prejudicial to the activity of the enzymes or the immunoreactants. Their main strength though is that they allow much better control than other methods over the stoichiometry of the conjugates, without the formation of homopolymers. The reason for this is that in one way or another the coupling is achieved through the intermediary of one or more thiol groups, which are present in limited amounts or are introduced in a controlled manner onto one or both moieties. In the case of the labeling of antibodies, it has proved very advantageous to use Fab' fragments rather than whole antibodies (for review see ISHIKAWA et al. 1983). These fragments bear approximately one free thiol group per molecule and lend themselves well to the preparation of 1 : 1 conjugates that have a very low level of nonspecific binding and hence a very high "useful signal : nonspecific binding ratio."

III. Labeling with Fluorescent Compounds

The use of a fluorescent signal associated with immunologic detection is of long standing, since as early as 1941 antibodies coupled to fluorescein have been employed in histology (COONS et al. 1941). Fluoroimmunoassays (FIA) have only been developed to any significant degree over the last 10 years or so. This trend can be explained by the detection sensitivity offered by fluorescent compounds and also by the possibilities they offer in terms of the development of homogeneous assays (for review see HEMMILÄ 1985).

1. Fluorescence

A fluorescent molecule is one that is capable of absorbing light and reemitting the absorbed energy in the form of photons of longer wavelength. The lower energy of emission is the consequence of nonradiative energy losses, the wavelength difference being referred to as the Stokes shift. In the case of fluorescence, this wavelength shift is rather small (30–50 nm). When electronic deexcitation occurs through the intermediary of a triplet state, the shift is greater ($\geqq 200$ nm), and in this case one speaks rather of phosphorescence.

A fluorescent molecule is also characterized by the following parameters: (a) its extinction coefficient; (b) its absorption spectrum; (c) its emission spectrum (shifted toward the long wavelength end of the spectrum as compared with the absorption spectrum); (d) its quantum yield, which is defined as the ratio between photons emitted and those absorbed; and (e) its fluorescent lifetime, which characterizes the decrease of photon emission.

Unlike radioactive disintegration, fluorescence emission is very sensitive to the physicochemical environment (temperature, pH, proteins, etc.). Although this dependence can be turned to advantage to perform homogeneous assays, it does nonetheless create several disadvantages, notably a drop in the detection sensitivity of the method.

The main problems encountered in the measurement of fluorescence are the following:

1. Scattering of the exciting or emitted light owing to the presence in solution of concentrated proteins or because of the use of a solid phase

2. The fluorescent background generated either by the proteins present (those of the serum in particular) or by the solid phase
3. Change in the environment of the fluorescent molecule or interaction with proteins, modifying its fluorescent characteristics (quantum yield, emission wavelength)
4. The presence in the vicinity of the fluorescent compounds of molecules absorbing either excitation or emission energy (e.g., another fluorescent molecule)

2. Organic Fluorescent Markers

A great many organic fluorescent markers have been used in fluoroimmunoassays for the labeling of antibodies, antigens, or haptens. The most common are fluorescein (FITC), rhodamine derivatives (TRMITC, RBITC), and dansyl chloride. All these compounds are characterized by high quantum yields (30%–85%). For further details about the structure of these compounds, as well as their fluorescent characteristics, see a review by HEMMILÄ (1985).

Each of these markers has its own optical properties which render them of varying interest in the design of heterogeneous and homogeneous assays. Because of the problems already described, the detection sensitivity of these markers (and hence of the corresponding tracers) falls in the range 10^{-9}–10^{-12} *M*, which makes them at best equal to ^{125}I.

3. Metal Chelates

Some rare earth chelates (europium or terbium) have really remarkable fluorescent properties, allowing very sensitive detection. In these chelates, the exciting light is absorbed by the ligand and the energy is transferred to a resonance level of the metal, which emits at its characteristic wavelength. This type of fluorescence is associated with a very marked Stokes shift (~250 nm), but also with a very long half-life (100–1000 μs), compared with those of other fluorescent compounds (1–100 ns).

Because the emitted radiation is of long wavelength (~600 nm for europium), interference due to scattering phenomena or to protein fluorescence is significantly reduced. By virtue of the prolonged half-life of these chelates, it is also possible to perform a delayed measurement of fluorescence emission (TRFIA) (SOINI 1984; DAKUBU et al. 1984), which allows great reduction in the other sources of fluorescence, whose decay is much faster. The combination of all these factors leads to a considerable reduction in the background fluorescence of the method, and an improvement in the detection sensitivity, which is around 10^{-18} mol. The employment of these chelates constitutes a remarkable advance and offers the possibility of synthesizing tracers of specific activity significantly higher than that of ^{125}I.

4. Heterogeneous Assays

A good many heterogeneous assays have been described that involve competition methods (labeling of hapten or antibody) or excess reagent methods (antibody labeling), and their principle is copied from their radioimmunoassay equivalent. Because of background fluorescence problems, these assays almost always involve a separation method using a solid phase with a low nonspecific binding capacity, so that the measurement is performed in a synthetic medium in the absence of interfering substances derived from biologic media.

For the reasons described previously, the performance of fluorescent assays involving organic markers does not exceed that of the corresponding radioimmunoassays (EKEKE et al. 1979; KOBAYASHI et al. 1982; KAPLAN et al. 1981; TSAY et al. 1980; SINOSICH and CHARD 1979; KARNES et al. 1981). In contrast, the use of rare earth chelates gives very good results. However, the preparation of tracers involving these chelates poses some problems:

1. It is difficult to obtain perfectly stable chelates, which remain bound to their immunologic partner in very dilute solutions containing other metals or other chelating agents.
2. Some chelates, and in particular those of europium, have a tendency to become hydrated and lose some of their fluorescent characteristics.

The most elegant solution to these difficulties is that provided by the dissociation-enhanced lanthanide fluorescence immunoassay (DELFIA) (HEMMILÄ et al. 1984), which has essentially been applied to the labeling of antibodies. In this method, the metal is bound very tightly to the antibody through the intermediary of powerful aminocarboxylic complexones (EDTA, EGTA, HEDTA). These complexes are of the stability required for participating in the immunologic reaction, but possess poor fluorescent characteristics.

After separation of the components of the immunologic reaction (solid-phase method), the metal is dissociated from the antibody (and hence from the solid phase) by the addition of a solution containing different chelating agents (β-diketone, trioctylphosphine) and a detergent (Triton X-100). In these conditions, one obtains a suspension of micelles containing the chelates, which exhibits optimal optical properties. This method, in association with TRFIA, has been applied successfully to some assays, which are characterized by high sensitivity and a wide working range (PETTERSON et al. 1983; SIITARI et al. 1983; TOIVONEN et al. 1986; JORONEN et al. 1986; VILPO et al. 1986).

5. Homogeneous Assays

Advantage has been taken of the sensitivity of fluorescence to the physicochemical environment to design many sorts of homogeneous fluoroimmunoassays. Most of these assays involve a competition in which the binding of antibody to the tracer modifies its fluorescence characteristics. We shall not deal in detail with these procedures, and interested readers may refer to the appropriate reviews (HEMMILÄ 1985) and original articles. Instead, we shall briefly cite the commonest of these methods.

Fluorescence enhancement immunoassays and fluorescence quenching immunoassays rely upon an increase or a reduction, respectively, of tracer fluorescence when immunoreaction takes place. The principle of these methods is closely analogous to that of the corresponding homogeneous EIA assays (EMIT), and like them they have essentially been applied to the assay of haptens (SMITH 1977; LI-

BURDY 1979; HANDLEY et al. 1979; SHAW et al. 1977). The indirect quenching immunoassays which involve antibodies directed against the fluorescent marker itself, can be applied to the assay of antigens (ZUK 1981; HASSAM et al. 1982).

The fluorescence excitation transfer immunoassay (FETIA) involves the labeling of two different molecules (a donor and an acceptor) with two different fluorescent probes (e.g., FITC-labeled antigen and TRMITC-labeled antibody). Because the emission and absorption spectra of the two fluorescent compounds overlap, emission from the donor will be quenched by the acceptor. This quenching effect being related to the distance between the two fluorescent probes, it will be maximum when antigen–antibody binding occurs. This method has been applied to both direct and indirect fluorescence quenching immunoassays (ULLMAN et al. 1976; MILLER et al. 1980; KHANNA and ULLMAN 1980; THAKRAR and MILLER 1982).

Fluorescence polarization immunoassays (FPIA) depend on the fact that tracer–antibody complexes have a more polarized emission than the free tracer, because of the greater size of the complex and its slower rotational motion (DANDLIKER et al. 1973; SPENCER et al. 1973; LU-STEFFES et al. 1982; JOLLEY et al. 1981; SPENCER 1981).

All these assays are rapid and simple. Most of them have been developed commercially. However, the sensitivity is often limited by the high fluorescent background produced by biologic samples, and also by the fact that the change induced in the fluorescence during the immunoreaction is invariably small. As with homogeneous enzyme immunoassays, these methods have mainly been employed for the assay of haptens present at high concentration in biologic fluids.

IV. Labeling with Luminescent Compounds

1. Luminescence

Luminescent compounds (whether chemi- or bioluminescent), like fluorescent compounds, are capable of emitting light upon the transition of an electron from an excited state to an orbital of lower energy. Unlike fluorescence, the excitation is not induced by electromagnetic radiation, but by a chemical oxidation reaction. Luminescence measurements are less subject to interference than conventional fluorescence measurements, first because biologic samples do not (or very rarely) contain luminescent products, but also because the problem of overlap between absorption and emission spectra encountered in fluorescence does not arise in luminescence. As a result, luminescent compounds can be detected with great sensitivity ($\sim 10^{-18}$ mol) (WEEKS et al. 1983a; DE LUCA 1978) and therefore allow preparation of tracers of high specific activity.

From a practical point of view, luminescence measurements suffer from two drawbacks:

1. These measurements should ideally be made with an automatic apparatus which includes a reagent addition device as well as the optical measuring system.

2. The measurement cannot be repeated as the luminescent product is irreversibly destroyed during the determination.

2. Chemiluminescence

Chemiluminescence reactions are distinguishable from bioluminescence reactions in that they do not necessarily involve biologic products. Quantum yields observed in chemiluminescence are generally low (<5%) because a significant portion of the energy stored during the chemical reaction is lost through nonradiative processes.

The commonest chemiluminescent markers are derivatives of luminol and isoluminol as well as acridinium esters (WOODHEAD et al. 1984; WOOD 1984; WEEKS et al. 1986). For all these compounds, the luminescence reaction is induced by the addition of hydrogen peroxide in alkaline conditions. In the case of luminol and isoluminol derivatives, the presence of a catalyst is necessary. These are most commonly enzymes (microperoxidase, pseudoperoxidase), which give the highest quantum yields. Other inorganic catalysts can however be used (SCHROEDER and YEAGER 1978). The reaction of acridinium esters, which is accompanied by cleavage of the molecule, does not require the presence of any catalyst (WEEKS et al. 1983 a).

These compounds have been used to label haptens (SCHROEDER et al. 1976 b; DE BOEVER et al. 1984; PAZZAGLI et al. 1982; KIM et al. 1982; BARNARD et al. 1981; ESHHAR et al. 1981) and antibodies (SIMPSON et al. 1979; WEEKS et al. 1983 a, b), and have enabled powerful assays to be developed. These are essentially heterogeneous assays using solid phases in order to minimize quenching phenomena that may occur during measurements.

The most interesting results for antigen assays have been obtained by excess reagent methods involving labeling of the antibody with acridinium ester derivatives. For example, the sensitivity obtained in the case of thyroid-stimulating hormone (WOODHEAD et al. 1984) is greater than that of the best radioimmunoassays available.

Two types of homogeneous chemiluminescent immunoassay (CLIA) have been described. The first type is based on an enhancement of chemiluminescence of isoluminol derivatives when antibody binding occurs (KOHEN et al. 1979, 1980 a, b). Like the corresponding enzymatic and fluorescent assays, they are characterized by poor sensitivity. The second type of assay involves the use of haptens or antigens labeled by an isoluminol derivative and antibodies labeled with fluorescein. When the antibody binds to the tracer, the luminescent signal is modified owing to a nonradiative energy transfer. The principle of this assay is rather similar to that of FETIA, and its sensitivity is given as equivalent to that of the corresponding radioimmunoassay (PATEL and CAMPBELL 1983).

3. Bioluminescence

As its name implies, bioluminescence involves the emission of luminescence by molecules of biologic origin. By far the commonest example is that of luciferin in the reduced state associated with the enzyme luciferase (EC 1.14.14.3). The luciferin–luciferase reaction which generates the luminescent emission is associated with other reactions that maintain the energetic balance of the system. These reactions involve the participation of other enzymes (glucose-6-phosphate dehydrogenase, pyruvate kinase, etc.) as well as certain enzymatic cofactors such as ATP,

NAD or NADP (DE LUCA 1978; WANNLUND and DE LUCA 1983; WHITEHEAD et al. 1979). Quantum yields observed in bioluminescent reactions are usually greater than those obtained in chemiluminescence (10%–100%) (DE LUCA 1978). The few bioluminescent immunoassays (BLIA) described are heterogeneous competition assays involving hapten–enzyme or antigen–enzyme conjugates. Depending on the assay, the enzyme involved in the conjugate is either luciferase (WANNLUND and DE LUCA 1983; JABLONSKI 1985) or one of the enzymes implicated in the auxiliary reactions (WANNLUND and DE LUCA 1983; WOOD 1984). Very recently, original homogeneous assays based on the use of solid-phase immobilized luciferase and FMN oxidoreductase have been developed (TÉROUANNE et al. 1986a, b; CARRIÉ et al. 1986). In this method, the bioluminescent reaction is restricted to the solid phase, thus allowing suppression of the separation step. Sensitivity comparable to that of the corresponding radioimmunoassay has been obtained. As far as we know, no system employs conjugates of luciferin. In fact, all these assays which are usually classified as BLIA depend rather on enzyme immunoassay methods using a luminescence detection technique (OLSSON et al. 1979; ARAKAWA et al. 1985; TAKAYASU et al. 1985).

Bioluminescence potentially offers the possibility of preparing tracers of very high specific activity by exploiting the amplification provided by the enzymes together with the luminescence detection sensitivity. Yet the assays published thus far have proved rather disappointing in term of sensitivity.

V. Other Labeling Methods

1. Labeling with Bacteriophages

Bacteriophages are viruses capable of infecting bacteria and inducing their death by cellular lysis. It has been shown that the infectious activity of these phages can be neutralized by antibodies which recognize determinants present at the surface of the virus. Hence, bacteriophage–antigen (HAIMOVICH and SELA 1969) and bacteriophage–hapten conjugates (HAIMOVICH et al. 1967; ANDRIEU et al. 1975) lose their infectious capacity in the presence of antibodies directed against these antigens of haptens. This effect has been used to develop homogeneous competition assays for antibodies, antigens, and haptens. In these tests, the immunologic reaction is performed and the infectious activity of the reaction medium is then measured by culture on plated bacteria. The final measurement consists in counting the lysis plaques in the Petri dishes.

The idea of using a viral probe is attractive a priori and should allow the preparation of tracers of very high specific activity since it is possible, in theory, to detect an individual viral particle by virtue of the amplifying power provided by bacterial culture techniques. This advantage is, moreover, confirmed by the assays published so far, which compare favorably with the corresponding radioimmunoassays (tritiated tracers) (ANDRIEU et al. 1975). Although these assays have been known for more than 15 years, they have still not enjoyed much success. This is most probably because the final measurement is too tiresome to perform.

2. Labeling with Metals

Apart from the rare earth chelates which we have discussed in Sect. C.III.3, metals have long been used for labeling immunologic reagants. In histochemistry, for example, ferritin labeling of antibodies has been used since 1959 (SINGER 1959). In this case, it was a question of indirect labeling since the iron atoms were bound to the antibody by the intermediary of an apoprotein. Methods have since been described for the direct labeling of antigens and haptens by metals. All these studies have been described in a review that the reader may usefully consult (CAIS 1983).

Two labeling processes have been used in metalloimmunoassay (MIA)

1. Direct introduction of the metal onto the hapten by formation of a carbon–metal bond (CAIS 1979)
2. Coupling of an organometallic compound or a coordination complex bearing an activatable chemical group (CAIS 1983; CAIS and TIROSH 1981; BROSSIER and MOISE 1982; WEBER and PURDY 1979; CAIS et al. 1977, 1978).

The second approach may be considered more valuable inasmuch as it can be used to develop more versatile labeling protocols.

Different analytic methods have been proposed for the determination of these metal tracers, the most common of which are electrochemical methods (WEBER and PURDY 1979; DOYLE et al. 1982) and atomic absorption spectroscopy (BROSSIER and MOISE 1984; CAIS 1983).

Most of the published methods relate to hapten and antigen assays and rely upon heterogeneous competition methods (BROSSIER and MOISE 1982; CAIS 1983). The development of these metal tracers owes a great deal to progress made in coordination chemistry, and is of undeniable value in a variety of areas of biology. However, it appears unlikely that they have a great future in the area of in vitro immunologic analysis. Indeed, the sensitivity attainable in the determination of these tracers (at best 10^{-13} mol) limits very seriously the sensitivity of the method, which requires, moreover, the employment of intensive analytic methods.

3. Labeling with Electroactive Compounds

Over the last few years, two types of assay have been described that depend on the preparation of electroactive tracers (WEHMEYER et al. 1983). In one type of assay, the tracer is a molecule of estriol labeled with nitro groups (WEHMEYER et al. 1982). This modification has the effect of increasing the redox potential of the molecule, which can be measured by differential pulse polarography with a mercury electrode at a potential where unmodified estriol is not reducible. It has been possible to develop a homogeneous competition assay based on the fact that the tracer loses its electroactive properties after reaction with a specific antibody.

In the second type of assay, human serum albumin is labeled by an electroactive metal (indium) and serves as a tracer in a heterogeneous competition assay (DOYLE et al. 1982). In this case, the metal is released in acid conditions before being determined by an anodic redissolving method. What is involved here in fact is a metalloimmunoassay employing an electrochemical detection method. These two electrochemical assays have the same drawbacks as metalloimmunoassays: too low a sensitivity and tedious experimental manipulation.

4. Labeling with Free Radicals

Free radicals possess unpaired electrons whose spins can be detected by resonance spectroscopy in an appropriate magnetic field. When stable free radicals are chemically coupled to haptens or antigens, this resonance can be modified by reaction with a specific antibody. Advantage has been taken of this effect in the development of homogeneous competition assays using these spin labels (STRYER and GRIFFITH 1965; LEUTE et al. 1972; HSIA et al. 1973; WEI and ALMIREZ 1975). These spin immunoassays (SIA) have proved to be applicable essentially to haptens present at high concentrations ($>10^{-7}$ M) in biologic fluids.

5. Labeling with Particles

Various nonisotopic labeling methods have been described that involve insoluble materials such as metallic colloidal suspensions, erythrocytes, or latex particles. Colloidal suspensions of gold, for instance, have long been employed either in immunologic tests (KOLMER and BOERNER 1945; LEUVERING et al. 1981, 1983) or immunohistochemical techniques (GEOGHEGAN and ACKERMAN 1977; HORISBERGER and VON LAUTHAN 1979; GOODMAN et al. 1979). Erythrocytes conjugated to antibodies or to antigens have also been used extensively in reliable and easily implemented agglutination tests (ALDER and LIU 1971).

However, it seems that the best results have been obtained by using latex particles as markers. It has been shown by some authors that it is possible to perform sensitive assays of antigens, haptens, or antibodies by using latex particles of well-defined size coated with one of the constituents of the immunoreaction (CAMBIASO et al. 1977; GRANGE et al. 1977; QUASH et al. 1978; MAGNUSSON and MASSON 1985; FAGNART et al. 1985; CHRISTIAN et al. 1958; ROBBIN et al. 1962). Some of these tests are known as particle counting immunoassays (PACIA). Depending on the assay, the agglutination reaction is followed by either a cell counter (CAMBIASO et al. 1977), by turbidimetry (DEZELIC et al. 1971), or by anistropic light scattering techniques (VON SCHULTHESS et al. 1976, 1980). Some of the published assays are of sensitivity comparable to that of the corresponding radioimmunoassays. These labeling procedures are easily applied to many molecules and form the basis of practical and automated tests. It may be that, within the next few years, they could be favorably compared with the best nonisotopic methods.

D. Conclusions

Throughout this chapter we have made a comparison of the respective merits of the different radioactive and nonradioactive labels. It appears today that the use of enzymatic, fluorescent, or luminescent probes allows the preparation of tracers having a specific activity significantly superior to that obtainable with ^{125}I. Such tracers which can be detected very sensitively (minimum detectable amount

$\sim 10^{-18}$ mol) may allow the determination of biologic molecules in the attomole range. Indeed, such performance constitutes significant progress compared with the possibilities classically offered by radioimmunoassays. This aim, however, will be achieved only if high affinity antibodies are used together with these high specific activity tracers. From this point of view, it seems that heterogeneous excess reagent methods (EKINS and JACKSON 1984) will provide better results because they allow a partial compensation of the limited affinity of antibodies by the use of concentrated reagents. Other experimental factors such as the quality of the separation method (mainly solid phase), which influence the level of nonspecific binding, also appear critical to ensure the quality of the assay. If the future development of nonisotopic methods is unquestionable, it is not possible, today, to make predictions about the prevalence of one of these methods over the others. It is much more likely, moreover, that there is a place for different labels appropriate to the analytic problem encountered and the experimental environment.

In addition, one should not believe that criticisms of radioimmunoassays will lead to their fast and inexorable decline. First, it should be recognized that some of these criticisms are exaggerated. For example, the risks of irradiation and contamination incurred by users of radioimmunoassay are in fact very small, bearing in mind the minimal amounts of radioactivity involved. These risks are probably equivalent to those associated with other toxic products (viruses, solvents, acids carcinogenic compounds, etc.) commonly employed in laboratories. Likewise, the high cost of automatic counters does not constitute a definitive argument since some of the most serious competitors of radioimmunoassay (luminescence, time-resolved fluoroimmunoassay) also use sophisticated and equally costly equipment.

Also, while the immutable nature of radioactive disintegration militates against the development of homogeneous assays, it does, in contrast, confer upon radioimmunoassay an increased insensitivity to certain types of interference due to biologic samples. Because of this reliability, radioimmunoassays are often employed as a reference for the validation of measurements provided by any alternative method.

Lastly, radioimmunoassays are firmly implanted in numerous research and analytic laboratories and have proved to be fully satisfactory, notably in endocrinology and neuroendocrinology. These laboratories will only abandon such methods, slowly since not only are they familiar, but the investment costs have already been incurred. For all these reasons, it is probably more appropriate to predict that radioimmunoassay will undergo a progressive regression rather than a rapid decline.

Acknowledgments. The authors wish to thank Mrs. Martine Ohana and Miss Helene Feller for typing the manuscript. Some parts of the work mentioned in this chapter as well as the redaction were made possible by financial support from Commissariat à l'Energie Atomique (J. G. and P. P.) and grants from PIRMED/INSERM (J. M.).

References

Abraham GE (1974) Radioimmunoassay of steroids in biological material. Proc symp radioimmunoassay relat proc med, Istambul. IAEA-SM 177/207 2:3–29

Al-Bassam MN, O'Sullivan MJ, Bridges JW, Marks V (1979) An improved methotrexate enzyme-immunoassay. Clin Chem 25:1448–1452

Alder FL, Liu C-T (1971) Detection of morphine by hemagglutination-inhibition. J Immunol 106:1684–1685

Alexander NM (1974) Oxidative cleavage of tryptophanyl peptide bonds during chemical and peroxidase-catalyzed iodination. J Biol Chem 249:1946–1952

Anderson GW, Zimmerman JE, Callahan FM (1964) The use of esters of N-hydroxysuccinimide in peptide synthesis. J Am Chem Soc 86:1839–1842

Andrieu JM, Mamas S, Dray F (1975) Viroimmunoassay of steroids: method and principles. In: Campbell EHD, Hiller SG, Griffiths K (eds) Steroid immunoassay. Proceedings of the fifth Tenovus workshop, Cardiff, April 1974. Alpha Omega, Cardiff, pp 189–198

Antonov PA, Pancheva RP, Ivanov IG (1985) Radiodination of naturally occurring phospholipids. Biochim Biophys Acta 835:408–410

Arakawa H, Maeda M, Tsuji A (1985) Chemiluminescence enzyme immunoassay for tyrosin with use of glucose oxydase and a bis (2,4,6-Trichlorophenyl) oxalate-fluorescent dye system. Clin Chem 31:430–434

Argentini M (1982) Labelling with iodine. A review of the literature. Federal Institute for Reactor Research, Wuerenlingen, Switzerland

Assoian RK, Blix PM, Rubenstein AH, Tager HS (1980) Iodotyrosylation of peptides using tertiary-butyloxycarbonyl-L-[^{125}I]iodotyrosine N-hydroxysuccinimide ester. Anal Biochem 103:70–76

Aubert ML (1971) Critical study of the radioimmunological assay for the dosage of the polypeptide hormone in plasma. Minerva Medica, Torino, p 1

Avrameas S (1969) Coupling of enzymes to proteins with glutaraldehyde. Immunochemistry 6:43–52

Avrameas S, Terninck T (1971) Peroxydase labelled antibody and Fab conjugates with enhanced intracellular penetration. Immunochemistry 8:1175–1179

Avrameas S, Terninck T, Guesdon JL (1978) Coupling of enzymes to antibodies and antigens. Scand J Immunol 8:7–23

Bacquet C, Twumasi DY (1984) A homogeneous enzyme immunoassay with avidine-ligand conjugates as the enzyme modulator. Anal Biochem 136:487–490

Barnard GJ, Collins WP, Kohen F, Linner HR (1981) The measurement of urinary estriol-16α-glucuronide by a solid phase chemiluminescence immunoassay. J Steroid Biochem 14:941–948

Bastiani RJ, Phillips RC, Schneider RS, Ullman EF (1973) Homogeneous immunochemical drug assays. Am J Med Technol 39:211–216

Bayly JR, Evans EA (1968) Storage and stability of compounds labelled with radioisotopes. Review 7. Radiochemical Centre, Amersham, p 1

Bayly JR, Evans EA, Glover JS, Rabinowitz JL (1976) Synthesis of labelled compounds. In: Tubis M, Wolf W (eds) Radiopharmacy. Wiley, New York, p 303

Berridge M, Jiang VW, Welch MJ (1979) Local effects in labeled molecules following iodine 125 decay. J Nucl Med 20:616

Berson SA, Yalow RS, Baumann A, Rotschild MA, Newerly K (1956) Insulin ^{131}I metabolism in human subjects: demonstration of insulin binding globulin in the circulation of insulin treated subjects. J Clin Invest 35:170–190

Bismuth J, Castay M, Lissitzky S (1971) Sephadex filtration of serum iodocompounds and protein binding of iodotyrosines. Clin Chim Acta 35:285–298

Bloch HS, Ray FE (1946) Organic radioiodo compounds for cancer research. J Natl Cancer Inst 7:61–66

Bogdanove EM, Strash AM (1975) Radioiodine escape is an unexpected source of radioimmunoassay error and chronic low level environmental contamination. Nature 257:426–427

Bolton AE (1977) Radioiodination techniques. Review 18. Radiochemical Centre, Amersham, England, p 1
Bolton AE, Hunter WM (1973) The labelling of proteins to high specific radioactivities by conjugation to a ^{125}I-containing acylating agent. Biochem J 133:529–539
Boorsma DM, Streefkerk JG (1979) Periodate or glutaraldehyde for preparing peroxydase conjugate. J Immunol Methods 30:245–255
Brossier P, Moise C (1982) Exemples d'utilisation d'un complexe organométallique comme marqueur de molécules thérapeutiques. In: Radioimmunoassay and related procedures in medicine, Vienna, IAEA SM-259/35:779–786
Brossier P, Moise C (1984) Dosage du fer dans le ferocène et quelques dérivés par spectrométrie d'absorption atomique sans flamme. Analysis 12:223–224
Brown BL, Reith WS (1967) A method of radio-iodination on a submicrogram scale. The preparation and stability of ^{131}I and ^{125}I-labelled 3-monoiodotyrosine and 3,5-diiodotyrosine of very high specific activity. Biochim Biophys Acta 148:423–434
Brunings KJ (1947) Preparation and properties of the iodohistidines. J Am Chem Soc 69:205–208
Burd JF, Wong RC, Feeney JE, Carrico RJ, Bogulasky RC (1977) Homogeneous reactant-labeled fluorescent immunoassay for therapeutic drugs exemplified by gentamicin determination in human serum. Clin Chem 23:1402–1408
Butler VP Jr (1973) Radioimmunoassay and competitive binding radioassay methods for the measurement of drugs. Metabolism 22:1145–1153
Butt WR (1972) The iodination of follicle stimulating and other hormones for radioimmunoassay. J Endocrinol 55:453–454
Cailla HL, Racine-Weisbuch MS, Delaage MA (1973) Adenosine 3′,5′ cyclic monophosphate assay at 10^{-15} mol level. Anal Biochem 56:394–407
Cais M (1979) Tests immunologiques à l'aide de complexes organométalliques. Un nouveau concept. Actual Chem 7:14–26
Cais M (1983) Metalloimmunoassay: Principles and practice. Methods Enzymol 92:445–458
Cais M, Tirosh N (1981) Metalloimmunoassay. IV. Manganese labelled metallohaptens via cymantrene derivatives. Bull Soc Chem Belg 90:27–35
Cais M, Dani S, Eden Y, Gandolfi O, Horn M, Isaacs EE, Josephy Y, Saar Y, Slouin E, Snarsky L (1977) Metalloimmunoassay. Nature 270:534–535
Cais M, Slouin E, Snarsky L (1978) Metalloimmunoassay. II. Iron-metallohaptens from estrogen steroids. J Organomet Chem 160:223–230
Cambiaso CL, Leek AE, De Steewinkel F, Billen J, Masson PL (1977) Particle counting immunoassay (PACIA) I. A general method for the determination of antibodies, antigens, and haptens. J Immunol Meth 18:33–44
Cardoso MT, Pradelles P (1982) Preparation of N(1-ethyl-2-pyrrolidyl-methyl) 2-methoxy-4-iodo-^{125}I-5-ethyl sulfonyl benzamide: a radioligand for the radioimmunoassay of sulpiride-related compounds. J Label Comp Radio 19:1103–1109
Carlson RA, White RM (1963) Formation of fragment ions from CH_3Te^{125} and $C_2H_5Te^{125}$ following the nuclear decays of $CH_3{}^{125}I$ and $C_2H_5{}^{125}I$. J Chem Phys 38:2930–2934
Carlsson J, Drevin H, Axen R (1978) Protein thiolation and reversible protein–protein conjugation. Biochem J 173:723–737
Caro RA, Ciscato VA, De Giacomini SMV, Quiroga S (1975) Labeling of proteins with ^{125}I and experimental determination of its specific activity. Int J Appl Radiat Isotopes 26:527–532
Carrico RJ, Cristner JE, Boguslaski RC, Yeung KK (1976) A method for monitoring specific binding reactions with cofactor labeled ligands. Anal Biochem 72:271–282
Carrié ML, Térouanne B, Brochu M, Nicolas JC, Crastes de Paulet A (1986) Bioluminescent immunoassay of progesterone: a comparative study of three different procedures. Anal Biochem 154:126–131
Castro A, Monji N (1981) Steric-hindrance enzyme immunoassay (SHEIA) using β-galactosidase as an enzyme label and maleimide derivative of hapten (or antigen) for enzyme coupling. Methods Enzymol 73:523–542

Christian CL, Mendes-Bryan R, Larson DL (1958) Latex agglutination test for disseminated lupus erythematosus. Proc Soc Exp Biol Med 98:820–823

Clark PMS, Price CP (1986) Enzyme-amplified immunoassay: a new ultrasensitive assay of thyrotropin evaluated. Clin Chem 32:88–92

Clinical Chemistry (1983) Product guide for radioassay and nonisotopic ligand assays. Clin Chem 29:890–986

Cohen Y (1976) Purity and stability of labeled compounds. In: Tubis M, Wolf W (eds) Radiopharmacy, Wiley, New York, p 379

Collignon F, Pradelles P (1984) Highly sensitive and specific radioimmunoassays for dihydroergotoxine components in plasma. Eur J Nucl Med 9:23–27

Commerford SL (1971) Iodination of nucleic acid in vitro. Biochemistry 10:1993–1999

Coons AH, Creech HJ, Jones RW (1941) Immunological properties of an antibody containing a fluorescent group. Proc Soc Exp Biol Med 47:200–202

Corric JET, Hunter WM (1981) ^{125}Iodinated tracers for hapten specific radioimmunoassays. Methods Enzymol 73:79

Crowl CP, Gibbons I, Schneider RS (1980) Recent advances in homogeneous enzyme immunoassays for haptens and proteins. In: Nakamura RM, Dito WR, Tucken ES III (eds) Immunoassays clinical laboratory techniques for the 1980s. Liss, New York, pp 89–126

Cvoric J (1969) Chemical forms of iodine in carrier free preparations of $Na^{125}I$. J Chromatogr 44:349–361

Dakubu S, Ekins RP, Jackson T, Marshall NJ (1984) High sensitivity, pulsed light time-resolved fluoroimmunoassay. In: Butt WR (ed) Practical immunoassay, the state of art. Dekker, New York, pp 71–101

Damle SR, Advani SH, Desai PB (1982) Non-isotopic analytical methods as alternative to radioimmunoassay in developing countries: a comparative study with reference to biological markers. In: Radioimmunoassay and related procedures in medicine. Vienna, IAEA-SM-259/84, pp 735–742

Dandliker WB, Kelly RJ, Dandliker J, Farquhar J, Levin J (1973) Fluorescence polarization immunoassay. Theory and experimental method. Immunochemistry 10:219–227

David GS, Reisfeld RA (1974) Protein iodination with solid state lactoperoxydase. Biochemistry 13:1014–1021

De Boever J, Kohen F, Vandekerckhove D, van Maele G (1984) Solid-phase chemiluminescence immunoassay for progesterone in unextracted serum. Clin Chem 30:1637–1641

De Luca MA (1978) Bioluminescence and chemiluminescence. Methods Enzymol 57:1

Dewanjee MK, Rao SA (1983) Principles of radioiodination and iodine-labeled tracers in biochemical investigation. In: Colombetti LG, Rayudu GVS (eds) Radiotracers for medical applications, volI. CRS, Boca Raton, p 1

Deželic G, Deželic N, Muic N, Pende B (1971) Latex particle agglutination in the immunochemical system human serum albumin-anti human serum albumin rabbit serum. Eur J Biochem 20:553–560

Donà V (1985) Homogeneous colorimetric enzyme inhibition immunoassay for cortisol in human serum with Fab anti-glucose 6-phosphate dehydrogenase as a label modulator. J Immunol Methods 82:65–75

Donabedian RK, Levine TA, Seligson D (1972) Micro-electrolytic iodination of polypeptide hormones for radioimmunoassay. Clin Chim Acta 36:517–520

Doran DM, Spar I-L (1980) Oxydative iodine monochloride iodination technique. J Immunol 38:155–163

Doyle MJ, Halsall HB, Heineman W (1982) Heterogeneous immunoassay for serum proteins by differential pulse anodic stripping voltammetry. Anal Chem 54:2318–2322

Dray F, Andrieu JM, Renaud F (1975) Enzyme immunoassay of progesterone at the picogram level using β-galactosidase. Biochim Biophys Acta 403:131–138

Edelhoch H (1962) The properties of thyroglobulin. J Biol Chem 337:2778–2787

Ekeke GI, Exley D, Abukneshar (1979) Immunofluorimetric assay of oestradiol-17β. J Steroid Biochem 11:1597–1600

Ekins RP (1973) The future development of immunoassay. In: Radioimmunoassay and related procedures in medicine, Istambul, IAEA SM-220/204, pp 241–275
Ekins RP (1978) Theoretical aspect of saturation analysis. In: Radioimmunoassay and related procedures in medicine, Berlin IAEA-SM-124/105, pp 325–353
Ekins RP, Jackson T (1984) Non isotopic immunoassay. An overview. In: Bizollon CA (ed) Monoclonal antibodies and new trends in immunoassays. Elsevier, Amsterdam, pp 149–163
Engvall E, Perlmann P (1971) Enzyme-linked immunosorbent assay (ELISA). Quantitative assay of immunoglobulin G. Immunochemistry 8:871–874
Erlanger BF (1973) Principles and methods for the preparation of drug-protein conjugates for immunological studies. Pharmacol Rev 25, p 271
Eshhar Z, Kim JB, Barnard G, Collins WP, Gilad S, Lindner HR, Kohen F (1981) Use of monoclonal antibody to pregnadiol-3α-glucuronide for the development of a solid-phase chemiluminescence immunoassay. Steroids 38:89–109
Fagnart OC, Mareschal JC, Cambiaso CL, Masson PL (1985) Particle counting immunoassay (PACIA) of pregnancy-specific β1-glycoprotein, a possible marker of various malignancies and Crohn's ileitis. Clin Chem 31:397–401
Finley PR, Williams RJ, Lichti DA (1980) Evaluation of a new homogeneous enzyme inhibitor immunoassay of serum thyroxine with use of bichromatic analyser. Clin Chem 26:1723–1726
Fraker PJ, Speck JC Jr (1978) Protein and cell membrane iodination with a sparingly soluble chloroamide 1,3,4,6-tetrachloro-3a,6a-diphenylglycoluril. Biochem Biophys Res Commun 80:849–857
Freedlender AE (1968) Practical and theoretical advantages for the use of I^{125} in radioimmunoassay. In: Margoulies M (ed) Protein and polypeptide hormones, part II. Excerpta Medica, Amsterdam, pp 351–354
Fromageot P, Pradelles P, Morgat JL, Levine H (1978) Radioactive labelling of peptide hormones. In: Gupta D, Voelter W (eds) Hypothalamic hormones. Chemistry, physiology and clinical applications. Verlag Chemie, Weinheim, p 59
Gandolfi C, Malvano R, Rosa V (1971) Preparation and immunoreactive properties of mono iodinated angiotensin labelled at high specific activity. Biochim Biophys Acta 251:254–261
Gemmill CL (1955) The apparent ionization constants of the phenolic hydroxyl groups of thyroxine and related compounds. Arch Biochem Biophys 54:359–367
Geoghegan WD, Ackerman GA (1977) Adsorption of horseradish peroxydase, ovomucoid and anti-immunoglobulin to colloidal gold for the indirect detection of concanavalin A, wheat germ agglutinin and goat anti-human immunoglobulin G on cell surfaces at the electron microscopic level: a new method, theory and application. J Histochem Cytochem 25:1187–1200
Gibbons I, Skold C, Rowley GL, Ullman EF (1980) Homogeneous enzyme immunoassay for proteins employing β-galactosidase. Anal Biochem 102:167–170
Goodman SL, Hodges GM, Trejdosiewicz LK, Livingston DC (1979) Colloidal gold probes – a further evaluation. Scan Electron Microsc 3:619–628
Grange J, Roch AM, Quash GA (1977) Nephelometric assay of antigens and antibodies with latex particles. J Immunol Methods 18:365–375
Greenwood FC, Hunter WM, Glover JS (1963) The preparation of ^{131}I labelled human growth hormone of high specific radioactivity. Biochem J 89:114–123
Grouselle D, Tixier-Vidal A, Pradelles P (1982) A new improvement of the sensitivity and specificity of radioimmunoassay for thyroliberin. Application to biological samples. Neuropeptides 3:29–44
Haimovich J, Sela M (1969) Protein-bacteriophage conjugates: application in detection of antibodies and antigens. Science 164:1279–1280
Haimovich J, Sela M, Drewdney JM, Batchelor FR (1967) Anti-penicilloyl antibodies: detection with penicilloylated bacteriophage and isolation with a specific immunoadsorbent. Nature 214:1369–1370
Handley G, Miller JN, Briges JW (1979) Development of fluorescence immunoassay method of drug analysis. Proc Anal Div Chem Soc 16:26–29

Harris CC, Yolken RH, Krokan H, Hsu IC (1979) Ultrasensitive enzymatic radio-immunoassay: application to detection of cholera toxin and rotavirus. Proc Natl Acad Sci USA 76:5336–5339
Hassam M, Landon J, Smith DS (1982) A novel non separation fluoroimmunoassay for thyroxine. J Immunoassay 3:1–15
Hayes CE, Goldstein IJ (1975) Radioiodination of sulphydryl-sensitive proteins. Anal Biochem 67:580–584
Helmkamp RW, Sears DA (1970) A label for the red cell membrane: diazotized diiodosulfanilic acid. Int J Appl Radiat Isol 21:683–684
Helmkamp RW, Goodland RL, Bale WF, Spar IL, Mutschler LE (1960) High specific activity iodination of γ-globulin with iodine 131 monochloride. Cancer Res 20:1495–1500
Helmkamp RW, Contreras MA, Bale WF (1967) ^{131}I-labeling of proteins by the iodine monochloride method. Int J Appl Radiat Isotopes 18:737–746
Hemmilä I (1985) Fluoroimmunoassays and immunofluorometric assays. Clin Chem 31:359–370
Hemmilä I, Dakubu S, Mukkala V-M, Siitari H, Lövgren T (1984) Europium as a label in time-resolved immunofluorometric assays. Anal Biochem 137:335–343
Hemmings WA, Redshaw M (1975) A biological test of damage caused to IgG by several methods of iodination. Int J Appl Radiat Isotopes 26:426–429
Horisberger M, van Lauthan M (1979) Fluorescent colloidal gold: a cytochemical marker for fluorescent and electron microscopy. Histochemistry 64:115–118
Horrocks DL (1981) Qualitative and quantitative measurements of radioiodines. J Radioanal Chem 65:307–320
Hsia JC, Wong LTL, Kalow W (1973) Homogeneous murine myeloma protein 315 and spin labeled DNP as a model system for spin-labeled hapten titration technique and spin immunoassay. J Immunol Methods 3:17–24
Hsu IC, Yolken RH, Harris CC (1981) Ultrasensitive enzymatic radioimmunoassay. Methods 73:383–394
Hugues WL (1957) The chemistry of iodination. Ann NY Acad Sci 70:3–18
Hung LT, Fermandjian S, Morgat JL, Fromageot P (1973) Peptide and protein labelling with iodine. Iodine monochloride reaction with aqueous solution of L-tyrosine, L-histidine, L-histidine-peptides and its effect on some simple disulfide bridges. J Label Compounds 10:3–21
Hunter WM (1982) Recent advances in radioimmunoassay and related procedures. In: Radioimmunoassay and related procedures in medicine, Vienna. IAEA-SM-259/101, pp 3–21
Hunter WM, Greenwood FC (1962) Preparation of iodine 131 labelled human growth hormone of high specific activity. Nature 194:495–496
Imagawa M, Yoshitake S, Ishikawa E, Endo Y, Ohtaki S, Kano E, Tsunetoshi Y (1981) Highly sensitive sandwich enzyme immunoassay of human IgE with β-D-galactosidase from Escherichia coli. Clin Chim Acta 117:199–207
Imagawa M, Ishikawa E, Yoshitake S, Tanaka K, Kan H, Inada M, Imura H, Kurosaki H, Tachibana S, Takagi M, Nishiura M, Nakazawa N, Ogawa H, Tsunetoshi Y, Nakajima K (1982) A sensitive and specific sandwich enzyme immunoassay for human thyroid-stimulating hormone. Clin Chim Acta 126:227–236
Imagawa M, Hashida S, Ishikawa E (1983) A highly sensitive sandwich enzyme immunoassay for insulin in human serum developed using capybara anti-insulin Fab'-horseradish peroxydase conjugate. Anal Lett 16:1509–1523
Inoue S, Hashida S, Ishikawa E, Mori T, Imura H, Ogawa H, Ichioka T, Nakajima K (1986) Highly sensitive enzyme immunoassay for human thyroid stimulating hormone (hTSH) in serum using monoclonal anti-TSH β-subunit IgG1-coated polystyrene balls and polyclonal anti-human chorionic gonadotropin Fab'-horseradish peroxydase conjugate. Anal Lett 19:845–861
Ishikawa E, Imagawa M, Hashida S, Yoshitake S, Hamaguchi Y, Ueno T (1983) Enzyme labeling of antibodies and their fragments for enzyme immunoassay and immunohistochemical staining. J Immunoassay 4:209–327

Jablonsky E (1985) The preparation of bacterial luciferase conjugates for immunoassay and application to rubella antibody detection. Anal Biochem 148:199–206
Jaklitsch AP, Schneider RS, Johannes RJ, Levine JE, Rosenberg GL (1976) Homogeneous enzyme immunoassay for T4 in serum. Clin Chem 22:1185
Janson JC (1967) Absorption phenomena on Sephadex. J Chromatogr 28:12–20
Jarvis RF (1979) The future outlook for enzyme immunoassays. Antibiot Chemother 26:105–117
Jeffcoate SL (1982) Use of ^{125}iodine tracers in steroid radioimmunoassays. In: Gupta D (ed) Radioimmunoassay of steroid hormones. Verlag Chemie, Weinheim, p 209
Jiang VW, Krohn KA, Welch MJ (1975) Intramolecular effects of radioiodine decay in O-iodophenol, a model for radioiodinated proteins. J Am Chem Soc 97:6551–6556
Jirousek L (1981) On the chemical nature of iodinating species. J Radioanal Chem 65:139–154
Johannsson A, Stanley CJ, Self CH (1985) A fast highly sensitive enzyme immunoassay system demonstrating benefit of enzyme amplification in clinical chemistry. Clin Chim Acta 148:119–124
Jolley ME, Stroupe SD, Wang C-HJ, Panas HN, Keegan CL, Schmidt RL, Schwenzer KS (1981) Fluorescent polarization immunoassay I. Monitoring aminoglycoside antibiotics in serum and plasma. Clin Chem 27:1190–1197
Joronen I, Hopsu-Havu VK, Manninen M, Rinne A, Järvinen M, Halonen P (1986) Detection of low molecular weight cysteine proteinase inhibitors by time-resolved fluoroimmunoassay. J Immunol Methods 86:243–247
Kaplan LA, Gau N, Stein EA, Fearn JA, Chen IW, Maxon HR, Volle C (1981) Comparison of two nonisotopic immunoassays with a radioimmunoassay for the analysis of serum thyroxine. In: Kaplan LA, Pesce AJ (eds) Nonisotopic alternatives to radioimmunoassay. Dekker, New York, pp 183–189
Karnes HT, Gudat JC, O'Donnel CM, Winefordner JD (1981) Double-antibody fluorescence immunoassay of tobramycin. Clin Chem 27:249–252
Kato K, Hamaguchi Y, Fukui H, Ishikawa E (1975) Enzyme-linked immunoassay II. A simple method for synthesis of the rabbit antibody-β-D-galactosidase complex and its general applicability. J Biochem 78:423–425
Kennedy JH, Kricka LJ, Wilding P (1976) Protein-protein coupling reactions and the application of protein conjugates. Clin Chim Acta 70:1–31
Khanna PL, Ullman EF (1980) 4′5′-dimethoxy-6-carboxy fluorescein. A novel dipole-dipole coupled fluorescence energy transfer acceptor for fluoroimmuno-assay. Anal Biochem 108:156–161
Kim JB, Barnard GJ, Collins WP, Kohen F, Lindner HR, Eshhar Z (1982) Measurement of plasma estradiol-17β by solid-phase chemiluminescence immunoassay. Clin Chem 28:1120–1124
King TP, Li Y, Kochoumian L (1978) Preparation of protein conjugates via intermolecular disulfide bond. Biochemistry 17:1499–1506
Kitagawa T, Aikawa T (1976) Enzyme coupled immunoassay of insulin using a novel coupling reagent. J Biochem 79:233–236
Kjeld JM, Kuku SF, Diamant L, Fraser TR, Joplin GF, Mashiter K (1975) Production and storage of (^{125}I)thyroxine and (^{125}I) triiodothyronine of high specific activity. Clin Chim Acta 61:381–389
Kobayashi Y, Yamata M, Watanabe I, Mikai K (1982) A solid-phase fluoroimmunoassay of serum cortisol. J Steroid Biochem 16:521–524
Kohen F, Pazzagli M, Kim JB, Lindner HR, Boguslasky RC (1979) An assay procedure for plasma progesterone based on antibody-enhanced chemiluminescence. FEBS Lett 104:201–205
Kohen F, Kim JB, Barnard G, Lindner HR (1980a) An assay for urinary estriol-16-glucuronide based on antibody-enhanced chemiluminescence. Steroids 36:405–419
Kohen F, Pazzagli M, Kim JB, Lindner HR (1980b) An immunoassay for plasma cortisol based on chemiluminescence. Steroids 36:421–437
Kolmer JA, Boerner F (1945) Procedure: Lange colloidal gold test. In: Approved laboratory technique. Appleton-Century, New York, p 585

Koshland ME, Engelberger FM, Erwin MJ, Gaddone SM (1963) Modification of amino acid residues in anti-*p*-azobenzene-arsonic acid antibody during extensive iodination. J Biol Chem 238:1343–1348
Krohn KA, Welch MJ (1974) Studies of radio-iodinated fibrinogen. II. Lactoperoxydase iodination of fibrinogen and model compounds. Int J Appl Radiat Isot 25:315–323
Langone JJ (1981) Radioiodination by use of the Bolton-Hunter and related reagents. Methods Enzymol 73:112
Leute R, Ullman EF, Goldstein A (1972) Spin immunoassay of opiate narcotics in urine and saliva. JAMA 221:1231–1234
Leuvering JHW, Thal PJHM, Van Der Waart M, Schuurs AHWM (1981) A sol particle agglutination assay for human chorionic gonadotropin. J Immunol Methods 45:183–191
Leuvering JHW, Tal JHM, Schuurs AHWM (1983) Optimisation of a sol particle immunoassay for human chorionic gonadotropin. J Immunol Methods 62:175–184
Li TM, Benovic JL, Buckler RT, Burd JF (1981) Homogeneous substrate-labeled fluorescent immunoassay for theophylline in serum. Clin Chem 27:22–26
Liburdy RP (1979) Antibody induced fluorescence enhancement of an N-(3-pyrene) maleimide conjugate of rabbit anti-human immunoglobulin G: quantitation of human IgG. J Immunol Meth 28:233–242
Ling N, Leppaluoto J, Vale W (1976) Chemical biochemical and immunological characterization of mono and diiodo thyrotropin releasing factor. Anal Chem 76:125–133
Lu-Steffes M, Pittluck GW, Jolley ME, Panas HN, Olive DL, Wang GHJ, Nystrom DD, Keegan CL, Davis TP, Stroupe SD (1982) Fluorescence polarization immunoassay IV. Determination of phenytoin and phenobarbital in human serum and plasma. Clin Chem 28:2278–2282
Magnusson CGM, Masson PL (1985) Immunoglobulin E assayed after pepsin digestion by an automated and highly sensitive particle counting immunoassay: application to human cord blood. J Allergy Clin Immunol 75:513–524
Mani RS (1983) Reactor-produced radionuclides. In: Helus F, Colombetti LG (eds) Radionuclides production, vol 2. CRC, Boca Raton, p 1
Marchalonis JJ (1969) An enzymatic method for the trace iodination of immuno-globulins and other proteins. Biochem J 113:299–305
Marks V, Mould GP, O'Sullivan MJ, Teale JD (1980) Monitoring of drug disposition by immunoassay. In: Bridges JW, Chasseaud LF (eds) Progress in drug metabolism. Wiley, New York, vol 5, p 255
Matkovics B, Rakonczay Z, Rajki SE (1971) Steroids XII – Iodination of aromatic steroids by peroxydases. Steroidologia 2:77–79
Mc Conahey PJ, Dixon FJ (1980) Radioiodination of proteins by the use of the chloramine T method. Methods Enzymol 70:210
McFarlane AS (1958) Efficient trace-labelling of protein with iodine. Nature 182:53
McFarlane AS (1968) In vivo behavior of ^{131}I fibrinogen. J Clin Invest 42:346–361
Markwell MA (1982) A new solid-state reagent to iodinate proteins. Anal Biochem 125:427–432
Mersel M, Benenson A, Doljanski F (1976) Lactoperoxydase-catalyzed iodination of surface membrane lipids. Biochem Biophys Res Com 70:1166–1171
Miles LEM, Hales CN (1973) Immunoradiometric assay procedures: new developments. In: Radioimmunoassay and related procedures in medicine, Istambul. IAEA-SM 124/107, pp 483–490
Miller JN, Lim CS, Bridges JN (1980) Fluorescamine and fluoresceine as labels in energy transfer immunoassay. Analyst 105:91–92
Miyachi Y, Vaitukaitis JL, Nieschlag E, Lipsett MB (1972) Enzymatic radioiodination of gonadotropins. J Clin Endocrinol Metab 34:23–28
Morgat JL, Hung LT, Fromageot P (1970) Preparation of highly labelled (^{3}H) angiotensin II. Biochim Biophys Acta 207:374–376
Morrison M (1980) Lactoperoxydase-catalysed iodination as a tool for investigation of proteins. Methods Enzymol 70:214

Morton RK (1957) The kinetics of hydrolysis of phenyl phosphate by alkaline phosphatase. Biochem J 65:674–682
Nakane PK, Kawoi A (1974) Peroxydase-labeled antibody a new method of conjugation. J Histochem Cytochem 22:1084–1091
Neddermeyer PA, Rogers LB (1968) Gel filtration behavior of inorganic salts. Anal Chem 40:755–762
Nelander B (1969) U.V. absorption spectra of complexes between some disulfides and iodine. Acta Chem Scand 23:2136–2148
Ngo TT, Lenhoff HM (1980) Enzyme modulators as tools for the development of homogeneous enzyme immunoassay. FEBS Lett 116:285–288
O'Beirne AJ, Cooper HR (1979) Heterogeneous enzyme immunoassay. J Histochem Cytochem 27:1148–1162
Oellerich M (1984) Enzyme-immunoassay: a review. J Clin Chem Clin Biochem 22:895–904
Olsson T, Brunius G, Carlsson HE, Thore A (1979) Luminescence immunoassay (L.I.A.): a solid-phase immunoassay monitored by chemiluminescence. J Immunol Methods 25:127–135
O'Sullivan MJ (1984) Enzyme immunoassay. In: Butt WR (ed) Practical immunoassay. Dekker, New York
O'Sullivan MJ, Marks V (1981) Methods for the preparation of enzyme-antibody conjugates for use in enzyme immunoassay. Methods Enzymol 73:147–166
Patel A, Campbell AK (1983) Homogeneous immunoassay based on chemiluminescence energy transfer. Clin Chem 29:1604–1608
Pazzagli M, Serio M, Munson P, Rodbard D (1982) A chemiluminescent immunoassay (L.I.A.) for testosterone. In: Radioimmunoassay and related procedures in medicine, Vienna. IAEA-SM-25, pp 747–755
Pettersson K, Siitari H, Hemmilä I, Soini E, Lövgren T, Hänninen V, Tanner P, Stenman UH (1983) Time-resolved fluoroimmunoassay of human choriogonadotropin. Clin Chem 29:60–64
Porath J, Flodin P (1959) Gel filtration: a method for desalting and group separation. Nature 183:1657–1659
Porstmann B, Porstmann T, Nugel E, Evers U (1985) Which of the commonly used marker enzymes gives the best results in colorimetric and fluorometric enzyme immunoassay: horseradish peroxydase, alkaline phosphatase or β galactosidase? J Immunol Methods 79:27–37
Pradelles P (1977) Contribution à l'étude de la conformation du TRF (facteur de libération de l'hormone thyréotrope) par dichroïsme circulaire. Etudes physico-chimiques, marquages radioactifs et application biologiques. Thesis, University of Paris VII
Pradelles P, Gros C, Rougeot C, Bepoldin O, Dray F, Llorens-Cortes C, Pollard H, Schwartz JC, Fournier-Zaluski MC, Cracel G, Roques BP (1978) Dosage radioimmunologique des enképhalines. In: Radioimmunoassay and related procedure in medicine 1977, vol 2. IAEA-SM 220/67, Vienna, p 495
Pradelles P, Grassi J, Maclouf J (1985) Enzyme immunoassays of eicosanoids using acetylcholine-esterase as label: an alternative to radioimmunoassay. Anal Chem 57:1170–1173
Pressman D, Eisen HN (1950) The zone of localization of antibodies. V. An attempt to saturate antibody-binding site in mouse kidney. J Immunol 64:273–279
Quash GA, Roch AM, Niveleau A, Grange J, Keolouangkhot T, Huppert J (1978) The preparation of latex particles with covalently bound polyamines, IgG and measles agglutinins and their use in visual agglutination tests. J Immunol Methods 22:165–174
Rae PA, Schimmer BP (1974) Iodinated derivatives of adrenocorticotropic hormone. J Biol Chem 249:5649–5653
Ramachandran LK (1956) Protein-iodine interaction. Chem Rev 56:199–218
Redshaw MR, Lynch SS (1974) An improved method for the preparation of iodinated antigens for radioimmunoassay. J Endocrinol 60:527–528
Regoeczi E (1984) Iodine labelled plasma protein. CRS, Boca Rato, p 1

Riad-Fahmy D, Read GF, Joyce BG, Walker RF (1981) Steroid immunoassays in endocrinology. In: Voller A, Bartlett A, Bidwell D (eds) Immunoassays for the 80s. MIT Press, Lancaster, p 205

Robbin JL, Hill GA, Carle BN, Calquist JH, Marcus C (1962) Latex agglutination reactions between human chorionic gonadotropin and rabbit antibody. Proc Soc Exp Biol Med 109:321–325

Rosa U, Scassellati A, Pennisi F, Riccioni N, Giagnoni P, Giordoni R (1964) Labelling of human fibrinogen with ^{131}I by electrolytic iodination. Biochim Biophys Acta 86:519–526

Rosenberg R, Murray TM (1979) The mechanism of methionine oxidation concomitant with hormone radioiodination. Comparative studies of various oxidants using a simple new method. Biochim Biophys Acta 584:261–269

Rotman B (1961) Measurement of activity of single molecules of β-D-galactosidase. Proc Natl Acad Sci USA 47:1981–1991

Rubenstein KE, Schneider RS, Ullman EF (1972) Homogeneous enzyme-immunoassay. A new immunochemical technique. Biochem Biophys Res Commun 47:846–874

Rudinger J, Ruegg U (1973) Preparation of *N*-succinimidyl 3-(4-hydroxyphenyl) propionate. Biochem J 133:538–539

Rowley GL, Rubenstein KE, Huisjen J, Ullman EF (1975) Mechanism by which antibodies inhibit hapten-malate dehydrogenase conjugates. An enzyme-immunoassay for morphine. J Biol Chem 250:3759–3766

Salacinski PR, Mc Leau C, Sykes JE, Clement-Jones VY, Lowry PJ (1981) Iodination of proteins, glycoproteins and peptides using a solid-phase oxiding agent, 1,3,4,6-tetrachloro-3α,6α-diphenyl glycoluril (iodogen). Anal Chem 117:136–146

Samols E, Williams HS (1961) Trace labelling of insulin with iodine. Nature 190:1211–1212

Schall RF, Tenoso HJ (1981) Alternatives to radioimmunoassay: labels and methods. Clin Chem 27:1157–1164

Schneider RS, Lindquist P, Wong ET, Rubenstein KE, Ullmann EF (1973) Homogeneous enzyme immunoassay for opiate in urine. Clin Chem 19:821–825

Schroëder HR, Yeager FM (1978) Chemiluminescence yields and detection limit of some isoluminol derivatives in various oxydation systems. Anal Chem 50:1114–1120

Schroëder HR, Carrico RJ, Boguslaski RC, Christner JE (1976) Specific binding reactions monitored with ligand-cofactor conjugates and bacterial luciferase. Anal Biochem 72:283–292

Schroëder HR, Vogelhut PO, Carrico RJ, Boguslaski RC, Buckler RT (1976) Competitive protein binding assay for biotin monitored by chemiluminescence. Anal Chem 48:1933–1937

Schuurs AHWM, Van Veemen BK (1977) Enzyme-immunoassay. Clin Chim Acta 81:1–40

Schuurs AHWM, van Veemen BK (1980) Enzyme-immunoassay: a powerfull analytical tool. J Immunoassay 1:229–249

Sefton BM, Vickus GG, Burge BW (1973) Enzymatic iodination of Sindbis virus proteins. J Virol 11:730–735

Seidah NS, Dennis M, Corvol P, Rochemont J, Chretien M (1980) A rapid high performance liquid chromatography purification method of iodinated polypeptide hormones. Anal Chem 109:185–191

Shalev A, Greenberg AH, Mc Alpine PJ (1980) Detection of attograms of antigens by a high-sensitivity enzyme-linked immunosorbent assay (H-S ELISA) using a fluorogenic substrate. J Immunol Meth 38:125–139

Shaposhnikov JD, Zerov YP, Ratavitski EA, Ivanov SD, Bobrov YF (1976) In vitro RNA iodination with aid of chloramine T. Anal Chem 75:234–240

Shaw EJ, Watson RAA, Landon J, Smith DS (1977) Estimation of serum gentamicin by quenching fluoroimmunoassay. J Clin Pathol 30:526–531

Shechter Y, Burstein Y, Patchornik A (1975) Selective iodination of methionine residues in proteins. Biochemistry 14:4497–4503

Shima K, Sawazaki N, Tanaka R, Tarui S, Nishikawa M (1975) Effect of an exposure to chloramine T on the immunoreactivity of glucagon. Endocrinology 96:1254–1260
Siitari H, Hemmilä I, Soini E, Lövgren T, Koistinen V (1983) Detection of hepatitis B surface antigen using time-resolved fluoroimmunoassay. Nature 301:258–260
Sikorav JL, Grassi J, Bon S (1984) Synthesis "in vitro" of precursors of the catalytic subunits of acetylcholinesterase from Torpedo marmorata and Electrophorus electricus. Eur J Biochem 145:519–524
Simpson JSA, Campbell AK, Ryall MET, Woodhead JS (1979) A stable chemiluminescent-labeled antibody for immunological assay. Nature 279:646–647
Singer SJ (1959) Preparation of an electron-dense antibody conjugate. Nature 183:1523–1524
Sinosich MJ, Chard T (1979) Fluoroimmunoassay of alphafoetoprotein (AFP) in amniotic fluid. Ann Clin Biochem 16:334–336
Smith DS (1977) Enhancement fluoroimmunoassay of thyroxine. FEBS Lett 77:25–27
Soini E (1984) Pulsed light, time resolved fluorometric immunoassay. In: Bizollon CA (ed) Monoclonal antibodies and new trends in immunoassays. Elsevier, Amsterdam, pp 197–208
Sorimachi K, Ui N (1975) Ion exchange chromatography analysis of iodothyronines. Anal Chem 67:157–165
Spencer RD (1981) Applications of fluorescent polarization in clinical assays. In: Kaplan LA, Pesce A (eds) Nonisotopic alternatives to radioimmunoassays. Dekker, New York, pp 143–169
Spencer RD, Toledo FB, Williams BT, Yoss NL (1973) Design, construction, and two applications for an automated flow-cell polarization fluorometer with digital read out: enzyme-inhibitor (antitrypsin) assay and antigen-antibody (insulin-insulin antiserum) assay. Clin Chem 19:838–844
Stagg BH, Temperley JM, Rochman H, Morley JS (1970) Iodination and the biological activity of gastrin. Nature 228:58–59
Stanley CJ, Paris F, Plumb A, Webb A, Johannsson A (1985a) Enzyme amplification: a new technique for enhancing the speed and sensitivity of enzyme immunoassays. Int Biotechnol Lab 3:46–51
Stanley CJ, Johansson A, Self CH (1985b) Enzyme amplification can enhance both the speed and the sensitivity of immunoassays. J Immunol Meth 83:89–95
Stryer L, Griffith OH (1965) A spin-labelled hapten. Proc Natl Acad Sci USA 54:1785–1791
Tack BF, Wilder RL (1981) Tritiation of proteins to high specific activity: application to radioimmunoassay. Methods Enzymol 73:138
Takayasu S, Maeda M, Tsujii A (1985) Chemiluminescent enzyme immunoassay using β-D-galactosidase as the label and the bis(2,4,6-trichlorophenyl)oxalate-fluorescent dye system. J Immunol Methods 83:317–325
Taylor DM (1981) The radiotoxicology of iodine. J Radioanal Chem 65:195–208
Térouanne B, Carrié ML, Nicolas JC, Crastes de Paulet A (1986a) Bioluminescent immunosorbent for rapid immunoassays. Anal Biochem 154:118–125
Térouanne B, Nicolas JC, Crastes de Paulet A (1986b) Bioluminescent immunoassay for α-fetoprotein. Anal Biochem 154:132–137
Thakrar H, Miller JN (1982) New developments in fluorescence immunoassay. Anal Proc 19:329–330
Thorell JI, Johansson BG (1971) Enzymatic iodination of polypeptides with ^{125}I to high specific activity. Biochim Biophys Acta 251:363–369
Tijssen P, Kurstak E (1984) Highly efficient and simple methods for the preparation of peroxydase and active peroxydase-antibody conjugates for enzyme immunoassays. Anal Biochem 136:451–457
Toivonen E, Hemmilä I, Marniemi J, Jørgensen PN, Zeuthen J, Lövgren T (1986) Two-site time-resolved immunofluorometric assay of human insulin. Clin Chem 32:637–640
Tsang VCW, Hancock K, Maddison SE (1984) Quantitative capacities of glutaraldehyde and sodium m-periodate coupled peroxydase-anti-human IgG conjugates in enzyme-linked immunoassays. J Immunol Meth 70:91–100

Tsay YG, Wilson L, Keefe E (1980) Quantitation of serum gentamicin concentration by solid-phase immunofluorescence method. Clin Chem 26:1610–1612

Tuskynki GP, Knight L, Piperno JR, Walsh PN (1980) A rapid method to removal of (^{125}I) iodide following iodination of protein solutions. Anal Chem 106:118–122

Ullman EF, Schwarzberg M, Rubenstein KE (1976) Fluorescence excitation transfer immunoassay. A general method for determination of antigens. J Biol Chem 251:4172–4178

Ullman EF, Yoshida RA, Blakemore JI, Maggio E, Leute R (1979) Mechanism of inhibition of malate dehydrogenase by thyroxine derivatives and reactivation by antibodies. Homogeneous enzyme-immunoassay for thyrosine. Biochim Biophys Acta 567:66–74

Van Der Waart M, Schuurs AHWM (1976) Enzymoimmunoassay. Towards the development of a radioenzyme-immunoassay (REIA). Z Anal Chem 279:142

Van Veemen BK, Schuurs AHWM (1971) Immunoassay using antigen-enzyme conjugates. FEBS Lett 15:232–236

Van Veemen BK, Schuurs AHWM (1972) Immunoassay using hapten-enzyme conjugates. FEBS Lett 24:77–81

Van Veemen BK, Schuurs AHWM (1975) The influence of heterologous combinations of antiserum and enzyme-labelled estrogen as the characteristics of estrogen enzyme-immunoassays. Immunochemistry 12:667–690

Vigny M, Bon S, Massoulié J, Leterrier F (1978) Active-site catalytic efficiency of acetylcholinesterase molecules forms in Electrophorus, Torpedo, rat and chicken. Eur J Biochem 85:317–323

Vilpo JA, Rasi S, Suvanto E, Vilpo LM (1986) Time-resolved fluoroimmunoassay of 5-methyl-2′-deoxycytidin. Anal Biochem 154:436–440

Voller A (1982) Application of immunoassays in parasitic diseases. In: Radioimmunoassay and related procedures in medicine, Vienna. IAEA-SM-259/108, p 689

Von Schulthess GK, Cohen RJ, Benedek GB (1976) Laser light scattering spectroscopic immunoassay in the agglutination-inhibition mode for human chorionic gonadotropin (hCG) and luteinizing hormone (hLH). Immunochemistry 13:963–966

Von Schulthess GK, Giglio M, Cannel DS, Benedek GB (1980) Detection of agglutination reactions using anisotropic light scattering: an immunoassay of high sensitivity. Mol Immunol 17:81–92

Walker WHC (1977) An approach to immunoassay. Clin Chem 23:384–402

Wannlund J, De Luca MA (1983) Bioluminescent immunoassays. Methods Enzymol 92:426–432

Weber SG, Purdy WC (1979) Homogeneous voltametric immunoassay: a preliminary study. Anal Litt 12:1–9

Weeks I, Beheshti I, Mc Capra F, Campbell AK, Woodhead JS (1983a) Acridinium esters as high-specific-activity labels in immunoassay. Clin Chem 29:1474–1479

Weeks I, Campbell AK, Woodhead JS (1983b) Two site-immunochemiluminometric assay for human α_1 fetoprotein. Clin Chem 29:1480–1483

Weeks I, Sturgess ML, Woodhead JS (1986) Chemiluminescence immunoassay: an overview. Clin Sci 70:403–408

Wehmeyer KR, Halsall HB, Heineman WR (1982) Electrochemical investigation of hapten-antibody interactions by differential pulse polarography. Clin Chem 28:1968–1972

Wehmeyer KR, Doyle MJ, Halsall HB, Heineman WR (1983) Immunoassay by electrochemical technique. Methods Enzymol 92:432–444

Wei R, Almirez R (1975) Spin immunoassay of progesterone. Biochem Biophys Res Commun 62:510–516

Whitehead TP, Kricka LJ, Carter TJN, Thorpe HG (1979) Analytical luminescence: its potential in the clinical laboratory. Clin Chem 25:1531–1546

Wilson MB, Nakane PK (1978) Recent development in the periodate method of conjugating horseradish peroxydase (HRPO) to antibodies. In: Knapp W, Holuber K, Wick G (eds) Immuno-fluorescence and related staining techniques. Elsevier, Amsterdam, pp 215–224

Wilzbach KE (1957) Tritium-labelling by exposure of organic compounds to tritium gas. J Am Chem Soc 79:1013
Wood FT, Wu MN, Gerhart JC (1975) The radioactive labeling of proteins with an iodinated amidination reagent. Anal Chem 69:339–349
Wood WG (1984) Luminescence immunoassays: problems and possibilities. J Clin Chem Clin Biochem 22:905–918
Woodhead JS, Sturgess M, Jones MK, Week I (1984) Chemiluminescent labels in immunoassay. In: Bizollon CA (ed) Monoclonal antibodies and new trends in immunoassays. Elsevier, Amsterdam, pp 165–174
Worah D, Yeung KK, Ward FE, Carrico RJ (1981) A homogeneous fluorescent immunoassay for human immunoglobulin. M Clin Chem 27:673–677
Yalow RS (1980) Radioimmunoassay. Am Rev Biophys Bioeng 9:327–345
Yalow RS, Berson SA (1968) General principles of radioimmunoassays. In: Hayer RL, Goswitz FA, Murphy BEP (eds) Radioisotopes in medicine: in vitro analysis. Atomic Energy Commission Symposium Series 11, Oak Ridge, p 7
Yoshitake S, Yamada Y, Ishikawa E, Masseyeff R (1979) Conjugation of glucose oxydase from *Aspergillus niger* and rabbit antibodies using N-hydroxysuccinimide ester of *N*-(4-carboxycyclohexylmethyl)-maleimide. Eur J Biochem 101:395–399
Zuk RF (1981) Fluorescence-protection immunoassay applied to the measurement of serum proteins. In: Kaplan LA, Pesce A (eds) Nonisotopic alternatives to radioimmunoassay. Dekker, New York, pp 83–95

CHAPTER 6

Strategies for Developing Specific and Sensitive Hapten Radioimmunoassays

E. EZAN, S. MAMAS, C. ROUGEOT, and F. DRAY

A. Introduction

Small molecules (molecular weight <1000) are poorly immunogenic or nonimmunogenic per se, but become immunopotent when they are attached in sufficient number to a macromolecular carrier (LANDSTEINER 1946). The first application of this principle was made for steroids (ERLANGER et al. 1957; GOODFRIEND and SEHON 1958; LIEBERMAN et al. 1959), and opened the way for the production of antibodies against any small molecule of biologic interest such as hormones, vitamins, mediators, metabolites, or drugs.

Several factors can influence the specificity of antibodies produced against haptenic determinants (epitopes). The most important of them are the site of fixation on the small molecule and the structure of the bridge which binds it to the carrier. Figure 1 shows the examples where the small molecules, prostaglandin $PGF_{2\alpha}$, and the neuropeptide, luteinizing hormone-releasing hormone (LHRH), are directly coupled to the carrier, whereas in the case of 17β-estradiol, the conjugation was performed through a succinyl or carboxymethoxime group possibly involving a functional group of the original hapten.

It can be predicted that the specificity will be reduced against some related analogs when a functional group is engaged in the link to the carrier and that the heterogeneity of antibody populations will be increased by the existence of antibridge antibodies. This chapter will analyze some aspects of specificity of antihapten antibodies and strategies to be developed to improve this specificity.

B. Hapten–Carrier Conjugation

The choice of carrier will depend on the immunogenic character, the degree of purity, the solubilization properties, the commercial availability, and the cost. For these reasons, bovine serum albumin (BSA) has been widely used. However, a higher titer of antiserum has sometimes been reported with other proteins like immunoglobulin, thyroglobulin, or keyhole limpet hemocyanin (KLH). All these proteins present available groups which are reactive under various conditions:

α-Amino groups of NH_2 terminal amino acids
ε-Amino groups of lysine residues
Phenolic hydroxyl groups of tyrosine residues
Hydroxyl groups of serine and threonine residues
Sulfhydryl groups of cysteine residues

No bridge introduced

Prostaglandin $F_{2\alpha}$

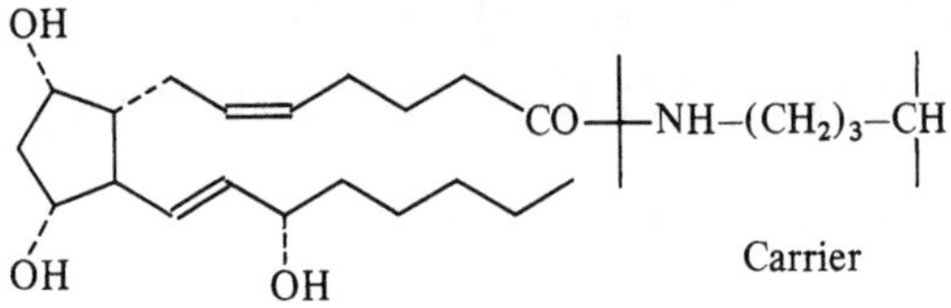

Luteinizing–Hormone Releasing–Hormone

pGLU–HIS–TRP–SER–TYR–GLY–LEU–ARG–PRO–GLY–NH_2

↑

Carrier

Bridge introduced

Estradiol–17β

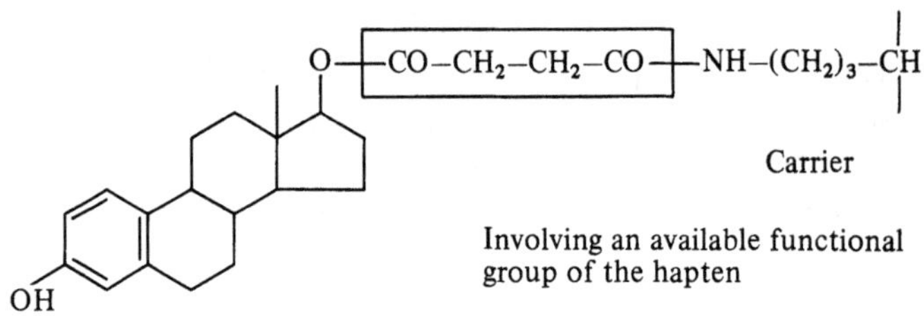

Involving an available functional group of the hapten

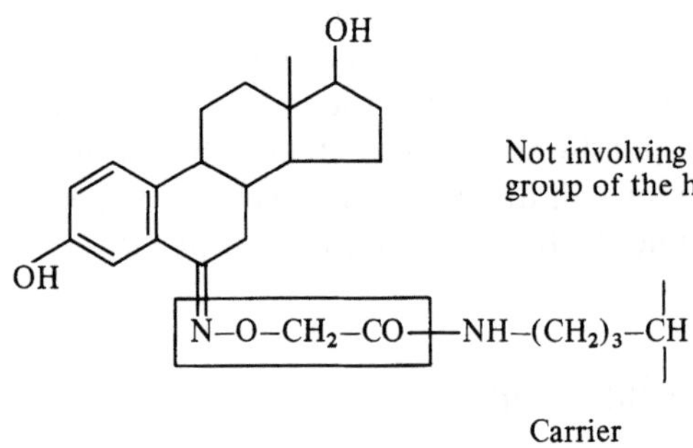

Not involving an available functional group of the hapten

Fig. 1. Heterogeneity introduced into haptens through conjugation to carrier

Imidazole groups of histidine residues
α-Carboxy groups of COOH terminal amino acids
Carboxyl groups (lateral chain) of aspartate or glutamate residues

The functional groups of haptens can involve one or more of these groups and in some instances, unusual groups like ketones or aldehydes. The chemical procedure of coupling reactions have already been reviewed (ERLANGER 1973; KABAKOFF 1980). We will describe the main cross-linking reagents used according to the chemical groups involved in the reaction.

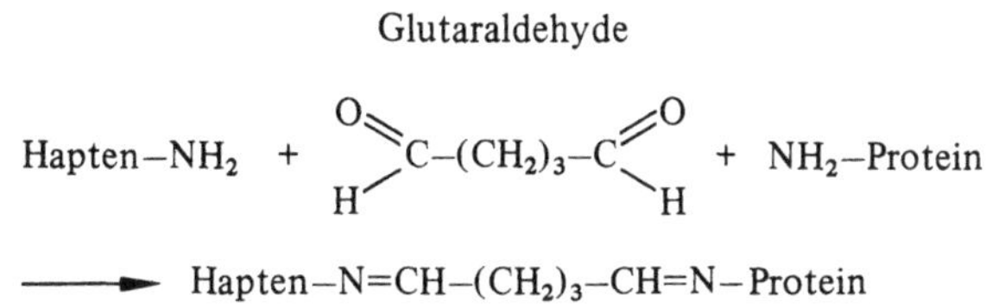

Fig. 2. One of the possible mechanisms for hapten–protein coupling via glutaraldehyde

I. Haptens with Amino Groups

The chemical cross-links involved in this case are homobifunctional reagents since they react mainly with amino groups of the carrier protein.

1. Glutaraldehyde

Glutaraldehyde (GA) was initially used for protein–protein conjugation. The prevalent idea is that this dialdehyde forms a bridge through Schiff base formation between primary amino groups on both protein and hapten (Fig. 2). In fact, the coupling mechanism appears not to be completely clarified (for review see PESCE 1976). Coupling with GA is performed under mild conditions near neutral pH with a wide range of buffers, except those containing amino groups like glycine or Tris buffers. In some cases, the Schiff base needs to be reduced by sodium cyanoborohydride or sodium borohydride and the free aldehyde groups need to be quenched by external amines (lysine, ethanolamine). Two-step procedures are often effective in reducing intramolecular cross-linking. The length of the linking chain allows free rotation of the hapten, thus avoiding possible steric hindrance by the carrier.

2. Benzoquinone

p-Benzoquinone (BQ) has been reported for protein–polysaccharide and protein–protein conjugations (TERNYNCK and AVRAMEAS 1977; BRANDT et al. 1975). This procedure has been applied to produce peptide immunogens. The quinone is first activated by nucleophilic attack to yield a monosubstituted hydroquinone (Fig. 3a). Hydrogen is then eliminated by reaction with a second molecule of BQ to give a monosubstituted quinone (Fig. 3b). A nucleophilic group in the protein can react with the hapten-bound quinone which results in a 2,5-substituted hydroquinone (Fig. 3c). This coupling is performed in a two-step procedure. Protein polymerization is avoided by a proper choice of reagent concentrations. It has been shown that BQ can react with amino or thiol groups and with the hydroxyl group of tyrosine, serine, or threonine.

3. Difluorodinitrobenzene

Difluorodinitrobenzene (DFDNB) has been used for coupling peptides (glucagon, bradykinin) to BSA with high coupling efficiency (TAGER 1976). Since it can react with α-NH_2, ε-NH_2, imidazole nitrogen, and aromatic hydroxyl groups, the

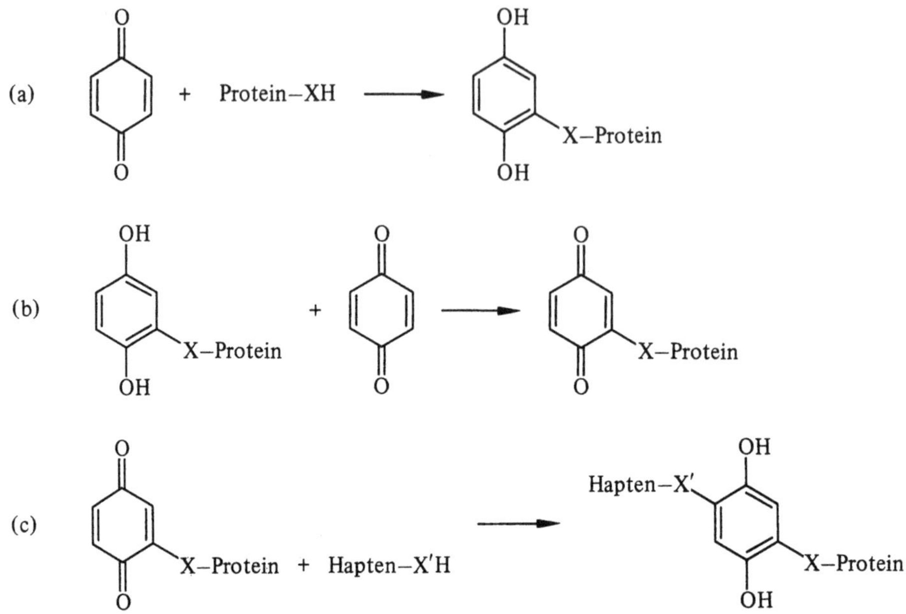

Fig. 3. Probable mechanism for the activation of a protein by *p*-benzoquinone and subsequent coupling to the hapten. (TERNYNCK and AVRAMEAS 1977)

bifunctional reagent concentration has to be optimized to avoid intramolecular coupling.

4. Toluene-2,4-diisocyanate

This reagent reacts mainly with α- and ε-amino groups of lysine and has been used in one- or two-stage reactions. This method has been described for coupling angiotensin to polylysine (HABER et al. 1965) and bradykinin to BSA (SPRAGG et al. 1966).

5. Diazotization Procedure

Aromatic amino groups or reduced aromatic nitro groups can be conjugated to proteins by the classical diazotization procedure of LANDSTEINER. A representative example of this method was the production of antibodies by a chloramphenicol–protein conjugate (HAMBURGER 1966).

6. *m*-Maleimidobenzoic acid *N*-Hydroxysuccinimide Ester

m-Maleimidobenzoic acid *N*-hydroxysuccinimide ester is used as a heterofunctional reagent to introduce maleimide residues into haptens containing amino groups. Then, the maleimide is coupled to thiol groups from protein cysteine residues or resulting from reduction of disulfide bounds of protein cystine residues. Specific antibiotic antibodies have been successfully obtained by using such hapten–protein conjugates (KITAGAWA 1981).

II. Haptens with Carboxyl Groups

This class represents a large number of haptens since many chemical methods are available for introducing activatable carboxylic groups when they are not present in the hapten. For instance, succinylation may be achieved on hydroxyl groups of steroid hormones by incubation with succinic anhydride to form hemisuccinates (ABRAHAM 1974). Haptens with keto or aldehyde groups can be converted to *O*-(carboxymethyl)oxime by adding them to *O*-(carboxymethyl)hydroxylamine (ABRAHAM and GROVER 1971). Introduction of a thiocarboxymethyl group on the steroid carbon skeleton is easily effected by addition of thiocarboxylic acid (BAUMINGER et al. 1974). Preparation of 7-carboxymethyl derivatives has been reported for testosterone and progesterone (DUVAL et al. 1980) and for cortisol, corticosterone, and desoxycorticosterone (DUVAL et al. 1985). These treatments yield carboxylated derivatives which can be coupled to the amino groups of proteins via one of the following coupling agents.

1. Carbodiimides

Carbodiimides form an amide linkage with amino groups of proteins through the initial formation of an activated *O*-acylisourea (Fig. 4). The coupling takes several hours and the pH must remain below 5 during the first hour of reaction. The conjugation can be carried out in aqueous solution with 1-cyclohexyl-3-(2-morpholinoethyl)carbodiimidemetho-*p*-toluenesulfonate (CMCI) or with 1-ethyl-3-(3-dimethylaminopropyl)carbodiimide hydrochloride (EDCI) while dicyclohexylcarbodiimide (DCC) has been used in solvent media.

2. *N*-Hydroxysuccinimide Ester

This method is derived from the carbodiimide method. In the presence of *N*-hydroxysuccinimide ester and DCC and in nonaqueous solvents the acidic group of the hapten forms an active ester. This latter is isolated and then hydrolyzed in alkaline phosphate buffer to form a peptide bond with the protein (DRAY and GROS 1981). This procedure is successful with poorly reactive carboxylic groups.

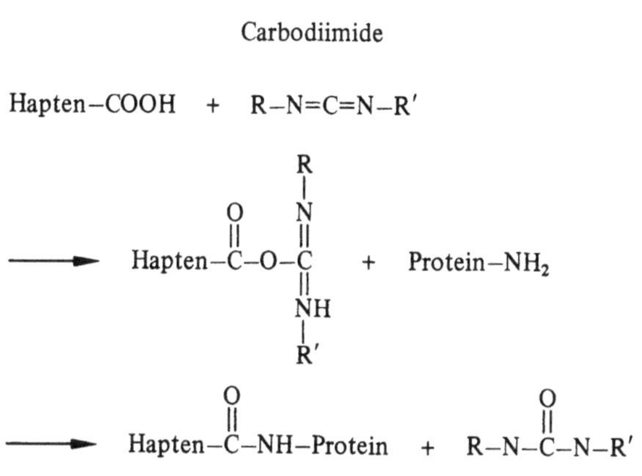

Fig. 4. Reaction of carbodiimide with carboxylic groups

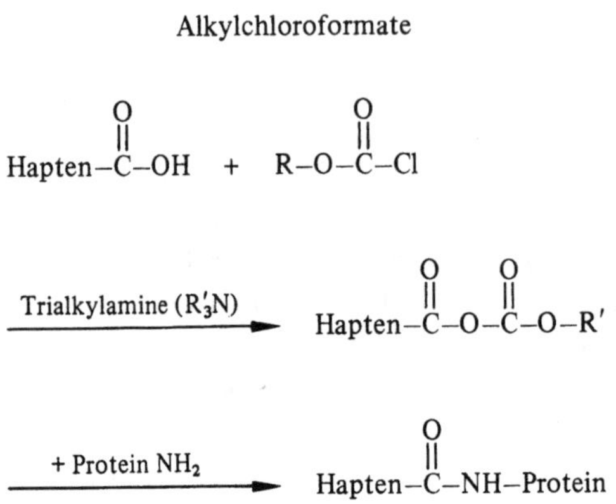

Fig. 5. The mixed anhydride method

3. Mixed Anhydride Method

Mixed anhydrides are formed by reaction between an acid, an alkyl chloroformate, and a trialkylamine, at low temperature in an inert organic solvent (Fig. 5; VAUGHAN and SATO 1952; ERLANGER et al. 1957, 1959). Free carboxyl groups converted into acid anhydrides are then added to a protein solution to react with amino and hydroxyl residues. Isobutylchloroformate, tri-*n*-butylamine, triethylamine, or *N*-methylmorpholine are the most common reagents for this activation.

4. *N*-Ethoxycarbonyl-2-ethoxy-1,2-dihydroquinoline

N-Ethoxycarbonyl-2-ethoxy-1,2-dihydroquinoline (EEDQ) acts on haptenic carboxyl group by forming a mixed anhydride which gives a stable amide bond with primary amines (BOSCHETTI 1978) (Fig. 6). This reagent presents many advantages: stable pH during coupling, short incubation time, and possible use in water–ethanol mixture with slightly water-soluble haptens.

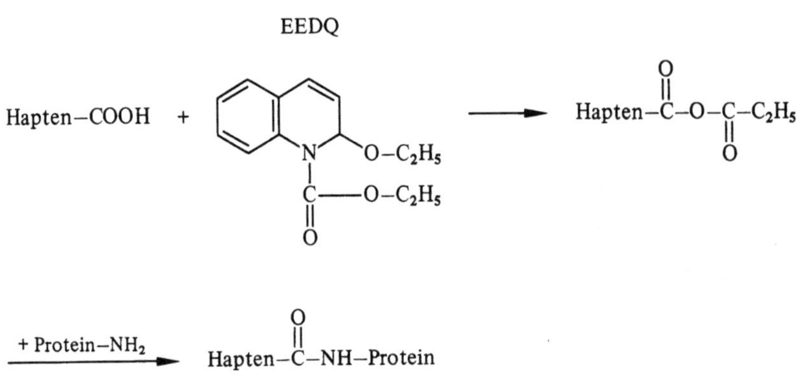

Fig. 6. Reaction scheme of the *N*-ethoxycarbonyl-2-ethoxy-1,2-dihydroquinoline (EEDQ) method

III. Haptens with Other Functional Groups

1. Phenolic Groups

Bisdiazotized benzidine has been used as a coupling agent between haptens and proteins through tyrosine residues. The covalent link involves the carbons in the *ortho* position of the hydroxyphenols (BASSIRI and UTIGER 1972).

2. Vicinal Hydroxyl Groups

Vicinal glycol of carbohydrate residues can be converted by the addition of sodium metaperiodate to yield dialdehydes which can react with amines by reductive alkylation under mildly alkaline conditions (Fig. 7). This procedure has been used to produce digoxin immunogen (BUTLER and CHEN 1967).

3. Hydroxyl Groups

These groups can be derivatized by succinylation or maleiation to form carboxyl groups. They can also react with phosgene to yield the highly reactive chlorocarbonate derivative which reacts directly with the amino groups of the protein in the presence of bicarbonate (ERLANGER et al. 1957). Additionally, EDCI has been used for coupling free hydroxyl groups of haptens (like ecdysone; HUNG et al. 1980) to a protein carrier which had been previously succinylated to increase the number of COOH groups (HABEEB 1967).

4. Sulfhydryl Groups

N-Succinimidyl-3-(2-pyridyldithio)propionate is a heterobifunctional reagent containing one *N*-hydroxysuccinimide ester moiety reacting with amino groups

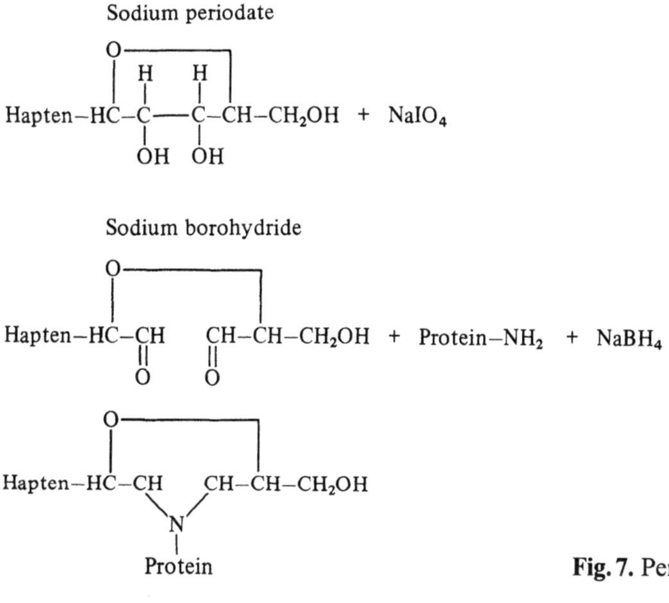

Fig. 7. Periodate oxidation method

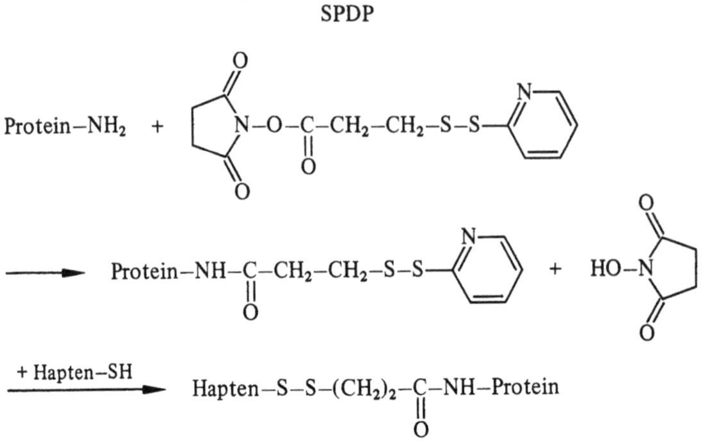

Fig. 8. Reaction involving *N*-succinimidyl-3-(2-pyridyldithio)propionate (SPDP) for hapten–protein conjugation

of proteins to give a stable amino bond (Fig. 8; CARLSSON et al. 1978). This intermediate compound reacts with aliphatic thiols to form a disulfide bond. The conjugation is performed under mild conditions in aqueous media and proceeds in two steps: introduction of the protected thiol groups in the protein and production of the conjugate with the sulfhydryl group of the hapten.

5. Indole Nitrogen

Preparation of hapten–protein conjugate has been reported by a Mannich addition involving formaldehyde condensation. This reaction allows a bridge between amino groups of a protein and compounds containing one or more reactive hydrogens as the indole nitrogen groups (Fig. 9). This method has been used for serotonin, melatonin (GROTA and BROWN 1974), and lysergic acid (RATCLIFF et al. 1977) immunogen preparation.

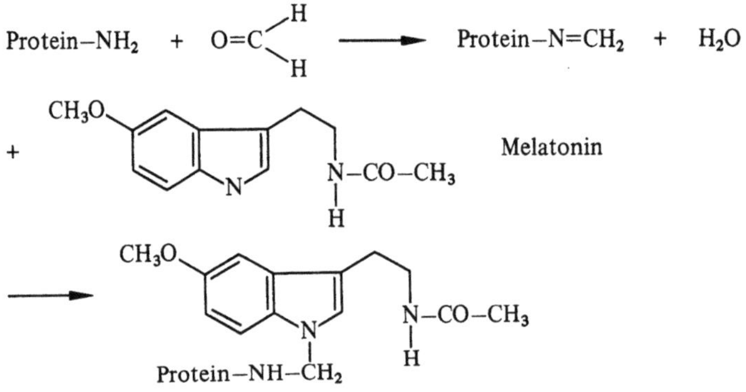

Fig. 9. Preparation of a melatonin–protein conjugate by the Mannich addition. (BESSELIEVRE et al. 1980)

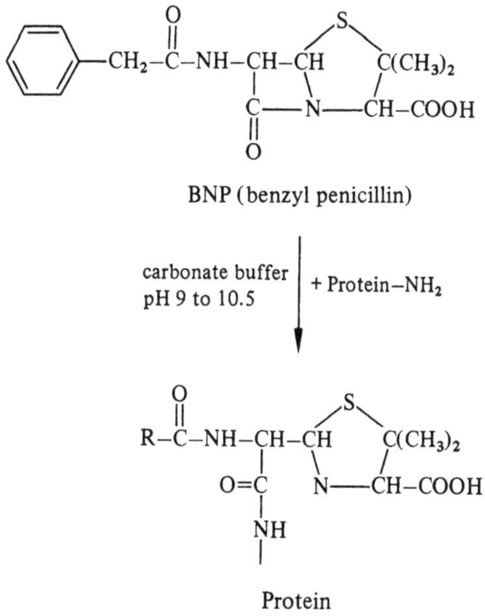

Fig. 10. Formation of a benzylpenicilloyl conjugate

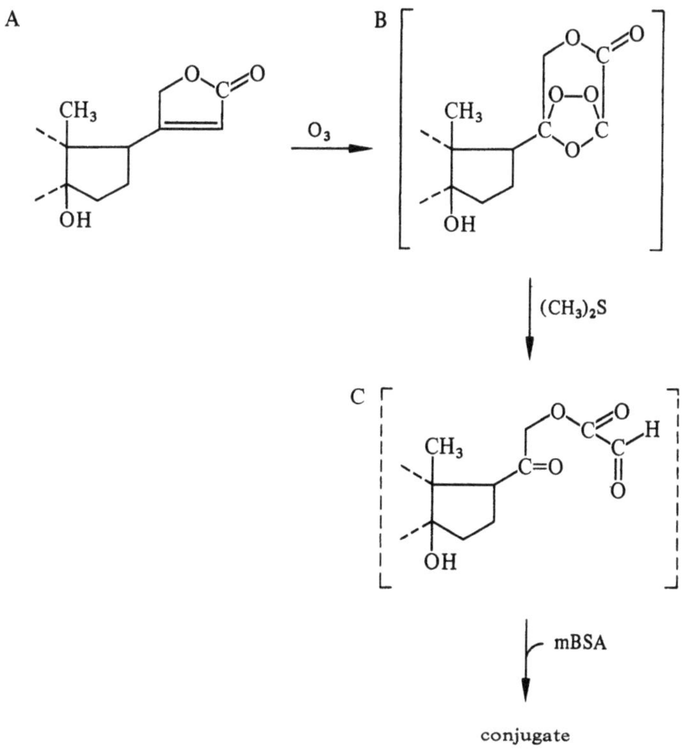

Fig. 11. Reaction scheme of the reductive ozonolysis of the lactone ring of digoxin. (THONG et al. 1985)

6. β-Lactam Ring

Benzylpenicilloyl immunogens have been prepared according to a method using the property of penicillin to open its β-lactam ring spontaneously in basic medium and thus to link NH_2 groups of a protein covalently (Fig. 10; MAMAS and DRAY 1979).

7. γ-Lactone Ring

Recently, a novel digoxin conjugate has been obtained by coupling through the lactone ring of the steroid moiety (THONG et al. 1985). The steps of the reaction (Fig. 11) involve: formation of an ozonide on the lactone ring (*A*), reduction with dimethylsulfide (*B*), and coupling to methylated BSA at alkaline pH in the presence of sodium cyanoborohydride (*C*).

C. Strategy for Increasing Specificity

I. General Considerations: From Hapten Size to Epitope Size

An ideal situation can exist for some haptens which fit exactly with the antibody-binding site, thus resulting in a restricted heterogeneity of antibody population. For small haptens (molecular weight <400), the participation of the bridge which links the original hapten to the carrier may become significant and diversified, owing on the one hand to the eventual existence of a side chain introduced into the hapten after derivatization, and on the other hand to the nonequivalence of the lysine residues of the protein carrier, which are often involved in the link. The small size of the original hapten will result in low response whereas the anti-bridge responses may predominate. The elimination of such antibody populations through affinity chromatography may improve the specificity, but the resulting anti-hapten antibodies usually have low affinities. In fact, the recent introduction of clonal selection through the hybridoma technique should overcome these difficulties (KOHLER and MILSTEIN 1975). For large haptens (molecular weight >1000), the multiplicity of epitopes will generate a large diversity of antibody populations, and partial recognition of the original hapten. In this case, one can take advantage of the respective properties of polyclonal antibodies (broad spectrum of recognition), and of the monoclonal antibody, to develop specific sandwich techniques.

II. Steroids

1. Problems Related to the Choice of the Steroid–Protein Complex

Steroids are good models for studying antibody specificity since their basic structure is common to them all. They all contain a cyclopentanophenanthrene nucleus (Fig. 12) with differences in the nature, number, and location of oxidized groups, the existence of double bonds, the spatial configuration of *A* and *B* rings, and the stereoisomerism of the alcohols.

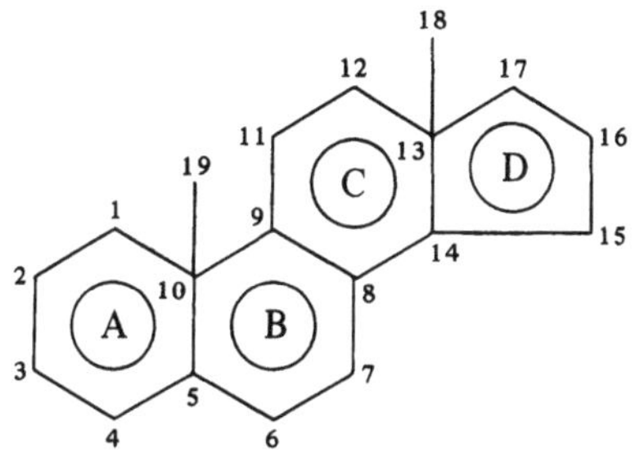

Fig. 12. Basic structure of a steroid molecule

Therefore, there are a lot of related compounds which can be more or less recognized by the antibody-binding sites. A major aspect of the strategy for eliciting highly specific antibodies is to keep intact the natural functional groups of the steroid, but to place them far from the bridge which links the hapten to the protein. This problem is illustrated by the study of the cross-reactivities of various anti-testosterone antisera (Fig. 13).

Dihydrotestosterone (DHT) differs from testosterone by reduction through the action of 5α-reductase of the double bond between C-4 and C-5. DHT-cross-reactivity can be classified according to the coupling position of immunogen:

1. C_3, C_6, and C_7: masking of DHT identity which results in high cross-reactivity.
2. C_1, C_{11}, and C_{19}: reduced carrier hindrance to the bond between C-4 and C-5 produces slight DHT recognition.
3. C_{15}: low cross-reactivity comes from an attachment point far from the *A* ring.

Progesterone differs from testosterone by substitution of the β-OH by a CH_3–CO group. The specificity of testosterone antibodies for this steroid is inverted. Coupling testosterone to the *A, B,* or *C* rings provides conjugates eliciting antisera with no cross-reactivity for progesterone whereas antibodies to *D* ring testosterone conjugates show higher progesterone recognition. Consequently, the choice of the testosterone–protein immunogen will depend on the nature of the biologic medium since possible cross-reactants may be found in various proportions from one medium to another.

Another example comes from the production of antisera against 5α-androstane-3α,17β-diol (A-3α-diol) and 5α-androstane-3β,17α-diol (A-3β-diol) (Kouznetzova and Dray 1977). Figure 14 shows the specificity of such antisera obtained after immunization by various androstanediol–BSA conjugates via a carboxymethyl group (Condom and Emiliozzi 1974).

Antibodies obtained by using the A-3α-diol coupled in position 1α were much less specific than those obtained when the steroid was coupled at position 15α. All antibodies against the 1 conjugate cross-reacted to a great extent with steroids with modification of the *A* ring. However, the presence of a ketone group at the C-17 position in androsterone induces an important change which affects the degree of cross-reactivity. On the other hand, the anti-A-3α-diol-15–BSA antisera possessed a very high specificity, except for androsterone. This could be explained by the proximity of the bridge in position 15α to the OH group in position 17, which may result in steric hindrance. Another possible reason for this cross-reactivity would be that the immunogen was contaminated with small amounts of an-

Steroid	Coupling position C1	C3	C19	C6	C7	C11	C15	C17
Testosterone 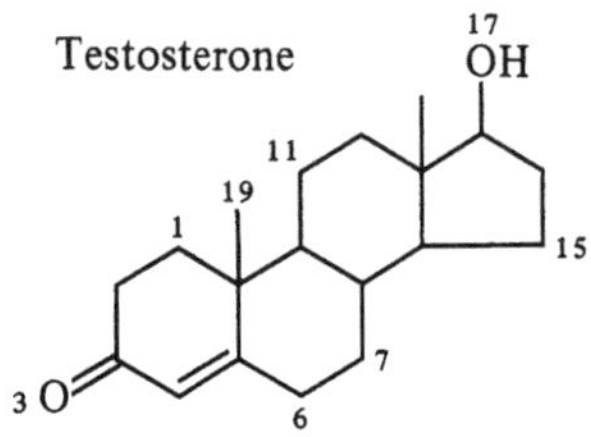	100	100	100	100	100	100	100	100
Progesterone 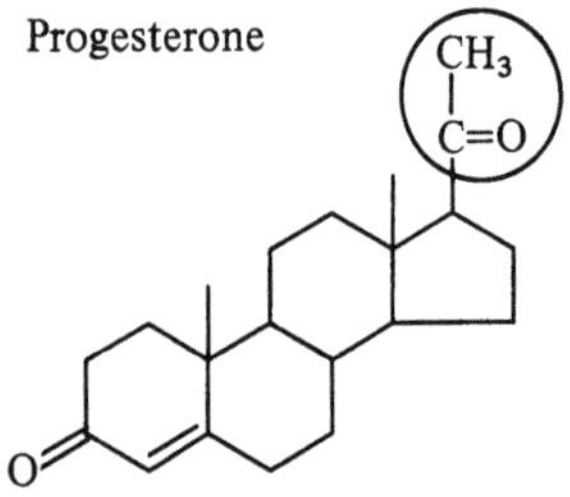		<0.1	0.01	0.01	<0.01		1.5	55
5α-Dihydro-Testosterone 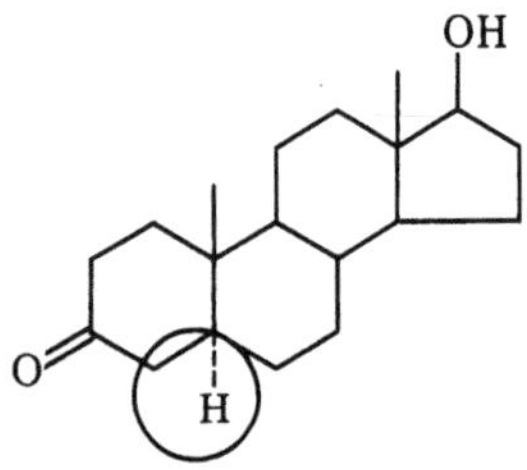	8	46	10	47	42	15	2.2	

Fig. 13. Cross-reactivity of antisera against different testosterone conjugates. *Circles* indicate those portions of the molecule which differ in structure from that found in testosterone. References: C-1 (KOHEN et al. 1975); C-3 (HARMAN et al. 1980); C-6, C-15, and C-19 (MALVANO 1983); C-7 (WEINSTEIN et al. 1972); C-11 (HILLER et al. 1973); C-17 (NISWENDER and MIDGLEY 1970)

drosterone-15α-carboxymethyl, this compound being the penultimate stage in the synthesis of the immunogen A-3α-diol-15-carboxymethyl–BSA.

Comparison of the specificity of both types of anti-A-3β-diol sera shows that the cross-reactions are heterogeneous. The difference in specificity relating to the position of coupling is much less clear than in the case of anti-A-3α-diol. However, when the coupling is through a carbon atom of the *B* ring, there is an increase in the specificity of antibodies.

2. Stereospecificity

The purity and identity of both original haptens and their derivatives have to be controlled, e.g., by high pressure liquid chromatography (HPLC) or gas chroma-

Steroid	Immunogen			
	A-3α, 17β -diol-1-BSA	A-3α, 17β -diol-15-BSA	A-3β, 17β -diol-1-BSA	A-3β, 17β -diol-7-BSA
A-3α-17β-diol 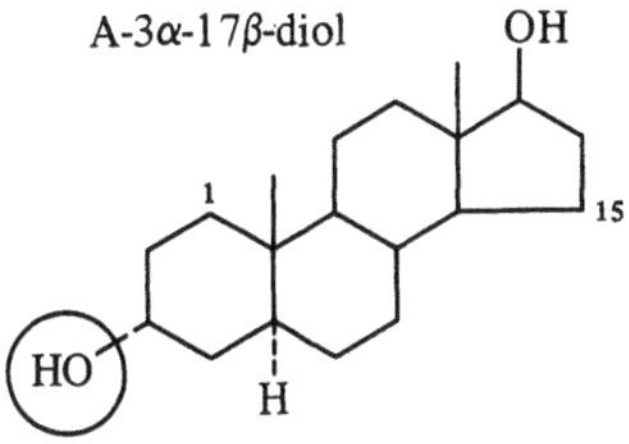	100	100	10.9	1.3
A-3β-17β-diol 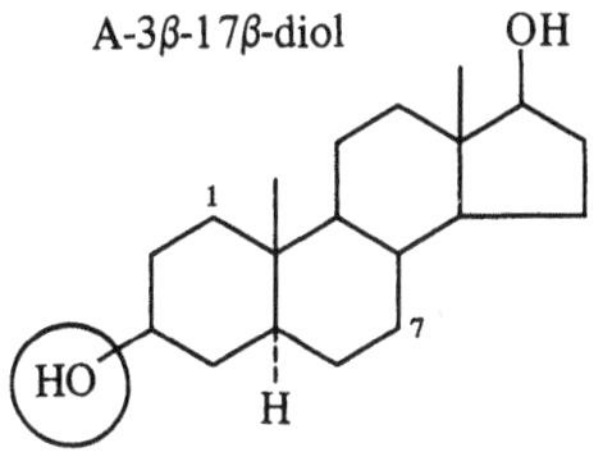	38	<0.36	100	100
Testosterone 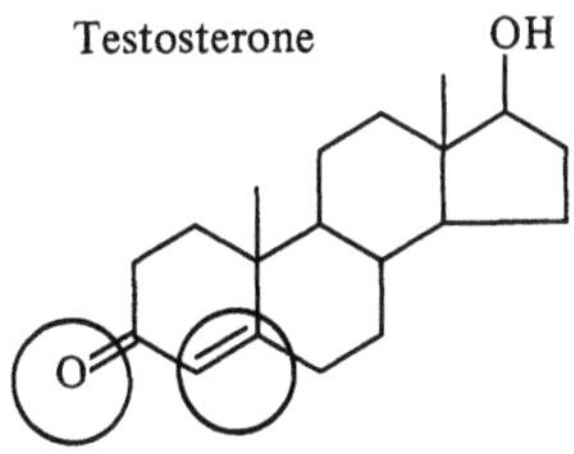	53	2.20	22	1.7
5α-Dihydro-Testosterone 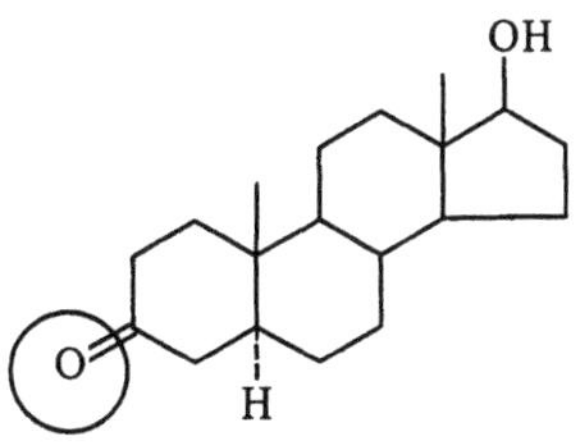	64	2.5	50	3.9
Androsterone 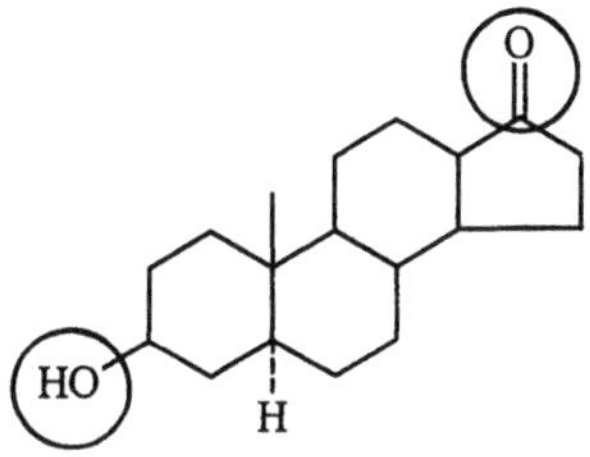	0.4	12	0.9	0.3

Fig. 14. Cross-reactivity of antisera raised against androstanediol. *Circles* indicate those portions which differ in structure from that found in A-3α,17β-diol or A-3β-17β-diol

tography–mass spectrometry (GC–MS) before their use for coupling to protein carriers. In the case of isomeric forms, a racemic mixture of the hapten is generally used for conjugation of the carrier, resulting in the production of stereospecific antibodies. As in racemic mixtures of drugs, the enantiomers do not share the same metabolic fate; the analysis of biologic samples containing arbitrary proportions of isomers cannot be correctly determined from a reference curve established with the racemic mixture. These difficulties may be overcome by using monoclonal antibodies, as shown for such haptens as abscisic acid, a plant hormone which naturally exists in the *cis* (+) form (PONTAROTTI et al. 1983; DELAAGE et al. 1984), and thymic hormone (SAVINO et al. 1982).

Another source of isomers are the oxims derivatives *(syn* and *anti)* when a keto group is used for the derivatization of the original hapten, as in steroid series. Recently, it has been shown that the 7α- and 7β-carboxymethyl derivatives of cortisol, corticosterone, and desoxycorticosterone, coupled to BSA, elicited highly specific polyclonal antibodies. In all immunized rabbits, the antisera obtained with the 7α derivative had a higher affinity and a narrower specificity than the antiserum obtained with the 7β derivative (DUVAL et al. 1985).

3. Shortened Incubation Time

There seems to be general agreement that steroid RIAs should operate better at equilibrium. However, the commonly used method for estimating the time required for the label to reach equilibrium grossly underestimates the time required. It has been shown by VINING et al. (1981) that the percentage of label bound in the presence of a cross-reacting competitor will approach a plateau much more slowly than in the absence of that competitor. Label and competitors all bind to the antiserum equally rapidly, but the competitors dissociate faster, and some of their places will be filled by the more slowly dissociating labeled hormone. It was suggested that, to ensure that an assay system is operating at its maximum possible specificity, the incubation time should be at least several times as long as the dissociation half-life of the steroid–antibody complex. We think that these results, limited to few steroids, should have wide application for hapten immunoassay.

4. Induction of Immunotolerance

The difficulty in improving the specificity of anti-testosterone and anti-DHT antibodies led some investigators (TATEISHI et al. 1980; DUPRET et al. 1984) to induce specific immunotolerance toward one of these molecules. The key step in these attempts was a preimmunization using the interfering cross-reactive compound (testosterone or DHT) which was linked with the same carboxylic side chain as that employed in the corresponding hapten (DHT or testosterone, respectively) and coupled covalently to a D-glutamic acid–D-lysine copolymer which confers tolerogenic properties on the linked compound. Further immunization with the corresponding hapten coupled to an antigenic protein led to antisera showing a decreased cross-reactivity with the interfering compound used in the preimmunization. The results obtained with small numbers of animals suggest nevertheless that there is some advantage in employing haptens and tolerogens bearing the intermediate link (C-17 or C-15) remote from the antigenic determinants toward which the specificity needs to be improved. The better results obtained with 17β-hemisuccinamido haptens suggest to DUPRET et al. (1984) that the structure of the hapten exerts a strong influence on the induction of immunotolerance.

III. Other Small Haptens

Although the specificity of steroid antisera has been widely studied, other relevant examples in the strategy of immunogen preparation are described in the literature. Cyclic adenosine monophosphate (cAMP) antisera elicit a poor specificity when the immunogen is obtained by coupling NH_2 of adenine to protein. The phosphoribose common to phosphonucleosides is indeed recognized by antibodies. On the contrary, when a succinyl group is linked to C-2′ of ribose of adenosine, specific anti-cAMP sera are obtained (Fig. 15; CAILLA 1973a).

Extensive studies of radioimmunoassays for narcotic alkaloid compounds have demonstrated how the choice of the coupling position can be important in determining specificity (FINDLAY et al. 1981). Narcotic molecules like morphine or codeine are subjected to metabolic modifications and the choice of the immunogen will depend on the purpose of the assay. On one hand, urine screening programs for detection of drug abuse should include immunoassays that are subject to cross-reactivity by as many metabolites of the drugs of interest as possible. On the other hand, for drug disposition or pharmacokinetic studies, the specificity of the radioimmunoassay has to be limited to the drug of interest.

A similar problem has been reported for assay of digoxin, a steroid–carbohydrate conjugate (Fig. 16; THONG et al. 1985). Digoxin metabolism leads to a sequential loss of glycosidic units and/or saturation of the steroid lactone ring. The relative importance of both these routes of biotransformation may vary from one subject to another. Conjugates of digoxin coupled to BSA through periodate oxidation on the carbohydrate moiety give antisera with a high cross-reactivity with the glycosidic metabolites. However, immunogens obtained by reductive ozonolysis of the lactone ring and subsequent coupling to BSA elicit antibodies which discriminate between changes in the carbohydrate region of digoxin, but lack specificity with dihydrodigoxin (the digoxin derivative with a saturated lactone ring).

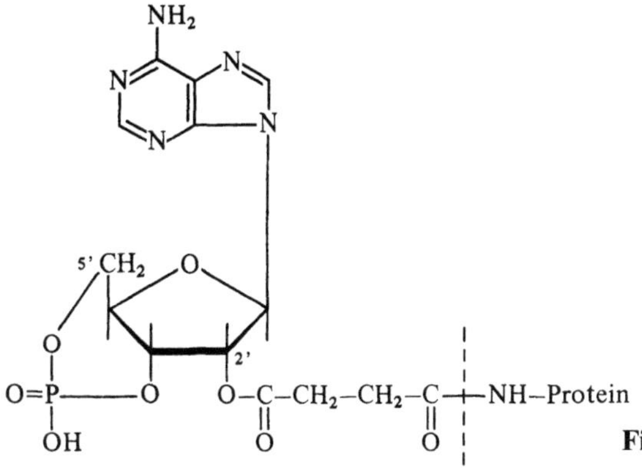

Fig. 15. Structure of cAMP immunogen

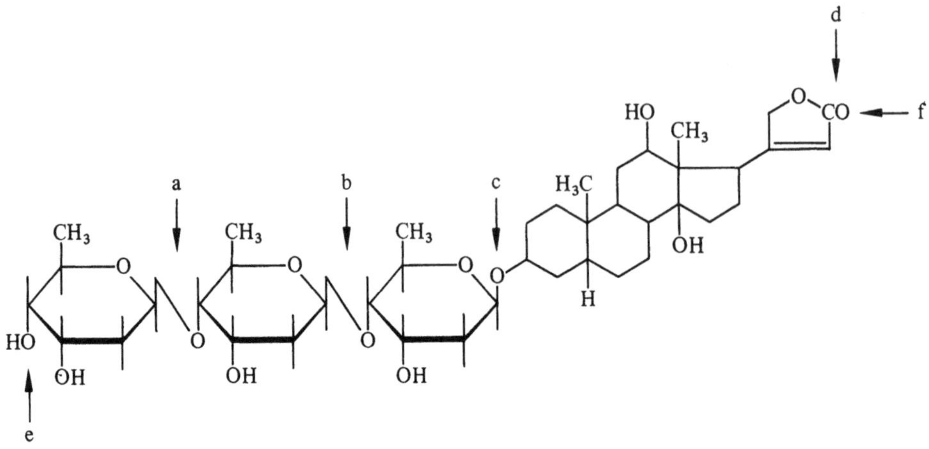

Fig. 16. Structure of digoxin. Positions *a, b, c,* and *d* show metabolism routes. Positions *e* and *f* refer to possible coupling for immunogen preparation via periodate oxidation (*e*) or reductive ozonolysis (*f*)

IV. Peptides

The use of homopolymers of amino acids or peptide proteins as antigens has established that the epitopic determinant is composed of four to six amino acids which contribute to binding with the antibody-combining site. However, a unideterminant antigen is not capable by itself of initiating an immune response. The immunogenic activity declines sharply with peptides smaller than the undecapeptide. An uninterrupted sequence of at least seven amino acid residues appears to be required for immunogenicity. The immune response of synthetic antigen is strongly enhanced by the incorporation of aromatic amino acids into the antigen. Therefore, the size, the conformation, and the components of the determinant are the main factors in the determination of the immunopotency.

In all cases, anti-peptide antibodies can be raised when the peptide is conjugated by covalent or electrostatic binding with a carrier protein.

Several groups of amino acid peptide sequences are activatable, for covalent coupling:

1. Amino groups have similar properties whether they are localized in the principal chain (α-NH_2) or in the lateral peptide chain (ε-NH_2 of lysine). They can be alkylated with dinitrofluorobenzene,acylated with various activated acids, esters (*N*-hydroxysuccinimide ester), and maleimide derivativized (MBSE) or carbamylated with phenylisothiocyanate derivatives. They also form addition compounds with aldehyde derivatives (like glutaraldehyde). The guanidyl group of arginine residues can also react with these reagents, but with a much lower reactivity.

2. The lateral carboxyl group from aspartate and glutamate amino acids or α-carboxyl groups can be esterified with various dialkylcarbodiimides (EDCI, DCC) or mixed anhydride reagents, to form a substituted amide with the amino groups of the carrier.

3. The phenolic group of tyrosine has a great intrinsic reactivity by virtue of the presence of a hydroxyl group and can involve an electrophilic substitution with halogenated reagent like bisdiazotized benzidine.

4. The imidazole nucleus of histidine can also be acylated, like amino groups, and *ortho* substituted, like the phenolic group of tyrosine, with halogenated reagents.

5. The thioester radical of methionine is less reactive, but can be alkylated, like amino groups, to give a sulfonium ion.

6. The hydroxyl groups of serine, threonine, and tyrosine can react like amino groups.

We note that several groups present in the peptide chain can be activated with the same reagent. This property decreases the specificity of the link between the hapten and the carrier and consequently reduces the specificity of the antibodies produced.

In fact, the different groups do not have the same reactivity, either because of their different intrinsic reactivity (the guanidium ion is less reactive than the α-NH_2 terminal group) or because of their different environment in the intrapeptide chain (involvement of ionizable groups in ionic or hydrogen binding). An alternative consists in protecting the reactive group by blocking the α-NH_2 and α-COOH groups in metallic complexes. The commonly used and easily eliminated protecting reagents are the acyl radicals and the alkyloxycarbonyls like benzyloxycarbonyl (BOC) or the *t*-butoxycarbonyl (*t*BOC).

1. Arginine Vasopressin

Arginine vasopressin (AVP) is a cyclic nonaamino acid peptide. Its structure (Fig. 17) is very similar to that of oxytocin, differing in two amino acids (isoleucine instead of phenylalanine in position 3, and leucine instead of arginine in position 8). Oxytocin and AVP are transported to the posterior pituitary from the

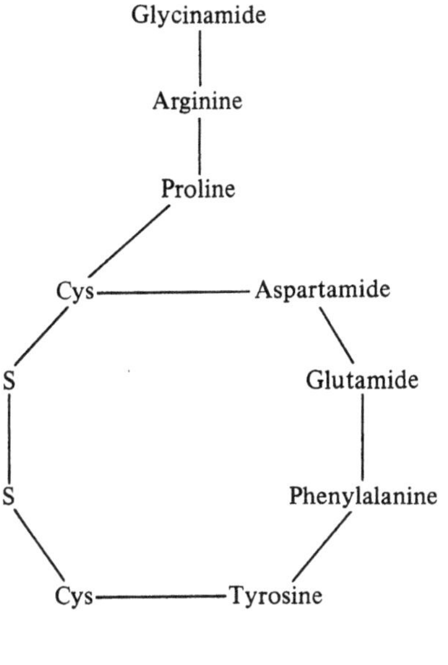

Fig. 17. Structure of arginine vasopressin

paraventricular and supraoptic nuclei, respectively. From this region these peptides are secreted into the peripheral circulation where they are found at a similar concentration (about 1 pg/ml). Therefore, the antibody used in the assay must distinguish one peptide from the other for measurements in the posterior pituitary, in plasma, or in urine. In this last biologic sample, the immunoassay must not measure the biologically inactive metabolites of vasopressin (1–6 peptide + glycinamide) (WALTER and BROWMAN 1973). For these reasons, the majority of antibodies produced must bind predominantly the tripeptide tail region, hence distinguishing AVP from oxytocin and from its metabolites in urine samples. Thus, this tripeptide tail structure must not be masked by the procedure of linking with the carrier. This may be accomplished by activation of AVP with a carbodiimide reagent which creates a link between the α-NH_2 of the peptide and the activated carboxyl group of the carrier (BAYLIS and HEATH 1977). Another specific procedure consists in linking the specific tyrosine residue of AVP to tyrosine or histidine residues of the carrier with bisdiazotized benzidine.

2. Methionine Enkephalin

Methionine enkephalin (the 61–65 sequence of LPH, β-lipotropin hormone), and leucine enkephalin, are endogenous pentapeptides which were discovered in the brain (HUGHES 1975). β-Endorphin (61–91 LPH) is a COOH terminal extension of Met-enkephalin, and is produced in the anterior pituitary gland. β-Lipotropin (1–91) is a pituitary hormone, preproopiomelanocortin (MAINS et al. 1977). Two copies of the Met-enkephalin sequence possess COOH terminal extensions of two or three amino acids, giving rise to the octapeptide Met-enkephalin-Arg-Gly-Leu and the heptapeptide Met-enkephalin-Arg-Phe (ROSSIER et al. 1980; KILPATRICK et al. 1981). All these neuropeptides are distributed throughout the central and peripheral nervous systems and exhibit opioid activity. Proteolytic enzymes quickly destroy enkephalins in tissue, and a specific aminopeptidase, enkephalinase, releases the Tyr-Gly-Gly from enkephalin peptides.

A highly specific Met-enkephalin immunoassay is required to distinguish Met-enkephalin from active opioid precursors and inactive metabolites. The assays using antisera raised against Met-enkephalin conjugated to the carrier with glutaraldehyde distinguish the different opioid peptides from Met-enkephalin (MILLER et al. 1978; CHILDERS et al. 1977).

In fact, only the primary amino group is accessible by the glutaraldehyde, whereas the COOH terminal group stays free. The resulting antibodies distinguish modifications in this part; e.g., leucine instead of methionine in Leu-enkephalin and the amide bridge in β-endorphin, octapeptide, LPH, and heptapeptide, instead of free carboxyl.

Another coupling reaction may be considered with dialkylcarbodimide in aqueous solution at pH 5. Actually, since the carboxylate ion is stable in these conditions, the carboxyl group of Met-enkephalin is less reactive than the amino group. However, as the carboxyl groups of the carrier are present at high concentrations, a small-proportion will react with the amino groups of Met-enkephalin (GROS et al. 1978). Figure 18 shows the various coupling possibilities used for Met-enkephalin.

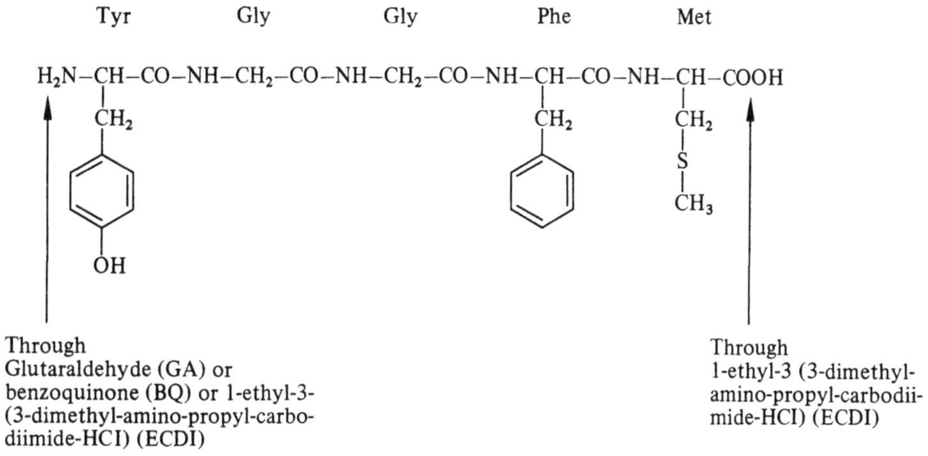

Fig. 18. Various coupling points for Met-enkephalin immunogen preparation

V. Transformation of Immunogen

The coupling reaction may introduce heterogeneity in the hapten through chemical modifications. One of the difficulties in raising specific PGE_2 antibodies arises from this. However, severe coupling conditions allow one to control hapten transformation during coupling (DRAY et al. 1982). Nevertheless, even coupled to a carrier, PGs of the E series (PGE) are unstable, because of the existence of the β-hydroxyketone moiety (Fig. 19).

In serum of all species, enzymes like dehydrase and isomerases, convert PGEs into stable PGBs. Almost all animals immunized against PGE coupled through a peptidic bond to a carrier give an antibody response generally negative for anti-PGE. The antisera recognize essentially PGA and PGB. In a few cases, the results were apparently positive, i.e., the serum of such animals contained relatively specific PGE antibodies that were convenient for developing the corresponding radioimmunoassay (RIA). However, these assays showed that PGB antibodies were in the majority (Table 1).

Since 1981, we have tested 362 animals of various species to select one without dehydrase. Animals with low enzymatic rate gave a low anti-PGE_2 response and a high anti-PGB_2 recognition. The only animal found without dehydrase activity, a rabbit, was immunized against 6-keto-PGE_1-BSA, along with nine other rabbits. As expected, this animal was the only one to produce a specific 6-keto-PGE_1 antiserum with a high binding affinity (cross-reactivity $<0.1\%$ with the other PGs, final titer 1:240000, inhibition concentration at 50% of initial binding 1 pg per tube).

The difficulties encountered were the same for PGDs as for PGEs. In order to estimate PGD_2 degradation, we have incubated tritiated PGD_2 with normal rabbit serum, mice ascites, and sheep lymph. After solvent extraction, samples were run through an HPLC column. All the radioactive peaks (8–12, depending on the biologic medium) except the first eluted showed binding capacity with the only PGD_2 antiserum previously obtained in our laboratory by immunizing animals with a PGD_2–BSA conjugate. Moreover, an important percentage of nonex-

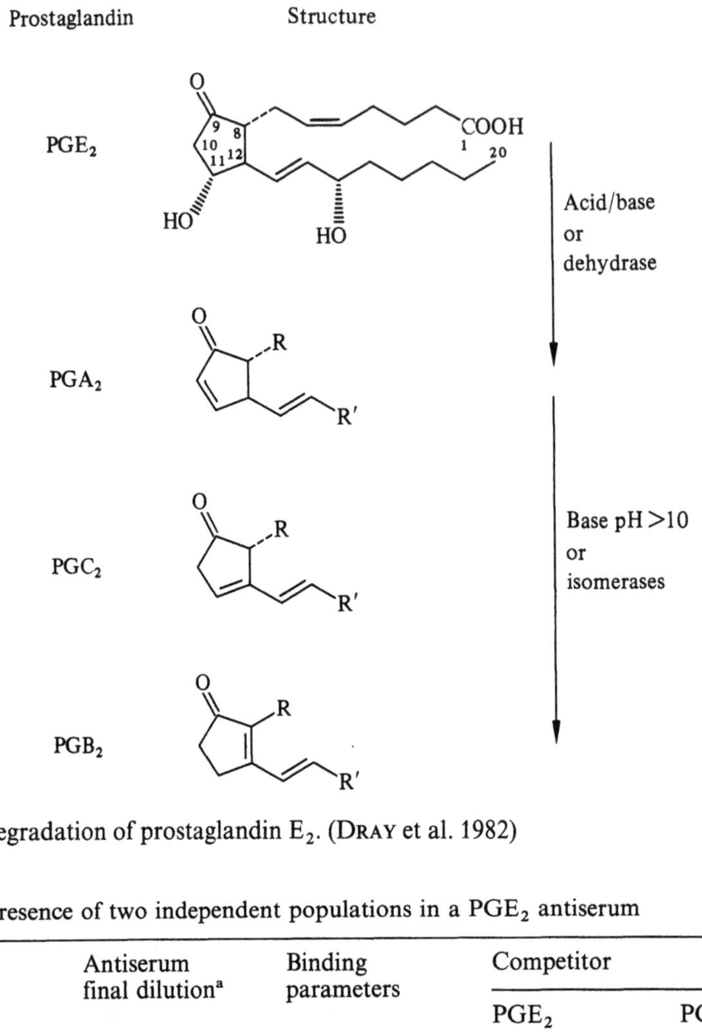

Fig. 19. Degradation of prostaglandin E_2. (DRAY et al. 1982)

Table 1. Presence of two independent populations in a PGE_2 antiserum

Tracer	Antiserum final dilution[a]	Binding parameters	Competitor	
			PGE_2	PGB_2
PGE_2 ^3H	1: 6000	IC_{50}	11	$>2\times10^5$
		K_a	8.2×10^{10}	
		CR%	100	<0.01
PGB_2 ^3H	1:12000	IC_{50}	$>2\times10^5$	29
		K_a		1.8×10^9
		CR%	<0.01	100

[a] Final dilution to obtain an initial binding of 40%.
IC_{50} pg ligand corresponding to 50% of initial tracer binding; K_a affinity constant (M^{-1}); CR% percentage cross-reactivity.

tractable degradation products has not been tested for immunoreactivity. These results show that this PGD_2 antiserum cross-reacts or is directed against several degradation products, and explain the difficulties in obtaining anti-PGD_2.

Another case of enzymatic degradation is observed with dehydroepiandrosterone (DHA) sulfate. When coupled to a carrier with a carboxymethyl bridge in

Table 2. Cross-reactivity (%) of dehydroepiandrosterone sulfate and dehydroepiandrosterone with anti-DHA sulfate serum

Tracer	Competitor	
	DHA sulfate	DHA
DHA sulfate ^3H	100	0.01
DHA ^3H	70	100

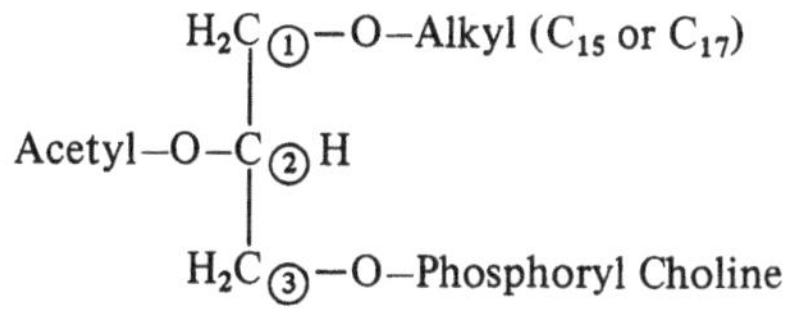

Fig. 20. Structure of PAF. Degradation enzymes act on the three chains; tritiated tracers can be obtained by introduction of tritium on C-1, C-2, or C-3 chains, proteins or reactive groups can be bound to lyso-PAF (hydroxyl group at C-2) and iodinated

the 7β position (CHARPENTIER 1982) and injected into rabbits, the hapten is probably partially hydrolyzed by a sulfatase. The resulting antisera reflect this transformation (Table 2) They exhibit DHA sulfate recognition without DHA cross-reactivity when tested with tritiated DHA sulfate, but also DHA recognition when tested with tritated DHA.

The problem of obtaining platelet-activating factor (PAF) antibodies is also not yet resolved. PAF alone, adsorbed or linked (Fig. 20) to a carrier, is very quickly degraded when injected into animals. PAF is a strong hypotensive and aggregant mediator, and hyperventilation is necessary to avoid death of animals just after PAF injection. Positive results obtained by testing platelet anti-aggregation activity of antisera are due to enzymatic effects. Thus, for biologic tests and for RIAs, precautions should be taken to eliminate all interference, positive or negative. In particular, purified immunoglobulins, obtained by a prior immunoprecipitation or by ion exchange chromatography, are required to test anti-PAF activity.

In order to obviate the problem of hapten degradation, some recommendations are to be followed:

1. Chemical stabilization of hapten, tracer, and ligand (GRANSTROM et al. 1982)
2. Selection of animals without enzyme
3. Production of monoclonal antibodies in vivo or in vitro.

This solution has been adopted in our laboratory to obtain PGE_2 antibodies (DAVID et al. 1985).

D. Immunoassay Sensitization

Three major factors limit the sensitivity of immunoassays: (a) the binding affinity between antigen and antibody; (b) the level of detection of the label; and (c) the type of immunologic reaction. High affinity antibodies display affinity constants of about 10^9–10^{12} M^{-1}, resulting in sensitivities of about 10^{-9}–10^{-12} M. Since the methods for detecting radioisotopes (such as γ-emitters) or enzymes can detect less than 0.2 fmol, the affinity between antigen and antibody is probably the limiting factor in immunoassays that use the most sensitive label detection methods. Theoretically, noncompetitive immunoassays performed in the presence of excess reagent at each step of the procedure should be more sensitive than competitive immunoassays. This type of assay allows the maximum reaction between antigen and antibody, thereby permitting measurement of lower antigen concentrations, as shown for ACTH 1–24 (ROUGEOT et al. 1983). An advantage of enzyme immunoassay over RIA in terms of sensitivity is the possible amplification of the antibody–antigen signal, chemically achieved by catalytic, cycling, or multiplication mechanisms.Although the development of noncompetitive assays and the use of nonisotopic labels have attracted great interest in the past decade, we will focus our attention on the predominant type of assay, i.e., the competitive binding RIA.

I. Increase in Tracer Specific Activity

1. Theoretical Considerations

During the first few years of immunoassay development, labeling with tritium (^3H) was widely used in competitive binding RIA. Yet this label suffers from a number of disadvantages and efforts have been made to introduce γ-emitting radioisotopes such as iodine (^{125}I). The use of radioiodinated compounds of high specific activity has many advantages: simplification of work, shortening of counting time, and increase of assay sensitivity.

Competitive RIA are based on competition between unlabeled L and labeled L* molecules of the ligand for the same antibody sites R. At equilibrium, the association constants are

$$K_a = \frac{[LR]}{[L]\,[R]}$$

for the unlabeled system and

$$K_a^* = \frac{[L^*R]}{[L^*]\,[R]}$$

for the labeled system.

In the case of tritiated tracers, there is no difference between the structure of the competitors and $K_a = K_a{}^*$. The sensitivity of the assay will be directly dependent both on K_a and on the specific activity of the tracer. In the case of iodinated tracers, the structures of the competitors are different and are recognized differently by the same antibody sites, so $K_a \neq K_a{}^*$.

On the one hand, blockade of the carboxyl group by a peptide bond increases the structural analogy between the tracer and immunogen as depicted in Fig. 21.

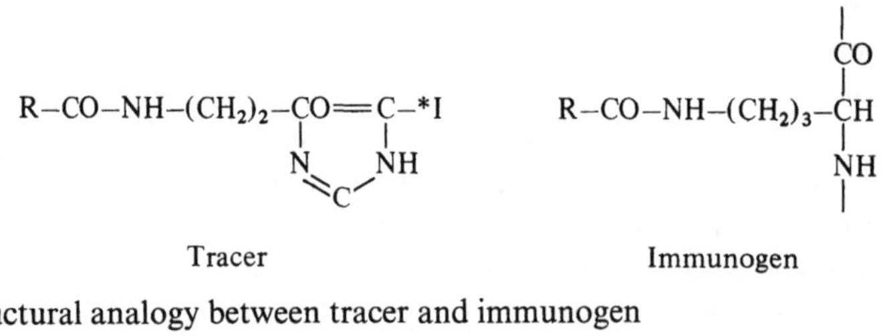

Fig. 21. Structural analogy between tracer and immunogen

The immunologic recognition of the iodinated tracer may be increased, and more unlabeled ligand R–COOH is necessary for competition. This will result in less sensitive assays and a shallow dose–response curve.

On the other hand, the high specific radioactivity obtained with ^{125}I makes the weight of tracer added in the assay negligible and requires less antiserum for binding. Increase in dilution of the antiserum may involve antibody populations with low capacity but high affinity; the effect will be an increase in sensitivity. Finally, we are in the presence of contradictory effects and the consequence of this on sensitivity cannot be predicted: it can only be observed by analyzing dose–response curves.

To circumvent this bridge effect and the subsequent decrease in assay sensitivity, other approaches have been reported. The ideal strategy would be to realize a homologous situation where both labeled and unlabeled ligands are structurally identical to the haptenic determinant. However, for chemical or practical considerations, this approach has not been developed until now.

2. Homologous Situation

A way of increasing the sensitivity of competitive assays with iodinated tracer is to block the carboxyl group of the ligand, in order to mimic the structure of the tracer, as illustrated for PGs of the E series (Table 3). The sensitivity of the system using radioactive ligand was compared when using as inhibitors the unmodified PGE_1 or PGE_2, or their derivatives synthesized by adding histamine (PG-His), lysine (PG-Lys), or methyl ester (PG-ME) derivatives. The carboxyl group of the ligand was conjugated through a peptide bond, either to the antigenic carrier, for the preparation of the immunogen, or to iodinatable substances, for tracer synthesis. The sensitivity was improved in both PGE_1 and PGE_2 RIAs when a peptide bound was formed. Blocking the carboxyl group in a peptide bond with carbodiimide would be the best solution. However, this transformation is incomplete and time-consuming when applied to a large number of samples. The homolo-

Table 3. Comparison of the sensitivity[a] of PGE_1 and PGE_2 radioimmunoassay with different tracers and inhibitors (DRAY 1982)

Tracer	Inhibitor	RIA sensitivity (fmol/ml)	
		PGE_1	PGE_2
PG ^3H	PG	300	51
PG–His ^3H	PG	4510	43
PG–ME ^3H	PG	4610	47
PG–ME ^3H	PG–ME	1410	50
PG–His ^{125}I	PG	170	25
PG–His ^{125}I	PG–His	110	24
PG–His ^{125}I	PG–Lys	60	19
PG–His ^{125}I	PG–Me	90	28

[a] Expressed as 50% of initial tracer binding.

Table 4. Comparison of progesterone assay performance with homologous bridge between tracer and immunogen (pg ligand corresponding to 50% of initial tracer binding)

Immunogen bridge	Tritiated tracer	Iodinated tracer bridge	
		Glucuronide	Succinyl
Glucuronide	80	160	
Succinyl	160		1280

gous situation may however result in a sensitive assay when the bridge structure is poorly recognized by the antibodies, as exemplified with a glucuronide group at the C-11 position of progesterone (Table 4; CORRIE et al. 1981).

A similar approach in competing ligand modification is shown with cAMP. The strategy in immunogen preparation led to antibodies possessing good specificity, but lacking sensitivity when used with an iodinated tracer obtained by coupling histamine ^{125}I to 2′-*O*-succinyl cAMP. To circumvent this problem, both standard and sample can be converted into 2′-*O*-succinyl cAMP in aqueous medium. The resulting homology between tracer and ligand enhances the sensitivity of the system by a factor of 100. Furthermore, this method offers many advantages: ease of technical manipulation, succinylation yield of 100%, and high reproducibility (CAILLA et al. 1973b).

More recently, an RIA kit for histamine has been developed with homologous ligand and tracer. Because of its small size, histamine is coupled to the carrier via a succinyl–glucinamide linkage (DELAAGE et al. 1984). Therefore, the weak affinity of the native molecule is compensated by incorporating a similar linkage to tracer and sample.

A different problem with a bridge effect has been reported with a small peptide: thyrotropin-releasing hormone (TRH) (GROUSSELLES et al. 1982). TRH was linked to sunflower protein by means of the bisdiazotized benzidine method and ^{125}I was introduced directly without a bridge on the histidine residue. In order to compensate the enhanced steric fit of the tracer to the antibody-binding sites, standard and samples were submitted to the same iodination treatment. This chemical modification allowed a 250% increase of sensitivity over the unmodified system.

Another alternative was brought about by selective removal of bridge antibodies, as shown in an immunoassay of cotinine, one of the major metabolites of nicotinine (KNIGHT et al. 1985). Rabbit anti-cotinine antibodies were adsorbed by nicotine bound to hemocyanin via an identical bridge. The resulting modified assay provided a steeper standard curve and a considerable increase in sensitivity, although the low nicotinine cross-reactivity (0.3%–0.4%) remained unchanged.

3. Heterologous Situation

An alternative approach is the deliberate reduction of affinity of label for antibody by the use of heterologous systems in which either the hapten, the site of the linkage, or the nature of the bridge are altered. Changing the hapten used in preparing the immunogen will prevent the bridge effect, as demonstrated for a melatonin RIA with the use of 5-methoxytryptamine for both immunogen and tracer preparations (TIEFENAUER and ANDRES 1984). The chemical linkage for coupling the carrier and for introducing the iodinated moiety is similar to the terminal group of melatonin. In this case, the absence of a bridge problem is demonstrated by an equal affinity between the tritiated and the iodinated tracers, and results in a substantial gain in sensitivity (aproximately eight-fold based on the 50% inhibition of tracer binding).

Different coupling positions in both immunogen and tracer generate many situations which are unpredictable a priori: increase of sensitivity, nonrecognition of the tracer, or strong alteration of specificity. However, introducing a subtle change in the chemical bridge is a more interesting alternative, as demonstrated for an androstenedione assay with heterologous combination of an ether or ester bridge in tracer and immunogen (Table 5; NORDBLOM et al. 1980).

To increase the sensitivity of progesterone and estradiol RIA, we have tested several iodinated tracers with anti-progesterone-11-hemisuccinate–BSA and anti-estradiol-6-carboxymethyloxime–BSA sera (Table 6, Fig. 22). Antisera titers re-

Table 5. Androstenedione assay sensitization by modification of the bridge structure (pg ligand for 50% initial tracer binding inhibition)

Immunogen bridge	Tritiated tracer	Tracer bridge	
		Ester	Ether
Ester	110	625	30
Ether	110	420	82

Table 6. Comparison of sensitivities of two steroid RIAs with heterologous (1, 2, 4, 5) and homologous (3, 6) tracers. Numbers correspond to Fig. 22

Tracer		Antiserum final dilution[a]	Delayed tracer incubation	IC_{50}[b]
		Anti-E_2-6-CMO–BSA		
E_2 ^3H	(1)	1: 12000	No	45
16α-E_2 ^{125}I	(2)	1: 3750	No	115
E_2-6-CMO-His ^{125}I	(3)	1:600000	No	2471
			Yes	206
		Anti-P-11-HS–BSA		
P ^3H	(4)	1: 30000	No	70
P-11α-Gluc-Tyr ^{125}I	(5)	1: 3000	No	13.5
P-11α-HS–His ^{125}I	(6)	1:600000	No	>3000
			Yes	334

[a] Final dilution to obtain an initial binding of 40%.
[b] Picograms ligand corresponding to 50% of initial tracer binding.
BSA, bovine serum albumin; CMO, carboxymethyloxime; E_2, estradiol; Gluc, glucuronide; HS, hemisuccinate; P, progesterone.

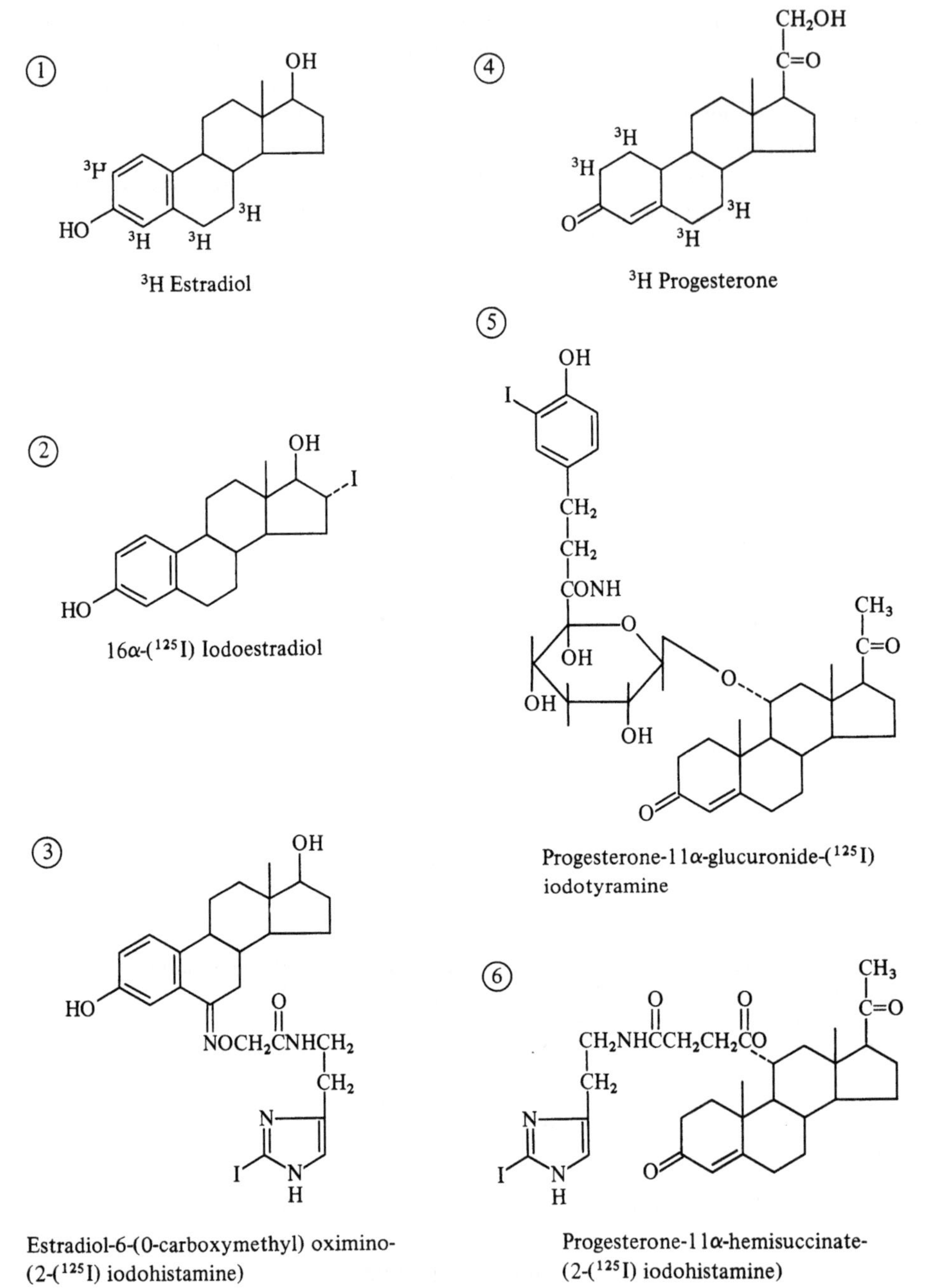

Fig. 22. Tritiated and iodinated tracers. (From Amersham Research Products, England.) Numbers correspond to Table 6

flect the antibody activities for these tracers. Homologous tracers with the same bridge as in the immunogen can involve very high titers, but provide a low assay sensitivity. This latter is, however, slightly increased by preincubation of ligand with antibodies. On the other hand, a significant gain in sensitivity is obtained with heterologous tracers.

II. Modification of Reagent Concentrations and Assay Procedures

1. Reagent Concentrations

The objective of changing reagent concentrations is to achieve maximal accuracy of measurement in the standard range of concentration and to lower the minimal detectable dose. In the case of a competitive RIA, the only variable of the system is the ligand concentration, whereas tracer and antibody concentrations are susceptible to modification. Optimal choice of reagents can be made by analyzing several parameters: slope of the dose–response curve, minimal detectable dose, and standard deviation of replicates. Several authors have dealt with this problem and shown the importance of statistical analysis in assessing assay performance (YALOW and BERSON 1971; EKINS 1979).

2. Assay Procedures

a) pH of Incubation Medium: Prostaglandin Systems

The introduction of radioiodinated tracers in prostaglandin RIA does not systematically improve assay performance. As in the case of steroid systems, the increase in tracer specific activity may be counterbalanced by increase in antibody–tracer affinity, which results in loss of assay sensitivity.

In the $PGF_{2\alpha}$ system, the introduction of an iodinated tracer leads to a considerable decrease in sensitivity in comparison with a tritiated tracer at the same pH. However, a pH change from 7.4 to 5 allows one to suppress this decrease (Table 7). The same pH effect is demonstrated by using an iodinated tracer with 6-keto-$PGF_{1\alpha}$. In the case of other compounds, like PGE_2 and thromboxane B_2

Table 7. Radioimmunoassay with three prostaglandin $F_{2\alpha}$ antisera; tritiated and iodinated tracer at pH 7.4 or pH 5

	Tracer			
	$PGF_{2\alpha}$ ^3H		His–$PGF_{2\alpha}$ ^{125}I	
pH of incubation medium	7.4	5	7.4	5
Antiserum 1				
Final dilution[a]	1:15000		1:300000	1:300000
IC_{50}[b]	36		139	40
Affinity constant (M^{-1})	0.3×10^{10}			
Antiserum 2				
Final dilution	1:9000	1:9000	1:19800	1:23100
IC_{50}	4.3	8.6	43.5	6.3
Affinity constant (M^{-1})	1.2×10^{10}			
Antiserum 3				
Final dilution	1:3000	1:3000	1:9900	1:12480
IC_{50}	11.5	16.1	61.6	10.9
Affinity constant (M^{-1})	0.8×10^{10}			

[a] Final dilution to obtain an initial binding of 40%.
[b] Picograms ligand corresponding to 50% of initial tracer binding.

Table 8. Radioimmunoassay with different anti-prostaglandin antisera, iodinated tracer at pH 7.4 and 5

Anti-prostaglandin	pH of incubation medium	
	7.4	5
Anti-6-keto-PCF$_{1\alpha}$		
Final dilution[a]	1:120000	1:120000
IC_{50}[b]	11.4	6.8
Anti-PGE_2		
Final dilution	1:60000	1:60000
IC_{50}	2.6	7.1
Anti-TXB_2		
Final dilution	1:60000	1:60000
IC_{50}	10.5	50

[a] Final dilution to obtain an initial binding of 40%.
[b] Picograms ligand corresponding to 50% of initial tracer binding.

Table 9. Cross-reactions (%) at pH 7.4 and 5 of various prostaglandins in the 6-keto-PGF$_{1\alpha}$ assay (iodinated tracer)

Prostaglandin	pH of incubation medium	
	7.4	5
6-Keto-PGF$_{1\alpha}$	100	100
6-Keto-PGE$_1$	39	20
$PGF_{2\alpha}$	24	10
PGE_2	34	0.6
2,3-Dinor-6-keto-PGF$_{1\alpha}$	134	100

(TXB_2), sensitivity is better at pH 7.4 than pH 5 (Table 8). One can assume that there is an optimal pH which allows every reactant ligand, tracer or antibody, to be in the best operational status. In these optimal conditions, the antibody affinity for tracer is the same or slightly lower than for ligand.

It can be observed that the specificity may be affected by this pH effect, as demonstrated for the binding properties of 6-keto-PGF$_{1\alpha}$ (Table 9). In the case of PGE_2 cross-reactivity, two phenomena can be considered (steric modification and chemical alteration), while for other PGs the lower effect could be due only to a steric modification.

b) Sequential Incubation of Ligand and Tracer

Assay sensitization based on delayed tracer incubation has been described for numerous peptide systems. This effect relies on the reduced dissociation rate of the preformed "cold immunocomplexes." Thus, the first reaction is nearly irreversible and the equilibrium can hardly be dissociated by tracer addition resulting in

Table 10. Assay sensitization by sequential tracer incubation

Peptide	Tracer from beginning	Tracer addition delayed [b]
Rat GHRF	160 [a]	30
CRF	325	100
Somatostatin	30	11
Vasopressin	250	100
ACTH	60	15
LHRH	25	12

[a] Picograms ligand corresponding to 50% of initial tracer binding.
[b] Sequential incubations 24+24 h.

an incomplete competitive system. For peptide RIAs developed in our laboratory, marked improvements in sensitivity have been obtained by allowing premixing of standard and antiserum (Table 10).

This approach is not efficient with smaller haptens like steroids since their higher dissociation rate allows them to reequilibrate too quickly for delayed tracer addition to be effective (MALVANO 1983). However, when homologous radioiodinated tracers are used, the sequential incubation may be effective because of the competitive imbalance between native and labeled ligand (see Table 6).

E. Validation

Validation of the RIA is necessary to render the system suitable for biologic analysis. In most cases, the problem comes from the presence of nonspecific materials inhibiting the specific ligand–antibody binding. This inhibition may be caused by various factors, depending on the nature of the antigen and the biologic medium. The first factor concerns the presence of antibody-like substances, such as plasma protein binding, endogenous antibodies, lipid complexes, or tracer-degrading enzymes. A second factor is the presence of antigen-like substances. Molecular heterogeneity, as encountered with steroids or prostaglandins, will be a severe handicap for the immunoassay. If structurally related forms are present in the sample, they will be detected in various proportions, depending on their antigenic similarity. An identical situation occurs in pharmacokinetic experiments when the metabolic fate of drugs results in minor structural differences which may be recognized by antibodies. Weak cross-reactions with metabolites may cause relative errors which have no relevance to the determination of the parent drug. Common antigenic structures are also found between peptides and their precursors, polymers, or fragments. In order to discover and to reduce nonspecific interference, the assayist must be methodical and take considerable care to avoid these problems.

I. Nonantigenic Materials

The interference of nonantigenic material can be assessed by various tests: tracer binding in the presence of increasing quantities of analyte-free material, analytic recovery of added analyte in the same material, or parallelism between sample serial dilutions and the standard curve. The tests will give the dilution necessary to eliminate interference in order to get an unbiased result. However, this alternative will be possible only when the substance to be assayed is present at a high concentration since sample dilution reduces assay sensitivity.

Theoretically, the best solution would be the incorporation of an identical amount of nonspecific material both in the standard curve and in the sample. However, since the range of sample concentrations is sometimes difficult to estimate, several standard curves with various quantities of nonspecific substances have to be constructed. Furthermore, the variation of interference from one sample to another may raise unpredictable errors in the assay.

Another approach has been provided by microencapsulation of antibodies in a nylon membrane which prevents the passage of substances of molecular weight greater than 20000. Thus, the antibody–antigen reaction remains free from protein and high molecular weight substance interactions. This technique was described for digoxin (HALPERN and BORDENS 1979a), thyroxine (HALPERN and BORDENS 1979b), cortisol (BORDENS and HALPERN 1980), and 17-hydroxyprogesterone (WALLACE and WOOD 1984). In the case of a progesterone assay, interfering steroid hormone binding proteins have been eliminated by incubating samples with cortisol. This causes the release of progesterone from its binding globulin (HAYNES et al. 1980). This solution was possible since the amount of cortisol required did not cross-react with the specific progesterone antibodies used.

In many cases, extraction of the substance from the sample will be the way to reduce nonspecific effects. The choice of the organic solvent will depend on the nature of interfering materials and physicochemical properties of the substance. However, each method presents practical problems since the extracting solvent has to be removed by evaporation or lyophilization and a nonnegligible fraction of the substance to be assayed is coeluted with nonspecific materials. This may be counterbalanced by possible sample concentration and apparent gain in assay sensitivity.

II. Antigen-like Materials

The assessment of the importance of cross-reacting materials in the measurement requires a specific purification step which may be performed with various methods, like thin layer chromatography, silicic acid column chromatography or HPLC. The choice of one of these methods depends on the properties of the substances, the recovery of the purified material, and the resolution between structurally related compounds. In some cases, sequential purification steps are necessary to control the specificity.

III. Example of Validation: Peptide Radioimmunoassay

We report here the development of a procedure for the validation of the RIA of D-Trp^6-LHRH, an analog of LHRH (Fig. 23). Tracer was obtained by direct introduction of ^{125}I by the iodogen method, and immunogen by coupling the peptide to BSA via glutaraldehyde (GA-anti-D-Trp^6-LHRH) or via benzoquinone (BQ-anti-D-Trp^6-LHRH).

The first step concerns the assessment of assay characteristics: dose–response parameters, evaluation of nonspecifically interfering materials, and estimation of specificity (Tables 11–13). The RIA with BQ-anti-D-Trp^6-LHRH has good specificity, but shows high interference with human plasma. Furthermore, its sensitivity is not suitable for pharmacologic studies since very low hormone concentrations are obtained. Before choosing GA-anti-D-Trp^6-LHRH, we have undertaken an assay validation for measurements in unextracted plasma.

LHRH

pGLU–HIS–TRP–SER–TYR–GLY–LEU–ARG–PRO–GLY–NH_2

D–Trp^6–LHRH

pGLU–HIS–TRP–SER–TYR–DTRP–LEU–ARG–PRO–GLY–NH_2

↑ Tracer (TYR) ↑ Immunogen (ARG)

Fig. 23. Structure of LHRH and D-Trp^6-LHRH

Table 11. Assay sensitivity[a]

	BQ-Anti-D-Trp^6-LHRH	GA-Anti-D-Trp^6-LHRH
Tracer from beginning	45	13
Tracer addition delayed[b]	23	4

[a] Expressed as picogram amounts corresponding to 50% of initial tracer binding.
[b] Sequential incubation of 24 + 24 h.

Table 12. Evaluation of non-specific materials: analytic recovery of added D-Trp^6-LHRH in unextracted human plasma

Plasma dilution	1:1	1:2	1:4	1:8	1:16	1:64
GA Recovery (%)	164	132	117	103	96	104
BQ Recovery (%)	206	158	135	123	108	98

Table 13. Cross-reaction (%) of various peptides with D-Trp^6-LHRH antiserum

	BQ-Anti-D-Trp^6-LHRH	GA-Anti-D-Trp^6-LHRH
D-Trp^6-LHRH	100	100
LHRH	20	< 0.01
(1–5) LHRH	< 0.02	< 0.02
(1–7) LHRH	< 0.02	< 0.02
(1–8) LHRH	< 0.02	< 0.02
(1–9) LHRH	< 0.02	< 0.02
(3–10) LHRH	17	< 0.02
(4–10) LHRH	7	< 0.02
(5–10) LHRH	13	< 0.02
(7–10) LHRH	17	< 0.02

The nature of nonspecific interference was assessed by gel filtration of hormone-free plasma and measurement of apparent immunoreactivity in collected fractions. Interfering materials appeared to be proteins, with a molecular weight >60000 and which were not peptide-degrading enzymes. These substances compete with the specific antibodies in their binding to peptides and lead to measurement of an apparently high plasma level. As the sample dilution increases, the recovery becomes quantitative (Table 12). However, the reduction of the assay sensitivity (detection limit 10 pg/ml in buffer vs 80 pg/ml in unextracted plasma at 1:8 dilution) requires the elimination of proteins for low hormone concentration samples. Therefore, we have tested various procedures for protein extraction which have been previously described for peptide RIAs. These methods involve protein precipitation by organic solvent: methanol/acetic acid (HANDELSMAN et al. 1984), ethanol/acetic acid (PEETERS et al. 1981), methanol (NETT and ADAMS 1977), acetone/ether (AMICO et al. 1979), or selective elution through a C_{18} silica cartridge Sep-Pak (Waters Associates) (EZAN et al. 1986). The efficiency of various solvents for extracting D-Trp^6-LHRH from human plasma was studied as follows (Table 14):

Table 14. Efficiency of D-Trp^6-LHRH extraction from human plasma

Extraction procedure	Tracer[a] recovery	Free plasma[b]	Free plasma[c] +20 pg	Free plasma[c] +200 pg
No extraction		80	117	103
Methanol	60	70	96	66
Methanol/acetic acid	79	<10	71	84
Ethanol/acetic acid	47	40	67	65
Acetone	76	<10	69	79
Sep-Pak	85	<10	79	85

[a] %. 100000 cpm added in hormone-free plasma.
[b] Apparent immunoreactivity (pg/ml) in hormone-free plasma.
[c] Recovery (%) of added D-Trp^6-LHRH.

1. ^{125}I-labeled D-Trp^6-LHRH was added to plasma and extracted. Among the solvent procedures examined, acetone and methanol/acetic acid gave the largest number of counts in the supernatant obtained after centrifuging the precipitated proteins. Tracer recovery after Sep-Pak elution was about 85%.

2. Subsequently, 1 ml plasma containing 0, 20, or 200 pg D-Trp^6-LHRH was extracted or radioimmunoassayed. Apparently, high quantitative recovery was found in methanol extract, probably owing to the presence of nonprecipitated proteins as shown by the high blank value (70 pg/ml). Except in the case of ethanol, the other procedures showed disappearance of apparent immunoreactive material from free plasma.

Among the five methods described, Sep-Pak chromatography combines effective recovery (75%–85%), elimination of interference (blank value <10 pg/ml), and high reproducibility (coefficient of variation 5%).

Assay specificity was then estimated as follows (Ezan et al. 1986):

1. Injection of 100 µg D-Trp^6-LHRH into humans and sampling at various time intervals
2. Sample extraction by Sep-Pak column
3. HPLC of selected samples

The HPLC was calibrated by determining retention time and analytic recovery of D-Trp^6-LHRH. Plasma samples fractionated using this system contained a major immunoreactive peak which corresponded to D-Trp^6-LHRH. Therefore, the majority of measured material in direct plasma assay with GA-anti-D-Trp^6-LHRH appeared to be the intact polypeptide form. Furthermore, calculation of the peptide half-life after subcutaneous injection gave identical results when samples were assayed either in unextracted plasma, after Sep-Pak extraction, or after extraction plus HPLC purification. A similar procedure has to be developed for the validation of RIA systems in order to have an idea of the magnitude and type of error in the final data. This approach emphasizes the importance of RIA sensitivity and specificity to obviate the possible problem of interfering materials and to avoid purification steps and time-consuming assays.

References

Abraham GE (1974) Radioimmunoassay of steroids in biological materials. In: Radioimmunoassay and related procedures in medicine, vol 2. Proceedings of a symposium on related procedures in clinical medicine and research, Istanbul, 1973. International Atomic Energy Agency, Vienna, pp 3–29

Abraham GR, Grover PK (1971) Covalent linkage of steroid hormones to protein carrier for use in radioimmunoassay. In: Odell WD, Daughaday WH (eds) Principles of competitive protein-binding assays. Lippincott, Philadelphia, pp 140–157

Amico JA, Seif SM, Robinson AG (1979) Oxytocin in human plasma-correlation with neurophysin and stimulation with estrogen. J Clin Endocrinol Metab 52:988–993

Bassiri S, Uttiger RD (1972) The preparation of antibody to thyrotropin releasing hormone. Endocrinology 80:722–729

Bauminger S, Kohen F, Lindner MR (1974) Antiserum to 5-hydrotestosterone: production, characterization and use in radioimmunoassay. Steroids 24:477–482

Baylis PH, Heath DA (1977) Plasma arginine vasopressin response to insulin induced hypoglycemia. Lancet II:428–430

Besselievre R, Lemaitre BJ, Husson HP, Hartman L (1980) Structural immunochemistry of melatonin-BSA binding, model of amino and indole groups cross-linking. Biomed 33:226–228

Bordens RW, Halpern EP (1980) Microencapsulated antibodies in radioimmunoassay. III. Determination of cortisol. Clin Chem 26:633–634

Boschetti E, Corgier M, Garelle R (1978) Immobilization of ligands for affinity chromatography. A comparative study of two condensation agents: 1-cyclohexyl-3-(2-morpholinoethyl)-carbodiimide-metho-p-toluene sulfonate (CMC) and *N*-ethoxycarbonyl-2-ethoxy-1,2-dihydroquinoline (EEDQ). Biochim 60:425–427

Brandt J, Andersson LO, Porath J (1975) Covalent attachment to polysaccharide carriers by means of benzoquinone. Biochem Biophys Acta 386:196–202

Butler VP, Chen JP (1967) Digoxin specific antibodies. Proc Natl Acad Sci USA 57:71–78

Cailla HL, Cros GS, Jolu EJ, Delaage MA, Depieds RC (1973a) Comparison between rat and rabbit anticyclic AMP antibodies. Specificity towards acyl derivatives of cyclic AMP. Anal Biochem 56:383–393

Cailla H, Racine-Weisbuch M, Delaage MA (1973b) Adenosine 3′,5′ cyclic monophosphate assay at 10^{-15} mole level. Anal Biochem 56:394–407

Carlsson J, Drevin H, Axen R (1978) Protein thiolation and reversible protein-protein conjugation. Biochem J 173:723–737

Charpentier B (1982) Préparation d'haptènes et ligands en série stéroide. Application dans les dosages par immunoprécipitation et dans la purification de protéines par chromatographie d'affinité. Thèse de Doctorat de 3ème cycle – Université de Nice – UER Domaine Méditerranéen

Childers SR, Rabi Simanton, Snyder SH (1977) Enkephalin: radioimmunoassay and radioreceptor assay in morphine dependant rats. Eur J Pharmacol 46:289–293

Condom R, Emiliozzi R (1974) Preparation of steroid antigens through positions of the steroid not bearing functional groups. Steroid 23:483–498

Corrie JET, Hunter WM, MacPherson JS (1981) A strategy for radioimmunoassay of plasma progesterone with use of a homologous-site ^{125}I-labeled radioligand. Clin Chem 27:594–599

David F, Somme G, Provost-Wisner A, Roth C, Astoin M, Dray F, Theze J (1985) Characterization of monoclonal antibodies against prostaglandin E_2: fine specificity and neutralization of biological effects. Mol Immunol 22:339–346

Delaage M, De Veyrac B, Morel A, Cailla H, Drocourt JL (1984) Monoclonal antibodies to haptens. In: Bizollon CA (ed) Monoclonal antibodies and new trends in immunoassays. Elsevier, Amsterdam, pp 53–58

Dray F (1982) Iodinated derivatives as tracers for eicosanoid radioimmunoassays. Methods Enzymol 86:297–306

Dray F, Gros C (1981) Competitive enzyme immunoassay of progesterone. In: Ishikawa E, Kawai T, Miyai T (eds) Enzyme immunoassay. Igaku-Shoin, Tokyo, pp 146–156

Dray F, Mamas S, Bizzini B (1982) Problems of PGE antisera specificity. Methods Enzymol 86:258–269

Dupret J, Grenot C, Rolland De Ravel M, Mappus E, Cuilleron CY (1984) Improvement of specificity of anti-testosterone and anti-5-α-dihydrotestosterone rabbit antibodies by immunotolerance techniques. J Steroid Biochem 20:1345–2352

Duval D, Desfosses B, Emiliozzi R (1980) Preparation of dehydroepiandrosterone, testosterone and progesterone antigens through 7-carboxymethyl derivatives: characteristics of the antisera to testosterone and progesterone. Steroids 35:235–249

Duval D, Predinet J, Charpentier B, Emiliozzi R (1985) Synthesis of 7α- and β-carboxymethyl derivatives of cortisol, corticosterone, deoxycorticosterone and cortisone. Imunogenic properties of cortisol, corticosterone and deoxycorticosterone derivatives. J Steroids Biochem 22:67–78

Ekins R (1979) Assay design and quality control. In: Bizollon CA (ed) Radioimmunology 1979, proceedings of IVth international symposium on radioimmunology. Elsevier, Amsterdam, pp 239–255

Erlanger BF (1973) Principles and methods for the preparation of drug protein conjugates for immunological studies. Pharmacol Rev 25:271–280

Erlanger BF, Borek F, Beiser SM, Lieberman S (1957) Steroid-protein conjugate. I. Preparation and characterization of conjugates of bovine serum albumin with testosterone and cortisone. J Biol Chem 228:713–727

Erlanger BF, Borek F, Beiser SM, Lieberman S (1959) Steroid-protein conjugates. II. Preparation and characterization of conjugates of bovine serum albumin with deoxycorticosterone and estrone. J Biochem 52:196–202

Ezan E, Drieu K, Chapelat M, Rougeot C, Dray F (1986) Radioimmunoassay of D-Trp^6-LHRH: its application to animal pharmacokinetic studies after single injection and long-acting formulation administration. Regul Peptides 14:155–167

Findlay JWA, Butz RF, Jones EC (1981) Relationships between immunogen structure and antisera specificity in the narcotic alkaloid series. Clin Chem 27:1524–1535

Goodfriend L, Semon AM (1958) Preparation of an estrone-protein conjugate. Can J Biochem 36:1177–1184

Granström E, Fitzpatrick FA, Kindahl H (1982) Radioimmunologic determination of 15-keto-13,14-dihydro-PGE_2: a method for its stable degradation product, 11-deoxy-15-keto-13,14-dihydro-11β, 16ε-cyclo-PGE_2. Methods Enzymol 86:306–320

Gros C, Pradelles P, Rougeot C, Bepoldin O, Dray F, Fournie Zauski MC, Roques BP, Pollard H, Llorens Cortes C, Shwartz JC (1978) Radioimmunoassay of methionine and leucine enkephalins in regions of rat brain and comparison with endorphins estimated by a radioreceptor assay. J Neurochem 31:29–39

Grota LJ, Brown GM (1973) Antibodies to indolealkylamines, serotonin and melatonin. Can J Biochem 52:196–202

Grousselles D, Tixier-Vidal A, Pradelles P (1982) A new improvement of the sensitivity and specificity of radioimmunoassay for thyroliberin, application to biological samples. Neuropeptides 3:29–44

Habeeb DFSA (1967) Quantitation of conformational changes on chemical modification of proteins: use of succinylated proteins as a model. Arch Biochem Biophys 121:652–664

Haber E, Page LB, Jacoby FA (1965) Synthesis of antigen in branch-chain copolymers of angiotensin and poly-L-lysine. Biochemistry 4:693–698

Halpern EP, Bordens RW (1979a) Microencapsulated antibodies in radioimmunoassay. I. Determination of digoxin. Clin Chem 25:860–861

Halpern EP, Bordens RW (1979b) Microencapsulated antibodies in radioimmunoassay. II. Determination of free thyroxine. Clin Chem 25:1561–1563

Hambuger RN (1966) Chloramphenicol-specific antibody. Science 152:203–205

Handelsman DJ, Jansen RPS, Boylan LM, Spaliviero JA, Turtle JR (1984) Pharmacokinetics of gonadrotropin-releasing hormone: comparison of subcutaneous and intravenous routes. J Clin Endocrinol Metab 59:739–746

Harman SM, Tsitouras PD, Kowatch MA, Kowarski AA (1980) Advantage of florisol over charcoal separation in a mechanized testosterone radioimmunoassay. Clin Chem 26:1613–1616

Haynes SP, Corcoran JM, Eastman CJ, Doy FA (1980) Radioimmunoassay of progesterone in unextracted serum. Clin Chem 26:1607–1609

Hillier SG, Brownsey BXC, Cameron END (1973) Some observations on the determination of testosterone in plasma by radioimmunoassay using antisera raised against testosterone-3-BSA and testosterone-11α-BSA. Steroids 21:735–754

Hughes JT (1975) Isolation of an endogenous compound from the brain with the pharmacological properties similar to morphine. Brain Res 88:295

Hung DT, Benner SA, Williams CM (1980) Preparation of an ecdysone immunogen for radioimmunoassay work. J Biol Chem 255:6047–6048

Kabakoff S (1980) Chemical aspects of enzyme-immunoassay. In: Maggio TE (ed) Enzyme-immunoassay. CRC, Boca Raton, pp 71–104

Kilpatrick DL, Jones BN, Kojima K, Udenfriends S (1981) Identification of the octapeptide (Met) Enkephalin-Arg^6-Gly^7-Leu^8 in extracts of bovine adrenal medulla. Biochem Biophys Res Commun 103:698–705

Kitagawa T (1981) Competitive enzyme immunoassay of antibiotics. In: Ishikawa E, Kawai T, Miyai K (eds) Enzyme immunoassay. Igaku-Shoin, Tokyo, pp 136–145
Knight GJ, Wylfe P, Holman MS, Haddow JE (1985) Improved ^{125}I radioimmunoassay for cotinine by selective removal of bridge antibodies. Clin Chem 31:118–121
Kohen F, Bauminger S, Lindner HR (1975) Preparation of antigenic steroid-protein conjugates. In: Cameron EHD, Hillier SG, Griffiths K (eds) Steroid immunoassay. Alpha Omega, Cardiff, pp 11–32
Kohler G, Milstein C (1975) Continuous culture of fused cells secreting antibody of defined specificity. Nature 256:495–497
Kouznetzova B, Dray F (1978) Specific antisera to androstandiols: influence of the type of bridge and position of coupling. J Steroid Biochem 9:349–352
Landsteiner F (1946) Specificity of serological reactions. Harvard University Press, Cambridge
Lieberman S, Erlanger BF, Bieser SM, Agate FJ (1959) Steroid-protein conjugates: their chemical, immunochemical and endocrinological properties. Recent Progr Horm Res 15:165–200
Mains RE, Eipper BA, Ling N (1977) Common precursor to corticotropins and endorphins. Proc Natl Acad Sci USA 74:3014–3018
Malvano R (1983) Radioimmunoassay of steroids. In: Odell WD, Franchimont P (eds) Principles of competitive protein-binding assays. Lippincott, Philadelphia, pp 161–352
Mamas S, Dray F (1979) Viroimmunoenzymoassay for detection of antipenicilloyl antibodies and penicilloyl residues: comparison of results obtained by radio-, viro-, and enzymoimmunoassay. Ann Immunol Institut Pasteur) 130C:407–418
Miller RJ, Kwenjen C, Cooper B, Cuatrecasas G (1978) Radioimmunoassay and characterization of enkephalins in rat tissues. J Biol Chem 253.2:531–538
Nett TN, Adams TE (1977) Further studies on the radioimmunoassay of gonadotropin-releasing hormone: effect of radioiodination, antiserum and unextracted serum on levels of immunoreactivity in serum. Endocrinology 101:1136–1144
Niswender GD, Midgley AR (1970) Hapten radioimmunoassay for steroid result. In: Peron FG, Caldwell BV (eds) Immunological methods in steroid determination. Appleton-Century-Crofts, New York, pp 149–173
Nordblom GD, Webb R, Counsell RE, England BG (1980) A chemical approach to solving phenomenon in hapten radioimmunoassay. Clin Chem 26:1007
Peeters TL, Depraetere Y, Vantrappen GR (1981) Simple extraction method and radioimmunoassay for somatostatin in human plasma. Clin Chem 27:888–891
Pesce A (1976) Nature of the bond formed by glutaraldehyde as a protein-protein coupling reagent. In: Van weemen BK (ed) Enzyme-immunoassay, Organon-Teknik. Oss, pp 57–67
Pontarotti PA, Leborgne de Kaouel C, Verrier M, Cupo AA (1983) Monoclonal antibodies against Met Enkephalin as proved in the central nervous system. J Neuroimmunol 4:47–59
Ratcliff WA, Fletcher SM, Moffat AC (1977) Radioimmunoassay of lysergic acid diethylamide (LSD) in serum and urine by using antisera of different specifities. Clin Chem 23:169–174
Rossier J, Audigier Y, Ling N, Gros J, Udenfriend S (1980) Met enkephalin-Arg^6-Phe^7 present in high amounts in brain of rat, cattle and man, is an opioid agonist. Nature 288:88–90
Rougeot C, Trivers ET, Harris CC, Dray F (1983) ACTH immunoassays: development of ultrasensitive and specific procedure for the measurement in biological materials. In: Avrameas (ed) Immunoenzymatic techniques. Elsevier, Amsterdam, pp 179–191
Savino W, Dardenne M, Papiernik M, Bach JF (1982) Thymic hormone. Hormone containing cells. Characterization and localization of serum thymic factor in young mouse thymus studied by monoclonal antibodies. J Exp Med 156:628–633
Spragg J, Austen KT, Haber E (1966) Production of antibodies against bradykinin: demonstration of specificity by complement fixation and radioimmunoassay. J Immunol 96:865–871

Tager MS (1976) Coupling of peptides to albumin with difluorodinitrobenzene. Anal Biochem 71:367–375
Tateishi K, Hamaoka T, Takatsu K, Hayashi C (1980) A novel immunization procedure for production of anti-testosterone and anti-5α-dihydrotestosterone antisera of low cross-reactivity. J Steroid Biochem 13:951–959
Ternynck T, Avrameas S (1977) Conjugation of p-benzoquinone treated enzymes with antibodies and Fab fragments. Immunochem 14:767-774
Thong B, Soldin SJ, Lingwood CA (1985) Lack of specificity of current anti-digoxin antibodies and preparation of a new, specific polyclonal antibody that recognizes the carbohydrate moiety of digoxin. Clin Chem 31:1625–1631
Tiefenauer LX, Andres RY (1984) Prevention of bridge effects in haptenic immunoassay systems exemplified by an iodinated radioimmunoassay for melatonin. J Immunol Meth 74:293–298
Vaughan JR, Sato RL (1952) Preparation of peptides using mixed carbonic-carboxylic anhydrides. J Am Chem Soc 74:676–678
Vining RF, Compton P, McGinley R (1981) Steroid radioimmunoassay. Effect of shortened incubation time on specificity. Clin Chem 27:910–913
Wallace AM, Wood DA (1984) Development of simple procedure for the preparation of semipermeable antibody containing microcapsules and their analytical performance in a radioimmunoassay for 17-hydroxyprogesterone. Clin Chim Acta 140:203–212
Walter R, Browman RH (1973) Mechanism of inactivation of vasopressin and oxytocin by the isolated perfused rat kidney. Endocrinology 92,1:189–193
Weinstein A, Lindner HR, Friedlander A, Bauminger S (1972) Antigenic complexes of steroid hormones formed by coupling to protein through position 7: preparation from Δ^4-3-oxosteroids and characterization of antibodies to testosterone and androstenedione. Steroids 20:789–812
Yalow RS, Berson SA (1971) Introduction and general considerations. In: Odell WD, Daughaday WH (eds) Principles of competitive protein-binding assays. Lippincott, Philadelphia, pp 1–24

CHAPTER 7

How to Improve the Sensitivity of a Radioimmunoassay

G. CIABATTONI

A. Introduction

Definitions of sensitivity fall into at least two major groups; one is specified by the definition proposed by Ekins and Newman, the second is derived from the "classical" theory developed by Berson and Yalow.

According to EKINS and NEWMAN (1970), the sensitivity of an assay may be defined as "the precision of measurement of a zero quantity." Defined in this way, sensitivity is equal to the detection limit of the assay, i.e., the lowest concentration of unlabeled antigen which can be distinguished from the absolute absence of substance. If we assume that the precision of a radioimmunoassay (RIA) reflects the reproducibility of the measurement, and that this reproducibility may be represented by the standard deviation (SD) of replicate estimates (assuming that they are normally distributed about the mean), then the detection limit is represented by the concentration (read from the standard curve)

$$B_0 + 2\,\mathrm{SD}\,(B_0),$$

where B_0 represents the binding with no unlabeled antigen present. The minimal detectable concentration of substance in a given assay will be represented by the detection limit of the standard curve multiplied by the dilution of the sample in the assay, (i.e., the ratio of sample volume to assay volume).

According to BERSON and YALOW (1973), the sensitivity is related to the slope of the standard dose–response curve. In this case, the concept of detection limit is extended to the minimal detectable difference between two points located anywhere on the standard curve, actually, the minimal change in concentration that is detectable from any point on the curve. In this case, we must assume that a linear relationship exists between the B/F ratio (or B/B_0 ratio) and the concentration of unlabeled material. If we suppose a linear relationship between these two variables, then it will be of the following type

$$y = ax + b,$$

where a represents the slope of the curve. If x' is a point two standard deviations away from x and $x' > x$, then

$$y' = ax' + b = y + 2\,\mathrm{SD}\,(y)$$

and

$$x' = \frac{y' - b}{a}$$

the detection limit will be given by $x-x'$ and

$$x=\frac{y-b}{a}$$

$$x'=\frac{y'-b}{a}=\frac{(y+2\,\mathrm{SD}(y))-b}{a}$$

$$x-x'=\frac{y-b}{a}-\frac{y+2\,\mathrm{SD}(y)-b}{a}$$

$$=\frac{y-b-(y+2\,\mathrm{SD}(y))-b}{a}=\frac{2\,\mathrm{SD}(y)}{a}.$$

It is evident that the minimal detectable difference between two concentrations on the standard curve is proportional to the slope a of the curve. Assuming that the experimental error is constant over different RIA determinations, then increasing the slope of the dose–response curve results in a reduction of the minimal detectable quantity, i.e., an increase of the sensitivity.

B. Basic Considerations

The theory underlying RIA and other radioligand assay methods has been extensively developed by several groups of workers (ELKINS and NEWMAN 1969, 1970; BERSON and YALOW 1959; YALOW and BERSON 1970; RODBARD et al. 1971; FELDMAN and RODBARD 1971). BERSON and YALOW (1959) have considered the equilibrium theory of RIA in the following cases: (a) monovalent antigen, monovalent antibody; (b) divalent antigen, monovalent antibody; and (c) monovalent antigen and two "orders" of antibody sites. In the present analysis, we will refer to the theory developed by these two authors, assuming that labeled and unlabeled antigen behave identically in the immune system and that the reaction follows the law of mass action. The simplest system is assumed, i.e., the bimolecular reaction between a homogeneous monovalent antigen and a single order of antibody-combining sites. These conditions, however, may not be completely satisfied in any RIA system, as a given antiserum generally contains more than one class of immunoglobulins with different association constants K. Nevertheless, when working at appropriate dilution of antisera, the antibodies with the highest association constant predominate and, from a practical point of view, a single order of combining sites may be assumed to be responsible for the antigen–antibody reaction.

The immunoassay reaction is described by the Berson–Yalow equation

$$B/F = K([\mathrm{Ab}^0] - B), \tag{1}$$

where B represents the molar concentration of bound antigen or of bound antibody-combining sites, F the molar concentration of uncoupled antigen, and $[\mathrm{Ab}^0]$ the molar concentration of total antibody-combining sites (BERSON and YALOW 1973). If b is the fraction of antigen that is bound, so that $B = b\,[\mathrm{Ag}]$, where [Ag] is the total antigen concentration, Eq. (1) becomes

$$\frac{B}{F}=\frac{b}{1-\mathrm{b}}=K([\mathrm{Ab}^0]-b[\mathrm{Ag}])\,. \tag{2}$$

From Eq. (2) it is evident that B/F decreases as the total amount of antigen in the RIA system increases. A further mathematical analysis of the Berson-Yalow equation has been proposed by SAMOLS and BARROWS (1978)

$$b[\mathrm{Ag}]=[\mathrm{Ab}^0]-\frac{b}{1-b}K^{-1}$$

or

$$[\mathrm{Ag}]=\frac{[\mathrm{Ab}^0]}{b}-\frac{K^{-1}}{1-b}\,. \tag{3}$$

In the original assumption of Berson and Yalow, the labeled and unlabeled antigen compete for binding sites with the same energy. Thus, in Eq. (3), we must consider that the total antigen concentration is represented by the unlabeled substance plus the labeled material. The concentration of unlabeled substance in an RIA system is therefore given [from Eq. (3)] by

$$[\mathrm{Ag}]_{\mathrm{unlabeled}}=\frac{[\mathrm{Ab}^0]}{b}-\frac{K^{-1}}{1-b}-[\mathrm{Ag}]_{\mathrm{labeled}}\,. \tag{4}$$

A different equation to describe the antigen–antibody reaction at equilibrium has been developed by FELDMAN and RODBARD (1971).

A divergence from the hyperbola represented by the "classical" equation of Berson and Yalow will occur if a second order of antibody-binding sites of lower affinity constant is present in a given antiserum and participates in the total binding of the labeled material. This problem is minimized for monoclonal antibodies, but is present when conventional antisera are employed. Antibodies suitable for RIA purposes are rarely represented by a single order of combining sites: in other words, the total IgG antibodies which are present in a given antiserum are inhomogeneous. The participation of antibody-binding sites of lower affinity constant K in the total binding of a given antigen leads to a sigmoid shape of the dose–response relationship, rather than a true hyperbola asymptotic to a straight line at high antigen concentrations (SAMOLS and BARROWS 1978).

In the original kinetic analysis, Berson and Yalow assumed that the concentration of labeled antigen was negligible compared with the concentration of antibody-binding sites. From this point of view, an RIA system generally works at extreme antibody excess with respect to labeled tracer. However, when working at high dilutions of antisera which possess high affinity constants, the amount of antigen and the amount of antibody-binding sites are comparable. Under these conditions, $[\mathrm{Ab}^0]\sim[\mathrm{Ag}]$. This means that, as the labeled antigen concentration approaches zero, the limit of the sensitivity that can be achieved with any given antiserum will be represented by the affinity constant K characterizing the reaction of the predominating antibodies. In fact, the Berson–Yalow equation may be recognized as a specific form of the equation

$$y = ax + b$$

expressing a linear relationship between an independent (B) and a dependent variable (B/F) known as the Scatchard plot. It is easy to demonstrate that the slope of this line is equal to $-K$ and, therefore, the sensitivity of this system is regulated by the affinity constant.

Finally, Berson and Yalow did not distinguish between radiolabeled antigen and unlabeled antigen. As a consequence, their analysis requires equilibrium conditions which are not achieved in many RIA systems, owing to the short incubation time. Moreover, it is unable to describe the situation when labeled and unlabeled antigen are added at different times (RODBARD et al. 1971). In this chapter, we will analyze the different variables regulating the sensitivity of the RIA system by an experimental approach, referring the reader to Chap. 8 for a more extensive and detailed mathematical analysis.

C. The Influence of Labeled Tracer

It is evident from the previous discussion that lowering the antigen concentration results in an increase in the slope of the standard dose–response curve and, therefore, in a higher sensitivity of the assay. If the concentrations of antibody and tracer are low, then only a small amount of unlabeled substance is required to displace the tracer from the antibody. An example of the behavior of the sensitivity under these conditions is given in Fig. 1. Three different standard curves were ob-

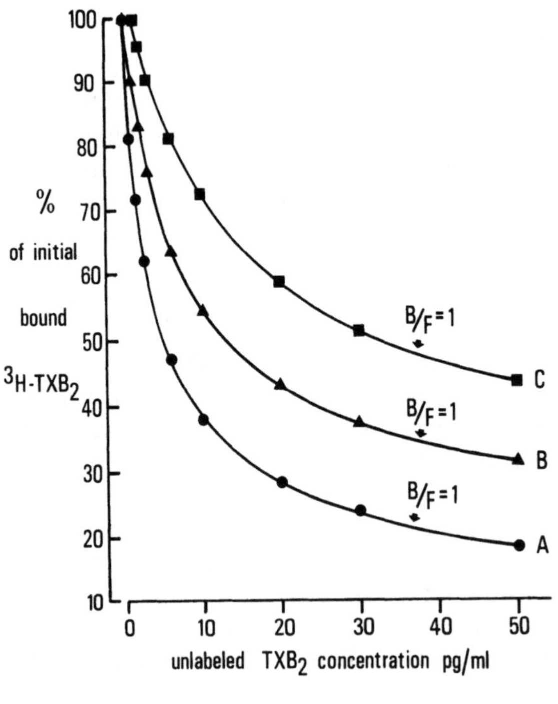

Fig. 1. Standard dose–response curves obtained by incubating an anti-thromboxane B_2 serum with increasing amounts of ^3H-labeled thromboxane B_2: curve *A* 5000 dpm per tube; curve *B* 10000 dpm per tube; curve *C* 20000 dpm per tube. Final antiserum dilution: curve *A* 1:900000; curve *B* 1:500000; curve *C* 1:250000. Ordinate: percentage of initial bound labeled thromboxane B_2; abscissa: standard thromboxane B_2 concentrations

tained by incubating an anti-thromboxane B_2 serum with increasing amounts of labeled thromboxane B_2. Curve *A* was obtained by incubating the antiserum with 5000 dpm ^3H labeled thromboxane B_2 per tube, at a final dilution of 1:900000; curve *B* by incubating a double amount of radioactive material (10000 dpm per tube) under the same conditions with a final dilution of antiserum of 1:500000; and curve *C* after incubation of 20000 dpm per tube with antiserum diluted 1:250000. In any case, the B/F ratio in the absence of unlabeled thromboxane B_2 was approximately 1.0. It is obvious that to maintain $B/F = 1$ at $[Ag]_{unlabeled} = 0$, the concentration of antibody-binding sites $[Ab^0]$ must increase with increase in $[Ag]_{labeled}$. As $[Ab^0]$ increases, the product $K[Ab^0]$ in Eq. (2) increases, and a less sensitive dose–response curve is obtained. As a consequence, on increasing the radioactivity, the slope of the curve decreases. Conversely, reducing the total amount of radioactivity toward zero (i.e., $[Ag]_{labeled} \rightarrow 0$) and concomitantly increasing the antiserum dilution would increase the sensitivity of the immune reaction. The limit of this approach is represented by the possibility of measuring a very low radioactivity with satisfactory precision. The problem is even more complex for β-emitters, which require liquid scintillation counters with a lower efficiency compared with γ-counters. In any case, a compromise must be achieved among several opposite requirements, i.e., optimal sensitivity, optimal precision, low radioactive concentration, low counting efficiency, low experimental error, and, in particular, excessively prolonged counting times. The last problem represents a serious limitation when a large number of samples are to be assayed. In any case, the lowest amount of radioactivity which can be detected with satisfactory precision by the counting equipment should be employed.

As for the specific activity of the labeled tracer, it is obvious that the higher the degree of incorporation of the radioactive isotope into the antigen molecule, the lower will be the concentration required to yield a given number of dpm or cpm in the assay. Increasing the specific activity, if the total dpm in each tube remains constant, leads to a reduction of $[Ag]_{labeled}$ in Eq. (4). If the fraction b of antigen that is bound to antibody is maintained constant, a lower concentration of antibody-binding sites $[Ab^0]$ is required to give the same B/F value and the sensitivity of the system improves. When considering the specific activity of a labeled tracer, two specific problems should be analyzed. A low specific activity may be due either to a low degree of incorporation of the radioactive isotope into the molecule, or to the presence in the labeled tracer of labeled and unlabeled molecules in variable proportions. In the first case, an increase in the specific activity may be achieved only by increasing the number of atoms of the radioactive isotope bound to or substituted within the molecular structure of the tracer. The limit of this approach is represented either by the maximum number of radioactive atoms which can be introduced into a single molecule, or by the emission spectrum of the particular isotope which is employed. In the second case, i.e., contamination of the labeled tracer with unlabeled substance, a sensitive increase in the specific activity may be achieved by increasing the efficiency of the labeling procedure, up to a limit represented by 100% of labeled molecules within the tracer.

Finally, the radiochemical purity of the tracer represents an important factor in limiting the sensitivity of the RIA system. The presence of labeled material, which is not represented by the labeled antigen, but is due to degradation products, introduces into the assay system a variable which cannot be distinguished from $[Ag]_{labeled}$. The RIA only measures the binding of a labeled substance to a given antiserum and the displacement of this binding caused either by the standard or the endogenous substance. If we introduce damaged labeled antigen into the system, we will have an apparent loss of sensitivity, because the total labeled antigen concentration will not be equal to the radioactivity present in the system, but

$$[Ag]_{labeled} = \text{total radioactivity} - \text{damaged } [Ag]_{labeled}.$$

The B/F value is therefore apparently decreased because of the higher value of F, which does not correspond to the true unbound $[Ag]_{labeled}$. As a consequence, a higher antibody concentration will increase the B/F value, thus allowing the RIA system to work in a relative excess of antibody-binding sites $[Ab^0]$. Therefore, the final result of using a tracer with low radiochemical purity is represented by a loss of sensitivity and precision of the assay.

D. The Influence of Antiserum Dilution

The extensive literature dealing with the mathematics of competitive binding radioassay agrees on the convenience of using an antiserum dilution that yields $B/F \leqq 1$ when tracer alone is present without added unlabeled antigen. In the original Berson–Yalow equation, B/F is regulated by both antigen and antibody-binding site concentrations [Ag] and $[Ab^0]$ in Eq. (2). A further mathematical analysis performed by Berson and Yalow demonstrated that sensitivity is maximal at any value of b [Eq. (2)] when [Ag] approaches zero, and is greatest at $b = {}^1/_3$ (BERSON and YALOW 1973). If the fraction of bound antigen is ${}^1/_3$ (33%), then $B/F = 0.5$. At any given concentration of $[Ag]_{labeled}$, it is possible to find an antiserum dilution yielding 33% binding of total labeled tracer. Figure 2 shows the different degrees of sensitivity which can be achieved with a fixed amount of radioactivity with different dilutions of the same antiserum. About 12000 dpm per tube of ^3H-labeled thromboxane B_2 (NEN, 113 Ci/mmol) was incubated with increasing dilutions of anti-thromboxane B_2 serum, yielding 50% (1:600000), 40% (1:800000), and 33% (1:1000000) binding, respectively, in the absence of unlabeled thromboxane B_2. The three standard curves obtained at these different antibody dilutions are represented in Fig. 2 by: curve A (B/F in the absence of unlabeled material = 1.0); curve B ($B/F = 0.66$); and curve C ($B/F = 0.5$), respectively. There is a significant loss of sensitivity with the use of increasing concentrations of antiserum, although the dilutions employed in the present experiment yield B/F ratios confined within a narrow range, which is usually considered optimal for the sensitivity of the RIA system. It is clear that an excess of antibody results in a loss of sensitivity, since a higher number of binding sites must be occupied before the unlabeled antigen can compete with and displace the labeled molecules. From this point of view, the condition of minimal sensitivity of the system is represented by a nearly total binding of labeled tracer to antibody at equilibrium, i.e., a b value in Eq. (2) approaching 1. As it is not possible to obtain total (100%) binding of the labeled tracer without using a very high antibody concen-

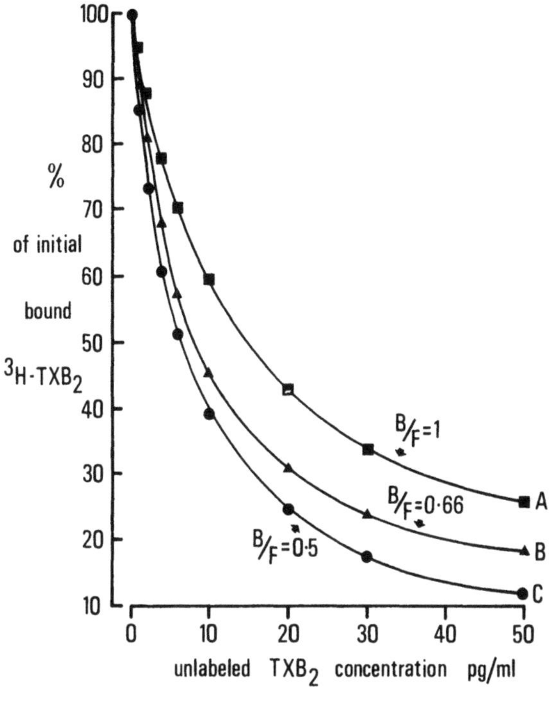

Fig. 2. Variation of a thromboxane B_2 assay sensitivity for a fixed amount of radioactivity, but different dilutions of the antiserum. ^3H-labeled thromboxane B_2: 12000 dpm per tube for all the standard dose–response curves. Final anti-thromboxane B_2 serum dilution: curve *A* 1:600000; curve *B* 1:800000; curve *C* 1:1000000. Ordinate: percentage of initial bound labeled thromboxane B_2; abscissa: standard thromboxane B_2 concentrations

tration, a value of B/F greater than 1 (50% binding) generally represents a relative excess of antibody, with more free binding sites available in the immune reaction. On the other hand, the limit of increasing antiserum dilution indefinitely is represented by excessively poor binding, with a relative excess of labeled tracer in the assay system. Under these conditions, sensitivity and precision decrease. A reasonable limit may be experimentally determined and is generally represented by about 30% binding of labeled tracer in the absence of unlabeled material.

E. The Role of Incubation Volume

From Eqs. (2) and (3), it is evident that at equilibrium B/F, or the fraction of antigen that is bound, is dependent on the concentrations of both antigen and antibody, including in the former both labeled and unlabeled material. From Eq. (2) we can also deduce that if we keep the total amount of labeled antigen constant, we can maintain the B/F ratio constant when modifying the incubation volume by increasing (or decreasing) the antibody concentration by an amount which is dependent on K. Since the slope of the standard dose–response curve is dependent on antigen concentration (as shown in Fig. 1), increasing the incubation volume, but maintaining the total radioactivity constant causes a decrease of antigen concentration in the assay system and, therefore, a steeper slope of the standard curve in the linear region of the dose–response relationship. This means an increase in the sensitivity of the RIA system.

Three ways may be used to reduce labeled tracer concentration in the RIA system: lowering the total radioactivity, increasing the specific activity of the tracer, or increasing the assay volume. The first and second possibilities have been previously discussed. An increase of incubation volume, maintaining the total amount of radioactivity constant, also causes a reduction of antigen concentration. Therefore, to achieve the highest sensitivity of the system, we should use the lowest detectable amount of tracer in the highest convenient volume. A limitation of this approach is represented by the technical problem of counting large volumes of aqueous solutions. This restriction is harder for β-emitters, which require a large supply of scintillation cocktails. The time to reach equilibrium is lengthened proportionally to the increase of incubation volume. Moreover, in many RIA kits, antibody, tracer, and standard solutions are used at fixed volumes, allowing quick handling of samples and reduced cost of the disposable material. However, for research purposes, or in any case to improve both sensitivity and precision, a larger incubation volume is required. For radioactive isotopes which emit γ-rays, a suitable incubation volume is 2–2.5 ml. This allows more precise handling of reagents and the improved sensitivity allows a higher dilution of the samples in the assay. Such requirements are of crucial importance when working with unextracted biologic fluids, such as plasma, serum, synovial fluid, gastric juice, or urine. For labeled tracers which require β-counting equipment, the major limitations are represented by: the size of the vial which contains the mixture with the scintillation cocktail;the quenching of the radioactivity caused by high amounts of buffer solution; and the cost of the liquid scintillator. However, an incubation volume of 1.5 ml can easily be used with tritium-labeled tracers, without marked quenching of the samples and with satisfactory sensitivity and precision (PATRONO et al. 1982).

F. Temperature and pH Effects

The equilibrium constant for the reaction of antigen with antibody is temperature dependent. Studies performed by Berson and Yalow in the early days of RIA show that higher B/F ratios are generally observed when mixtures are equilibrated at 4 °C than at room temperature (20°–25 °C). Temperature may influence the rate of antigen–antibody reaction, the ratio of products (i.e., antigen–antibody complexes) at equilibrium, the binding of free antigen to the adsorbing agents, and the stability of the labeled and unlabeled antigen. The time required to reach equilibrium is longer when an immunologic reaction is carried out at low temperature. To shorten this incubation time, RIA systems are often incubated at room temperature for a few hours. However, this procedure is not recommended. The advantage of reducing the overall time required to perform the RIA is generally counterbalanced by a reduction of sensitivity, precision, and accuracy of the assay. Moreover, incubating the RIA system for a short time may result in disequilibrium or nonequilibrium at the time of separation of antibody-bound from free antigen, with a possible reduction of sensitivity (see Sect. G). It should also be remarked that some dissociation of complexes may occur if mixtures are incubated at 4 °C and the antibody-bound and free antigen are separated at room temperature. For this reason, it is important to maintain incubation mixtures at a constant temperature during addition of specific adsorbent (e.g., in an ice–water bath) and to use a refrigerated centrifuge.

As for the pH of the incubation mixture, it is relevant that antigen–antibody complexes are dissociated by extremes of pH. However, studies performed by Berson and Yalow show that peptide hormone systems do not exhibit a significant pH dependency in the range 7.0–8.5. The pH may be relevant for substances which show limited chemical stability in aqueous solution and may, at extreme values of pH, be converted into hydration, dehydration, or oxidation products which could not be recognized by antisera. A classical example is given by the 15-keto-13,14-dihydro metabolite of prostaglandin E_2, which undergoes spontaneous dgradation in aqueous solution, particularly at high or low pH. The dehydration product is represented by the 15-keto-13,14-dihydroprostaglandin A_2. However, at higher pH, a bicyclic compound, 11-deoxy-13,14-dihydro-15-keto-11,16-cycloprostaglandin E_2 (bicyclo-PGE_2),is formed (GRANSTRÖM et al. 1980). The latter compound is stable and radioimmunologic determination of the systemic metabolite of prostaglandin E_2 has been performed by converting the unstable 15-keto-13,14-dihydro metabolite into the biclycic compound (GRANSTRÖM and KINDAHL 1980).

G. Disequilibrium Conditions

The previously reported analysis of the antigen–antibody reaction refers to equilibrium conditions. It has been clearly demonstrated that if labeled and unlabeled antigen and antibody are added to the incubation mixture at the same time, the sensitivity of the RIA system is maximal when the reaction is carried to equilibrium (RODBARD et al. 1971; EKINS and NEWMAN 1970). It should be stressed that "equilibrium" means a condition which is achieved when the duration of incubation approaches infinity ($t \rightarrow \infty$). This means that a few hours of incubation are generally not sufficient to achieve the condition of maximal sensitivity. In some cases, RIAs have been performed with incubation times of several days, up to 1 week (BERSON and YALOW 1968).

When labeled and unlabeled antigen are added simultaneously to the antibody at time zero, the fraction of labeled antigen bound [b in Eq. (2)] is equal to the fraction of unlabeled antigen bound. If the assay is terminated prior to attainment of equilibrium, the fraction of both labeled and unlabeled antigen bound will be below the equilibrium level. This means that: (a) the number of counts for antigen bound is decreased, with a concomitant increase in counting error; (b) the slope of the standard dose–response curve is lower than the equilibrium curve in the low dose region, i.e., the assay becomes less sensitive and precise; and (c) the unequilibrated system is more sensitive to variations when bound and free antigen are separated, with a consequent increase in the experimental error.

Different conditions may be realized if unlabeled antigen is incubated with antibody until equilibrium is reached and then the labeled antigen is added. This and the following conditions have been extensively investigated by RODBARD et al. (1971). If, after the antibody and unlabeled antigen reaction has reached equilibrium, we add the labeled tracer, there is no labeled antigen–antibody complex present at time zero, but as time progresses there will be dissociation of the unlabeled complex and formation of the labeled antigen–antibody complex. Thus, as time progresses, the fraction of bound labeled antigen incrases progressively, while the fraction of unlabeled antigen progressively decreases, reaching the equilibrium value as time approaches infinity. However, when the assay is terminated prior to attainment of equilibrium, the standard dose–response curve is shifted to the left and its slope becomes steeper in the low dose region. The sensitivity of the assay system, therefore, increases. If the opposite conditions are realized, i.e., if labeled antigen reacts with antibody to equilibrium and then the unlabeled substance is added, the dose–response curves will be shifted to the right compared with the equilibrium curve, because we are approaching the equilibrium value from above the final value, rather than from below. Thus, while all RIA systems at equilibrium give the same dose–response curve, irrespective of the initial conditions, when we are dealing with "nonequilibrium" assays, we can improve or decrease the sensitivity, depending on the kinetics of the system. Some criticism of this approach has been expressed by BERSON and YALOW (1973), since this two-stage RIA results in a significant enhancement of sensitivity only if the rate constant for dissociation of the antigen–antibody complex is small compared with the rate constant for association. However, for labeled peptide substances which are particularly susceptible to progressive damage on prolonged incubation, this procedure may prevent an excessive degree of incubation damage (see Sect. H).

H. Other Factors Affecting Sensitivity

The "incubation damage" of labeled antigen, leading to the formation of "damaged tracer" in the RIA system, represents a serious problem which has been extensively investigated by Berson and Yalow (see STRAUS and YALOW 1977 for detailed discussion). The damage of the labeled antigen during incubation will decrease the fraction of bound tracer and increase the nonspecific counts in the incubation mixture. Thus, the sensitivity and the reliability of the RIA procedure is decreased and, if the extent of damage is serious, the results may be completely invalidated. A method of calculating the amount of damaged antigen has been proposed by BERSON and YALOW (1973). It consists in a control mixture containing the labeled antigen and either the unknown sample or the diluent used for the standards, but without antiserum. In this way it is possible to evaluate the differential damage occurring during the incubation period. This method, however, cannot be used in every case (STRAUS and YALOW 1977). In some cases, the only way to reduce the degree of incubation damage of the tracer is to shorten the incubation time. This, however, contrasts with the prolonged time required to reach equilibrium when antibody and tracer are highly diluted. A possible solution, at least in part, may be represented by the two-stage RIA procedure, as described by RODBARD et al. (1971).

Finally, an apparent reduction of the sensitivity of the assay may be seen when a secondary, nonspecific binding substance interferes with the assay system. This is the case with proteins, such as albumin or other plasma proteins, which may be present at high concentrations in the measured samples (such as plasma, serum, synovial fluid) and act synergistically with antiserum. Generally, the presence of this phenomenon is revealed by a binding of labeled tracer higher than the maximum binding in the absence of unlabeled material. If we add a constant amount of the sample, deprived of the endogenous substance which is specifically recognized by the antibody, to the standard curve, we will observe a shift to the right of the dose–response curve. This loss of sensitivity is due to a secondary binding system. The nonspecific affinity constant of molecules like albumin is generally two or three orders of magnitude lower than the K value of antiserum. Thus, increasing dilutions of the sample progressively lower the nonspecific binding until it completely disappears.

References

Berson SA, Yalow RS (1959) Quantitative aspects of the reaction between insulin and insulin-binding antibody. J Clin Invest 38:1996–2016

Berson SA, Yalow RS (1968) Radioimmunoassay of ACTH in plasma. J Clin Invest 47:2725–2751

Berson SA, Yalow RS (1973) Radioimmunoassay. General. In: Berson SA, Yalow RS (eds) Methods in investigative and diagnostic endocrinology. North-Holland, Amsterdam, vol 2A, pp 84–120

Ekins RP, Newman GB (1969) The optimisation of precision and sensitivity in the radioimmunoassay method. In: Margoulies M (ed) Protein and polypeptide hormones. Excerpta Medica International Congress Series, no 161, pp 329–331

Ekins RP, Newman GB (1970) Theoretical aspects of saturation analysis. In: Diczfalusy E (ed) Steroid assay by protein binding. Acta Endocrinol 64 Suppl 147:11–36

Feldman H, Rodbard D (1971) Mathematical theory of radioimmunoassay. In: Odell WD, Daughaday WH (eds) Competitive protein binding assay. Lippincott, Philadelphia, pp 158:203

Granström E, Kindahl H (1980) Radioimmunologic determination of 15-keto-13,14-dihydro-PGE_2: a method for its stable degradation product, 11-deoxy-15-keto-13,14-dihydro-11 beta, 16-cyclo-PGE_2. Adv Prostaglandin Thromboxane Res 6:181–182

Granström E, Hamberg M, Hansson G, Kindahl H (1980) Chemical instability of 15-keto-13,14-dihydro-PGE_2: the reason for low assay reliability. Prostaglandins 19:933–957

Patrono C, Pugliese F, Ciabattoni G, Patrignani P, Maseri A, Chierchia S, Peskar BA, Cinotti GA, Simonetti BM, Pierucci A (1982) Evidence for a direct stimulatory effect of prostacyclin on renin release in man. J Clin Invest 69:231–239

Rodbard D, Ruder HJ, Vaitukaitis J, Jacobs HS (1971) Mathematical analysis of kinetics of radiologand assays: improved sensitivity obtained by delayed addition of labeled ligand. J Clin Endocrinol Metab 33:343–355

Samols E, Barrows GH (1978) Automated data processing and radioassays. Semin Nucl Med, vol 8 2:163–179

Straus E, Yalow RS (1977) Specific problems in the identification and quantitation of neuropeptides by radioimmunoassay. In: Gainer H (ed) Peptides in neurobiology. Plenum, New York, pp 39–60

Yalow RS, Berson SA (1970) General aspects of radioimmunoassay procedures. In: Proceeding of the symposium on "in vitro" procedures with radioisotopes in clinical medicine and research. Vienna, 8–12 September 1969. International Atomic Energy Agency, Vienna, SM-124/106, pp 455–472

CHAPTER 8

Statistical Aspects of Radioimmunoassay

D. RODBARD, V. GUARDABASSO, and P. J. MUNSON

A. Introduction

The logit-log method (RODBARD and LEWALD 1970) remains the most popular method for RIA data reduction in use today (Fig. 1). It has the virtue of simplicity (RODBARD 1979). The method can be implemented graphically, using special logit-log graph paper. It can be implemented using a small programmable hand-held calculator or computer (DAVIS et al. 1980). Even the smallest microcomputers can easily perform this method with proper weighting and detailed statistical analysis. A weighted linear regression can even be performed by special "macro" programs for popular spreadsheet programs. The logit-log method has been incorporated into numerous commercial systems (β- and γ-counters). When the logit-log method "works" (which is probably about 90%–95% of the time) everything is fine. Unfortunately, in about 5%–10% of assays, the logit-log method fails to provide an adequate description of the RIA dose-response curve. What should the assayist do when the logit-log method fails? Since numerous alternatives are available, each with its own advantages and limitations, the assayist is often faced with a bewildering situation (RODBARD 1979). Are we to try all possible methods, using a trial-and-error approach? When will we be able to say that we have a "good" method, an "adequate" or "optimal" method? Is there a systematic approach? Are we to be restricted to the curve-fitting methods implemented in commercial, turnkey, "black box" approaches to data analysis? This chapter will attempt to provide a simple, rational approach to the question, "what should I do when the logit-log method fails?"

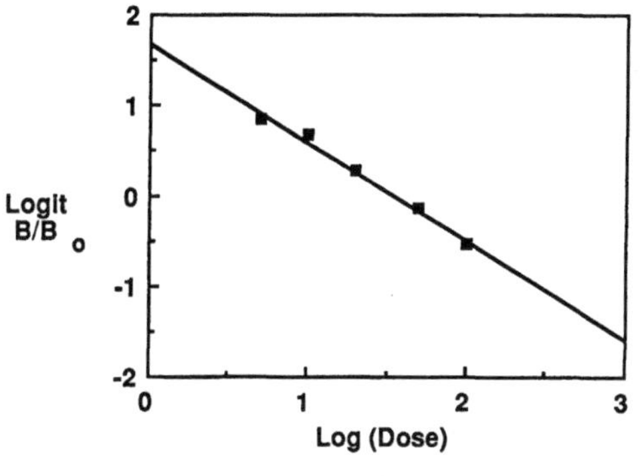

Fig. 1. Logit-log representation of RIA dose-response curve. Linear relationship between logit Y and log X. The dose resulting in $B/B_0 = 0.50$ is termed IC_{50} or ED_{50}, and may also be designated c. The simplicity of the straight line relationship facilitates curve fitting and dose interpolation. However, appropriate weighting must be used to compensate for the systematic nonuniformity of variance introduced by the logit transformation

B. The Logit-Log Method

The logit-log method (RODBARD and LEWALD 1970) is based on a transformation of data defined as

$$\operatorname{logit}(Y)=\log_e\left(\frac{Y}{1-Y}\right) \tag{1}$$

The transformed variable must be $Y=B/B_0$: counts of bound labeled antigen (above nonspecific) divided by counts in the absence of unlabeled antigen (above nonspecific). It reduces a hyperbolic curve (counts versus dose) to a straight line (logit B/B_0 versus log dose) characterized by just two parameters, slope and intercept (Fig. 1). This method and its use in RIA data analysis have been extensively described (RODBARD 1980).

C. Is the Logit-Log Method Failing?

Sometimes, one may think that there is a "failure", when in fact there is no adequate reason to discard the method. Often, the method is implemented incorrectly. One must apply the logit transformation to B/B_0, not to the bound-to-total ratio, B/T. The B/B_0 ratio must be corrected for nonspecific binding N, both in numerator and denominator, i.e.,

$$B/B_0=\left(\frac{\text{counts}-\bar{N}}{\bar{B}_0-\bar{N}}\right), \tag{2}$$

where $\bar{N}$ = mean counts for nonspecific binding; and $\bar{B}_0$ = mean counts for zero dose of unlabeled antigen. Further, the estimates of $\bar{B}_0$ and $\bar{N}$ must be precise and accurate. To reduce the uncertainty in $\bar{B}_0$ and $\bar{N}$, it is desirable that both $\bar{B}_0$ and $\bar{N}$ be measured on the basis of multiple replicates (e.g., quadruplicates). An outlier among the measurements of counts for B_0 or N could result in nonlinearity. The measurement of N can be problematic. Sometimes, in the context of a double-antibody RIA, N is measured simply by omitting the specific (first) antiserum. This changes the protein concentration, and may affect the reaction with the second antibody. As a result, the observed counts may differ considerably from the mean response for "infinite" dose. In such cases, it may be preferable to measure N by addition of a large excess of ligand or antigen, when this is economically feasible (for inexpensive ligands such as thyroxine, cortisol, digoxin, other drugs, etc.). Problems with the measurement of $\bar{B}_0$ or $\bar{N}$ may account for a large portion of the apparent failures of the logit-log method. If these cannot be corrected by modification of the assay procedure, then the four parameter logistic method (see Sect. D) can be used to adjust and improve the estimates of $\bar{B}_0$ and $\bar{N}$ (designated as a and d in the four-parameter model).

One good way to test the adequacy of the logit-log method is to test for linearity. This is easily accomplished by a computer program (e.g., RIAPROG). We can fit two straight lines, one to the first half and one to the second half of the points (Fig. 2a). Then we can use a t-test to evaluate whether the slopes for the

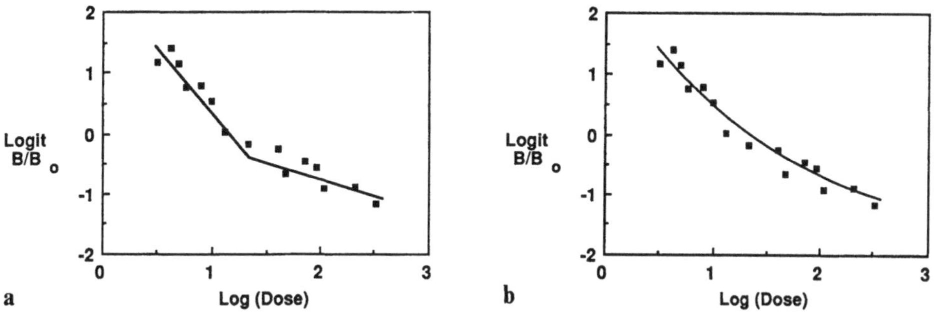

Fig. 2 a, b. Tests for linearity. One should test whether more complicated models, e.g., **a** two separate straight line segments or **b** a parabola, result in an improved fit, in terms of the average size of a deviation (RMS error), or in terms of the randomness of the residuals (deviations of the points from the line)

two line segments are compatible. Alternatively, one can test the two line segments for identity, both in terms of slope and intercept. A closely related method is to fit a straight line to the entire data set, and then fit a parabola (Fig. 2b). If the parabola gives a "statistically significant" better fit than the straight line, then one has evidence of curvature. If the curvature is small, and if the parabola gives a good fit for the entire working range of the assay, then one can use the parabolic or quadratic logit-log method as the basis for dose interpolation. This has been used in a number of systems (RODBARD et al. 1985). However, the parabolic logit-log should be regarded, at best, as a temporary substitute. The parabolic method loses the simplicity, stability, and many of the theoretical advantages of the linear logit-log method. Instead, the parabolic method should be regarded primarily as a test of the linear model. If the linear model fails, then one can use either the symmetric or asymmetric versions of the logistic equation (Sects. D and F.II.3–4). These methods are slightly more computationally demanding than the parabolic logit-log method, but they retain the virtue of being "monotonic," i.e., either increasing or decreasing, as appropriate in most RIA systems. In contrast, the parabola will "wrap around" on itself.

What should we do if the parabolic test or the linear segment test indicates curvature? First, we should determine whether or not it is a consistent finding. Perhaps, one assay will be concave upward and the next assay will be concave downward. If there is no consistent effect, then we may be dealing with random fluctuations, outliers, or minor technical difficulties. If the improvement in terms of goodness of fit (RMS error) is small, we should consider retaining the use of the linear model. Also, it is desirable to perform an analysis of variance, to evaluate whether the error variance for replicates is compatible with the variance of dose-means around the regression line. Larger variance of dose-means around the predicted value than expected on the basis of errors for replicates may imply one of two things: (a) systematic lack of fit of the model; or (b) non-independence of the errors in the observations for the replicates for any given dose level. If there is a discrepancy between the within-dose and between-dose variance, one should use the latter as the basis for testing curvature (comparing the parabola with the straight line). This is tantamount to using only the means at each dose to perform the linear and parabolic regression. This results in a major loss of degrees of freedom. Hence, one may do well to use fewer replicates (e.g., duplicates) at more dose levels. If the curvature is consistently in the same direction, no matter how small, one has strong evidence for lack of fit and the need for a more elaborate, nonlinear model.

D. The Four-Parameter Logistic

When the logit-log model fails, the first recourse should be to try the four-parameter logistic method (Fig. 3; RODBARD and HUTT 1974). The logistic equation has four parameters, a, b, c, and d, corresponding respectively to expected response for $X = 0$, slope factor, ED_{50}, and expected response for infinite dose. This method will solve the problem in most cases. As a matter of fact, the four-parameter logistic method is superior to the logit-log method, both theoretically and in practice, and really should be regarded as the primary, not the secondary method (RODBARD 1974, 1979). However, we began this chapter with consideration of the logit-log method, owing to its popularity, simplicity, historical precedence, and for didactic reasons. The four-parameter logistic method is a generalization of the logit-log method. If the results of the logit-log curve fitting are graphically displayed in the original coordinate system (counts versus log dose) rather than as logit B/B_0 versus log dose, then we see a smooth symmetric sigmoidal curve with an upper and lower plateau. In the logit-log method, the plateaus are defined by $\bar{B}_0$ and $\bar{N}$, the means of the triplicate or quadruplicate counts for zero dose of unlabeled antigen and for nonspecific binding. However, these values could be in error due to an outlier, or due to a technical problem of some kind. Suppose that the B_0 or N tubes were accidentally dropped on the floor, or left out of the assay. Some kit manufacturers even recommend the omission of B_0 and N, to reduce cost in terms of reagents and time. Without B_0 or N, the B/B_0 ratio and logit B/B_0 cannot be calculated. Of course, we might look at the curve, and make an intelligent guess as to where B_0 and/or N ought to be. There may be several doses with responses on (or very near) the upper plateau (Fig. 4a). If we regard the curve in terms of counts versus arithmetic dose (Fig. 4b), we see that we are trying to extrapolate to $X = 0$. Our best estimate of the y-intercept is not just the point at

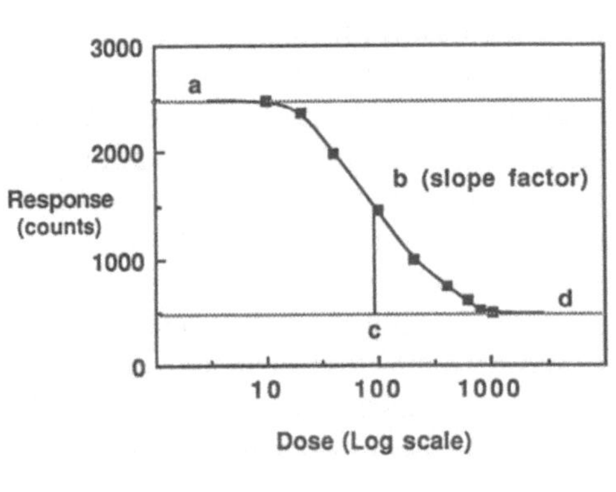

Fig. 3. Schematic representation of the four-parameter logistic model. Note the smooth, symmetric sigmoidal curve of response versus log dose. Parameters: a = expected response when $X = 0$; b = slope factor, corresponding to the slope of a logit-log plot, or the pseudo-Hill coefficient; $c = ED_{50}$ or IC_{50}, i.e., dose with an expected response halfway between the upper and lower plateaus: response = $y = (a + d)/2$; d = expected response for infinite dose

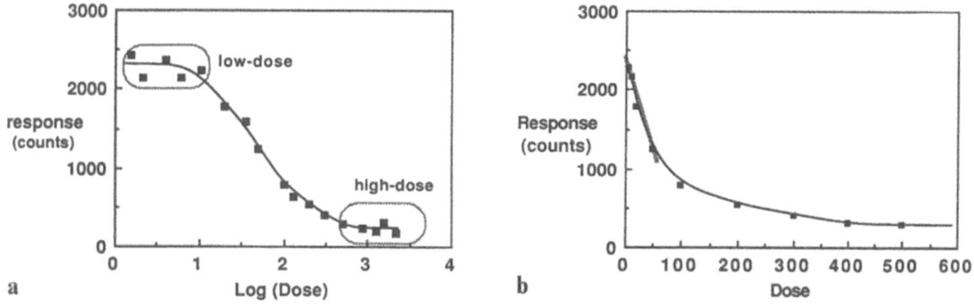

Fig. 4. a All of the values in the low dose region (on the upper plateau) contribute information regarding the parameter *a*. Likewise, all doses in the high dose region contribute information regarding the parameter *d*. **b** With dose shown on a linear scale, we see that all of the low doses in the linear region help to provide an estimate of *a*. This is especially true when *a* must be extrapolated because no observations with $X = 0$ are present

$X = 0$. Instead, we can fit a line (or curve) to all of the points in the vicinity of $X = 0$ (Fig. 4b), and use the combined information to estimate initial binding (now designated *a*). This approach can provide an estimate of *a* even if there was no measured point at $X = 0$.

Likewise we can use the results from all of the "very high dose" points, and extrapolate to obtain an estimate of the high dose plateau, designated *d* (Fig. 4a). Sometimes this will differ appreciably from our observed measurement at $X = \infty$. Of course, it is very difficult to get an "infinite" amount of anything into a small test tube. We never really do. One often encounters solubility problems. For instance, one might have to use a solvent, like ethanol, to get 10^{-3} *M* antigen or ligand into the tube. The ethanol (or salts, or other contaminants) may interfere with the binding reaction. So, it may be better to use the results from 10^{-7} to 10^{-10} *M* (for example), rather than carrying the curve too far to the right. In the four-parameter logistic method, we program the computer to estimate *a* and *d* (corresponding to $\bar{B}_0$ and $\bar{N}$) on the basis of the entire curve, considering all dose levels, not just the zero and "infinite dose" levels. Thus, the four-parameter logistic can be regarded as an "endpoint-adjusted" logit-log.

In a simple, elegant method of calculation developed in this laboratory (cf. Appendix A of Rodbard et al. 1978a) one first performs a logit-log regression. Then, one adjusts the values for the endpoints, also using a linear regression of observed counts versus predicted B/B_0. Then, the logit-log regression is recalculated, using the revised estimates of *a* and *d*. This iterative process is repeated until the values for all four parameters have stabilized, and the (weighted) deviations of points around the curve can no longer be improved, i.e., until the method has converged. This "2 + 2" algorithm first fits two of the parameters (*b, c*), and then fits the other two (*a, d*).

This method is a simple way to conceptualize the fitting of the four-parameter logistic. Also, it is easy to implement in a program, and only requires use of weighted linear regression. However, a more efficient method is to use a generalized nonlinear least-squares curve-fitting method. The problem is regarded as nonlinear, not because the fitted relationship (*y* versus *X* or *y* versus log *X*) is curvilinear, but because the parameters (b, c), appear in a nonlinear form in the equation

$$y = \frac{a-d}{1+(X/c)^b} + d. \quad (3)$$

Programs are available that use the logistic equation for fitting RIA standard curves (NIHRIA, SIGMOID) or families of dose–response curves (ALLFIT) (DELEAN et al. 1978).

Once we have a general program to perform nonlinear regression, then we can deal with curves of almost any shape and degree of complexity. The theoretical bases for such curve fitting cannot be included here, for reasons of space. In essence, one requires a computer algorithm to adjust the parameters so as to optimize the goodness of fit, e.g., so as to obtain the minimum discrepancy between observed and predicted values. However, nonlinear least-squares regression is well treated in many textbooks (MAGAR 1972; BARD 1974; DRAPER and SMITH 1966). The necessary computer programs have been implemented on small inexpensive microcomputers. Programs are available in FORTRAN and BASIC, with or without matrix operators (MCINTOSH 1984; RODBARD 1984). Again, programs can even be developed using the "macro" language of spreadsheets for microcomputers. Suffice it to say that we can simply specify the function, a set of initial estimates of parameters (perhaps obtained from the logit-log method) and a description of the nature of the errors in response (e.g., variance of y proportional to y), and the curve-fitting program will provide: (a) the best estimates of the parameters in order to minimize the weighted sum of squares of deviations; (b) standard errors of the parameters; and (c) evaluation of the goodness of fit by a variety of criteria.

E. Examples of Problems

In some assays, it may be impossible to obtain a measurement of the response for infinite dose (Fig. 5a). The working region of the assay may be substantially below the point where the curve seems to reach a plateau or asymptote and may not even reach the IC_{50} or inflection point. Hence, the logit-log method may not be applicable – we have little or no information regarding N or d. Yet, the four parameter logistic may still suffice: the program can, in effect, try all possible values for d, and select the one that enables it to fit the best smooth symmetric logistic to the entire curve. Sometimes this will work just fine – provided one can make a fairly good initial "guesstimate" of d. Other times, the program may find that d is quite indeterminate. This is especially likely to be the case when a large or even an infinite number of combinations of b, c, and d will result in about the same degree of goodness of fit for the observed region of the curve (Fig. 5a). We would say that the problem is "ill-conditioned": we do not have enough information to estimate all four parameters. In this case, we would like to reduce the complexity of the curve. For example, we might set the slope factor $b = 1$. Or we might set the parameter d to be equal to its highest plausible value, perhaps equal to the total number of counts in the assay tube, and hence the limiting physical

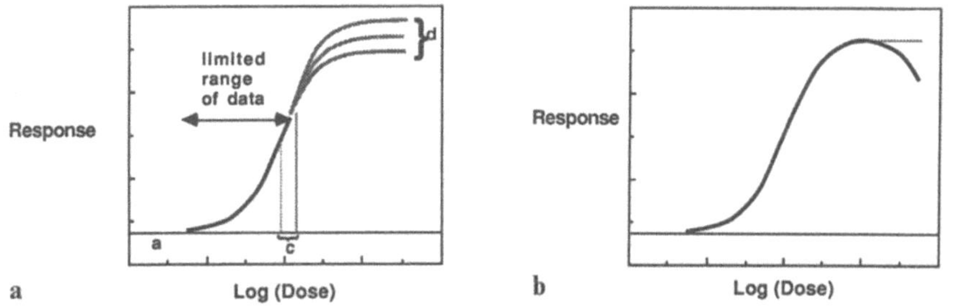

Fig. 5. **a** When observations are obtained over a limited range of dose levels, one or more of the parameters may be ill determined. Here, any of a large number of combinations of b, c, and d might give rise to a satisfactory fit over the observed range. One can often obtain a satisfactory fit by setting d equal to an arbitrary constant well above the highest value, by setting $b = 1$, or by setting $a = 0$. **b** The curve is nonmonotonic. However, a logistic curve can provide an adequate approximation for a major portion of the upstroke

value. By introducing one or another constraint ($b = 1$, or $d =$ *maximum counts or* $a = 0$), we reduce the number of fitted parameters to only three. Then, the available data may be adequate to permit the program to provide a unique best set of estimates of parameters. These parameters can be used for purposes of curve fitting and dose interpolation, but may or may not correspond to the "true" underlying values.

A somewhat more complicated problem is shown in Fig. 5b. Here, the curve is "nonmonotonic", i.e., first increasing and then decreasing. This kind of curve is frequently encountered in "sandwich" assays or "two-site" IRMAs, ELISAs, and ultrasensitive enzyme RIAs (USERIAs). Indeed, this shape can be predicted theoretically (RODBARD et al. 1978b). Later in this chapter, we shall consider better ways to handle this problem. However, for now, we shall suggest a simple expedient which is adequate for many purposes, though certainly not optimal. If we are only interested in the major upstroke of the curve, we may (temporarily) ignore the peak and the decline. The four-parameter logistic may still provide an adequate description of the upstroke. We would need to truncate the range of the data, and then fit the best three- or four-parameter logistic model. After truncation, there may or may not be enough information remaining to obtain a uniquely defined value for d. If not, we may have to introduce a constraint – as in the previous example – to reduce the number of parameters and permit convergence to a reasonable estimate of parameters.

I. Evaluating Goodness of Fit

How can we evaluate the goodness of fit of the four-parameter logistic model or of other nonlinear models? We can no longer simply phrase this in terms of linearity versus nonlinearity, as in the case of the logit-log method (see Fig. 2; RODBARD 1984).

1. Magnitude of Scatter

The magnitude of the scatter of the points around the fitted curve is evaluated in terms of the (weighted) sum of squares of deviation (WSS), the mean square (MS = WSS divided by degrees of freedom) or residual variance, or in terms of the root mean square (RMS = square root of MS). In the unweighted case, the RMS corresponds to the best estimate of the standard deviation of a point around the curve in the vertical direction (in terms of the response variable). This can be compared with the observed mean standard deviation of response for replicates (pooled for all standards or possibly for all standards and unknowns in the assay). It should also be compared with the average RMS in all previous (or perhaps the 20 most recent) assays.

In the case of the use of weighted curve fitting, the RMS has an entirely different meaning. Here, it is the magnitude of the observed error, relative to that predicted on the basis of a specified weighting model and its weighting coefficients. For example, if weights were calculated on the assumption that all of the observations were subject to a 1% error, and if the RMS error were 5, this would imply that the scatter, on the average, corresponded to 5% error (given the assumption that a constant percentage error were indeed applicable; RODBARD et al. 1983).

2. Distribution of Residuals

A large scatter (or RMS) may be due to a single bad point or outlier, it may be due to a larger degree of random scatter than expected, or it may be due to a systematic departure of the points from the fitted curve. How can we discriminate among these three possibilities? The best way is to construct a residuals plot (Fig. 6a), i.e., a plot of the deviations of the points from the curve, versus position on the curve. The deviation may be in absolute terms (e.g., counts or optical density),

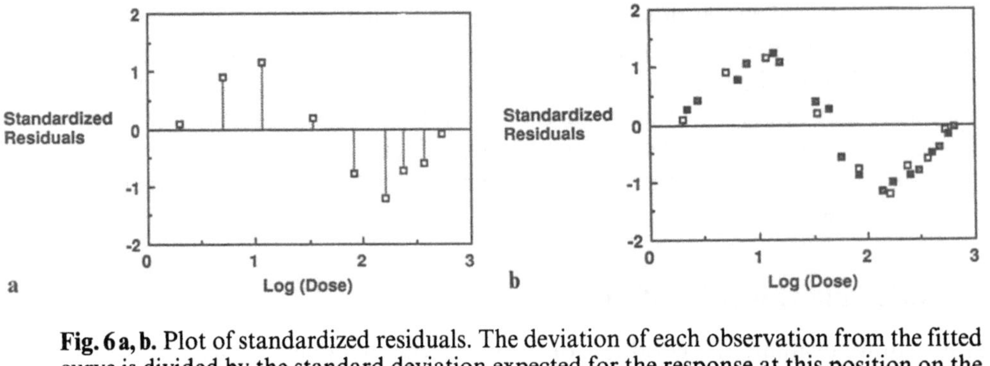

Fig. 6a, b. Plot of standardized residuals. The deviation of each observation from the fitted curve is divided by the standard deviation expected for the response at this position on the curve. If the observations were independent, the residuals should be randomly distributed. The presence of a systematic pattern indicates significant lack of fit (**a**). If the same pattern were observed in several consecutive assays, we would have an indication that the model being used is inappropriate or inadequate. By performing this kind of analysis for the pooled results from several assays (**b**), one may be able to determine the type of modification needed to improve the model.

or expressed relative to the error expected for this position on the curve. The latter is termed the standardized residual (SR). The horizontal axis can be in terms of dose, log dose, the predicted *Y* value, or simply the "standard point number." If the observations were independent, subject to a gaussian distribution, then the residuals plot should have a random appearance. We can also make a histogram of the residuals, and test for consistency with a normal or gaussian distribution. Of course, a display of all of these analyses routinely might lead to an excessive degree of output. Instead, the computer should make this histogram "in its mind," "look at it" (i.e., examine it numerically, not visually), and decide if everything is "OK." If it is "OK," the program can print: "Normality of standardized residuals: OK," or "Normality test: passed." Or it can print nothing. It is only when the normality test is *not* passed that we need to have a warning to the user, perhaps with some additional information, such as: "Normality test not passed, owing to presence of an outlier, point number 14." The program should then recalculate the curve fit without the outlier, and retest the residuals for outliers. This philosophy is similar to that of the BRIGHT STAT-PACK (RODBARD et al. 1983).

The residuals should also be randomly distributed. How can we test for randomness? Again, an experienced assayist can do this by eye. But visual inspection can be misleading. We need objective, statistically valid methods. We recommend the runs test, together with the mean square successive difference (MSSD) test or the serial correlation of residuals (BENNETT and FRANKLIN 1959).

a) Runs test

We can consider each positive deviation as a (+), or as a "head" in a coin toss. Each negative deviation can be termed a (−) or a "tail." Now, we can look at the total number of (+)s and (−)s. If we have 20 points, we would not expect 19 "heads" and only 1 "tail." More sensitive than the total number of (+)s and (−)s, is the number of "runs." For example, we would not expect 10 "heads" in a row. We can define a run as a series of observations of consecutive (+)s, or of consecutive (−)s. We count the total number of runs observed. This can be compared with the number predicted, assuming that the observations were independent. A probability or *p* value can be assigned. Thus, the program can decide if these residuals are or are not compatible with a random series at any desired probability level (e.g., $p < 0.01$). The runs test is excellent when we have a large number of dose levels (e.g., 20). It is not seriously perturbed by one or two outliers or by departure from a normal distribution: it is "robust." However, in the typical RIA setting, where we have only 5 or 10 dose levels, the runs test does *not* perform well; it is not sensitive when aplied to only a few observations. So we turn to one of two alternative but equivalent methods: the MSSD and the serial correlation of residuals.

b) Mean Square Successive Difference Test

The MSSD test evaluates whether the overall or global scatter of the replicates (measured by the standard deviation of all residuals) is consistent with the scatter of the points in local regions of the curve. If the points are oscillating wildly above and below the curve, the local variability (between any two adjacent points) may be larger than the average overall variability. Conversely, if the local variability is much smaller than the overall scatter, this implies a gradual trend or slow oscillation of the points around the curve (Fig. 6a).

c) Serial Correlation Method

This method evaluates the data in terms of the same hypothesis (randomness) from a slightly different point of view. Here, we ask whether knowing the size and direction of a deviation of a point tells us anything about the deviation (or standardized residual) for the two neighboring points. The calculations are very simple. We can plot the deviation for one point on the X-axis, and the deviation for the next point on the Y-axis. This is repeated for all points. Under the null hypothesis, the correlation (and hence the slope of the best fitting line) should be indistinguishable from zero. Accordingly, we simply calculate the slope and its standard error (SE), and the ratio $t = \text{slope}/\text{SE}$. If this is significantly different from zero ($p < 0.05$ or < 0.01), the program provides a warning. A positive correlation implies a slow trend of the points around the fitted curve. A negative correlation implies a "bouncing ball" pattern or rapid oscillation.[1] A computer program for RIA curve fitting should make these tests "internally," and present the results to the user only when there is evidence that an assumption has been violated.

We stress the importance of the routine use of these tests of goodness of fit. They apply to virtually all methods for curve fitting. They should be used to help determine whether any particular model or method is adequate.

II. Pooling of Information Over Assays

Any one assay may show a small degree of lack of fit, but this may not be statistically significant by any of the criteria mentioned so far, because the number of data points is limited. Accordingly, we would like to pool information about randomness of residuals from several assays. We can display the standardized residuals for all recent assays, superimposed (Fig. 6b).

By using "the law of averages" to reduce random scatter, any consistent underlying pattern is more likely to emerge. We can then apply the runs test, MSSD test, and/or serial correlation to the *mean* standardized residual obtained for each dose level in the assay. Any large changes in the positions of the curves for several assays have been stripped away by the individual (e.g., logistic) curve fits. We examine only what is left over, i.e., looking for small, subtle, but systematic departures from the model. This approach can be very effective in routine RIAs, where the curves are likely to be constructed with the same protocol (and dose levels), time after time.

[1] Technical comment. The MSSD test and the serial correlation of standardized residuals are essentially equivalent. The serial correlation involves a more familiar concept and mode of computation, and can be more easily implemented by many workers.

F. Strategy to Deal with Failure of Logistic Models

What should we do if the four-parameter logistic method fails to provide a good fit? Suppose that the criteria we have discussed indicate failure of the logistic model.

I. Does it Matter?

First, we might ask if it matters, i.e., "is it meaningful?" For example, with modern automated assay methods, the error variance can often be reduced dramatically, perhaps (to exaggerate a bit) to a level of $\pm 1\%$ in terms of response. Then, a tiny systematic departure from the model (e.g., $\pm 1.5\%$) might become detectable, especially if we pool results from many assays or curves. But such a small error might have no measurable detrimental effect on the use of the model for dose interpolation or on the use of the assay.

There is a paradox here: as the measurement or between-replicate (within-dose) error becomes smaller and smaller, i.e., as we make the assay more precise, our ability to detect very subtle departures from the model improves, perhaps dramatically (Fig. 7). For example, with a 5%–10% error in the assay, the ability to detect departures from the logistic model may be negligible. In contrast, with 1% –2% errors, very small, systematic, but still not very important departures from the model may be detected relatively easily and frequently (Raab 1983).

Even if the systematic departure is small, we should explore ways to handle it. First, is it an indication of an artifact or technical problem with the assay? Perhaps there was a flaw in the way the standard curve was constructed. Perhaps an inadvertent error was made in the serial dilution of a stock solution of standards, so that results for all concentrations between 10^{-9} and 10^{-7} M were shifted by 10% relative to dose levels between 10^{-7} and 10^{-5} M. Careful scrutiny of the curve and review of procedures can often detect such small errors or accidents.

A careful search should be made to find any such sources of departure from the ideal curve. When all possible technical artifacts have been excluded, then we would be forced to conclude that there is significant lack of fit. The logistic curve

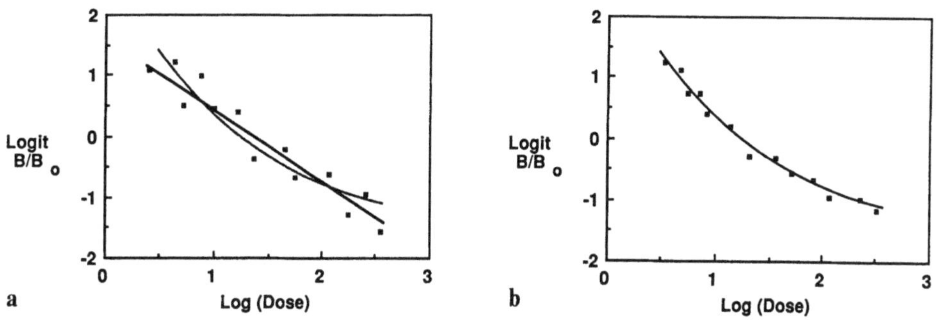

Fig. 7. a Large random errors can obscure significant systematic departures from the model (nonlinearity is present in this case). **b** Conversely, when random experimental errors are small, even very subtle and possibly inconsequential departures from the model become detectable

is, in theory, a *close approximation* to the ideal RIA dose–response curve. But, it is an approximation (FELDMAN and RODBARD 1971; RODBARD 1979; RAAB 1983; FINNEY 1983) and indeed small departures from the model can be expected in theory and are observed in practice.

II. Choice of Other Methods

Later in this chapter, we will consider various ways to enhance the logistic methods, to enable them to deal with asymmetry, nonmonotonicity, or other types of lack of fit. Before doing so, we shall consider two other families of methods: those based on the mass action law and empirical methods.

1. Mass Action Law Models

One can have a perfectly good RIA obeying the mass action law, which does *not* give a perfect fit with the logistic method. This applies, even if the antiserum were monoclonal and the antigen were homogeneous (FELDMAN and RODBARD 1971; RAAB 1983; FINNEY 1983), the reaction reaches equilibrium, and there is perfect separation of bound and free antigen or ligand.

So, if the logistic model is rejected, one should try the mass action law methods. All one has to do is to construct a Scatchard plot, of B/F versus B

$$\begin{aligned} B/F &= (B/T)/(1-B/T) \\ B &= (B/T)(X+X^*), \end{aligned} \tag{4}$$

where B = bound ligand concentration (specific binding only), assuming that labeled and unlabeled ligands have the same B/T ratio; B/F = bound to free ratio for the labeled ligand; B/T = bound to total ratio for labeled ligand (corrected for nonspecific binding); X = concentration of unlabeled ligand; and X^* = concentration of labeled ligand.

For a homogeneous univalent ligand reacting with a homogeneous antibody, this should result in a linear relationship between B/F and B. The slope should be $-K$, where K is the equilibrium constant of association (affinity), and the extrapolated intercept on the B axis should be the maximum binding capacity. If a straight line is obtained with satisfactory goodness of fit, then a simple mass action law model is likely to be satisfactory.

The mass action law offers the advantages of theoretical justification and simplicity. There are only two major parameters – the binding capacity [Ab^0], and the affinity constant K. There are also two minor parameters, i.e., nonspecific binding N and concentration of labeled ligand X^*. As a rule, X^* cannot simply be set equal to zero – usually, more than an infinitesimal concentration is present. If the experimentalist knows the specific activity, then the concentration of labeled ligand can be calculated, and set eqαal to a constant.

One could use a Scatchard plot to estimate K and [Ab^0]. This can even be done graphically, or by use of simple linear regression. These two approaches are, however, only satisfactory if the data are essentially free of error (e.g. $<1\%$ error).

In the presence of an appreciable amount of scatter, one should use nonlinear least-squares curve fitting to estimate K and $[Ab^0]$ (MUNSON and RODBARD 1980) and one should use weighting. This avoids the statistical problems of the Scatchard plot, which arise from the presence of errors in both the B/F ratio and B, and the fact that these errors are highly correlated (RODBARD et al. 1980; MUNSON and RODBARD 1984).

The same kind of nonlinear least-squares curve-fitting algorithm used for the four-parameter logistic model can also be used to fit the parameters of the mass action law models. Since we are using nonlinear regression, we can also regard N and X^* as parameters to be estimated. WILKINS et al. (1978) have advocated the use of a four-parameter mass action law model (fitting K, $[Ab^0]$, N, X^*), and have shown that it works extremely well – sometimes even better than the four-parameter logistic – for a wide range of assays.

This laboratory has used a different "four-parameter" mass action model. We have usually been content to leave X^* and $\bar{N}$ at their measured values. Instead, we provide for the presence of two independent classes of binding sites. Hence, we fit K_1, $[Ab^0]_1$, K_2, and $[Ab^0]_2$. We can fit nonspecific binding N simultaneously as well (RODBARD and TACEY 1978). In principle, the mass action law models could be extended to consider more complex cases, e.g., involving three classes of binding sites, cooperativity, nonidentity of labeled and unlabeled ligands, incomplete bindability[2] or heterogeneity of labeled ligand, and/or systematic errors in the separation of bound and free. Program LIGAND (MUNSON and RODBARD 1980) could be used for curve fitting for several (though not all) of these complex cases. However, for routine RIA data processing, if the simple two- and four-parameter mass action law models do not consistently provide a good fit, then we would recommend the use of another family of methods.

2. Empirical Methods

If the four-parameter logistic and two-, three-, or four-parameter mass action models fail to provide a satisfactory fit, then we would turn to the empirical methods (RODBARD 1979). In their most primitive form, these methods simply connect adjacent points by straight line segments. Many assayists use this kind of approach as their first and only method. It is, after all, a "universal" method, and it is simple. There is only one problem – it gives almost no information about the quality of the fit! The "line" always goes exactly through the points. All of the "residuals" are zero, the RMS is zero, and we cannot test the residuals for randomness. The curve will make an arbitrarily large detour for an outlier. For any local region of the curve, we are using only two points – in effect we are throwing away all of the information which might have been available from the other points on the curve. We also lose the information that, in theory, the curve is supposed to be smooth, continuous, and monotonically increasing or decreasing.

Use of polynomials or spline functions is slightly better than linear segments: these methods provide "smoothness." Both of these methods are fairly easy to

---

[2] The fraction of tracer that can be bound with "infinite excess" of antibody.

compute with small programs readily adapted to microcomputers (RODBARD 1979). One should still use weighting, as in the case of the logistic and mass action law models. Choice of coordinate system is important: should we fit Y versus X, Y versus log X, $1/Y$ versus X, $1/Y$ versus log X, logit Y versus log X, B/F versus B, etc? Each of these coordinate systems has its advantages and limitations, cf. Appendix C of RODBARD et al. (1978 a).

A cubic spline is simply a way to describe local segments of the curve by a cubic polynomial, i.e., $y = a + bx + cx^2 + dx^3$. These curvilinear segments are tied together at "knots," with the constraint that the curve and its first two derivatives must be continuous at the "knot." Some types of splines automatically place a knot at each dose level. Other types of splines require the user to specify the location of the knots. One can force the spline to pass exactly through the data points (an interpolating spline). Or, one can try to find a compromise between closeness to the points and smoothness (smoothing splines). The amount of smoothing should depend on the magnitude of the errors. Improper choice of a smoothing factor can lead to oscillations or overshoot of the curve between the points. There are many algorithms for splines (e.g., WOLD 1974), and one must acquire some experience with them. Properly used, they can still yield quite a bit of statistical information about the assay system, e.g. ED_{50}, effective slope for a tangent to the central segment of the curve, confidence limits for an observation around the curve, and standard errors or confidence limits for potency estimates. A major feature of spline functions is that they will fit nonmonotonic curves. This is an advantage if the curve is *biphasic,* a drawback if the curve is truly monotonic.

3. The Asymmetric Logistic

When the four-parameter logistic model shows a systematic lack of fit owing to asymmetry of the data curve, one can use a model involving just one more parameter, an asymmetry parameter, designated m (Appendix A of RODBARD et al. 1978 a)

$$y=\frac{a-d}{[1+(X/c)^{b}]^{m}}+d. \qquad (5)$$

If $m = 1$, then this model becomes identical to the four-parameter logistic. If $m \neq 1$, then various degrees of mild to moderate asymmetry will be introduced. This model, introduced by PRENTICE (1976) in another context, has been successfully applied to RIA (RAAB 1983). We shall designate it as the five-parameter logistic. Since the parameters b and m can be highly interdependent, we suggest the following modus operandi:

1. Only use this method when you have dose–response curves with a large number of dose levels (e.g., ten or more).
2. Fit the four-parameter logistic (set $m = 1$).
3. Then fit all five parameters (a, b, c, d, m), using the values for a, b, c, d from step 2 as initial estimates.
4. Test whether the five-parameter fit is statistically significantly better than the four-parameter model, by the "extra sum of squares principle" (DRAPER and

SMITH 1966). This is nearly, but not exactly, equivalent to performing a t-test on m, to see whether it is significantly different from unity

$$t=\frac{m-1}{\mathrm{SE}(m)}. \tag{6}$$

If the five-parameter fit is significantly better than the four-parameter fit, we have an indication that the four-parameter model is inadequate.

5. Collect values of m (and their standard errors) for standard curves from 10 or 20 consecutive assays. Test them for homogeneity, e.g., using a χ^2 test. Then, calculate the weighted average of m, and its standard error. Test whether this mean value for m is statistically different from unity. If so, then there is a consistent departure from the four-parameter logistic. If the standard error of m is small, then one can regard the mean value of m as characteristic for the assay system.

6. In future assays, one can use the five-parameter logistic model to fit the curve, with the parameter m "fixed" (constrained) to be constant. This reduces the number of fitted parameters and eliminates the problem of interaction between b and m. In some cases, it may be desirable to constrain b, possibly at unity, and then fit only m. Constraining both b and m corresponds to "freezing" the shape of the curve, so that only the vertical limits (a, d) and the horizontal position (c) are to be adjusted.

7. Periodically, one can and should update and adjust the values of m and/or b.

The five-parameter logistic method shows the flexibility which can be obtained by introduction of just one more parameter. The five-parameter logistic will take care of mild degrees of asymmetry. It is preferable to the use of a parabolic logit-log method, since it retains monotonicity. Comparison with the five-parameter model provides another way to test the adequacy of the four-parameter logistic mode. However, even the five-parameter model will not handle multistep or nonmonotonic curves, or severely asymmetric curves. For these cases, we turn to the next two options.

4. The Two-Slope Logistic

If the four- and five-parameter logistics do not suffice, we could, in principle, continue to add parameters (and complexity) until we get a satisfactory fit. All of this can be done, at the risk of having too many parameters and an "ill-conditioned" curve fit. We should try to select our additional parameters in such a manner that they can be readily estimated from the data which are available (or likely to become available).

We return to consider a logit-log plot. One obvious type of departure from linearity would be to have *two* distinct linear segments (Fig. 2a). Here, we might use one linear relationship to describe the curve in one region, and another straight line to describe the curve in a second region. We could, in effect, process the data for the two regions entirely separately.

We have previously tried to fit two different straight lines to two halves of the data, as a simple test of linearity (Fig. 2a). Of course, the two segments do not necessarily each contain exactly half of the points. So, we would like to permit the breakpoint to occur at any arbitrary dose level (Appendix A of RODBARD et al. 1978a).

With two linear segments, we would have four parameters – two slopes (b_1, b_2) and two intercepts (see Fig. 2a). However, if we require the two lines to come together at a single point, this introduces a constraint which allows us to eliminate one of the intercepts. If the dose at the breakpoint (g) is specified, then one intercept and the two slopes for the two segments will fully define the relationship. If the position of the break in the slope is also unknown, one might try to fit g as well.

The transition from one line segment to the other line segment may be very abrupt – essentially instantaneous – or it may be more gradual. We can allow the slope to shift from b_1 to b_2 by any arbitrary monotonic function. For simplicity, we can make it a smooth logistic function. We would like to be able to control the rate of this transition, and we can do so by introducing yet another parameter (f). Now we have introduced a total of five parameters [b_1, b_2, f, g, and ED_{50} or c from Eq. (3)] to describe essentially a hyperbolic relationship in terms of the logit-log plot. We still have to define the upper and lower plateaus (a, d) corresponding to the upper (100%) and lower (0%) plateau levels before applying the logit transformation [also corresponding to a, d in Eq. (3)]. This makes for a total of seven parameters: a, b_1, b_2, c, d, f, g

$$Y=\frac{a-d}{1+(X/c)^{b_1}\left[\frac{1+(X/g)^f}{1+(c/g)^f}\right]^{[(b_2-b_1)/f]}}+d\,, \qquad (7)$$

where a, b_1, c, d correspond to a, b, c, d of the original four-parameter logistic, b_2 defines the slope of the second segment, g defines the dose at the breakpoint, and the rate of transition from b_1 to b_2 is controlled by the parameter f.

Given sufficient data with small errors, one could, in principle, fit all seven of the parameters to a single curve from a single assay. However, owing to the large number of highly interactive parameters, instability can become very problematical. Accordingly, we would use the kind of approach described for the five-parameter asymmetric logistic [Eq. (5)], i.e., trying to introduce one parameter at a time, and pooling information from several curves or assays. Equation (7) might be used with two, one, or no parameters fixed (constrained to be constant), and with five, six, or seven parameters to be fitted, respectively. Accordingly, the two-slope logistic may also be designated as the "five-, six-, seven-parameter logistic," in contradistinction to the "four-parameter logistic." We have developed a computer program to facilitate this type of analysis (V. Guardabasso, P. J. Munson, and D. Rodbard, in preparation).

If we can set two or three of the parameters constant, based on our previous knowledge and experience with the 10–20 most recent assays, then we will only have to fit a few (perhaps four) parameters for each new assay calibration curve. No one would want to fit a complex seven-parameter model to a family of curves for each assay. But, if we can do this just once, hold some of the parameters constant, and then return to the use of a relatively simple program for routine use, then the improved flexibility of the model describing the curve shape may be worthwhile. In any particular assay, we might only need to find a, d, c, and perhaps b_1. We could also test whether the use of the seven-parameter model provided any improvement relative to the simpler four- or five-parameter logistic models.

If asymmetry is present, it may be desirable to alter the experimental design, to facilitate more refined characterization of the curve shape. For example, one might use ten closely spaced doses in duplicate, rather than seven widely spaced doses in triplicate. Or, one might widen the range of doses, even beyond the limits customarily needed for assay purposes, to help to characterize the shape of the curve. Having done so, one could later return to the use of a larger number of replicates at fewer doses, located in the range of greatest interest.

The two-slope logistic model also turns out to be useful to characterize at least some nonmonotonic curves. The values of b_1 and b_2 can have different signs: one can be positive, the other negative. This property can be useful to characterize curves for those ELISAs and IRMAs in which a high dose "hook effect" is present (RODBARD et al. 1978b), and cooperative immunoassays (EHRLICH and MOYLE 1983). There is one significant restriction: the curve shape is not infinitely flexible, and the left- and right-hand horizontal plateaus must be at the same level.

III. A "Universal" Approach

The methods described should be adequate for about 99% of immunoassays. But surely we would like a universally applicable method, combining the flexibility of the splines with the statistical properties of the constrained four-parameter logistic, in terms of its ability to combine information from multiple curves or to test curves for similarity of shape (congruence) and slope (parallelism). We would like to incorporate the virtues of the asymmetric logistic models, in terms of their ability to identify and characterize more complex curve shapes. However, we do not want to have to be faced with equations of ever increasing complexity. We have previously outlined the principles of such an approach (cf. Appendix C of RODBARD et al. 1978a).

We have recently developed a method (four-parameter constrained smoothing spline) and computer program (GUARDABASSO et al. 1987) to characterize curves of arbitrary shape and complexity. The dose–response curves from several assays can be analyzed simultaneously, allowing for normalization in the vertical direction (in terms of maximal and minimal responses, corresponding to parameters a and d), superimposing the curves by shifting them horizontally according to their relative potency (c). A fourth parameter (b) can be used to "stretch" the horizontal axis, i.e. to adjust the slopes of curves that are not parallel. The computer program can find the best estimates for the four parameters (a, b, c, d), for each curve. The common shape for all curves is described by a constrained spline function. The program provides for weighted curve fitting, if necessary.

This method may be ideal for many assays (e.g., IRMA, ELISA, USERIA, BIOIRMA, and cooperative immunoassay) with complex shapes for the dose–response curves. It combines the virtues of the empirical methods (such as the splines) with the advantages of modeling. As such, it represents a new and powerful method, not just for immunoassay, but for analysis of all kinds of dose–response curve.

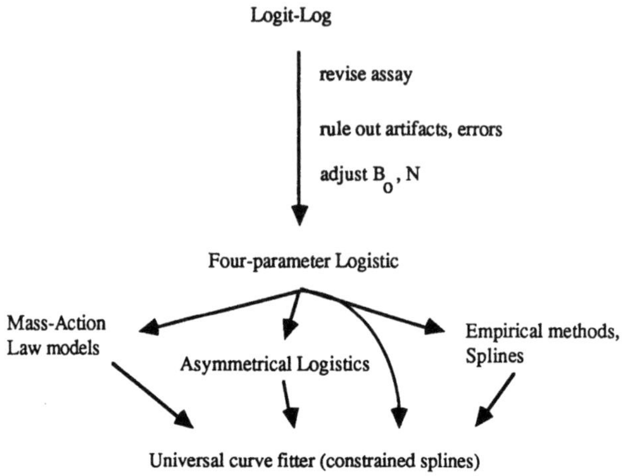

Fig. 8. Strategy to deal with a failure of the logit-log method. Many alternatives are available. Usually, the four-parameter logistic method is adequate. When this fails, we turn to mass action law models (Sect. F.II.1), empirical methods (Sect. F.II.2), two types of asymmetric extension of the logistic model (Sects. F.II.3 and 4). Finally, the four-parameter constrained splines approach provides a definitive, general method to fit curves of arbitrary shape and degree of complexity when simpler methods fail. This method combines the advantages of both the empirical methods (e.g., smoothing splines) and the model-based approaches (e.g., four-parameter logistic)

G. Concluding Remarks

The logit-log method, and the closely related four-parameter logistic method, are usually sufficient and optimal for routine RIA data processing. We have compared the advantages and limitations of these approaches. We have also provided a strategy to deal with the relatively rare cases where the logit-log and logistic methods fail (Fig. 8). This requires consideration of the mass action law models, spline methods, some extensions of the logistic, and a new method, combining the virtues of the spline and the logistic families of methods.

Acknowledgment. Mrs. Annette Kuo and Mrs. Mary Keffer provided excellent secretarial assistance.

References

Bard Y (1974) Nonlinear parameter estimation. Academic, New York
Bennett CA, Franklin NL (1954) Statistical analysis in chemistry and the chemical industry. Wiley, New York
Davis SE, Jaffe ML, Munson PJ, Rodbard D (1980) Radioimmunoassay data processing with a small programmable calculator. J Immunoassay 1 (1):15–25
DeLean A, Munson PJ, Rodbard D (1978) Simultaneous analysis of families of sigmoidal curves: applications to bioassay, radioligand assay and physiological dose-response curves. Am J Physiol 235:E97–E102
Draper NR, Smith H (1966) Applied regression analysis. Wiley, New York

Ehrlich PH, Moyle WR (1983) Cooperative immunoassays: ultrasensitive assays with mixed monoclonal antibodies. Science 221:279–281

Feldman HA, Rodbard D (1971) Mathematical theory of radioimmunoassay. In: Odell WD, Daughaday WH (eds) Competitive protein binding assays. Lippincott, Philadelphia, chap 7, pp 158–203

Finney DJ (1983) Response curves for radioimmunoassay. Clin Chem 29:1762–1766

Guardabasso V, Rodbard D, Munson PJ (1987) A model-free approach to estimation of relative potency in dose-response curve analysis. Am J Physiol: Endocrinology and Metabolism 252(15):E357–E364

Magar ME (1972) Data analysis in biochemistry and biophysics. Academic, New York

McIntosh JEA (1984) Overview of mathematical modeling with computers in endocrinology. In: Rodbard D, Forti G (eds) Computers in endocrinology. Raven, New York, p 37

Munson PJ, Rodbard D (1980) Ligand: characterization of ligand binding systems: a versatile computerized approach. Anal Biochem 107:220–239

Munson PJ, Rodbard D (1984) Computerized analysis of ligand binding data: basic principles and recent developments. In: Rodbard D, Forti G (eds) Computers in endocrinology. Raven, New York, pp 117–145

Prentice RL (1976) A generalization of the probit and logit methods for dose-response curves. Biometrics 32:761–768

Raab GM (1983) Comparison of a logistic and a mass-action curve for radioimmunoassay. Clin Chem 29:1757–1761

Rodbard D (1974) Statistical quality control and routine data processing for radioimmunoassays (RIA) and immunoradiometric assays (IRMA). Clin Chem 20:1255–1270

Rodbard D (1979) Dose interpolation for radioimmunoassays: an overview. In: Natelson S, Pesce AJ, Dietz AA (eds) Clinical immunochemistry, chemical and cellular bases and applications in disease, vol 3. American Association for Clinical Chemistry, Washington, DC, pp 477–494

Rodbard D (1980) Mathematics and statistics of immunoassay. In: Langan J (ed) Proceedings of a symposium on the use of non-isotopic methods in immunoassay. Masson, New York, chap 3, pp 45–101

Rodbard D (1984) Lessons from the computerization of radioimmunoassays: an introduction to the basic principles of modeling. In: Rodbard D, Forti G (eds) Computers in endocrinology. Raven, New York, pp 75–99

Rodbard D, Hutt DM (1974) Statistical analysis of radioimmunoassays and immunoradiometric (labeled antibody assays: a generalized, weighted, iterative least squares method for logistic curve fitting. In: Proceedings, symposium on radioimmunoassay and related procedures in medicine. International Atomic Energy Agency, Vienna, pp 165–192 (Unipub, New York)

Rodbard D, Lewald JE (1970) Computer analysis of radio-ligand assay and radioimmunoassay data. Karolinsky symposia, research methods reproductive endocrinology, Geneva, March 1970. Acta Endocr 64, Suppl 147:79–103

Rodbard D, Tacey RL (1978) Radioimmunoassay dose interpolation based on the mass action law with antibody heterogeneity. Anal Biochem 90:13–21

Rodbard D, Munson PJ, DeLean A (1978a) Improved curve-fitting, parallelism testing, characterization of sensitivity and specificity, validation, and optimization for radioligand assays. In: Radioimmunoassay and related procedures in medicine. International Atomic Energy Agency, Vienna, vol 1, pp 496–504 (Unipub, New York)

Rodbard D, Feldman Y, Jaffe ML, Miles LEM (1978b) Kinetics of two-site immunoradiometric ("sandwich") assays. II. Studies on the nature of the "high dose hook effect." Immunochemistry 15:77–82

Rodbard D, Munson PJ, Thakur AK (1980) Quantitative characterization of hormone receptors. Cancer 46(12):2907–2918

Rodbard D, Cole BR, Munson PJ (1983) Development of a friendly, self-teaching, interactive statistical package for analysis of clinical research data: the BRIGHT STAT-PACK. In: Dayhoff RE (ed) Proceedings of the seventh annual symposium on computers in medical care. IEEE Computer Society, Silver Spring, MD, pp 701–704

Rodbard D, Foy J, Munson P, Zweig M, Lewis T (1985) A personal-computer-based system for acquisition and analysis of radioimmunoassay data. Proceedings, symposium on computer applications in medical care. IEEE Computer Society Press, Silver Spring, MD, pp 455–459

Wilkins TA, Chadney DC, Bryant J, Palmstrom SH, Winder RL (1978) Nonlinear least-squares curve-fitting of a simple theoretical model to radioimmunoassay dose-response data using a minicomputer. In: Radioimmunoassay and related procedures in medicine, vol 1. International Atomic Energy Agency, Vienna, pp 399–423 (Unipub, New York)

Wold S (1974) Spline function in data analysis. Technometrics 16:1–11

CHAPTER 9

Validation Criteria for Radioimmunoassay

C. PATRONO

A. Introduction

From a pharmacologist's point of view, radioimmunoassay (RIA) represents a highly sophisticated form of bioassay, substituting the classical smooth muscle strip with a soluble antibody as the biological reactant. Both assay methods involve indirectly assessing the concentration of the unknown by comparing its effects on a measurable read-out of the reaction evoked (changes in radioactivity, tension, optical density, etc.) with those of known concentrations of a standard. Many bioassay techniques involve measuring biological responses that are mediated by agonist–receptor interactions. Such interactions usually occur in accordance with the law of mass action, similarly to antigen–antibody reactions. The presence of a radioactive tracer in RIA greatly enhances the sensitivity of the detection by limiting the mass of the antigen involved, to an extent inversely related to its specific activity. The specificity is also potentially increased in RIA vis-à-vis bioassay by virtue of a smaller number of possible reagents being present in a highly diluted antiserum than in isolated cell preparations or whole tissue fragments. Obviously, biological activity of the unknown is inherently measured by the latter, but not necessarily by the former. Thus, what is measured by RIA is immunochemical behaviour which may or may not be related to parts of the molecule responsible for biological activity (YALOW 1982).

To those interested in basic or clinical pharmacology, RIA and bioassay may offer unique opportunities, though both methods suffer from limitations. Thus, RIA has the potential for easily detecting biologically inactive metabolites of drugs or endogenous autacoids. However, the chemical instability of some biologically active substances may preclude their detection by RIA. It was through classical bioassay techniques that many fundamental discoveries were made in the field of prostaglandins and related eicosanoids, although it is through RIA and gas chromatography–mass spectrometry (GC–MS) that some understanding is being gained of their pathophysiologic significance in humans (see Chap. 18).

Because of the similarities outlined above, it must be appreciated that the biological read-out of RIA, i.e., variably reduced binding of the labelled antigen to a presumably "specific" antiserum as a function of increasing concentrations of one or more unlabelled antigens, is liable to many forms of artifact potentially affecting bioassay. Thus, it is important to establish validation criteria for the critical assessment of RIA measurements, with the understanding that these should be employed whenever a new technique is developed or a different type of biological fluid is being analysed by an established technique.

B. Conditions Necessary, but not Sufficient for Establishing the Validity of RIA Measurements

An important requirement of all assay methods is that a known amount of the compound of interest added to the biological fluid to be assayed be recovered quantitatively. In the case of RIA, this would simply imply verifying that the exogenous antigen is not being degraded by enzymes or modified chemically, and that the antigen is not being prevented from binding to the antibodies by other binding proteins present in the biological sample (e.g., plasma proteins, endogenous antibodies, hormone-binding proteins). When dealing with substances that are chemically modified in plasma, a possible solution is represented by the development of antibodies to the chemically stable derivative. Such an approach is described by PESKAR et al. (p. 460) for the bicyclic derivative of 13,14-dihydro-15-keto-PGE_2 and by NYBERG et al. (p. 235) for Met-enkephalin sulfoxide.

Proteolytic degradation of some peptide hormones can be prevented or reduced by drawing blood samples into tubes containing protease inhibitors such as aprotinin. The problem of interfering steroid hormone-binding proteins is discussed by PAZZAGLI and SERIO (p. 367).

A similarly necessary but insufficient condition for validating RIA measurements is represented by immunochemical identity of standard and unknown (YALOW 1985). This requires that the apparent concentration of the measured antigen be independent of the dilution at which it is assayed. A classical way of demonstrating immunochemical identity of standards and unknowns is represented by superposability of their dilution curves over at least a 100-fold range of concentrations (YALOW 1985). Failure to use an adequate dilution range and/or a linear plot of the data may obscure lack of superposability (YALOW 1985). Nonparallelism of dilution curves obtained with standard and unknown may reflect a variety of nonspecific as well as specific sources of interference with the immune reaction. These include: (a) the variable effect of pH, ionic strength and chemical composition (including the presence of anticoagulants such as heparin) of the incubation medium of standards and unknowns; (b) variable degradation of labelled and unlabelled antigen and/or antibody during the incubation period; (c) the use of heterologous peptide standards; (d) the presence of more than one immunoreactive form of many peptide hormones in plasma and tissue extracts, possibly reflecting the existence of precursors and/or metabolites of the biologically active moiety; and (e) the presence of structurally related substances in the biological fluid of interest, showing varying cross-reactivities with the particular antiserum employed (YALOW 1985). The influence of these various factors on the antigen–antibody reaction is discussed by CIABATTONI (p. 58) and by EZAN et al. (p. 172).

Attention should be drawn to the fact that all of the sources of nonspecific or specific interference with the immune reaction are related to one or both of the following situations: (a) the same molecular species is present in standards and unknowns, but it reacts with the antibody in a different milieu; (b) the antibody is confronted with a single antigen or hapten in the solution containing the standard, whereas it reacts with a complex mixture of closely related antigens in the solution containing the unknown. While the former can be adequately dealt with by ensuring that standards and unknowns are reacted with antibody in the same incubation medium, the latter requires a diversified strategy essentially centered on separation (by extraction and chromatographic procedures) of the immunoreactive species of interest from the heterogeneous materials present in the same biological fluid. No rigid requirement should be defined a priori, but rather the degree of sample purification prior to RIA should be individually evaluated depending upon the particular substance to be measured, the particular antiserum employed and the nature of the biological fluid.

Measurement of the same substance in different biological fluids with the same antiserum may require entirely different work-up procedures prior to RIA. Thus, measurement of thromboxane (TX) B_2 in serum (range of concentrations 200–400 ng/ml) can be performed in highly diluted (1:5000 to 1:15000) unextracted samples because of the relative abundance of this eicosanoid vis-à-vis other potentially cross-reacting arachidonate metabolites (PATRONO et al. 1980). On the other hand, measurement of TXB_2 in urine (range of concentrations 30–100 pg/ml) requires extraction and silicic acid column chromatography prior to RIA (CIABATTONI et al. 1979) because of the presence in urine of a 10- to 100-fold excess of at least 20 different TXB_2 metabolites (ROBERTS et al. 1981).

Validation of the TXB_2 RIA for the assay of unextracted serum samples should not be extrapolated to measurements of the same compound in unextracted urine, as the latter will grossly overestimate the true concentrations because of cross-reacting metabolites.

Demonstrating that a known amount of standard added to the biological fluid of interest is recovered quantitatively and that displacement curves obtained with serial dilutions of standard and unknown can be superposed does not necessarily establish chemical identity of the two. Thus, additional validation criteria should be used whenever these basic requirements can be met.

C. Appropriate Biological Behaviour of the Measured Immunoreactivity

In the early days of RIA, when these techniques were being developed for the measurement of peptide hormones in plasma, one important criterion for the validity of these measurements was that the immunoreactive material be undetectable at an appropriate interval following surgical ablation of the secreting gland (YALOW 1985). The later recognition of multiple sites of synthesis of the same hormone has somewhat limited the usefulness of this criterion. The increase or decrease of the circulating or urinary immunoreactivity following appropriate physiologic stimuli or pharmacologic interventions was also frequently used to validate RIA measurements of a particular hormone or autacoid. Thus, the opposite effects of hyperglycemia on pancreatic versus gut glucagon were used initially to characterize the nature of plasma glucagon-like immunoreactivity (C. Patrono and R. S. Yalow 1970, unpublished work). Limitations inherent in this approach are represented by the limited availability of highly selective physiological or pharmacological manoeuvres specifically affecting the synthesis and release of a particular substance.

The use of cyclooxygenase inhibitors (e.g. indomethacin, aspirin) to demonstrate suppression of PG-like or TX-like immunoreactivity in plasma or urine is often quoted as a useful validation criterion for RIA of various eicosanoids originating through the cyclooxygenase pathway of arachidonate metabolism. While failure of the measured PG-like or TX-like immunoreactivity to decrease after administration of indomethacin or aspirin (in variable dosage, depending upon the cellular target involved) would strongly argue for the measured compound or compounds being unrelated to cyclooxygenase activity, the reverse, i.e. the expected reduction, does not necessarily validate the measurement as being specific for PGX or PGY, but simply indicates that the immunoreactive material is likely to be derived from arachidonate via the cyclooxygenase pathway. Thus, as an example of the limitation of this particular criterion, serum 6-keto-$PGF_{1\alpha}$-like immunoreactivity is profoundly suppressed by aspirin and is enhanced by a TX synthase inhibitor, such as dazoxiben, because of increased availability of platelet-derived PG-endoperoxides (PATRIGNANI et al. 1984). Although consistent results were obtained with four different anti-6-keto-$PGF_{1\alpha}$ sera (PATRIGNANI et al. 1984), direct comparison of RIA and GC–MS measurements (PEDERSEN et al. 1983) revealed a substantial quantitative artifact in estimates of 6-keto-$PGF_{1\alpha}$-like immunoreactivity, possibly due to combined cross-reactivities of PGE_2, PGD_2, and $PGF_{2\alpha}$. While this type of problem would not invalidate the main point of the observation, i.e. enhanced PGI_2 synthesis by white blood cells as a consequence of platelet TX synthase inhibition by the drug, it may misrepresent the quantitative aspects of this rather interesting phenomenon.

Interestingly enough, even methodological artifacts occasionally exhibit up- and downregulation by allegedly physiological mechanisms, as reported for the effect of dietary sodium content on plasma PGA-like immunoreactivity in normal humans (ZUSMAN et al. 1973). The use of different pharmacological manoeuvres in vitro to demonstrate "appropriate behaviour" of cyclic nucleotide measurements is discussed by PARKER (p. 513).

D. Limited Cross-Reactions of Structurally Related Substances

In describing a new RIA technique it is necessary to include information on the cross-reactivity of the particular antiserum employed with substances structurally related to the primary antigen or hapten. Limited cross-reactions (1% or lower) of a list of 5–10 different compounds is often used as an argument to claim "specificity" of the RIA measurements. However, it is important to realize that such a list of compounds tested is usually far from being complete simply because some structurally related compounds may not have yet been discovered or made available. Thus, historically, the first papers to describe RIA measurements of $PGF_{2\alpha}$ and PGE_2 (JAFFE et al. 1973) in the early 1970s usually listed cross-reactivity data with compounds, such as $PGF_{1\alpha}$, PGE_1, PGA_1, or PGA_2, which are either not formed to any substantial amount in vivo or merely represent artifacts. The later discovery of TXA_2 (HAMBERG et al. 1975) and PGI_2 (GRYGLEWSKI et al. 1976) and the corresponding stable hydration products and enzymatic derivatives (reviewed

Table 1. Immunological specificity of antisera directed against 6-keto-$PGF_{1\alpha}$ (PATRONO et al. 1982)

Substance measured	Relative cross-reaction (%)			
	AS 1	AS 2	AS 3	AS 4
6-keto-$PGF_{1\alpha}$	100	100	100	100
6-keto-PGE_1	1.9	16.6	0.8	18.1
PGE_2	0.3	1.0	0.2	2.1
PGD_2	0.2	0.3	0.2	1.0
13,14-Dihydro-6,15-diketo-$PGF_{1\alpha}$	0.1	0.7	8.7	3.3
6,15-Diketo-$PGF_{1\alpha}$	0.06	0.8	3.5	1.4
$PGF_{2\alpha}$	0.05	2.0	0.3	8.0
2,3-Dinor-6-keto-$PGF_{1\alpha}$	0.006	0.004	18.5	14.6
TXB_2	0.003	0.006	0.004	0.03

by FITZGERALD et al. 1983) made it clear that these were much more relevant compounds to monitor for potential cross-reactions.

Unfortunately, immunological characterization of anti-eicosanoid sera is often limited to primary PGs and TXB_2, with limited information, if any, on the cross-reactivity of major plasma or urinary metabolites. Such limited information may be entirely misleading, as outlined in Table 1 describing the immunological specificity of four different antisera raised against 6-keto-$PGF_{1\alpha}$ in different laboratories (PATRONO et al. 1982). Thus, if one looks at relative cross-reactions of AS 3 with 6-keto-PGE_1, PGE_2, PGD_2, $PGF_{2\alpha}$, and TXB_2, one might be induced to characterize this particular antiserum as being "highly specific" for 6-keto-$PGF_{1\alpha}$. If, on the other hand, the relevant plasma and urinary metabolites are examined for cross-reactivity, then one has a completely different picture, i.e., that of a relatively poor immunological specificity. Obviously the same antiserum may allow perfectly valid measurements of 6-keto-$PGF_{1\alpha}$ when applied to the incubation medium of vascular fragments or to the culture medium of endothelial cells, because only limited formation of those metabolites occurs in vitro. However, when applied to measurement of 6-keto-$PGF_{1\alpha}$ in urine, this antiserum will grossly overestimate its true concentrations because of the 18.5% cross-reaction with 2,3-dinor-6-keto-$PGF_{1\alpha}$ i.e. a major enzymatic metabolite of PGI_2 excreted in a 5- to 10-fold excess vis-à-vis the parent compound (PATRONO et al. 1982). Inasmuch as urinary 6-keto-$PGF_{1\alpha}$ largely reflects renal PGI_2 synthesis, while 2,3-dinor-6-keto-$PGF_{1\alpha}$ is a reflection of extrarenal vascular PGI_2 production (FITZGERALD et al. 1983), the use of incompletely characterized antisera may generate quantitatively as well as qualitatively misleading information on drugs differentially affecting cyclooxygenase activity in the renal versus extrarenal vascular compartments.

Thus, such methodological problems will obscure the effects of selective cyclooxygenase inhibitors, e.g. sulindac (CIABATTONI et al. 1984), low dose aspirin (PATRIGNANI et al. 1982) and sulphinpyrazone (CATELLA et al. 1984) on the kidney

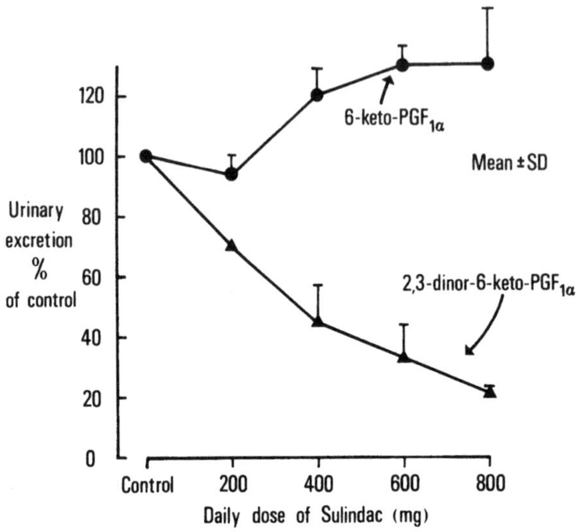

Fig. 1. Urinary excretion of 6-keto-PGF$_{1\alpha}$ and 2,3-dinor-6-keto-PGF$_{1\alpha}$ before and during chronic dosing with sulindac (200–800 mg/day, each dose given for 7 days in successive weeks) in seven healthy subjects. Mean (±SD) percentage changes versus control measurements are depicted for both compounds. (Data from CIABATTONI et al. 1987a)

by virtue of the reduced excretion of cross-reacting systemic metabolites masking the continued synthesis and excretion of renal prostaglandins (PATRONO and DUNN 1987). The long-term administration of sulindac reduces dose-dependently the urinary excretion of 2,3-dinor-6-keto-PGF$_{1\alpha}$ in healthy subjects (CIABATTONI et al. 1987) with no appreciable effects on urinary 6-keto-PGF$_{1\alpha}$ throughout the whole dose range (Fig. 1). A remarkably similar pattern of differential inhibition is obtained with sulphinpyrazone (CATELLA et al. 1984; PEDERSEN and FITZGERALD 1985). The selective sparing of renal cyclooxygenase activity by both redox prodrugs would be partially or totally obscured by improperly validated RIA measurements because of these considerations.

E. Use of Multiple Antisera

Conceptually, the use of multiple antisera in order to enhance the likelihood of specifically recognizing a given substance is similar to the cascade superfusion of several assay organs arranged in series, as originally developed by FERREIRA and VANE (1967), whereby the specificity of the biological recognition is enhanced as a function of the unique responses of the different reactants. The pattern of immunological specificity of each antiserum is sufficiently unlikely to be reproduced by chance as to allow its use as a unique biological probe. Very similar concentrations of substance X measured by three or four different antisera in a given biological fluid are more likely to be explained by the same substance being recognized homogeneously than by a variable mixture of substance X plus cross-reacting substances W, Y, Z giving the same global signal with different antisera. As with other criteria described already, however, unexpected results will ring a bell while the expected fulfilment of this criterion cannot be taken to prove the validity of RIA measurements of substance X. Furthermore, this approach may yield different results when applied to the same biological fluid obtained under different experimental conditions, as illustrated by our studies of urinary 6-keto-PGF$_{1\alpha}$-

like immunoreactivity measured by four different antisera before and after furosemide injection in healthy women (PATRONO et al. 1982). Thus, while 2- to 6-fold higher excretory values were measured by AS 2, AS 3, and AS 4 as compared with AS 1 (Table 1) in basal urine collections, quite similar results were obtained by the four different antisera in the highly diluted urine obtained after furosemide injection (PATRONO et al. 1982).

The use of three different anti-parathyroid hormone sera allowed, for the first time, characterization of its immunochemical heterogeneity in plasma (BERSON and YALOW 1968). The nature of the hormonal forms responsible for the observed heterogeneity of plasma and tissue parathyroid hormone was subsequently elucidated by fractionation studies (SILVERMAN and YALOW 1971).

F. Identical Chromatographic Behaviour of Standards and Unknowns

Following the discovery that insulin is synthesized by way of a higher molecular weight precursor, proinsulin (STEINER and OYER 1967), it was shown that total circulating immunoreactive insulin is comprised of at least two components (ROTH et al. 1968). Thus, when plasma samples obtained from healthy subjects were filtered on Sephadex G-50 columns, the endogenous immunoreactive insulin was recovered in two peaks. One peak, quite discrete, appeared at 0.45 of the distance between the protein and salt peaks (little insulin); the other peak (big insulin) was detected halfway between little insulin and the plasma protein peak and accounted for 0%–50% of the total immunoreactivity (ROTH et al. 1968). The discovery that plasma immunoreactive insulin was not a homogeneous substance soon raised questions about the nature of other circulating hormones and led to the recognition that many, if not all, peptide hormones are found in more than one form in plasma and in tissue extracts (reviewed by YALOW 1985). While the 6000-dalton peptide with full biological activity is the predominant form of insulin-like immunoreactivity in the circulation of virtually all subjects in the stimulated state, this is not often the case with other peptide hormones (YALOW 1985). Thus, precursor forms and/or metabolic fragments, with or without biological activity and with variable turnover times, have been characterized for parathyroid hormone, gastrin, adrenocorticotropic hormone and growth hormone (YALOW 1985).

The use of Sephadex gel filtration or starch gel electrophoresis has played a pivotal role in allowing characterization of the heterogeneity of peptide hormones, by separating different molecular forms on the basis of size and/or charge. A similar approach has been adopted subsequently, using different chromatographic techniques such as thin layer chromatography (TLC) or high pressure liquid chromatography (HPLC), to characterize the nature of steroid-like or eicosanoid-like immunoreactivity. Again, demonstration of identical chromatographic behaviour of standards and unknowns cannot be taken as final proof of chemical identity, inasmuch as closely related haptens may comigrate with a particular solvent system. However, characterization of the heterogeneity of the measured endogenous immunoreactivity on TLC and/or HPLC will represent an important clue to define the chromatographic procedures prior to RIA or to evaluate different antisera. Thus, when characterizing the nature of urinary 6-keto-PGF$_{1\alpha}$-like immunoreactivity on TLC, we found that only AS 1 (Table 1) recognized a single peak of immunoreactive material comigrating with authentic 6-keto-PGF$_{1\alpha}$, whereas AS 2, AS 3, and AS 4 also revealed the presence of increasing amounts of immunoreactivity in both more and less polar zones of the plate (Fig. 2), thus

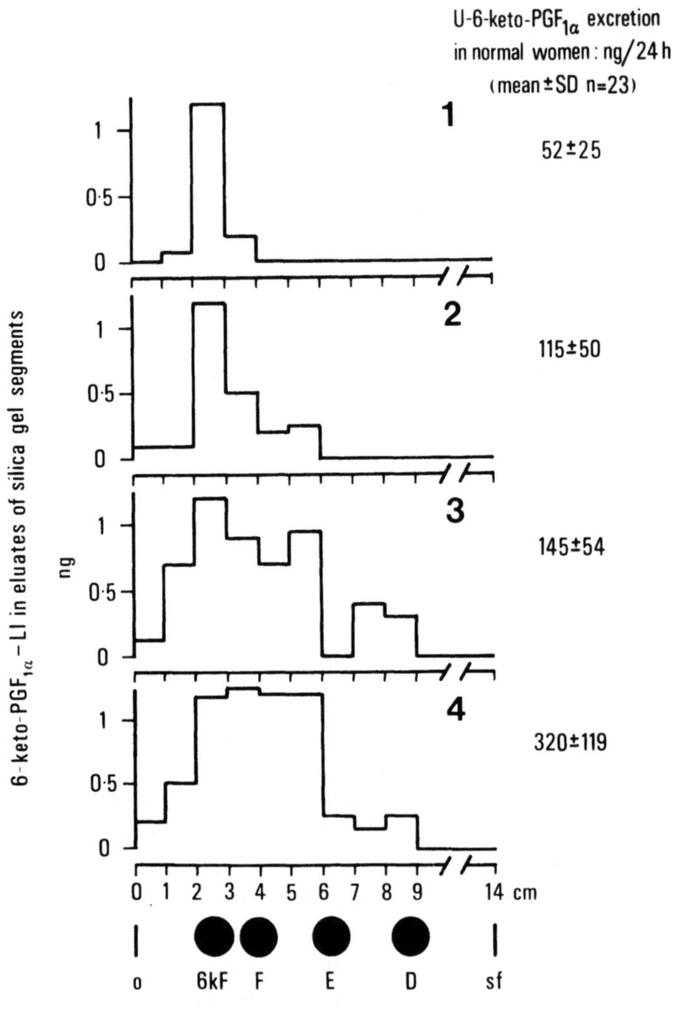

Fig. 2. Heterogeneity of urinary 6-keto-PGF$_{1\alpha}$-like immunoreactivity (LI). Purified extracts, prepared from urine obtained under basal conditions, were subjected to thin layer chromatography. The whole lane corresponding to one particular extract was divided into 1-cm segments, the silica gel was scraped off and eluted with methanol. All the eluates were assayed for 6-keto-PGF$_{1\alpha}$-LI with four anti-6-keto-PGF$_{1\alpha}$ sera of different specificities (as detailed in Table 1). The *vertical marks* on the abscissa indicate the origin (*o*) and the solvent front (*sf*) of the plate. The *dots* indicate the location of cochromatographed authentic PGs: 6kF 6-keto-PGF$_{1\alpha}$; F PGF$_{2\alpha}$; E PGE$_2$; D PGD$_2$. Mean ($\pm$SD) excretion rates measured by the 4 different antisera in the urine of 23 healthy women are also indicated. (PATRONO et al. 1982. Reproduced from *The Journal of Clinical Investigation,* 1982, 69, 231–239, by copyright permission of The American Society for Clinical Investigation)

indicating the presence of specifically interfering substances coeluting with 6-keto-$PGF_{1\alpha}$ from silicic acid columns (PATRONO et al. 1982). It is interesting to note that all the antisera employed satisfied the basic requirements of RIA, as outlined here, i.e. quantitative recovery of added standard and parallelism of standard and unknown. It can be easily appreciated from the data presented in Fig. 2 that reliable measurements of 6-keto-$PGF_{1\alpha}$ would be obtained by AS 1 in urinary extracts subjected to silicic acid column chromatography without a TLC step prior to RIA. In contrast, the use of any of the other antisera would necessitate the inclusion of a TLC step in order to separate 6-keto-$PGF_{1\alpha}$ from other cross-reacting substances. A similar approach has been adopted in order to characterize the nature of the enhanced TXB_2-like immunoreactivity in the urines of patients with systemic lupus erythematosus (PATRONO et al. 1985b).

We have exploited the high cross-reactivities (50%–100%) of the 2,3-dinor derivatives of TXB_2 and 6-keto-$PGF_{1\alpha}$ with antisera raised against the parent compounds to develop RIA methods employing homologous (PATRONO et al. 1985b) or commercially available heterologous tracers (CIABATTONI et al. 1987b). GC–MS-validated measurements were obtained by including a TLC step prior to RIA, thereby separating 2,3-dinor-TXB_2 or 2,3-dinor-6-keto-$PGF_{1\alpha}$ from cross-reacting TXB_2 and 6-keto-$PGF_{1\alpha}$, respectively (PATRONO et al. 1985b; CIABATTONI et al. 1987b). Thus, perfectly valid measurements can be obtained even using antisera of limited immunological specificity, provided that adequate chromatographic steps are introduced prior to RIA in order to separate structurally related cross-reacting substances.

The use of HPLC or some other fractionation system is discussed by NYBERG et al. (p. 231) for the validation of RIA measurements of opioid peptides; by PARKER (p. 513) for cyclic nucleotides; and by ZWEERINK et al. (p. 485) for leukotrienes.

G. Comparison with an Independent Method of Analysis

A direct comparison of RIA measurements with those performed by an independent assay method in the relevant biological fluid or fluids should be sought whenever such a method exists and is accessible to the investigator. This comparison should preferably be performed blind, and involve a reasonable number of samples (20–40) spanning the whole range of concentrations of potential pathophysiological and pharmacological interest (Fig. 3).

In the field of peptide hormones, a readily available comparison is with radioreceptor assays as discussed by NYBERG et al. (p. 235) for opioid peptides. Some discrepancies are likely to be encountered because of the existence of multiple forms of many peptide hormones, variably recognized by different antisera and displaying varying degrees of biological activity. In the case of the parathyroid hormone assay, this is a test seldom properly validated because of the differences between standards and unknowns and the consequent difficulty of interpretation (YALOW 1985). The existence of two different hormonal forms with similar biological activity and markedly different turnover time, such as the 34 and 17 amino acid peptide forms of gastrin (see WALSH and WONG, p. 316), may further

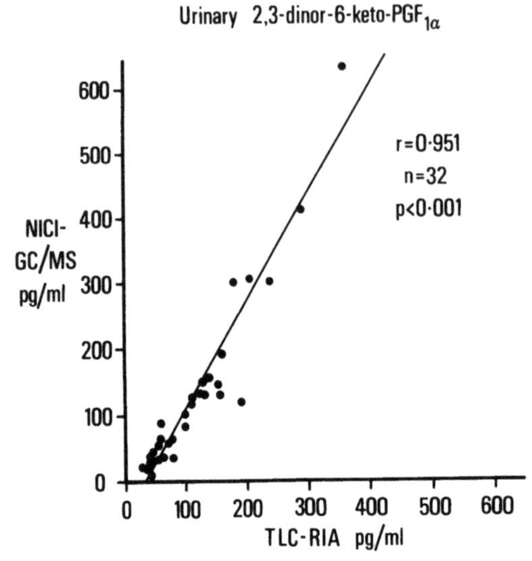

Fig. 3. Correlation of measurements of urinary 2,3-dinor-6-keto-PGF$_{1\alpha}$ obtained by thin layer chromatography–radioimmunoassay (TLC–RIA) and by negative ion chemical ionization gas chromatography–mass spectrometry (NICI GC–MS). (Data from Ciabattoni et al. 1987b)

require knowledge of the circulating hormonal forms in order to interpret RIA data in different clinical situations (Yalow 1985).

For the validation of thyroid hormone RIA measurements (see Bartalena et al., p. 416), reference is made to a direct, absolute equilibrium technique, i.e. equilibrium dialysis followed by RIA of free thyroid hormones in the dialysate and in the dialysant.

In the field of eicosanoids (see Peskar et al., p. 456) as well as in the field of steroid hormones (see Pazzagli and Serio, p. 369), comparison with GC–MS measurements represents the ultimate validation criterion. GC–MS techniques of adequate sensitivity to match that of RIA are now available, such as negative ion chemical ionization (NICI) GC–MS (Fig. 3). Direct comparisons for the major urinary metabolites of TXB_2 (Patrono et al. 1986) and PGI_2 (Ciabattoni et al. 1987b) have been reported recently. It has been the experience of our group that eicosanoid RIAs satisfying all the previously described requirements, including identical behaviour of standards and unknown on TLC, usually provide measurements consistent with GC–MS verification.

H. Are We Measuring the Right Compound, in the Right Compartment?

Although not strictly inherent in the problem of validating RIA measurements, but perhaps equally relevant to the task of monitoring endogenous mediators of pharmacological interest, is the question of what to measure and where. Obviously, a quite different approach applies to autacoids as compared with classical hormones. As for the latter, measurement of the biologically active form (or forms) in the peripheral circulation often provides all the relevant information. In contrast, meaningful measurements of autacoids often require getting at or near the relevant site (or sites) of synthesis and action. This implies obtaining blood from the venous drainage of a specific area of interest, provided that this is readily accessible and that blood cells are not capable of synthesizing the same substance. Alternatively, it requires measuring the compounds of interest in a different compartment, e.g. the cerebrospinal fluid, gastric juice, or bile. The issue of measuring opioid peptides in cerebrospinal fluid is discussed by NYBERG et al. (p. 239). Measurement of leukotrienes in human synovial fluid is discussed by ZWEERINK et al. (p. 495).

The field of prostaglandin and thromboxane measurement has perhaps suffered more from such conceptual inadequacies than from methodological issues (see PESKAR et al., p. 455). This can be best exemplified by the problems related to measuring platelet TXB_2 synthesis in humans. Serum TXB_2, i.e. TXB_2 synthesized by platelets ex vivo during whole blood clotting at 37 °C (PATRONO et al. 1980) provides a readily accessible capacity index related to platelet cyclooxygenase and TX synthase activity (ALESSANDRINI et al. 1985). This has allowed the biochemical pharmacology of inhibitors of these enzymes to be established in humans (PATRONO et al. 1985a; FITZGERALD et al. 1985). However, the capacity of platelets to release and oxygenate arachidonate in response to appropriate stimuli ex vivo greatly exceeds the actual synthetic rate in vivo. Thus, 1 ml human whole blood allowed to clot at 37 °C for 60 min will generate roughly 300–400 ng TXB_2 per ml serum, i.e. a quantity approximating the hourly total body production of 420 ng (PATRONO et al. 1986). This fact has obvious relevance to the reliability of TXB_2 measurements in peripheral or coronary sinus plasma. Thus, platelet activation during and after sampling by as little as 0.1% of their maximal capacity would largely account for the putative "circulating" concentrations of TXB_2 reported by many investigators, i.e. 10–300 pg/ml (FITZGERALD et al. 1983). It is important to emphasize that at least some of these measurements were obtained with properly validated RIA techniques or GC–MS. Similar considerations also apply to the measurement of platelet-derived proteins in the circulation (see PEPPER, p. 520). The recent characterization of long-lived enzymatic metabolites of TXB_2 in the human circulation (LAWSON et al. 1986) and development of appropriate analytical techniques promises to bypass the problem of platelet activation during and after sampling.

Whenever the thing in question is not so much the identity of the compound being measured, but rather its actual presence in the circulating blood prior to sampling, one should refer to studies measuring the endogenous secretion and plasma clearance rates of a given substance, thereby obtaining a maximal estimate of its circulating concentration. Studies of this nature have been performed with both PGI_2 (FITZGERALD et al. 1981) and TXB_2 (PATRONO et al. 1986). Unfortunately, this information is not available for platelet-derived proteins.

J. How to Evaluate an RIA Kit Critically

Although a sizeable proportion of investigators in this area usually develop their own reagents for RIA or perhaps only purchase the radioactive tracer, an increasing number of potential users, particularly in the pharmaceutical industry, are now relying upon commercially available kits. Because it is commercially available, a kit tends to be viewed as a magic box necessarily generating valid and reliable measurements. Although this may well be the case under certain circumstances, one has to develop and apply a critical attitude towards assessing the validity of RIA data, whatever the source of the reagents may be. Thus, the specificity data of an anti-TXB_2 serum supplied by the manufacturer may be perfectly adequate for assessing the feasibility of measuring this compound in serum or platelet-rich plasma, but are absolutely insufficient to evaluate the feasibility of urinary measurements inasmuch as none of the relevant urinary metabolites of TXB_2 have been examined. Surprisingly, this may be so despite an impressive table describing the cross-reactivity of 28 different compounds. Moreover, validation of RIA data thus obtained in a particular biological fluid is often desumed from the literature. This may be quite misleading for many good reasons, as outlined in this chapter.

Short of obtaining confirmation from a reliable independent method of analysis, the use of a single validation criterion will not be sufficient to validate RIA measurements, although it may alert the careful investigator to the existence of problems. However, the combined application of several validation criteria will greatly enhance the likelihood of measuring the right amount of the right compound.

Acknowledgements. During preparation of this manuscript, the author was Visiting Professor in the Cardiovascular Unit of the Royal Postgraduate Medical School and in the Pharmacology Department of King's College, University of London. The author's sabbatical was supported by the British Heart Foundation and the British Council. The author wishes to express his gratitude to Pat Tierney for expert editorial assistance.

References

Alessandrini P, Avogaro P, Bittolo Bon G, Patrignani P, Patrono C (1985) Physiologic variables affecting thromboxane B_2 production in human whole blood. Thromb Res 37:1–8

Berson SA, Yalow RS (1968) Immunochemical heterogeneity of parathyroid hormone. J Clin Endocrinol 28:1037–1047

Catella F, Pugliese F, Patrignani P, Filabozzi P, Ciabattoni G, Patrono C (1984) Differential platelet and renal effects of sulfinpyrazone in man (Abstract.) Clin Res 32:239A

Ciabattoni G, Pugliese F, Cinotti GA, Stirati G, Ronci R, Castrucci G, Pierucci A, Patrono C (1979) Characterization of furosemide-induced activation of the renal prostaglandin system. Eur J Pharmacol 60:181–187

Ciabattoni G, Cinotti GA, Pierucci A, Simonetti BM, Manzi M, Pugliese F, Barsotti P, Pecci G, Taggi F, Patrono C (1984) Effects of sulindac and ibuprofen in patients with chronic glomerular disease. N Engl J Med 310:279–283

Ciabattoni G, Boss AH, Patrignani P, Catella F, Simonetti BM, Pierucci A, Pugliese F, Filabozzi P, Patrono C (1987 a) Effects of sulindac on renal and extra-renal eicosanoid synthesis in man. Clin Pharmacol Ther 41:380–383

Ciabattoni G, Boss AH, Daffonchio L, Daugherty J, FitzGerald GA, Catella F, Dray F, Patrono C (1987 b) Radioimmunoassay measurement of 2,3-dinor-metabolites of prostacyclin and thromboxane in human urine. Adv Prostaglandin, Thromboxane Leukotriene Res (in press)

Ferreira SH, Vane JR (1967) Prostaglandins: their disappearance from and release into the circulation. Nature 216:868–873

FitzGerald GA, Brash AR, Falardeau P, Oates JA (1981) Estimated rate of prostacyclin secretion into the circulation of normal man. J Clin Invest 68:1272–1276

FitzGerald GA, Pedersen AK, Patrono C (1983) Analysis of prostacyclin and thromboxane biosynthesis in cardiovascular disease. Circulation 67:1174–1177

FitzGerald GA, Reilly IA, Pedersen AK (1985) The biochemical pharmacology of thromboxane synthase inhibition in man. Circulation 72:1194–1201
Gryglewski RJ, Bunting S, Moncada S, Flower RJ, Vane JR (1976) Arterial walls are protected against deposition of platelet thrombi by a substance (prostaglandin X) which they make from prostaglandin endoperoxides. Prostaglandins 12:685–700
Hamberg M, Svensson J, Samuelsson B (1975) Thromboxanes: a new group of biologically active compounds derived from prostaglandin endoperoxides. Proc Natl Acad Sci USA 72:2994–2998
Jaffe BM, Behrman HR, Parker CW (1973) Radioimmunoassay measurement of prostaglandins E, A and F in human plasma. J Clin Invest 52:398–407
Lawson JA, Patrono C, Ciabattoni G, FitzGerald GA (1986) Long-lived enzymatic metabolites of thromboxane B_2 in the human circulation. Anal Biochem 155:198–205
Patrignani P, Filabozzi P, Patrono C (1982) Selective cumulative inhibition of platelet thromboxane production by low-dose aspirin in healthy subjects. J Clin Invest 69:1366–1372
Patrignani P, Filabozzi P, Catella F, Pugliese F, Patrono C (1984) Differential effects of dazoxiben, a selective thromboxane-synthase inhibitor, on platelet and renal prostaglandin-endoperoxide metabolism. J Pharmacol Exp Ther 228:472–477
Patrono C, Dunn MJ (1987) The clinical significance of inhibition of renal prostaglandin synthesis. Kidney Int, in press
Patrono C, Ciabattoni G, Pinca E, Pugliese F, Castrucci G, De Salvo A, Satta MA, Peskar BA (1980) Low dose aspirin and inhibition of thromboxane B_2 production in healthy subjects. Thromb Res 17:317–327
Patrono C, Pugliese F, Ciabattoni G, Patrignani P, Maseri A, Chierchia S, Peskar BA, Cinotti GA, Simonetti BM, Pierucci A (1982) Evidence for a direct stimulatory effect of prostacyclin on renin release in man. J Clin Invest 69:231–239
Patrono C, Ciabattoni G, Patrignani P, Pugliese F, Filabozzi P, Catella F, Davi G, Forni L (1985a) Clinical pharmacology of platelet cyclo-oxygenase inhibition. Circulation 72:1177–1184
Patrono C, Ciabattoni G, Remuzzi G, Gotti E, Bombardieri S, Di Munno O, Tartarelli G, Cinotti GA, Simonetti BM, Pierucci A (1985b) Functional significance of renal prostacyclin and thromboxane A_2 production in patients with systemic lupus erythematosus. J Clin Invest 76:1011–1018
Patrono C, Ciabattoni G, Pugliese F, Pierucci A, Blair IA, FitzGerald GA (1986) Estimated rate of thromboxane secretion into the circulation of normal humans. J Clin Invest 77:590–594
Pedersen AK, FitzGerald GA (1985) Cyclooxygenase inhibition, platelet function and metabolite formation during chronic sulfinpyrazone dosing. Clin Pharmacol Ther 37:36–42
Pedersen AK, Watson M, FitzGerald GA (1983) Inhibition of thromboxane synthase in serum: limitations of the measurement of immunoreactive 6-keto-$PGF_{1\alpha}$. Thromb Res 33:99–103
Roberts LJ II, Sweetman BJ, Oates JA (1981) Metabolism of thromboxane B_2 in man. Identification of twenty urinary metabolites. J Biol Chem 256:8384–8393
Roth J, Gorden P, Pastan I (1968) "Big insulin": a new component of plasma insulin detected by immunoassay. Proc Natl Acad Sci USA 61:138–145
Silverman R, Yalow RS (1971) Heterogeneity of parathyroid hormone: clinical and physiological implications. J Clin Invest 52:1958–1971
Steiner DF, Oyer PE (1967) The biosynthesis of insulin and a probable precursor of insulin by a human islet cell adenoma. Proc Natl Acad Sci USA 57:473–480
Yalow RS (1982) The limitations of radioimmunoassay (RIA). Trends Anal Chem 1:128–131
Yalow RS (1985) Radioimmunoassay of hormones. In: Wilson JD, Foster DW (eds) Williams textbook of endocrinology, 7th edn. Saunders, Philadelphia, pp 123–132
Zusman RM, Spector D, Caldwell BV, Speroff L, Schneider G, Mulrow PJ (1973) The effect of chronic sodium loading and sodium restriction on plasma prostaglandin A, E, and F concentrations in normal humans. J Clin Invest 52:1093–1098

CHAPTER 10

Measurement of Opioid Peptides in Biologic Fluids by Radioimmunoassay

F. NYBERG, I. CHRISTENSSON-NYLANDER, and L. TERENIUS

A. Introduction

During the past decade, a large family of peptides with opioid activity has been described. These peptides, also termed endorphins (an acronym for endogenous morphine) have attracted interest as alleged neurotransmitters, as neurohormones, and as hormones, particularly within the hypothalamic – pituitary – adrenal axis. These compounds have chemical homology in the NH_2 terminal sequence Tyr-Gly-Gly-Phe-(Met or Leu), which forms the part of the molecule which is critical for opioid activity, and chemical heterogeneity in the COOH terminal sequences, which may contribute to receptor affinity and selectivity of interaction between opiate receptor types. A rather bewildering chemical complexity among these peptides is due to the fact that they derive from three distinct prohormones, each with a unique distribution in different cell types and neuronal pathways: proopiomelanocortin (with β-endorphin), proenkephalin A (with seven enkephalin core sequences), and proenkephalin B (with three different dynorphin peptides); (Table 1 see also Sect. B).

Immunologic techniques have been instrumental in confirming or refuting a functional role of these peptides in various organs and under various conditions. Immunohistochemical visualization of their anatomic distribution has defined

Table 1. Structure of major opioid peptides

Precursor/peptide	Structures
Proopiomelanocortin	
β-Endorphin	Tyr-Gly-Gly-Phe-Met-Thr-Ser-Glu-Lys-Ser-Gln-Thr--Pro-Leu-Val-Thr-Leu-Phe-Lys-Asn-Ala-Ile-Ile-Lys--Asn-Ala-Tyr-Lys-Lys-Gly-Glu
Proenkephalin A	
Leu-enkephalin	Tyr-Gly-Gly-Phe-Leu
Met-enkephalin	Tyr-Gly-Gly-Phe-Met
Met-enkephalin-Arg^6Phe^7	Tyr-Gly-Gly-Phe-Met-Arg-Phe
Proenkephalin B	
Dynorphin A	Tyr-Gly-Gly-Phe-Leu-Arg-Arg-Ile-Arg-Pro-Lys-Leu--Lys-Trp-Asp-Asn-Gln
Dynorphin B (rimorphin)	Tyr-Gly-Gly-Phe-Leu-Arg-Arg-Gln-Phe-Lys-Val--Val-Thr
α-Neoendorphin	Tyr-Gly-Gly-Phe-Leu-Arg-Lys-Tyr-Pro-Lys

their site of formation and radioimmunoassay (RIA) has given useful quantitative information. RIA analysis of these peptides in body fluids has been used both in experimental animals and in humans, either to study normal physiologic relationships or altered activity in disease. The problems of chemical heterogeneity have already been mentioned, these problems are amplified in analysis of body fluids since the peptides may derive from various tissues and there may be contributions from several prohormones. This emphasizes the need for specific analytic procedures. The structural homologies between the peptides are so extensive that it is frequently necessary to introduce a preseparation step prior to RIA. Another matter of consideration is the metabolic instability of the peptides in body fluids, which requires special precautions.

B. General Aspects

I. Distribution and Processing of Opioid Peptides

Each of the three opioid peptide systems has a clearly distinct regional distribution. Peptides derived from proopiomelanocortin are mainly produced in the pituitary gland (Bloom et al. 1977; Pelletier et al. 1977). Neuronal cell bodies containing β-endorphin are relatively few. They are for instance found in the region of the arcuate nucleus of the medial basal hypothalamus (Bloch et al. 1978; Bloom et al. 1978; Pelletier 1980; Watson and Akil 1979, 1980). These neurons project extensively to many areas of the limbic system and the brain stem. Peptides produced by proenkephalin A (enkephalins) are widely distributed in CNS, peripheral nervous system, and endocrine tissues (Elde et al. 1976; Hökfelt et al. 1977; Pickel et al. 1980; Sar et al. 1978; Uhl et al. 1979; Watson et al. 1982). In CNS, enkephalin-containing elements are found at every level of the neuraxis, from cells in the cortex all the way to cells in the spinal cord. Peripherally, peptides related to proenkephalin A are found in the adrenal medulla, the gastrointestinal tract, and also in several other structures. Neurons with dynorphin- (proenkephalin B)-related peptides are relatively few. They form a striatonigral pathway, are present in the hypothalamus, in certain other brain nuclei and in spinal interneurons (Maysinger et al. 1982; Vincent et al. 1982; Watson et al. 1982a, b). Dynorphin peptides have also been shown immunohistochemically to coexist with vasopressin in magnocellular neurons of the hypothalamus (Watson et al. 1982; Weber et al. 1982).

It has already been emphasized that each of the endorphin prohormones is pluripotent, i.e., capable of generating several biologically active fragments. The principal active peptides are flanked by pairs of basic amino acid residues (Lys/Arg); a common route of processing is a trypsin-like cleavage, followed by "trimming" of the COOH terminal basic amino acid by a carboxypeptidase C-like enzyme. These processes are probably partly statistical and a whole family of peptides, differing in length, may be generated. The minimum sequence for significant opioid activity is Tyr-Gly-Gly-Phe-X, where X is Leu or Met; opioid activity is however maintained or even enhanced by COOH terminal elongation such as in β-endorphin. Another enzymatic pathway, acetylation of the NH_2 terminal

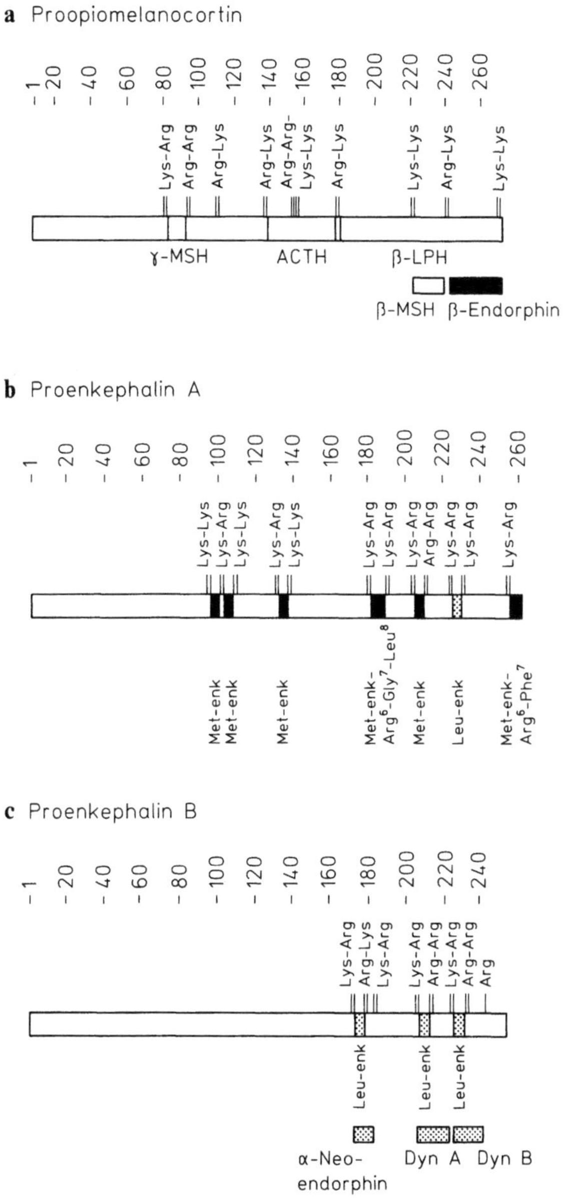

Fig. 1 a–c. Schematic structures of **a** proopiomelanocortin, **b** proenkephalin A and **c** proenkephalin B. *Black boxes* represent opioid peptides with Met-enkephalin sequence, *grey boxes* with Leu-enkephalin, *white boxes* nonopioid peptides. Dyn dynorphin; enk enkephalin

Tyr leads to loss of opioid activity. These general principles will be exemplified below.

Processing of proopiomelanocortin (NAKANISHI et al. 1979) releases β-endorphin from its COOH terminus (Fig. 1 a). β-Endorphin constitutes the COOH terminus of a previously characterized hormone, β-lipotropin (β-LPH; LI et al.

1965). β-Endorphin may be processed further to its α (1–16), γ (1–17), or δ (1–27) fragments, or by acetylation of its NH_2 terminus (ZAKARIAN and SMYTH 1979). Proenkephalin A (GUBLER et al. 1982) contains no less than six Met-enkephalin core sequences and one Leu-enkephalin sequence, and may generate a variety of enkephalin-related peptides (Fig. 1 b). The posttranslational processing of this precursor is not yet fully clarified. In addition to the enkephalin pentapeptides sequential trypsin, carboxypeptidase C action leads to Met-enkephalin-Arg^6Phe^7 (STERN et al. 1979), -$Arg^6Gly^7Leu^8$ (KILPATRICK et al. 1981), and two COOH terminally amidated enkephalin-containing sequences, metorphamide (WEBER et al. 1983) and amidorphin (SEIZINGER et al. 1984). In proenkephalin B (KAKIDANI et al. 1982), there are three unique Leu-enkephalin-related peptides: α-neoendorphin (KANGAWA et al. 1979), dynorphin A (GOLDSTEIN et al. 1981; TACHIBANA et al. 1982), and dynorphin B or rimorphin (FISCHLI et al. 1982; KILPATRICK et al. 1982). Other Leu-enkephalin-containing peptides consisting of fragments or combinations of these three peptides have also been characterized (Fig. 1 c). Generation of Leu-enkephalin from this precursor was once thought to be prohibited; more recent work, however, shows that it happens, for instance in the striatonigral dynorphin pathway (ZAMIR et al. 1984). Thus, Leu-enkephalin can originate from any of the two proenkephalin systems and can therefore not serve as a specific marker. The extent of processing and other posttranslational enzymatic events may markedly affect not only opioid activity, but also the receptor profile, there being several opioid receptors.

II. Characterization of Opioid Receptors

The concept of multiple opioid receptors with different properties derives from a wealth of biochemical and pharmacologic data. It has thus been possible to distinguish at least three main types of opioid receptors named μ, δ, and κ (KOSTERLITZ 1979; KOSTERLITZ et al. 1982). Each of these receptors has been characterized by its affinity for specific ligands. The receptor preference of the endogenous opioids is highly structure dependent (PATERSON et al. 1983). It appears, for example, that the enkephalins are mainly ligands for δ receptors, whereas dynorphin and other COOH terminally elongated Leu-enkephalin-containing peptides bind quite selectively to κ receptors. Met-enkephalin-Arg^6-Phe^7 and -Arg^6-Gly^7-Leu^8 have about equal affinity for both δ and κ sites. Interestingly, metorphamide is unique in being highly μ-selective, i.e., pharmacologically similar to morphine (WEBER et al. 1983).

The CNS distribution of opioid receptors has been thoroughly examined. In general, there appears to be an enrichment of μ receptors within areas involved in pain processing (the periaqueductal grey, the raphe nuclei, and the dorsal horn of the spinal cord), whereas a high density of δ receptors is found in parts of the limbic system (CHANG et al. 1979; BONNET et al. 1981; NINKOVIC et al. 1981; SNYDER and GOODMAN 1980). κ receptors are also present in limbic structures, in the cerebral cortex, and in the dorsal spinal cord. Interestingly, analgesia can be induced via all three receptors at the spinal level.

Opioid receptors are also present in several tissues outside the CNS. Most thoroughly studied are those in intestinal nerve plexuses which are mainly of the

μ-type and in the intestinal mucosa layer which mainly are of the δ type (NORTH and EGAN 1983). Receptors of the δ type seem to exist in bronchial tissue and lungs (HOLADAY 1983). In general, relatively little is known about opioid receptors in nonneuronal tissues and the target organs for the endocrine secretion of opioid peptides are poorly defined.

C. Methodological Aspects

I. Sampling Protocol

It is important to standardize the sampling protocol as far as possible. In our protocol for CSF analysis, the patient gives a sample of 12.5 ml lumbar CSF early in the morning before rising after an overnight fast. In general, a 22-gauge spinal needle inserted in the lumbar region at L-3 to L-4 is used. The CSF is collected in polyethylene or polypropylene tubes placed on ice. The sample is subsequently mixed, centrifuged in the cold, frozen in suitable aliquots, and then stored at −90 °C until assay. Plasma samples should preferably be collected in EDTA tubes followed by the same storage conditions. EDTA also serves as a protease inhibitor. Other protease inhibitors may also be added, but there are few controlled studies in the literature of the effectiveness of these additions. Extreme precautions against degradation have to be taken in enkephalin determinations. For instance, CLEMENT-JONES et al. (1980c) collected plasma on acid in aprotinin-treated tubes which were immediately frozen in liquid nitrogen.

II. Sample Workup (Preseparation)

Prior to RIA, a separation step is frequently needed so that greater specificity can be achieved. Separation of the opioid peptides can be obtained by several techniques, such as ion exchange or reverse phase chromatography. Furthermore, preseparation of cross-reacting components extends the application of RIA to the measurement of individual components in a single sample. In the analysis of series of samples it is necessary to have access to a procedure which allows rapid and effective separation. It is particularly advantageous to use several parallel separation systems cheap enough that they are used only once. This is possible, e.g., with mini ion exchange columns (BERGSTRÖM et al. 1983), which give complete separation of enkephalin penta- and hexapeptides and longer peptides such as dynorphin A (1–8) and dynorphin A. Preseparation on reverse-phase cartridges has been utilized by several groups. If sufficient material is available, structure confirmation of the measured activity should preferentially be used by combining the RIA with HPLC. Several laboratories have also used gel filtration techniques to settle the identity of the analyzed immunoreactivity, at least by molecular weight.

III. Radioimmunoassay of Opioid Peptides from Individual Systems

1. β-Endorphin and Related Peptides

a) RIA Procedures

The presence of β-endorphin immunoreactivity in CSF and plasma has been demonstrated in many laboratories. Chromatographic procedures have revealed that the major part of this activity present in CSF is identical with β-endorphin, whereas the remaining part derives from the precursor β-lipotropin (CLEELAND et al. 1984; JEFFCOATE et al. 1978; MCLOUGHLIN et al. 1980). A β-endorphin fragment (1–27) has also been identified in rat CSF (JACKSON et al. 1985). In plasma β-lipotropin is the predominant of the two peptides (NAKAO et al. 1978; CAHILL et al. 1983; KIKUCHI et al. 1984; DETRICK et al. 1985). Most antibodies raised against β-endorphin recognize the COOH terminal region of the molecule, but NH_2 terminal-directed β-endorphin antiserum has in fact been used in a few studies. However, in these cases it has been difficult to discriminate between the activity due to β-endorphin or its fragments and the contribution from other opioid peptides (EKMAN 1985).

In the assays for β-endorphin, the iodinated peptide (labeled with ^{125}I or ^{131}I) is generally used as tracer and, in the separation of bound and free peptide, charcoal adsorption or double-antibody procedures have frequently been used. Incubations are normally performed at 4 °C in plastic (polystyrene or polypropylene) vials or siliconized glass tubes, with antibodies, tracer, and unknown sample or standard in a final volume of 200–500 µl phosphate buffer, pH 7.4–7.5, containing albumin (HÖLLT et al. 1978; ROSS et al. 1979; WARDLAW and FRANTZ 1979; WIEDEMANN et al. 1979). In order to minimize enzymatic degradation of tracer and peptide adsorption to the walls, many procedures also include EDTA and detergent (ROSS et al. 1979; GHAZAROSSIAN et al. 1980; NYBERG and TERENIUS 1982). In most β-endorphin RIAs, samples and antibodies are allowed to preincubate for 16–24 h before the addition of tracer followed by additional 24 h incubation. In the charcoal adsorption procedure (HÖLLT et al. 1978; ROSS et al. 1979; NYBERG and TERENIUS 1982), unbound peptide is removed by adding 1 ml charcoal slurry followed by centrifugation within 5 min. A defined aliquot of the supernatant solution is then counted in a γ counter.

A critical point of the charcoal adsorption technique is to determine the optimal charcoal concentration. It may vary for every dilution and from batch to batch of the antiserum and, furthermore, it also seems to depend on whether samples are extracts from plasma or CSF. Double-antibody or second antibody precipitation procedures (FACCHINETTI and GENAZZANI 1979; HO et al. 1980; WARDLAW and FRANTZ 1979; WIEDEMANN et al. 1979; CAHILL et al. 1983) are more time-consuming, but in general less complicated to handle. The antigen–antibody complexes are precipitated by addition of goat or sheep anti-rabbit immunoglobulin. Following centrifugation, the pellet is counted in a γ counter.

b) Generation of Antisera

Most radioimmunoassays for β-endorphin utilize antibodies raised against the thyroglobulin complex with the authentic peptide standard. However, owing to the sensitivity of the methionine residue in position 5 to oxidation, some laboratories have also used its Leu^5 analog, both as tracer and for raising antibodies (Ross et al. 1979; Ghazarossian et al. 1980). The detection limits reported for β-endorphin RIAs vary from below 1 fmol per tube up to 10 fmol per tube.

c) Preseparation

The reported CSF and plasma levels of β-endorphin vary between different laboratories. On average, the CSF concentration in normal subjects is around 20 fmol/ml, whereas that of plasma is about 10 fmol/ml. The discrepancy in levels between different laboratories may partly result from the different procedures used for CSF and plasma extraction. Many investigators have used silicic acid adsorption and desorption with acidic acetone (Nakao et al. 1978; Höllt et al. 1979; Facchinetti et al. 1983), while others have used talc tablets (Wardlaw and Frantz 1979) or Sep-Pak C_{18} cartridges (Cahill and Akil 1982). In silicic acid extraction, plasma samples (3–6 ml) are normally added to polypropylene or siliconized glass tubes containing 100–300 mg silicic acid followed by vortexing for a few seconds or by mixing in the cold for about 30 min (Höllt et al. 1978; Nakao et al. 1978; Ross et al. 1979; Gambert et al. 1981; Facchinetti et al. 1983). After centrifugation, the silicic acid is washed out with water or buffer before peptide desorption with acidic acetone. The acetone extract is evaporated and the final residue redissolved in the appropriate volume of RIA buffer.

In some laboratories, the plasma samples are acidified prior to extraction (Ross et al. 1979). Acidic acetone extraction of plasma β-endorphin without silicic acid adsorption has also been described (Ho et al. 1980; Detrick et al. 1985). Pickar et al. (1980), who studied β-endorphin in plasma from psychiatric patients, used aluminum oxide for plasma extraction. The talc tablet extraction procedure (Wardlaw and Frantz 1979) has been used in several clinical studies of plasma β-endorphin. In this procedure β-endorphin and β-lipotropic hormone are extracted, generally from 2–20 ml plasma, and subsequently eluted with acidic acetone. The extracts are evaporated and reconstituted fractions assayed directly for total β-endorphin-like immunoreactivity. The recovery of β-endorphin reported with these extraction procedures varies between 50% and 80%. An alternative approach used for extraction of β-endorphin immunoreactivity in human CSF is based on antibody adsorption (Gerner and Sharp 1982). The peptide was extracted by circulating the CSF through gel beads coupled to β-LPH antibodies prior to elution with acetic acid/HCl.

In our studies, the CSF β-endorphin has been recovered by fractionation on Sephadex G-10 (Nyberg et al. 1986) or an RP-8 reverse-phase silica gel cartridge (Lyrenäs et al. 1987). In plasma analysis, the sample was passed through an ion exchanger (DEAE–Sepharose) before extraction on the reverse-phase cartridge. In recovery studies, these extraction procedures show high efficiency, with over 90% recovered when synthetic peptide is present in picomole concentrations. However, when adding β-endorphin in low femtomole levels, the recovery does not exceed 50%. None of these techniques separates β-endorphin from its cross-reacting precursor β-LPH. To remove contributions from β-LPH, an additional separation step is needed. In many studies, this is accomplished by gel filtration (NAKAO et al. 1978; MCLOUGHLIN et al. 1980; FACCHINETTI et al. 1983). In a recent study of plasma β-endorphin levels during labor, the peptide was recovered from the acidified plasma sample by reverse-phase HPLC (RÄISÄNEN et al. 1984). The fraction containing β-endorphin was collected with a programmable fraction collector. By this procedure plasma β-endorphin was purified from other plasma components and completely separated from β-LPH. The mean recovery of the procedure for the synthetic peptide added to plasma samples was about 60%. However, when these approaches for plasma extraction are combined with gel filtration or HPLC, the overall procedures become time-consuming and fairly complicated. The development of new rapid techniques with high and reliable recovery of β-endorphin and applicable to series of samples should therefore have high priority in this area of research.

In certain pathologic cases, β-endorphin may rise to comparatively high levels. For instance, in plasma of patients with Addisons's disease, Cushing's disease, or ectopic ACTH syndrome, β-endorphin was elevated up to 1000 fmol/ml (SMITH et al. 1981). Parallel increases in plasma ACTH and β-lipotropin were also observed in these patients.

2. Peptides Derived from Proenkephalin A

In the proenkephalin A system, Met-enkephalin has been most extensively studied. Reported CSF levels range from approximately 5 fmol (AKIL et al. 1978a) up to several pmol/ml (BURBACH et al. 1979). The values at the pmol level derive from some early studies, performed without any preseparation of the CSF sample, and the detected activity may include contributions from cross-reacting components present in the fluid. Excluding these extremes, the average CSF concentration of Met-enkephalin is around 30 fmol/ml (CLEMENT-JONES et al. 1980d; LO et al. 1983; NYBERG et al. 1986). The corresponding value calculated from plasma determinations is significantly higher (100–200 fmol/ml) (CLEMENT-JONES et al. 1980d; SHANKS et al. 1981). In CSF studies of Met-enkephalin, both gel filtration (YAKSH et al. 1983; JACKSON et al. 1985; NYBERG et al. 1986) and reverse-phase cartridge adsorption (LO et al. 1983) have been used for peptide extraction. Reverse-phase systems have also been useful for enkephalin extraction from plasma samples (CLEMENT-JONES et al. 1980c; PANERAI et al. 1983). Some laboratories have tried acidic ethanol or acetone for the extraction from plasma of enkephalins or other peptides derived from proenkephalin A (CHOU et al. 1983).

In the radioimmunoassay, the tritium-labeled peptide has been used as tracer in many laboratories. In conformity with the β-endorphin RIA, incubations are in general performed at 4 °C and allowed to continue overnight. Other conditions such as buffers, separation of bound and free peptides, are almost identical with the conditions already described for β-endorphin. In general, the enkephalin RIAs are less sensitive than those of, say, β-endorphin or dynorphin. Detection limits in the range of 10–100 fmol/per tube have been reported. Most RIA data for Met-enkephalin may be relatively unreliable, owing to cross-reaction between the antibodies and other enkephalin-related peptides, e.g., Leu-enkephalin. Another problem with the RIA for Met-enkephalin arises from the fact that the peptide is unstable and undergoes spontaneous oxidation at its methionine residue to methionine sulfoxide during storage, extraction, and assay. To overcome this difficulty, antibodies have also been raised to the Met-enkephalin sulfoxide, Met-*O*-enkephalin, and the oxidized peptide was also used as tracer (CLEMENT-JONES et al. 1980c). All samples were then oxidized by hydrogen peroxide before assay and the modified RIA was reported to be highly sensitive and specific and could reliably distinguish between Met-enkephalin and Leu-enkephalin, for example.

The introduction of the iodinated peptide as tracer and the separation of bound and free by second antibody precipitation have also contributed to increasing the sensitivity of the RIA (LO et al. 1983). A similar RIA has been developed for another Met-enkephalin-containing opioid peptide, Met-enkephalin-Arg^6-Phe^7 (BOARDER et al. 1982a). Very few studies of Leu-enkephalin in CSF or plasma have been reported. Some laboratories have reported CSF Leu-enkephalin concentrations of 100–200 fmol/ml (NEUSER et al. 1984; KLEINE et al. 1984), whereas others have been negative (WAY et al. 1984). In the former studies, no chromatographic structure confirmation was reported and contributions from cross-reacting components could therefore not be excluded. We have observed approximately fivefold lower CSF concentration of Leu-enkephalin as compared with Met-enkephalin (NYBERG et al. 1986). However, it has also appeared from our studies that the enkephalins are not the major endogenous opioids in human CSF (NYBERG and TERENIUS 1982; NYBERG et al. 1983a; NYBERG et al. 1986). By use of radioreceptorassay in combination with electrophoresis and HPLC we could tentatively identify Met-enkephalin-Lys^6 and Met-enkephalin-Arg^6Phe^7 as major opioid components in the fluid. The presence of Met-enkephalin-Lys^6 in the CSF specimens was later confirmed by RIA (TERENIUS et al. 1984).

In a recent study (NYBERG et al. 1986), we used a substantial number of RIA methods directed against peptides considered to be of importance in all three opioid peptide systems (see Table 2). From data obtained, it is evident that the enkephalyl hexapeptides were present in human CSF in significantly higher concentration than Met- or Leu-enkephalin themselves. The predominance of enkephalyl hexapeptides over the pentapeptides has also been observed in spinal perfusate from cats (YAKSH et al. 1983; NYBERG et al. 1983b). Evidence for the presence of COOH terminal Met-enkephalin extensions in plasma has also been documented. The heptapeptide Met-enkephalin-Arg^6Phe^7 has been identified and quantified in human plasma (CHOU et al. 1983). Its extraction was carried out with acidic ethanol and reverse-phase Sep-Pak cartridge, and structure identity was settled by HPLC. The RIA for Met-enkephalin-Arg^6Phe^7 used the ^{125}I-analog as radiolabel and the charcoal adsorption procedure was used to separate free and bound antigen. The plasma levels of the heptapeptide in normal subjects ranged between 10 and 20 fmol/ml. The characterization of a 8.5 kdalton proenkephalin A-derived peptide circulating in bovine plasma was recently reported (BAIRD et al. 1984). The presence of large proenkephalin-derived structures in human CSF will be discussed in the following paragraphs.

Variations in the CSF and plasma levels of Met-enkephalin have been observed in certain clinical samples. For example, in patients with renal failure, the plasma concentrations of Met-enkephalin were found to be about ten times higher than in those of controls (SMITH et al. 1981). High levels of the pentapeptide have also been recorded in CSF and plasma following acupuncture-like treatment of addicts and chronic pain patients (CLEMENT-JONES et al. 1979; KISER et al. 1983). The latter study used the specific RIA procedure developed earlier by CLEMENT-JONES et al. (1980d), for measurement of plasma Met-enkephalin; levels went up to approximately 300 fmol/ml. An approximately threefold increase of Met-enkephalin in cat CSF after somatic stimuli was previously observed in our laboratory (YAKSH et al. 1983).

In studies of the enkephalins in plasma as well as CSF, it is important to consider their relatively low stability in these fluids. As mentioned already (and as will be discussed later), high potency of enkephalin-degrading activity has been found in both fluids and sampling should preferentially include the addition of protease inhibitors.

3. Peptides Derived from Proenkephalin B

Very few reports dealing with proenkephalin B-related peptides in plasma and CSF are found in the literature. This may reflect the fact that these endorphins are not detectable or only present in very low concentrations. In fact, recent studies performed in our laboratory (NYBERG et al. 1986) have indicated that the CSF levels of both dynorphin A and B, are below 10 fmol/ml (Table 2), dynorphin A being the major component. Furthermore, the material cross-reacting with the antiserum against dynorphin A was chemically heterogeneous. The main activity was associated with components of molecular weight exceeding that of the authentic heptadecapeptide. The presence of dynorphin A-like immunoreactivity in human CSF was first suggested by a study using an RIA for the dynorphin A fragment (1–13) (WAHLSTRÖM and TERENIUS 1980). We have later observed an in-

Table 2. Levels of opioid peptides in human CSF determined by RIA. Mean values of data obtained from 5–6 individual patients with suspected but unconfirmed neurologic deficits; they are not corrected for recovery. (Data from NYBERG et al. 1986)

Peptide	Concentration (fmol/ml)
β-Endorphin	21
Met-enkephalin	51
Met-enkephalin-Lys^6	170
Leu-enkephalin[a]	11
Dynorphin A	9.3
Dynorphin A (1–8)	< 3
Dynorphin B	< 1
Leu-enkephalin-Arg^6	41

[a] May derive from both proenkephalin A and B.

crease in RIA-detectable dynorphin A in cat CSF during sciatic nerve stimulation (YAKSH et al. 1983). The identity of the measured dynorphin A immunoreactivity and the synthetic peptide was confirmed by gel filtration (Sephadex G-10) and electrophoresis (NYBERG et al. 1983b).

The existence of immunoreactive dynorphin A in human CSF was also reported by TOZAWA et al. (1984). To some extent the characteristics of their immunoreactive material differed from our observations. They described much lower proportions of the high molecular weight dynorphin activity than we found in the fluid (NYBERG et al. 1986); on the other hand, they found an additional component tentatively identified as dynorphin A (1–13). This is explainable since their antiserum, in contrast to ours, had no specificity toward the COOH terminus of the dynorphin A molecule. A recent report describes the presence of the dynorphin A (1–8) fragment in human CSF (ZHANG et al. 1985). The CSF level of the peptide ranged from an average of 90 fmol/ml (in schizophrenic patients) up to 130 fmol/ml (controls). However, no separation of the CSF specimens prior to assay was reported in this study. Using a radioreceptorassay in combination with HPLC and electrophoresis, we had previously observed a CSF component among the so-called fraction I endorphins, which was chromatographically indistinguishable from dynorphin A (1–8) (NYBERG et al. 1983a). With radioimmunoassay, however, only trace amounts of this fragment were detected (NYBERG et al. 1986). Our studies have also been focused on the dynorphin A (1–6) fragment, i.e., Leu-enkephalin-Arg^6. This is the common NH_2 terminal sequence of all opioid peptides derived from proenkephalin B. Its levels in CSF greatly exceed those of both dynorphin A and dynorphin B (see Table 2). This observation is compatible with the presence of a Leu-enkephalin-Arg^6-generating endopeptidase in the CSF, as recently reported (NYBERG et al. 1985). The enzyme was found to specifically cleave dynorphin A, dynorphin B, and α-neoendorphin at the Arg^6–Arg^7 or Arg^6–Lys^7 bonds, respectively.

The difficulty in quantitating proenkephalin B peptides in plasma is typified by the study of plasma dynorphin reported by HOWLETT et al. (1984). They found that the apparent dynorphin A activity detected by RIA in the fluid was artifactual and caused by enzymatic degradation of tracer. With a Sep-Pak C_{18} extraction procedure, no dynorphin A activity was recovered from acidified plasma. However, the use of an affinity chromatography procedure for nonacidified plasma (the antibody was coupled to CNBr-activated Sepharose) resulted in reliable extraction of a dynorphin A immunoreactive component. This component, on the other hand, was chromatographically different from the heptadecapeptide. The presence of Leu-enkephalin-containing structures in human plasma has also been suggested by other studies (BOARDER et al. 1982b). We have recently identified a large Leu-enkephalin-Arg^6-containing entity circulating in the plasma of a schizophrenic patient (NYBERG et al., submitted).

Most radioimmunoassays for dynorphin-related peptides use the iodinated peptide as tracer. Separation of bound and free peptide is achieved by the second antibody precipitation procedure (CHRISTENSSON-NYLANDER et al. 1985) or by the charcoal adsorption technique (HOWLETT et al. 1984). The addition of detergent, e.g., Triton X-100, is necessary to prevent adsorption of peptide on the walls of test tubes in these assays. Preseparation of plasma or CSF samples is in general accomplished by extraction with reverse-phase cartridges, gel filtration, or ion exchange chromatography.

4. Sequential Enzymatic Radioimmunoassay Procedures for Proenkephalin-Derived Peptides

Procedures to detect all possible opioid peptides derived from proenkephalin A or B have been described by several laboratories. Most investigators have worked with the enzymatic method of Udenfriend and collaborators (Lewis et al. 1978) with sequential trypsin, carboxypeptidase treatment followed by enkephalin pentapeptide RIAs. In our CSF studies (Nyberg and Terenius 1985; Nyberg et al. 1986), we have used a modification of this method. Thus, in accordance with Cupo et al. (1984), our approach is to omit carboxypeptidase treatment, and requires only one enzymatic step which provides better recovery. Moreover, trypsin treatment of any enkephalin precursor (with COOH terminal extension of at least two residues) would release products maintaining their identity of origin, Leu-enkephalin-Arg^6 being uniquely formed from proenkephalin B and Met-enkephalin-Arg^6/Lys^6 and Leu-enkephalin-Lys^6 from proenkephalin A. By selective assays of these peptides, it is theoretically possible to discriminate between all possible opioid hexapeptides generated from the two enkephalin systems. However, a problem has been that hexapeptides released from proenkephalin A cross-react in the RIA for Leu-enkephalin-Arg^6. Therefore, in the presence of peptides from both proenkephalin A and B, evaluation of data requires correction for cross-reaction. The validity of this approach has been confirmed by HPLC analysis. To minimize the contribution of cross-reacting NH_2 terminal extensions of the enkephalin hexapeptides, which may be present in the trypsin digest, the degradation mixture is separated on an SP-Sephadex ionexchanger (Bergström et al. 1983) prior to RIA analysis.

Using this sequential trypsin degradation/RIA procedure, we were able to identify several Leu- and Met-enkephalin-containing polypeptides in human CSF

Table 3. Multiple precursors to enkephalin peptides of various sizes in human CSF. (Data from Nyberg et al. 1986)

Estimated molecular weight[a]	Peptides derived from	
	Proenkephalin A (equiv./ml × 10^{15})	Proenkephalin B (equiv./ml × 10^{15})
20–25000	30	2
15000	23	
9–10000	26	4
5000		9
3000		10
1.5–2000		13
1300	9	
700	25	5

[a] Molecular weight for enkephalins 500.

(NYBERG and TERENIUS 1985; NYBERG et al. 1986). It appeared that proenkephalin A-related structures predominate in the high molecular weight region (> 5 kdalton), whereas Leu-enkephalin-containing peptides showed highest concentrations in the intermediate 1–3 kdalton range (Table 3). Furthermore, the large structures were detected in concentrations comparable to those found for smaller peptides, e.g., β-endorphin and enkephalins (cf. Tables 2 and 3). Polypeptides derived from proenkephalin A have also been identified in rat CSF by use of an RIA procedure for the COOH terminal heptapeptide sequence of the precursor, Met-enkephalin-Arg^6Phe^7 (JACKSON et al. 1985).

Opioid active tryptic fragments have also been observed in studies of human plasma (BOARDER et al. 1982b). The study involved trypsin treatment of gel-filtered plasma fractions and the released activity was monitored by receptor assay. In a similar study on plasma from a schizophrenic patient, we were able to identify tryptic fragments cross-reacting in the RIA for Leu-enkephalin-Arg^6 (NYBERG et al., 1987). Judging from their chromatographic behavior, some of these enkephalin hexapeptide-containing fractions were of comparatively high molecular weight (5–10 kdalton). Further studies of those components are in progress.

D. Functional Aspects

I. Opioid Peptides in Biologic Fluids

Most clinical studies of opioid peptides have focused on their levels in CSF or plasma. Since CSF is in direct contact with the CNS, it seems likely that the CSF levels reflect neuropeptide activity in CNS regions more adequately than that in the plasma. Another point in favor of analysis of CSF components is the low concentrations of total protein. However, precautions may have to be taken against enzymatic degradation, a phenomenon not generally acknowledged, despite reports of enzymes in CSF (HAZATO et al. 1983; SCHWEISFURTH and SCHIÖBERG-SCHIEGNITZ 1984; NYBERG et al. 1985; LANTZ and TERENIUS 1985). The limited accessibility of CSF is another problem; even if one tap can be considered as clinical routine, repeated sampling is rarely possible. On the other hand, plasma sampling can be carried out very frequently without greatly affecting the patient. This makes plasma analysis ideal for longitudinal studies within the same patient. Limitations with endorphin analysis in plasma samples have several origins, some of which may be difficult to overcome. First, plasma levels of endogenous opioids mainly represent contributions from peptides produced outside the blood–brain barrier, for instance in the pituitary and adrenal glands or in paracrine tissue of the gastrointestinal tract. In fact, it has been shown that CSF levels of, say, β-endorphin do not parallel those in plasma, thus suggesting a different origin of the peptide in the two fluids. Thus, plasma analysis rarely reflects endorphin activity in CNS directly. Second, proteases are ubiquitous in plasma and vascularized tissues, leading to rapid and extensive degradation.

Quantitative analysis of opioid peptides in CSF or plasma may be of interest for two main reasons. First, they may serve as markers of functional activity. Second, the compounds as measured in the two compartments may be of relevance as mediators of hormonal action.

1. Markers of Functional Activity

It seems clear that changes in opioid peptide levels in CSF or plasma may be a reflection of activities in neuronal or nonneuronal cells expressing these peptide systems. For instance, immunoreactive β-endorphin in human ventricular CSF rises after focal electrical brain stimulation for treatment of intractable pain (AKIL et al. 1978b; HOSOBUCHI et al. 1979). Increased CSF endorphins have also been observed in cats and rats after sciatic nerve stimulation (YAKSH et al. 1983). However, in this study, the peptide species showing most consistent increases did not correspond to major peptides common in brain tissue, which were used as markers. This is in agreement with the observation that major opioid peptides in human CSF are distinct from "common" peptides and represent higher molecular weight forms (NYBERG et al. 1986). One likely explanation is differential metabolic stability favoring peptides of somewhat higher molecular weight. Markers of activity in any of the three opioid peptide systems may also be sought among sequences not having the opioid message. Thus, NH_2 terminal segments of proenkephalin A (syn-enkephalin, LISTON and ROSSIER 1984) or proopiomelanocortin (CHAN et al. 1983) have been measured by specific RIAs. The latter group reported elevated level NH_2 terminal fragments in patients with Cushing's disease, Nelson's syndrome, and other disorders of the pituitary–adrenal axis or in chronic renal failure. The NH_2 terminal fragments may be useful markers of gene activity, both because they are metabolically stable and because there may be less potential for cross-reaction with other peptide sequences; however, they cannot reflect processing events involving the opioid, i.e., the message sequences.

2. Hormones

Opioid peptides may have an endocrine role, both within the CNS and in the conventional sense via circulation in plasma. Within the CNS, there are fibers with proopiomelanocortin which terminate in ventricular walls, and which may release β-endorphin in a neuroendocrine fashion. These peptides may interact with CNS opioid receptors as previously discussed. With opioid peptides released into plasma, there is uncertainty as to possible target organs. This is partly due to the relatively meager data on peripheral effects of opioid peptides given systemically. Their putative role in stress and shock (see Sect. E.II) suggests that possible target organs are: vascular beds, particularly in the lung; nerve plexi in the gastrointestinal tract and sympathetic ganglia (which however, are probably protected by nerve–blood barriers); and probably endocrine tissues. A direct diuretic effect of dynorphin on κ-receptors in renal tubules has been described (SLIZGI et al. 1984). Apparently, opioid peptides may have actions quite different from those typical of classical opiates; the definition of these actions is still incomplete and remains important.

E. Clinical Aspects

In a number of studies, opioid peptides as measured in body fluids have been taken as an index of functional activity. The following survey is by no means exhaustive.

I. Pain

The presence of endorphins and opioid receptors in the regions of the CNS concerned with pain transmission supports their involvement in endogenous pain suppression mechanisms (for review see TERENIUS 1982). Of particular interest in the context of this chapter is the involvement of endorphins in stimulation-produced analgesia (TERENIUS 1981, 1982; AKIL and WATSON 1984). Stimulation-produced analgesia has been shown to be partially reversed by naloxone, a potent narcotic antagonist, in rats (AKIL et al. 1976), in cats, and humans (HOSOBUCHI et al. 1977; AKIL et al. 1978b). It has also been shown that there is concurrent increase of endorphins in the CSF in rats and cats (CESSELIN et al. 1982; YAKSH and ELDE 1981; YAKSH et al. 1983). In patients with pain, the release of endogenous opioids into the CSF by clinically effective stimulation has been demonstrated (AKIL et al. 1978b; CLEMENT-JONES et al. 1980a; HOSOBUCHI et al. 1979; SJÖLUND et al. 1977; VON KNORRING et al. 1978).

Another phenomenon which associates the endorphins with pain is the paradigm of stress-induced analgesia. It is known that stress-induced analgesia has both opioid and nonopioid components (BODNAR et al. 1978; LEWIS et al. 1980). The work of Liebeskind's group has shown that the footshock stressor with different parameters and duration can lead to either nonopioid-mediated or opioid-mediated stress-induced analgesia (LEWIS et al. 1980). Footshock stress, under conditions which produce naloxone-reversible analgesia, brings about a significant depletion of β-endorphin-like material from the anterior pituitary and an increase in the plasma concentration of β-endorphin (GUILLEMIN et al. 1977; AKIL and WATSON 1982). Evidence suggesting that dynorphin levels in hypothalamus and pituitary are altered by stress-induced analgesia has also been reported (AKIL et al. 1984).

II. Stress, Physical Exercise, and Shock

Since proopiomelanocortin is the prohormone common to the stress hormone ACTH and to β-endorphin, it is logical to assume that β-endorphin secretion occurs as part of the stress response. Likewise, the enkephalin peptides synthesized in the adrenal medulla are present in the same cells as the catecholamines and stress hormones, and are likely to be released in concert. A more detailed discussion of endogenous opioids and stress is given by AKIL et al. (1984).

Recent interest in the endocrine consequences of long-distance running and other exhausting sport activities, has been extended to studies of endorphins. Thus, moderately exhaustive or long-distance running increases plasma levels of β-endorphin and ACTH (GAMBERT et al. 1981; JANAL et al. 1984). There were accompanying increases in pain tolerance and subjective effects of mental well-being, the "runner's high." Increases in plasma β-endorphin were recently reported to be even more important in the early postexercise period (DE MEIRLEIR et al. 1985).

A related area of study is shock. The functional relationships and the cardiovascular effects of opioids have recently been reviewed (HOLADAY 1983). However, very little is known about which opioid peptides are released into plasma or other body fluids during shock.

III. Narcotic Dependence

The role of opioid receptors and endorphins in opioid tolerance and dependence has been reviewed by TERENIUS (1984). Several investigators have analyzed opioid peptides in body fluids of addicts. For instance, increased β-endorphin levels were observed in CSF and plasma of addicts in withdrawal (CLEMENT-JONES et al. 1979; EMRICH et al. 1983). Evidence for changes of receptor-active CSF endorphins during withdrawal with a minimum level at 26 h followed by a rebound has also been reported (HOLMSTRAND et al. 1981; O'BRIEN et al. 1983). However, measurements of Met-enkephalin gave no evidence of changes (CLEMENT-JONES et al. 1979). Furthermore, an intravenous dose of naloxone in addicts previously maintained on methadone and in a detoxification period failed to raise plasma levels of proopiomelanocortin-derived peptides, while in healthy volunteers it more than tripled the level (GOLD et al. 1981). Progress in the understanding of the biosynthesis of the endorphins has provided new directions for studies of the opioid peptide systems in drug dependence. For example, it was shown that an observed decrease in pituitary β-endorphin in morphine-treated rats was caused by a decrease in the biosynthesis of proopiomelanocortin (HÖLLT et al. 1981; GIANOULAKIS et al. 1982).

IV. Psychiatry

In recent years, extensive investigation has been focused on the role of opioid peptides in psychiatric diseases (for reviews see WATSON et al. 1983; TERENIUS and NYBERG 1986). The anatomic distribution of endogenous opioid peptides with dense innervation in limbic structures suggests involvement in regulation of mood and behavior. It is also known that the endorphins interact with central catecholamines that are implicated in psychiatric disorders. The first biochemical evidence in support of a link between endorphins and psychosis was reported by TERENIUS et al. (1976). Increased levels of radioreceptor-active opioid activity in the CSF of schizophrenic and manic depressive subjects were observed. This study and the observation of an antihallucinatory effect of naloxone in certain cases of schizophrenia (GUNNE et al. 1977) have been followed up and extended in many different laboratories (BLOOM et al. 1976; DOMSCHKE et al. 1979; EMRICH et al. 1979; GERNER and SHARP 1982; LINDSTRÖM et al. 1978, 1982; LO et al. 1983; NABER et al. 1981; NAKAO et al. 1980; PICKAR et al. 1982; RIMON et al. 1980; WATSON et al. 1983). In most of these studies, individual opioid peptides such as β-endorphin or enkephalins have been quantitated by RIA, although certain studies have used radioreceptorassay (RRA). The strongest correlation between elevated endorphins and the presence and severity of psychiatric disorder arises from studies based on measurements of proenkephalin A-derived peptides. Most studies focused on β-endorphin appear negative, whereas data from proenkephalin B-derived peptides is still needed. However, ZHANG et al. (1985) reported dynorphin A (1–8) to be lower in patients with schizophrenia than in controls. It has been suggested that the major pool of opioid peptides recognized by receptor assay is related to the proenkephalin A family (NYBERG et al. 1983 a).

V. Neuroendocrinology and Endocrine Tumors

The very high concentration of opioid peptides in the hypothalamus suggests that they may be important in neuroendocrine regulation. In fact, several indices of endorphin activity are neuroendocrine. It has been shown that Met-enkephalin and β-endorphin are potent stimulators of prolactin and growth hormone secretion in rats and in humans (BRUNI et al. 1977; STUBBS et al. 1978; REID et al. 1981 a, b). It seems likely that in both species endorphins control prolactin release predominantly through an interaction with hypothalamic dopamine (RAGAVAN and FRANTZ 1981; VAN LOON et al. 1980). For growth hormone the situation appears more complex with evidence for multiple neurotransmitter interactions in rats (KATAKAMI et al. 1981; RIVIER et al. 1977; TERRY et al. 1982) and a possible cholinergic link in humans (MAYER and KOBBERLING 1981; DELITALA et al. 1983). An inhibitory effect of exogenous opioids has been observed for thyrotropin (TSH) secretion in rats and humans, and this effect is easily reversed by naloxone (BRUNI et al. 1977; GROSSMAN et al. 1981 b). It has been suggested that the opioid control of TSH secretion is mediated via an inhibition of hypothalamic dopamine, as is the situation with prolactin (DELITALA et al. 1981 a, b; SHARP et al. 1981). However, the magnitude of the opioid effect does not indicate that endorphins are likely to be important in the physiologic control of TSH in humans.

The effects of opioids on the gonadotropins in humans are broadly similar to those observed in rats; with opioid peptides decreasing and naloxone increasing their levels (CICERO 1980; KINOSHITA 1980, 1982; MORLEY 1980). These changes appear to be due to opioid modulation of the pulsatile release of luteinizing hormone-releasing hormone (LHRH) (DROUVA et al. 1981; DELITALA et al. 1981 c; GROSSMAN et al. 1981 b). In rats, opioid modulation of LHRH release may be important in the initiation of puberty and the control of sexual behavior (WILKINSON and BHANOT 1982; SIRINATHSINGHJI et al. 1983), whereas in humans the clearest opioid control of gonadotropins is exemplified by the change in LH pulsatility associated with hyperprolactinemia (GROSSMAN et al. 1982).

There is clear evidence that hypothalamic opioid peptide systems suppress the pituitary–adrenal axis in both humans and rats (STUBBS et al. 1978; DELITALA et al. 1981 a; FERRI et al. 1982), giving a decrease in circulating ACTH, LPH, and cortisol. For instance, in Addison's disease there is a loss of steroidal feedback on ACTH release, and the effect of exogenous opioids in suppressing circulating ACTH is very much magnified (GAILLARD et al. 1981; ALLOLIO et al. 1982). It is likely that the effect of opioids on the corticotrophs is mediated through inhibition of corticotropin-releasing factor (CRF) release. Opioids are probably involved in the release of CRF in response to stress (BUCKINGHAM 1982), but the precise details of this control mechanism are unknown.

An inhibitory effect of opioids on the release of posterior pituitary peptide hormones has also been demonstrated. Thus, there is increasing evidence that opioid peptides are potent inhibitors of vasopressin (AZIZ et al. 1981; IVERSEN et al. 1980; GROSSMAN et al. 1980; REID et al. 1981 b) and oxytocin release (HALDAR and BADE 1981; LUTZ-BUCHER and KOCH 1980). However, the receptors mediating this response are relatively insensitive to naloxone (LIGHTMAN et al. 1982). It therefore seems likely that the opioid effects on the posterior pituitary are mediated through sites with poor affinity for μ-ligands, e.g., δ- or κ-receptors. This would be compatible with the high concentrations of Leu-enkephalin and dynorphin found in this region of the gland (WATSON et al. 1982).

Several tumors of both endocrine and nonendocrine origin associated with disturbed opioid peptide involvement have been found in humans. Among the endocrine tumors, the β-endorphin/ACTH-producing pituitary adenomas connected with Cushing's disease and Nelson's syndrome are common (GILLIES et al. 1980; SUDA et al. 1979, 1980). Moreover, β-endorphin-like immunoreactivity has been observed in several nonendocrine tumors associated with the ectopic ACTH syndrome (ORTH et al. 1978; HASHIMOTO et al. 1980; PULLAN et al. 1980). These tumors were also found to contain Met-enkephalin immunoreactive material. High concentrations of Met-enkephalin and its high molecular weight precursor have been detected in human adrenal medullary pheochromocytomas (CLEMENT-JONES et al. 1980 b). These tumors have been used for the isolation of Met-enkephalin from human tissue (CLEMENT-JONES et al. 1980 b). The clinical sequelae of the opioid peptides found in endocrine and nonendocrine tumors, however, are not yet fully clarified.

F. Concluding Remarks

Although this chapter is by no means exhaustive, it illustrates the broad biologic and clinical context of RIA analysis of opioid peptides in body fluids. The explosive growth in interest in the endorphin field which occurred during the last decade can explain some of the conceptual errors and methodological flaws which abound in the literature. For instance, one critical issue which is often overlooked is that of using the proper compartment (fluid). There is no evidence that opioid peptide secretion into CSF and plasma is coordinated and there is restricted passage of opioid peptides from plasma into CSF. Far too many studies take the stand that opioid peptides as measured in plasma can be used as an index of CNS activity. CSF levels are likely to be a much better choice. Also relevant to functional indications is the selection of the proper peptide for measurement. The most frequent choice is β-endorphin, whereas the proenkephalin-derived peptides, which are more abundant, have received much less attention. This is partly on methodological grounds; for instance, commercial RIA kits are available for β-endorphin measurements, but not until recently for other opioid peptides.

The chemical relevance of measured activity needs continued attention and there is much room for methodological development. The sampling protocol is also critical, particularly in plasma analysis, and needs to be meticulously standardized. Authors and referees have the responsibility of ensuring that papers describing RIA analysis of opioid peptides in body fluids have adequate detail in the description of sampling protocol, preseparation procedures, and assay conditions. There is still a need to identify the chemical nature of opioid peptides in body fluids and to establish functional relationships, i.e., suitable markers for measurement.

The complexity of the opioid peptide systems is baroque, both in chemical and anatomic/physiologic terms. It will require considerable unraveling to identify functional roles in this variety of systems. RIA analysis of body fluids will remain important, particularly in studies of the normal physiology and pathophysiology in humans.

Acknowledgments. The authors are supported by the Swedish Medical Research Council (Grant 21X-5095).

References

Akil H, Watson S (1982) Opioid peptides and pain. In: Fink G, Whalley L (eds) Neuropeptides: basic and clinical aspects. Churchill Livingstone, London, pp 50–58

Akil H, Madden J, Patrick RL, Barchas JD (1976) Stress-induced increase in endogenous opiate peptides: concurrent analgesia and its partial reversal by naloxone. In: Kosterlitz H (ed) Opiates and endogenous opioid peptides. Elsevier, Amsterdam, pp 63–70

Akil H, Watson S, Sullivan S, Barchas JD (1978 a) Enkephalin-like material in human CSF: measurements and levels. Life Sci 23:121–126

Akil H, Hughes J, Richardson DE, Barchas JD (1978 b) Enkephalin-like material elevated in ventricular cerebrospinal fluid of pain patients after analgetic focal stimulation. Science 201:463–465

Akil H, Watson S, Young E, Lewis M, Khachaturian H, Walker M (1984) Endogenous opioids: biology and function. Annu Rev Neurosci 7:223–255

Allolio B, Winkelmann W, Hipp FX, Kaulen D, Miles R (1982) Effects of a Met-enkephalin analog on adrenocorticotropin (ACTH), growth hormone, and prolactin in patients with ACTH hypersecretion. J Clin Endocrinol Metab 55:1–7

Aziz IA, Forsling ML, Wolf CJ (1981) The effects of intracerebroventricular injections of morphine on vasopressin release in the rat. J Physiol (Lond) 311:401–409

Baird A, Klepper R, Ling N (1984) In vitro and in vivo evidence that the C-terminus of preproenkephalin-A circulates as an 8500-dalton molecule. Proc Exp Biol Med 175:304–308

Bergström L, Christensson I, Folkesson R, Stenström B, Terenius L (1983) An ion-exchange chromatography and radioimmunoassay procedure for measuring opioid peptides and substance P. Life Sci 33:1613–1619

Bloch B, Bugnon C, Fellman D, Lenys D (1978) Immunocytochemical evidence that the same neurons in the human infundibular nucleus are stained with antiendorphins and antisera of other related peptides. Neurosci Lett 10:147–152

Bloom F, Segal D, Ling N, Guillemin R (1976) Endorphins: profound behavioral effects in rats suggest new etiological factors in mental illness. Science 194:630–632

Bloom FE, Battenberg E, Rossier J, Ling N, Leppaluoto J, Vargo TM, Guillemin R (1977) Endorphins are located in the intermediate and anterior lobes of the pituitary gland, not in the neurohypophysis. Life Sci 20:43–48

Bloom FE, Rossier J, Battenberg E, Bayon A, French E, Henricksen SJ, Siggins GR, Segal D, Browne R, Ling N, Guillemin R (1978) Betaendorphin: cellular localization, electrophysiological and behavioral effects. Adv Biochem Psychopharmacol 18:89–109

Boarder M, Lockfeld J, Barchas J (1982a) Measurement of methionine-enkephalin (Arg^6, Phe^7) in rat brain by specific radioimmunoassay directed at methionine sulphoxide enkephalin- (Arg^6, Phe^7). J Neurochem 38:299–304
Boarder M, Erdelyi E, Barchas J (1982b) Opioid peptides in human plasma: evidence for multiple forms. J Clin Endocrinol Metab 54:715–720
Bodnar RJ, Kelly DD, Spiaggia A, Ehrenberg C, Glusman M (1978) Dose dependent reductions by naloxone of analgesia induced by cold-water stress. Pharmacol Biochem Behav 8:661–666
Bonnet KA, Grotin J, Gioamnini T, Cortes M, Simon EJ (1981) Opiate receptor heterogeneity in human brain regions. Brain Res 221:437–440
Bruni JF, van Vugt DA, Marshall S, Meites J (1977) Effects of naloxone, morphine and methionine enkephaline on serum prolactin, luteinizing hormone, follicle-stimulating hormone, thyroid stimulating hormone and growth hormone. Life Sci 21:461–466
Buckingham JC (1982) Secretion of corticotropin and its releasing factor in response to morphine and opioid peptides. Neuroendocrinology 35:111–116
Burbach JPH, Loeber JG, Verhof J, deKloet ER, van Ree JM, de Wied D (1979) Schizophrenia and degradation of endorphins in cerebrospinal fluid. Lancet II:480–481
Cahill CA, Akil H (1982) Plasma betaendorphin-like immunoreactivity, self reported pain perception and anxiety levels in women during pregnancy and labor. Life Sci 31:1879–1882
Cahill CA, Matthews JD, Akil H (1983) Human plasma beta-endorphin-like peptides: a rapid, high recovery extraction technique and validation of radioimmunoassay. J Clin Endocrinol Metab 56:992–997
Cesselin F, Olivieras JL, Bourgoin S, Sieralta F, Michelit R, Besson JM, Hamon M (1982) Increased levels of Met-enkephalin-like material in the CSF and anaesthetized cells after tooth pulp stimulation. Brain Res 239:325–330
Chan JSD, Seidh NG, Chrétien M (1983) Measurement of N-terminal (1-76) of human proopiomelanocortin in human plasma: correlation with adrenocorticotropin (ACTH). J Clin Endocrinol Metab 56:791–796
Chang K-J, Cooper BR, Hazum E, Cuatrecasas P (1979) Multiple opiate receptors: different regional distribution in the brain and differential binding of opiates and opioid peptides. Mol Pharmacol 16:91–104
Chou J, Tang J, Costa E (1983) Met^5-enkephalin-Arg^6-Phe^7 content of human and rabbit plasma. Life Sci 32:2589–2595
Christensson-Nylander I, Nyberg F, Ragnarsson U, Terenius L (1985) A general procedure for analysis of proenkephalin B derived opioid peptides. Regul Pept 11:65–76
Cicero TJ (1980) Effects of exogenous and endogenous opiates on the hypothalamic-pituitary-gonadal axis in the male. Fed Proc 39:2551–2554
Cleeland CS, Shachman S, Dahl JL, Orrison W (1984) CSF beta-endorphin and the severity of pain. Neurology 34:378–380
Clement-Jones V, McLoughlin L, Lowry PJ, Besser GM, Rees LH, Wen HL (1979) Acupuncture in heroin addicts: changes in met-enkephalin and beta-endorphin in blood and cerebrospinal fluid. Lancet II:380–383
Clement-Jones V, McLoughlin L, Tomlin S, Besser GM, Rees LH, Wen HL (1980a) Increased beta-endorphin but not met-enkephalin levels in human cerebrospinal fluid after acupuncture for recurrent pain. Lancet II:946–949
Clement-Jones V, Corder R, Lowry PJ (1980b) Isolation of human met-enkephalin and two groups of putative precursors (2K-pro-met-enkephalin) from an adrenalmedullary tumour. Biochem Biophys Res Commun 95:665–673
Clement-Jones V, Lowry PJ, Rees LH, Besser GM (1980c) Met-enkephalin circulates in human plasma. Nature 283:295–297
Clement-Jones V, Lowry PJ, Rees LH, Besser GM (1980d) Development of a specific extracted radioimmunoassay for methionine enkephalin in human plasma and cerebrospinal fluid. J Endocrinol 86:231–243
Cupo A, Pontarotti PA, Jarry T, Delaage M (1984) A new immunological approach to the detection and the quantitation of the Met^5-enkephalin precursors in rat brain. Neuropeptides 4:375–387

Delitala G, Devilla L, Canessa A, D'Astra F (1981 a) On the role of dopamine receptors in the central regulation of human TSH. Acta Endocrinol 98:521–527
Delitala G, Grossman A, Besser GM (1981 b) Changes in pituitary hormone levels induced by Met-enkephalin in man – the role of dopamine. Life Sci 29:1537–1544
Delitala G, Devilla L, Arata L (1981 c) Opiate receptors and anterior pituitary hormone secretion in man. Effect of naloxone infusion. Acta Endocrinol 97:150–156
Delitala G, Grossman A, Besser GM (1983) Opiate peptides control growth hormone through a cholinergic mechanism in man. Clin Endocrinol 18:401–405
De Meirleir K, Naaktgeboren N, van Steirteghem A, Block P (1985) Plasma beta-endorphin levels after exercise. J Endocrinol Invest 8:89
Detrick JM, Pearson JW, Frederickson RCA (1985) Endorphins and parturition. Obstet Gynecol 65:647–651
Domschke W, Dichschas A, Mitznegg P (1979) CSF betaendorphin in schizophrenia. Lancet I:1024
Drouva SV, Epelbaum J, Tapia-Arancibia L, Laplante E, Kordon C (1981) Opiate receptors modulate LHRH and SRIF release from mediobasal hypothalamic neurons. Neuroendocrinology 32:163–167
Ekman R (1985) Radioimmunoassay of neuropeptides in the CSF. Problems and pitfalls. Nord Psykiatr Tidsskr 39 [Suppl 11]:31–34
Elde R, Hökfelt T, Johansson O, Terenius L (1976) Immunohistochemical studies using antibodies to leucine enkephalin: initial observations on the nervous system of the rat. Neuroscience 1:349–351
Emrich HM, Höllt V, Kissling W, Fischler M, Laspe H, Heinemann H, von Zerssen D, Hertz A (1979) Beta-endorphin-like immunoreactivity in cerebrospinal fluid of patients with schizophrenia and other neuropsychiatric disorders. Pharmakopsychiatrie 12:269–276
Emrich HM, Nusselt L, Gramsch C, John S (1983) Beta-endorphin immunoreactivity in plasma increases during withdrawal. Pharmakopsychiatrie 16:93–96
Facchinetti F, Genazzani AR (1979) Simultaneous radioimmunoassay of beta-LPH and beta-EP in human plasma. In: Albertini A, Da Prada M, Peskar BA (eds) Radioimmunoassay of drugs and hormones in cardiovascular medicine. North-Holland, Amsterdam, pp 347–354
Facchinetti F, Bagnoli R, Petraglia F, Parrini D, Sardelli S, Genazzani AR (1983) Fetomaternal opioid levels and parturition. Obstet Gynecol 62:764–768
Ferri S, Candeletti S, Cochi D, Giagnoni G, Rossi T, Scoto G, Spampinato S (1982) Interplay between opioid peptides and pituitary hormones. Adv Biochem Psychopharmacol 33:109–115
Fischli W, Goldstein A, Hunkapiller MW, Hood LE (1982) Isolation and amino acid sequence analysis of a 4000-dalton dynorphin from porcine pituitary. Proc Natl Acad Sci USA 79:5435–5437
Gaillard RC, Grossman A, Smith R, Rees LH, Besser GM (1981) The effects of a Met-enkephalin analogue on ACTH, beta-LPH, beta-endorphin and Met-enkephalin in patients with adrenocortical disease. Clin Endocrinol 14:471–478
Gambert S, Garthwaite T, Ponter C, Cook E, Tristani F, Duthie E, Martinson D, Hagen T, McCarty D (1981) Running elevates plasma beta-endorphin immunoactivity and ACTH in untrained human subjects. Proc Soc Exp Biol Med 168:1–4
Gerner R, Sharp B (1982) CSF beta-endorphin immunoreactivity in normal, schizophrenic, depressed, manic and anorexic subjects. Brain Res 237:244–247
Ghazarossian VE, Dent RR, Otsu K, Ross M, Cox B, Goldstein A (1980) Development and validation of a sensitive radioimmunoassay for naturally occurring beta-endorphin-like peptides in human plasma. Anal Biochem 102:80–87
Gianoulakis C, Drouin J-N, Seidah NG, Kalant H, Chrétien M (1982) Effect of chronic morphine treatment on beta-endorphin biosynthesis by the rat neurointermediate lobe. Eur J Pharmacol 72:313–321
Gillies C, Ratter S, Grossman A, Gaillard R, Lowry PJ, Besser GM, Rees LH (1980) Secretion of ACTH, LPH and beta-endorphin from human pituitary tumours in vitro. Clin Endocrinol 13:197–205

Gold MS, Pottash ALC, Extein I, Martin DA, Finn LB, Sweeney KR, Kleber HD (1981) Evidence for an endorphin dysfunction in methadone addicts: lack of ACTH response to naloxone. Drug Alcohol Depend 8:257–262

Goldstein A, Fischli W, Lowney LI, Hunkapiller M, Hood L (1981) Porcine pituitary dynorphin: complete aminoacid sequence of the biologically active heptadecapeptide. Proc Natl Acad Sci USA 78:7219–7223

Grossman A, Besser GM, Milles JJ, Baylis PH (1980) Inhibition of vasopressin release in man by an opiate peptide. Lancet II:1108–1110

Grossman A, Stubbs WA, Gaillard RC, Delitala G, Rees LH, Besser GM (1981 a) Studies on the opiate control of prolactin, GH and TSH. Clin Endocrinol 14:381–386

Grossman A, Moult PJA, Gaillard RC, Delitala G, Toff WD, Rees LH, Besser GM (1981 b) The opioid control of LH and FSH release: effects of a Met-enkephalin analogue and naloxone. Clin Endocrinol 14:41–47

Grossman A, West S, Williams J, Evans J, Rees LH, Besser GM (1982) The role of opioid peptides in the control of prolactin in the puerperium, and TSH in primary hypothyroidism. Clin Endocrinol 16:317–320

Gubler U, Seeburg P, Hoffman BJ, Gage LP, Udenfriend S (1982) Molecular cloning establishes proenkephalin as precursor of enkephalin-containing peptides. Nature (Lond) 295:206–208

Guillemin R, Vargo T, Rossier J, Minick S, Ling N, Rivier C, Vale W, Bloom F (1977) Beta-endorphin and adrenocorticotropin are secreted concomitantly by the pituitary. Science 197:1367–1369

Gunne LM, Lindström L, Terenius L (1977) Naloxone-induced reversal of schizophrenic hallucinations. J Neural Transm 40:13–19

Haldar J, Bade V (1981) Involvement of opioid peptides in the inhibition of oxcytocin release by heat stress in lactating mice. Proc Soc Exp Biol Med 168:10–14

Hashimoto K, Takahara J, Ogawa N, Yumoki S, Tadashi O, Arata A, Kanda S, Terada K (1980) Adrenocorticotropin, beta-lipotropin, beta-endorphin, and corticitropin-releasing factor-like activity in an adrenocorticotropin-producing nephroblastoma. J Clin Endocrinol Metab 50:461–465

Hazato T, Shimamura M, Katayama T, Kasama A, Nishioka S, Kaya K (1983) Enkephalin degrading enzymes in cerebrospinal fluid. Life Sci 33:443–448

Ho WKK, Wen HL, Ling N (1980) Beta-endorphin-like immunoactivity in the plasma of heroin addicts and normal subjects. Neuropharmacology 19:117–120

Hökfelt T, Elde R, Johansson O, Terenius L, Stein L (1977) The distribution of enkephalin-immunoactive cell bodies in rat central nervous system. Neurosci Lett 5:25–31

Hökfelt T, Ljungdahl LA, Terenius L, Elde RP, Nilsson G (1977) Immunohistochemical analysis of peptide pathways possibly related to pain and analgesia: enkephalin and substance P. Proc Natl Acad Sci USA 74:3081–3085

Holaday JW (1983) Cardiovascular effects of endogenous opiate systems. Annu Rev Pharmacol Toxicol 23:541–594

Höllt V, Przewlocki R, Hertz A (1978) Radioimmunoassay of beta-endorphin. Basal and stimulated levels in extracted rat plasma. Naunyn Schmiedebergs Arch Pharmacol 303:171–174

Höllt V, Müller OA, Fahlbusch R (1979) Beta-endorphin in human plasma: basal and pathologically elevated levels. Life Sci 25:37–44

Höllt V, Haarmann I, Hertz A (1981) Long-term treatment of rats with morphine reduces the activity of messenger ribonucleic acid coding for the beta-endorphin/ACTH precursor in the intermediate pituitary. J Neurochem 37:619–626

Holmstrand J, Gunne LM, Wahlström A, Terenius L (1981) CSF-endorphins in heroin addicts during methadone maintance and during withdrawal. Pharmacopsychiatria 14:126–128

Hosobuchi Y, Adams J, Linchitz R (1977) Pain relief by electric stimulation of the central grey matter in humans and its reversal by naloxone. Science 197:183–186

Hosobuchi Y, Rossier J, Bloom FE, Guillemin R (1979) Stimulation of human periaqueductal gray for pain relief increases immunoreactive beta-endorphin in ventricular fluid. Science 203:279–281

Howlett TA, Walker J, Besser GM, Rees LH (1984) "Dynorphin" in plasma: enzymatic artifact and authentic immunoreactivity. Regul Pept 8:131–140
Iversen LL, Iversen SD, Bloom FE (1980) Opiate receptors influence vasopressin release from nerve terminals in rat neurohypophysis. Nature (Lond) 284:350–353
Jackson SR, Corder R, Kiser S, Lowry PJ (1985) Proenkephalin peptides possessing Met-enkephalin-Arg^6-Gly^7-Leu^8-immunoreactivity in rat CSF, striatum and adrenal gland. Peptides 6:169–178
Janal MN, Colt EWD, Clark WC, Glusman M (1984) Pain sensitivity, mood and plasma endocrine levels in man following long-distance running: effects of naloxone. Pain 19:13–25
Jeffcoate WJ, McLoughlin L, Rees LH, Lowry JP, Hope J, Besser GM (1978) Beta-endorphin in human cerebrospinal fluid. Lancet II:119–121
Kakidani H, Furutani Y, Takahashi H, Noda M, Morimoto Y, Hirose T, Asai M, Inayama S, Nakanishi S, Numa S (1982) Cloning and sequence analysis of cDNA for porcine beta-neo-endorphin/dynorphin precursor. Nature (Lond) 298:245–249
Kangawa K, Matsuo H, Igarashi M (1979) Alpha-neo-endorphin: a "big" Leu-enkephalin with potent opiate activity from porcine hypothalami. Biochem Biophys Res Commun 86:153–160
Katakami H, Kato Y, Matsushita N, Shimatsu A, Imura H (1981) Possible involvement of gamma-aminobutyric acid in growth hormone release induced by a Met-enkephalin analog in conscious rats. Endocrinology 109:1033–1036
Kikuchi K, Tanaka M, Abe K, Yamaguchi K, Kimura S, Adachi I (1984) Rapid and specific radioimmunoassays for beta-endorphin and beta-lipotropin in affinity-purified human plasma. J Clin Endocrinol Metab 59:287–292
Kilpatrick DL, Jones BN, Kojima K, Udenfriend S (1981) Identification of the octapeptide Met-enkephalin-Arg^6-Gly^7-Leu^8 in extracts of bovine medulla. Biochem Biophys Res Commun 103:698–705
Kilpatrick DL, Wahlström A, Lahm HW, Blacher R, Udenfriend S (1982) Rimorphin, a unique, naturally occurring (Leu)enkephalin containing peptide found in association with dynorphin and alpha-neo-endorphin. Proc Natl Acad Sci USA 79:6480–6483
Kinoshita F, Nakai Y, Katakami H, Kato Y, Yajima H, Imura H (1980) Effect of beta-endorphin on pulsatile luteinizing hormone release in conscious castrated rats. Life Sci 27:843–846
Kinoshita F, Nakai Y, Katakami H, Imura H (1982) Suppressive effect of dynorphin-(1-13) on luteinizing hormone release in conscious castrated rats. Life Sci 30:1915–1919
Kiser RS, Jackson S, Smith R, Rees LH, Lowry PJ, Besser GM (1983) Endorphin-related peptides in rat cerebrospinal fluid. Brain Res 288:187–192
Kleine TO, Klempel K, Fünfgeld EW (1984) Methionine-(Met)-enkephalin and leucine-(Leu)-enkephalin in CSF of chronic schizophrenics: correlation with psychopathology. Protides Biol Fluids 32:175–178
Kosterlitz HW (1979) Interaction of endogenous opioid peptides and their analogs with opiate receptors. In: Bonica JJ, Liebeskind JC, Albe-Fessard DG (eds) Advances in pain research and therapy. Raven, New York, pp 377–384
Kosterlitz HW, Paterson SJ, Robson LE (1982) Opioid peptides and their receptors. In: Fink G, Whalley LJ (eds) Neuropeptides: basic and clinical aspects. Churchill Livingstone, Edinburgh, pp 3–11
Lantz I, Terenius L (1985) High enkephalyl peptide degradation due to angiotensin converting enzyme-like activity in human CSF. FEBS Lett 193:31–33
Lewis JW, Cannon JT, Liebeskind JC (1980) Opioid and non-opioid mechanisms of stress analgesia. Science 208:623–625
Lewis RV, Stein S, Gerber LD, Rubinstein M, Udenfriend S (1978) High molecular weight opioid-containing proteins in striatum. Proc Natl Acad Sci USA 75:4021–402
Li CH, Barnafi L, Chrétien M, Chung D (1965) Isolation and amino-acid sequence of beta-LPH from sheep pituitary glands. Nature 208:1093–1094
Lightmann SL, Langdon N, Todd K, Forsling M (1982) Naloxone increases the nicotine-stimulated rise of vasopressin secretion in man. Clin Endocrinol 16:353–358

Lindström LH, Widerlöv E, Gunne LM, Terenius L (1978) Endorphins in human cerebrospinal fluid: clinical correlation to some psychotic states. Acta Psychiatr Scand 57:153–164

Lindström LH, Besev G, Gunne LM, Sjöström R, Terenius L, Wahlström A, Wistedt B (1982) Cerebrospinal fluid content of endorphins in schizophrenia. In: Shah NS, Donald AG (eds) Endorphins and opiate antagonists in psychiatry. Plenum, New York, pp 245–256

Liston D, Rossier J (1984) Distribution of syn-enkephalin immunoreactivity in the bovine brain and pituitary. Regul Pept 8:79–87

Lo CW, Wen HL, Ho WKK (1983) Cerebrospinal fluid (Met^5)enkephalin level in schizophrenics during treatment with naloxone. Eur J Pharmacol 92:77–81

Lutz-Bucher B, Koch B (1980) Evidence for a direct inhibitory effect of morphine on the secretion of posterior pituitary hormones. Eur J Pharmacol 66:375–378

Lyrenäs S, Nyberg F, Lutsh H, Lindberg B, Terenius L (1987) Cerebrospinal fluid dynorphin and β-endorphin at term pregnancy and six months after delivery. No influence of acupuncture treatment. Acta Endocrinol, in press

Mayer G, Kobberling J (1981) Effect of pimozide and cyproheptadine on FK 33-824-induced hGH and prolactin secretion in normal man. Acta Endocrinol 97 [Suppl 243]: Abstract 41

Maysinger D, Höllt V, Seizinger BR, Mehraein P, Pasi A, Hertz A (1982) Parallel distribution of immunoreactive alpha-neo-endorphin and dynorphin in rat and human tissues. Neuropeptides 2:211–225

McLoughlin L, Lowry PJ, Ratter S, Besser GM, Rees LH (1980) Beta-endorphin and beta-MSH in human plasma. Clin Endocrinol 12:287–292

Morley JE, Baranetsky NG, Wingert TD, Carlson HE, Hershman JM, Melmed S, Levin SR, Jamison KR, Weitzman R, Chang RJ, Varner AA (1980) Endocrine effects of naloxone-induced opiate receptor blockade. J Clin Endocrinol Metab 50:251–257

Naber D, Pickar D, Post RM, van Kammen DP, Waters RN, Ballenger JC, Goodwin FK, Bunney WE Jr (1981) Endogenous opioid activity and beta-endorphin immunoreactivity in CSF of psychiatric patients and normal volunteers. Am J Psychiatr 138:1457–1462

Nakanishi S, Inoue A, Kita T, Nakamura M, Chang ACY, Cohen SN, Numa S (1979) Nucleotide sequence of cloned cDNA for bovine corticotropin-beta-lipotropin precursor. Nature 278:423–427

Nakao K, Nakai Y, Oki S, Horii K, Imura H (1978) Presence of beta-endorphin in normal human plasma: a concomitant release of beta-endorphin with adrenocorticotropin after metyrapone administration. J Clin Invest 62:1395–1398

Nakao K, Oki S, Tanaka I, Horii K, Nakai Y, Furui T, Fukushima M, Kawayama A, Kageyama N, Imura H (1980) Immunoreactive beta-endorphin and adrenocorticotropin in human cerebrospinal fluid. J Clin Invest 66:1383–1390

Neuser D, Lesch KP, Stasch JP, Przuntek H (1984) Beta-endorphin-, leucine enkephalin-, and methionine enkephalin-like immunoreactivity in human cerebrospinal fluid. Eur Neurol 23:73–81

Ninkovic M, Hunt SP, Emson P, Iversen LL (1981) The distribution of multiple opiate receptors in bovine brain. Brain Res 214:163–167

North A, Egan TM (1983) Actions and distribution of opioid peptides in peripheral tissues. Br Med Bull 39:71–75

Nyberg F, Terenius L (1982) Endorphins in human cerebrospinal fluid. Life Sci 31:1737–1740

Nyberg F, Terenius L (1985) Identification of high molecular weight enkephalin precursor forms in human cerebrospinal fluid. Neuropeptides 5:537–540

Nyberg F, Wahlström A, Sjölund B, Terenius L (1983a) Characterization of electrophoretically separable endorphins in human CSF. Brain Res 259:267–274

Nyberg F, Yaksh TL, Terenius L (1983b) Opioid activity released from cat spinal cord during sciatic nerve stimulation. Life Sci 33 [Suppl 1]:17–20

Nyberg F, Nordström K, Terenius L (1985) Endopeptidase in human cerebrospinal fluid which cleaves proenkephalin B opioid peptides at consecutive basic amino acids. Biochem Biophys Res Commun 131:1069–1074

Nyberg F, Nylander I, Terenius L (1986) Enkephalin-containing polypeptides in human cerebrospinal fluid. Brain Res 371:278–286

Nyberg F, Cappelen C, Terenius L (1987) Hemodialysis decreases circulating opioid active material in a schizophrenic patient responding to naloxone treatment. Submitted

O'Brien CP, Terenius L, Wahlström A, McLellan AT, Krivoy W (1983) Endorphin levels in opioid dependent human subjects: a longitudinal study. Ann NY Acad Sci 398:377–383

Orth DN, Guillemin R, Ling N, Nicholson WE (1978) Immunoreactive endorphins, lipotropins and adrenocorticotropins in a human nonpituitary tumor: evidence for a common precursor. J Clin Endocrinol Metab 46:849–852

Panerai AE, Martini A, Di Guilio AM, Fraioli F, Vegni C, Pardi G, Marini A, Mantegazza P (1983) Plasma beta-endorphin, beta-lipotropin, and Met-enkephalin concentrations during pregnancy in normal and drug-addicted women and their newborn. J Clin Endocrinol Metab 57:537–543

Paterson SJ, Robson LE, Kosterlitz HW (1983) Classification of opioid receptors. Br Med Bull 39(1):31–36

Pelletier G (1980) Ultrastructural localization of a fragment (16K) of the common precursor for adrenocorticotropin and beta-LPH in the rat hypothalamus. Neurosci Lett 16:85–90

Pelletier G, Leclerc R, LaBrie F, Cote J, Chretien M, Lis M (1977) Immunohistochemical localization of beta-LPH hormone in the pituitary gland. Endocrinolgy 100:770–776

Pickar D, Cutler NR, Naber D, Post RM, Pert CB, Bunney WE (1980) Plasma opioid activity in manic-depressive illness. Lancet II:937

Pickar D, Cohen MR, Naber D, Cohen RM (1982) Clinical studies of the endogenous opioid systems. Biol Psychiatr 17:1243–1276

Pickel VM, Sumal KK, Beckley SC, Miller RJ, Reis DJ (1980) Immunocytochemical localization of enkephalin in the neostriatum of rat brain: a light and electron microscopic study. J Comp Neurol 189:721–740

Pullan PT, Clement-Jones V, Corder R, Lowry PJ, Rees GM, Rees LH, Besser GM, Macedo MM, Galvao-Teles A (1980) Ectopic production of methionine enkephalin and beta-endorphin. Br Med J 280:758–763

Ragavan VV, Frantz AG (1981) Opioid regulation of prolactin secretion: evidence for a specific role of beta-endorphin. Endocrinology 109:1769–1771

Räisänen I, Paatero H, Salminen BA, Laatikainen T (1984) Pain and plasma beta-endorphin level during labor. Obstet Gynecol 64:783–786

Reid RL, Hoff D, Yen SS, Li CH (1981 a) Effects of exogenous beta-endorphin on pituitary hormone secretion and its disappearance rate in normal human subjects. J Clin Endocrinol Metab 52:1179–1184

Reid RL, Yen SS, Artman H, Fischer DA (1981 b) Effects of synthetic beta-endorphin on release of neurohypophyseal hormones. Lancet I:1169–1170

Rimon R, Terenius L, Kampman R (1980) Cerebrospinal endorphins in schizophrenia. Acta Psychiatr Scand 61:395–403

Rivier C, Brown M, Vale W (1977) Stimulation in vivo of the release of prolactin and growth hormone by beta-endorphin. Endocrinology 100:751–754

Ross M, Berger P, Goldstein A (1979) Plasma beta-endorphin immunoactivity in schizophrenia. Science 205:1163–1164

Sar M, Stumpf WE, Miller RJ, Chang KJ, Cuatrecasas P (1978) Immunohistochemical localization of enkephalin in rat brain and spinal cord. J Comp Neurol 182:17–37

Schweisfurth H, Schiöberg-Schiegnitz S (1984) Assay and biochemical characterization of angiotensin I converting enzyme in cerebrospinal fluid. Enzyme 32:12–19

Seizinger BR, Liebisch DC, Gramsch C, Hertz A, Weber E, Evans CJ, Esch FS, Bohlen P (1984) Isolation and structure of a novel C-terminally amidated opioid peptide, amidorphin, from bovine adrenal medulla. Nature 313:57–59

Shanks MF, Clement-Jones V, Linsell CJ, Mullen PE, Rees LH, Besser GM (1981) A study of 24-hour profiles of plasma Met-enkephalin in man. Brain Res 212:403–409

Sharp B, Morley JE, Carlson HE, Gordon J, Briggs J, Melmed S, Hershman JM (1981) The role of opiates and endogenous opioid peptides in the regulation of rat TSH secretion. Brain Res 219:335–344

Sirinathsinghji DJ, Whittington PE, Audsley A, Fraser HM (1983) Beta-endorphin regulates lordosis in female rats by modulating LH-RH release. Nature 305:62–64

Sjölund B, Terenius L, Eriksson M (1977) Increased cerebrospinal fluid levels of endorphins after electroacupuncture. Acta Physiol Scand 100:382–384

Slizgi GR, Taylor CJ, Ludens JH (1984) Effects of the highly selective kappa opioid, U-50,488, on renal function in the anesthetized dog. J Pharm Exp Ther 230:641–645

Smith R, Grossman A, Gaillard R, Clement-Jones V, Ratter S, Mallinson J, Lowry PJ, Besser GM, Rees LH (1981) Studies on circulating Met-enkephalin and beta-endorphin in normal subjects and patients with renal and adrenal disease. Clin Endocrinol 15:291–300

Snyder S, Goodman RR (1980) Multiple neurotransmitter receptors. J Neurochem 35:5–15

Stern A, Lewis RV, Kimura S, Rossier J, Gerber L, Brink L, Stein S, Udenfriend S (1979) Isolation of the opioid heptapeptide Met-enkephalin (Arg^6, Phe^7) from bovine adrenal medullary granules and striatum. Proc Natl Acad Sci USA 76:6680–6683

Stubbs WA, Delitala G, Jones A, Jeffcoate WJ, Edwards CRW, Ratter JS, Besser GM, Bloom SR, Alberti KGMM (1978) Hormonal and metabolic responses to an enkephalin analogue in normal man. Lancet II:1225–1227

Suda T, Abe Y, Demura H, Demura R, Shizume K, Tamahashi N, Sasano N (1979) ACTH, beta-LPH, and beta-endorphin in pituitary adenomas of the patients with Cushing's disease: activation of beta-LPH conversion to beta-endorphin. J Clin Endocrinol Metab 49:475–477

Suda T, Demura H, Demura R, Jibiki K, Tozawa F, Shizume K (1980) Anterior pituitary hormones in plasma and pituitaries from patients with Cushing's disease. J Clin Endocrinol Metab 51:1048–1051

Tachibana S, Araki K, Ohya S, Yoshida S (1982) Isolation and structure of dynorphin, an opioid peptide, from porcine duodenum. Nature 295:339–340

Terenius L (1981) Endorphins and pain. Front Horm Res 8:162–177

Terenius L (1982) Endorphins and modulation of pain. Adv Neurol 33:59–64

Terenius L (1984) Opiate tolerance and dependence: roles of receptors and endorphins. Res Adv Alcohol Drug Probl 8:1–21

Terenius L, Nyberg F (1986) Opioid peptides in CSF of psychiatric patients. In: Van Ree JM, Matthysse S (eds) Progress of brain research, vol 65. Elsevier, Amsterdam, pp 207–219

Terenius L, Wahlström A, Lindström LH, Widerlöv E (1976) Increased CSF levels of endorphins in chronic psychosis. Neurosci Lett 3:157–162

Terenius L, Nyberg F, Wahlström A (1984) Opioid peptides in human cerebrospinal fluid. In: Hughes J, Collier HOJ, Rance MJ, Tyers MB (eds) Opioids: past, present and future. Taylor and Francis, London, pp 179–191

Terry LC, Crowley WR, Johnson MD (1982) Regulation of episodic growth hormone secretion by the central epinephrine system. Studies in the chronically cannulated rat. J Clin Invest 69:104–112

Tozawa F, Suda T, Tachibana S, Tomori N, Demura H, Shizume K (1984) Presence of immunoreactive dynorphin in human cerebrospinal fluid. Life Sci 35:1633–1637

Uhl GR, Goodman RR, Kuhar MJ, Childers SR, Snyder SH (1979) Immunocytochemical mapping of enkephalin containing cell bodies, fibers and nerve terminals in the brain stem of the rat. Brain Res 116:75–94

Van Loon GR, Ho D, Kim C (1980) Beta-endorphin-induced decrease in hypothalamic dopamine turnover. Endocrinology 106:76–80

Vincent SR, Hökfelt T, Christensson I, Terenius L (1982) Dynorphin-immunoreactive neurons in the central nervous system of the rat. Neurosci Lett 35:185–190

Von Knorring L, Almay B, Johansson F, Terenius L (1978) Pain perception and endorphin levels in cerebrospinal fluid. Pain 5:359–366

Wahlström A, Terenius L (1980) Chemical characteristics of endorphins in human cerebrospinal fluid. FEBS Lett 118:241–244

Wardlaw S, Frantz A (1979) Measurement of beta-endorphin in human plasma. J Clin Endocrinol Metab 48:176–180

Watson SJ, Akil H (1979) Presence of two alpha-MSH positive cell groups in rat hypothalamus. Eur J Pharmacol 58:101–103
Watson SJ, Akil H (1980) Alpha-MSH in rat brain: occurrence within and outside brain beta-endorphin neurons. Brain Res 182:217–223
Watson SJ, Akil H, Berger PA, Barchas J (1979) Some observations on the opiate peptides and schizophrenia. Arch Gen Psychiatry 36:35–41
Watson SJ, Akil H, Fischli W, Goldstein A, Zimmerman E, Nilaver G, van Wimersma-Greidanus TB (1982a) Dynorphin and vasopressin: Common localization in magnocellular neurons. Science 216:85–87
Watson SJ, Khachaturian H, Akil H, Coy D, Goldstein A (1982b) Comparison of the distribution of dynorphin systems and enkephalin systems in brain. Science 218:1134–1136
Watson SJ, Albala E, Berger E, Akil H (1983) Peptides in psychiatry. In: Krieger D, Brownstein M, Martin J (eds) Brain peptides. Wiley, New York, pp 351–368
Way WL, Hosobuchi Y, Johnson BH, Eager EI, Bloom FE (1984) Anesthesia does not increase opioid peptides in cerebrospinal fluid of humans. Anesthesiology 60:43–45
Weber E, Roth KA, Barchas JD (1982) Immunocytochemical distribution of alpha-neo-endorphin/dynorphin neuronal systems in rat brain: evidence for colocalization. Proc Natl Acad Sci USA 79:3062–3066
Weber E, Esch FS, Bohlen P, Paterson S, Corbett AD, McKnight AT, Kosterlitz HW, Barchas JD, Evans CJ (1983) Metorphamide: isolation, structure and biologic activity of an amidated opioid octapeptide from bovine brain. Proc Natl Acad Sci USA 80:7362–7366
Wiedemann E, Saito T, Linfoot JA, Li CH (1979) Specific radioimmunoassay of human β-endorphin in unextracted plasma. J Clin Endocrinol Metab 49:478–486
Wilkinson M, Bhanot R (1982) A puberty-related attenuation of opiate peptide-induced inhibition of LH secretion. Endocrinology 110:1046–1048
Yaksh TL, Elde RP (1981) Factors governing release of methionine enkephalin-like immunoreactivity from mesencephalon and spinal cord of the cat in vivo. J Neurophysiol 46:1056–1075
Yaksh TL, Terenius L, Nyberg F, Jhamandas K, Wang JY (1983) Studies on the release by somatic stimulation from rat and cat spinal cord of active materials which displace dihydromorphine in an opiate-binding assay. Brain Res 264:119–128
Zakarian S, Smyth D (1979) Distribution of active and inactive forms of endorphins in rat pituitary and brain. Proc Natl Acad Sci USA 76:5972–5976
Zamir N, Weber E, Palkovits M, Brownstein M (1984) Differential processing of prodynorphin and proenkephalin in specific regions of the rat brain. Proc Natl Acad Sci USA 81:6886–6889
Zhang AZ, Zhou GZ, Xi GF, Gu NF, Xia ZY, Yao JL, Chang JK, Webber R, Potkin S (1985) Lower levels of dynorphin(1-8) immunoreactivity in schizophrenic patients. Neuropeptides 5:553–556

CHAPTER 11

Radioimmunoassay of Pituitary and Hypothalamic Hormones

D. COCCHI

A. Introduction

Radioimmunoassay (RIA) systems have been developed to quantitate virtually every hormone available in pure form. This exquisitely sensitive technique has revolutionized the fields of endocrine physiology and clinical endocrinology. Bioassay techniques which have been employed for many years are not sufficiently sensitive to measure accurately all the anterior pituitary hormones in plasma; the development of RIAs in biologic fluids and tissues has permitted studies which have greatly expanded our knowledge of the factors involved in anterior pituitary hormone synthesis, metabolism, and action. A chapter on the general principles of RIAs for anterior pituitary hormones would have the disadvantage of being repetitive, several excellent reviews on this topic being already available in the literature.

In view of these points, this chapter, in addition to quoting many papers from the literature describing the technical procedures of pituitary hormone RIAs in several animal species, will focus on some aspects thought to be of peculiar interest. More space will be given to the second part of the chapter, on the RIA detection of hypophysiotropic neurohormones. This is an expanding field in endocrinology, particularly after the recent recognition of corticotropin-releasing factor (CRF) and growth hormone-releasing hormone (GHRH). Besides a description of the general problems related to the assay of hypophysiotropic peptides and a critical assessment of available techniques, the significance of determinations of these peptides in brain areas or biologic fluid as an index of neuronal function will be considered. For general considerations on RIA techniques and description of the basic principles of antigen–antibody interaction, the reader is referred to Chaps. 1 and 2.

B. General Principles of Determination

I. The Antibody

Chapters 3 and 4 deal with problems related to the production of the antibody, the agent playing a key role in the RIA method. The sensitivity of the assay is directly related to the avidity of the antibody for the antigen, and its specificity depends on the reaction between the determinants of the antigen and the sites of the antibody.

While in RIA methods polyclonal antibodies are generally employed, monoclonal antibodies produced by hybrid myelomas (see Chap. 4) are valuable reagents for use in immunoradiometric assays (IRMA; see Sect. B.V) in which the antibody is the labeled component. Once a suitable antibody has been obtained, it may be produced in large quantities with constant characteristics. It may be easily purified from other immunoglobulins for preparation of solid-phase and labeled antibody reagents. Moreover, it is possible to produce different antibodies reacting with distinct determinants in a given molecule, which can find application in two-site assays (see Sect. B.V).

The major problem related to the use of monoclonal antibodies in RIA is the difficulty of obtaining one with a sufficiently high affinity in order to achieve maximum sensitivity. High affinity may be less important for IRMAs in which antibody may be added in excess, although the kinetics of antibody binding may still be a factor in relation to convenience of assay protocol. Antisera can be stored frozen almost indefinitely since they rarely deteriorate. Deterioration of antiserum is not the usual cause of problems in RIAs.

II. The Labeled Antigen

All the problems related to radioiodination and purification of labeled antigens are discussed in Chap. 5. Detailed procedures for iodination will be described for some particular hormones. In contrast to antisera, most purified antigens for iodination, kept in solution (even frozen) for more than 1 month, may lose their ability to bind well with antiserum. In case of such difficulty, an aliquot of antigen should be weighed and solubilized before each iodination.

III. Reference Preparations

The problem of reference preparations is very important in the RIA of most anterior pituitary hormones, which must be obtained from pituitary extracts. Establishment of international standards has enabled investigators to report results either in international units or in terms of the reference material, and has made it possible to compare directly figures obtained by different laboratories. Reference preparations of pituitary hormones are now available from two major organizations: the National Institute of Arthritis, Diabetes, Digestive and Kidney Diseases (NIADDK) (see Table 1) and the National Institute for Biological Standards and Controls, Holly Hill, Hampstead, London NW3 6RB.

For each standard, it is specified the species of animal from which the hormone is derived, the source, and the type of assay (RIA or bioassay) for which the standard is suitable. The potency of the standard preparation is expressed in international units (IU). It is advisable that each author, when describing the RIA method, reports the type of reference preparation used and possibly its biologic potency. This procedure will facilitate comparison between results.

IV. Separation of Bound from Free Labeled Hormone

Separation of unreacted free labeled hormone from the antibody-bound label in RIAs can be achieved by many methods which utilize the difference in physicochemical properties between the antibody and the test hormone. They include differential solubility in organic solvents (THOMAS and FERIN 1968), molecular size (HABER et al. 1965), and electrophoretic mobility (YALOW and BERSON 1959). For routine checking of binding, organic solvents, with or without salts, are extremely useful and rapid. However, when serum samples are to be assayed, the results do not appear to be satisfactory. This could possibly be due to coprecipitation of the free label along with the antigen–antibody complex in the presence of excess protein. Solid-phase RIA systems employ the antibody coated directly on the inner wall of test tubes (CATT and TREGEAR 1967) or covalently bound to a solid support immunoadsorbent (CATT et al. 1966). The major advantage of a solid-phase antibody is to facilitate the separation of antibody-bound from free labeled antigens. The method is simple, but has the drawback of low sensitivity because the immunoreactivity of the antibody is decreased through coating. The use of a double-antibody (antibody to the immunoglobulin of the animal in which the hormone antibody was prepared) to precipitate the soluble hormone–antibody complex formed during the first incubation is a widely employed technique (UTIGER et al. 1962). Verification of the quality of each batch of anti-immunoglobulin serum is extremely important. This is done by adding increasing dilutions of the serum to an incubation mixture containing the specific antiserum and the labeled antigen in the absence of cold antigen. The radioactivity curve of the labeled antigen should show a plateau before falling off at high dilution. The presence of serum or plasma may cause a delay in the precipitation; this interference is due to serum complement (MORGAN et al. 1964). The inhibition of precipitation is prevented by addition of EDTA (0.005–0.01 *M*) or heparin (1000 IU/ml).

Other separation methods remove the free antigen from the incubation mixture. They are based on the high affinity of free antigen for materials such as charcoal, silica, and some ion exchange resins, to which the antigen–antibody complex binds poorly. These methods are generally useful for rather small polypeptides (glucagon, ACTH, hypophysiotropic peptides) which bind strongly to the adsorbing material. In the case of charcoal, it has been implied that coating with an appropriately sized molecule (different for each hormone) is necessary to provide good separation of bound and free hormone, by allowing the free hormones to penetrate to the adsorbent through pores (between the coating molecules) that are too small to permit the hormone–antibody complexes to pass. HERBERT et al. (1965) reported that both dextran and Ficoll, a nonionic polymeric alcohol (average molecular weight 400000) are useful coating agents. Dextran-coated charcoal is available commercially and is used extensively.

V. Future Development of RIAs

In addition to development of alternative methods which do not employ radioactive materials such as fluoroimmunoassays and luminiscent immunoassays, the efforts of qualified researchers are inclined to development of assays of greater sensitivity, specificity, and speed than RIAs. A method which displays performance characteristics which are superior to the past generation of RIAs is that employing labeled antibodies. The technical constraints on the use of such assays have rested, in the past, on the problems associated with the production of relatively large amounts of labeled antibody of appropriate specificity. Labeled antibody techniques need much greater amounts of antibody than labeled antigen techniques; moreover, the extraction, labeling, and purification of labeled antibody from antisera raised by conventional in vivo immunization technique are both time-consuming and wasteful. The advent of monoclonal antibody production techniques clearly offers the possibility of circumventing these practical difficulties. These techniques (see Chap. 4) offer the possibility of the synthesis, in relatively pure or easily purifiable form, of large amounts of antibody.

IRMA, which involves preparing a radiolabeled derivative of the protein to be assayed by reacting it with labeled antibody and separating the derivative from excess labeled antibody, was first described by Miles and Hales (1968). There are two processes in all IRMA methods: (a) the interaction of antigen with labeled antibody; and (b) the reaction either of the unreacted labeled antibody with solid-phase linked antigen or of the labeled antibody–antigen complex with solid-phase linked antibody (the two-site variant).

Several different methods of labeling and types of solid phase have been successfully used in two-site assays. The immobilization of antibodies by simple adsorption to the walls of plastic incubation tubes is a particularly simple technique. This allows easy assessment and comparison of several different antibodies, and provides a basis for a rapid and convenient separation step in the assay itself. Many two-site IRMA methods have been developed for pituitary hormones and will be quoted in the appropriate sections.

C. Radioimmunoassay of Anterior Pituitary Hormones

The RIA technique for determination of anterior pituitary hormones in biologic fluids and tissue extracts is now very well established. Many high quality commercial kits are available for detection of human hormones and many manufacturers also sell purified antigens, labeled or unlabeled, and antibodies for RIAs of pituitary hormones in experimental animals, namely rats.

The problems of obtaining antisera and purified antigens for researchers who do not want to use commercial products or who determine hormones in animal species other than the rat (dog, mouse, etc.) have been overcome by the ready availability of these materials from the Hormone Distribution Program of the NIADDK. Table 1 lists the research materials procured and distributed to qualified investigators by the NIADDK. Everybody can easily find the complete, up-to-date list of reagents in certain issues of *Endocrinology* and/or *Journal of Clinical Endocrinology and Metabolism.*

Table 1. National Institute of Arthritis, Diabetes, Digestive and Kidney Diseases. Research materials available to qualified investigators[a]

	Human RIA vials			Rat RIA vials			Ovine RIA vials			Bovine RIA vials		Porcine RIA vials	
	H	AS	RP	H	AS	RP	H	AS	RP	H	AS	H	AS
GH	1[a]	1	1	1[b]	1	1	1[c]	1		1[d]		1[d]	
PRL	1[e]	1	1	1[b]	1	1	1[b]	1		1[d]		1[d]	
FSH	1[e]	1	1	1[e]	1	1	1[e]	1	1	5[f]		5[f]	
LH	1[b]	1	1	1[e]	1	1				1[c]		1[d]	
TSH	1[b]	1	1	1[e]	1	1	1[g]	1		1[g]	1		
ACTH	1[g]	1				1							

Subunits				Miscellaneous[j]	
Human (h), ovine (o), rat (r) kits				Canine	
oα-LH	1	hα-LH	1[g]	GH	1 kit
oβ-LH	1	hβ-LH	1[e]	PRL	1 kit
				Rabbit	
hαCG	1[g]	rα-LH	antiserum only	GH	1 kit
hβ-CG	1[g]	rβ-LH	1	PRL	1 kit
hα-TSH	1[g]	rβ-TSH	1	LH	1 kit
hβ-TSH	1[h]	hα-CG	2–5 mg	FSH	1 kit
hβ-FSH	1[g]	hβ-CG	2–5 mg	TSH	1 kit
				Murine	
				GH (RIA)	0.1 mg
				PRL	1 kit
				TSH RP only	1 vial

H, Hormone; AS, antiserum; RP, reference preparation.
[a] 2 mg; [b] 400 μg; [c] 500 μg; [d] 1 mg; [e] 100 μg; [f] 25 μg; [g] 50 μg; [h] 200 μg.
[j] These materials are distributed by Dr. A. F. Parlow, Director, Pituitary Hormones and Antisera Center, Harbor-UCLA Medical Center, 100 West Carson Street, Torrance, CA 90509, USA, through his own program.
Address request to: National Hormone and Pituitary Program, Suite 501-9, 210 West Fayette Street, Baltimore, MD 21201, USA.

I. Prolactin

There has been a plethora of studies over the past decade dealing with various facets of prolactin (PRL) and its actions. The endocrinologic literature of the last 15 years is replete with studies of the chemistry, mechanism of action, regulation, and function of PRL in both laboratory animals and humans. The impetus for the striking increase in research activity in these areas was provided partially by the demonstration that PRL exists as a discrete pituitary hormone in humans and is elevated in a significant number of women with altered menstrual and other endocrinologic functions. In addition, as PRL fulfills several functions in animals, possible analogies of function in a variety of mammals, including humans, have been sought (Jaffe 1981).

1. Human Prolactin

Many heterologous and homologous RIAs for human (h) PRL have been developed since the early 1970s and for details the reader is referred to the reviews of Felber (1974) and Jacobs (1979). The most important problem concerning hPRL RIA is the extreme susceptibility of the antigen to iodination damage. Human PRL has a strong tendency for aggregation during iodination, giving problems with binding to antibody or with displacement from it. Iodination by the lactoperoxidase method (Thorell and Johansson 1971) is preferable to the chloramine-T procedure (Hunter and Greenwood 1962) since it results in a tracer which retains substantial degrees of biologic as well as immunologic activity (Hughes et al. 1982).

An attempt to solve the problem of damage to hPRL during iodination has been reported by Vander Laan et al. (1983). It entails the use of a synthetic 13 amino acid analog of the NH_2 terminus of hPRL in which tyrosine is substituted for valine at position 13. The peptide was used for iodination and displacement curves were found to be parallel when human PRL and the synthetic peptide were compared as standards. This RIA has the advantage of purity of the synthetic peptide in its roles as hapten and as labeled peptide, of ease of iodination of the peptide, and of its stability after iodination. Some of the most commonly employed RIAs for hPRL have been extensively reviewed (Jacobs 1979). Upper limits of normal plasma PRL levels in most laboratories are between 15 and 20 ng/ml. Any elevation of PRL is suspect, and pituitary adenomas have been found associated with PRL concentrations as low as 23 ng/ml (Chang et al. 1977). The physiologic and patophysiologic states with alterations of plasma PRL levels have been reviewed by Jaffe (1981).

2. RIA Determination of Prolactin as an Index of Biologic Activity

Clinicians have long been aware that RIA measurements of serum PRL do not always correlate with clinical findings. For example, some patients with high RIA PRL levels have normal menstrual function and no galactorrhea, whereas other patients with normal RIA PRL values have marked physiologic disturbances.

Since the introduction of RIA for measurement of pituitary hormones, investigators have examined the validity of the assays by comparison with bioassays. There has been good agreement in the majority of cases. There are reports, however, which indicate that at times the immuno- and bioassays did not agree. In these cases, the bioactivity was always greater than the immunoreactivity. Nicoll (1975) has written an excellent review of this topic, covering both GH and PRL. His conclusion is that the RIA of circulating levels of the hormones does not give a physiologically meaningful result.

Discrepancies between RIA and bioassay were noted when PRL secretion in response to the suckling stimulus was measured (Nicoll et al. 1976). This was also noted in rats treated with perphenazine or bearing a PRL-secreting tumor (Leung et al. 1978). The investigators calculated that the RIA measures only about 25% of the activity detected by the mammary gland organ culture bioassay.

Studies have been carried out to identify the forms of PRL in the pituitary gland which react differently in bioassay and RIA (for review see LEWIS et al. 1985). PRL is not a single entity, but a family of proteins that differ in size and charge, and most likely play different physiologic roles. The possibility exists that these different forms of PRL are present in the circulation in varying proportions among individuals and may react differently in RIA and biologic assays.

Until the recent introduction of the Nb2 rat lymphoma cell bioassay for PRL (TANAKA et al. 1980) there had been no practical method for the routine assessment of PRL bioactivity in human serum. Previous methods, such as the pigeon crop sac assay (NICOLL 1967) and the radioreceptor assay (FRIESEN and MCNEILLY 1977), have to some degree lacked the sensitivity and specificity necessary for the study of human serum samples.

Many studies have been performed comparing serum PRL values determined by RIA and by Nb2 lymphoma cell bioassay. TANAKA et al. (1980, 1983) showed that there was no significant difference in estimates of serum PRL values obtained by bioassay or RIA in serum from subjects with PRL levels in the normal range, or in patients with elevated levels of the hormone. Slightly different data have been obtained by SUBRAMANIAN et al. (1984) who found that in hyperprolactinemic women RIA values were significantly higher than the bioassay results, and the values correlated better in women who were hyperprolactinemic with obvious menstrual cycle disturbances than in hyperprolactinemic women without menstrual cycle disturbances.

3. Prolactin RIA in Other Animal Species

Availability of antigens and antisera from NIADDK or from the Pituitary Hormones and Antisera Center to qualified investigators has eliminated most problems in the development of homologous PRL RIAs in most species of laboratory animals (see Table 1). For the technical procedures to be followed in RIA determinations of nonhuman species, the reader can refer to: rat PRL (NISWENDER et al. 1969a); mouse PRL (SINHA et al. 1972a; MARKOFF et al. 1981); canine PRL (RICHARDS et al. 1980; DE COSTER et al. 1983); ovine PRL (MCNEILLY and ANDREWS 1974); bovine PRL (MACHLIN et al. 1974); rabbit PRL (FUCHS and WAGNER 1981); and hamster PRL (SOARES et al. 1983; BORER et al. 1982).

II. Growth Hormone

Methods for accurate determination of circulating GH concentrations in humans have been available since 1963 (GLICK et al. 1963; SCHALCH and PARKER 1964). The hGH RIA is highly specific. Addition of GHs prepared from various nonprimate species fails to displace the binding of the labeled hGH to anti-hGH serum. However, GH from primates significantly cross-reacts with the hGH assay, and this assay may be used for estimations of plasma GH in monkeys. Successful performance of the hGH RIA is dependent upon the quality of the protein reagents used in the assay, a problem now overcome by the availability of reagents supplied by NIADDK. Antibodies obtained from this source, as well as those of most commercial kits, show negligible cross-reactivity with other pituitary hormones, particularly PRL. The chloramine-T method can be employed for radioiodination and, prior to use in each assay, an aliquot of iodinated hGH is further purified by chromatography on a Sephadex G-100 column. Careful descriptions of possible RIA procedures for hGH can be found in the literature (Peake et al. 1979; Webb et al. 1984).

1. Evaluation of Human Growth Hormone Levels in Blood

Basal levels of GH in fasting adult subjects are in the range 0–5 ng/ml. Investigations in which GH levels have been detected by RIA after sample extraction and concentration on Sepharose columns with adsorbed anti-GH serum have shown that maximal basal values were 0.8 ng/ml (Drobny et al. 1983). According to this view, values of 1–2.5 ng/ml might be considered high levels. GH levels are elevated in the fetus and during the first days of life (Gluckman 1982). Spontaneous secretory episodes are present at all ages (Martin et al. 1978), but are more evident in adolescence; after puberty, an increase in the amplitude and frequency of secretory peaks becomes evident, as well as an increase in mean GH values (Miller et al. 1982). It is well known that GH secretion is a highly labile function, responsive to even minor changes in the environment. A variety of metabolic stress and endocrine factors may affect its release. Classically, it has been postulated that at least three major control mechanisms exist: one concerned with maintaining circulating energy–substrate levels in the presence of a decrease in the substrate for intracellular energy production (glucoprivation, fasting, lowering of nonesterified fatty acids, etc.); the second associated with changes in aminoacidemia; and the third related to a variety of physical and physiologic stimuli. The hormone exhibits a circadian periodicity of secretion with peak levels during the initial episode of slow wave sleep (for review see Müller 1979).

2. Heterogeneity of Human Growth Hormone in the Circulation

It is now clear that GH is not a single substance. Knowledge of the chemical nature of GH in blood is of considerable biologic importance because of the largely unexplained discrepancy between GH-like bioactivity in plasma and GH concentrations as measured by RIA. The bioactivity in plasma has been reported to exceed immunoassayable GH by as much as 50- to 200-fold (Ellis et al. 1978). Since this cannot be attributed solely to the presence of somatomedins in plasma (Ellis et al. 1978), the possibility of highly bioactive and/or poorly immunoreactive forms of GH as part of the overall growth-promoting activity of plasma must be considered.

In the early 1970s, it was shown that fractionation of plasma on Sephadex yields two to four (usually three) immunoreactive hGH species with molecular weights of about 20000, 45000, and >60000 (Yalow 1974). Since no information existed about the structure of these species, they were named "little," "big," and "big-big" hGH. A method was developed for examining serum for multiple forms of hGH (Bauman et al. 1983).

The previous interpretation of the nature of little and big hGH was essentially confirmed: these species represent monomeric and dimeric hGH. In addition, big-

big hGH was recognized as a heterogeneous mixture of hGH oligomers. BAUMANN et al. (1983), by using physicochemical techniques applied to plasma extracts by immunoaffinity chromatography, showed that the predominant 22 K form of GH accounted for about 85% of the total hGH, the second most prevalent form was 20 K (10% of the total), and only 5% of total immunoreactive hGH could be attributed to one or more unidentified acidic forms.

It is presently not clear what the biologic role of the various forms of hGH is. As assessed by radioreceptor assay, monomeric 22 K and some of its cleaved congeners have the highest intrinsic potency, although this seems not to be the only important parameter for overall bioactivity (BAUMANN 1984). It could be possible that some fragments, generated in vivo from the parent hormone, are capable of maintaining the biologic activity while losing immunoreactivity toward anti-GH serum. From all these data, it would appear that detection of very low radioimmunoassayable plasma GH in some patients (i.e., short children) cannot exclude a normal biologic activity of the hormone, suggesting other factors as the cause of the short stature.

3. Growth Hormone RIA in Other Animal Species

Although most of the experimental studies on the regulation of GH secretion have been performed in the rat, this animal does not represent a suitable model for GH studies. Its major drawback is the discrepancy among results obtained by the bioassay "tibia test" and those obtained by RIA. Most of the conditions such as cold exposure, insulin hypoglycemia, or prolonged fasting which are found to be associated with reduced levels of pituitary GH, when measured by bioassay, displayed no effect on pituitary GH concentrations or the circulating GH levels when measured by RIA (MÜLLER 1973). Despite this drawback, the rat is a very useful experimental animal for studies on the mechanism of action of GHRH and somatostatin, the hypophysiotropic hormones which dually regulate GH synthesis and release (MÜLLER 1979; THORNER and CRONIN 1985). RIA for rat GH was first described by SCHALCH and REICHLIN (1966). An animal model more similar to humans as far as the regulation of GH secretion is concerned, is the dog. In contrast to rats, dogs show a very stable basal GH secretion, and, in contrast to primates, they are scarcely susceptible to sampling-related stress. Dogs exhibit a GH response very similar to humans to most neuropharmacologic stimuli.

Several RIAs for dog GH have been described (COCOLA et al. 1976; HAMPSHIRE et al. 1975; LOVINGER et al. 1974; EIGEMANN and EIGEMANN 1981). Monkeys are considered to be ideal hosts for immunization procedures with canine GH as well as for other nonprimate species. The reason is that primate and mammalian GHs (dog, ox, pig, rat) exhibit appreciable structural differences with virtually no immunologic cross-reactivity between the two classes (WILHELMI 1974). As for hGH, repurification of ^{125}I-labeled canine GH before each assay is important for reproducible assay sensitivity (HAMPSHIRE et al. 1975; EIGEMANN and EIGEMANN 1981); use of tracer without repurification, although giving straight line standard curves, results in a loss of sensitivity in the lower range of the curve.

Increasing interest in GH secretion in mice is principally due to the existence of a mutant dwarf mouse (dw/dw, Snell dwarf mouse) endowed with a defective pituitary gland which is lacking in acidophils (and consequently having undetectable circulating GH levels; Sinha et al. 1975). These mice represent a very suitable experimental model for studying the action of agents (such as GHRH) to be used for the therapy of human dwarfism. Although homologous RIAs for mouse GH have been described (Sinha et al. 1972 b), in our laboratory we use the heterologous RIA for rat GH with excellent results (utilizing reagents provided by NIADDK). In addition to the previously mentioned RIAs for rat, dog, and mouse GH, homologous assays for rabbit and ovine GH have been reported (Chihara et al. 1984; Davis et al. 1977).

III. Glycoprotein Hormones

1. Gonadotropin RIA

The cross-reaction of gonadotropin antisera across species barriers was recognized many years ago (Moudgal and Li 1961). This property was successfully exploited by Niswender et al. (1968 b) to develop heterologous RIA systems for measuring luteinizing hormone (LH) in many species by using a universal antiserum against ovine LH (Niswender et al. 1969 b). Improved RIAs of the homologous type have been described by various authors for gonadotropins of several animal species (see Moudgal et al. 1979 for review; Yen et al. 1968, 1970; Kosasa et al. 1976 for human gonadotropins; Neill et al. 1977 for simian LH; Niswender et al. 1968 a; Daane and Parlow 1972 for rat gonadotropins; L'Hermite et al. 1972 for ovine FSH).

Basal (tonic) gonadotropin secretion, which is characterized by pulsatile, rhythmic circhoral discharges of LH and follicle stimulating hormone (FSH) is controlled by the classical negative feedback action of estradiol and the initiation of the preovulatory gonadotropin surge is the consequence of a stimulatory effect of the same steroid when its concentration in serum surpasses a given threshold for a well-defined period of time (for review see Knobil 1974). The signal which initiates the circhoral pulsatile discharges of gonadotropins originates in the central nervous system and is mediated by a pulsatile release of gonadotropin-releasing hormone (LHRH) (Carmel et al. 1976).

a) Gonadotropin RIA in the Rat

Although primates represent the experimental animals closest to humans as far as gonadotropin regulation is concerned, the rat is extensively used. Variable levels of LH and FSH are reported in the literature, depending on the potency of the reference preparation available. For instance, rat LH RIA reference preparation NIADDK rLH-RP-2 (the one available at present) is 61 times as potent as the previously distributed reference preparation NIADDK rLH-RP-1. Therefore, expression of serum LH values in terms of NIADDK-rLH-RP-2 gives values (ng/ml) which are 61 times lower than previously. For rat FSH, the new reference preparation (RP-2) is 45 times more potent than RP-1. In the following paragraphs, we will describe an RIA for rat LH currently performed in our laboratory. The RIA is carried out with materials obtained from the NIADDK, using minor changes from the printed format supplied with those materials.

Each assay tube contains 200 µl of either rabbit anti-rat LH (initial dilution 1:10000) or buffer (PBS 0.01 *M*, EDTA 0.05 *M*, 3% normal rabbit serum, pH 7.6, for detection of nonspecific binding) and 200 µl standard or unknown. NIADDK rat LH-I-6 is iodinated by the chloramine-T method (2.5 µg LH in 10 µl PBS is iodinated with 1 mCi ^{125}I by the addition of 25 µg chloramine-T). The reaction is allowed to proceed for 45 s and is terminated by the addition of 50 µg sodium metabisulfite and 50 µl 0.5% KI.

Purification of the labeled LH is carried out using a 5-ml plastic disposable syringe packed with Sephadex G-50 medium grade in PBS. Labeled LH (100 µl), containing approximately 15000 CPM per tube, is added to all assay tubes and these are incubated at room temperature for 24 h. Thereafter, 200 µl goat anti-rabbit IgG is added and the samples incubated overnight. To maximize the sensitivity of the assay, when very low plasma LH levels have to be determined, disequilibrium incubation is used. The radioimmunoassay of rat FSH routinely performed in our laboratory is very similar to that employed for LH. In order to improve the sensitivity of the assay, the reagents are always preincubated prior to addition of the tracer.

b) Biologic and Immunologic Activity of Human Gonadotropins

The introduction of a simple method for the measurement of biologically active LH (based on testosterone production by dispersed interstitial cells of the rat testis; Dufau et al. 1974) have allowed one to obtain a reliable measure of LH present in human plasma. The profiles of LH biologic activity in normal subjects were found to be identical with those measured by RIA. However, a difference between the two methods was found when LH levels were very high (postmenopausal women, and after LHRH administration; Dufau et al. 1976), suggesting that circulating LH molecules exhibit a relatively higher biologic activity during states of increased biosynthesis and release of gonadotropins.

In accordance with the two-pool theory proposed by Hoff et al. (1977), it could be supposed that a low basal gonadotropin secretion rate may induce the release of LH only from the "early" pituitary (readily releasable) pool, consisting of molecules with a low bioassay: immunoassay ratio. On the other hand, when the pituitary is maximally stimulated by LHRH, the "later" (newly synthetic) pool may also be utilized, leading to the secretion of a more bioactive form of LH. This theory could also explain the greater responsiveness for bioactive LH of the pubertal pituitary, when compared with that observed in adult men (Montanini et al. 1984).

2. Thyrotropin RIA

Among the pituitary hormones, thyrotropin (TSH) is the most finely controlled by the periphery. Thyroid hormones represent the primary control mechanism for TSH secretion (for review see Morley 1981). For this reason, measurements of basal serum TSH concentration are extremely valuable in the diagnosis and management of hypothyroidism or thyrotoxicosis. Detection of TSH levels in newborns allows one to screen those affected by congenital hypothyroidism who will develop irreversible mental retardation unless immediately treated with thyroid hormones.

Thyrotropin has considerable chemical and thus immunologic similarities to pituitary and chorionic gonadotropins. All these hormones have two polypeptide subunits, one of them (the α-subunit) is similar or identical in structure in the four hormones. The other subunit (β) differs substantially and confers the biologic specificity to the hormone. For this reason anti-TSH sera contain anti-gonadotropin antibodies as well as anti-TSH antibodies (Jacquet et al. 1971). The reactivity of anti-TSH serum with gonadotropins can be reduced by absorption of anti-TSH serum with human chorionic gonadotropin (hCG) (Binoux et al. 1974), a procedure restricting the population of antibodies directed against α-subunits. Absorption is easily performed by incubating the anti-TSH serum with hCG for 48 h at 4 °C. Attempts have been made to increase the specificity of the assay by using antibodies directed against TSH β-subunit. However anti-β-subunit sera have a very low cross-reactivity for intact TSH (lower than 1%); therefore, they cannot be used to determine the hormone. By standard RIA techniques performed with polyclonal TSH antibodies, a low sensitivity (0.8–1 μU/ml) can be reached, making it very difficult to distinguish undetectable TSH levels from low-normal values. A high percentage of normal subjects show low TSH values indistinguishable from those really undetectable in hyperthyroid patients. In all these cases, a stimulatory test with thyrotropin-releasing hormone (TRH) is mandatory for evaluation of hyperthyroidism (for review see Scanlon et al. 1983).

Development of anti-TSH monoclonal antibodies has in part solved these problems. They are endowed with remarkable specificity (they do not cross-react with LH, FSH, and hCG) and sensitivity capable of distinguishing low-normal levels from undetectable values in hyperthyroid patients (Rifgway et al. 1982). Monoclonal antibodies are mainly used in IRMAs which show a higher sensitivity than RIAs (see Sect. B.V). For reviews of the methodological problems related to human TSH determination the reader can refer to Utiger (1979).

Two methods for quantitation of TSH have been described. The first (Eriksson et al. 1984) does not require centrifugation or decantation steps. It involves coupling of a second antibody to agarose spheres which also contain entrapped crystals of bismuth oxide in order to attenuate the radiation from the bound fraction of the ^{125}I-labeled antigen. The system is designed so that the samples may be measured in a counter 15 min after the spheres have been added, and give data with a good correlation with data obtained by conventional methods.

The second method (Soos et al. 1984) is a two-site IRMA using two monoclonal antibodies, one of which is adsorbed on plastic tubes and the other labeled with ^{125}I. The optimized assay covers a working range of 0.17–10 ng/ml TSH in a 4 h, single-incubation protocol, with no significant interference from other glycoprotein hormones at their maximum physiologic or pathologic concentrations. For screening of congenital hypothyroidism at birth, methods have been developed for determination of TSH in filter paper disks impregnated by cord blood (the earliest convenient sample available from the infant) (Foley et al. 1977). Thyrotropin RIAs have been developed for many experimental animals: homologous for rat (Kieffer et al. 1974) and dog (Quinlan and Michaelson 1981); heterologous for ox (Chin et al. 1981) and mouse (Gershengorn 1978).

3. RIA of α- and β-Subunits

As already mentioned, the glycoprotein hormones, TSH, LH, FSH, and hCG are composed of two covalently bound subunits, α and β. The α-subunits are virtually identical in structure and are almost indistinguishable by RIA. The β-subunit, which is unique to each, confers specificity. Separately, neither subunit exerts biologic activity (for review see PIERCE and PARSONS 1981). Elevated levels of the free subunits are found in the blood of persons with increased levels of the complete hormones, e.g., TSH in primary hypothyroidism, LH and FSH in post menopausal women, and hCG in pregnancy (KOURIDES et al. 1975). Elevated subunit levels have been found in patients with pituitary tumors (KOURIDES et al. 1976; MACFARLANE et al. 1980, 1982). In some of these patients, elevated subunit levels were reduced by surgical removal of adenoma tissue and external pituitary irradiation, indicating that measurement of serum subunits may be useful as a marker of residual pituitary tumor.

Higher concentrations of β-TSH are present in patients with idiopathic central hypothyroidism due to the secretion of TSH with reduced biologic activity (FAGLIA et al. 1983). Since this is the only pathologic condition with a high β:α ratio, this finding might be a useful diagnostic tool for this disorder. Reagents for developing RIAs of α- and β-subunits can be obtained from NIADDK. The cross-reactivity among different α-subunits (LH, TSH, hCG) is almost complete.

Intact LH, FSH, and TSH, on the contrary, have variable cross-reactivity in most homologous subunit RIAs (up to 15%–20%). The high circulating levels of glycoprotein hormones (postmenopausal women, hypothyroid patients) can interfere with the measurement of the free α-subunit, masking the small amounts of ectopically produced subunit. Less important is the cross-reactivity displayed by intact hormones or α-subunits in the different β-subunit RIAs (1% or less). The same is true for cross-reactivity among different β-subunits, confirming the immunologic specificity of the individual β-subunits. For a detailed description of RIA procedures for α- and β-subunits, the reader can refer to KOURIDES et al. (1974), MACFARLANE et al. (1980), and FAGLIA et al. (1983), for TSH subunits; BEITINS et al. (1977), for α- and β-LH and FSH, and PAPAPETROU et al. (1980), PAPAPETROU and ANAGNOSTOPOULOS (1985), for α- and β-hCG.

IV. Adrenocorticotropin, Endorphins, and Related Peptides

Adrenocorticotropin (ACTH) is a 39 amino acid peptide with a molecular weight of approximately 4500. The pituitary is also known to contain such hormones as β-lipotropin (β-LPH) and melanotropin (α-MSH) which share several amino acids with ACTH. In addition to these hormones, β-endorphin (β-EP), the COOH terminal portion of the β-LPH molecule, is contained in the pituitary and hypothalamus. It is known that all these peptides are derived from the common precursor molecule, pro-ACTH-β-LPH (NAKANISHI et al. 1979). At the COOH terminus of preproopiomelanocortin (POMC) there exists β-LPH; next to β-LPH there is a sequence of two amino acids, a possible cleavage site by proteolytic enzymes, and then there is ACTH.

The common precursor can be processed differently in the pituitary gland, thus giving rise to different peptides. In the rat, POMC is processed differently from ACTH, β-EP, β-LPH-, and α-MSH in the anterior lobe and differently from β-EP and α-MSH in the intermediate lobe (Rossier et al. 1979; Mains and Eipper 1979). The pituitary of adult humans does not exhibit a pars intermedia. The fetal pituitary contains a large amount of α-MSH (Silman et al. 1976) while in the adult only trace amounts of α-MSH are present (Scott et al. 1973). According to the data reported in the rat, in humans, the major products of POMC present in the pituitary (anterior), are ACTH and β-LPH and only small amounts of β-EP have been found (Liotta et al. 1978). In human pituitaries, ACTH and β-LPH are present in equimolar concentrations and are concomitantly secreted under different physiologic conditions and following stimulation (Krieger et al. 1980). It is obvious from the foregoing that the mechanism which controls the secretion of ACTH from the corticotropic cells must be the same as that controlling the release of POMC-related peptides contained in the same cells. Concomitant reduction of plasma levels after glucocorticoid administration, and the striking increase after adrenalectomy or after blockade of cortisol biosynthesis, support the concept that ACTH, β-EP, and β-LPH share a common mechanism of regulation (for review see Imura et al. 1982).

In addition to the pituitary, immunoreactivity for ACTH has also been detected in the hypothalamus, CSF, and other brain areas of the rat. Also, the other POMC-derived peptides (α-MSH, β-LPH, and β-EP) have been localized in several brain areas: limbic system, mesencephalon, medulla, pons, cortex, and several hypothalamic nuclei (for review see Grossman et al. 1985). POMC-related peptides have also been detected in human placenta (Krieger et al. 1980), in male reproductive organs (Tsong et al. 1982), and in the gastrointestinal system (see Grossman et al. 1985).

1. Adrenocorticotropin RIA

The estimation of basal circulating levels of ACTH (20–60 pg/ml) by direct RIA requires an antibody of very high affinity, and such antibodies are not readily available. Therefore, the vast majority of ACTH RIAs employ antibodies of only moderate affinity and utilize a preliminary extraction step involving adsorption on powdered glass to concentrate the analyte from relatively large volumes of plasma in order to obtain the necessary assay sensitivity.

A second problem has been the lack of specificity of ACTH RIAs. Thus, the cross-reaction of the precursor molecules, fragments resulting from peptidase activity on the intact molecule, and other peptides with sequence homology in the RIA has led to discrepancies between estimates obtained by bioassay and RIA and difficulties in interpreting the latter. General problems with RIAs of ACTH and related peptides are similar to those encountered for hypophysiotropic peptides (see Sect. D.II).

a) Iodination of Adrenocorticotropin and Related Peptides

The hormone used for iodination and standard must be as pure as possible. If the antibodies are directed toward a portion of the 1–24 sequence of ACTH, any peptide which includes the 1–24 sequence – (1–24) ACTH or any species of natural or synthetic (1–39) ACTH – should be a satisfactory standard. Human ACTH can be used as tracer and standard for measuring ACTH in tissue extracts of all other animal species.

Two alternative methods can be used: the first is a modification of the chloramine-T method of HUNTER and GREENWOOD (1962) described by ORTH (1979), the other, which avoids the use of the Sephadex column, is described in detail by REES et al. (1971). Iodination by the first procedure is carried out with 1 mCi ^{125}I and 2 μg highly purified or synthetic ACTH (provided by the National Pituitary Agency). The iodination mixture is purified on a Sephadex G-50 fine gel column.

According to the other method (REES et al. 1971), the separation of damaged from undamaged ^{125}I-labeled ACTH is accomplished by adsorption on QUSO glass (Corning). The same procedures can be followed for radioiodination of other POMC-derived peptides. Alternatively, iodination with lactoperoxidase can be performed (OGAWA et al. 1979; PANERAI et al. 1983).

b) Antisera

For antibody production, the antigen (ACTH) has to be conjugated to heterologous serum albumin, thyroglobulin, or other protein carriers for the hapten. It is important to couple the polypeptide hapten to as many of the available exposed protein-reactive carboxy and/or amino groups as possible. Thus, a large molar excess of hapten is used in the reaction. Two methods suitable for binding (1–24) ACTH or other ACTH preparations to albumin with carbodiimide or glutaraldehyde have been described in detail by ORTH (1979).

Once the antiserum has been obtained after immunization, it has to be characterized, with particular reference to its specificity. For RIA purposes, only the specificity of the antibodies in the antiserum that bind labeled tracer are relevant. The antisera have to be characterized by defining the portion of the ACTH molecule to which antibodies are directed and the cross-reactivity with α-MSH, and particularly with γ-LPH and β-LPH, since these two hormones possess a common heptapeptide sequence with ACTH and are secreted in parallel with it (KRIEGER et al. 1980). The high degree of cross-reactivity of antibodies also represents the major problem for RIAs of other POMC-related peptides, i.e., β-EP and β-LPH. β-EP antibodies cross-react up to 100% with β-LPH, acetylated β-EP, POMC, and, to a lesser degree, with α- and γ-EP (SUDA et al. 1978; PANERAI et al. 1982). More specific antibodies (anti-β-EP) have been obtained by some authors (WILKES et al. 1980; WIEDEMAN et al. 1979), making possible the direct measurement of such peptides in tissues. The antiserum for detection of βEP + βLPH is generally raised against human β-EP, that for determination of these peptides in rats is raised against camel β-EP since SEIDAH et al. (1978) have shown that rat, ovine, and camel β-EP are identical. In view of the poor selectivity of the antibodies, the assay has to be preceded by separation of the different peptides or, alternatively, several antibodies have to be used (WARDLAW and FRANTZ 1979; GENAZZANI et al. 1981; PANERAI et al. 1982). Affinity chromatography (see Sect. D.2) has been extensively employed to separate POMC-related peptides, overcoming the problem of poorly specific antibodies by using more than one (SHIOMI and Akil 1982; Cahill et al. 1983). The RIA described by the latter authors are very sensitive and can measure concentrations of β-EP-like peptides in 2–4 ml human plasma (sensitivity 5–15 fmol per tube).

c) IRMA of Adrenocorticotropin

A nonextracted "two-site" immunoradiometric assay for corticotropin has been developed by Hodgkinson et al. (1984). As mentioned before, IRMAs are potentially more sensitive and specific, and are quicker and easier to perform than RIAs. The assay described by Hodgkinson et al. (1984) is based on the simultaneous addition of ^{125}I-labeled sheep anti-(NH_2 terminal ACTH) IgG antibodies and rabbit anti-(COOH terminal ACTH) antiserum to standards and unknowns (0.5 ml followed by 18 h incubation). The use of solid-phase reagents is avoided in order to minimize nonspecific effects and the time required for reactants to reach equilibrium. Instead, the separation of ACTH-bound from free labeled antibody is achieved by the addition of sheep anti-rabbit antiserum, which precipitates bound labeled antibody by complex formation with rabbit anti-ACTH antibodies which are also hormone-bound. By this RIA, a level of ACTH as low as 8 pg/ml can be detected directly in plasma.

D. Radioimmunoassay of Hypothalamic Hypophysiotropic Hormones

I. Premises

The ultimate mediators of the neuroregulation of anterior pituitary function are the so-called releasing and inhibiting hormones produced by the neurosecretory neurons of the medial basal hypothalamus. The chemical structure of most of them (six of seven) has now been identified (Table 2). It has been known for many years that GH is under a dual regulatory mechanism by two hypothalamic hypophysiotropic hormones, one inhibitory, and the other excitatory. The inhibitory component is somatostatin (SS), while recent identification of growth hormone-

Table 2. Identified hypophysiotropic neurohormones

Hypothalamic hormone	Abbreviation	Structure
Growth hormone- (GH)-	GHRH-40	H-40 aa –OH
-releasing hormones	GHRH-44	H-44 aa $-NH_2$
Somatostatins	SS-14	H-14 aa –OH
	SS-28	H-28 aa –OH
Prolactin-inhibiting factor	GAP	H-56 aa –OH
Thyrotropin- (TSH)-releasing hormone	TRH	H-3 aa $-NH_2$
Luteinizing hormone- (LH)- and follicle stimulating-hormone- (FSH)-releasing hormone	LHRH	H-10 aa $-NH_2$
Corticotropin- (ACTH)-releasing factor	CRF	H-41 aa $-NH_2$

aa, amino acids; GAP, gonadotropin-releasing hormone associated peptide.

releasing hormone (GHRH) from a pancreatic tumor makes the hypothalamic counterpart of this molecule the most likely candidate for the exicitatory component. Studies leading to the characterization of GHRH from a human pancreatic tumor were motivated by the observation of few patients with bronchial carcinoid and acromegaly (for review see FROHMAN 1984a). The tumors contained a GHRH-like activity as assessed on primary cultures of rat pituitary cells.

Purification of GHRH from human pancreatic tumors that caused acromegaly was accomplished independently by GUILLEMIN et al. (1982) and RIVIER et al. (1982). The primary structure of GHRH determined by Guillemin and colleagues was that of a 44 amino acid peptide; the structure obtained by RIVIER et al. (1982) was superimposable, the only exception being the lack of the NH_2 terminal peptide. The localization of GHRH neurons seems to be mainly confined to the medial basal hypothalamus, with very few neurons in extrahypothalamic areas (SAWCENKO et al. 1985).

The hypothalamic hormone responsible for the inhibitory influence on GH release was identified and synthesized as a tetradecapeptide, somatostatin, by Guillemin and his group (BRAZEAU et al. 1973). In addition to the hypothalamus, SS-like immunoreactivity is present in some amygdaloid nuclei, zona incerta, hippocampus, cortex, other CNS areas, gastrointestinal tract, and pancreas (for review see REICHLIN 1983). The juxtaposition of SS-containing D cells to insulin and glucagon-containing B and A cells within pancreatic islets, and the ability of SS to inhibit the release of insulin and glucagon indicate that SS acts as a local (paracrine) regulator of islet A and B cell function. Subsequently, an NH_2 terminal extended form of SS-14 comprised of 28 amino acids (SS-28) was first identified in the intestine and characterized in the hypothalamus (ESCH et al. 1980).

In 1971, SCHALLY et al. reported for the first time the isolation, amino acid composition, and chemical and biologic properties of a nonapeptide, which had both luteinizing hormone (LH) and follicle-stimulating hormone (FSH) releasing activity. In a subsequent paper, the same group reported the definitive structure of a pig neurohormone, a decapeptide containing, in addition to the nine amino acid residues, one tryptophan residue. A similar structure has been reported for human, ovine, bovine, and rat LHRH (for review see SCHALLY et al. 1979).

The isolation, elucidation of structure, and synthesis of porcine and ovine thyrotropin-releasing hormone (TRH) represented a real breakthrough in neuroendocrinology, since TRH was the first hypophysiotropic hormone to be identified and characterized (BÖLER et al. 1969; BURGUS et al. 1970). In addition to the hypothalamus, TRH is localized in several extrahypothalamic areas such as the spinal cord and limbic forebrain where TRH may function as a neurotransmitter or modulator (JACKSON 1983). TRH has also been detected in the gastrointestinal tract and pancreas of the rat.

As early as 1955, SAFFRAN and SCHALLY provided evidence for the existence of a factor present in the hypothalamus that could stimulate ACTH release. However, until recently, several methodological difficulties hindered the biochemical indentification of corticotropin-releasing factor (CRF), a hypothalamic regulatory peptide identified and characterized later than TRH, LHRH, and SS. In 1981, VALE et al. reported the structure of ovine CRF, a 41 amino acid peptide. Like the other hypophysiotropic peptides, CRF functions not only as a regulator of the release of a specific hormone, but also as a neuromodulator mediating several neuroendocrine and autonomic responses. The characteristic distribution of CRF in the CNS, in close association with a large number of autonomic centers, suggests that CRF is an important regulatory peptide mediating several stress responses (Vale et al. 1983). The availability of hypophysiotropic peptides in pure form led to the development of specific RIAs for these substances and allowed several studies to be performed aimed at describing their localization and, possibly, functions.

II. General Principles of Determination

There are three main technical problems in the determination of hypophysiotropic as well as other peptides: (a) the degradation of their amino acid chain by tissue peptidases; (b) the low concentrations in tissues which demand highly sensitive assays; and (c) the specificity of antibodies used to detect them by RIA.

1. Inactivation of Hypophysiotropic Peptides in Biologic Fluids and Tissues

One of the major problems connected with the assay of peptides in biologic fluid and tissues is that, in general, their immunologic as well as biologic activity is rapidly destroyed. Plasma and tissues are the most active in this sense, while destruction by urine is generally much slower and cerebrospinal fluid (CSF) is minimally active in this regard.

Proteolytic enzymes in serum or tissue extracts have at least two effects on the RIA of peptides:

1. Degradation of the peptide in the sample would lead to underestimation of its actual level in the unknown if the antiserum recognizes the entire peptide molecule, or possibly, to overestimation of the actual quantity of the peptide if the degraded fragments retain immunologic activity.

2. Degradation of the radioiodinated peptide would prevent its binding to antiserum and would be interpreted falsely as inhibition of binding of the radioiodinated peptide (i.e., overestimation of unknown in the sample).

Inactivation of the tripeptide TRH immunoreactivity by plasma or serum at 37 °C is rapid, occurring at a rate of 50% or more per minute in vitro (Utiger and Bassiri 1973; Eskay et al. 1976). Loss of activity by plasma proteolytic enzymes is also a feature of LHRH (Nett and Adams 1977), somatostatin, and GHRH (Frohman and Downs 1986).

Unlike plasma, CSF inactivates peptide immunologic activity only slowly. For instance, TRH in CSF scarcely loses its immunologic properties, even at 37 °C (Shambaugh et al. 1975). Homogenates of most tissues destroy the bioactivity and immunoreactivity of the hypophysiotropic peptides, and the rate of such destruction is comparable to that occurring in plasma (Bassiri and Utiger 1974; Griffiths et al. 1975; Patel and Reichlin 1979).

To overcome these problems, it is recommended that the hypophysiotropic peptide be extracted from the biologic sample in which it is to be measured using a system which eliminates macromolecules (i.e., proteins which bind radioiodinated peptides nonspecifically and enzymes which degrade the peptide itself).

Furthermore, a very useful method for inhibiting peptide degradation by tissue is that of boiling extracts or performing extraction with acids. As an example, incubation for 24 h at 4 °C of ^{125}I-labeled Tyr-SS with homogenates of cerebral cortex in buffer at pH 7.5 induced a 50% loss of the label, evaluated by chromatoelectrophoresis. The label, however, appears to be completely stable in boiled extracts (pH 7.5) or in unboiled acetic acid extracts (PATEL and REICHLIN 1979). These finding suggest that the damaging effect of tissue enzymes on SS can be prevented by prior boiling of tissue extracts or by acidifying tissues. The latter procedure, however, can be insufficient for extracts containing high concentrations of acid proteases (stomach). Alternatively to boiling, peptide degradation in ex vivo dissected tissues (from rats) can be blocked by killing animals by microwave irradiation, a procedure which is known to stop any enzymatic activity almost instantaneously (HERCHL et al. 1977). The massive degradation of peptides in plasma can be overcome by the use of enzyme inhibitors such as aprotinin, bacitracin, *O*-phenantroline, or EDTA, and by sample acidification (HONG et al. 1977; CLEMENT-JONES et al. 1980).

A series of experiments must be performed in order to verify that the extraction conditions maintain the immunologic properties of the peptide and allow a good recovery. Detection of the peptide in the different extracts gives a measure of the loss of peptide during the procedures and allows one to choose the technique giving the lowest rate of decay. For recovery studies, different amounts of synthetic peptides are added to tissues before starting all the procedures connected with extraction. Recovery must be constant for peptide concentrations in the range of those endogenously present in the tissue.

2. Specificity of Antibodies

An important problem is connected with the specificity of antibodies used to identify and quantify neuropeptides either by RIA or by immunoenzymatic and immunohistochemical methods. Antibodies, in fact, either polyclonal or monoclonal, base their specificity on short amino acid sequences of the antigen (see Chap. 2). These sequences are very seldom specific to a single neuropeptide; they are often contained in the sequence of other known or still unknown neuropeptides or proteins. It follows that an antibody recognizing (even in a highly specific way, as for monoclonal antibodies) a part of such a common sequence, will recognize all other peptides or proteins containing the same sequence.

From the foregoing, the necessity arises of identifying single peptides by non-RIA methods and using the latter exclusively to quantify peptides previously identified on physicochemical bases.

Several methods can be applied to separate different peptides: gel filtration, ion exchange chromatography, electrophoresis, electrofocusing, thin layer chromatography, high pressure liquid chromatography (HPLC), and high affinity chromatography. Gel filtration is certainly the most utilized method for peptide separation. Although this technique is suitable for getting separation of molecules in a very restricted range of molecular weights (i.e., 0–700), it does not separate molecules with very similar molecular weights, but differing in some amino acids.

High pressure liquid chromatography (SNYDER and KIRKLAND 1979; AKIL et al. 1981) allows quick resolution of structurally similar peptides. This method is more rapid than gel chromatographic techniques since one sample can be run in a relatively short time and from each sample it is possible to separate several peptides in a single run.

Affinity chromatography (CUATRECASAS and ANFISEN 1971) is based on the use of a solid phase bearing active groups (antibodies or receptors) directed against the substance to be purified. This technique, owing to the inherent limitation of antibody specificity, is suitable for concentrating samples with a low content of the peptide or for preparative processes, allowing one to achieve a rough separation of different peptides.

III. Features Peculiar to the Determination of a Single Hypophysiotropic Hormone

1. Somatostatin

a) RIA Procedure

For detailed descriptions of RIA for SS, the reader can refer to PATEL and REICHLIN (1979), GERICH et al. (1979), and HERMANSEN et al. (1979). Anti-SS antibodies are generally obtained in rabbits immunized with glutaraldehyde conjugates of synthetic peptide with bovine serum albumin (BSA) (REICHLIN et al. 1968), or carbodiimide conjugates with bovine thyroglobulin (SKOWSKY and FISHER 1972). Since SS does not contain tyrosine, tyrosinated analogs of the peptide are used as antigens for iodination (lactoperoxidase method). The iodination mixture can be purified by gel chromatography (OGAWA et al. 1977) or by HPLC. Either the double-antibody method or BSA-coated charcoal can be used for separation of bound from free labeled peptide.

b) Determination of Somatostatin in Plasma

The RIA of SS in plasma is subject to several difficulties: (a) the well-known general capacity of unextracted plasma to degrade both native and tyrosylated analogs of the peptide used as tracers; (b) the presence in plasma of SS-binding proteins (KRONHEIM et al. 1979) and substances that interfere with antibody binding (LUNDQUIST et al. 1979); and (c) the heterogeneity of plasma SS-like immunoreactivity with only a small proportion of it being attributable to a 1600 dalton molecule corresponding to SS-14 (KRONHEIM et al. 1978).

Methods for separation of SS from proteins and other components of plasma that interfere with SS RIA have relied upon protein precipitation by acid or organic solvent (ARIMURA et al. 1978; LUNDQUIST et al. 1979) or extraction on glass (PENMAN et al. 1979). Gel filtration of plasma before RIA alleviates degradation of tracer and interference with antibody binding, providing a better (near complete) recovery of SS than the latter procedures (MACKES et al. 1981). In addition, it permits determination of the SS-like immunoreactivity in plasma due to the 1600 dalton molecule. Briefly, human plasma (5 ml) is applied to a Sephadex G-25 column; previously equilibrated with PBS 0.01 *M*, pH 7.4, 1% serum albumin, 0.05 *M* EDTA. Two peaks of SS-like immunoreactivity are found, one eluting in the void volume (>5000 dalton), the other peak coeluting with synthetic SS in the 1600 dalton region. SS values detected in human plasma by this method ranged between 40 and 100 pg/ml ($3–6 \times 10^{-11}$ *M*). However, chromatographic separation requires considerable time and large samples.

A simple and rapid method for the extraction of SS from plasma on octadecylsilylsilica (ODS) cartridges has been reported by VASQUEZ et al. (1982). ODS cartridges (Sep-Pak C_{18}) are prepared by washing with 5 ml acetonitrile followed by 5 ml water. Plasma (3 ml) is applied, weakly bound plasma components eluted with 5 ml water followed by 5 ml 0.1% trifluoroacetic acid (TFA) and SS is eluted with 3 ml 80:20 (v/v) acetonitrile: 0.1% TFA. The eluate is frozen, lyophilized and stored at −20 °C until subsequent assay. The rapidity of this method allows one to process 60–80 plasma samples per day. The range of basal SS measured in ODS extracts of human plasma is similar to that obtained following chromatographic separations.

2. Growth Hormone-Releasing Hormone

In our laboratory, we routinely perform GHRH (1–44) RIA according to the method described by FROHMAN and DOWNS (1986), slightly modified. The synthetic peptide is radioiodinated by the chloramine-T method (0.2 mCi Na ^{125}I, 1 µg peptide, and 10 µg chloramine-T). The iodination mixture is purified by HPLC by application to a C_{18} column and elution with a 30%–40% acetonitrile gradient in 0.01 *M* TFA (FROHMAN 1984b). Alternatively, the mixture can be separated by gel chromatography (Sephadex G-50). The antiserum we use (kindly donated by Dr. L.A. Frohman, Cincinnati) has been raised in rabbits against GHRH (1–40) coupled to hemocyanin by the glutaraldehyde technique, and exhibits considerable cross-reactivity with various forms and fragments of human GHRH, including GHRH (1–44). The RIA is carried out in phosphate buffer 50 m*M*, 10 m*M* EDTA, pH 7.2 + 0.3% BSA + 0.05% Triton, in a total volume of 800 µl for 48 h at 4 °C. The bound/free antigen separation is accomplished by adding a plasma–dextran-coated charcoal suspension. Rat and human GHRH have considerable sequence heterogeneity in the COOH terminal region. Consequently, antisera raised against human peptides exhibit a very low cross-reactivity for rat GHRH. For RIA determination of rat GHRH (1–43), we routinely use an iodination procedure for the synthetic peptide identical to that described for human GHRH. In the RIA procedure, hypothalamic extracts or standards are incubated in HCl 0.1 *M* + Triton 0.1%; the tracer and the antiserum are diluted in phosphate buffer 0.15 *M*, pH 7.4 (total volume 300 µl); free/bound antigen separation is accomplished by adding a plasma–dextran-coated charcoal mixture. In our hands, this method of separation gives lower nonspecific binding than the double-antibody procedure. The anti-rat GHRH antiserum was kindly donated by Dr. W. B. Wehrenberg (Milwaukee) and does not cross-react with GHRH isolated from other species (WEHRENBERG et al. 1984).

As for other peptides, measurement of GHRH immunoreactivity in unextracted plasma has been shown to be unreliable, owing to interference from non-GHRH material (FROHMAN 1984b). Thus, plasma samples must be extracted in order to obtain accurate GHRH estimates. According to the method described by Frohman and Downs (1986), acidified plasma samples are extracted by adsorption onto hydrophobic C_{18} Sep-Pak cartridges. The Sep-Pak is washed with TFA 0.1 *M* to remove interfering substances and GHRH is eluted with 80% acetonitrile/20% 0.01 *M* TFA. The eluate is frozen, lyophilized, and reconstituted in assay buffer. This procedure shows a recovery of more than 90%. Extraction of plasma on Vycor glass has been employed by Penny et al. (1984) in their very sensitive RIA for circulating GHRH (sensitivity 7 pg/ml).

3. Gonadotropin-Releasing Hormone

The fundamental paper describing detailed methods for LHRH was published by Nett et al. (1973) and almost all the methods published thereafter were slight modifications of their procedure. Nett and Niswender (1979) wrote a chapter in which they carefully described methodologies for the conjugation of LHRH to BSA, the production of antisera, and the preparation and purification of radioiodinated LHRH.

If one wants to produce antisera having specificity for the COOH terminus of LHRH, then LHRH conjugated to a carrier molecule via internal amino acids or via the NH_2-terminus should be used as immunogen. Conversely, if antisera having specificity for the NH_2-terminus of LHRH are desired, then LHRH conjugated to a carrier molecule via internal amino acids or via the COOH terminus should be used as immunogen. Antisera with the highest specificity, with respect to fragments of LHRH, have been produced when LHRH conjugated to a carrier molecule via internal amino acid residues was used as the immunogen.

Measurement of LHRH in peripheral plasma of sheep, humans, and rats (several papers quoted in Jonas et al. 1975) gave widely disparate levels. Jonas et al. (1975) made a comparison between plasma levels of LHRH in sheep, either unextracted or extracted with methanol. They found LHRH levels ranging from 0 to 400 pg/ml when RIA was performed in unextracted plasma; however, this immunoreactivity had no resemblance to synthetic LHRH added to ovine plasma. No endogenous LHRH could be detected in methanol extracts, while it was possible to determine LHRH exogenously administered.

The same extraction procedure was used to determine LHRH in portal blood collected from sheep (Clarke and Cummins 1982). These authors found peripheral LHRH concentrations ranging from 10 to 30 pg/ml. A different method for extracting LHRH from plasma has been reported by Miyake et al. (1980). Plasma was applied to a carboxymethylcellulose column eluted with ammonium acetate buffer, gradient pH 4.6. LHRH plasma levels reported by these authors in women varied during the menstrual cycle from 4–5 pg/ml to 40–50 pg/ml.

a) LHRH Analogs

Many analogs of LHRH have been synthesized to study the structure–activity relationship and obtain potent and long-acting agonists and antagonists of LHRH. D-Ser(*t*-Bu)^6-des-Gly^{10}-Pro^9-*NH*-Et-LHRH (buserelin) is an LHRH agonist that has been used in the treatment of various diseases, including prostate cancer, breast cancer, endometriosis, and precocious puberty (Bex and Corbin 1984). In order to determine the levels of buserelin in plasma and urine of patients under treatment, an RIA has been developed by SAITO et al. (1985). This assay does not show any appreciable cross-reaction with native LHRH and its analogs.

4. Thyrotropin-Releasing Hormone

An RIA for TRH was developed by BASSIRI and UTIGER (1972a). For a careful description of the main problems encountered in determining TRH in tissues and biologic fluids, the reader can refer to BASSIRI and UTIGER (1972b) and BASSIRI et al. (1979). Antisera prepared by immunization with TRH–BSA conjugates according to BASSIRI and UTIGER (1972a) have been used successfully in several laboratories (SAITO et al. 1975; ESKAY et al. 1976). An alternative method for coupling TRH to protein has been described by VISSER and KLOOTWIJSK (1981). TRH is first reacted with excess 1,5-difluoro-2,4-dinitrobenzene (DFDNB); after removal of unreacted DFDNB, the fluorodinitrophenyl-TRH intermediate is subsequently reacted with protein, yielding a TRH-fluorodinitrophenyl conjugate. This procedure seems to give anti-TRH sera with the highest affinity. Similarly to LHRH, TRH cannot be assayed in unextracted plasma. Interference in the assay also occurs with several incubation media (Locke, Krebs-Ringer) (JOSEPH-BRAVO et al. 1979). Simple extraction with methanol can be used; affinity chromatography can conveniently be employed to extract and concentrate the peptide from biologic fluids (MONTOYA et al. 1975; EMERSON and UTIGER 1975).

TRH immunoreactivity was detected in ethanol extracts of rat pituitary stalk blood after HPLC resolution (SHEWARD et al. 1983). The immunoreactivity resolved into three components, the first of which had an identical retention time to the synthetic tripeptide. The absolute amount of TRH released into stalk blood (1.667 ng/h) was in the same range as the values reported previously (FINK et al. 1982) and is very large relative to the amount of LHRH and SS release.

5. Corticotropin-Releasing Hormone

An RIA for ovine CRF (1–41) was developed by VIGH et al. (1982) for measuring CRF concentrations in rat hypothalami. The authors used two different iodinated analogs of ovine CRF as tracers Tyr(O)-CRF (1–41) and Tyr(35)-CRF (36–41), three different antibodies directed against the native peptide, and two fragments of it, CRF (37–41) and CRF (22–41), coupled to BSA. In their opinion, this method was suitable for measuring endogenous rat CRF. VALE et al. (1983) used an antiserum directed against the NH_2 terminal region of CRF, obtained by immunizing rabbits with Tyr^{22}, Gly^{23}-CRF (1–23), to detect CRF-like immunoactivity in sheep, dogs, rats, monkeys, and humans. On the contrary, immunization of animals with immunoglobulin-coupled CRF (1–41) produces antibodies directed toward the middle or COOH terminal region of CRF. These sera read sheep CRF well, but detect rat and human CRF, which differs from ovine CRF in several amino acids, poorly (VALE et al. 1983).

SKOFITSCH and JACOBOWITZ (1985) made a comparison between two antisera directed against synthetic ovine CRF and two antisera directed against rat/human CRF. They evaluated the CRF-like immunoreactivity in the rat brain and found that the two kinds of antisera recognize different forms of CRF-like immunoreactivity. Anti-ovine CRF serum recognized a peptide, prominent in the medulla and the spinal cord, which was not detected by anti-rat/human CRF serum.

Iodination of CRF is generally performed by the chloramine-T methodology carefully described by Nicholson et al. (1983), with one exception due to Linton and Lowry (1982) who used the Bolton–Hunter reagent. Immunoreactive CRF has been detected in human unextracted plasma (Nicholson et al. 1983; Stalla et al. 1984) from subjects receiving exogenous synthetic ovine CRF. The authors analyzed plasma immunoreactive CRF by gel exclusion chromatography and found that it coeluted with ^{125}I-labeled ovine CRF, indicating that most of the activity was similar in size to intact ovine CRF. Plasma levels of immunoreactive ovine CRF before the subjects received the peptide were always undetectable (<80 pg/ml). Instead, CRF-like activity was determined in portal blood of rats after extraction by adsorption onto C_{18} cartridges. Extracted portal plasma gave a response parallel to that of either synthetic ovine CRF or highly purified rat CRF (Gibbs and Vale 1982). Based upon ovine standard, about 0.5 ng/ml CRF was found in portal blood.

IV. Significance of Hypophysiotropic Peptides in Biologic Fluids

1. Determination in the Peripheral Blood

Measurement of hypophysiotropic peptides in hypothalamic tissue has been possible with all assays developed to date, and most investigators have observed good agreement between levels of the peptides measured by bioassay and those determined by RIA (Table 3). It has been proposed that measurement of hypothalamic releasing hormones in the systemic circulation might be used as a direct index of hypothalamic function; however, the wide anatomic distribution of these peptides and the dilution of portal vessel blood in the systemic circulation makes it unlikely that measurement of peptide levels in body fluids can be used to infer dysfunction in the hypothalamus. Different plasma levels have been reported by several laboratories, ranging from undetectable or a few picograms per milliliter to several nanograms per milliliter. Using radioactive microspheres, Nett et al. (1974) have

Table 3. Hypophysiotropic peptide concentrations in rat hypothalamus and pituitary portal blood

	Hypothalamus	Pituitary portal blood
Somatostatin-LI (SS-14, SS-28)	40–50 ng/MBH[a]	200–400 pg/ml[d]
Growth hormone-releasing hormone-LI	1–2 ng/MBH[a]	200–800 pg/ml[e]
Thyrotropin-releasing hormone-LI	8–9 ng/hypothalamus[b]	3–6 ng/ml[b]
Gonadotropin-releasing hormone-LI	3–5 ng/MBH[a]	30–100 pg/ml[f]
Corticotropin-releasing factor-LI	1–2 ng/MBH[c]	400–500 pg/ml[g]

LI, like immunoreactivity; MBH, medial basal hypothalamus.
[a] Author's results; [b] Fink et al. (1983); [c] Vale et al. (1983); [d] Millar et al. (1983); [e] Plotsky and Vale (1985); [f] Eskay et al. (1977); [g] Gibbs and Vale (1982).

estimated in the sheep that portal blood is diluted approximately 500-fold by the time it enters the external jugular vein. Since neurohormone levels in pituitary portal blood are generally below 1 ng/ml, maximum levels of hypothalamus-derived peptide in the peripheral circulation would not be expected to exceed a few picograms per milliliter.

The presence of TRH in human peripheral blood is controversial, with some workers finding it to be undetectable and others detecting significant quantities of TRH-like immunoreactivity. The concentrations described cannot be reconciled with a known central or peripheral source. Interestingly, the neonatal rat has high levels of authentic TRH in the systemic circulation (ENGLER et al. 1981) as well as a low TRH-degrading activity (NEARY et al. 1978). In these animals, the peptide contained in plasma appears to derive from extra-CNS structures (pancreas or gastrointestinal tract) since they are not affected by encephalectomy (ENGLER et al. 1981). SS is present in high concentration in the stomach, duodenum, jejunum, and pancreas. Therefore, these sources may be mainly responsible for maintaining circulating concentrations of the peptide (100 pg/ml). It has been postulated that circulating SS might play a potential hormone role. At concentrations similar to those present in blood, SS can affect pituitary as well as gastrointestinal and pancreatic endocrine function. Moreover, in the dog, systemic administration of SS antiserum has been shown to alter gastrointestinal function (SCHUSDZIARRA et al. 1980).

Determination of plasma levels of GHRH has resulted in a very useful method of diagnosing possible ectopic secretion of the neurohormone from bronchial or pancreatic tumors. Basal GHRH levels in normal subjects are generally undetectable, as were those of a large population of acromegalic subjects (THORNER and CRONIN 1985). However, high levels of the peptides are present in patients with ectopic GHRH syndrome. By using a highly sensitive RIA, SOPWITH et al. (1985) found GHRH levels of 7–15 pg/ml in plasma of normal subjects after an overnight fast. GHRH concentrations rose to higher levels (20–77 pg/ml) when measured in the same subjects sampled 120 min after a mixed breakfast. These data would suggest that the predominant source of the GHRH detected in plasma is extrahypothalamic, possibly from upper gut or pancreas. In view of the poor correlation between circulating GHRH and the hypothalamic neurons, detection of low plasma levels of GHRH in acromegaly does not exclude the possibility of an enhanced activity of GHRH-producing neurons among the etiopathogenetic factors of the disease.

2. Determination in the Cerebrospinal Fluid

Determination of releasing and inhibiting hormones in the CSF would appear a more reliable index of hypothalamic neuronal activity than their determination in peripheral blood, the biologic fluid most accessible to investigators in humans. These water-soluble peptides, released from the neurons into the extracellular space, readily pass into the CSF. However, as already mentioned, most hypophysiotropic hormones are widely distributed in extrahypothalamic regions of the CNS, suggesting a role for these peptides in neuronal function. Therefore, peptides coming from extrahypothalamic areas can readily pass into the CSF, adding to those coming from the hypothalamus; this also implies that detection of CSF levels of hypophysiotropic hormones can by no means be taken as an index of hypothalamic neuronal activity.

The ability of peptides, which are in general low molecular weight and water-soluble substances, to traverse the blood–brain barrier would appear severely restricted, with the exception of the "leaky" areas between blood and brain which occur at the level of circumventricular organs (Weindl 1973). This is of importance for excluding a contribution from peptides coming from the gastrointestinal tract and peripheral nervous system to increase CSF levels. However, since some neural peptides occur not only in the hypothalamus and brain, but also in the spinal cord, caution must be exercised in the interpretation of peptide levels measured in lumbar CSF, the most commonly applied sampling technique. It is also possible that significant changes could occur in the concentrations of a peptide in the third ventricle without producing alterations in the levels elsewhere in the ventricular system. Despite these caveats, many data have been produced on the measurement of endogenous hypophysiotropic peptides in the CSF.

First RIA quantitation of CSF TRH-like immunoreactivity was performed by Oliver et al. (1974) in CSF obtained by cisternal puncture from patients undergoing neuroradiologic examination. The levels reported ranged from 60 to 290 pg/ml. In another study performed on 17 neurologic patients undergoing pneumoencephalography, the CSF TRH levels were somewhat lower (4.9 ± 2.6 pg/ml; Liira et al. 1978).

Immunoreactive CRF was detected in human CSF by Suda et al. (1983) by using a specific ovine CRF RIA and immunoaffinity chromatography in order to concentrate the peptide (levels reported 7.4 ± 1.1 fmol/ml). Most reports agree that the levels of LHRH in the CSF of different animal species, including humans (see Jackson 1984) are low or undetectable. Gunn et al. (1974) reported that LHRH was absent from 23 of 26 samples determined, while in the other 3 samples, levels of 20–120 pg/ml were measured, but dilutions showed nonparallelism, thus creating doubts about identity of the immunoreactivity. SS is readily detectable in the CSF where it appears to be present in concentrations in the same range as those reported for plasma.

Many studies give support to the hypothesis that SS in the CSF does not derive solely from the hypothalamus. In the study by Berelowitz et al. (1982) performed in the rhesus monkey, intravenous infusion of human GH lowered CSF SS levels. Since in vivo GH administration stimulates hypothalamic SS release, CSF SS is likely derived from extrahypothalamic brain secretion which responds in an opposite way to GH feedback. The same conclusions have been drawn in the human by Sorensen et al. (1981). These authors collected serial spinal fluid aliquots and found a reduction in the protein content when comparing initial and subsequent samples, owing to protein increase from the cerebral ventricles over the cerebral cisterns and in the lumbar sac. In contrast, CSF SS did not decrease, indicating that it is released diffusely from the CNS including the spinal cord. In addition, they found no correlation between basal arginine-provoked plasma GH and the CSF content of SS. In all, these and the previously mentioned results do not support the idea that CSF SS may be taken to denote the activity or tone of hypothalamic SS release. The presence of immunoreactive GHRH in CSF has been detected by an antiserum specific to the COOH terminal portion of hGHRH (1–44)-NH_2 (KASHIO et al. 1985). CSF samples were collected in volumes of 2–10 ml and concentrated by antibody affinity chromatography.

Immunoreactive GHRH was measurable in the CSF of all control subjects with no endocrinopathy (mean concentrations 29.3 ± 2.0 pg/ml). GHRH was not detectable in any of the four patients with hypothalamic germinoma which invaded the medial basal hypothalamus, destroying the neurons containing the hypothalamic hormones.

These results suggest that the main source of CSF GHRH lies in the medial basal hypothalamus, as already demonstrated by immunohistochemical studies showing that most of the GHRH-containing neurons in the human brain are located around the third ventricle (BLOCH et al. 1983). Therefore,it would appear that, in contrast to SS, assay of GHRH levels in the CSF would be a reliable index of hypothalamic neuronal function of the releasing hormone.

References

Akil H, Ueda Y, Lin HL, Watson SJ (1981) A sensitive coupled HPLC/RIA technique for separation of endorphins: multiple forms of β-endorphin in rat pituitary intermediate and anterior lobe. Neuropeptides 1:429–446

Arimura A, Lundquist G, Rotman J, Change R, Fernandez-Durango R, Elde R, Coy DH, Meyer C, Schally AV (1978) Radioimmunoassay of somatostatin. Metabolism 27:1139–1143

Bassiri RM, Utiger RD (1972a) The preparation and specificity of antibody to thyrotropin releasing hormone. Endocrinology 90:722–727

Bassiri RM, Utiger RD (1972b) Serum inactivation of the immunological and biological ativity of thyrotropin. Endocrinology 91:657–664

Bassiri RM, Utiger RD (1974) Thyrotropin-releasing hormone in the hypothalamus of the rat. Endocrinology 94:188–197

Bassiri RM, Dvorak J, Utiger RD (1979) Thyrotropin-releasing hormone. In: Jaffe BM, Behrman HR (eds) Methods of hormone radioimmunoassay. Academic, New York, pp 45–56

Baumann G (1984) Heterogeneity of human growth hormone in the circulation. In: Labrie F, Proulx L (eds) Endocrinology. Excerpta Medica, Amsterdam, pp 749–752

Baumann G, MacCart JG, Amburn K (1983) The molecular nature of circulating growth hormone in normal and acromegalic man: evidence for a principal and minor monomeric form. J Clin Endocrinol Metab 56:946–951

Beitins Z, Derfel RL, O'Laughlin K, Mc Arthur JW (1977) Immunoreactive luteinizing hormone, follicle stimulating hormone and their subunits in human urine following gel filtration. J Clin Endocrinol Metab 44:149–159

Berelowitz M, Perlow MJ, Frohman LA (1982) The effect of intravenous growth hormone infusion on cerebrospinal fluid somatostatin levels in the rhesus monkey. Brain Res 241:355–358

Bex FJ, Corbin A (1984) LHRH and analogs: reproductive pharmacology and contraceptive and therapeutic utility. In: Martin L, Ganong WF (eds) Frontiers in neuroendocrinology. Raven, New York, pp 85–151

Binoux M, Pierce JG, Odell WD (1974) Radioimmunological characterization of human thyrotropin and its subunits: applications for the measurement of human TSH. J Clin Endocrinol Metab 38:674–682

Bloch B, Brazeau P, Ling N, Böhlen P, Esch F, Wehrenberg WB, Benoit R, Bloom F, Guillemin R (1983) Immunohistochemical detection of growth hormone releasing factor in brain. Nature 301:607–608

Böler J, Enzmann F, Folkers K, Bowers CY, Schally AV (1969) The identity of chemical and hormonal properties of the thyrotropin releasing hormone and pyroglutamil-histidylproline-amide. Biochem Biophys Res Comm 37:705–710
Borer KT, Kelch RP, Corley K (1982) Hamster prolactin: physiological changes in blood and pituitary concentrations as measured by a homologous radioimmunoassay. Neuroendocrinology 35:13–21
Brazeau P, Vale W, Burgus R, Ling N, Butcher M, Rivier J, Guillemin R (1973) Hypothalamic polypeptide that inhibits immunoreactive pituitary growth hormone. Science 179:77–79
Burgus R, Dunn T, Desiderio DV, Ward DN, Vale W, Guillemin R (1970) Characterization of ovine hypothalamic hypophysiotropic TSH-releasing factor. Nature 226:321–325
Cahill C, Matthews JD, Akil H (1983) Human plasma β-endorphin-like peptides: a rapid, high recovery extraction technique and validation of radioimmunoassay. J Clin Endocrinol Metab 56:992–997
Carmel PW, Araki S, Ferin M (1976) Pituitary stalk portal blood collection in rhesus monkey: evidence for pulsatile release of gonadotropin-releasing hormone (GnRH). Endocrinology 99:243–248
Catt KJ, Tregear GW (1967) Solid-phase radioimmunoassay in antibody-coated tubes. Science 158:1570–1572
Catt JK, Niall HD, Tregear GW (1966) Solid-phase radioimmunoassay of human growth hormone. Biochem J 100:31C–33C
Chang RJ, Keye WR Jr, Young JR, Wilson CB, Jaffe RB (1977) Detection, evaluation and treatment of pituitary microadenomas in patients with galactorrhea and amenorrhea. Am J Obstet Gynecol 128:356–363
Chihara K, Kashio Y, Abe M, Minamitami N, Kaji H, Kita T, Fujita T (1984) Noradrenergic modulation of human pancreatic growth hormone-releasing factor-induced growth hormone release in conscious male rabbits: involvement of endogenous somatostatin. Endocrinology 114:1402–1406
Chin WW, Maloof F, Martorana MA, Pierce JG, Ridgway EC (1981) Production and release of TSH and its subunits by monolayer cultures containing bovine anterior pituitary cells. Endocrinology 108:387–394
Clarke IJ, Cummins JT (1982) The temporal relationship between gonadotropin releasing hormone (GnRH) and luteinizing hormone (LH) secretion in ovariectomized ewes. Endocrinology 111:1737–1739
Clement-Jones V, Lowry PJ, Rees LH, Besser GM (1980) Development of a specific extracted radioimmunoassay for methionine enkephalin in human plasma and cerebrospinal fluid. J Endocrinol 86:231–243
Cocola F, Udeschini G, Secchi C, Panerai AE, Neri P, Müller EE (1976) A rapid radioimmunoassay method of growth hormone in dog plasma. Proc Soc Exp Biol Med 151:140–145
Cuatrecasas P, Anfisen CB (1971) Affinity chromatography. Meth Enzym 22:345–378
Daane TA, Parlow F (1971) Serum FSH and LH in constant light induced persistent estrus: short term and long term studies. Endocrinology 88:653–663
Davis SL, Sasser RG, Thacker DL, Ross RH (1977) Growth rate and secretion of pituitary hormones in relation to age and chronic treatment with thyrotropin-releasing hormone in prepubertal dairy heifers. Endocrinology 100:1394–1402
De Coster R, Beckers JF, Beerens D, De Mey J (1983) A homologous radioimmunoassay for canine prolactin: plasma levels during the reproductive cycle. Acta Endocrinol 103:473–478
Drobny EC, Amburn K, Baumann G (1983) Circadian variation of basal plasma growth hormone in man. J Clin Endocrinol Metab 57:524–528
Dufau ML, Mendelson C, Catt KJ (1974) A highly sensitive in vitro bioassay for luteinizing hormone and chorionic gonadotropin. Testosterone production by dispersed Leydig cells. J Clin Endocrinol Metab 39:610–618
Dufau ML, Pock R, Neubauer A, Catt KJ (1976) Bioassay of luteinizing hormone in human serum. J Clin Endocrinol Metab 42:958–963

Eigemann JE, Eigemann RV (1981) Radioimmunoassay of canine growth hormone. Acta Endocrinol 96:514–520
Ellis S, Vodian MA, Grindeland RE (1978) Studies on bioassayble growth hormone-like activity of plasma. Recent Prog Horm Res 34:213–238
Emerson CH, Utiger RD (1975) Plasma thyrotropin-releasing hormone concentration in the rat: effect of thyroid excess and deficiency and cold exposure. J Clin Invest 56:1564–1570
Engler D, Scanlon MF, Jackson IMD (1981) Thyrotropin-releasing hormone in the systemic circulation of the neonatal rat is derived from the pancreas and other extraneural tissues. J Clin Invest 67:800–808
Eriksson H, Mattiasson B, Thorell J (1984) Combination of solid-phase second antibody and internal sample attenuator counting techniques. Radioimmunoassay of thyroid-stimulating hormone. J Immunol Methods 71:117–125
Esch F, Böhlen P, Ling N, Benoit R, Brazeau P, Guillemin R (1980) Primary structure of ovine hypothalamic somatostatin-28 and somatostatin-25. Proc Natl Acad Sci USA 77:6827–6831
Eskay RL, Oliver C, Warberg J, Porter JC (1976) Inhibition of degradation and measurement of immunoreactive thyrotropin releasing hormone concentration in rat and plasma. Endocrinology 98:269–277
Eskay RL, Mical RS, Porter JC (1977) Relationship between luteinizing hormone-releasing hormone concentration in hypophyseal portal blood and luteinizing hormone release in intact, castrated and electrochemically stimulated rats. Endocrinology 100:263–270
Faglia G, Beck-Peccoz P, Ballabio M, Nava C (1983) Excess of β-subunit of thyrotropin (TSH) in patients with idiopathic central hypothyroidism due to the secretion of TSH reduced biological activity. J Clin Endocrinol Metab 56:908–914
Felber JP (1974) Radioimmunoassay of polypeptide hormones and enzymes. Methods Biochem Anal 22:1–94
Fink G, Koch Y, Ben Aroya N (1982) Release of thyrotropin releasing hormone into hypophysial portal blood is high relative to other neuropeptides and may be related to prolactin secretion. Brain Res 243:186–189
Fink G, Koch Y, Ben Aroya N (1983) TRH in hypophysial portal blood: characteristics of release and relationship to thyrotropin and prolactin secretion. In: Griffiths EC, Bennett GW (eds) Thyrotropin-releasing hormone. Raven, New York, pp 127–143
Foley TP, Klein AH, Agustin AV (1977) Adaptation of TSH filter paper method for regionalized screening for congenital hypothyroidism. J Lab Clin Med 90:11–17
Friesen HG, McNeilly AS (1977) Nature of prolactin and its measurement. In: Martini L, Besser GM (eds) Clinical neuroendocrinology. Academic, London, pp 309–317
Frohman LH (1984a) Ectopic hormone production by tumors: growth-hormone releasing factor. In: Müller EE, MacLeod RM (eds) Neuroendocrine perspectives, vol 3. Elsevier, Amsterdam, pp 201–224
Frohman LA (1984b) Growth hormone-releasing factor: a neuroendocrine perspective. J Lab Clin Med 103:819–832
Frohman LA, Downs TR (1986) Measurement of growth hormone-releasing factor. In: Conn PM (ed) Neuroendocrine peptides. Methods Enzymol 124:370–389
Frohman LA, Thominet JL, Webb CB, Vance ML, Underman H, Rivier J, Vale W, Thorner MO (1984) Metabolic clearance and plasma disappearance rates of human pancreatic tumor growth hormone releasing factor in man. J Clin Invest 73:1304–1311
Fuchs AR, Wagner G (1981) Effect of mating on plasma levels of oxytocin and prolactin in male and female rabbits. J Endocrinol 90:245–253
Genazzani AR, Facchinetti F, Parrini D (1981) β-Lipotropin and β-endorphin plasma levels during pregnancy. Clin Endocrinol (Oxf) 14:409–418
Gerich J, Greene K, Hara M, Rizza R, Patton G (1979) Radioimmunoassay of somatostatin and its application in the study of pancreatic somatostatin secretion in vitro. J Lab Clin Med 93:1055–1061
Gershengorn MC (1978) Regulation of thyrotropin production by mouse pituitary thyrotropic tumor cells in vitro by physiological levels of thyroid hormones. Endocrinology 102:1122–1128

Gibbs DM, Vale W (1982) Presence of corticotropin releasing factor-like immunoreactivity in hypophysial portal blood. Endocrinology 111:1418–1420

Glick SM, Roth J, Yalow RS, Berson SA (1963) Immunoassay of human growth hormone in plasma. Nature 199:784–787

Gluckman PD (1982) The hypothalamic-pituitary unit: the maturation of the neuroendocrine system in the fetus. In: Besser GM, Martini L (eds) Clinical neuroendocrinology, vol 2. Academic, New York, pp 31–43

Griffiths EC, Hooper KC, Jeffcoate SL, White N (1975) Peptidases in the rat hypothalamus inactivating thyrotropin-releasing hormone (TRH). Acta Endocrinol 79:209–216

Grossman A, Clement-Jones V, Besser GM (1985) Clinical implications of endogenous opioid peptides. In: Müller EE, MacLeod RM, Frohman LA (eds) Neuroendocrine perspectives. Elsevier, Amsterdam, vol 4, pp 243–295

Guillemin R, Brazeau P, Böhlen P, Esch F, Ling N, Wehrenberg WB (1982) Growth hormone-releasing factor from a human pancreatic tumor that caused acromegaly. Science 218:585–587

Gunn A, Fraser HM, Jeffcoate SL, Holland DT, Jeffcoate WJ (1974) CSF and release of pituitary hormones. Lancet I:1057

Haber E, Page LB, Richards FF (1965) Radioimmunoassay employing gel filtration. Anal Biochem 12:163–172

Hampshire T, Altszuler N, Steele R, Greene LJ (1975) Radioimmunoassay of canine growth hormone: enzymatic radioiodination. Endocrinology 96:822–827

Herbert V, Lan KS, Gottlieb CW, Bleicher SJ (1965) Coated charcoal immunoassay of insulin. J Clin Endocrinol Metab 25:1375–1384

Herchl R, Havlicek V, Rezek M, Kroeger E (1977) Cerebroventricular administration of somatostatin (SRIF): effect on central levels of cyclic AMP. Life Sci 20:821–826

Hermansen K, Christensen SE, Ørskoo H (1979) Characterization of somatostatin release from the pancreas. The role of calcium and acetylcholine. Diabetologia 16:261–275

Hodgkinson C, Allolio B, Landon J, Lowry PJ (1984) Development of a non-extracted "two site" immunoradiometric assay for corticotropin utilizing extreme aminoacid carboxy-terminally directed antibodies. Biochem J 218:703–711

Hoff JD, Lasley BL, Yen SSC (1977) The two pools of pituitary gonadotropin: regulation during the menstrual cycle. J Clin Endocrinol Metab 44:302–313

Hong JS, Yang HYT, Costa E (1977) On the location of methionine enkephalin neurons in rat striatum. Neuropharmacology 16:541–548

Hughes JP, Tanaka T, Gout PW, Beer CT, Noble RL, Friesen HG (1982) Effect of iodination on human growth hormone and prolactin: characterized by bioassay radioimmunoassay, radioreceptor assay and electrophoresis. Endocrinology 111:827–832

Hunter WM, Greenwood FC (1962) Preparation of iodine-131 labelled human growth hormone of high specific activity. Nature 194:495–496

Imura H, Nakai Y, Nakao K, Oki S, Tanaka I (1982) Control of biosynthesis and secretion of ACTH, endorphins and related peptides. In: Müller EE, MacLeod RM (eds) Neuroendocrine perspectives. Elsevier, Amsterdam, vol 1, pp 137–167

Jackson JMD (1983) Thyrotropin releasing hormone (TRH): distribution in mammalian species and its functional significance. In: Griffiths EC, Bennett GW (eds) Thyrotropin-releasing hormone. Raven, New York, pp 3–18

Jackson JMD (1984) Neuropeptides in the cerebrospinal fluid. In: Müller EE, MacLeod RM (eds) Neuroendocrine perspectives, vol 3. Elsevier, Amsterdam, pp 121–159

Jacobs LS (1979) Prolactin. In: Jaffe BM, Behrman HR (eds) Methods of hormone radioimmunoassay. Academic, New York, pp 199–222

Jaffe RB (1981) Physiologic and pathophysiologic aspects of prolactin production in humans. In: Jaffe RB (ed) Prolactin. Elsevier, New York, pp 181–217

Jaquet P, Ketelslegers JM, Jakubowski H, Hellman G, Franchimont P (1971) Dosage radio-immunologique de l'hormone thyreotrope humaine (H-TSH). I. Etude de la cummunatè antigènique entre TSH humaine, bovine, porcine et les autres hormone glycoprotèiques (HFSH, HLH, HCG). Ann Endocrinol 32:423–495

Jonas HA, Burger HG, Cumming JA, Findlay JK, Kretser DM (1975) Radioimmunoassay for luteinizing hormone releasing hormone (LHRH): its application to the measurement of LHRH in ovine and human plasma. Endocrinology 96:384–393

Joseph-Bravo P, Charli JL, Palacios JM, Kordon C (1979) Effect of neurotransmitters on the in vitro release of immunoreactive thyrotropin-releasing hormone from rat medial basal hypothalamus. Endocrinology 104:801–806
Kashio Y, Chihara K, Kaji H, Minamitani N, Kita T, Okimura Y, Abe H, Iwasaki J, Fujita T (1985) Presence of growth hormone-releasing factor-like immunoreactivity in human cerebrospinal fluid. J Clin Endocrinol Metab 60:396–398
Kieffer JD, Weintraub BD, Baigelman W, Leeman S, Maloof F (1974) Homologous radioimmunoassay of thyrotropin in rat plasma. Acta Endocrinol 76:495–505
Knobil E (1974) On the control of gonadotropin secretion in the rhesus monkey. Recent Prog Horm Res 30:1–46
Kosasa TS, Thompson IE, Byer WB, Taymor ML (1976) A comparison between gonadotropin radioimmunoassays with the use of 48 hour and six day incubation methods. Am J Obstet Gynecol 124:116–125
Kourides IA, Weintraub BD, Levko MA, Maalof F (1974) Alpha and beta subunits of human thyrotropin: purification and development of specific radioimmunoassays. Endocrinology 94:1411–1420
Kourides IA, Weintraub BD, Ridgway EC, Maloof F (1975) Pituitary secretion of free α and β subunit of human thyrotropin in patients with thyroid disorders. J Clin Endocrinol Metab 40:872–885
Kourides IA, Weintraub BD, Rosen SW, Ridgway EC, Kliman B, Maloof F (1976) Secretion of α subunit of glycoprotein hormones by pituitary adenomas. J Clin Endocrinol Metab 43:97–106
Krieger DT, Liotta AS, Brownstein MJ, Zimmerman EA (1980) ACTH, β lipotropin and related peptides in brain, pituitary and blood. Recent Prog Horm Res 36:277–343
Kronheim S, Berelowitz M, Pimestone B (1978) The characterization of somatostatin-like immunoreactivity in human serum. Diabetes 27:523–529
Kronheim S, Sheppard M, Shapiro B, Pimestone B (1979) Ultracentrifugation evidence for a somatostatin-binding protein in serum. Biochem Biophys Acta 586:568–573
Leung FC, Russell SM, Nicoll CS (1978) Relationship between bioassay and radioimmunoassay estimates of prolactin in rat serum. Endocrinology 103:1619–1628
Lewis VJ, Sinha YN, Markoff E, Vander Laan WP (1985) Multiple forms of prolactin: properties and measurement. In: Müller EE, MacLeod RM, Frohman LA (eds) Neuroendocrine perspectives, vol 4. Elsevier, Amsterdam, pp 43–57
L'Hermite M, Niswender GD, Reichert LE Jr, Midgley AR Jr (1972) Serum follicle-stimulating hormone in sheep as measured by radioimmunoassay. Biol Reprod 6:325–329
Liira J, Leppaluoto J, Hyppa MT (1978) Hypothalamic and pituitary hormones in human CSF. Acta Neurol Scand 57:272–273
Linton EA, Lowry PI (1982) Radioimmunoassay and chromatographic characterization of CRF-41-like immunoreactivity in hypothalami from several species. Neuropeptides 3:45–50
Liotta AS, Toshiro S, Krieger DT (1978) β-lipotropin is the major opioid-like peptide of human pituitary and rat pars distalis. Proc Natl Acad Sci USA 75:2950–2952
Lovinger R, Boryczka AT, Schackelford R, Kaplan SL, Ganong WF, Grumbach MM (1974) Effect of synthetic somatotropin release inhibiting factor on the increase in plasma growth hormone elicited by L-dopa in the dog. Endocrinology 95:943–946
Lundquist G, Gustavsson S, Elde R, Arimura A (1979) A radioimmunoassorbent assay for plasma somatostatin. Clin Chim Acta 101:15–25
MacFarlane IA, Beardwell CG, Shalet SM, Darbyshire PJ, Hayward E, Sutton ML (1980) Glycoprotein hormone α subunit secretion by pituitary adenomas: influence of external irradiation. Clin Endocrinol 13:215–222
MacFarlane IA, Beardwell CG, Shalet SM, Ainslie G, Rankin E (1982) Glycoprotein hormone α-subunit secretion in patients with pituitary adenomas: influence of TRH, LRH and bromocriptine. Acta Endocrinol 99:487–492
Machlin LJ, Jacobs LS, Cirulis N, Kimes R, Miller R (1974) An assay for growth hormone and prolactin-releasing activities using a bovine pituitary cell culture system. Endocrinology 95:1350–1358
Mackes K, Itoh M, Greene K, Gerich J (1981) Radioimmunoassay of human plasma somatostatin. Diabetes 30:728–734

Mains RE, Eipper BA (1979) Synthesis and secretion of corticotropins, melanotropins and endorphins by rat intermediate pituitary cells. J Biol Chem 254:7989–7994
Markoff E, Colosi P, Talamantes F (1981) Homologous radioimmunoassay for secreted mouse prolactin. Life Sci 28:203–211
Martin JB, Brazeau P, Tannenbaum GS, Willoughby JO, Epelbaum J, Terry LC, Durand D (1978) Neuroendocrine organization of growth hormone regulation. In: Reichlin S, Baldessarini RJ, Martin JB (eds) The hypothalamus. Raven, New York, pp 329–357
McNeilly AS, Andrews P (1974) Purification and characterization of caprine prolactin. J Endocrinol 60:359–366
Miles LEM, Hales CN (1968) Labelled antibodies and immunological assay systems. Nature 219:186–189
Millar RP, Sheward WJ, Wegener J, Fink G (1983) Somatostatin-28 is an hormonally active peptide secreted into hypophysial portal vessel blood. Brain Res 260:334–337
Miller JD, Tannenbaum GS, Colle E, Guyda HJ (1982) Daytime pulsatile growth hormone secretion during childhood and adolescence. J Clin Endocrinol Metab 110:540–550
Miyake A, Kawamura Y, Aono T, Kurachi K (1980) Changes in plasma LRH during the normal menstrual cycle in women. Acta Endocrinol 93:257–263
Montanini V, Celani MF, Baraghini GF, Carani C, Marrama P (1984) Effects of acute stimulation with luteinizing hormone-releasing hormone on biologically active and immunoreactive luteinizing hormone in pubertal boys. Acta Endocrinol 107:289–294
Montoya E, Seibel MJ, Wilber JF (1975) Thyrotropin releasing hormone secretory physiology: studies by radioimmunoassay and affinity chromatography. Endocrinology 96:1413–1418
Morgan CR, Sorenson RL, Lazarow A (1964) Studies of an inhibitor of the two antibody immunoassay system. Diabetes 13:1–5
Morley JE (1981) Neuroendocrine control of thyrotropin secretion. Endocr Rev 2:396–436
Moudgal NR, Li CH (1961) An immunochemical study of sheep pituitary interstitial cell stimulating hormone. Arch Biochem Biophys 95:93–98
Moudgal NR, Muralidhar K, Madhwa Raj HG (1979) Pituitary gonadotropins. In: Jaffe BM, Behrman HR (eds) Methods in radioimmunoassay. Academic, New York, pp 173–198
Müller EE (1973) The control of growth hormone secretion. Neuroendocrinology 11:338–369
Müller EE (1979) The control of somatotropin secretion. In: Li CH (ed) Hormonal proteins and peptides, vol VII. Academic, New York, pp 123–204
Nakanishi S, Inone A, Kita T, Nakamura M, Chang ACY, Cohen SN, Imura S (1979) Nucleotide sequence of cloned DNA for bovine, corticotropin β-lipotropin precursor. Nature 278:423–427
Neary JT, Kieffer JD, Nakamura C, Mover H, Soodak M, Maloof F (1978) The developmental pattern of thyrotropin-releasing hormone degrading activity in the plasma of rats. Endocrinology 103:1849–1854
Neill JD, Parton JM, Dailey RA, Tsou RC, Tindall GT (1977) Luteinizing hormone releasing hormone (LHRH) in pituitary stalk blood of rhesus monkeys: relationship to level of LH release. Endocrinology 101:430–434
Nett TM, Adams TE (1977) Further studies on the radioimmunoassay of gonadotropin-releasing hormone: effect of radioiodination, antiserum and unextracted serum on levels of immunoreactivity in serum. Endocrinology 101:1135–1144
Nett TM, Akbar AM, Niswender GD, Hedlund MT, White WF (1973) A radioimmunoassay for gonadotropin-releasing hormone (Gn-RH) in serum. J Clin Endocrinol Metab 36:880–885
Nett TM, Akbar AM, Niswender GD (1974) Serum levels of luteinizing hormone and gonadotropin-releasing hormone in cycling castrated and anestrous ewes. Endocrinology 94:713–718
Nett TM, Niswender GD (1979) Gonadotropin-releasing hormon. In: Jaffe BM, Behrman HR (eds) Methods of hormone radioimmunoassay. Academic, New York, pp 57–75.

Nicholson WE, De Cherney GS, Jackson RV, De Bold CR, Underman H, Alexander AN, Rivier J, Vale W, Orth DN (1983) Plasma distribution, disappearance half-time, metabolic clearance rate and degradation of synthetic ovine corticotropin-releasing factor in man. J Clin Endocrinol Metab 57:1263–1269

Nicoll CS (1967) Bioassay of prolactin. Analysis of the pigeon crop sac response to local prolactin injection by an objective and quantitative method. Endocrinology 80:641–655

Nicoll CS (1975) Radioimmunoassay and radioreceptor assays for prolactin and growth hormone: a critical appraisal. Am Zool 15:881–903

Nicoll CS, Mena F, Nichols CW Jr, Green SH, Tai M, Russell SM (1976) Analysis of suckling-induced changes in adenohypophyseal prolactin concentration in the lactating rat by three assay methods. Acta Endocrinol 83:512–521

Niswender GD, Midgley AR Jr, Monroe SE, Reichert LE (1968a) Radioimmunoassay of rat LH with anti-ovine LH serum and ovine LH^{131}. Proc Soc Exp Biol Med 128:807–811

Niswender GD, Midgley AH, Reichert LE Jr (1968b) Radioimmunological studies with murine, bovine, ovine, porcine luteinizing hormones. In: Rosemberg E (ed) Gonadotropins. Geron-X Inc, Los Altos, pp 299–315

Niswender GD, Chen CL, Midgley AR Jr, Meites OJ, Ellis S (1969a) Radioimmunoassay for rat prolactin. Proc Soc Biol Med 130:533–536

Niswender GD, Reichert LE, Midgley AR Jr, Nalbandov AV (1969b) Radioimmunoassay for bovine and ovine luteinizing hormone. Endocrinology 84:1166–1173

Ogawa N, Thompson T, Friesen HG, Martin JB, Brazeau P (1977) Properties of soluble somatostatin-binding protein. Biochem J 165:269–277

Ogawa N, Panerai AE, Lee S, Forsbach G, Havlicek V, Friesen HG (1979) β-Endorphin concentration in the brain of intact and hypophysectomized rats. Life Sci 25:317–326

Oliver C, Charvet JP, Codaccioni JL, Vague J, Porter JC (1974) TRH in human CSF. Lancet I:873

Orth DN (1979) Adrenocorticotropic hormone (ACTH). In: Jaffe BM, Behrman HR (eds) Methods of hormone radioimmunoassay. Academic, New York, pp 245–284

Panerai AE, Martini A, De Rosa A, Salerno F, Di Giulio AM, Mantegazza P (1982) Plasma β-endorphin and met enkephalin in physiological and pathological conditions. Adv Biochem Pharmacol 33:139–149

Panerai AE, Martini A, Di Giulio Am, Fraioli F, Vegni C, Pardi G, Marini A, Mantegazza P (1983) Plasma β-endorphin, β-lipotropin and met-enkephalin concentrations during pregnancy in normal and drug-addicted women and their newborn. J Clin Endocrinol Metab 57:537–543

Papapetrou PD, Anagnostopoulos NI (1985) A gonadotropin and α-subunit suppression test for the assessment of the ectopic production of human chorionic gonadotropin and it subunits after the menopause. J Clin Endocrinol Metab 60:1187–1195

Papapetrou PD, Sakarebou NP, Braonzi H, Fessas PH (1980) Ectopic production of human chrorionic gonadotropin (hCG) by neoplasms: the value of measurements of immunoreactive hCG in the urine as a screening procedure. Cancer 45:2583–2589

Patel YC, Reichlin S (1979) Somatostatin. In: Jaffe BM, Behrman HR (eds) Methods of hormone radioimmunoassay. Academic, New York, pp 77–99

Peake GT, Morris J, Buckman MT (1979) Growth hormone. In: Jaffe BM, Behrman HR (eds) Methods of hormone radioimmunoassay. Academic, New York, p 223

Penman E, Wass JAH, Lund A, Lowry PJ, Stewart J, Dawson AM, Besser GM, Rees LH (1979) Development and validation of a specific radioimmunoassay for somatostatin in human plasma. Ann Clin Biochem 16:15–25

Penny ES, Penman E, Price P, Rees LH, Sopwith AM, Wass JAH, Lytras N, Besser GM (1984) Circulating growth hormone releasing factor concentrations in normal subjects and patients with acromegaly. Br Med J 289:453–455

Pierce JG, Parsons TF (1981) Glycoprotein hormones: structure function. Annu Rev Biochem 50:465–495

Plotsky PM, Vale W (1985) Patterns of growth hormone-releasing factor and somatostatin secretion into the hypophyseal-portal circulation of the rat. Science 230:461–463

Quinlan WJ, Michaelson S (1981) Homologous radioimmunoassay for canine thyrotropin: response of normal and X-irradiated dogs to propilthiouracil. Endocrinology 108:937–942

Rees LH, Cook DM, Kendall JW, Allen CF, Kramer RM, Ratcliffe JG, Knight RA (1971) A radioimmunoassay for rat plasma ACTH. Endocrinology 89:254–261

Reichlin M, Schumize JJ, Vance VK (1968) Induction of antibodies to porcine ACTH in rabbits with non steroidogenic polymers of BISA and ACTH. Proc Soc Exp Biol Med 128:347–350

Reichlin S (1983) Somatostatin. N Engl J Med 309:1945–1500

Richards GE, Holland FJ, Aubert ML, Ganong WF, Kaplan SL, Grumbach MM (1980) Regulation of prolactin and growth hormone secretion. Neuroendocrinology 30:139–143

Rifgway EC, Ardisson LJ, Meskell MJ, Mudgett-Hunter M (1982) Monoclonal antibody to human thyrotropin. J Clin Endocr Metab 55:44–48

Rivier J, Spiess J, Thorner M, Vale W (1982) Characterization of a growth hormone-releasing factor from a human pancreatic islet tumor. Nature 300:276–278

Rossier J, Vargo TM, Mimick S, Ling N, Bloom FE, Guillemin R (1979) Regional dissociation of βEP and met enk content in brain and pituitary. Proc Natl Acad Sci USA 11:5163–5165

Saffran M, Schally AV (1955) The release of corticotropin by anterior pituitary tissue in vivo. Can J Biochem Physiol 33:408–415

Saito S, Musa K, Yamamoto S, Oshima S, Funato T (1975) Radioimmunoassay of thyrotropin releasing hormone. Endocrinology Jpn 22:303–309

Saito S, Saito H, Yamasaki E, Hosoi E, Komatsu M, Iwahana H, Maeda T (1985) Radioimmunoassay of an analog of luteinizing hormone-releasing hormone [D-Ser (*t* Bu)]^6 des-Gly-NH_2^{10} ethylamide (Buserelin). J Immunol Methods 79:173–183

Sawcenko PE, Swanson LW, Rivier J, Vale W (1985) The distribution of growth-hormone releasing factor (GRF) immunoreactivity in the central nervous system of the rat: an immunohistochemical study using antisera directed against rat hypothalamic GRF. J Compar Neurol 237:100–115

Scanlon MF, Peters J, Food S, Dieguez C, Hall R (1983) The clinical relevance of TRH in diagnosis and investigation. In: Griffiths EC, Bennet GW (eds) Thyrotropin releasing hormone. Raven, New York, pp 303–314

Schalch DS, Parker ML (1964) A sensitive double antibody immunoassay for human growth hormone in plasma. Nature 203:1141–1142

Schalch DS, Reichlin S (1966) Plasma growth hormone concentrations in the rat determined by radioimmunoassay: influence of sex, pregnancy lactation, anesthesia hypophysectomy and extrasellar pituitary transplant. Endocrinology 79:275–280

Schally AV, Arimura A, Baba Y, Nair RMG, Matsuo H, Redding TW, Debeljuk L, White WF (1971) Isolation and properties of the FSH and LH-releasing hormone. Biochem Biophys Res Commun 43:393–399

Schally AV, Coy DH, Meyers CA, Kastin AJ (1979) Hypothalamic peptide hormones: basic and clinical studies. In: Li CH (ed) Hormonal proteins and peptides. Academic, New York, pp 1–54

Schusdziarra V, Zyznar E, Rouiller D, Boden G, Brown J, Arimura A, Unger R (1980) Splanchnic somatostatin: a hormonal regulator of nutrient homeostasis. Science 207:530–532

Scott AP, Radcliffe JG, Rees LH, Landon J (1973) Pituitary peptides. Nature 244:65–67

Seidah NG, Gianoulakis G, Crine P, Lis M, Benjannet S, Routhier R, Chretien M (1978) In vitro biosynthesis and chemical characterization of β-lipotropin, lipotropin and β endorphin in rat pars intermedia. Proc Natl Acad Sci USA 75:3153–3157

Shambaugh GE, Wibber JF, Montoya E, Ruder H, Blonsky ER (1975) Thyrotropin releasing hormone (TRH): measurement in human spinal fluid. J Clin Endocrinol Metab 41:131–134

Sheward WJ, Harmar AJ, Fraser HM, Fink G (1983) Thyrotropin-releasing hormone in rat pituitary stalk blood and hypothalamus: studies with high performance liquid chromatography. Endocrinology 113:1865–1869

Shiomi H, Akil H (1982) Pulse-chase studies of the pituitary POMC/β-endorphin system in normal and stressed rats. Life Sci 31:2185–2187
Silman RE, Chard T, Lowry PJ, Smith I, Young IM (1976) Human foetal pituitary peptides and parturition. Nature 260:716–718
Sinha YN, Selby FW, Lewis VJ, Vanderlaan WP (1972a) Studies of prolactin secretion in mice by a homologous radioimmunoassay. Endocrinology 91:1045–1053
Sinha YN, Selby FW, Lewis UJ, Vanderlaan WP (1972b) Studies of GH secretion in mice by a homologous radioimmunoassay for mouse GH. Endocrinology 91:784–792
Sinha YN, Salocks CB, Vanderlaan WP (1975) Pituitary and serum concentrations of prolactin and GH is Snell dwarf mice. Proc Soc Exp Biol Med 150:207–210
Skofitsch G, Jacobowitz D (1985) Distribution of corticotropin releasing factor-like immunoreactivity in the rat brain by immunohistochemistry and radioimmunoassay: comparison and characterization of ovine and rat/human CRF antisera. Peptides 6:319–336
Skowsky WR, Fisher DA (1972) The use of thyroglobulin to induce antigenicity to small molecules. J Lab Clin Med 80:134–144
Snyder LR, Kirkland JJ (1979) Introduction to modern liquid chromatography. Wiley, New York
Soares MJ, Colosi P, Talamantes F (1983) Development of a homologous radioimmunoassay for secreted hamster prolactin. Proc Soc Exp Biol Med 172:379–381
Soos M, Taylor SJ, Gard T, Siddle K (1984) A rapid, sensitive two-site immunometric assay for TSH using monoclonal antibodies: investigation of factors affecting optimisation. J Immunol Methods 73:237–249
Sopwith AM, Penny ES, Besser GM, Rees LH (1985) Stimulation by food of peripheral plasma immunoreactive growth hormone releasing factor. Clin Endocrinol 22:337–340
Sorensen KU, Christensen SE, Hansen AP, Ingerslev J, Pederson E, Orskov H (1981) The origin of cerebrospinal fluid somatostatin: hypothalamic or disperse central nervous system secretion. Neuroendocrinology 32:335–338
Stalla GK, Hartiwimmer J, von Werder K, Müller OA (1984) Ovine (o) and human (h) corticotrophin releasing factor (CRF) in man: CRF-stimulation and CRF-immunoreactivity. Acta Endocr 106:289–297
Subramanian MG, Spirtos NJ, Moghissi KS, Magyar DM, Hayes MF, Gala RR (1984) Correlation and comparison of NO_2 lymphoma cell bioassay with radioimmunoassay for human prolactin. Fertil Steril 42:870–874
Suda T, Liotta AS, Krieger DT (1978) β-Endorphin is not detectable in plasma from normal human subjects. Science 202:221–223
Suda Y, Tozawa F, Mouri T, Demura H, Shizume K (1983) Presence of immunoreactive corticotropin-releasing factor in human cerebrospinal fluid. J Clin Endocrinol Metab 57:225–226
Tanaka T, Shiu RPC, Gout PW, Beer CT, Noble RR, Friesen HG (1980) A new sensitive and specific bioassay for lactogenic hormones: measurement of prolactin and growth hormone in human serum. J Clin Endocrinol Metab 51:1058–1063
Tanaka T, Shishiba Y, Gout PW, Beer CT, Noble RL, Friesen HG (1983) Radioimmunoassay and bioassay of human growth hormone and human prolactin. J Clin Endocrinol Metab 56:18–20
Thomas K, Ferin J (1968) A new rapid radioimmunoassay for HCG (LH, ICSH) in plasma using dioxane. J Clin Endocrinol Metab 28:1667–1670
Thorell JJ, Johansson BG (1971) Enzymatic iodination of polypeptides with ^{125}I to high specific activity. Biochem Biophys Acta 251:363–365
Thorner MO, Cronin MJ (1985) Growth hormone releasing factor: clinical and basic studies. In: Müller EE, MacLeod RM, Frohman LA (eds) Neuroendocrine perspectives, vol 4. Elsevier, Amsterdam, pp 95–144
Tsong SD, Phyllips D, Halmi N, Liotta AS, Margioris A, Bardin CW, Krieger DT (1982) ACTH and β-endorphin-related peptides are present in multiple sites in the reproductive tract of the male rat. Endocrinology 110:2204–2206

Utiger RD (1979) Thyrotropin. In: Jaffe BM, Berham HR (eds) Methods of hormone radioimmunoassay. Academic, New York, pp 315–325
Utiger RD, Bassiri RM (1973) Thyrotropin-releasing hormone (TRH) radioimmunoassay. In: Gual C, Rosemberg E (eds) Hypothalamic hypophysiotropic hormones. Int Congr Ser No 263. Excerpta Medica, Amsterdam, pp 146–153
Utiger R, Parker ML, Daughaday WH (1962) Studies on human growth hormone. A radioimmunoassay for human growth hormone. J Clin Invest 41:254–261
Vale W, Spiess J, Rivier C, Rivier J (1981) Characterization of a 41-residue ovine hypothalamic peptide that stimulates secretion of corticotropin and beta-endorphin. Science 213:1394–1397
Vale WW, Rivier C, Spiess J, Rivier J (1983) Corticotropin releasing factor. In: Krieger DT, Brownstein MJ, Martin JB (eds) Brain peptides. Wiley, New York, pp 961–974
Vander Laan WP, Ling N, Sigel MB, Vander Laan EF (1983) Radioimmunoassay of human prolactin based on a 13 amino acid synthetic analog of the amino terminus. Biochem Biophys Res Commun 115:346–350
Vasquez B, Harris V, Unger RH (1982) Extraction of somatostatin from human plasma on octadecylsilyl silica. J Clin Endocrinol Metab 55:807–809
Vigh S, Merchenthaler I, Torres-Aleman I, Sueiras-Diaz J, Coy DH, Carter WH, Petrusz P, Schally AV (1982) Corticotropin releasing factor (CRF): immunocytochemical localization and radioimmunoassay (RIA). Life Sci 31:2441–2448
Visser TJ, Klootwijsk W (1981) Approaches to a markedly increased sensitivity of the radioimmunoassay for thyrotropin releasing hormone by derivatization. Biochem Biophys Acta 673:454–466
Wardlaw SL, Frantz AG (1979) Measurement of β-endorphin in human plasma. J Clin Endocrinol Metab 48:176–180
Webb CB, Vance ML, Thorner MO, Perisutti G, Thominet J, Rivier J, Vale W, Frohman LA (1984) Plasma growth hormone responses to constant infusions of human pancreatic growth hormone releasing factor: intermittent secretion or response attenuation. J Clin Invest 74:96–103
Wehrenberg WB, Bloch B, Phyllips BJ (1984) Antibodies to growth hormone-releasing factor inhibit somatic growth. Endocrinology 115:1218–1220
Weindl A (1973) Neuroendocrine aspects of circumventricular organs. In: Ganong WF, Martin L (eds) Frontiers in neuroendocrinology. Oxford University Press, London, pp 1–32
Wiedeman E, Saito T, Linfoot JA, Li CH (1979) Specific radioimmunoassay of human β-endorphin in unextracted plasma. J Clin Endocrinol Metab 49:478–480
Wilhelmi AE (1974) Chemistry of growth hormone. In: Greep RO, Astwood EB (eds) Handbook of physiology section 7: Endocrinology vol IV. The pituitary gland and its neuroendocrine control, part 2. Williams and Wilkins, Baltimore, pp 59–78
Wilkes MM, Steward RD, Bruni JF, Quigley ME, Yen SSC, Ling N, Chretien M (1980) A specific homologous radioimmunoassay for human β-endorphin: indirect measurement in biological fluids. J Clin Endocrinol Metab 50:309–312
Yalow RS (1974) Heterogenity of peptide hormones. Recent Prog Horm Res 30:597–634
Yalow RS, Berson SA (1959) Assay of plasma insulin in human subjects by immunological methods. Nature 194:1648–1649
Yen SSC, Llerena O, Little B, Pearson LH (1968) Disappearance rates of endogenous luteinizing and chorionic gonadotropin in man. J Clin Endocrinol Metab 28:1763–1767
Yen SSC, Llerena O, Person LH, Littell AS (1970) Disappearance rates of endogenous follicle-stimulating hormone in serum following surgical hypophysectomy in man. J Clin Endocrinol Metab 30:325–329

CHAPTER 12

Radioimmunoassay of Nonpituitary Peptide Hormones

G. NILSSON

A. Hormones Involved in Regulation of Carbohydrate Metabolism

I. Insulin

1. Chemistry and Physiology

Insulin is a polypeptide consisting of two chains of amino acids linked by two disulfide bridges (Fig. 1). Only minor differences in amino acid composition exist between species (Table 1). These differences are generally not sufficient to influence the biologic activity in heterologous species, but may make the insulin antigenic.

Insulin is produced in the endoplasmic reticulum of the beta cells of the pancreatic islets. It is secreted as proinsulin from the normal pancreas following prolonged stimulation (Fig. 2) and also from some islet cell tumors. The peptide segment (the C-peptide) that connects the A and B chains of insulin is normally detached in the granules of the beta cells before secretion and is then released in

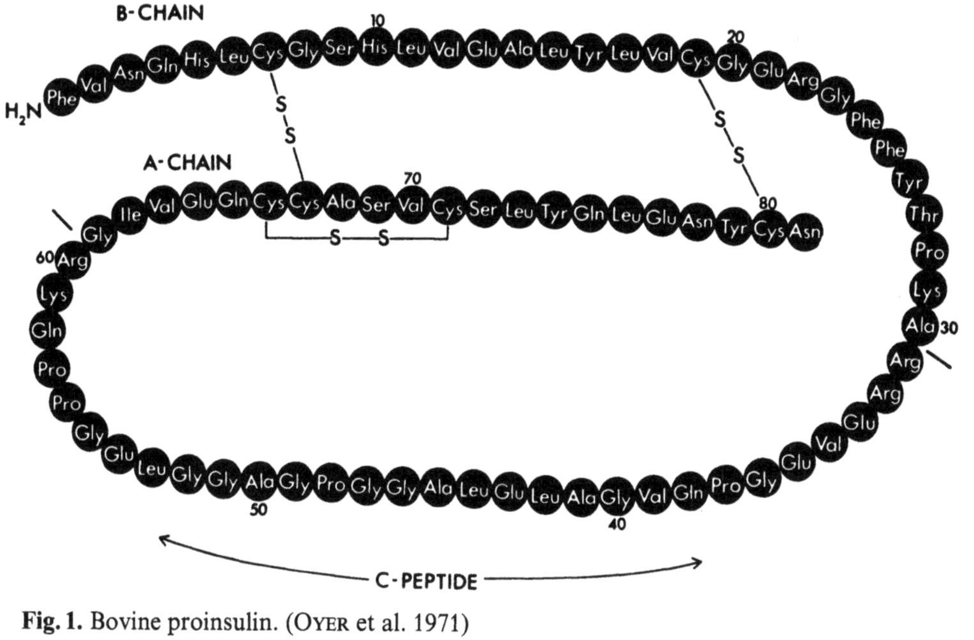

Fig. 1. Bovine proinsulin. (OYER et al. 1971)

Table 1. Discrepancies from amino acid sequence in human insulin

Species	A chain position 8 9 10	B chain position 30
Pig, dog, sperm whale	Thr-Ser-Ile	Ala
Rabbit	Thr-Ser-Ile	Ala
Cattle, goat	Ala-Ser-Val	Ala
Sheep	Ala-Gly-Val	Ala
Horse	Thr-Gly-Ile	Ala
Sei whale	Ala-Ser-Thr	Ala

equimolar amounts, together with the two-chained polypeptide of insulin (HORWITZ et al. 1975).

Insulin is secreted in response to a great number of stimulants. For example, cholinergic and β-adrenergic stimulation of nerves release insulin whereas α-adrenergic excitation inhibits the release of insulin. Humoral factors, such as gastrin, secretin, glucagon (see Fig. 7), and in particular GIP have been shown to cause stimulation following exogenous administration, whereas somatostatin inhibits the insulin release. In addition, derivates of carbohydrates, proteins (Fig. 3), and fat evoke release of insulin from the pancreas. The secretion of insulin in response to most stimulants occurs in two phases. Initially, there is a re-

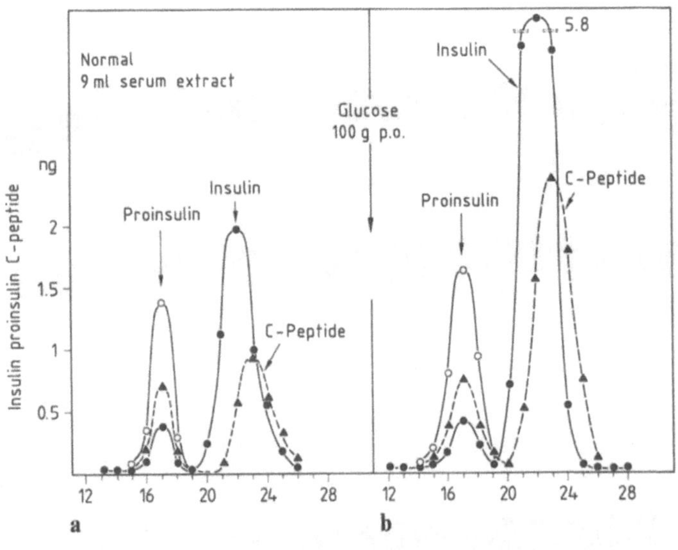

Fig. 2a, b. Serum samples from a healthy subject following fasting (**a**) and 60 min after oral glucose administration (**b**) after filtration on a Bio-Gel P-30 column. Fractions were assayed by an insulin assay with insulin (*full circles*) and proinsulin standards (*open circles*) and by a C-peptide assay with C-peptide standard (*full triangles*). (RUBINSTEIN et al. 1973)

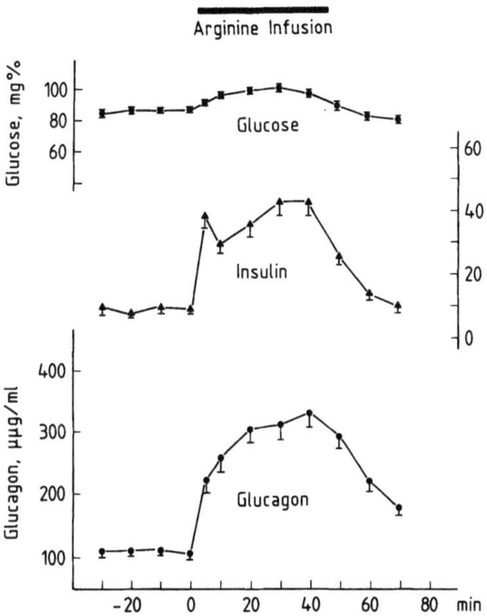

Fig. 3. Plasma glucose, insulin, and glucagon concentrations following infusion of arginine (11.7 mg kg^{-1} min^{-1}) in healthy subjects. (UNGER et al. 1970)

sponse that reaches its maximum a few minutes after the start of stimulation, whereas the second phase reaches its peak value considerably later. Figure 4 illustrates the secretion of insulin to the blood in dogs following vagus nerve stimulation induced by sham feeding or by a protein-rich test meal. In Table 2 factors that influence the secretion of insulin are summarized. Insulin is not bound to any transporting proteins in plasma and its turnover rate is rapid with a half-life less than 5 min (HORWITZ et al. 1975). A considerable portion of the released insulin is extracted from the circulation by the liver. The secretion of insulin varies from one individual to another and is also related to the amount and sort of food that

Table 2. Factors influencing insulin secretion

Stimulators	Inhibitors
Glucose	Somatostatin
Mannose	2-Deoxyglucose
Amino acids	Mannoheptulose
β-Keto acids	α-Adrenergic stimulating agents (epinephrine, norepinephrine)
Gastrointestinal hormones (GIP, gastrin, GRP, CCK, glucagon)	β-Adrenergic blocking agents (propranolol)
Cyclic AMP and different cyclic AMP-generating substances	Diazoxide
	Thiazide diuretics
Acetylcholine	K^+ depletion
β-Adrenergic stimulating agents	Phenytoin
Theophylline	Alloxan
Sulfonylureas	Microtubule inhibitors
	Insulin

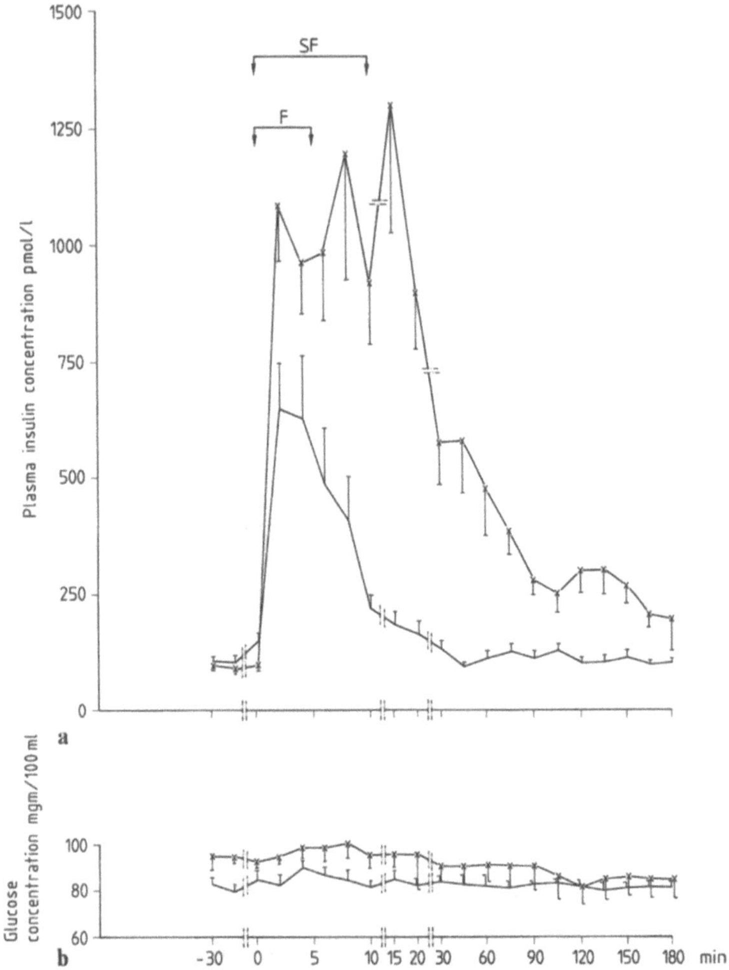

Fig. 4a, b. Plasma insulin (**a**) and glucose concentrations (**b**) before, during, and after 10 min of sham feeding (SF, *lower curves*) and a test meal of minced boiled liver (F, *upper curves*) in dogs. (NILSSON and UVNÄS-WALLENSTEN 1974)

is consumed. In particular, stimulation with a high content of carbohydrates will raise the concentration of insulin in plasma. The insulin response to a carbohydrate meal in humans is illustrated in Fig. 5. The insulin secretion is also related to the amount of body fat. Thus, both basal and stimulated insulin levels are greater in individuals with obesity.

Insulin principally exerts its physiologic effects at three locations in the body, i.e., in the adipose tissue, in the muscles, and in the liver. For example, the transport of glucose into the cells is facilitated. Within the cells lipogenesis as well as the synthesis of glycogen is promoted. The uptake of amino acids into the cells and intracellular synthesis of protein are also stimulated. Inhibition of lipolysis

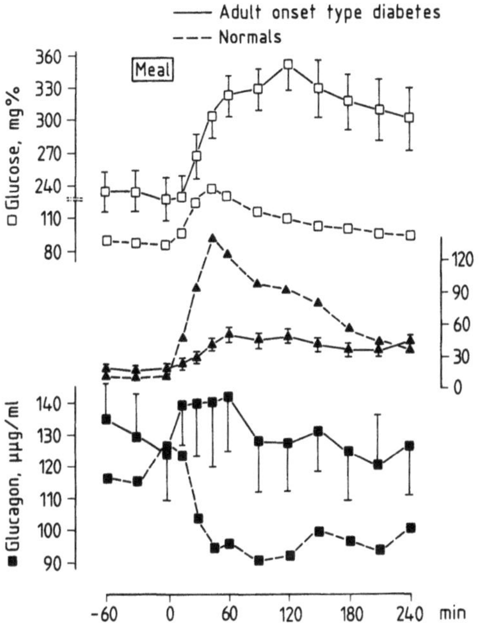

Fig. 5. Plasma glucose, insulin, and glucagon concentrations following a large carbohydrate meal in normal subjects and adult-type diabetics. (MÜLLER et al. 1970)

Table 3. Effect of insulin on adipose tissue, striated muscles, and liver

Adipose tissue
Increased glucose entry
Increased fatty acid synthesis
Increased glycerol phosphate synthesis
Increased triglyceride deposition
Activation of lipoprotein lipase
Inhibition of hormone-sensitive lipase
Increased K^+ uptake
Muscle
Increased glucose entry
Increased glycogen synthesis
Increased amino acid uptake
Increased protein synthesis in ribosomes
Decreased protein catabolism
Decreased release of gluconeogenic amino acids
Increased ketone uptake
Increased K^+ uptake
Liver
Decreased cyclic AMP
Decreased ketogenesis
Increased protein synthesis
Increased lipid synthesis
Decreased glucose output due to decreased gluconeogenesis and increased glycogen synthesis

and gluconeogenesis are other principal anabolic actions exerted by insulin. Table 3 summarizes the effects of insulin on adipose tissue, striated muscles, and liver. In insulin deficiency, glucose uptake into the cells will be reduced and catabolism of proteins and lipolysis will increase, leading to the complex metabolic abnormalities found in diabetes.

2. Pathophysiology

In the juvenile form of diabetes, requiring pharmacologic treatment with insulin, the secretion of insulin may be considerably reduced. In less severe forms of diabetes, secretion of insulin is moderately reduced. Great individual variations in secretory behavior of insulin have been noted in patients with suspected or established diabetes. According to some authors (CERASI and LUFT 1967; CERASI et al. 1972), individuals with a prediabetic disposition may be detected by determination of insulin concentrations in plasma on stimulation by an intravenous glucose infusion. In the prediabetic individual, the early phase of insulin secretion is reduced by such stimulation. It has been estimated that 20%–25% of the population belong to this group of low insulin responders. It is, however, still too early to conclude whether a reduced insulin response combined with normal glucose levels and glucose tolerance tests indicates a risk of the individual patient developing diabetes later in life.

Hypersecretion of insulin may occur. Such hypersecretion can be caused by insulin-producing tumors of the pancreatic beta cells, and is associated with hypoglycemia. Sometimes, hyperinsulinemic states occur by the secretion of insulin or insulin-like material from non-beta cell tumors. Such secretion may take place from retroperitoneal sarcomas and from tumors of the lung.

3. Radioimmunoassay and Its Clinical Application

Insulin was the first hormone to be measured by radioimmunoassay (YALOW and BERSON 1960). Compared with many other peptide assays, few problems are connected with the determination of insulin in blood. Estimation of insulin in urine, on the other hand, is complicated by the presence of substances in urine that interfere with the antigen–antibody reaction.

With human or porcine insulin, immunization is preferably performed in guinea pigs since the insulin in these animals differs from the human or porcine hormone in 17 amino acids (SMITH 1966). Rabbits, on the other hand, are not suitable for immunization since guinea pig insulin is identical with the rabbit insulin, except for the terminal amino acids of the insulin B chain (SMITH 1966). Commercial insulin, available from a great number of companies, can be used without previous coupling to a larger protein. Most immunized animals will respond with high antibody titers following 3–5 immunizations given biweekly.

Labeling of insulin can be performed with ^{125}I (COFFEY et al. 1974; KAGAN 1975) using the chloramine-T method (HUNTER and GREENWOOD 1962). Several techniques have been applied for the purification of the labeled material. For example, starch gel electrophoresis, gel filtration on Sephadex G-50, or purification on a cellulose column. The methods mentioned have been reviewed by YALOW and BERSON (1973). The most stable radioiodinated insulin is usually obtained by

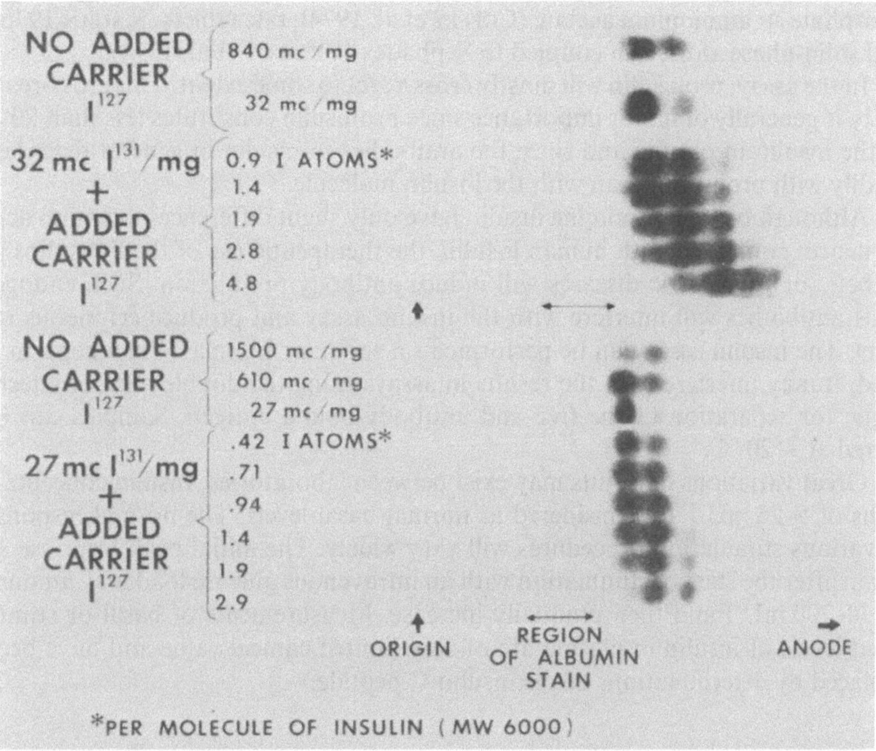

Fig. 6. Autoradiograph of starch gel electrophoresis of ^{131}I-labeled insulin preparations. The numbers given for iodine atoms per molecule of insulin indicate the average value for the preparation; they are calculated as the product of the starting ratios and the percentage iodination. The molecular weight of the bovine insulin monomer was taken as 6000. (BERSON and YALOW 1966)

the starch gel electrophoresis technique (Fig. 6). After some time, the labeled insulin purified by starch gel electrophoresis may have to be repurified by gel filtration on Sephadex.

Assays are generally carried out in 1:10 to 1:25 dilutions of plasma, depending on the sensitivity of the assay. At these dilutions, incubation damage is negligible and no protective agents have to be used. As diluents, barbital–HSA (KAGAN 1975; YALOW and BERSON 1973, p. 864), phosphate–BSA (ALBANO et al. 1972; COFFEY et al. 1974; HALES and RANDLE 1963), phosphate–HSA (HEDING 1972a; VELASCO et al. 1974), and borate–BSA (NAKAGAWA et al. 1973; SOELDNER and SLONE 1965) have been used. The choice of buffer system seems to be of little significance for the sensitivity or precision of the assay. It should, however, be noted that batches of bovine serum albumin (BSA) have to be selected since some batches may have impurities that will interfere with the assay. For separation, a large number of methods have been applied: the double-antibody technique (HALES and RANDLE 1963; SOELDNER and SLONE 1965), dextran-coated charcoal (ALBANO et al. 1972; HERBERT et al. 1965), uncoated charcoal (NILSSON and UV-

NÄS-WALLENSTEN 1977), polyethylene glycol (NAKAGAWA et al. 1973), zirconyl phosphate + ammonium acetate (COFFEY et al. 1974), talc tablets (KAGAN 1975), and solid-phase antiserum coupled to Sephadex (VELASCO et al. 1974).

In the assay, proinsulin will mostly cross-react to some extent. This cross-reactivity is generally of minor importance since proinsulin constitutes less than 20% of the insulin in plasma and since the antibodies to insulin in general react less readily with proinsulin than with the insulin molecule.

Although bovine or porcine insulin have only slight differences in amino acid sequences compared with human insulin, the therapeutic use of these insulins in diabetic or psychiatric diseases will induce antibody production. Such endogenous antibodies will interfere with the insulin assay and produce erroneous results. The insulin assay can be performed on serum or plasma. When heparin is used, it may interfere with the results in assays using the double-antibody technique for separation of the free and antibody-bound antigen. Samples can be stored at −20 °C.

Great variations in results may exist between laboratories. Insulin concentrations of 5–25 mU/l are considered as normal basal levels. The normal response to various stimulation procedures will vary widely. The initial peak response 3–6 min after the start of stimulation with an intravenous glucose load may amount to 50–200 mU/l and then gradually increase. Measurements of basal or stimulated levels of insulin in plasma are of very limited clinical value and have been replaced by determinations of the insulin C-peptide.

II. C-Peptide of Insulin

The C-peptide of insulin (Fig. 1) is released from the beta cells in equimolar concentrations with insulin (HORWITZ et al. 1975), and along with about 5% of intact proinsulin (STEINER 1972). Insulin is metabolized by the liver whereas the C-pep-

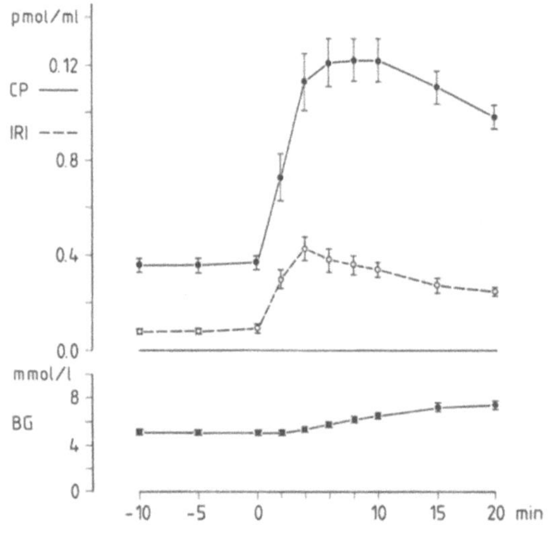

Fig. 7. Plasma glucose (BG), insulin (IRI), and C-peptide (CP) concentrations before and after the intravenous administration of 1 mg glucagon in normal subjects. (HOEKSTRA et al. 1982)

tide is not. The half-lives of insulin and C-peptide in humans have been estimated as 4.8 and 11.1 min, respectively (HORWITZ et al. 1975). When measured in peripheral blood, C-peptide levels will reflect the beta cell secretory activity (HORWITZ et al. 1975). No physiologic effects have been attributed to the C-peptide.

1. Radioimmunoassay and Its Clinical Application

There are wide species differences in the amino acid composition of the C-peptide (RUBENSTEIN et al. 1970). For this reason, only the human C-peptide can be used for antibody production, labeling, and standard in assays that will be applied on the study of human C-peptide. The C-peptide per se is a weak antigen and has therefore to be coupled to a large protein to produce antibodies (MELANI et al. 1970). There is no tyrosine residue in the C-peptide molecule so it has to be incorporated before labeling by the method of HUNTER and GREENWOOD (1962), as modified by FREYCHET et al. (1971).

Antisera to the C-peptide cross-react to some extent with proinsulin (because the C-peptide is part of the molecule) and with degradation products of the C-peptide. Under normal conditions, the concentration of C-peptide in serum is about ten times the concentration of proinsulin. On a molar basis, proinsulin then contributes less than 3% of the measured C-peptide-like concentration (HORWITZ et al. 1978). There are, however, certain conditions in which proinsulin levels in serum are elevated and may influence the concentration of C-peptide immunoreactivity. Such elevated levels of proinsulin have been observed in severe insulin-deficient diabetes (GORDON et al. 1974), in islet cell tumors (TURNER and HEDING 1977), and in obese diabetic subjects (DUCKWORTH et al. 1972). Also, familial hypersecretion of proinsulin has been described (GABBAY et al. 1976). Proinsulin levels may also be elevated in subjects treated with insulin who develop antibodies to the exogenous insulin. Since proinsulin to some degree cross-reacts with such antibodies, proinsulin will remain in the circulation and its concentration increases (HEDING 1978).

Several techniques have been proposed by which proinsulin can be separated from the C-peptide. For example, separation can be performed according to molecular size on Bio-Gel or Sephadex (MELANI et al. 1970; ROTH et al. 1968). Others have removed the proinsulin from serum by binding it to insulin antibodies coupled to Sepharose before determination with C-peptide antibodies (HEDING 1975). C-peptide estimations have also been performed on serum samples from diabetics after removal of insulin antibodies and proinsulin bound to such antibodies by treatment with polyethylene glycol (KUZUYA et al. 1977). The C-peptide immunoreactivity in serum or plasma samples will decrease considerably with time (KUZUYA et al. 1977) and depending on the temperature at which the samples are stored. Samples stored at −20 °C will thus lose great amounts of immunoreactivity whereas no change seems to occur when samples have been stored for 3 months at −70 °C (GARCIA-WEBB and BONSER 1979). Plasma concentrations of C-peptide in normal and diabetic subjects during fasting and following stimulation are shown in Table 4. Radioimmunologic determination of the C-peptide in blood has provided important information concerning the natural history and pathophysiology of insulin-dependent diabetes and the investigation of

Table 4. Plasma concentrations of C-peptide in normal subjects and in short-term insulin-dependent diabetic subjects during fasting and after intravenous stimulation with 1 mg glucagon. (HOEKSTRA et al. 1982)

	C-peptide concentration (ng/ml)	
	Normal subjects	Diabetic subjects
Fasting	1.08 (0.78–1.89)	0.54 (0.18–1.23)
After stimulation	3.84 (2.73–5.64)	1.02 (0.24–2.34)

Table 5. Clinical applications of the C-peptide assay

Hypoglycemic states
Diagnosis of insulinoma (suppression test)
Diagnosis of factitious hyperinsulinemia
Diagnosis of insulinoma in insulin-dependent diabetic subjects
Hyperglycemic states
Follow-up after pancreatectomy
Establishing endogenous insulin production and the need for insulin therapy in diabetic subjects treated with insulin

Table 6. Conditions associated with increased and decreased C-peptide levels. (HOEKSTRA et al. 1982)

Condition	C-peptide level	Remarks
Insulinoma	Increased	
Beta cell hyperplasia	Increased	
Nesidioblastosis	Increased	
Obesity	Increased (slightly)	Insulin resistance
Renal failure	Increased	Impaired renal excretion
Presence of insulin antibodies and proinsulin	Increased (spuriously)	Cross-reaction with proinsulin bound to circulating insulin antibodies
Beta cell failure	Decreased	
Exogenous hyperinsulinism	Decreased	Feedback inhibition

insulin feedback mechanisms in hypoglycemia in humans. Besides that, the clinical application of the assay is limited and essentially concerns the investigation of certain conditions with hypo- and hyperglycemia. A summary of clinical applications of the C-peptide assay is given in Table 5. Studies describing various kinds of suppression tests (ASHBY and FRIER 1981; SERVICE et al. 1977; TURNER and HEDING 1977; TURNER and JOHNSON 1973), diagnostic investigations of factitious hyperinsulinemia (SAFRIT and YOUNG 1977; SERVICE et al. 1975; STELLON and TOWNELL 1979), insulinoma in insulin-dependent diabetic subjects (SANDLER et al. 1975), and hyperglycemic states after pancreatectomy (KARESEN et al. 1980) are available in the literature. C-peptide measurements have also been performed to assess residual pancreatic beta cell function in patients with insulin-dependent diabetes (FABER and BINDER 1977) to estimate the need for insulin therapy in diabetic subjects already treated with insulin (Table 6).

III. Glucagon

1. Chemistry and Physiology

Human glucagon is a linear polypeptide with a molecular weight of 3485 (Table 7). It contains 29 amino acids and has the same structure as porcine and bovine glucagon. Glucagon is produced by the alpha cells in the pancreatic islets of Langerhans. Some glucagon is also present in the gastrointestinal mucosa. In addition, several enteroglucagons seem to exist in the mucosa of the gut. One of these enteroglucagons is glicentin which has been purified by SUNDBY et al. (1976). Glicentin also seems to be present in the alpha cells of the pancreatic islets (ALUMETS et al. 1978; LARSSON and MOODY 1980; SMITH et al. 1977). If so, it is found in the peripheral portion of the same cells. Glicentin has some glucagon activity. However, the relation between these two peptides in the pancreas is not known.

A great number of factors may induce release of glucagon from the pancreas. Both cholinergic and β-adrenergic stimulation cause release of glucagon. Amino acids, and in particular the glucogenic amino acids alanine, serine, glycine, cysteine, and threonine, are stimulators of glucagon secretion. Figure 3 shows glucagon release following the intravenous administration of the amino acid arginine. Glucagon release is also evoked by excercise, infections, and other stresses. In addition, it has been reported that CCK, gastrin, and cortisol may cause release of glucagon. Inhibition of glucagon release can be accomplished by α-adrenergic stimulation, glucose (see Fig. 5), free fatty acids, and ketones. Insulin, somatostatin, and secretin also reduce the output of glucagon from the pancreas. A more extensive summary of factors influencing glucagon release is presented in Table 8.

Glucagon exerts glucogenolytic, gluconeogenic, and lipolytic actions. The lipolytic effect in turn will lead to an increased ketogenesis. The glucogenolytic effect of glucagon takes place in the liver. No such activity occurs in muscles. The

Table 7. Amino acid sequence of human, bovine, and porcine glucagon

1	2	3	4	5	6	7	8	9	10	11	12	13	14	15	16	17	18	19	20
His-	Ser-	Gln-	Gly-	Thr-	Phe-	Thr-	Ser-	Asp-	Tyr-	Ser-	Lys-	Tyr-	Leu-	Asp-	Ser-	Arg-	Arg-	Ala-	Gln-
21	22	23	24	25	26	27	28	29											
Asp-	Phe-	Val-	Gln-	Trp-	Leu-	Met-	Asn-	Thr											

Table 8. Factors influencing glucagon release

Stimulators	Inhibitors
Amino acids	Glucose
CCK, gastrin	Ketones
Cortisol	Somatostatin
Exercise	Secretin
Infections	Insulin
Other stresses	Phenytoin
Acetylcholine	α-Adrenergic stimulators
β-Adrenergic stimulators	
Theophylline	

blood glucose concentration will be raised because glucagon stimulates adenylate cyclase of the liver cells. This will lead to the activation of phosphorylase and a subsequent breakdown of glycogen.

Gluconeogenesis is stimulated by glucagon from several substrates, and the effect takes place at different biochemical sites. Liver uptake of lactate is stimulated as well as the conversion of lactate and pyruvate to glucose. Glucagon also stimulates the liver uptake of circulating amino acids and their conversion to glucose. In addition, glucagon stimulates the hepatic catabolism of proteins. Plasma values of glucagon in humans, determined under fasting, various types of meal, and infusion of arginine, are shown in Tables 9 and 10. In summary, glucagon

Table 9. Plasma glucagon concentration in normal subjects during fasting and stimulation. (FALOONA 1973)

	Plasma glucagon concentration (pg/ml)	
	Mean	Range
Fasting	75	30–210
Carbohydrate meal maximum decrease	45	20–120
Protein meal maximum increase	102	50–180
Arginine infusion maximum increase	239	104–590

Table 10. Plasma glucagon concentrations in diabetic subjects during fasting and following stimulation. (UNGER 1973)

	Plasma glucagon concentration (pg/ml)	
	Mean	Range
Fasting	119	40– 240
Carbohydrate meal maximum ↓ decrease	8	0– 40
Protein meal maximum increase	75	0– 150
Arginine infusion maximum increase	365	108–1040

mobilizes energy stores and therefore has an opposite effect to insulin, which principally acts as a hormone of energy storage. The concentration of glucagon in blood will vary according to the level of glucose in the blood and the nutritional state. Also, the molar ratio of insulin and glucagon in the circulation will fluctuate considerably owing to the conditions preceding the collection of blood. For example, the insulin : glucagon molar ratio on a balanced diet is approximately 2.3. Following 3 days of starvation, the ratio will fall to 0.4, whereas the ratio after the intake of a large carbohydrate meal may increase to 70 (GANONG 1985).

2. Pathophysiology

As indicated in Table 8, the concentration of glucagon in plasma is dependent on a variety of circumstances. The tests show that the normal alpha cells of the pancreas will respond by a decreased output of glucagon when the glucose concentration in plasma is raised and by an increased secretion of glucagon in response to amino acids, or when glucose in plasma is low. Some disease states have been described in which abnormal function of the alpha cells will result in deficiency or excess of glucagon. Insufficient function may be a consequence of an isolated absence of alpha cells with normal function of the remainder of the endocrine and exocrine pancreatic tissue (LEVY et al. 1970). It may also be due to inflammatory disease of the pancreas (UNGER 1973), neoplastic replacement of the pancreas, or its surgical removal.

In diabetes mellitus, hypersecretion of glucagon may be found under several conditions, despite the raised concentration of glucose in the blood, and it has been suggested (UNGER et al. 1970) that the hypersecretion of glucagon plays a role in the metabolic disturbance of ketoacidosis. It may do so by exaggerating the metabolic consequencies of the lack of insulin. More insulin will therefore be required to overcome the ketoacidosis. The somewhat paradoxical effects of glucagon secretion from an adult-type diabetic are illustrated in Fig. 5. Table 10 shows glucagon concentrations in plasma in diabetes mellitus. These values should be compared with the concentrations of glucagon seen in normal subjects and illustrated in Table 9.

Hypersecretion of glucagon may be found in patients with glucagonoma, hypercalcemia, acute pancreatitis, or diabetes mellitus (PALOYAN 1967; PALOYAN et al. 1966; UNGER 1973; UNGER et al. 1970). Glucagonomas can be suspected at glucagon levels of 500 pg/ml or more following an overnight fast. Hypersecretion of glucagon in connection with hypercalcemia has been reported in connection with hyperparathyroidism or with the Gelhorn–Phimpton syndrome and leukemia (PALOYAN 1967). Excess secretion of glucagon has also been noted in animal experiments in which acute pancreatitis has been induced (PALOYAN et al. 1966). Such hypersecretion of glucagon may explain both the hyperglycemia and the hypocalcemia that may be associated with this disorder.

3. Radioimmunoassay and Its Clinical Application

Radioimmunoassay of glucagon has been hampered by the following problems: (a) difficulties in producing antisera of sufficient titer and affinity to permit the detection of glucagon in body fluids; (b) considerable cross-reactivity with peptides of nonpancreatic origin; and (c) damage of iodinated and endogenous glucagon by proteolytic enzymes present in blood. A number of methods for the production of glucagon antisera have been described. Several of these techniques have been summarized by HEDING (1969, 1972b).

The problem of degradation of the iodinated and endogenous glucagon in the test tube can be overcome by the addition of 500 units per milliliter plasma of the proteinase inhibitor aprotinin. Labeling of glucagon can be carried out by the chloramine-T method (HUNTER and GREENWOOD 1962). Purification of the labeled product can be accomplished by gel filtration on Sephadex G-25. All eluates should then be collected in test tubes prepared with aprotinin. A quick and practical method of purifying the labeled glucagon is to adsorb the labeled material on a cellulose column and to elute it with acid–ethanol in tubes that contain the standard diluent with 500 units aprotinin per milliliter.The method has been described for purification of labeled insulin by YALOW and BERSON (1973, p. 864), but can just as well be applied for the purification of iodinated glucagon.

As standard, crystalline bovine or porcine glucagon can be used. The glucagon is dissolved in dilute HCl, which is then diluted to the appropriate concentration by the addition of the buffer solution used as diluent; a 0.2 *M* glycine buffer solution with 0.25% human serum albumin and 1% normal sheep serum at pH 8.8 has been used (FALOONA 1973) as well as a 0.02 *M* barbital buffer solution at pH 8.6, supplemented with 1% control guinea pig or rabbit serum (depending on the species used for immunization) and 0.25% human serum albumin (R. S. YALOW and S. A. BERSON 1971, personal communication). At incubation, the incubation mixture should contain 500 units aprotinin per milliliter. Tubes are incubated for 4–6 days at 4 °C.

As for other polypeptide substances, several adsorption techniques have been used to separate the free from the antibody-bound antigen. Separation can be done by a suspension of charcoal (Norit A) with (LECLERCQ-MEYER et al. 1970) or without (NILSSON and UVNÄS-WALLENSTEN 1977) 0.25% dextran 80. Before charcoal is added to tubes containing standards or diluted plasma, a suitable volume of outdated plasma is added to the tubes to make the protein concentration in tubes as near equal as possible. Separation can also be carried out by precipitating the glucagon–antibody complex by Na_2SO_4 (UNGER et al. 1963), immunoprecipitation (CHESNEY and SCHOFIELD 1969; HAZZARD et al. 1968), or by ethanol (EDWARDS et al. 1970; HEDING 1971). The antigen can also be adsorbed on a cellulose powder (NONAKA and FOÀ 1969) or Amberlite G-400 (WEINGES 1968), or separated from the antibody complex by paper chromatography (UNGER et al. 1961) or chromatoelectrophoresis (ASSAN et al. 1967).

In summarizing the clinical application of the glucagon radioimmunoassay, it can be stated that there are several situations where glucagon concentrations in blood may be abnormal. A clinical need for the glucagon radioimmunoassay is thus established when glucagonomas are suspected and in cases of isolated glucagon deficiency. Fasting plasma concentrations in excess of 500 pg/ml are highly suggestive of glucagonoma in the absence of diabetes mellitus. In isolated glucagon deficiency, infusion of arginine will cause hypoglycemia due to insulin release and the glucagon concentration in the blood will not rise.

B. Hormones Regulating Calcium Homeostasis

I. Physiology of Calcium Homeostasis

The regulation of calcium homeostasis is essentially accomplished by parathyroid hormone (PTH), calcitonin, and vitamin D (1,25-dihydroxycholecalciferol). Variations in the concentration of free or ionized calcium present in the extracellular fluid will initiate the acute phases of this homeostasis. Following a reduction of the Ca^{2+} concentration in the blood, there is a rapidly occurring increase of PTH in the circulation. The released PTH will cause an increase in the blood concentration of Ca^{2+}. This increase is attained by an elevated rate of osteocytic osteolysis in the bone and by an increase in the net rate of Ca^{2+} resorption from the bone. Also, the absorption of Ca^{2+} from the upper small intestine increases and the excretion of Ca^{2+} from the kidneys will be reduced. Simultaneously, there is an increased urinary excretion of phosphate. Phosphate metabolism also contributes to the homeostatic regulation of calcium inasmuch as free calcium and phosphate exist in equilibrium in the blood and the solubility product of calcium phosphate contributes to the relative concentration of both free calcium and phosphate in biologic tissues; i.e., when the concentration of phosphate is reduced the concentration of Ca^{2+} is increased. At a raised concentration of Ca^{2+}, the secretion of PTH is lowered and the secretion of calcitonin is stimulated. Under the influence of the latter hormone, the rate of bone resorption is inhibited, resulting in a decreased calcium concentration in the blood. Vitamin D influences the actions of PTH on the bone tissue by contributing to the mobilization of Ca^{2+} from the bone.

II. Parathyroid Hormone

1. Chemistry

PTH is produced in the dark chief cells of the parathyroid glands. As with most peptide hormones, several molecular forms have been detected (Berson and Yalow 1968). The most abundant form that is stored in the glands and that which constitutes the primary secretion product under normal conditions has a molecular weight of 9500. Besides that, other forms, of molecular weight 7000 and 4500 also exist (Habener et al. 1971; Mayer et al. 1979; Tanaka et al. 1975). The former, which is biologically inactive, primarily constitutes the carboxyl end of the PTH molecule (Mayer et al. 1977) whereas the biologically active 4500 molecular weight form is considered to be the amino end of the molecule (Canterbury and Reiss 1972). The 9500 molecular weight form represents approximately 50% of the PTH peptides present in the circulation (Habener and Potts 1977; Habener et al. 1971). The amino acid composition of this molecular form from a number of species is illustrated in Table 11. The 7000 and 4500 molecular weight forms are created by peripheral enzymatic conversion of the 9500 molecular weight form. The half-life in blood for the 9500 molecular weight form has been estimated to be very short, for the NH_2 terminal portion somewhat longer, and for the COOH terminal portion of the PTH molecule still longer. Metabolic studies of PTH have been reviewed by Christensen (1979). On decrease in serum

Table 11. Amino acid sequences of rat (R), human (H), bovine (B), and porcine (P) PTH. *Dashes* indicate residues identical to those in the human molecule. (AURBACH et al. 1985)

Species									10		
R	Ala	–	–	–	–	–	–	–	–	–	–	–
H	Ser	Val	Ser	Glu	Ile	Gln	Leu	Met	His	Asn	Leu	Gly
B	Ala	–	–	–	–	–	Phe	–	–	–	–	–
P	Ser	–	–	–	–	–	Phe	–	–	–	–	–
								20				
R	–	–	–	Ala	–	Val	–	–	Met	Gln	–	–
H	Lys	His	Leu	Asn	Ser	Met	Glu	Arg	Val	Glu	Trp	Leu
B	–	–	–	Ser	–	Met	–	–	–	–	–	–
P	–	–	–	Ser	–	Leu	–	–	–	–	–	–
						30						
R	–	–	–	–	–	–	–	–	–	–	–	Ser
H	Arg	Lys	Lys	Leu	Gln	Asp	Val	His	Asn	Phe	Val	Ala
B	–	–	–	–	–	–	–	–	–	–	–	–
P	–	–	–	–	–	–	–	–	–	–	–	–
				40								
R	–	–	Val	Gln	Met	–	Ala	–	Glu	Gly	Ser	Tyr
H	Leu	Gly	Ala	Pro	Leu	Ala	Pro	Arg	Asp	Ala	Gly	Ser
B	–	–	–	Ser	Ile	–	Tyr	–	–	Gly	Ser	–
P	–	–	–	Ser	Ile	Val	His	–	–	Gly	–	–
		50										60
R	–	–	–	Thr	–	–	–	–	–	–	–	–
H	Gln	Arg	Pro	Arg	Lys	Lys	Glu	Asp	Asn	Val	Leu	Val
B	–	–	–	–	–	–	–	–	–	–	–	–
P	–	–	–	–	–	–	–	–	–	–	–	–
										70		
R	Asp	Gly	Asn	Ser	–	–	–	–	–	Gly	–	–
H	Glu	Ser	His	Glu	Lys	Ser	Leu	Gly	Glu	Ala	Asp	Lys
B	–	–	–	Gln	–	–	–	–	–	–	–	–
P	–	–	–	Gln	–	–	–	–	–	–	–	–
								80				84
R	–	–	–	Asp	–	–	Val	–	–	–	–	–
H	Ala	Asp	Val	Asn	Val	Leu	Thr	Lys	Ala	Lys	Ser	Gln
B	–	–	–	Asp	–	–	Ile	–	–	–	Pro	–
P	–	Ala	–	Asp	–	–	Ile	–	–	–	Pro	–

Ca^{2+} concentration, PTH is essentially secreted as the 9500 molecular weight form.

The PTH concentration in plasma is subject to a diurnal variation, with peak concentrations in the evening and with the lowest concentrations in the morning. Samples for determination of hypo- or hypersecretion conditions are usually collected in the morning. The normal concentration of PTH in serum ranges between 120 and 160 pmol equiv./l (1.1–2.5 ng/ml). Problems exist with overlap in the values of normal individuals and of those that suffer from hyperparathyroidism.

2. Radioimmunoassay and Its Clinical Application

The first radioimmunologic method for determination of PTH was described by BERSON and YALOW (1963). Since then, a great number of assays have been reported. Porcine PTH has been used for antiserum production in guinea pigs (ARNAUD et al. 1971) and bovine PTH in rabbits (DEFTOS 1974) or chicken (REISS and CANTERBURY 1968). Several problems are associated with the radioimmunologic determination of PTH. For example, the use of bovine or porcine PTH for immunization has produced antisera where the immunologic reactivity of human PTH is only 20%–50% of the bovine or porcine PTH against which the antisera have been raised. The heterogeneity of PTH in blood will contribute to variations of results between laboratories. Thus, antibodies that are raised by bovine PTH in particular react against the 9500 molecular weight component in blood. Antisera that have been raised by porcine PTH, on the other hand, primarily react with the 4500 molecular weight form of PTH. Other problems related to the PTH assay are the great ability of PTH to be adsorbed on glass surfaces which has to be counteracted by high proportions of serum in the standard diluent. PTH also has a great vulnerability to enzymatic destruction. The latter problem can be handled by the presence of aprotinin (500 units/ml) at collection of plasma samples and in the incubation medium and by rapid freezing of collected plasma samples.

Labeling of PTH may be carried out by the chloramine-T method (HUNTER and GREENWOOD 1962) by ^{125}I (ALMQVIST et al. 1975; CHRISTENSEN 1976; DEFTOS 1974; SCHOPMAN et al. 1970; TANAKA et al. 1975) and the labeled hormone purified by QUSO (YALOW and BERSON 1973, p. 971). Labeling with ^{125}I has also been described by the lactoperoxidase method, with subsequent separation on Bio-Gel P-10 (THORELL and LARSSON 1978). As diluent, phosphate–BSA (CHRISTENSEN 1978; SCHOPMAN et al. 1970), barbital buffer (DEFTOS 1974), barbital–BSA (REISS and CANTERBURY 1968), barbital–guinea pig serum (ARNAUD et al. 1978), and barbital–HSA (TANAKA et al. 1975) have been used.

Incubation to reach a state of equilibrium of the antigen–antibody reaction is most commonly carried out for 96 h at 4 °C. At separation, dextran-coated charcoal (PALMIERI et al. 1971) can be added before centrifugation and removal of the supernatant from the charcoal pellet. In other cases, plasma is added to the incubation volume before the subsequent addition of uncoated charcoal suspended in barbital buffer (PALMIERI et al. 1971). Other methods employed for separation are electrophoresis on cellulose acetate (REISS and CANTERBURY 1968), paper chromatography (CHRISTENSEN 1976), and the double-antibody technique (TANAKA et al. 1975). The radioimmunologic determination of PTH is mainly used to differentiate between various causes of hypercalcemia. A number of such causes are listed in Table 12.

In primary hyperparathyroidism, the secretion of PTH is elevated. This kind of hypersecretion has successfully been differentiated from others by determining the 7000 molecular weight form in the blood. Also in hyperparathyroid patients who have normal blood values of calcium, the concentration of PTH may be too high compared with the calcium level. Levels of PTH in patients with hypersecretion of calcium have been extensively studied by ARNAUD et al. (1973). If the hypercalcemia is due to causes other than hyperparathyroidism, the patients will

Table 12. Disorders associated with hypercalcemia. (HAMILTON 1975)

Neoplastic	*Nonneoplastic*
Primary hyperparathyroidism	Sarcoidosis
Parathyroid adenoma	Vitamin D intoxication
Parathyroid hyperplasia	Milk–alkali syndrome
Parathyroid carcinoma	Addison's disease
Nonendocrine malignancies	Thiazide diuretics
Bronchogenic carcinoma	
Breast carcinoma	
Lymphoma	
Multiple myeloma	
Leukemia	
Direct malignant bone involvement	

have levels of PTH that are below the detection limit for the hormone. In cases of chronic renal failure, the 7000 molecular weight component seems to be elevated whereas the 9500 molecular weight form is only occasionally raised in this patient group. Raised levels of the 7000 molecular weight form have also been found in patients with primary hyperparathyroidism. According to the conclusion of ARNAUD et al. (1973), elevated concentrations of the 7000 molecular weight component in the blood will reflect a state of chronic hypersecretion of PTH, as in hyperparathyroidism. Blood concentrations of the 9500 molecular weight form, on the other hand, reflect acute changes in the secretory state of the parathyroid gland.

For localization of parathyroid adenomas at surgery, amino terminal-specific antisera have been successfully used (TANAKA et al. 1975). Such antisera are supposed to reflect the acute secretion of PTH in the region of parathyroid adenomas. Lowered concentrations of PTH will be found in hypercalcemia, independent of PTH hypersecretion. Such hypercalcemia exists in connection with malignant tumors as well as in other conditions listed in Table 12 (HAMILTON 1975; LAFFERTY 1966; RODMAN and SHERWOOD 1978; STEWART et al. 1980).

Lately, attempts have been made to produce antibodies directed against the intermediate portion of the PTH molecule (amino acids 53–68). Assays using such antibodies (Cambridge Medical Diagnostics, England) seem to be of better diagnostic value than assays using antibodies directed against the COOH or NH_2 termini in conditions with primary hyperparathyroidism, in preoperative searches for PTH-secreting adenomas, and in hypoparathyroidism. The assay can also be used for diagnostic purposes and to follow secondary hyperparathyroidism in renal insufficiency.

III. Calcitonin

1. Chemistry and Physiology

Calcitonin is a polypeptide consisting of 32 amino acids (molecular weight 3400), characterized by a 1–7 disulfide bridge at the NH_2 terminus of the molecule and by the existence of proline as a COOH terminal amino acid. Calcitonin has been isolated from a large number of species. Differences in structure of the calcitonin molecule in these species are illustrated in Table 13. The biologic activity seems to require the contribution of the whole polypeptide molecule. As for other polypeptide hormones, higher molecular weight forms have been reported although not chemically identified. In humans, calcitonin has been demonstrated in the C cells, which are small groups of interstitial and parafollicular cells in the thyroid. Such cells have also been found in the parathyroid glands and in the thymus.

Following an elevated concentration of Ca^{2+} (>9 mg per 100 ml) in the blood, secretion of calcitonin is initiated. The hormone is considered to act directly on the bone tissue by decreasing the rate of remodeling of the bone. In cases of Paget's disease of bone tissue, which is characterized by increased bone destruction, increased bone formation, and an abnormal architecture in the newly formed bone, treatment with calcitonin seems to be effective.

Table 13. Amino acid sequence of calcitonin in various species. *Dashes* indicate residues identical to those in the human molecule. (AURBACH et al. 1985)

Species		2		4		6		8		10	
Eel	–	Ser	–	–	–	–	–	Val	–	–	Lys
Salmon I	–	Ser	–	–	–	–	–	Val	–	–	Lys
Salmon II	–	Ser	–	–	–	–	–	Val	–	–	Lys
Salmon III	–	Ser	–	–	–	–	–	Met	–	–	Lys
Human	Cys	Gly	Asn	Leu	Ser	Thr	Cys	Met	Leu	Gly	Thr
Rat	–	–	–	–	–	–	–	–	–	–	–
Porcine	–	Ser	–	–	–	–	–	Val	–	Ser	Ala
Bovine	–	Ser	–	–	–	–	–	Val	–	Ser	Ala
Ovine	–	Ser	–	–	–	–	–	Val	–	Ser	Ala
	12		14		16		18		20		22
Eel	Leu	Ser	–	Glu	Leu	His	–	Leu	Gln	–	Tyr
Salmon I	Leu	Ser	–	Glu	Leu	His	–	Leu	Gln	–	Tyr
Salmon II	Leu	Ser	–	–	Leu	His	–	Leu	Gln	–	–
Salmon III	Leu	Ser	–	–	Leu	His	–	Leu	Gln	–	–
Human	Tyr	Thr	Gln	Asp	Phe	Asn	Lys	Phe	His	Thr	Phe
Rat	–	–	–	–	Leu	–	–	–	–	–	–
Porcine	–	Trp	Arg	Asn	Leu	–	Asn	–	–	Arg	–
Bovine	–	Trp	Lys	–	Leu	–	Asn	Tyr	–	Arg	–
Ovine	–	Trp	Lys	–	Leu	–	Asn	Tyr	–	Arg	Tyr
		24		26		28		30		32	
Eel	–	Arg	–	Asp	Val	–	Ala	–	Thr	–	
Salmon I	–	Arg	–	Asn	Thr	–	Ser	–	Thr	–	
Salmon II	–	Arg	–	Asn	Thr	–	Ala	–	Val	–	
Salmon III	–	Arg	–	Asn	Thr	–	Ala	–	Val	–	
Human	Pro	Gln	Thr	Ala	Ile	Gly	Val	Gly	Ala	Pro–NH_2	
Rat	–	–	–	Ser	–	–	–	–	–	–	
Porcine	Ser	Gly	Met	Gly	Phe	–	Pro	Glu	Thr	–	
Bovine	Ser	Gly	Met	Gly	Phe	–	Pro	Glu	Thr	–	
Ovine	Ser	Gly	Met	Gly	Phe	–	Pro	Glu	Thr	–	

2. Radioimmunoassay and Its Clinical Application

Synthetic human calcitonin is commercially available from several companies. Antisera to human calcitonin have been produced in both guinea pigs and rabbits (TASHJIAN and VOLLKEL 1974). Labeling of calcitonin with ^{125}I using both the chloramine-T method (ALMQVIST et al. 1975; CLARK et al. 1969; DEFTOS 1971; SILVA et al. 1974) and the lactoperoxidase method (THORELL and LARSSON 1978) has been described. Purification of the labeled hormone can be accomplished by QUSO as for PTH. As diluent, phosphate–HSA (CLARK et al. 1969), phosphate–BSA (ALMQVIST et al. 1975; HEYNEN and FRANCHIMONT 1974), TRIS buffer with human serum (TASHJIAN and VOLLKEL 1974) or borate–HSA (SILVA et al. 1974) have been used in different laboratories. For separation, methods using dextran-coated charcoal (ALMQVIST et al. 1975; CLARK et al. 1969; DEFTOS 1971; HEYNEN and FRANCHIMONT 1974; TASHJIAN and VOLLKEL 1974), polyethylene glycol (SILVA et al. 1974), or dioxane precipitation (DEFTOS 1971) have been proposed.

Serum as well as plasma samples can be used for the radioimmunoassay determination. Normal values of calcitonin in plasma are 0.02–0.04 ng/ml. Calcitonin is normally present in urine in concentrations lower than 1 ng/ml. In patients with medullary carcinoma of the thyroid, concentrations ranging between nanogram and microgram levels can be found. In urine, concentrations as high as 3 μg per milliliter urine have been noted in patients with medullary carcinoma.

Medullary carcinoma of the thyroid starts as a malignancy of the thyroid C cells. Determination of calcitonin has been shown to be useful as an early indication of such malignancy and for follow-up studies on such patients (ALMQVIST et al. 1975; HEYNEN and FRANCHIMONT 1974). Determination of calcitonin also seems to be of great value in screening relatives of patients with a hereditary tendency to medullary carcinoma. Pathologic levels of calcitonin have also been found in patients with chronic hypocalcemia and pseudohypoparathyroidism as well as in carcinoid tumors.

References

Albano JD, Edkins RP, Maritz G, Turner RC (1972) A sensitive, precise radioimmunoassay of serum insulin relying on charcoal separation of bound and free hormone moieties. Acta Endocrinol 70:487

Almqvist S, Hjern B, Wästhed B (1975) The diagnostic value of a radioimmunoassay for parathyroid hormone in human serum. Acta Endocrinol 78:493

Alumets J, Håkansson R, O'Dorisio T, Sjölund K, Sundler F (1978) Is GIP a glucagon cell constituent? Histochemistry 58:253

Arnaud CD, Tsao HS, Littledike T (1971) Radioimmunoassay of human parathyroid hormone in serum. J Clin Invest 50:21

Arnaud CD, Goldsmith RS, Sizenore GW, Oldham SB (1973) Studies on the characterization of human parathyroid hormone in hyperthyroid serum; practical considerations. In: Frame B, Parfitt AM, Duncan H (eds) Clinical aspects of metabolic bone disease. Excerpta Medica, Amsterdam, p 207

Ashby JP, Frier BM (1981) Circulating C-peptide: measurement and clinical application. Ann Clin Biochem 18:125

Assan R, Rosslin G, Dolais J (1967) Effets sur la glucagonémie des perfusions et ingestions d'acide aminés. Journees annuelles de de Diabétologie de l'Hotel – Dieu, 7:25

Aurbach GD, Marx SJ, Spiegel AM (1985) Parathyroid hormone, calcitonin and the calciferols. In: Wilson JD, Foster DW (eds) Textbook of endocrinology. Saunders, Philadelphia, p 1137
Berson SA, Yalow RS (1966) Iodoinsulin used to determine specific activity of iodine-131. Science 152:205
Berson SA, Yalow RS (1968) Immunochemical heterogeneity of parathyroid hormone in plasma. J Clin Endocr 28:1037
Berson SA, Yalow RS, Aurbach GD, Potts JT Jr (1963) Immunoassay of bovine and human parathyroid hormone. Proc Natl Acad Sci USA 49:613
Canterbury JM, Reiss E (1972) Multiple immunoreactive forms of parathyroid hormone in human serum. 140:1393
Cerasi E, Luft R (1967) Plasma insulin response to glucose infusion in healthy subjects and in diabetes mellitus. Acta Endocrinol 55:278
Cerasi E, Luft R, Efendic S (1972) Decreased sensitivity of the pancreatic beta cells to glucose in prediabetic and diabetic subjects. Diabetes 21:224
Chesney TMcC, Schofield JG (1969) Studies on the secretion of pancreatic glucagon. Diabetes 18:627
Christensen MS (1976) Sensitive radioimmunoassay of parathyroid hormone in human serum using a specific extraction technique. Scand J Clin Lab Invest 36:313
Christensen MS (1979) Radioimmunoassay of human parathyroid hormone. Dan Med Bull 26:157
Clark MB, Byfield PHG, Boyd OW, Foster VU (1969) A radioimmunoassay for human calcitonin. Lancet 2:74
Coffey IW, Nagy CF, Lenusky R, Hansen HI (1974) A radioimmunoassay for plasma insulin using zirconyl phosphate gel. Biochem Med 9:54
Deftos LJ (1971) Immunoassay for human calcitonin. I. Method. Metabolism 20:1122
Deftos LJ (1974) Parathyroid hormone. In: Jaffe BM, Behrman HR (eds) Methods of hormone radioimmunoassay. Academic, New York, p 231
Duckworth WC, Kitabchi AE, Heinemann M (1972) Direct measurement of plasma proinsulin in normal and diabetic subjects. Am J Med 53:418
Edwards JC, Howell SL, Taylor KW (1970) Radioimmunoassay of glucagon released from isolated guinea pig islets of Langerhans incubated in vitro. Biochem Biophys Acta 215:297
Faber OK, Binder C (1977) C-Peptide response to glucagon. A test for the residual beta cell function in diabetes mellitus. Diabetes 26:605
Faloona GR (1973) Radioimmunoassay: Glucagon and GLI. In: Berson SA, Yalow RS (eds) Methods in investigative and diagnostic endocrinology. North-Holland, p 919
Freychet P, Roth J, Neville DM (1971) Monoiodoinsulin: demonstration of its biological activity and binding to fat cells and liver membranes. Biochem Biophys Res Commun 43:400
Gabbay KH, Deluca K, Fisher JN Jr, Mako ME, Rubenstein AH (1976) Familial hyperproinsulinemia: an autosomal dominant defect. N Engl J Med 294:911
Ganong WF (1985) Review of medical physiology, 12th edn. Lange Medical, Los Altos
Garcia-Webb P, Bonser A (1979) Decrease in measured C-peptide immunoreactivity on storage. Clin Chem Acta 95:139
Gordon P, Hendricks CM, Roth J (1974) Circulating proinsulin-like component in man – increased proportion in hypoinsulinemia states. Diabetologia 11:469
Habener JF, Potts JT Jr (1977) Chemistry, biosynthesis, secretion and metabolism of parathyroid hormone. In: Aurbach GD (ed) Handbook of physiology, Sect. 7. Endocrinology, vol VII. Williams and Wilkins, Baltimore, p 313
Habener JF, Powell D, Murray TM, Mayer GP, Potts JT Jr (1971) Parathyroid hormone: secretion and metabolism in vivo. Proc Natl Acad Sci 68:2986
Hales CN, Randle PJ (1963) Immunoassay of insulin with insulin-antibody precipitate. Biochem J 88:137
Hamilton CR (1975) Radioimmunoassay of the calcium regulating hormones. In: Rothfeld B (ed) Nuclear medicine in vitro. Lippincott, Philadelphia, p 124

Hazzard WR, Crockford PM, Buchanan KD, Vance JE, Chen R, Williams RH (1968) A double antibody assay for glucagon. Diabetes 17:179
Heding LG (1969) The production of glucagon antibodies in rabbits. Hormone Metabol Res 1:87
Heding LG (1971) Radioimmunological determination of pancreatic and gut glucagon in plasma. Diabetologia 7:10
Heding LG (1972a) Determination of total serum insulin (IRI) in insulin-treated diabetic patients. Diabetologia 8:260
Heding LG (1972b) Immunologic properties of pancreatic glucagon: antigenicity and antibody characteristics. In: Lefebvre PJ, Unger RH (eds) Glucagon. Pergamon, Oxford, p 187
Heding LG (1975) Radioimmunological determination of human C-peptide in serum. Diabetologia 11:541
Heding LG (1978) Insulin, C-peptide and proinsulin in non-diabetics and insulin-treated diabetics: characterization of the proinsulin in insulin-treated diabetics. Diabetes 27 (Suppl):178
Herbert V, Lau K, Gottlieb CW, Bleicher CJ (1965) Coated charcoal immunoassay of insulin. J Clin Endocrin Metab 25:1375
Heynen G, Franchimont P (1974) Human calcitonin radioimmunoassay in normal and pathological conditions. Eur J Clin Invest 4:213
Hoekstra JBL, van Rijn HJM, Erkelens DN, Thijssen JHH (1982) C-peptide. Diabetes Care 5:438
Horwitz DL, Starr JI, Marks ME, Blackard WG, Rubenstein AH (1975) Proinsulin, insulin and C-peptide concentrations in human portal and peripheral blood. J Clin Invest 55:1278
Horwitz DL, Rubenstein AH, Steiner DF (1978) Proinsulin and C-peptide in diabetes. Med Clin North Am 62:723
Hunter WM, Greenwood FC (1962) Preparation of iodine-131-labelled growth hormone of high specific activity. Nature 194:495
Kagan A (1975) Radioimmunoassay of insulin. Semin Nucl Med 5:183
Karesen R, Tronier B, Aune S (1980) Immunoreactive glucagon, insulin and C-peptide in man after resection of the pancreas and total pancreatectomy. Am J Surg 140:272
Kuzuya H, Blix P, Horwitz DL, Steiner DF, Rubenstein AH (1977) Determination of free and total insulin and C-peptide in insulin-treated diabetics. Diabetes 26:22
Lafferty FW (1966) Pseudohyperparathyroidism. Medicine (Baltimore) 45:247
Larsson LI, Moody AJ (1980) Glicentin and gastric inhibitory polypeptide immunoreactivity in endocrine cells of the gut and pancreas. J Histochem Cytochem 28:925
Leclercq-Meyer V, Miahle P, Malaisse WJ (1970) Une méthode de dosage radioimmunologique du glucagon comportant une separation par le charbon-dextran. Diabetologia 6:121
Levy LJ, Spergel G, Bleicher SJ (1970) Proc Endocrin Soc 134 (Abstr)
Mayer GP, Keaton JA, Hurst JG, Habener JF (1977) Effects of plasma calcium concentration on the relative proportion of hormone and carboxyl fragment in parathyroid venous blood. Endocrinology 100s:234 [Abstr] 355
Melani F, Rubenstein AH, Oyer PE, Steiner DF (1970) Identification of proinsulin and C-peptide in human serum by specific immunoassay. Proc Natl Acad Sci USA 67:148
Müller WA, Faloona GR, Aguilar-Parada E, Unger RH (1970) Abnormal alpha-cell function in diabetes. Response to carbohydrate and protein ingestion. N Engl J Med 283:109
Nakagawa S, Nakayama H, Sasaki T, Yoshino K, Ying Yu Y, Shinozaki K, Aoki S, Mashimo K (1973) A simple method for the determination of serum free insulin levels in insulin-treated patients. Diabetes 22:590
Nilsson G, Unväs-Wallensten K (1974) Effect of teasing, sham feeding and feeding on plasma insulin concentrations in dogs. In: Radioimmunoassay: methodology and applications in physiology and in clinical studies. Commemorative issue for Solomon A. Berson. Horm Metab Res 91
Nilsson G, Uvnäs-Wallensten K (1977) Effect of teasing and sham feeding on plasma glucagon concentration in dogs. Acta Physiol Scand 100:298

Nonaka K, Foà (1969) A simplified glucagon immunoassay and its use in a study of incubated pancreatic islets. Proc Soc Exp Biol Med 130:330
Oyer PE, Cho S, Peterson JD, Steiner DF (1971) Studies on human proinsulin. J Biol Chem 246:1375
Palmieri GMA, Yalow RS, Berson SA (1971) Adsorbent techniques for the separation of antibody-bound from free peptide hormones in radioimmunoassay. Horm Metab Res 3:301
Paloyan E (1967) Recent developments in the early diagnosis of hyperparathyroidism. Surg Clin North Am 47:61
Paloyan D, Paloyan E, Worobec R, Ernst K, Deininger E, Harper PV (1966) Serum glucagon levels in experimental acute pancreatitis in the dog. Surg Forum 17:348
Reiss E, Canterbury JM (1968) Experience with radioimmunoassay of parathyroid hormone in human sera. Proc Soc Exp Biol Med 128:501
Rodman JS, Sherwood LM (1978) Disorders of mineral metabolism in malignancy. In: Avioli LA, Krane SM (eds) Metabolic bone disease. Academic, New York, vol II, p 578
Roth J, Gordon P, Pastan I (1968) "Big insulin": a new component of plasma insulin detected by immunoassay. Proc Natl Acad Sci USA 61:138
Rubenstein AH, Welbourne WP, Maks M, Melani F, Steiner DF (1970) Comparative immunology of bovine, porcine and human proinsulins and C-peptides. Diabetes 19:546
Rubenstein AH, Melani F, Steiner DF (1973) Proinsulin and C-peptide in human serum. In: Berson SA, Yalow RS (eds) Methods in investigative and diagnostic endocrinology. North-Holland, Amsterdam, p 870
Safrit HF, Young CW (1977) Factitious hypoglycemia. N Engl J Med 298:515
Sandler R, Horwitz DL, Rubenstein AH, Kuzuya H (1975) Hypoglycemia and endogenous hyperinsulinism complicating diabetes mellitus. Am J Med 59:730
Schopman W, Hackeng HL, Leguin RM (1970) A radioimmunoassay for parathyroid hormone in man. I. Development of a radioimmunoassay for bovine PTH. Acta Endocrinol 63:643
Service FJ, Horwitz DL, Rubenstein AH (1977) C-peptide suppression test for insulinoma. J Lab Clin Med 90:180
Service FJ, Rubenstein AH, Horwitz DL (1975) C-peptide analysis in diagnosis of factitial hypoglycemia in an insulin-dependent diabetic. Mayo Clin Proc 50:697
Silva OL, Snider RH, Becker KL (1974) A radioimmunoassay of calcitonin in human plasma. Clin Chem 20:337
Smith LF (1966) Species variation in the amino acid sequence of insulin. Am J Med 40:662
Smith PH, Merchant FW, Johnson DG, Fujimoto WY, Williams RH (1977) Immunocytochemical localization of a gastric inhibitory polypeptide-like material within A-cells of the endocrine pancreas. Am J Anat 149:585
Soeldner JS, Slone D (1965) Critical variables in the radioimmunoassay of serum insulin using the double antibody technique. Diabetes 19:1
Steiner DF (1972) Proinsulin. Triangle 11:51
Stellon A, Townell NH (1979) C-peptide assay for factitious hyperinsulinism. Lancet 2:148
Stewart AF, Horst R, Deftos LJ, Cadman EC, Lang R, Broadus AE (1980) Biochemical evaluation of patients with cancer-associated hypercalcemia. New Engl J Med 303:1377
Sundby F, Jacobsen H, Moody AJ (1976) Purification and characterization of a protein from porcine gut with glucagon-like immunoreactivity. Horm Metab Res 8:366
Tanaka M, Abe K, Adachi I, Yamaguchi K, Miyakawa S, Hirakawa H, Kumasaka S (1975) Radioimmunoassay specific for amino (N) and carboxyl (C) terminal portion of parathyroid hormone. Endocrinol Jpn 22:471
Tashjian AH, Vollkel EF (1974) Human calcitonin: application of affinity chromatography. In: Jaffe BM, Behrman HR (eds) Methods of hormone radioimmunossay. Academic, New York, p 199
Thorell JI, Larsson SM (1978) Radioimmunoassay and related techniques. Mosby, Saint Louis, Appendix 17

Turner RC, Heding LG (1977) Plasma proinsulin, C-peptide and insulin in diagnostic suppression tests for insulinomas. Diabetologia 13:351

Turner RC, Johnson PC (1973) Suppression of insulin release by fish-insulin-induced hypoglycemia with reference to the diagnosis of insulinomas. Lancet 2:1483

Unger RH (1973) Clinical evaluation of glucagon excess and deficiency. In: Berson SA, Yalow RS (eds) Methods in investigative and diagnostic endocrinology. North-Holland, Amsterdam, p 937

Unger RH, Eisentraut AM, McCall MS, Madison LL (1961) Glucagon antibodies and an immunoassay for glucagon. J Clin Invest 40:1280

Unger RH, Eisentraut AM, Madison LL (1963) The effects of total starvation upon levels of circulating glucagon and insulin. J Clin Invest 42:1031

Unger RH, Ohneda E, Aguilar-Parada E, Eisentraut AM (1970) Studies of pancreatic alfa cell function in normal and diabetic subjects. J Clin Invest 49:837

Velasco CA, Cole HS, Damerino-Davalos PA (1974) Radioimmunoassay of insulin. Clin Chem 20:700

Weinges KF (1968) Labelled proteins. In: Tracer studies. European Atomic Energy Community Euroatom, p 271

Yalow RS, Berson SA (1960) Immunoassay of endogenous plasma insulin in man. J Clin Invest 39:1157

Yalow RS, Berson SA (1973) Radioimmunoassay. In: Berson SA, Yalow RS (eds) Methods in investigative and diagnostic endocrinology. North-Holland, Amsterdam, p 864

CHAPTER 13

Radioimmunoassay of Gastrointestinal Polypeptides

J. H. WALSH and H. C. WONG

A. Introduction

The gastrointestinal hormones comprise a group of peptides produced by endocrine cells that line the stomach and intestine. The pancreatic hormones insulin, glucagon, and pancreatic polypeptide are often included in this category. Several other peptides are produced in the gut exclusively by nerves or by both nerves and endocrine-type cells. These endocrine peptides and neuropeptides exert a wide variety of biologic effects on target organs in the gastrointestinal tract. They serve as endocrine and/or neurocrine regulators of many physiologic functions. They also produce clinical disease, most dramatically when there is massive overproduction by a hormone-secreting tumor.

Measurement of gastrointestinal peptides has developed greatly during the past two decades. The first radioimmunoassay developed by Yalow and Berson was used to measure insulin. Other assays were developed to measure glucagon and later gastrin. During the past decade, it has become routine to develop radioimmunoassays to measure new peptides almost as soon as they are discovered and characterized. Region-specific antisera are often utilized to permit more subtle characterization of immunoreactive peptides, and the use of two or more region-specific assays may allow complete identification of a molecular form of hormone present in blood or tissue extracts without the necessity of chromatographic separation.

In this chapter, we will present the methods used in our laboratory for the measurement of several gastrointestinal peptides. Most of these are classified as hormones, but a few such as vasoactive intestinal polypeptide and the bombesin/gastrin-releasing peptides are exclusively neuropeptides. We will discuss the physiologic and pathologic roles of these peptides and their chemistry only briefly. For more comprehensive reviews of these subjects readers are referred to WALSH (1983) and WALSH (1987). We will present the methods that we find useful, characterize our own antisera, and discuss briefly the assay methods reported by others. It should be recognized that normal values depend greatly on the antibody, assay condition, and standards employed in different laboratories. Furthermore, the presence of multiple molecular forms of many of these peptides creates a more complicated situation that cannot be described adequately by a single number. We will present the normal range obtained in our laboratory, recognizing that it may differ from values obtained in other laboratories. Whenever possible, we measure all biologically active forms of peptide in a single assay, and this value is the one that is presented. For some peptides, such as neurotensin, the biologically inactive fragment is most commonly reported. Therefore, we will give normal values for the biologically active fragments and for the biologically inactive fragment.

B. Gastrin

I. Physiologic Relevance

Gastrin is a hormone produced in the antral portion of the stomach that is responsible for regulation of food-stimulated gastric acid secretion (WALSH and GROSSMAN 1975). Gastrin also has a trophic effect on the mucosa of the acid-secreting portion of the stomach (JOHNSON 1976). Hypersecretion of gastrin may be caused by a gastrin-secreting tumor, or gastrinoma, or by primary hyperfunction of antral gastrin cells (WALSH and GROSSMANN 1975). Hypergastrinemia also commonly occurs in patients with very low or absent gastric acid secretion (ARNOLD et al. 1982). Primary hypergastrinemia due to antral gastrin cell hyperplasia is a rare cause of peptic ulcer disease (LEWIN et al. 1984).

The concentrations of gastrin in the blood are sufficient to account for a major part of the gastric acid secretory response to an amino acid meal (FELDMAN et al. 1978). There is a high correlation between serum gastrin and acid responses to graded concentrations of amino acids and peptides (LAM et al. 1980). The concentrations of gastrin found during graded stimulations of acid secretion caused by gastric perfusion with increasing concentrations of peptone corresponded closely with the gastrin values expected for the observed acid secretory rates (MAXWELL et al. 1984).

II. Chemical Composition

Gastrin consists of two major molecular forms, a 34 amino acid "big gastrin" (G34) and a 17 amino acid "little gastrin" (G17) (WALSH and GROSSMANN 1975). These two peptides differ in clearance rates, with the smaller form being cleared 4–8 times more rapidly than the larger form (WALSH et al. 1976). The tyrosine side chain on gastrin may be sulfated or nonsulfated. Approximately equal amounts of both forms are found in tissues and circulation. Sulfation does not affect the acid stimulatory activity of gastrin (GREGORY 1979).

Despite earlier evidence that big gastrin was less active than little gastrin, recent studies suggest that large and small forms have approximately equal potency in the blood (EYSSELEIN et al. 1984 a). Therefore, measurement of the biologically active carboxy terminal portion of gastrin can give a good reflection of circulating acid stimulatory activity due to gastrin. Postsecretory processing of gastrin may result in biologically active forms containing less than 17 amino acid residues. Gastrins containing 14 amino acids, 6 amino acids, and 5 amino acids have been isolated from tissues (GREGORY et al. 1979, 1983; EYSSELEIN et al. 1984 b).

III. Characteristics of Antibodies and Labeled Peptides

The most useful antibodies for gastrin radioimmunoassay are those directed against the biologically active carboxy terminal tetrapeptide amide region. They should not distinguish between sulfated and nonsulfated gastrins. By now, many satisfactory gastrin radioimmunoassays have been described and a number of antibodies have been characterized (ROSENQUIST and WALSH 1980). A number of additional antibodies have been produced that react with other portions of the gastrin gene product. The sequence for cDNA that codes for gastrin has been published (YOO et al. 1982).

IV. Other Published Assays

Immunization of rabbits with gastrin conjugated to carrier protein has resulted in production of antibodies with K_a as high as 10^{12} M^{-1} and specific either for the carboxy terminus or amino terminus of G17 (REHFELD et al. 1972) or, rarely, of the entire G17 molecule (DOCKRAY and TAYLOR 1977). Some examples of other antibody specificities are antibodies directed at the amino terminus of G17 (DOCKRAY and WALSH 1975), at the amino terminus of big gastrin (DOCKRAY 1980), and at carboxy terminal extended forms of gastrin (SUGANO and YAMADA 1985). Antibodies have also been prepared against the carboxy terminus of preprogastrin (DESMOND et al. 1985). Antibodies can also recognize single amino acid residue substitutions (REHFELD and MORLEY 1983). An antibody specific for the amino terminus of G34 was used to demonstrate amino terminal fragments of G34 in human plasma (PAUWELS and DOCKRAY 1982).

The most common gastrin radioimmunoassays utilize antibodies specific for the biologically active carboxy terminus of gastrin that react approximately equally well with G34, G17, and G14, do not distinguish between sulfated and nonsulfated gastrins, and exhibit minimal cross-reactivity with cholecystokinin peptides (ROSENQUIST and WALSH 1980). Specificity for sulfated gastrins may cause unexpected results if not recognized (REHFELD et al. 1981). Normal basal gastrin concentrations are of the order of 10–20 pM (20–40 pg/ml G17 equivalent). Further characterization of molecular forms of gastrin in plasma or tissue extracts can be done without further separation by use of G17-specific antibody (DOCKRAY et al. 1978; TAYLOR et al. 1979). Alternatively, molecular forms of gastrin can be separated by gel filtration (REHFELD et al. 1974) or by a combination of affinity chromatography and gel filtration (LAMERS et al. 1982), and different molecular forms of gastrin can be determined by their characteristic elution positions and their reactions with region-specific antisera (DOCKRAY et al. 1975a, b; DOCKRAY and WALSH 1975; HOLMQUIST et al. 1979).

C. Cholecystokinin

Cholecystokinin (CCK) is a hormone and a neuropeptide. The endocrine cells that produce CCK are located in the mucosa of the upper half of the small intestine, and the nerve cell bodies are located primarily in the myenteric plexus lining in the muscle layers of the small and large intestine. CCK has many biologic effects, including stimulation of gallbladder emptying, stimulation of pancreatic enzyme secretion, inhibition of food intake, inhibition of gastric acid secretion, inhibition of gastric emptying, and stimulation of small intestinal motility (WALSH 1987). At least the effects on gallbladder contraction and on pancreatic enzyme secretion are felt to be physiologic. There are no diseases known to be associated with hypersecretion of CCK. A suitable bioassay for plasma CCK has been developed by LIDDLE et al. (1985). This assay utilizes release of pancreatic enzymes from isolated acini and has a sensitivity equivalent to that of a highly sensitive radioimmunoassay. LIDDLE et al. (1985) have found the basal levels of CCK to be approximately 1 p*M* with responses to meals in the range 5–10 p*M*.

I. Published Radioimmunoassays

Development of sensitive and specific radioimmunoassays for CCK has been difficult. REHFELD (1978) prepared suitable tracer by conjugation with 125 I-labeled hydroxyphenylpropionic acid succinimide ester. Others have labeled 99% pure CCK33 (GO et al. 1971; THOMPSON et al. 1975) or have utilized CCK39 in order to label the free amino terminal tyrosine residue (SCHLEGEL et al. 1977). Antibodies directed at the carboxy terminal region of CCK often have a high degree of cross-reactivity with gastrin. Antibodies directed against midportion or amino terminal antigenic determinants in porcine CCK may be species specific (GO et al. 1971; STRAUS and YALOW 1978). An additional potential problem with amino terminal antisera is failure to measure small, biologically active carboxy terminal fragments of CCK. Estimates of fasting and food-stimulated CCK concentrations in humans have ranged widely in different assay systems. REHFELD (1984) has discussed the problems with CCK radioimmunoassay in detail.

Use of gel filtration combined with a nonspecific carboxy terminal gastrin/CCK antibody permits easy identification of plasma activity in the region of CCK8, but larger forms are "hidden" in the more abundant gastrin peaks (WALSH et al. 1982).

Several radioimmunoassays have been developed that use antibodies with high specificity for the biologically active sulfated carboxy terminal region of CCK and only minimal cross-reactivity with gastrin (JANSEN and LAMERS 1983; BYRNES et al. 1981; IZZO et al. 1984). Three different assays of this type have produced modest variation in basal CCK immunoreactivity, ranging from around 1 p*M* to around 15 p*M*. Responses to mixed meals and to fat have been increases of about 10 p*M*.

D. Secretin

Secretin was originally described as the first peptide hormone by BAYLISS and STARLING (1902). It is responsible for stimulation of pancreatic bicarbonate secretion during duodenal acidification (WALSH 1987). The secretin cells are located in the mucosa of the upper half of the small intestine. Secretin release is activated by a decrease in the pH of the intestinal contents below a threshold between pH 4 and 3. Secretin is also released in small amounts by bile acids and may be released by certain spices. The concentrations of secretin in blood during and after the ingestion of a protein meal or an acidified meal are sufficient to explain most of the pancreatic bicarbonate response to the meal. There are no known diseases caused by excessive release of secretin. There may be increased blood levels of secretin in conditions characterized by gastric acid hypersecretion. Defective secretin release is found in diseases affecting the intestinal mucosa such as celiac sprue. The decreased secretin release in sprue may account in part for diminished pancreatic secretion.

I. Chemistry

Secretin is structurally related to other gastrointestinal peptides, including glucagon, vasoactive intestinal polypeptide, and gastric inhibitory peptide. Only one molecular form is known to circulate. Human secretin appears to be slightly different from porcine secretin, but specific immunoreactivity and biologic activity studies have not been performed with human secretin.

II. Antibody Production

Antibodies to secretin were readily produced in rabbit by synthetic porcine secretin pentaacetate (E. R. Squibb, Princeton, New Jersey) conjugated to keyhole limpet hemocyanin (Calbiochem, San Diego, California) using 1-ethyl-3-(3-dimethylaminopropyl)carbodiimide hydrochloride (ICN Pharmaceuticals, Plainview, New Jersey) as a coupling agent. To 4 mg secretin, dissolved in 6 ml 0.05 *M* sodium phosphate buffer, pH 7.4, 200 μl saline-dialyzed keyhole limpet hemocyanin (5 mg), 40 mg carbodiimide hydrochloride, and 10000 cpm ^{125}I-labeled secretin were added. The mixture was stirred at room temperature for 2–3 h and dialyzed against 2 l 0.15 *M* NaCl at 4 °C. The efficiency of conjugation was estimated by measuring the radioactivity of ^{125}I-labeled secretin before and after dialysis.

Six 6- to 8-week-old New Zealand white female rabbits were used for immunization; 50–70 μg conjugate was used per rabbit in each injection. The conjugate was diluted in saline so that the amount to be injected was contained in 1 ml. The conjugate was emulsified with an equal amount of complete Freund's adjuvant (Difco Laboratory, Detroit, Michigan) and injected into 20 intradermal sites in the shaved area at the dorsal part of the rabbit. In the initial immunization, 0.5 ml pertussis toxoid (Eli Lilly, Indianapolis, Indiana) was injected intramuscularly in each rabbit. A booster injection was given 6–8 weeks after the first immunization. Useful antibodies developed after the first booster. Rabbits were bled by the ear artery or vein 7–10 days after the booster injection and then checked every 1–2 weeks for antibodies. In our immunization program, four of six rabbits used produced useful secretin antibodies. Antibody 7842 was used at a 1 : 200000 dilution in our radioimmunoassays.

III. Radioiodination

In a small glass tube, containing 5 μg synthetic secretin pentaacetate (Research Plus, Denville, New Jersey), dissolved in 100 μl, 0.25 *M* sodium phosphate buffer, pH 7.4, were added in order: 5 μl Na ^{125}I (Amersham, England) 100 mCi/ml in NaOH, and 10 μl chloramine-T, 2 mg/ml in phosphate buffer. The reaction mixture was mixed gently for exactly 60 s and the reaction was terminated by the addition of 20 μl sodium metabisulfite, 5 mg/ml in phosphate buffer. The iodination mixture was first purified on a 1.0 × 10 cm Sephadex G-10 column prepared in 0.1 *M* sodium acetate buffer, pH 5.2, with 0.15 *M* NaCl and 0.5% bovine serum albumin. Enough 1-ml fractions were collected to include both the labeled peptide and free iodide. Radioactivity was measured in each tube in order to estimate the percentage incorporation of radioiodide into the peptide. The peptide peak usually contained 60%–70% of total radioactivity.

The labeled peptide (first peak) from the Sephadex G-10 column was applied to a 1.0 × 30 cm SP Sephadex C-25 column equilibrated with 0.1 *M* sodium acetate buffer, pH 5.2, with 0.15 *M* NaCl and 0.5% bovine serum albumin (Fig. 1) and 1-ml fractions were collected. Peaks from the descending slope of the labeled secretin were pooled, divided into aliquots, and frozen for use in radioimmunoassay. Specific activity was estimated by radioimmunoassay (Fig. 2). A dilution of antibody was chosen that produced a bound/free (*B*/*F*) ratio of approximately 1.0 with trace amounts of labeled peptide of approximately 1000 cpm per milliliter incubation mixture. Two types of inhibition curve were constructed. In one type of curve, unlabeled secretin was added in graded increments to determine the concentration of peptide that produced 50% inhibition of the initial *B*/*F* (ID_{50}). In the other type of curve, doubling increments of labeled secretin were added over the range 500–100000 cpm per 2 ml incubation volume. The ID_{50} for labeled secretin was determined as the cpm/ml that produces a 50% decrease in *B*/*F*. From these two numbers, the specific activity could be calculated as the ratio of ID_{50} for labeled/unlabeled secretin and expressed in cpm/pg. Such tests of specific activity have revealed that labeled secretin obtained from the descending slope of the SP Sephadex column had a specific activity of 1500 μCi/nmol, while labeled peptide from the ascending slope had a specific activity as low as 100 μCi/nmol.

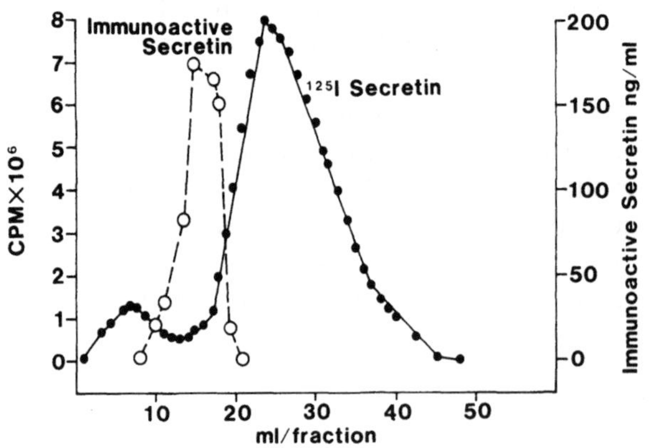

Fig. 1. Secretin label purification on SP Sephadex column

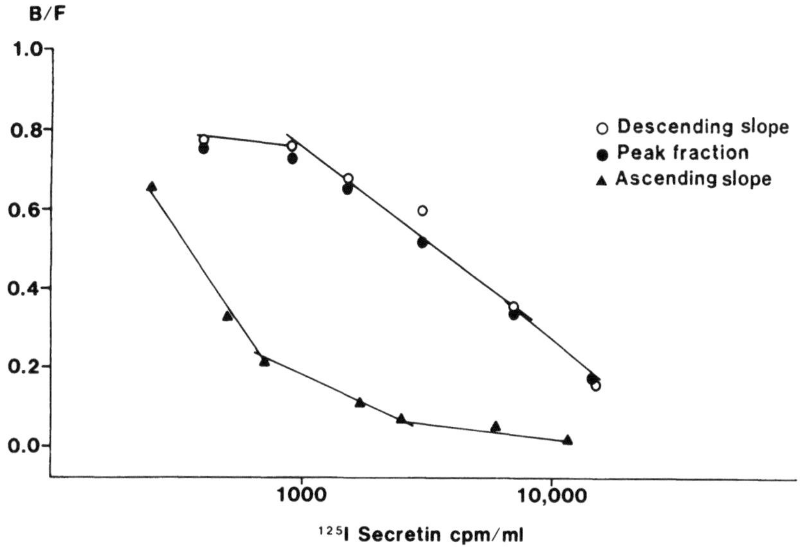

Fig. 2. Secretin label evaluation from SP Sephadex column; antibody 7842 used at 1:200000 dilution

IV. Radioimmunoassay Procedure

All pipetting procedures should be carried out in an ice bath. Samples, standards, antiserum, and labeled secretin are diluted in 0.1 *M* sodium acetate, pH 4.5, containing 2% serum bovine albumin, 2500 kIU aprotinin (FBA Pharmaceuticals, New York, New York) per milliliter buffer, and 0.02 *M* EDTA. Standards are prepared first by diluting the standards to contain 1000, 100, and 10 pg/ml. Each of these standards is pipetted in amounts of 200, 100, 50, and 20 µl, producing ten different concentrations with two points of overlap. Standards and serum samples or other unknowns are diluted to contribute a volume of 1 ml to the reaction mixture. For assays of serum specimen it is desirable to add charcoal-stripped serum to the standard samples so as to correct for nonspecific interference by serum protein (Fig. 3). Aliquots of unknown serum samples 200 and 50 µl are diluted to 1 ml with standard buffer to give a final concentration of 1/10 and 1/40 in the reaction mixture.

Labeled secretin and antibody are added to each tube to give a final incubation volume of 2 ml. To each assay tube, 2000 cpm labeled secretin is added, plus diluted antibody (predetermined to bind 50% of the label). The nonspecific binding controls contain the diluted label and standard buffer instead of antibody. Tubes are incubated for 24–72 h at 4 °C. Separation of bound and free labeled peptide is performed with dextran-coated charcoal. The separation mixture contains 20 mg activated charcoal (Mallinckrodt, Paris, Kentucky), 20 mg dextran T-70 (Pharmacia, Piscataway, New Jersey), and 20 µl 5% bovine serum albumin in a final volume of 0.2 ml. Tubes should be kept on ice during the separation procedure. After thorough mixing, the tubes are centrifuged at 3000 rpm (~2000 g) for 10–15 min and the supernatant solutions are removed by pouring off into a separate tube. Both the pellet containing the free secretin and the supernatant

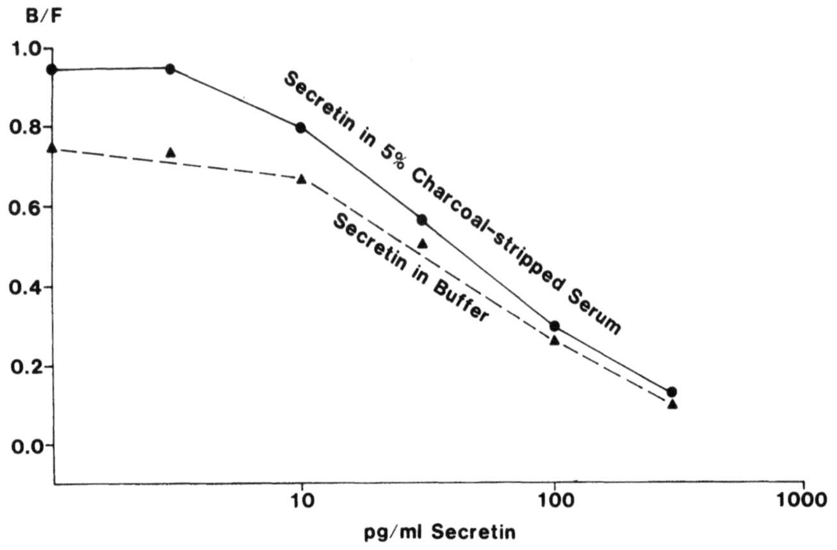

Fig. 3. Serum effect in secretin radioimmunoassay

containing the antibody-bound secretin are counted in a gamma scintillation spectrometer for a minimum of 2 min. Calculations of unknown concentrations are performed by comparing the B/F ratio of the unknown to the B/F ratio of the standard curve.

V. Characterization of Antibodies

Antibody titers were defined as the final dilution of antiserum that would bind 50% of labeled secretin added in a concentration of 1000 cpm/ml after incubation at 4 °C for 48 h. Titers obtained with our secretin-immunized rabbits ranged between 1:50000 and 1:300000. The most useful secretin antibody 7842 is used in a final concentration of 1:300000 in radioimmunoassay.

Antibodies were characterized for avidity and specificity. Sensitivity was expressed as the final concentration of peptide that caused 50% inhibition of the initial B/F ratio (ID_{50}). Sensitivity could be altered by varying the conditions of incubation time, buffer used, specific activity of labeled secretin, antibody dilution, and by addition of aprotinin. Under the optimal conditions, ID_{50} for antibody 7842 for secretin was 12.5 pM (50 pg/ml). No cross-reactivity was found with gastrin, gastric inhibitory peptide, vasoactive intestinal polypeptide, cholecystokinin, or other related peptides tested at contrations of 10 ng/ml (Fig. 4).

VI. Sample Preparation for Radioimmunoassay

Blood was collected in ice-chilled EDTA Vacutainer tubes (Becton Dickinson, Rutherford, New Jersey) containing 2500 kIU aprotinin per milliliter blood. Blood was mixed and centrifuged at 4 °C for 10 min. The plasma was frozen until use. Charcoal-stripped plasma was prepared by adding 5 g activated charcoal to

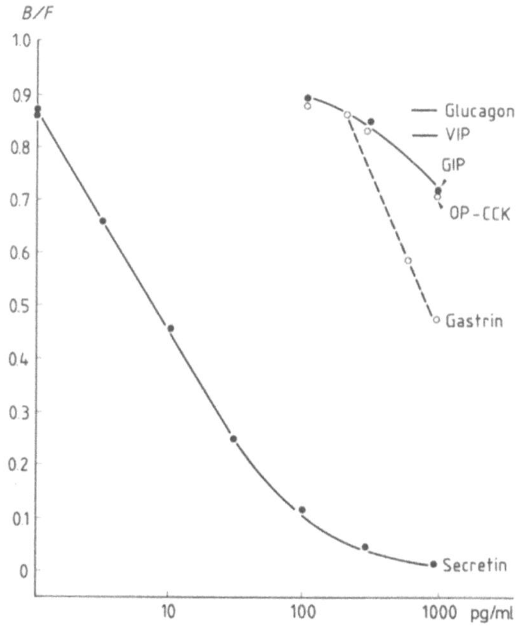

Fig. 4. Secretin antibody 7842 characterization; antibody used at 1:200000 dilution

100 ml normal human plasma. The mixture was stirred overnight at room temperature, and was centrifuged at 10000 *g* for 30 min. The supernatant was saved and filtered through a 0.45-μm filter membrane (Millipore, Bedford, Massachusetts). Charcoal-stripped plasma was then divided into aliquots and stored at −20 °C until use.

VII. Other Published Radioimmunoassays

Immunization of rabbits with either unconjugated or conjugated secretin has produced high affinity secretin antibodies (FAHRENKRUG et al. 1976). Most reported antibodies have had little cross-reactivity with other related gut peptides, possibly because they seem to be directed at antigenic sites in the carboxy terminal portion of the molecule (FAHRENKRUG et al. 1976; BODEN and CHEY 1973; TAI and CHEY 1978). Substitution by tyrosine at the amino terminus of the secretin molecule appears to result in less loss of immunoreactivity than substitution in position 6 (YANAIHARA et al. 1976). Purification of label by ion exchange chromatography improves assay performance (SCHAFFALITZKY DE MUCKADELL et al. 1981; KLEIBEUKER et al. 1984).

Secretin radioimmunoassay has been used successfully to measure increases in plasma immunoreactivity following acid instillation into the duodenum (BODEN and CHEY 1973; RAYFORD et al. 1976). After extraction of plasma to remove interfering factors, fasting secretin concentrations in normal humans were found to average less than 5 p*M* (SCHAFFALITZKY DE MUCKADELL and FAHRENKRUG 1977; TAI and CHEY 1978). Extraction with alcohol or with small C_{18} reverse-phase columns produces similar results. Increased sensitivity may be obtained by inclusion of aprotinin in the incubation and by delayed addition of labeled secretin (KLEIBEUKER et al. 1984).

E. Somatostatin

Somatostatin, originally isolated from the hypothalamus, is widely distributed in the gastrointestinal tract, pancreas, and peripheral nerves (LUFT et al. 1978). Although a number of radioimmunoassays (ARIMURA et al. 1979; PENMAN et al. 1979b; KRONHEIM et al. 1979; VINIK et al. 1981; PIMSTONE et al. 1978) have been reported to measure somatostatin in the human circulation, considerable problems in the methodologies still exist. These include in particular peptide disintegration and interference by plasma protein when standard buffers of neutral pH are used (VINIK et al. 1981; NEWGAARD and HOLST 1981) and the need in most assays to extract somatostatin from plasma before measurement.

I. Antibody Production

To 5 mg somatostatin (Penisula Lab, San Carlos, California), dissolved in 3 ml 0.05 *M* sodium phosphate buffer, pH 7.4, 50 mg 1-ethyl-3-(3-dimethylaminopropyl)carbodiimide (ICN Pharmaceuticals, Plainview, New Jersey) and 500 µl (12.5 mg) dialyzed keyhole limpet hemocyanin slurry (Calbiochem, San Diego, California) were added. The mixture was stirred for 5 h at room temperature and was then dialyzed against 2 l 0.15 *M* NaCl for 24 h. The efficiency of conjugation, estimated by the incorporation of a trace amount of radioiodinated peptide, was 72%.

Five 8-week-old New Zealand white rabbits were immunized at intervals of 6–10 weeks by multiple intradermal injections of an emulsion containing equal parts of complete Freund's adjuvant (Difco Laboratory, Detroit, Michigan) and approximately 50–70 µg conjugated peptide. Useful antibodies 8401 and 8402 were obtained after the second booster injection.

II. Radioiodination

Radioactively labeled Tyr-1-somatostatin (Peninsula Lab, San Carlos, California) was prepared by a modification of the chloramine-T method. To a small 12 × 75 mm glass tube, containing 3 µg Tyr-1-somatostatin dissolved in 100 µl 0.25 *M* sodium phosphate buffer, pH 7.4, were added in order: 0.5 mCi $Na^{125}I$ in 5 µl solvent and 10 µl chloramine-T 1 mg/ml 0.25 *M* sodium phosphate buffer, pH 7.4. The reaction was terminated at 12 s by addition of 20 µl sodium metabisulfite, 5 mg/ml, 0.25 *M* sodium phosphate buffer, pH 7.4. Labeled peptide was separated from unreacted ^{125}I by chromatography on a 1 × 26 cm Sephadex G-25 fine column (Pharmacia, Piscataway, New Jersey) equilibrated with 0.1 *M* acetic acid and 0.1% heat-inactivated newborn bovine serum (Flow Lab, North Ryde, Australia).

Incorporation of radioiodine into the peptide peak averaged 60%–70%. This procedure provides reasonably good separation of labeled peptide from unreacted immunoreactive peptide, with the unreacted peptide eluted in an earlier fraction. The descending slopes of the labeled peptide peaks were used for radioimmunoassay. The labeled peptide retained full immunoreactivity for 2–3 weeks when stored frozen in small aliquots at −20 °C.

Specific activity of the labeled peptide was estimated by autoinhibition of binding in the radioimmunoassay system by serial dilutions of radioactivity in tubes obtained from various parts of the radioactive peak from a Sephadex G-25 fine column. Tubes from the descending slope had higher specific activity than tubes from the ascending slope. Radioiodinated peptide pooled from the descending slope had a specific activity of 1000 μCi/nmol based on comparision with unlabeled somatostatin, and exhibited parallel displacement curves.

III. Sampling and Treatment of Serum

Blood samples were collected into tubes containing 10% (v/v) aprotinin (10000 kIU/ml), FBA Pharmaceuticals, New York, New York). The blood was centrifuged and the serum was stored at −20 °C until use.

IV. Serum Extraction

C_{18} Sep-Pak cartridges (Waters Associates, Milford, Massachusetts) were rinsed first with 10 ml 100% acetonitrile and then with 10 ml 0.05 *M* ammonium acetate, pH 4.5. Serum was loaded onto the cartridge at a flow rate of 1 ml/min. The cartridge was rinsed with 10 ml 0.05 *M* ammonium acetate. The eluates containing somatostatin-like immunoreactivities were dried to the desired volume by rotary evaporation, and stored at −20 °C until use.

V. Radioimmunoassay Procedure

For radioimmunoassays, each tube contained 1500 cpm (3 fmol) labeled peptide, 0.2 ml unknown serum or standards prepared in charcoal-stripped serum over the range 1–1000 pg/ml using synthetic somatostatin (SRIF 14), somatostatin 28 (SRIF 28), and Tyr-1-somatostatin (Peninsula Lab, San Carlos, California), and antibody 8401 at a final dilution of 1:40000, made up to a final volume of 1.0 ml with 0.05 *M* ammonium acetate buffer, pH 4.5, containing 0.1% heat-inactivated newborn bovine serum and 2500 kIU/ml aprotinin. Tubes were set up in triplicate and incubated at 4 °C for 24 h. The whole pipetting procedure was carried out in an ice bath.

Bound and free labeled peptides were separated by the addition of 0.1 ml dextran-coated charcoal suspension (500 mg dextran T-70 and 10 g Norit-A activated charcoal in 200 ml 0.05 *M* ammonium acetate buffer). The mixtures were gently agitated and centrifuged at 2000 rpm for 5–10 min at 4 °C. Both the pellet (free) and the supernatant (bound) were counted in an automatic gamma scintillation spectrometer for 2–3 min. The concentrations of somatostatin-like immunoreactivity of the unknown sera were determined by comparing their bound/free ratios with the standard curves. Specificity was tested by the addition to separate

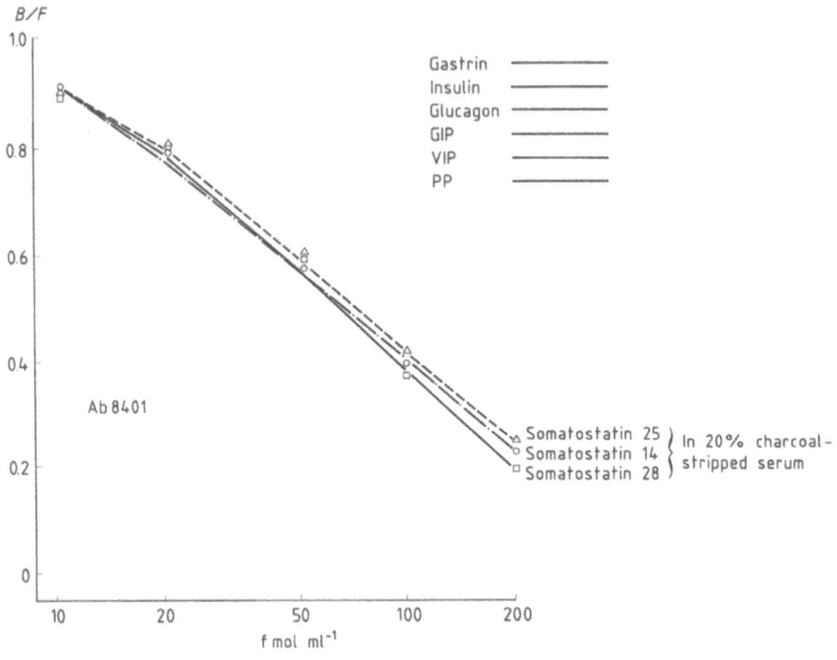

Fig. 5. Somatostatin antibody 8401 characterization. Standard curves represent inhibition of binding of labeled Tyr-1-somatostatin peptide to antibody at graded concentrations of somatostatin tetradecapeptide, somatostatin-28, and somatostatin-25, and lack of inhibition by other peptides listed

assay tubes of 1 nmol somatostatin 1–14, gastrin, insulin, glucagon, gastric inhibitory peptide, vasoactive inhibitory polypeptide, and pancreatic polypeptide (Fig. 5).

VI. Detection Limit

The smallest amount of somatostatin that could be differentiated from zero hormone concentration with 95% confidence was 1.5 fmol/ml, equivalent to 15 fmol/per milliliter serum. ID_{50} was 20 fmol/ml for Tyr-1-SRIF, 40 fmol/ml for SRIF 14, and 70 fmol/ml for SRIF 25 and SRIF 28. Somatostatin 1–14 and other peptides tested have less than 0.1% cross-reactivity. Somatostatin was unstable in serum, but degradation was substantially slowed by addition of aprotinin.

VII. Other Published Radioimmunoassays

A sensitive and specific radioimmunoassay for somatostatin was described by ARIMURA et al. (1978). After immunization with somatostatin, covalently bound to α-globulin by glutaraldehyde, rabbits produced antibodies that bound radioiodinated (Tyr-1) somatostatin with high affinity. This antiserum, R101, recognizes the 5th through 13th residues of somatostatin-14.

Antibodies with different specificities have been developed. They have been useful in studies of somatostatin metabolism. Analogs of somatostatin with substitutions of tyrosine for various residues in the molecule have been used to produce region-specific antibodies by selective conjugation through the tyrosine side chains and use of the same analog to prepare radioiodinated peptide (VALE et al. 1978). Antibodies specific for the amino terminus of somatostatin-14 react with labeled somatostatin in which a tyrosine residue is incorporated in the ring, at position 11 (VALE et al. 1978). Other antibodies have been raised that are specific for the amino terminus or dibasic 13–14 amino acids in somatostatin-28 (RUGGERE and PATEL 1985; BASKIN and ENSINCK 1984).

Radioimmunoassay for somatostatin in blood has presented some problems. Plasma may have nonspecific effects on antibody binding to labeled antigen and there are differences in region specificity among antibodies. Elimination of plasma interference can be achieved by extraction prior to assay (ARIMURA et al. 1978). Acetone or ethanol extraction produces almost complete recovery of somatostatin-14 and slightly lower recovery of somatostatin-28 (HILSTED and HOLST 1982). Extraction of plasma on small octadecylsilyl silica (Sep-Pak C_{18}) columns is also commonly used (COLTURI et al. 1984). An alternative approach is performance of assays at acid pH to prevent enzymatic damage to label. This type of assay requires an antibody that binds well at low pH (SEAL et al. 1982). Nonspecific effects of plasma could not be abolished by enzyme inhibitors or by use of somatostatin-free plasma for construction of standard curves, but good results were obtained after prior extraction of somatostatin from plasma onto silica glass (PENMAN et al. 1979a). Under these conditions, fasting immunoreactive somatostatin in human subjects ranged from 17 to 81 pg/ml with 78% recovery of somatostatin added to plasma. Prior extraction of rat plasma with acetone (ARIMURA et al. 1978) or with acid–ethanol (PATEL et al. 1980) also produced fasting concentrations in peripheral blood that were less than 50 pg/ml. These results are lower than fasting values of 274 pg/ml obtained in human serum that was not extracted (KRONHEIM et al. 1978). Affinity chromatography was used to concentrate large amounts of human plasma and to characterize normal circulating forms (BALDISSERA et al. 1983). Fasting concentrations were 11 p*M* (18 pg/ml) with approximately equal distribution into somatostatin-28 and somatostatin-14 peaks.

F. Motilin

Motilin is a peptide produced by endocrine cells in the mucosa of the upper small intestine that has major effects on intestinal motility (WALSH 1987). Spontaneous increase in the circulating concentrations of motilin are associated with the onset of the interdigestive migrating myoelectric complex of the intestine. This is a spontaneous wave of contraction that originates in the stomach and is propagated through the intestine to the cecum. It occurs approximately every 1.5–2 h in fasting dogs and humans. Intravenous infusion of motilin during the quiescent phase provokes premature appearance of this complex. The blood concentrations required are close to those that develop spontaneously during the normal contraction periods. Intravenous infusion of antibody to motilin diminishes or prevents the onset of these contractions. Motilin also has stimulatory effects on gallbladder emptying and on pancreatic secretion. There are no diseases known to be caused by overproduction of motilin, although it is found in certain endocrine tumors such as carcinoid tumors. Motilin dysfunction may be present in some patients with irritable bowel syndrome.

I. Chemistry

Motilin is a peptide of 22 amino acid residues. Canine motilin differs from porcine motilin in 5 positions, 8, 9, 13, 14, and 15. These differences cause some antisera raised against porcine motilin not to detect canine motilin. Human motilin has a structure identical to that of porcine motilin.

II. Antibody Production

Motilin antibodies have been raised in rabbits by motilin conjugated by 1-ethyl-3-(3-dimethylaminopropyl)carbodiimide to keyhole limpet hemocyanin. To 4 mg motilin, dissolved in 6 ml 0.05 *M* sodium phosphate buffer, pH 7.4, 200 ml saline-dialyzed keyhole limpet hemocyanin (5 mg), 40 mg carbodiimide hydrochloride, and 5000–10000 cpm ^{125}I-labeled motilin were added. The mixture was stirred at room temperature for 2–3 h and dialyzed against 2 l 0.15 *M* NaCl at 4 °C. The efficiency of conjugation was 40%, measured by the radioactivity of ^{125}I-labeled motilin before and after dialysis.

For immunization, six 6-week-old New Zealand white female rabbits were used; 50 µg conjugate was used per rabbit in each injection. A booster injection was given 6–8 weeks after the first immunization. Useful antibodies developed after the second injection. Rabbits were bled by the ear artery or vein 7–10 days after the booster injection and then checked every 1–2 weeks for the presence of antibodies. In our immunization program, three of six rabbits used produced antibodies to motilin. Their titer ranged between 1:10000 and 1:35000.

III. Radioiodination

In a small glass tube containing 4 µg porcine motilin (Peninsula Lab, San Carlos, California) dissolved in 50 µl 0.25 *M* sodium phosphate buffer, pH 7.4, the following reagents were added in order: 5 µl Na^{125}I, 100 mCi/ml in NaOH (Amersham, England) and 10 µl chloramine-T, 1 mg/ml in phosphate buffer. The reaction mixture was mixed gently for exactly 15 s and the reaction was terminated by the addition of 20 µl sodium metabisulfite, 5 mg/ml in phosphate buffer.

The iodination mixture was applied to a 1.0 × 10 cm Sephadex G-10 column prepared in 0.05 *M* ammonium acetate, pH 5.2, containing 0.2% bovine serum albumin. Enough 1-ml fractions were collected to include both the labeled peptide and free iodide. Radioactivity was measured in each tube in order to estimate the percentage incorporation of radioiodide into the peptide. The peptide peak usually contained 50% of total radioactivity.

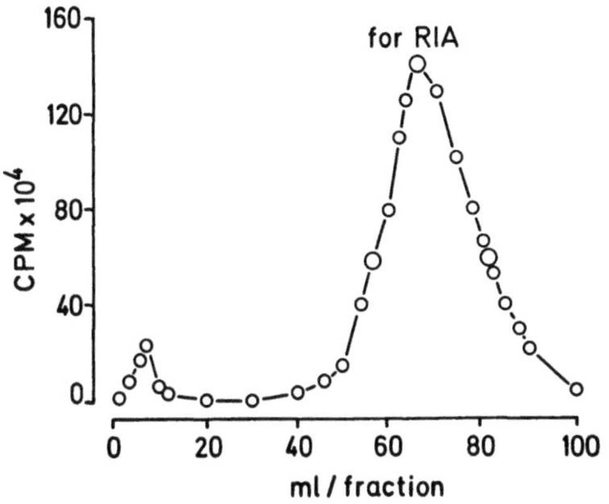

Fig. 6. Motilin label purification by CM Sephadex C-25 column. *Large circles* represent tubes taken for evaluation of specific activity

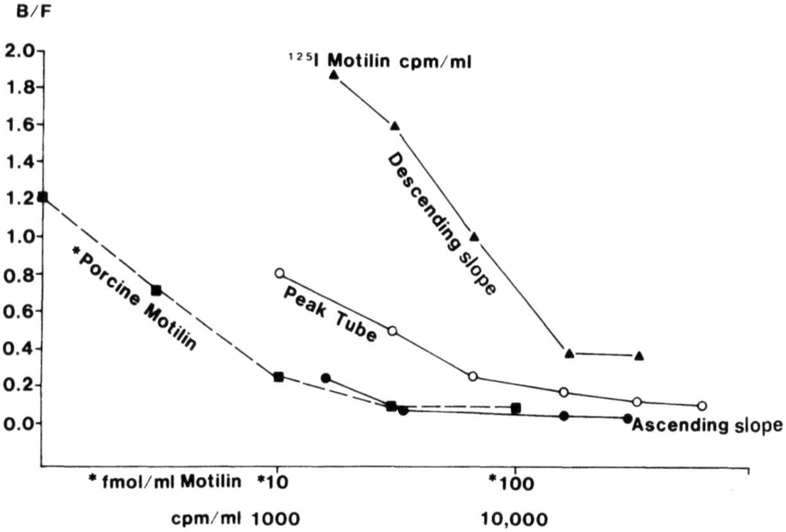

Fig. 7. Motilin label evaluation

Fractions from the first peak tube of the Sephadex G-10 column were applied to a 1.0 × 30 cm CM Sephadex C-25 column equilibrated with 0.05 *M* ammonium acetate, pH 5.2, containing 0.2% bovine serum albumin. Radioactive motilin was eluted from the column with a continuous gradient of ammonium acetate 0.05–1.0 *M*, pH 5.2, using a 50-ml mixing flask (Fig. 6). This step permitted partial separation of labeled motilin from unlabeled motilin.

Routinely, fractions from the descending slope of the eluted peak were pooled, divided into aliquots, and frozen at −40 °C or colder until used. Labeled motilin was stable for 6–8 weeks if kept frozen at −40 °C. Specific activity was estimated empirically by radioimmunoassay (see Sect. D). Labeled motilin obtained from the descending slope of the CM Sephadex C-25 column had a specific activity of 3000 cpm/fmol, while labeled peptide from the ascending slope had a specific activity lower than 500 cpm/fmol (Fig. 7).

IV. Radioimmunoassay Procedure

Samples and standards are diluted in 0.02 *M* sodium barbital buffer, pH 8.6, containing 0.15 *M* NaCl and 0.2% plasma protein (Travenol Laboratories, Glendale, California). Standards are prepared first by diluting the standards to contain 1000, 100, and 10 fmol/ml. Each of these standards is pipetted in amounts of 200, 100, 50, and 20 µl, producing ten different concentrations with two points of overlap. Standards and serum samples or other unknown are diluted to contribute a volume of 1 ml to the reaction mixture. For assays of serum specimen it is desirable to add charcoal-stripped serum to the standard samples so as to correct for nonspecific interference by serum protein (Fig. 8). Aliquots of unknown serum samples 200 and 50 µl are diluted to 1 ml with standard buffer to give a final concentration of 1/10 and 1/40 in the reaction mixture.

Labeled motilin and antibody are added to each tube to give a final incubation volume of 2 ml. To each assay tube, 2000 cpm labeled motilin is added, plus diluted antibody (predetermined to bind 50% of the label). The nonspecific binding controls contain the diluted label and standard buffer instead of antibody.

Tubes are incubated for 24–72 h at 4 °C. The optimal incubation time is determined empirically for each antibody. Separation of bound and free labeled peptide is performed with dextran-coated charcoal. The separation mixture contains 20 mg activated charcoal, 20 mg dextran T-70 (Pharmacia, Piscataway, New Jersey) and 20 µl plasma in a final volume of 0.2 ml. Tubes should be kept on ice during the separation procedure. After thorough mixing, the tubes are centrifuged at 3000 rpm for 10–15 min and the supernatant solutions are removed by pouring off into separate tubes. Both the pellet containing the free motilin and the supernatant containing the antibody-bound motilin are counted in a gamma

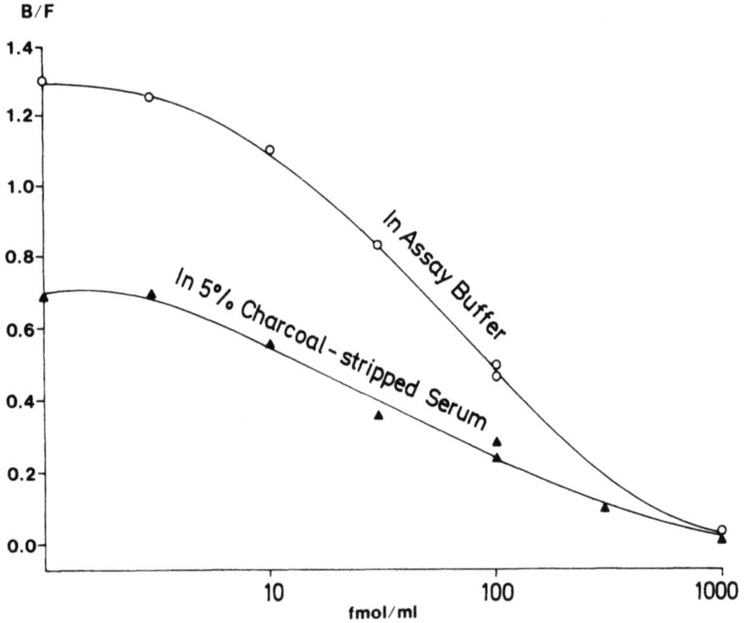

Fig. 8. Serum effect on motilin in radioimmunoassay

scintillation spectrometer for a minimum of 2 min, depending on the accuracy desired. Calculations of unknown concentrations are performed by standard methods.

V. Characterization of Antibodies

Antibody titers were defined as the final dilution of antiserum that would bind 50% of labeled motilin added in a concentration of 1000 cpm/ml after incubation at 4 °C for 48 h. Titers obtained with several rabbit antisera ranged between 1:5000 and 1:30000. The most useful antibody 7821 was used in a final concentration of 1:40000 in radioimmunoassays. Antibodies were characterized for avidity and specificity. Sensitivity was expressed as the final concentration of peptide that caused 50% inhibition of the initial B/F ratio (ID_{50}). Sensitivity could be altered by varying the conditions of incubation, buffer used, specific activity of labeled motilin, antibody dilution, incubation temperature, and incubation time. Under optimal conditions, ID_{50} for antibody 7821 was 5 pM. No cross-reactivity was found with gastrin, secretin, bombesin, vasoactive intestinal polypeptide, somatostatin, neurotensin, histamine-releasing peptide, cerulein, litorin, CCK 8, or Met-enkephalin at concentrations of 1 nmol/ml tested.

VI. Other Published Radioimmunoassays

Several motilin radioimmunoassays have been reported in some detail (BLOOM et al. 1976; DRYBURGH and BROWN 1975; ITOH et al. 1978; LEE et al. 1978). In each case, porcine motilin, usually conjugated to bovine serum albumin, was used for production of antisera, for labeling, and as the assay standard. In one assay (LEE et al. 1978), methanol extraction of plasma was used with recovery of about 70%. Regional specificity of motilin antibodies has been reported. One antibody was shown to be specific for a carboxy terminal fragment between 11 and 16 residues long (POITRAS et al. 1980). Antibodies specific for the amino terminal region have also been reported (CHANG et al. 1981).

G. Gastric Inhibitory Peptide

I. Antibody Production

We have produced gastric inhibitory peptide (GIP) antibodies in rabbits by GIP (gift from Dr. Viktor Mutt, Karolinska Institutet, Stockholm, Sweden) conjugated to keyhole limpet hemocyanin (Calbiochem, San Diego, California) by use of the carbodiimide condensation method (see Sect. E.I). In one immunization program, one of three rabbits developed useful antibody. Antibody 7831 was used at a final dilution of 1:10000 in radioimmunoassays.

For immunization, three 6- to 8-week-old New Zealand white female rabbits were used; 50–70 µg conjugate was used per rabbit. A booster injection was given 6–8 weeks after the first immunization. Animals were bled 5–7 days following immunization by ear artery or vein to check for the presence of antibodies.

II. Characterization of Antisera

Antibody titers were defined as the final dilution of antisera that would bind 50% of labeled GIP added in a concentration of 1000 cpm/ml after incubation at 4 °C for 48–72 h. Antibodies were checked for cross-reactivity with other structurally related peptides such as glucagon, secretin, and vasoactive intestinal polypeptide.

Our GIP antibody 7831 was shown not to cross-react with secretin, glucagon, gastrin, CCK, insulin, or vasoactive intestinal polypeptide at concentrations of up to 10 ng/ml. The sensitivity of the assay utilizing this antibody, expressed as the final concentration of peptide that caused 50% inhibition of initial B/F ratio (ID_{50}), was 200 pg/ml.

III. Radioiodination

Iodination was carried out in a shielded radiation hood at room temperature. In a small glass tube, the following were added in order: 2–3 µg GIP dissolved in 50 µl 0.25 M sodium phosphate buffer, pH 7.4, 5 µl $Na^{125}I$, 100 mCi/ml in NaOH (Amersham, England), and 10 µl chloramine-T, 1 mg/ml in phosphate buffer. The reaction mixture was gently mixed for 12–15 s and was terminated by the addition of 20 µl sodium metabisulfite, 5 mg/ml in phosphate buffer.

The iodination mixture was applied to a 1.0 × 10 cm Sephadex G-10 column prepared in 0.05 M ammonium acetate buffer, pH 5.2, with 0.2% bovine serum albumin and 2500 kIU aprotinin (FBA Pharmaceuticals, New York, New York) per milliliter buffer. Enough 1-ml fractions were collected to include both the labeled peptide and free iodide. Radioactivity was measured in each tube in order to estimate the percentage incorporation of radioiodide into the peptide. In this procedure, the peptide peak usually contained 40%–60% of the total radioactivity.

Labeled GIP fractions (first peak) from the G-10 column were applied to a 1.0 × 20 cm CM Sephadex C-25 column equilibrated with 0.05 M ammonium bicarbonate buffer, pH 8.6, containing 0.2% bovine serum albumin and 2500 kIU aprotinin per milliliter buffer. Radioactive GIP was eluted from the column with a continuous gradient of ammonium 0.05–1.0 M pH 5.2, with the same concen-

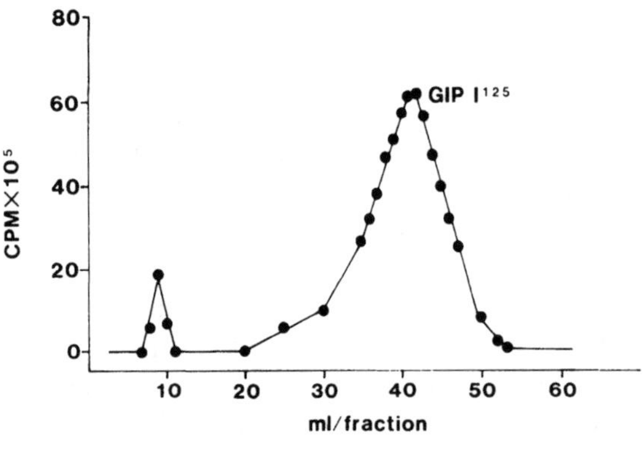

Fig. 9. GIP label purification by CM Sephadex C-25 column

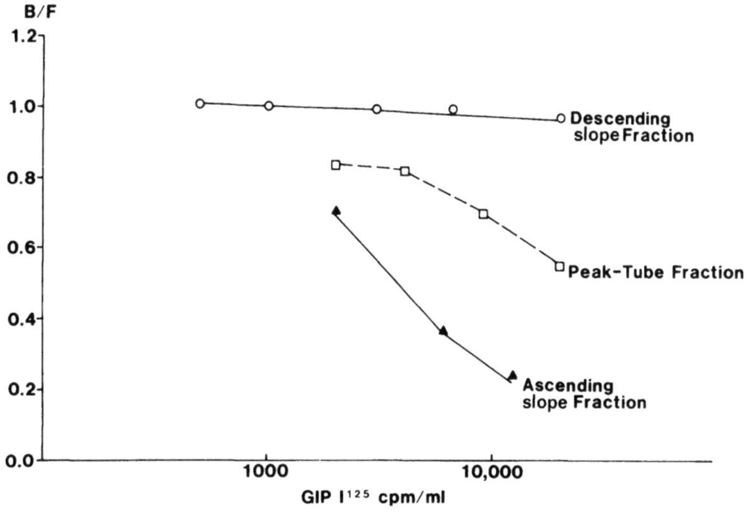

Fig. 10. GIP label evaluation; label was purified by CM Sephadex C-25 column

tration of bovine serum albumin and aprotinin, using a 50-ml mixing flask (Fig. 9).

Routinely, peak tubes from the descending slope of the labeled GIP were pooled, divided into aliquots, and stored at −70 °C for use in radioimmunoassays. Specific activity was estimated by radioimmunoassay. A dilution of antibody was chosen that produced a *B/F* ratio of approximately 1.0 with approximately 1000 cpm/ml labeled peptide in the incubation mixture. Labeled GIP obtained from the descending slope of the CM Sephadex C-25 column had a specific activity greater than 2000 cpm/fmol, while labeled peptide from the ascending slope had specific activity around 100 cpm/fmol (Fig. 10).

IV. Radioimmunoassay Procedure

Pipetting should be carried out in an ice bath. Samples and standards are diluted in 0.05 *M* sodium phosphate buffer, pH 6.5, with 5% bovine serum albumin and 2500 kIU aprotinin per milliliter buffer added. Standards are prepared first by diluting the standards to contain 100, 10, and 1 ng/ml. Each of these standards is pipetted in amounts of 100, 50, 20, and 10 µl, producing ten different concentrations with two points of overlap. Standards and serum samples or other unknowns are diluted to contribute a volume of 0.5 ml in the reaction mixture. For assays of serum specimen, it is desirable to add charcoal-stripped serum to the standard samples so as to correct for nonspecific interference by serum protein to the standards. Aliquots of unknown serum samples 100 and 20 µl are diluted to 0.5 ml with standard buffer to give a final dilution of 1/10 and 1/50 in the final 1 ml reaction mixture. Then, 1000–1500 cpm labeled GIP and antibody dilution (predetermined to bind 50% of the label) are added to each tube to give a final incubation volume of 1 ml. The nonspecific binding controls contain the diluted label and standard buffer instead of antibody.

Tubes are incubated for 48–72 h at 4 °C. Separation of bound and free labeled GIP is performed with dextran-coated charcoal. The separation mixture contains 20 mg activated charcoal (Mallinckrodt, Paris, Kentucky), 20 mg dextran T-70 (Pharmacia, Piscataway, New Jersey), and 20 µl 5% bovine serum albumin in a final volume of 0.2 ml. Tubes should be kept on ice during the separation procedure. After thorough mixing, tubes are centrifuged at 3000 rpm for 10–15 min at 4 °C. The supernatant solutions are removed by pouring off into separate tubes. Both the pellet containing the free GIP and the supernatant containing the antibody-bound GIP are counted in a gamma scintillation spectrometer for a minimum of 2 min. Calculations of unknown concentrations are performed by standard methods.

V. Other Published Radioimmunoassays

Specific radioimmunoassays have been developed for GIP (KUZIO et al. 1974; MAXWELL et al. 1980). Fasting normal human plasma GIP concentration averaged 9 p*M* and the peak concentration obtained after ingestion of glucose, fat, or a mixed meal was about 34 p*M* with one assay (SARSON et al. 1980). Higher stimulated values of 250–300 p*M* were reported by others (KUZIO et al. 1974; MORGAN et al. 1978). Gel filtration studies of plasma, combined with radioimmunoassay, reveal that immunoreactive GIP in the circulation consists of two major forms; 43 amino acid GIP with a molecular weight of approximately 5000 and another larger form. Both are increased during stimulation by a meal (BROWN et al. 1975; SARSON et al. 1980). Different antisera raised against porcine GIP produced markedly different estimates of circulating human GIP, but gave similar results for infused porcine GIP or for GIP release in pigs (JORDE et al. 1983, 1985; AMLAND et al. 1984). Human and porcine GIP must therefore differ structurally. Immunoreactive GIP appears to be relatively stable in plasma, and an enzyme inhibitor (aprotinin) was not necessary for preservation of immunoreactivity in plasma (SARSON et al. 1980).

H. Neurotensin

Neurotensin is found in neurons in the brain. In the gut it is found in the mucosa of the lower half of the small intestine in endocrine cells (WALSH 1987). The most potent neurotensin-releasing substances are the digestion products of fat. Neurotensin has inhibitory effects on gastric acid secretion, stimulatory action on pancreatic exocrine secretion, and some motility effects, but none of these actions is known to be physiologic. Neurotensin also has actions on blood pressure and regional blood flow. Carcinoid tumors of the intestine often produce neurotensin. It is not known if the flushing seen in carcinoid syndrome is due to release of neurotensin.

I. Chemistry

Neurotensin is a 13 amino acid peptide. It is degraded rapidly in the blood by peptidase activity to form two fragments: 1–8 and 9–13. Neither of these fragments has substantial biologic activity. The clearance of the 1–8 fragment is considerably slower than the clearance of the entire neurotensin molecule. Therefore, the 1–8 fragment accumulates in the blood. Antibodies specific for the amino terminal region of neurotensin measure predominantly the 1–8 fragment. The 9–13 fragment is cleared rapidly and does not accumulate in the blood. Therefore, measurements with antibodies specific for the carboxy terminal region of neurotensin give lower values that more accurately reflect the circulating biologically active molecule.

II. Preparation of Antigen

We have produced antibodies to neurotensin (Peninsula Lab, San Carlos, California) by immunization of New Zealand white rabbits with neurotensin conjugated to keyhole limpet hemocyanin (Calbiochem, San Diego, California) with carbodiimide (ICN Pharmaceuticals, Plainview, New Jersey) as a coupling agent. To 3 mg neurotensin, dissolved in 6 ml 0.05 *M* sodium phosphate buffer, pH 7.5, 300 mg 1-ethyl-3-(3-dimethylaminopropyl)carbodiimide, 150 μl keyhole limpet hemocyanin (4 mg), and 10000 cpm ^{125}I-labeled neurotensin were added. The mixture was stirred overnight at 4 °C and was then dialyzed against 2 l 0.14 *M* NaCl for 24 h at 4 °C. The efficiency of conjugation was 40%, measured by the radioactivity of ^{125}I-labeled neurotensin before and after dialysis.

III. Immunization of Animals

Six 8-week-old New Zealand white female rabbits were immunized at intervals of 6–8 weeks by multiple intradermal injections. In our most recent immunization program, antibodies 8551 and 8552 were obtained in two of the six rabbits used after the first booster injection.

IV. Radioiodination

The iodination of neurotensin was carried out in a shielded radioiodination hood at room temperature. To a small glass tube, containing 3–5 μg neurotensin dissolved in 50 μl 0.25 *M* sodium phosphate buffer, were added in order: 1.0 mCi Na^{125}I in 10 μl NaOH (Amersham, England) and 10 μl 2 mg/ml buffer solution of chloramine-T. The reaction mixture with mixed gently for exactly 15 s, and the oxidation reaction was terminated by the addition of 20 μl 5 mg/ml solution of sodium metabisulfite.

Labeled peptide was separated from the unreacted iodine by passing through a 1 × 10 cm Sephadex G-10 column equilibrated in 0.1% acetic acid containing 0.1% bovine serum albumin. Contents of the first radioactive peaks were pooled and then applied to a 1 × 20 cm column of CM Sephadex C-25 equilibrated with 0.1 *M* ammonium acetate, pH 5.2, containing 0.1% bovine serum albumin. Radioactive neurotensin was eluted from the column with a continuous gradient of ammonium acetate 0.1–1.0 *M*, pH 5.2, a 50-ml mixing flask. This step permitted partial separation of labeled from unlabeled neurotensin as shown in Fig. 11. The iodinated peptide eluted slightly behind the noniodinated peptide. Tubes from the

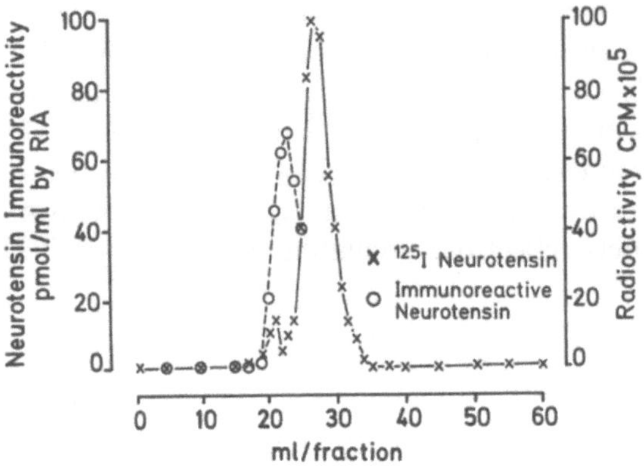

Fig. 11. Purification of neurotensin label on CM Sephadex C-25 column

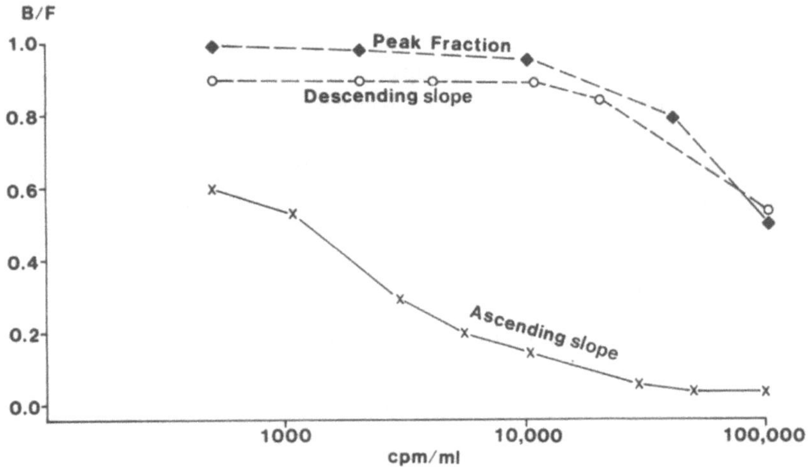

Fig. 12. Neurotensin label evaluation

descending peaks were pooled and stored at −70 °C for use in radioimmunoassays.

Specific activity of the labeled peptide was assessed by plotting label inhibition curves with material obtained from various regions in the peak of radioactivity obtained from the CM Sephadex C-25 column. Labeled neurotensin obtained from the ascending slope of the CM Sephadex C-25 columns had a specific activity of 500 μCi/nmol, while labeled peptide from the descending slope has a specific activity as high as 2200 μCi/nmol (Fig. 12). Labeled peptide retains immunoreactivity for at least 6 weeks when stored frozen at −70 °C.

1. Tissue Extracts

The neurotensin radioimmunoassay has been used successfully to measure neurotensin immunoreactivity in extracts prepared from the gut and brain of mammals and fish. Extraction can be done by boiling the frozen tissue for 5–10 min in 2%

–3% acetic acid. Strong acid produces some inhibition in the radioimmunoassay system; therefore, concentrated extracts must either be neutralized or lyophylized before assay.

2. Plasma

Blood samples were collected in lithium heparin tubes (Becton Dickinson, Rutherford, New Jersey) containing 500 kIU/ml aprotinin in an ice bath. The blood was centrifuged in a refrigerated centrifuge and the plasma was stored at −30 °C until use. The protein in the plasma was precipitated by the addition of ethanol. To 1 ml plasma, 2 ml absolute ethanol was added at room temperature. The mixture was vortexed and centrifuged.Duplicate aliquots of 1 ml supernatant (equivalent to 0.33 ml plasma) were placed in assay incubation tubes and dried under air streams in a water bath at 33 °C. The extracted plasma samples were reconstituted by adding 200 μl assay buffer before use.

V. Radioimmunoassay Procedure

Standard neurotensin is stored in aliquots containing 100 pmol/ml peptide in 0.05 *M* sodium phosphate buffer, pH 7.4. The standard solution is diluted 1 : 10, 1 : 100, and 1 : 1000 immediately prior to assay with 0.02 *M* sodium barbital buffer, pH 8.6, containing 2% bovine serum albumin and 100 kIU aprotinin/per milliliter buffer. Standard curves are prepared by pipetting 200, 100, 50, and 20 μl of each standard solution into duplicate tubes with a Micromedic automatic pipette and adjusting the final volume to 1.0 ml by addition of an appropriate amount of standard diluent. The range covered by the standard curve is 1–1000 fmol/ml with two points of overlap (10 and 100 fmol/ml) to test the accuracy of dilution of the peptide.

Unknown solutions are added in volumes of 200–20 μl, depending on the expected concentration of neurotensin immunoreactivity present and may be prediluted when neccessary. The final volume of unknown solutions is also brought up to 1.0 ml by addition of standard diluent. Labeled neurotensin is diluted in standard diluent to contain 2000 cpm per 0.8 ml and is added as a second step along with 0.2 ml diluted antibody (predetermined to bind 50% of the label). If desired, a final assay volume of 1 ml can be used. When assaying neurotensin concentration in plasma, standard curves containing the same protein concentration of the dried 1 : 3 alcohol extracts of charcoal-stripped plasma are used.

Tubes are incubated for 24–72 h at 4 °C. Separation of bound and free labeled peptide is performed with dextran-coated charcoal suspension. The separation mixture contains 20 mg Norit-A activated charcoal (Mallinckrodt, Paris, Kentucky), 20 mg dextran T-70 (Pharmacia, Piscataway, New Jersey), and 20 μl plasma protein fraction (Travenol Laboratories, Glendale, California) in a final volume of 0.2 ml. Tubes should be kept on ice during the separation procedure. After thorough mixing, tubes are centrifuged at 4 °C at 3000 rpm for 10–15 min. Supernatant solutions are removed by pouring off into separate tubes. Both the pellet containing the free neurotensin and the supernatant containing the antibody-bound neurotensin are counted in an automatic gamma scintillation spectrometer for a minimum of 2 min. Calculations of unknown concentrations are performed by comparing the B/F ratio with the B/F ratio of the standard used.

VI. Characterization of the Antibody

Antibody was characterized for avidity and specificity. Sensitivity of the assay could be altered by varying the conditions of incubation, buffer used, specific activity of labeled neurotensin, antibody dilution, assay volume, addition of the enzyme inhibitor aprotinin, incubation temperature, and incubation time. Specific-

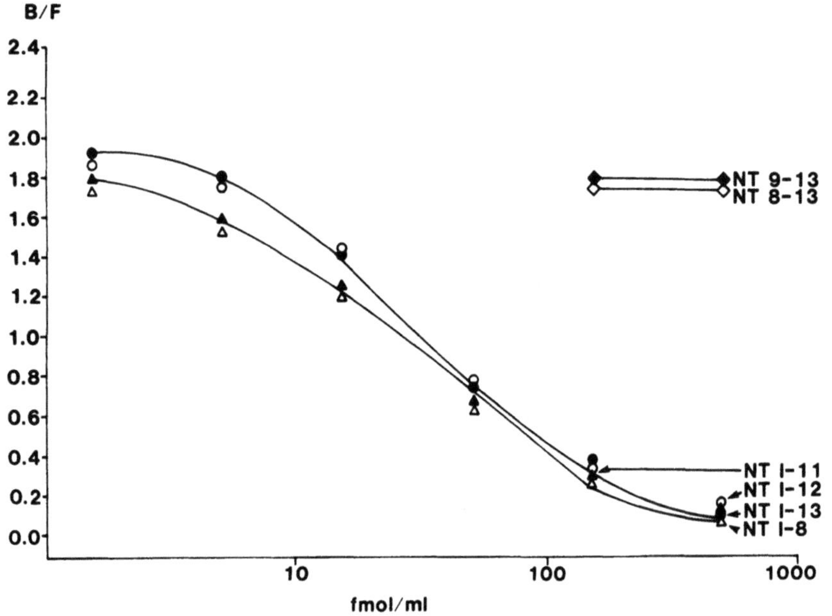

Fig. 13. Neurotensin antibody 8551 characterization; antibody used at 1:100000 dilution with neurotensin 1–13 label

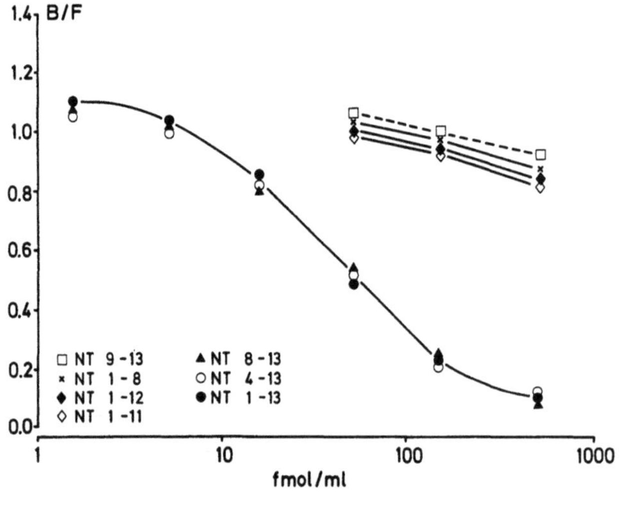

Fig. 14. Neurotensin antibody 8552 characterization; antibody used at 1:15000 dilution with neurotensin 4–13 label

ity was tested by addition to separate assay tubes of 1 nmol bombesin, bovine pancreatic polypeptide, vasoactive intestinal peptide, gastrin, gastric inhibitory polypeptide, substance P, and Met-enkephalin. No cross-reactivity was observed with these peptides tested with antibody 7852. Under optimal conditions, ID_{50} for antibody 7852 when used at 1:80000 dilution for neurotensin was 35 p*M*. It did not cross-react with carboxy terminal fragments of neurotensin. ID_{50} for antibody 8551, when used at 1:100000 dilution for neurotensin 1–13 was 30 p*M*. The antibody did not cross-react with the carboxy terminal portion of the neurotensin (Fig. 13). For antibody 8552, ID_{50} for neurotensin 4–13 is 45 p*M*. The antibody does not cross-react with amino terminal fragments of neurotensin (Fig. 14).

VII. Stability of Neurotensin

To determine the stability of neurotensin in plasma, a known concentration of pure peptide was added to fresh heparinized blood. Blood was left at room temperature for 1 h and at 4 °C for 8 h before being centrifuged and frozen. Similarly, plasma that had been immediately centrifuged and separated was left at room temperature for 1 h and at 4 °C for 8 h before being frozen at −30 °C. In addition, plasma was frozen and thawed under assay conditions 1–5 times prior to assaying. For each set of conditions, blood was taken in duplicate, one sample with and one without the added aprotinin (Mobay Chemical Corporation, New York, New York) 500 kIU/ml.

Recovery of neurotensin added to plasma prior to extraction over a range of 10–200 p*M* was $87.9 \pm 3.0\%$, $r = 0.994$. There was no difference between the buffer and the extracted charcoal-stripped plasma standard curves. Nonextracted charcoal-stripped plasma caused a substantial depression in binding and reduced the sensitivity of the measurement.

The within-assay coefficient of variation for a plasma sample measured ten times in a single assay was 7.8%, with a mean value of 20 p*M*, and 3.1% for a sample with a mean value of 75 p*M*. The coefficient of variation for a plasma sample with a mean value of 138 p*M* measured in eight assays over a 3-month period was 11%, and 12% for a sample with a mean value of 296 p*M*.In six assays, the coefficient of variation for a plasma sample with a mean value of 56 p*M* was 22%. Neurotensin was unstable in blood and plasma. Degradation was substantially slower at low temperature, but was prevented only slightly by addition of aprotinin.

VIII. Other Published Radioimmunoassays

Antibodies that recognize the whole neurotensin molecule and others with specificity for amino or carboxy terminal fragments have been produced by immunization of rabbits with conjugates prepared by coupling to different carrier molecules (Carraway and Leeman 1976). Use of antibodies with different regional specificities will produce different radioimmunoassay results under several circumstances. Amino terminal antisera markedly overestimate the concentration of biologically active neurotensin in the circulation because they detect biologically inactive neurotensin fragments that are the predominant circulating forms. Carboxy terminal-specific antisera must be used for any measurements that attempt to estimate biologic activity. However, large amounts of neurotensin-like carboxy terminal immunoreactivity can be generated from plasma by the action of pepsin-like enzymes at low pH (CARRAWAY 1984).

J. Vasoactive Intestinal Polypeptide

Vasoactive intestinal polypeptide (VIP) is a neuropeptide. Circulating VIP is important in the disease known as pancreatic cholera or VIPoma syndrome (WALSH 1983). The best-known effects of VIP that appear to be physiologic are relaxations of smooth muscle sphincters. The lower esophageal sphincter muscle, the internal anal sphincter, and the fundus of the stomach all contain relaxation reflexes that appear to be mediated by VIP. VIP also causes inhibition of gastric acid secretion. It has a strong stimulatory effect on intestinal secretion of electrolytes and water. This secretory effect accounts for the massive watery diarrhea found in patients with VIPoma syndrome. Measurement of plasma VIP is the most satisfactory method for diagnosing VIPoma syndrome.

I. Chemistry

VIP is structurally related to secretin, glucagon, GIP, and GHRH (growth hormone-releasing hormone). Only one major form is known to circulate in significant amounts, although fragments may be identified in the plasma of some patients with VIPoma.

II. Antibody Production

VIP antibodies have been successfully raised in rabbits by VIP–bovine serum albumin conjugates or VIP–keyhole limpet hemocyanin conjugates. To 2 mg VIP (Peninsula Lab, San Carlos, California) and 6 mg bovine serum albumin (Calbiochem, San Diego, California) in 2 ml 0.1 *M* ammonium acetate buffer, pH 7.0, 1 ml 0.02 *M* glutaraldehyde solution was added dropwise. The mixture was stirred gently at room temperature, in the dark, for 3–4 h and then dialyzed against 0.15 *M* NaCl for 24 h. The efficiency of conjugation can be estimated by inclusion of a trace amount of ^{125}I-labeled VIP in the reaction mixture and measurement of the radioactivity before and after dialysis and calculation of the unbound, dialyzable VIP from the difference in the radioactivity in the dialysates.

Keyhole limpet hemocyanin (Calbiochem, San Diego, California) was first dialyzed against 0.15 *M* NaCl for 24 h at room temperature. To 2 mg VIP, dissolved in 2 ml 0.05 *M* sodium phosphate buffer, pH 7.4, 100 µl (2.5 mg) dialyzed keyhole limpet hemocyanin and 20 mg 1-ethyl-3-(3-dimethylaminopropyl)carbodiimide hydrochloride (ICN Pharmaceuticals, Plainview, New Jersey) were added. The mixture was stirred at room temperature for 3–4 h and dialyzed against 0.15 *M* NaCl for 24 h at 4 °C.

For immunization, six 6-week-old New Zealand white female rabbits were used; 50–75 µg conjugate was used per rabbit in each injection. In our first immunization program, four of six rabbits produced useful antibodies specific to the amino terminus of the VIP molecule. Antibody 6 was used at 1:20000 dilution in radioimmunoassays. In our second immunization program, two of four rabbits produced useful antibodies to VIP. Antibody 7913 is used routinely at 1:1000000 dilution in radioimmunoassay. This antibody has carboxy terminal specificity.

III. Radioiodination

The methionine residue at position 17 of the VIP molecule is known to be very susceptible to oxidation damage. We have developed the following protocol for the iodination of VIP. Iodination is carried out in a shielded radioactive hood at room temperature. For each iodination, to 2–3 µg VIP in 200 µl 0.25 *M* sodium phosphate buffer, pH 7.4, in a small glass test tube, is added 5 µl $Na^{125}I$, 100 mCi/ml in NaOH (Amersham, England) and 10 µl chloramine-T, 1 mg/ml in phosphate buffer. The reaction is terminated in 10 s by the addition of 20 µl sodium metabisulfite, 5 mg/ml in phosphate buffer.

The iodination mixture is applied to a 1.0 × 10 cm Sephadex G-10 column prepared in 0.05 *M* ammonium bicarbonate buffer, pH 8.6, containing 1% bovine serum albumin and 2500 kIU aprotinin (FBA Pharmaceuticals, New York, New York) and 1.0-ml fractions are obtained. Enough fractions are collected to include both the labeled peptide and free iodide. Radioactivity is measured in each tube in order to estimate the percentage incorporation of radioiodide into the peptide. The peptide peak usually contains 40%–70% of total radioactivity.

The labeled peptide peak (first peak) from the Sephadex G-10 column is applied to a 1.0 × 20 cm column of CM Sephadex C-25 equilibrated with 0.05 *M* ammonium bicarbonate buffer, pH 8.6 containing 1% bovine serum albumin and 2500 kIU aprotinin. Radioactive VIP is eluted from the column with a gradient to 0.5 *M* ammonium bicarbonate containing the same concentration of bovine serum albumin and aprotinin, using a 50-ml mixing flask. This step permits partial separation of labeled from unlabeled VIP as shown in Fig. 15. Routinely, frac-

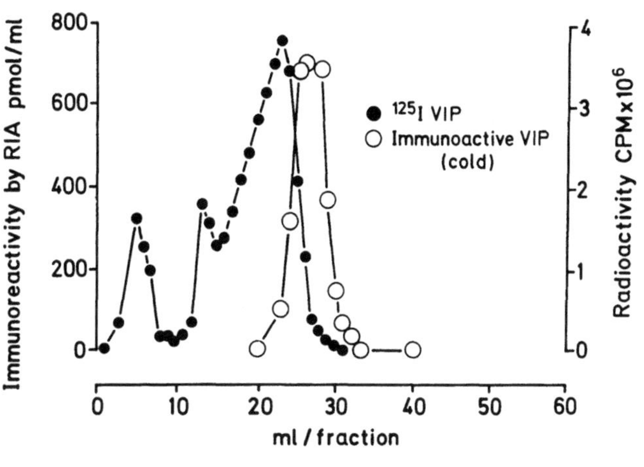

Fig. 15. VIP label purification on CM Sephadex C-25 column

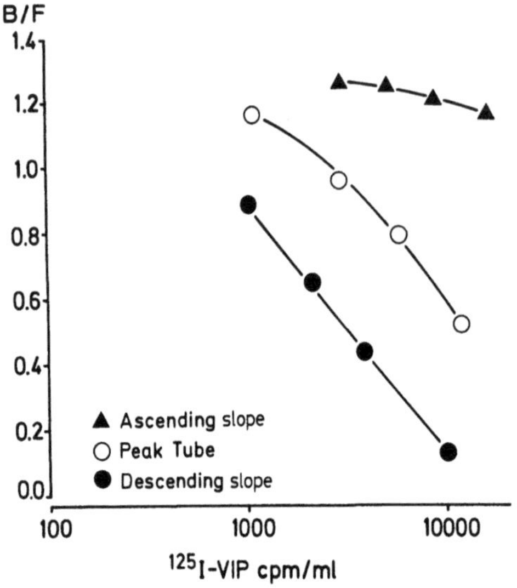

Fig. 16. VIP label evaluation from CM Sephadex C-25 column

tions from the ascending slope of the labeled VIP are pooled, divided into aliquots, and frozen for use in radioimmunoassay.

Labeled VIP obtained from the ascending slope of the CM Sephadex C-25 column has a specific activity of 5000 cpm/fmol, while labeled peptide from the descending slope has a specific activity as low as 500 cpm/fmol (Fig. 16). When it is desirable to maximize assay sensitivity, tubes are selected only from the midportion of the ascending slope in order to achieve greatest specific activity.

IV. Radioimmunoassay Procedure

The pipetting procedure should be carried out in an ice bath. Samples and standards are diluted in 0.02 *M* sodium barbital buffer, pH 8.6, containing 1% bovine serum albumin and 2500 kIU aprotinin per milliliter buffer. Standards are prepared first by diluting the standards to contain 1000, 100, and 10 fmol/ml. Each of these standards is pipetted in amounts of 200, 100, 50, and 20 µl, producing ten different concentrations with two points of overlap. Standards and plasma samples or other unknowns are diluted to contribute a volume of 1 ml to the reaction mixture. For assays of serum specimens it is desirable to add charcoal-stripped plasma to the standard samples so as to correct for nonspecific interference by serum protein. Aliquots of unknown samples, 200 and 50 µl, are diluted to 1 ml with standard buffer to give a final concentration of 1/10 and 1/40 in the reaction mixture. Labeled VIP and antibody are then added to each tube to give a final incubation volume of 2 ml. To each assay tube, 2000 cpm labeled VIP is added, plus diluted antibody (predetermined to bind 50% of the label). The nonspecific binding controls contain the diluted label and standard buffer instead of antibody.

Tubes are incubated for 24–72 h at 4 °C. The optimal incubation time is determined empirically for each antibody. Separation of bound and free labeled peptide is performed with dextran-coated charcoal. The separation mixture contains 20 mg activated charcoal, 20 mg dextran, and 20 µl plasma protein fraction (Travenol Laboratories, Glendale, California) in a final volume of 0.2 ml. Tubes should be kept on ice during the separation procedure. After thorough mixing, the tubes are centrifuged at 3000 rpm for 10–15 min and the supernatant solutions are removed by pouring off into separate tubes. Both the pellet containing the free VIP and the supernatant containing the antibody-bound VIP are counted in a gamma scintillation spectrometer for 2–5 min, depending on the accuracy desired. Calculations of unknown concentrations are performed by standard methods.

V. Characterization of Antibodies

Antibody titers were defined as the final dilution of antiserum that would bind 50% of labeled VIP added in a concentration of 1000 cpm/ml after incubation at 4 °C for 48 h. Titers obtained with several rabbit antisera ranged between 1 : 1000 and 1 : 2000 000. The most useful antibodies, 6 and 7913, were used in a final concentration of 1 : 20 000 and 1 : 2000 000, respectively.

Antibodies were characterized for avidity and specificity. Sensitivity of the assay was expressed as the final concentration of peptide that caused 50% inhibition of the initial B/F ratio (ID_{50}). Sensitivity could be altered by varying the conditions of incubation, buffer used, specific activity of labeled VIP, antibody dilution, incubation temperature, and incubation time. Under optimal conditions, ID_{50} for antibody 6 for VIP was 70 p*M*, and 12 p*M* for antibody 7913. The specificity of antibodies 6 and 7913 are indicated in Figs. 17 and 18. No cross-reactivity was found with gastrin, secretin, or other related peptides at concentrations up to 1 µ*M*.

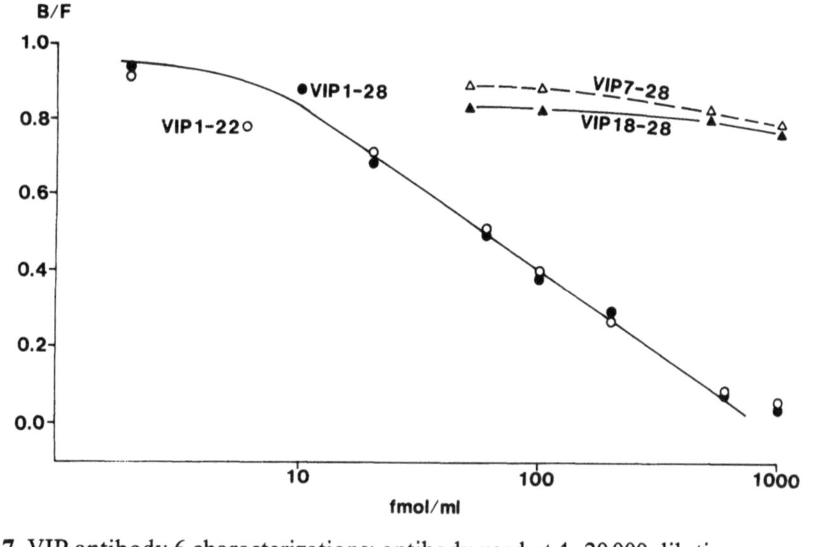

Fig. 17. VIP antibody 6 characterizations; antibody used at 1 : 20 000 dilution

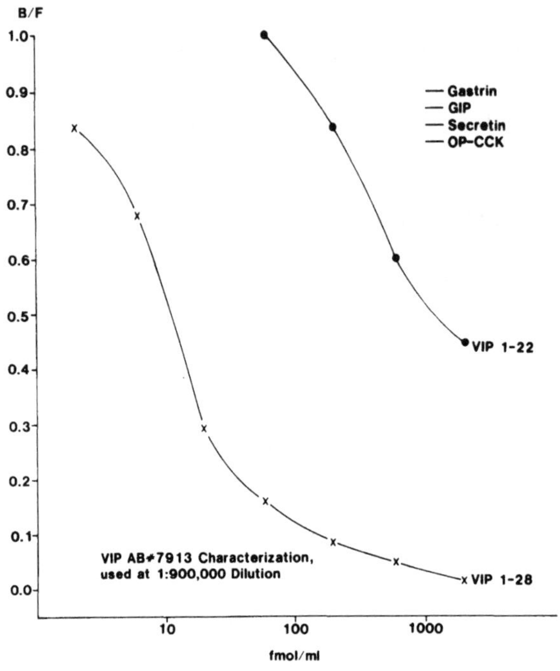

Fig. 18. VIP antibody 7913 characterization; antibody used at 1 : 900 000 dilution

VI. Sample Preparation for Radioimmunoassay

The VIP molecule contains two separate double basic amino acid sequences and can be rapidly destroyed by trypsin-like proteolytic enzymes. Considerable care is required in sample preparation for radioimmunoassay. Blood samples should be collected in ice-chilled heparinized tubes containing approximately 1000 kIU aprotinin per milliter whole blood, rapidly mixed, and centrifuged at 4 °C. The plasma should be free of hemolysis. Plasma should be kept frozen until use. Repeated thawing and freezing causes significant losses of hormonal activity.

VIP standards should be stored in small aliquots at −20 °C or colder. Aliquots of patient plasma with elevated known VIP levels should be used and run in each assay as interassay controls. Tissue extracts can be obtained by boiling a known amount of frozen tissue in 1%–3% acetic acid for 2–5 min. VIP is relatively stable at low pH. Strong acid produces some inhibition in the radioimmunoassay system; one should include a standard curve containing the same concentration of acetic acid as the unknown specimen and use it as a reference.

VII. Other Published Radioimmunoassays

The original radioimmunoassay method for VIP was developed by SAID and FALOONA (1975). Other useful assays have been described by MITCHELL and BLOOM (1978), FAHRENKRUG and SCHAFFALITZKY DE MUCKADELL (1977), and by GAGINELLA et al. (1978).

K. Pancreatic Polypeptide and Peptide YY

Pancreatic polypeptide (PP) was isolated from the pancreas, where it is present in endocrine cells. It consists of 36 amino acids and has an approximate molecular weight of 4200. Peptide YY is another gut endocrine peptide, found mainly in the ileum and colon. It also contains 36 amino acid residues, of which 18 are identical to those of PP.

The best-characterized biologic action of PP is the inhibition of pancreatic exocrine secretion. This may be a physiologically important action. Peptide YY inhibits both gastric and pancreatic secretion. Both peptides are released when fatty acids are present in the intestinal lumen. PP is also released by digested protein and amino acids. Atropine strongly inhibits the release of PP. The mose useful application of radioimmunoassay of either peptide has been the finding that PP may serve as a marker for hormone-secreting tumors of the pancreas. No clinical applications of peptide YY measurement have been described.

I. Pancreatic Polypeptide

Most measurements of mammalian PP have been performed with antibody raised to bovine PP by CHANCE et al. (1979) and supplied to different research laboratories (SCHWARTZ et al. 1978; TAYLOR et al. 1978a, b). This antibody has equal affinity for bovine and human PP and no cross-reactivity with other known gastrointestinal peptides. It does not detect avian PP and is not satisfactory for measurement of PP from rat pancreas. Antibody raised against the carboxy terminal hexapeptide appears to be suitable for measurement of rat PP (KIMMEL et al. 1984; TAYLOR and VAILLANT 1983). The PP concentration in fasting serum from dogs or normal human subjects is about 10–30 p*M*.

II. Peptide YY

Peptide YY is structurally related to PP. Several radioimmunoassays have been developed for measurement of peptide YY in plasma and tissue extracts (CHEN et al. 1984; TAYLOR 1985; ADRIAN et al. 1985). Antisera have been raised with or without conjugation to carrier protein. Specificities of antisera have been quite good, and only slight cross-reactivity with PP and neuropeptide Y have been found among all the peptides tested. Labeled peptide YY can be prepared by oxidative radioiodination, and the labeled peptide can be purified by ion exchange chromatography, gel filtration, or high pressure liquid chromatography. Specific activity in the range of 1–2 Ci/μmol has been achieved. Two assays are sufficiently sensitive to detect less than 2 p*M* changes in peptide YY concentration in the incubation tube (TAYLOR 1985; ADRIAN et al. 1985) and the other is about tenfold less sensitive. All three assays have been used successfully to measure peptide YY concentrations in human or canine plasma.

Region specificity of the antibody-binding site was evaluated by use of synthetic peptide fragments and tryptic peptides of natural peptide YY in one assay (ADRIAN et al. 1985). The amino terminal 1–19 fragment bound as well as intact peptide YY, but carboxy terminal fragments as long as 13–36 produced minimal inhibition of binding. The regional specificities of other antisera have not yet been reported. Basal plasma concentrations in humans averaged about 50 pg/ml in one assay (CHEN et al. 1984) and 9 p*M* (38 pg/ml) in another (ADRIAN et al. 1985). In dogs, basal plasma peptide YY concentrations averaged 62 p*M* (260 pg/ml) (PAPPAS et al. 1985).

References

Adrian T, Savage A, Sagor G, Allen J, Bacarese-Hamilton A, Tatemoto K, Polak J, Bloom S (1985) Effect of peptide YY on gastric, pancreatic, and biliary function in humans. Gastroenterology 89:494–499
Amland P, Jorde R, Revhaug A, Myhre E, Burhol P, Giercksky K-E (1984) Fasting and postprandial GIP values in pigs, rats, dogs, and man measured with five different GIP antisera. Scand J Gastroent 19:1095–1098
Arimura A, Lundquist G, Rothman J, Chang R, Fernandez-Durango R, Elde R, Coy DH, Meyers C, Schally AV (1978) Radioimmunoassay of somatostatin. Metabolism [Suppl 1] 27:1139–1144
Arimura A, Sato H, Coy D, Schally A (1979) Radioimmunoassay for GH-release inhibiting hormones. Proc Soc Exp Biol Med 148:784–789
Arnold R, Hulst M, Neuhof C, Schwarting H, Becker H, Creutzfeldt W (1982) Antral gastrin-producing G-cells and somatostatin-producing D-cells in different states of gastric acid secretion. Gut 23:285–291
Baldissera F, Munoz-Perez M, Holst J (1983) Somatostatin 1-28 circulates in human plasma. Regul Pept 6:63–69
Baskin D, Ensinck J (1984) Somatostatin in epithelial cells of intestinal mucosa is present primarily as somatostatin 28. Peptides 5:615–621
Bayliss WM, Starling EH (1902) The mechanism of pancreatic secretion. J Physiol (Lond) 28:325–333
Bloom S, Mitznegg P, Bryant M (1976) Measurement of human plasma motilin. Scand J Gastroenterol 11:47–52
Boden G, Chey WY (1973) Preparation and specificity of antiserum to synthetic secretin and its use in a radioimmunoassay (RIA). Endocrinology 92:1617–1624
Brown J, Dryburgh J, Ross S, Dupre J (1975) Identification and actions of gastric inhibitory polypeptide. Recent Prog Horm Res 31:487–532
Byrnes D, Henderson L, Borody T, Rehfeld J (1981) Radioimmunoassay of cholecystokinin in human plasma. Clin Chim Acta 111:81–89
Carraway R (1984) Rapid proteolytic generation of neurotensin-related peptide(s) and biologic activity during extraction of rat and chicken gastric tissues. J Biol Chem 259(16):10328–10334
Carraway R, Leeman SE (1976) Radioimmunoassay for neurotensin, a hypothalamic peptide. J Biol Chem 251:7035–7044
Chance R, Moon N, Johnson M (1979) Human pancreatic polypeptide (HPP) and bovine pancreatic polypeptide (BPP). In: Jaffe BM, Behrman HR (eds) Methods of hormone radioimmunoassay. Academic, New York, pp 657–670
Chang T-M, Wagner D, Chey W, Yajima H (1981) Motilin-antimotilin reaction reveals two antigenic determinants in motilin. Peptides 2:31–37
Chen M, Balasubramanian A, Murphy R, Tabata K, Fischer J, Chen I, Joffe S (1984) Sensitive radioimmunoassay for measurement of circulating peptide YY. Gastroenterology 87:1332–1338
Colturi T, Unger R, Feldman M (1984) Role of circulating somatostatin in regulation of gastric acid secretion, gastrin release, and islet cell function. Studies in healthy subjects and duodenal ulcer patients. J Clin Invest 74:417–423
Desmond H, Dockray G, Spurdens M (1985) Identification by specific radioimmunoassay of two novel peptides derived from the C-terminus of porcine preprogastrin. Regul Pept 11:133–142

Dockray G (1980) Immunochemical studies of big gastrin using NH_2-terminal specific antisera. Regul Pept 1:169–186
Dockray GJ, Taylor IL (1977) Heptadecapeptide gastrin: Measurement in blood by specific radioimmunoassay. Gastroenterology 71:971–977
Dockray G, Walsh J (1975) Amino terminal gastrin fragment in serum of Zollinger-Ellison syndrome patients. Gastroenterology 68:222–230
Dockray GJ, Debas HT, Walsh JH, Grossman MI (1975 a) Molecular forms of gastrin in antral mucosa and serum in dogs (38848). Proc Soc Exp Biol Med 149:550–553
Dockray G, Walsh J, Passaro E Jr (1975 b) Relative abundance of big and little gastrin in the tumours and blood of patients with the Zollinger-Ellison syndrome. Gut 16:353–358
Dockray G, Vaillant C, Hopkins C (1978) Biosynthetic relationships of big and little gastrins. Nature 273:770–772
Dryburgh J, Brown J (1975) Radioimmunoassay for motilin. Gastroenterology 68:1169–1176
Eysselein V, Maxwell V, Reedy T, Wunsch E, Walsh J (1984 a) Similar acid stimulatory potencies of synthetic human big and little gastrins in man. J Clin Invest 73:1284–1290
Eysselein V, Reeve J Jr, Shively J, Miller C, Walsh J (1984 b) Isolation of a large cholecystokinin precursor from canine brain. Proc Natl Acad Sci USA 81:6565–6568
Fahrenkrug J, Schaffalitzky De Muckadell O, Rehfeld J (1976) Production and evaluation of antibodies for radioimmunoassay of secretin. Scand J Clin Lab Invest 36:281–287
Fahrenkrug J, Schaffalitzky De Muckadell O (1977) Radioimmunoassay of vasoactive intestinal polypeptide (VIP) in plasma. J Lab Clin Med 89:1379–1388
Feldman M, Walsh JH, Wong HC (1978) Role of gastrin heptadecapeptide in the acid secretory response to amino acids in man. J Clin Invest 61:308–313
Gaginella T, Mekhjian H, O'Dorisio T (1978) Vasoactive intestinal peptide: quantification by radioimmunoassay in isolated cells, mucosa, and muscle of the hamster intestine. Gastroenterology 74:718–721
Go V, Ryan R, Summerskill W (1971) Radioimmunoassay of porcine cholecystokinin-pancreozymin. J Lab Clin Med 77:684–689
Gregory RA (1979) A review of some recent developments in the chemistry of the gastrins. Bioorg Chem 8:497–511
Gregory RA, Tracy HJ, Harris JI, Runswick MJ (1979) Minigastrin: Corrected structure and synthesis. Hoppe-Seyler's Z Physiol Chem 360:73–80
Gregory R, Dockray G, Reeve J Jr, Shively J, Miller C (1983) Isolation from porcine antral mucosa of a hexapeptide corresponding to the C-terminal sequence of gastrin. Peptides 4:319–323
Hilsted L, Holst J (1982) On the accuracy of radioimmunological determination of somatostatin in plasma. Regul Pept 4:13–31
Holmquist A, Dockray G, Rosenquist G, Walsh J (1979) Immunochemical characterization of cholecystokinin-like peptides in lamprey gut and brain. Gen Comp Endocrinol 37:474–481
Itoh Z, Takeuchi S, Aizawa I, Mori K, Taminato T, Seino Y, Imura H, Yanaihara N (1978) Changes in plasma motilin concentration and gastrointestinal contractile activity in conscious dogs. Dig Dis 23:929–935
Izzo R, Brugge W, Praissman M (1984) Immunoreactive cholecystokinin in human and rat plasma: correlation of pancreatic secretion in response to CCK. Regul Pept 9:21–34
Jansen J, Lamers C (1983) Radioimmunoassay of cholecystokinin in human tissue and plasma. Clin Chim Acta 131:305–316
Johnson LR (1976) The trophic action of gastrointestinal hormones. Gastroenterology 70:278–288
Jorde R, Burhol P, Schulz T (1983) Fasting and postprandial plasma GIP values in man measured with seven different antisera. Regul Pept 7:87–94
Jorde R, Amland P, Burhol P, Giercksky K-E (1985) What are "physiological" plasma GIP levels in man after intravenous infusion of porcine GIP? Scand J Gastroenterol 20:268–271

Kimmel J, Pollock H, Chance R, Johnson M, Reeve J, Taylor I, Miller C, Shively J (1984) Pancreatic polypeptide from rat pancreas. Endocrinology 114:1725–1731

Kleibeuker J, Eysselein V, Maxwell V, Walsh J (1984) Role of endogenous secretin in acid-induced inhibition of human gastric function. J Clin Invest 73:526–532

Kronheim S, Berelowitz M, Pimstone BL (1978) The characterization of somatostatin-like immunoreactivity in human serum. Diabetes 27(5):523–529

Kuzio M, Dryburgh J, Malloy K, Brown J (1974) Radioimmunoassay for gastric inhibitory polypeptide. Gastroenterology 66:357–364

Lam SK, Isenberg JI, Grossman MI, Lane WH, Walsh JH (1980) Gastric acid secretion is abnormally sensitive to endogenous gastrin released after peptone test meals in duodenal ulcer patients. J Clin Invest 65:555–562

Lamers C, Walsh J, Jansen J, Harrison A, Ippoliti A, Van Tongeren J (1982) Evidence that gastrin 34 is preferentially released from the human duodenum. Gastroenterology 83:233–239

Lee K, Chey W, Tai H, Yajima H (1978) Radioimmunoassay of motilin: validation and studies on the relationship between plasma motilin and interdigestive myoelectric activity of the duodenum of dog. Dig Dis Sci 23:789–795

Lewin K, Yang K, Ulich T, Elashoff J, Walsh J (1984) Primary gastrin cell hyperplasia: report of five cases and a review of the literature. Am J Surg Pathol 8:821–832

Liddle R, Goldfine I, Rosen M, Taplitz R, Williams J (1985) Cholecystokinin bioactivity in human plasma: molecular forms, responses to feeding, and relationship to gallbladder contraction. J Clin Invest 75:1144–1152

Luft R, Efendic S, Hokfelt T (1978) Somatostatin – both hormone and neurotransmitter? Diabetologia 14:13

Maxwell V, Shulkes A, Brown J, Solomon T, Walsh J, Grossman M (1980) Effect of gastric inhibitory polypeptide on pentagastrin-stimulated acid secretion in man. Dig Dis Sci 25:113–116

Maxwell V, Eysselein V, Kleibeuker J, Reedy T, Walsh J (1984) Glucose perfusion intragastric titration. Dig Dis Sci 29:321–326

Mitchell S, Bloom S (1978) Measurement of fasting and postprandial VIP in man. Gut 19:1043–1048

Morgan L, Morris B, Marks V (1978) Radioimmunoassay of gastric inhibitory polypeptide. Ann Clin Biochem 15(3):172–177

Newgaard C, Holst J (1981) Heterogeneity of somatostatin like immunoreactivity in extracts of porcine, canine and human pancreas. Acta Endocrinol (Copen) 98:564–572

Pappas T, Debas H, Goto Y, Taylor I (1985) Peptide YY inhibits meal-stimulated pancreatic and gastric secretion. Am J Physiol 248:G118–G123

Patel YC, Wheatley T, Fitz-Patrick D, Brock G (1980) A sensitive radioimmunoassay for immunoreactive somatostatin in extracted plasma: Measurement and characterization of portal and peripheral plasma in the rat. Endocrinology 107:306–313

Pauwels S, Dockray G (1982) Identification of NH2-terminal fragments of big gastrin in plasma. Gastroenterology 82:56–61

Penman E, Wass JA, Lund A, Lowry PJ, Stewart J, Dawson AM, Besser GM, Rees LH (1979a) Development and validation of a specific radioimmunoassay for somatostatin in human plasma. Ann Clin Biochem 16(1):15–25

Penman E, Wass J, Lund A, Lowry P, Stewart J, Dawson A, Besser G, Rees L (1979b) Development and validation of a specific radioimmunoassay for somatostatin in human plasma. Ann Clin Biochem 16:15–25

Pimstone B, Berelowitz M, Krnaold D, Shapiro B, Kronheim S (1978) Somatostatin-like immunoreactivity in human and rat serum. Metabolism [Suppl 1] 27:1145–1149

Poitras P, Steinbach J, Van Deventer G, Code C, Walsh J (1980) Motilin independent ectopic fronts of the interdigestive myoelectric complex in dogs. Am J Physiol 239:G215–G220

Rayford P, Curtis P, Fender H, Thompson J (1976) Plasma levels of secretin in man and dogs: validation of a secretin radioimmunoassay. Surgery 79:658–665

Rehfeld J (1978) Immunochemical studies on cholecystokinin. J Biol Chem 253:4016–4021

Rehfeld J (1984) How to measure cholecystokinin in plasma? Gastroenterology 87:434–438

Rehfeld J, Morley J (1983) Residue-specific radioimmunoanalysis: a novel analytical tool. Application to the C-terminus of CCK/gastrin peptides. J Biochem Biophys Meth 7:161–170

Rehfeld J, Stadil F, Rubin B (1972) Production and evaluation of antibodies for the radioimmunoassay of gastrin. Scand J Clin Lab Invest 30:221–232

Rehfeld J, Stadil F, Vikelsoe J (1974) Immunoreactive gastrin components in human serum. Gut 15:102–111

Rehfeld J, De Magistris L, Andersen B (1981) Sulfation of gastrin: effect on immunoreactivity. Regulatory Peptides 2:333–342

Rosenquist G, Walsh J (1980) Radioimmunoassay of gastrin. In: Jerzy Glass GB (ed) Gastrointestinal hormones, vol 33. Raven, New York, pp 769–795

Ruggere M, Patel Y (1985) Hepatic metabolism of somatostatin-14 and somatostatin-28: Immunochemical characterization of the metabolic fragments and comparison of cleavage sites. Endocrinology 117:88–96

Said S, Faloona G (1975) Elevated plasma and tissue levels of vasoactive intestinal polypeptide in the watery-diarrhea syndrome due to pancreatic bronchogenic and other tumors. N Engl J Med 293:155–160

Sarson D, Bryant M, Bloom S (1980) A radioimmunoassay of gastric inhibitory polypeptide in human plasma. J Endocrinol 85:487–496

Schaffalitzky De Muckadell O, Fahrenkrug J (1977) Radioimmunoassay of secretin in plasma. Scand J Clin Lab Invest 37(2):155–162

Schaffalitzky De Muckadell O, Fahrenkrug J, Nielsen J, Westphall I, Worning H (1981) Meal-stimulated secretin release in man: effect of acid and bile. Scand J Gastroent 16:981–988

Schlegel W, Raptis S, Grube D, Pfeiffer E (1977) Estimation of cholecystokinin-pancreozymin (CCK) in human plasma and tissue by a specific radioimmunoassay and the immunohistochemical identification of pancreozymin-producing cells in the duodenum of humans. Clin Chim Acta 80:305–316

Schwartz T, Holst J, Fahrenkrug J, Lindkaer Jensen S, Nielsen O, Rehfeld J, Schaffalitzky De Muckadell O, Stadil F (1978) Vagal, cholinergic regulation of pancreatic polypeptide secretion. J Clin Invest 61:781–789

Seal A, Yamada T, Debas H, Hollinshead J, Osadchey B, Aponte G, Walsh J (1982) Somatostatin-14 and -28 clearance and potency on gastric function in dogs. Am J Physiol 243:G97–G102

Straus E, Yalow R (1978) Species specificity of cholecystokinin in gut and brain of several mammalian species. Proc Natl Acad Sci USA 75:486–489

Sugano K, Yamada T (1985) Progastrin-like immunoreactivity in porcine antrum: identification and characterization with region-specific antisera. Bio Biophys Res Commun 126:72–77

Tai H-H, Chey W (1978) Rapid extraction of secretin from plasma by XAD-2 resin and its application in the radioimmunoassay of secretin. Anal Biochem 87:376–385

Taylor I (1985) Peptide YY radioimmunoassay. Gastroenterology 88:1094–1095

Taylor I, Vaillant C (1983) Pancreatic polypeptide-like material in nerves and endocrine cells of the rat. Peptides 4:245–253

Taylor I, Feldman M, Richardson C, Walsh J (1978 a) Gastric and cephalic stimulation of human pancreatic polypeptide release. Gastroenterology 75:432–437

Taylor I, Impicciatore M, Carter D, Walsh J (1978 b) Effect of atropine and vagotomy on pancreatic polypeptide response to a meal in dogs. Am J Physiol 235(4):E443–E447

Taylor I, Dockray G, Calam J, Walker R (1979) Big and little gastrin responses to food in normal and ulcer subjects. Gut 20:957–962

Thompson J, Fender H, Ramus N, Villar H, Rayford P (1975) Cholecystokinin metabolism in man and dogs. Ann Surg 182:496–504

Vale W, Rivier J, Ling N, Brown M (1978) Biologic and immunologic activities and applications of somatostatin analogs. Metabolism 27:1391–1401

Vinik A, Shapiro B, Glaser B, Wagner L (1981) Circulating somatostatin in primates. In: Bloom SR, Polak JM (eds) Gut hormones. Churchill Livingstone, Edinburgh, pp 371–375

Walsh J (1983) Gastrointestinal peptide hormones. In: Sleisenger MH, Fordtran JS (eds) Gastrointestinal disease, 3rd edn. Saunders, Philadelphia, pp 54–96

Walsh J (1987) Gastrointestinal hormones. In: Johnson L, Christensen J, Jackson M, Jacobson E, Walsh JH (eds) Physiology of the gastrointestinal tract, 2nd edn. Raven, New York, pp 181–254

Walsh J, Grossman M (1975) Medical progress: gastrin. N Engl J Med 292:1324–1332

Walsh J, Isenberg J, Ansfield J, Maxwell V (1976) Clearance and acid-stimulating action of human big and little gastrins in duodenal ulcer subjects. J Clin Invest 57:1125–1131

Walsh J, Lamers C, Valenzuela J (1982) Cholecystokinin-octapeptidelike immunoreactivity in human plasma. Gastroenterology 82:438–444

Yanaihara N, Sato H, Kubota M, Sakagami M, Hashimoto T, Yanaihara C, Yamaguchi K, Zeze F, Abe K, Kaneko T (1976) Radioimmunoassay for secretin using N-tyrosyl-secretin and [Tyr1]-secretin. Endocrinol Jpn 23(1):87–90

Yoo O, Powell C, Agarwal K (1982) Molecular cloning and nucleotide sequence of full-length cDNA colding for porcine gastrin. Proc Natl Acad Sci USA 79:1049–1053

CHAPTER 14

Radioimmunoassay of Atrial Peptide Blood and Tissue Levels

M. L. MICHENER, D. SCHWARTZ, M. G. CURRIE, D. M. GELLER, and P. NEEDLEMAN

A. Introduction

The observation (DEBOLD et al. 1981) that mammalian atria contain a peptide that is both natriuretic and diuretic has led to the discovery of a novel endocrine hormone (atriopeptin, AP) which regulates salt and water metabolism from the kidney (NEEDLEMAN et al. 1985). Atrial extracts were found to be both heat and acid stable (DEBOLD et al. 1981), and to possess spasmolytic activity on vascular smooth muscle (CURRIE et al. 1983). These bioassays facilitated the purification and sequence analysis of three related peptides, termed AP I, II, and III (CURRIE et al. 1984a). Several peptides of different lengths, but sharing a common core sequence to the APs, have been identified by different laboratories. These include cardionatrin (FLYNN et a. 1983), atrial natriuretic factor (ANF) (THIBAULT et al. 1983), atrial natriuretic polypeptide (ANP) (KANGAWA and MATSUO 1984), and auriculin (ATLAS et al. 1984). cDNA probes were constructed from the peptide sequence and used to determine the sequence for the AP gene (YAMANAKA et al. 1984; MAKI et al. 1984; OIKAWA et al. 1984; SEIDMAN et al. 1984). The AP gene codes for a 152 amino acid preprohormone which contains a hydrophobic 24 amino acid leader sequence (Fig. 1). Atrial myocytes store AP as a 126 amino acid polypeptide prohormone (AP126) which lacks both the leader sequence and two

Met-Gly-Ser-Phe-Ser-Ile-Thr-Lys-Gly-Phe-Phe-Leu-Phe-Leu-Ala-Phe-Trp-Leu-Pro-Gly-
His-Ile-Gly-Ala | (1) Asn-Pro-Val-Tyr-Ser-Ala-Val-Ser-Asn-(10) Thr-Asp-Leu | Met-Asp-Phe-Lys- (AP13-126 →)
Asn-Leu-Leu-(20) Asp-His-Leu-Glu-Glu-Lys-Met-Pro-Val-Glu-(30) Asp-Glu-Val-Met | Pro-Pro-Gln- (AP34-126 →)
Ala-Leu-Ser-(40) Glu-Gln-Thr-Asp-Glu-Ala-Gly-Ala | Ala-Leu-(50) Ser-Ser-Leu-Ser-Glu-Val-Pro- (AP48-67)
Pro-Trp-Thr-(60) Gly-Glu-Val-Asn-Pro-Ser-Gln-Arg | Asp-Gly-(70) Gly-Ala-Leu-Gly-Arg-Gly-Pro-
Trp-Asp-Pro-(80) Ser-Asp-Arg-Ser-Ala-Leu-Leu-Lys-Ser-Lys-(90) Leu-Arg-Ala-Leu-Leu-Ala-Gly-
Pro-Arg | (100) Ser-Leu-Arg-Arg-Ser-Ser-Cys-Phe-Gly-Gly-Arg-(110) Ile-Asp-Arg-Ile-Gly-Ala-Gln- (AP28)
Ser-Gly-Leu-(120) Gly-Cys-Asn-Ser-Phe-Arg-Tyr | Arg-Arg

Fig. 1. Amino acid sequence for rait AP gene product. The NH_2 terminal 24 amino acids represent the hydrophobic leader sequence. AP126, the atrial storage form, extends from Asn-1 to Tyr-126. AP34–126 represents the cyanogen bromide fragment of the prohormone. AP48–67 is the NH_2 terminal fragment (NTF) peptide used to develop the NTF RIA. AP28 is the biologically active circulating form of the hormone

COOH terminal Arg residues (KANGAWA et al. 1984). AP126 is relatively inactive in vitro and is cleaved in vivo to yield the active circulating hormone, AP28 (SCHWARTZ et al. 1985). AP126 is the predominant form in atrial extracts, however, only low molecular weight APs are found in the plasma (SCHWARTZ et al. 1985; THIBAULT et al. 1985) or in the perfusate from isolated hearts (CURRIE et al. 1984b; LANG et al. 1985a). This suggests that prohormone cleavage occurs either during or immediately following secretion.

AP is a potent diuretic and natriuretic in rats (CURRIE 1984a), dogs (YUKIMURA et al. 1984; WAKITANI et al. 1985a, b), and humans (RICHARDS et al. 1985). The diuretic and natriuretic effects of AP are probably due to a combination of direct and indirect effects on the kidneys. AP increases renal blood flow (WAKITANI et al. 1985a, b; HINTZE et al. 1985; HUANG et al. 1985), glomerular filtration (HUANG et al. 1985), and may decrease tubular reabsorption of sodium and water (BURNETT et al. 1984). In addition, AP indirectly increases sodium and water excretion by inhibiting both aldosterone secretion (ATARASHI et al. 1984; CAMPBELL and NEEDLEMAN 1985; CHARTIER et al. 1984), and vasopressin secretion (SAMSON 1985).

B. Development of Atriopeptin Radioimmunoassay

Radioimmunoassays for AP have been developed by several laboratories (SCHWARTZ et al. 1985; GUTKOWSKA et al. 1984; NAKAO et al. 1984; ARENDT et al. 1985; YAMAJI et al. 1985). This has made it possible to follow changes in plasma AP levels and begin to understand the physiology and pathophysiology of this endocrine system. In the following sections, we will review in detail our efforts to develop AP RIAs and how we and other laboratories have employed these RIAs for studying APs.

I. Immunization

The antisera to AP that we employ were raised in both guinea pigs and rabbits by various fragments and conjugates of AP126. A high molecular weight fragment of AP126 was generated by cyanogen bromide cleavage of partially purified rat prohormone. This fragment, derived from residues 34–126 of the prohormone, contains the biologically active COOH terminal portion of the prohormone and is itself biologically active. Two low molecular weight peptides were also used as antigens and were covalently coupled to bovine thyroglobulin. AP24, the COOH terminal, biologically active portion of the prohormone was synthesized and used for production of antisera that would recognize both the circulating form of the peptide and the storage form. A 20 amino acid peptide whose sequence corresponds to an NH_2 terminal fragment (NTF) of the prohormone (amino acids 48–67), was also synthesized and coupled to thyroglobulin. Antisera raised against this peptide were expected to recognize both the intact prohormone and NTFs of the prohormone, but not the COOH terminal low molecular weight APs. Each antigen was mixed with complete Freund's adjuvant and injected subcutaneously into the backs of guinea pigs and rabbits. Approximately 500 μg antigen was used in the initial injections followed by boosts of 100 μg in incomplete Freund's adjuvant 6 weeks later. Boosts were repeated in either 2- or 3-week cycles. Animals were then bled and titered for binding to either AP24 ^{125}I or *N*-Tyr-NTF48–67 ^{125}I.

II. Iodination of Peptides

AP24 and *N*-Tyr-NTF48–67 were iodinated using a modification of the chloramine-T procedure. Peptides (6 μg) were combined with chloramine-T (4 μg) and 1 mCi ^{125}I at pH 7.4 from 1 to 5 min at room temperature. To quench the reaction, reactants were separated by reverse-phase HPLC on a Bondapak C_{18} reverse-phase column using a linear gradient of acetonitrile in the presence of 0.05% trifluoroacetic acid. Quenching with metabisulfite resulted in ineffective radioligands, presumably owing to reductions of the cystine disulfide linkage of the AP. This HPLC system resolved both the monoiodo and diiodo peptides and free iodine from the unlabeled peptide. HPLC fractions containing the iodinated peptides were mixed 1:1 with 50 m*M* phosphate buffer (pH 7.4) that contained 0.25% BSA and stored at −20 °C.

III. Titering and Sensitivity

1. Atriopeptin RIA

Suitable antiserum for RIA of AP24 was obtained from a rabbit that had been immunized with the cyanogen bromide fragment, AP34–126. This antiserum was used at a final dilution of 1:6000 in 300 μl containing 10000 cpm AP24 ^{125}I, 20 μl goat anti-rabbit IgG (final dilution 1:750), with a buffer of 100 m*M* sodium phosphate (pH 7.4), 0.25% BSA, 10 m*M* sodium azide, and 3% polyethylene glycol. Assays were incubated overnight at 4 °C, diluted with 2 ml 0.25% BSA, centrifuged at 1500 *g* for 30 min, and the resulting pellet counted in a γ-counter. AP24 produces a half-maximal displacement of bound AP24 ^{125}I with 30 fmol (77 pg) (Schwartz et al. 1985; Manning et al. 1985). AP34–126, the cyanogen bromide fragment, and AP13–126, a biosynthetic prohormone construct, cross-reacted with this antiserum 100% and 40%, respectively. Both of these polypeptides contain intact COOH termini. The 20 amino acid peptide, NTF48–67, whose sequence does not overlap with AP24, did not cross-react. Atrial extracts produce parallel displacement curves to that of AP24. Unrelated peptides, such as vasopressin, angiotensin, and ACTH, do not cross-react with this antiserum.

2. NH_2 Terminal Fragment RIA

Sera from rabbits and guinea pigs immunized against the NTF peptide were analyzed at different dilutions following overnight incubation at 4 °C in 100 m*M* phosphate buffer (pH 6.8), 0.3% bovine γ-globulin, 10 m*M* EDTA, 0.05% sodium azide, and 10000 cpm *N*-Tyr-NTF ^{125}I. Antibody complexes were precipitated by addition of 1 ml 16% polyethylene glycol. These bleeds were compared for sensitivity in RIAs using the synthetic NTF peptide to displace NTF48–67 ^{125}I. GP71 antisera (final dilution 1:25000) displayed the greatest sensitivity, showing a half-maximal displacement with 60 fmol NTF (130 pg) per tube. AP34–126 and AP13–126, both of which contain the sequence of the NTF peptide, cross-reacted with GP71 with similar affinity to the NTF peptide. AT24, which does not share any sequence homology with the NTF peptide, did not cross-react in this assay.

C. Assay of Tissues and Plasma

I. Atrial Extracts

Acid extracts of rat atria yield microgram quantities of biologically active AP prohormone (CURRIE et al. 1983, 1984c; KANGAWA and MATSUO 1984). High yields are obtained when fresh atria are first quick-frozen and powdered under liquid nitrogen. The powder is heated to 100 °C for 10 min in 1 *M* acetic acid, homogenized at 0 °C, and centrifuged at 28000 *g* for 20 min. Purification of these extracts by solid-phase C_{18} (Baker ODS) extraction followed by reverse-phase HPLC on C_{18} columns demonstrates that the majority of the biologic activity and immunoreactivity reside in a single high molecular weight peak (MICHENER et al. 1986). This peak reacts equally well with either the AP RIA or the NTF RIA. Sequence analysis confirms that it represents the intact prohormone, AP126.

II. Plasma Immunoreactivity

1. Stimulation of Release by Pressor Agents

To detect AP immunoreactivity (APir) in rat plasma, chloral hydrate-anesthetized Sprague-Dawley rats were prepared with both jugular and carotid catheters. Blood was collected in 110 m*M* sodium citrate, the plasma separated and frozen at −70 °C until assayed. Basal plasma APir was found to be between 0.5 and 0.8 ng/ml (MANNING et al. 1985). Plasma produced displacement of AP24 ^{125}I parallel to that of AP24 in the AP RIA. In order to characterize the nature of the immunoreactivity properly and identify the circulating form of AP, sufficient amounts of plasma APir had to be obtained with which to obtain an amino acid sequence. We observed that arginine vasopressin and in particular the pressor analog of vasopressin, desamino-arginine vasopressin, caused a simultaneous and dose-dependent increase in mean arterial blood pressure and plasma APir (MANNING et al. 1985). Plasma from rats treated with desamino-arginine vasopressin also produced parallel displacement curves to AP24 when assayed in the AP RIA, but were shifted to the left of the displacement curves for control plasma.

2. Identification of Circulating Atriopeptin

The ability of pharmacologic doses of desamino-arginine vasopressin to stimulate AP release in vivo enabled us to collect enough plasma AP to characterize the circulating form of the peptide chemically. Plasma from rats treated with desamino-arginine vasopressin were collected in 0.44% EDTA, partially purified on octadecylsilane resins (Baker ODS), and eluted with 90% acetonitrile/0.1% trifluoroacetic acid. Eluates were lyophilized, reconstituted in 10% acetonitrile/1% trifluoroacetic acid, and separated by reverse-phase HPLC on a C_{18} column. A low molecular weight peak of APir, which was increased in plasma stimulated with desamino-arginine vasopressin, migrated in the vicinity of the AP24 and AP28 standards. When further separated by a shallower gradient on the HPLC, a large and a small peak of low molecular weight APir were revealed. Sequence analysis of these peaks identified the major peak as AP28 and the minor peak as AP24. Since the AP24 peak was about one-tenth the mass of the AP28 peak, we concluded that the major circulating form of AP in rats is AP28 and that AP24 represents a degradation product of AP28 (SCHWARTZ et al. 1985). The identity of the circulating form has now been confirmed by another laboratory (THIBAULT et al. 1985).

3. Mechanism of Atriopeptin Release

a) Right Atrial Stretch

The mechanism by which desamino-arginine vasopressin or other pressor agents caused release of AP in vivo was not clear. Other investigators have reported that distension of the atria, either by balloon inflation (LANG et al. 1985a) or by acute volume expansion (DIETZ 1984), stimulates AP release. We examined the relationship between both left and right atrial pressure and plasma levels of APir following administration of pressor agents (KATSUBE et al. 1985). Chloral hydrate-anesthetized rats were cannulated to record mean arteriolar blood pressure, right atrial pressure, and left ventricular end-diastolic pressure (an indication of left atrial pressure). Blood samples were collected in sodium citrate and the plasma assayed for APir. Rats were given intravenous bolus injections of either 1 μg arginine vasopressin, 1 μg desamino-arginine vasopressin, 30 μg phenylephrine, or 10 μg angiotensin II at time zero. Administration of both desamino-arginine vasopressin and arginine vasopressin resulted in pronounced increases in mean arterial blood pressure, right atrial pressure, and left ventricular end-diastolic pressure. The increases in plasma APir paralleled the observed pressure changes. Both phenylephrine and angiotensin II caused equal peak increases in mean arterial pressure, right atrial pressure, and left ventricular end-diastolic pressure, but these were more transient than those observed for the vasopressin analogs and the increases in plasma APir were smaller.

The differences in plasma APir responses could have been due either to an intrinsic difference in the mechanism (or mechanisms) of release between these agents or to the differences of the durations of the elevated atrial pressures. To address these possibilities, continuous infusions of these pressor agents were administered to rats to attain prolonged changes in pressure. Both desamino-arginine vasopressin (0.03 μg/min) and phenylephrine (10 μg/min) caused marked and sustained increases in mean arterial blood pressure, right atrial pressure, and plasma APir. In contrast, angiotensin II (3 μg/min) caused sustained elevations in mean arterial pressure and left ventricular end-diastolic pressure, but caused only transient elevations in right atrial pressure and plasma APir. These results suggested that pressor-induced release of AP was most dependent on right atrial pressure.

Plasma APir and right atrial pressure were examined under conditions of head-out water immersion (KATSUBE et al. 1985). Water immersion results in central hypervolemia, a rise in central venous pressure, atrial distension, and a natriuresis such as that seen following volume expansion. Chloral hydrate-anesthetized rats were immersed to the neck in a vertical position in 37 °C water. Control animals were placed in a similar position, but not immersed. Water immersion produced sustained elevations in both right atrial pressure and plasma APir, which returned to control values upon removal of the water (neither mean arterial pressure nor left ventricular end-diastolic pressure was significantly changed during immersion). Bilateral vagotomy, performed 10 min prior to immersion, had no effect on either right atrial pressure or on the accompanying increase in plasma APir.

b) α-Adrenergic Stimulation

Isolated retrogradely perfused rat hearts were used to characterize the regulation of AP release by adrenergic agents (CURRIE and NEUMAN 1986). In the isolated rat heart, values of APir release are elevated immediately after the initiation of perfusion and decline during the first 60 min. Release of AP was generally constant for the next 120 min of perfusion. The physiologic state of the heart was routinely assessed by measuring heart rate, left ventricular dP/dt, and perfusion pressure. Hearts that exhibited signs of failure tended to have spurious levels of AP release and were excluded from the study.

Constant infusion of norepinephrine (1 μM) and epinephrine (1 μM) caused a time-dependent increase of APir release as well as increases in heart rate, left ventricular dP/dt, and perfusion pressure. Isoproterenol (1 μM) did not affect APir release in spite of causing equivalent changes in heart rate and left ventricular dP/dt. Phenylephrine (a nonselective α-adrenergic agonist), but not BHT-920 (a selective α_2-adrenergic agonist) produced a concentration-dependent stimulation of APir release. The mixed α_1- and α_2-antagonist, phentolamine, antagonized the AP-stimulating effects of norepinephrine. These findings suggest that the adrenergic nervous system may play a role in the regulation of AP secretion from atrial myocytes through activation of an α_1-adrenergic receptor-mediated mechanism. How α_1-adrenergic stimulation is related to the ability of direct atrial stretch to stimulate AP release is unknown.

4. Human Plasma Immunoreactivity

Several radioimmunoassays have been developed for the detection of human AP (NAKAO et al. 1984; YAMAJI et al. 1985; ARENDT et al. 1985). Sequence analysis of the human AP gene revealed a high degree of homology between the rat and the human genes (SEIDMAN et al. 1984; OIKAWA et al. 1984; NAKAYAMA et al. 1984; NEMER et al. 1984; GREENBERG et al. 1984). The COOH terminal region of human AP prohormone is virtually identical to that of the rat with the exception of a methionine substitution for isoleucine at amino acid 110. Extracts of human plasma, chromatographed by both gel filtration and HPLC, have been found to contain a major circulating AP species which comigrated with the human AP28 (YAMAJI et al. 1985; MIYATA et al. 1985), analogous to the circulating species found in rats

(Schwartz et al. 1985). The actual sequence of this peptide has not been determined. The only other notable difference between rat AP and human AP was the discovery of a 56 amino acid antiparallel dimer of the 28 amino acid COOH terminal peptide (Kangawa et al. 1985) in small amounts in both atrial extracts and plasma. The significance of this dimer is unknown.

RIAs for human AP have been used to screen plasma from patients suffering from a variety of cardiovascular diseases which are characterized by alterations in fluid and electrolyte homeostasis. Basal human plasma APir levels have been reported to be 50 pg/ml or less and are elevated 2- to 10-fold in patients with various degrees of congestive heart failure (Shenker et al. 1985; Nakaoka et al. 1985; Hartter et al. 1985; Tikkanen et al. 1985 a), valvular and ischemic heart disease (Sugawara et al. 1985), and in pediatric patients with congenital heart defects (Lang et al. 1985 b) and renal failure (Rascher et al. 1985). The significance of altered AP levels in such patients, either as primary pathophysiologic events or compensatory adjustments, is not clear. The observed increases in plasma APir during congestive heart failure are probably not sufficient to maintain appropriate diuresis since congestive heart failure is characterized by frank fluid retention. A causal relationship between elevated levels of plasma APir and polyuria is postulated in patients diagnosed as having paroxysmal atrial tachycardia (PAT) (Schriffrin et al. 1985; Tikkanen et al. 1985 b), a condition which is often associated with excessive diuresis. The levels of plasma APir reported for patients with PAT-associated polyuria were consistent with the levels required for diuresis in response to AP infusion (Tikkanen et al. 1985 a).

III. Brain Atriopeptin

AP-containing neurons and fiber tracks habe been identified by immunohistochemical staining using anti-AP antibodies (Saper et al. 1985; McKenzie et al. 1985; Kawata et al. 1985). These neurons and fiber tracks appear in hypothalamic nuclei which have already been identified as being regulatory centers for cardiovascular control and fluid volume regulation. Fiber tracks have been seen in the periventricular preoptic nucleus, ventral and lateral stria terminalis, lateral hypothalamic area, paraventricular and arcuate nucleus. Cell bodies were found in the anteroventral periventricular preoptic nucleus, medial preoptic area, stria terminalis, ventral pallidum, parvocellular part of the paraventricular nucleus, and the periventricular nucleus. Extracts from these regions contain APir, with the hypothalamus showing the highest level (5 pmol/g) (Tanaka et al. 1984). HPLC separation of brain extracts reveals that the majority of the APir exists as low molecular weight peptide with some high molecular weight APir present (Glembotski et al. 1985). This differs from atrial extracts, where the predominant form of APir is the high molecular weight AP126. The significance of brain AP is unknown, although its localization leads one to suspect that it may participate in central cardiovascular and fluid volume control.

D. Processing of the Atriopeptin Prohormone

Proteolytic activation of AP126 was examined using the two selective RIAs for the prohormone. We expected that the 98 amino acid peptide that remains from AP126 after removal of AP28 would be coreleased from the atria. Rat plasma was examined for the presence of NTF immunoreactivity (NTFir) and increases in plasma NTFir 3 min after intravenous administration of 3 μg desamino-arginine vasopressin (Fig. 2). Basal plasma APir and NTFir levels were found to be 0.42 ± 0.05 ng/ml and 3.5 ± 0.4 ng/ml, respectively. Both APir and NTFir were elevated following desamino-arginine vasopressin administration, to 8.8 ± 2.6 ng/ml and 9.6 ± 1.4 ng/ml. Increases in plasma APir following desamino-arginine vasopressin administration parallels the transient increase in blood pressure, returning to baseline after 60 min, whereas plasma NTFir remains elevated for up to 2 h (MICHENER et al. 1986). The greater immunoreactive half-life of the circulating NTFir is also reflected by a basal plasma NTFir that is 5–8 times higher than the basal APir.

Differences in APir and NTFir half-lives suggested that the two RIAs were detecting separate metabolites of the same precursor. Plasma was collected from rats treated with desamino-arginine vasopressin, extracted on octadecylsilane resins, reconstituted, subjected to reverse-phase HPLC, and assayed for both APir and NTFir (MICHENER et al. 1986). APir was detected in a single low molecular weight peak that migrated in the vicinity of AP24 and AP28. No high molecular weight APir was detected. NTFir was detected as a series of high molecular weight peaks, the majority of which did not comigrate with AP126. These findings were consistent with the direct assays of plasma APir and NTFir in that distinct APir and NTFir species were identified.

Previous reports by this laboratory and by others have demonstrated that the predominant form of AP released from the isolated perfused heart is the low molecular weight AP (CURRIE et al. 1984b; LANG et al. 1985a). The immunoreactive HPLC profiles for both APir and NTFir from isolated rat heart perfusate were

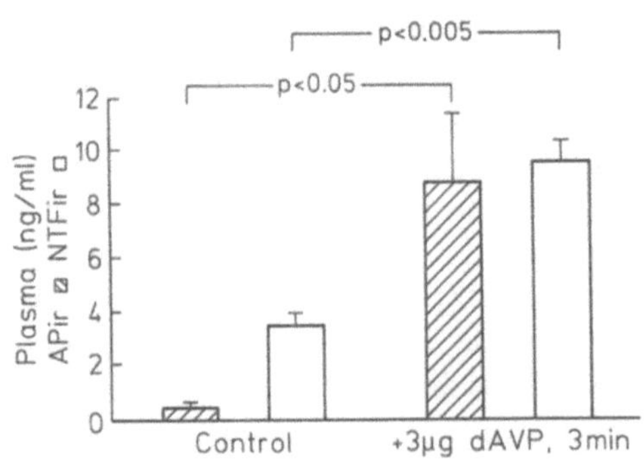

Fig. 2. Rat plasma APir and NTFir before and after intravenous bolus administration of 3 μg desamino-arginine vasopressin (dAVP). Male Sprague-Dawley rats were anesthetized with chloral hydrate, both jugular and carotid cannulae were inserted, and the animals equilibrated for 30 min. Then, 3 μg desamino-arginine vasopressin in saline was injected at time zero. Blood was collected in 110 mM sodium citrate (final citrate concentration 11 mM), centrifuged immediately, and the plasma stored at −70 °C until assayed

virtually identical to that seen for rat plasma (MICHENER et al. 1986). NTFir was found in multiple high molecular weight peaks which did not contain detectable APir. APir was detected only in low molecular weight peaks. This suggested that the isolated heart has the necessary enzymatic activity to process the prohormone fully. Whether this processing occurs before the peptide is released from the myocyte, or soon thereafter, is not clear.

E. Summary

AP has now been established as an important hormone for the regulation of vascular fluid volume and blood pressure. AP release in response to atrial stretch provides an ideal means of responding to changes in vascular volume. This response is composed of actions on the kidneys and vascular smooth muscle, and an integrated endocrine response through the inhibition of aldosterone and vasopressin secretion. The detection of AP in the cardiovascular control centers of the brain by both RIA and immunohistochemical staining suggests that AP may also play a role in the central regulation of the cardiovascular system. The development of RIAs for AP make it possible to measure stored and secreted AP and to begin to understand how changes in AP levels relate to physiologic and pathophysiologic conditions such as high blood pressure and congestive heart failure.

References

Arendt RM, Stangl E, Zahringer J, Liebsch DC, Herz A (1985) Demonstration and characterization of alpha-human atrial natriuretic factor in human plasma. FEBS Lett 189:57–61

Atarashi K, Mulrow PJ, Franco-Saenz R, Snajdar R, Rapp J (1984) Inhibition of aldosterone production by an atrial extract. Science 224:992–994

Atlas SA, Kleinert HD, Camargo MJ, Januszewicz A, Sealey JE, Laragh JH, Schilling JW, Lewicki JA, Johnson LK, Maack T (1984) Purification, sequencing and synthesis of natriuretic and vasoactive rat atrial peptide. Nature 309:717–719

Burnett JC Jr, Granger JP, Oppenorth TJ (1984) Effects of synthetic atrial natriuretic factor on renal function and renin release. Am J Physiol 247:F863–F866

Campbell WB, Needleman P (1985) Inhibition of aldosterone biosynthesis by atriopeptins in rat adrenal cells. Circ Res 57:113–118

Chartier L, Schriffrin E, Thibault G (1984) Effect of atrial natriuretic factor (ANF)-related peptides on aldosterone secretion by adrenal glomerulosa cells: critical role of the intramolecular disulphide bond. Biochem Biophys Res Commun 122:171–174

Currie MG, Neuman W (1986) Evidence for alpha-1 adrenergic receptor regulation of atriopeptin release from the isolated rat-heart. Biochem Biophys Res Commun 137:94–100

Currie MG, Geller DM, Cole BR, Boylan JG, YuSheng W, Holmberg SW, Needleman P (1983) Bioactive cardiac substances: potent vasorelaxant activity in mammalian atria. Science 221:71–73

Currie MG, Geller DM, Cole BR, Siegel NR, Fok KF, Adams SP, Eubanks SR, Galluppi GR, Needleman P (1984a) Purification and sequence analysis of bioactive atrial peptides (atriopeptins). Science 223:67–69

Currie MG, Sukin D, Geller DM, Cole BR, Needleman P (1984b) Atriopeptin release from the isolated perfused rabbit heart. Biochem Biophys Res Commun 124:711–717

Currie MG, Geller DM, Cole BR, Needleman P (1984c) Proteolytic activation of a bioactive cardiac peptide by in vitro trypsin cleavage. Proc Natl Acad Sci USA 81:1230–1233

DeBold AJ, Borenstein HB, Veress AT, Sonnenberg H (1981) A rapid and potent natriuretic response to intravenous injection of atrial myocardial extracts in rats. Life Sci 28:89–94

Dietz JR (1984) Release of natriuretic factor from heart lung preparation by atrial distension. Am J Physiol 247:R1093–R1096

Flynn TG, deBold ML, deBold AJ (1983) The amino acid sequence of an atrial peptide with potent diuretic and natriuretic properties. Biochem Biophys Res Commun 117:859–865

Glembotski CG, Wildey GM, Gibson TR (1985) Molecular forms of immunoactive atrial natriuretic peptide in the rat hypothalamus and atrium. Biochem Biophys Res Commun 129:671–678

Greenberg BD, Bencen GH, Seilhamer JJ, Lewicki JA, Fiddes JC (1984) Nucleotide sequence of the gene encoding human atrial natriuretic factor precursor. Nature 312:656–658

Gutkowska J, Thibault G, Januszewicz P, Cantin M, Genest J (1984) Direct radioimmunoassay of atrial natriuretic factor. Biochem Biophys Res Commun 122:593–601

Hartter E, Weissel M, Stummvoll HK, Woloszczuk W, Punzengruber C, Ludvik B (1985) Atrial natriuretic peptide concentrations in blood from right atrium in patient with severe right heart failure. Lancet II:93–94

Hintze TH, Currie MG, Needleman P (1985) Atriopeptins – renal-specific vasodilators in conscious dogs. Am J Physiol 248:H587–H591

Huang C-L, Lewicki J, Johnson LK, Cogan MG (1985) Renal mechanism of action of rat atrial natriuretic factor. J Clin Invest 75:769–773

Kangawa K, Matsuo H (1984) Purification and complete amino acid sequence of alpha-human atrial natriuretic polypeptide (alpha-hANP). Biochem Biophys Res Commun 118:131–139

Kangawa K, Tawaragi Y, Oikawa S, Mizuno A, Sakuragawa Y, Nakazato H, Fukuda A, Minamino N, Matsuo H (1984) Identification of rat gamma-atrial natriuretic polypeptide and characterization of the c-DNA encoding its precursor. Nature 312:152–155

Kangawa K, Fukuda A, Matsuo H (1985) Structural identification of beta- and gamma-human atrial natriuretic polypeptide. Nature 313:397–400

Katsube N, Schwartz D, Needleman P (1985) Release of atriopeptin in the rat by vasoconstrictors or water immersion correlates with changes in right atrial pressure. Biochem Biophys Res Commun 133:937–944

Kawata M, Nakao K, Morii N, Kiso Y, Yamashita H, Imura H, Sano Y (1985) Atrial natriuretic polypeptide: topographical distribution in the rat brain by radioimmunoassay and immunohistochemistry. Neuroscience 3:521–546

Lang RE, Tholken H, Ganten D, Luft FC, Rushoaho H, Unger T (1985a) Atrial natriuretic factor – a circulating hormone stimulated by volume loading. Nature 314:264–266

Lang RE, Unger T, Ganten D, Weil J, Bidlingmaier F, Dohlemann D (1985b) Alpha-atrial natriuretic peptide concentrations in plasma of children with congenital heart and pulmonary diseases. Br Med J 291:1241

Maki M, Takayanagi R, Misono KS, Pandey KN, Tibbetts C, Inagami T (1984) Structure of rat atrial natriuretic factor precursor deduced from cDNA sequence. Nature 309:722–724

Manning PT, Schwartz D, Katsube NC, Holmberg SW, Needleman P (1985) Vasopressin stimulated atriopeptin release: endocrine antagonists in fluid homeostasis. Science 229:395–397

McKenzie JC, Tanaka I, Misono KS, Inagami T (1985) Immunocytochemical localization of atrial natriuretic factor in the kidney, adrenal medulla, pituitary, and atrium of rat. J Histochem Cytochem 33:828–832

Michener ML, Gierse JK, Seetharam R, Fok KF, Olins PO, Mai MS, Needleman P (1986) Proteolytic processing of atriopeptin prohormone in vivo and in vitro (to be published)

Miyata A, Kangawa K, Toshimori T, Hatoh T, Matsuo H (1985) Molecular forms of atrial natriuretic polypeptides in mammalian tissues and plasma. Biochem Biophys Res Commun 129:248–255

Nakao K, Sugawara A, Morii N, Sakamoro M, Suda M, Soneda J, Ban T, Kihara M, Yamori Y, Shimokura M, Kiso Y, Imura H (1984) Radioimmunoassay for alpha-human and rat atrial natriuretic polypeptide. Biochem Biophys Res Commun 124:815–821
Nakaoka H, Imataka K, Amano M, Fuji J, Ishibashi M, Yamaji T (1985) Plasma levels of atrial natriuretic factor in patients with congestive heart failure. N Engl J Med 313:892–893
Nakayama K, Ohkubo H, Hirose T, Inayama S, Nakanishi S (1984) mRNA sequence for human cardiodilantin – atrial natriuretic factor precursor and regulation of precursor mRNA in rat atria. Nature 310:699–701
Needleman P, Adams SP, Cole BR, Currie MG, Geller DM, Michener ML, Saper CB, Schwartz D, Standaert DG (1985) Atriopeptins as cardiac hormones. Hypertension 7:469–482
Nemer M, Chamberland M, Sirois D, Argentin S, Drouin J, Dixon RA, Sivin RA, Condra JH (1984) Gene structure of human cardiac hormone precursor, pronatriodilatin. Nature 312:654–656
Oikawa S, Imai M, Ueno A, Tanaka S, Noguchi T, Nakazato H, Kangawa K, Fukuda A, Matsuo H (1984) Cloning and sequence analysis of cDNA encoding a precursor for human atrial natriuretic polypeptide. Nature 309:724–726
Rascher W, Tulassay T, Lang RE (1985) Atrial natriuretic peptide in plasma of volume-overloaded children with chronic renal failure. Lancet II:303–305
Richards AM, Ikram H, Yandle TG, Nicholls MG, Webster MWI, Espiner EA (1985) Renal haemodynamic, and hormonal effects of human alpha atrial natriuretic peptide in healthy volunteers. Lancet I:545–548
Samson WK (1985) Atrial natriuretic factor inhibits dehydration and hemorrhage-induced vasopressin release. Neuroendocrinology 40:277–279
Saper CB, Standaert DG, Currie MG, Schwartz D, Geller DM, Needleman P (1985) Atriopeptin-immunoreactive neurons in the brain: presence in cardiovascular regulatory areas. Science 227:1047–1049
Schriffrin EL, Gutkowska J, Kuchel O, Cantin M, Genest J (1985) Plasma concentration of atrial natriuretic factor in a patient with paroxysmal atrial tachycardia. N Engl J Med 312:1196–1197
Schwartz D, Geller DM, Manning PT, Siegel NR, Fok KF, Smith CE, Needleman P (1985) Ser-Leu-Arg-Arg-Atriopeptin. III. The major circulating form of atrial peptide. Science 229:397–400
Seidman CE, Duby AD, Choi E, Graham RM, Haber E, Homcy C, Smith JA, Seidman JG (1984) The structure of rat preproatrial natriuretic factor as defined by a complementary DNA clone. Science 225:324–326
Shenker Y, Sider RS, Ostafin EA, Grekin RJ (1985) Plasma levels of immunoreactive atrial natriuretic factor in healthy subjects and in patients with edema. J Clin Invest 76:1684–1687
Sugawara A, Nakao K, Morii N, Sakamoto M, Suda M, Shimokura M, Kiso Y, Kihara M, Yamori Y, Nishimura K, Soneda J, Toshihiko B, Imura H (1985) Alpha-human atrial natriuretic polypeptide is released from the heart and circulates in the body. Biochem Biophys Res Commun 129:439–446
Tanaka I, Misono KS, Inagami T (1984) Atrial natriuretic factor in rat hypothalamus, atria and plasma: determination by specific radioimmunoassay. Biochem Biophys Res Commun 124:663–668
Thibault G, Garcia R, Seidah NG, Lazure C, Cantin M, Chretien M, Genest J (1983) Purification of three rat atrial natriuretic factors and their amino acid composition. FEBS Lett 164:286–290
Thibault G, Lazure C, Schriffrin J, Gutkowska J, Chartier L, Garcia R, Seidah NG, Chretien M, Genest J, Cantin M (1985) Identification of a biologically active circulating form of rat atrial natriuretic factor. Biochem Biophys Res Commun 130:981–986
Tikkanen I, Metsarinne K, Fyhrquist F, Leidenius R (1985 a) Plasma atrial natriuretic peptide in cardiac disease and during infusion in healthy volunteers. Lancet II:66–69
Tikkanen I, Metsarinne K, Fyhrquist F (1985 b) Atrial natriuretic peptide in paroxysmal supraventricular tachycardia. Lancet II:40–41

Wakitani K, Cole BR, Geller DM, Currie MG, Adams SP, Fok KF, Needleman P (1985a) Atriopeptins – correlation between renal vasodilation and natriuresis. Am J Physiol 249:F49–F53

Wakitani K, Oshima T, Loewy AP, Holmberg SW, Cole BR, Adams SP, Fok KF, Currie MG, Needleman P (1985b) Comparative vascular pharmacology of the atriopeptins. Circ Res 56:621–627

Yamaji T, Ishibashi M, Takaku F (1985) Atrial natriuretic factor in human blood. J Clin Invest 76:1705–1709

Yamanaka M, Greenberg B, Johnson L, Seilhamer J, Brewer M, Friedemann T, Miller J, Atlas S, Laragh J, Lewicki J, Fiddes J (1984) Cloning and sequence analysis of the cDNA for the rat atrial natriuretic factor precursor. Nature 309:719–722

Yukimura T, Ito Y, Takenaga T, Yamamoto K, Kangawa K, Matsuo H (1984) Renal effects of a synthetic alpha-human atrial natriuretic polypeptide (alpha-hANP) in anesthetized dogs. Eur J Pharmacol 103:363–366

CHAPTER 15

Immunochemical Methods for Adrenal and Gonadal Steroids

M. PAZZAGLI and M. SERIO

A. Chemistry and Nomenclature

All adrenocortical and gonadal hormones are steroids. The term steroid is used to designate those compounds containing a four-ring structure, the cyclopentanoperhydrophenanthrene nucleus (Fig. 1). It includes many naturally occurring higher alcohols called sterols, such as cholesterol, ergosterol, the glycosides of digitalis, and many other substances without alcoholic hydroxyl groups, such as estrogens and progesterone.

The terminology for steroids is confusing because of the use of trivial names, chemical names, and different terms for the same substance. Much of this confusion arose in the early days of steroid chemistry. For instance the most abundantly secreted steroid from the adrenal cortex is commonly known as cortisol.

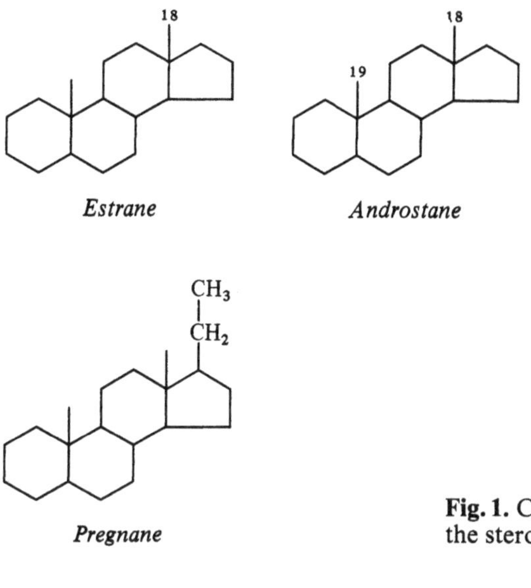

Fig. 1. Cyclic hydrocarbon structures related to the steroid hormones

Its chemical name is 11β-17,21-trihydroxypregn-4-ene-3, 20-dione. As it is a 17-hydroxy derivative of corticosterone, it has also been called 17-hydroxycorticosterone; likewise, its trivial name is hydrocortisone, implying it is a hydrogenated derivative of cortisone. However, the essentials of terminology are relatively simple.

The basic steroid nucleus, the cyclopentanoperhydrophenanthrene ring, contains 17 carbon atoms (Fig. 1). Each major steroid-secreting endocrine organ of the body produces a single major parent compound containing one, two, or four additional carbon atoms to this basic steroid nucleus. It has therefore become steroid shorthand to call those steroids secreted by the ovaries C_{18} steroids (estrane), those secreted by the testis C_{19} steroids (androstane), and those secreted by the adrenal cortex C_{21} (pregnane) (Fig. 1). The adrenal cortex secretes C_{19} steroids as a minor product, too.

All steroids derivatives from each of the three glands retain the parent compound name, modified by prefixes and suffixes designating the type and site of the alteration to the parent molecule. Only a limited number of alterations can be made. By convention, only one suffix may be used. The site of a double bond was formerly designated by placing the lower number of the carbon atom adjacent to the double bond symbol, e.g., Δ_4 indicates a double bond between carbon atoms 4 and 5. Recently, the Δ symbol has been abandoned in favor of placing the number in the parent compound before the designation of the bond, e.g., ene = one double bond, diene = two double bonds, etc. The position of a substituent group on the nucleus is indicated by the number of the carbon atom to which it is attached and its spatial arrangement below or above the plane of the nucleus may be designated by α or β. Hydroxyl groups are given as prefixes unless they are the only substituent, in which case they are designated as suffixes. To illustrate, the principal human adrenal corticosteroid has a trivial name cortisol or hydrocortisone, a C_{21} steroid. The old chemical name for this was 11β-17,21-trihyroxypregnene-3,20-dione. In the new terminology it is 11β-17,21-trihydroxypregn-4-ene-3,20-dione. This designates the three hydroxyl groups on the 11, 17, and 21 positions; the two carbonyl groups at the 3 and 20 positions; and one double bond between the fourth and fifth carbon atom. The most significant hormone secreted by the testis is a C_{19} steroid with the trivial name of testosterone, and is chemically designated androst-4-ene-17β-ol-3-one. The most significant secretion from the ovary is a C_{18} steroid, commonly called estradiol, which has the chemical name estra-1,3,5-triene-3,17β-diol. The systematic nomenclature of steroids is unnecessarily cumbersome for general use and trivial names of the common steroids are used instead.

B. Immunoassay Methods for Steroids

I. Basic Principles

Immunoassay is a term used to include the larger number of methods which are based on the specific binding of one reactant (the ligand) by an antibody. The ratio of the amount of ligand present in the bound and free fractions is usually determined by adding a small amount of labeled ligand as a tracer for the unlabeled ligand.

The development of a steroid immunochemical method requires the production of suitable antisera. Steroids by themselves are nonimmunogenic, and they must first be coupled covalently to protein carriers (ERLANGER et al. 1959; LIEBERMANN et al. 1959). Despite this early work, the first steroid radioimmunoassay was not published until 1969 (ABRAHAM 1969). Since then, however, radioimmunoassays have been developed for all major biologically active steroids and for many synthetic steroids (ABRAHAM et al. 1977; ABRAHAM 1974; CAMERON et al. 1975).

II. Synthesis of the Immunogen Derivative

The first step is to prepare an active derivative of the steroid that can be coupled to protein (ERLANGER 1981). Thus, the introduction of a free carboxyl group enables its coupling to the ε-amino group of lysine residues in the protein molecule. The technique used to prepare the steroid derivative depends on whether the steroid contains hydroxyl or keto groups. Hydroxyl groups can be esterified with succinic anhydride (Fig. 2). If the steroid has more than one such group, their different reactivities may still allow selective esterification. Keto groups can be reacted with *O*-(carboxymethyl)hydroxylamine hemihydrochloride to form oxime derivatives (Fig. 3). The major biologically active corticosteroids have two keto groups, and the production of a specific antiserum requires the selective formation of a monooxime derivative. The 3-oxime derivative can be produced under very mild alkaline conditions, as unsaturated ketones are more reactive than saturated ketones. However, this usually produces a mixture of mono- and dioximes, and a more satisfactory technique, in which a 3-enamine derivative is prepared, has been suggested by JANOSKI et al. (1974). Steroid glucuronides, however, can be coupled directly to proteins without further derivative formation (KELLIE 1975).

The specificity of the antibodies produced depends, to a large extent, on the site of conjugation (ABRAHAM 1975). In most biologically active steroids, the reactive groups are at the 3 and 17 positions. If the C-17 side chain is used to form

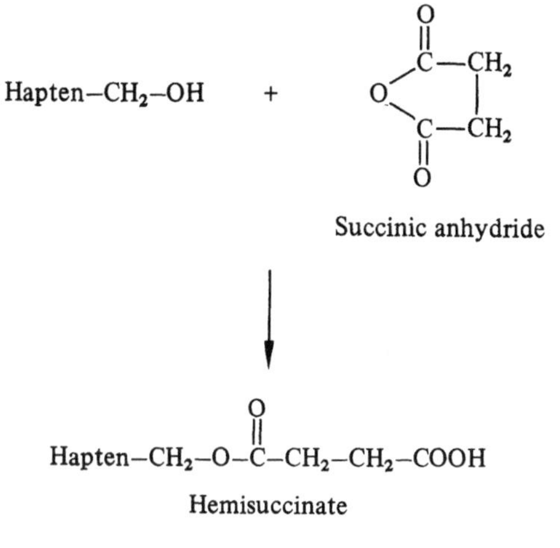

Fig. 2. Synthesis of a hemisuccinate derivative of a steroid containing a hydroxyl group

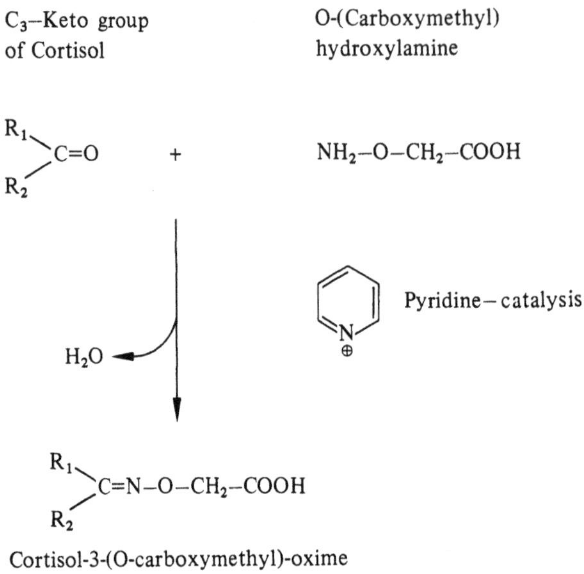

Fig. 3. Synthesis of a carboxymethyloxime derivative of a steroid (i.e., cortisol) containing a ketone group

the link between the steroid and protein, then the antiserum produced is usually specific for the end of the molecule furthest from the protein bridge, and is unable to distinguish between steroids with differences in structure that are located close to the site of conjugation. For this reason, several assays have now been developed using steroids that have additional reactive groups to those present in the natural steroids. Thus, antisera produced by immunization with progesterone-11 conjugates are more specific for progesterone than those resulting from immunization with progesterone-3 conjugates (MORGAN and COOKE 1972; YOUSSEFNEJADIAN et al. 1972).

The most commonly used method has been to couple the steroid derivative to the ε-amino group of lysine residues in either bovine or human serum albumin. Although there are 59 lysine residues and one terminal amino group present in the albumin molecule, the molar ratio of steroid to albumin rarely exceeds 35:1 (DAWSON et al. 1978). If the molar ratio achieved is below 10:1, the immunogenicity of the conjugate is likely to be poor (NISWENDER and MIDGLEY 1970). This subject has been well reviewed by ABRAHAM (1974) and NIESCHLAG et al. (1975).

III. The Radioactive Tracer

Two kinds of radioactive steroid are used in radioimmunoassay (RIA): tritium-labeled steroids with specific activities between 25 and 100 Ci/mmol; and ^{125}I-labeled steroid derivatives with specific activities between 2200 and 4400 Ci/mmol. Although it has been claimed that a greater sensitivity can be achieved with ^{125}I-labeled steroids, the sensitivity obtained in practice is generally in the same range as when using ^{3}H-labeled steroids (HUNTER et al. 1975). In fact, the attachment of iodine to the steroid molecule for a suitable assay has posed considerable technical problems.

Steroid RIA systems may incorporate iodinated tracers through a homologous bridge. The same chemical bond links the steroid to the peptide for both the immunogen and the tracer. In these homologous bridge systems, antibodies recognize the bridge and bind the labeled steroid with greater affinity than the native steroid, thereby causing loss of sensitivity in the assay (CORRIE and HUNTER 1981; CORRIE 1982). Sensitivity can be increased by using a heterologous bridge system in which the labeled steroid contains a bridge different in site or nature from the immunogen (NORDBLOM et al. 1979, 1981).

In conclusion, the iodinated radioligand is the most suitable tracer for laboratories without facilities for liquid scintillation counting. In other laboratories, the tritium-labeled steroids remain the best approach, as they are more stable, can be easily repurified by chromatography, and, lastly, can be used as internal standards for recovery calculation.

IV. Sample Preparation Before Immunoassay

As antibodies obtained against steroid hormones are not completely specific and react with related steroids, the use of these antisera as detection methods usually requires some purification of the steroid prior to the measurement. The degree of purification depends on: (a) the specificity of the antiserum: (b) the relative concentration of the interfering steroids in the biologic fluid; and (c) the nonspecific interference by plasma lipids and proteins or other biologic materials.

1. Direct Assay in Plasma

Plasma and serum are interchangeable since there is no difference in the levels of steroids measured in these two biologic fluids. If a steroid exists in a biologic fluid in a relatively large concentration compared with interfering steroids, this steroid can be measured directly. Dilution with assay buffer will minimize the nonspecific interference by proteins, lipids, or other biologic materials. Solid-phase methods are particularly suitable for this approach.

Until now, many procedures for direct immunoassays of steroids have been described (HAYNES et al. 1980; JURJENS et al. 1975; MCGINLEY and CASEY 1979; PRATT et al. 1975; RATCLIFFE 1983; ROLLERI et al. 1976), but a careful internal and external quality control program must be established before using them routinely.

2. Selective Solvent Extraction

If interfering steroids are greatly different in polarity from the steroid under measurement, the interfering steroid can be removed by selective solvent extraction. For example, corticosteroids in relation to their very high circulating levels interfere in the assay of progesterone. By extraction with a nonpolar solvent such as petroleum ether or hexane, progesterone can be correctly quantified because polar corticoids mostly remain in the plasma sample (MORGAN and COOKE 1972; DIGHE and HUNTER 1974).

3. Chromatographic Purification

Various chromatographic systems such as paper (FORTI et al. 1974), alumina columns (FURUYAMA et al. 1970), Sephadex LH-20 (CARR et al. 1971), celite columns (ABRAHAM et al. 1977), and, more recently, high pressure liquid chromatography (HPLC) (HEFTMANN and HUNTER 1979) have been employed to purify steroid hormones.

Celite chromatography is now the most widely used system since it is easy to prepare, is relatively inexpensive, nonspecific interference is negligible or absent, and it has a high capacity and high power of resolution. In addition, several steroids can be separated at the same time.

However, Sephadex LH-20 chromatography is also practicable and in some particular cases paper chromatography must still be used to obtain the necessary purification. On the contrary, in our experience the cost:benefit ratio of HPLC is too high for it to be suggested as routine procedure. Radioactive internal standards are obviously needed to calculate the recovery.

V. Bound/Free Separation Systems

Several separation methods have been used, as reviewed by COLLINS et al. (1975) and RATCLIFFE (1983). They are based on: (a) absorption of the free steroid (dextran-coated charcoal); (b) precipitation of the steroid–antibody complex (double-antibody, ammonium sulfate, and polyethylene glycol); and (c) solid-phase (antibody-coated polystyrene tubes and antibody covalently linked to beads; BOLTON and HUNTER 1973). One of the most practicable procedures is dextran-coated charcoal because it is simple to perform, inexpensive, rapid, and precise.

VI. Monoclonal Antibodies to Steroids

Monoclonal mouse antibodies to steroid hormones have been produced by several laboratories (GALFRE and MILSTEIN 1981). The affinity constant and the specificity of some monoclonal antibodies are sometimes lower than those of some traditional antisera. Now useful monoclonal antibodies to steroids can be produced using existing technology (ESHHAR et al. 1981; FANTL et al. 1981; KOHEN et al. 1982; WHITE 1983). In all probability, monoclonal antibodies will be produced with affinity and specificity similar to those of the best polyclonal antisera. Although it seems unlikely that such developments will make possible techniques until now unreliable with conventional antisera, the production of antibodies in the required quantities by means of hybridomas should enable the more widespread application of existing and future assay techniques which need high quality antibodies.

VII. Quality Assessment

Every laboratory must establish an internal quality control scheme to monitor the reproducibility of its assays, but it is equally important that every laboratory takes part in external quality control schemes to compare the performance of their assays, in terms of precision, specificity, and accuracy, with what is currently available elsewhere (JEFFCOATE 1981).

In the past few years, several external quality assessment schemes have been organized. They use samples in which the hormonal steroids have been measured by isotope dilution mass fragmentography, i.e., by very precise (coefficient of variation ~2%) and very specific methods which in some cases can be considered as definitive. Several such methods for steroid measurements have been published and they can be used as references (PATTERSON et al. 1984; SIEKMANN 1979).

VIII. Alternative Immunoassay Methods

1. Enzyme Immunoassay

a) Introduction

Enzymes have proven to be sensitive and versatile labels in various immunoassay systems for the measurement of plasma steroids (VAN WEEMAN et al. 1979; BLAKE and GOULD 1984). The sensitivity of detection is due to the amplification of the signal by prolonged incubation and catalytic turnover. Versatility is due to the modulation of enzymatic activity by a variety of factors. The factors affecting the choice of enzyme include: (a) the turnover number of the pure enzyme; (b) the purity of the enzyme preparation; (c) the sensitivity, ease, and speed of product detection; (d) the absence of interfering factors in the test fluid; (e) the presence of potentially reactive groups for coupling: (f) the stability of the enzyme conjugate; (g) the availability and cost; and (h) the suitability for homogeneous assay. The enzymes that have been used most extensively are horseradish peroxidase, alkaline phosphatase from calf intestinal mucosa, β-D-galactosidase from *Escherichia coli,* glucose oxidase from *Aspergillus niger,* urease, penicillinase, and luciferase in heterogeneous assay systems, and lysozyme, malate dehydrogenase, and glucose-6-phosphate dehydrogenase in homogeneous enzyme immunoassay (EIA). A variety of end points have been used in the development of EIA. For example, if a fluorescent product is produced after enzyme incubation, the method might be described as a fluorescence EIA.

In the case of EIA for steroids, two basic approaches have been described:

1. Antigen-labeled techniques: the procedure involves the competitive reaction between an enzyme-labeled steroid and the native steroid for a limited concentration of specific antibody (OGIHARA et al. 1977; RAJKOWSKI et al. 1977).

2. Antibody-labeled techniques, where the enzyme-labeling procedure occurred after the initial antibody-binding reaction to an immobilized antigen.

The amount of first antibody bound to the immunoabsorbant is measured by an enzymatic technique in which a heterologous bridging antibody and a soluble antibody–enzyme immune complex (rabbit anti-peroxidase–horseradish peroxidase) are added in sequence. Peroxidase activity is inversely proportional to the concentration of free antigen in the original sample (YORDE et al. 1976).

b) Limitations to Enzyme Immunoassay

The difficulties imposed by the analysis of biologic samples, however, have restricted the use of EIA. The technique has not yet been widely applied to the measurement of hormones and drugs, in plasma or saliva, for the following reasons:

1. Compared with radioisotopes the label has a high molecular weight which often interferes with the antibody-binding reaction by bridging effects or steric hindrance, and it may be necessary to develop assays which involve the use of heterologous combinations of reagents (VAN WEEMAN et al. 1979).
2. The synthesis of labeled antigen is difficult to control and the product needs to be characterized for enzyme and antibody-binding activities.
3. The homogeneous systems are relatively insensitive owing to nonspecific interference from the biologic sample.
4. The end point determination is more complex and time-consuming than for RIA.
5. The large size of the enzyme-labeled antigen limits the method of separation of the antibody-bound and free fractions to solid-phase or second antibody techniques.

2. Fluoroimmunoassay

a) Introduction

Fluorescence is one sort of luminescence, which is the emission of light resulting from the dissipation of energy from a substance in an electronically excited state. There are several forms of luminescence which differ only in the source of energy involved in exciting the electrons to a higher energy level. These include: (a) radioluminescence, where excitation is effected by a radioisotope; (b) chemiluminescence, where excitation is effected by a chemical reaction; (c) bioluminescence, where the chemical reaction is mediated by enzymes in a biologic system; (d) thermoluminescence (incandescence); (e) electroluminescence (e.g., neon light); and (f) fluorescence or photoluminescence, where excitation is effected by photons of infrared, visible, or ultraviolet light. The light emitted occurs in the form of a photon of longer wavelength than that of the exciting photon.

Developments in immunoassay procedures with fluorescent labels have been reviewed (SOINI and HEMMILA 1979; SMITH et al. 1981; EXLEY 1983; HEMMILA et al. 1984; SOINI and KOJOLA 1983). These can be divided into fluorescence immunoassay (FIA), where labeled antigen is used, and immunofluorimetric assay (IFMA) using labeled antibodies. Each can be further subdivided into homogeneous (nonseparation) (KOBAYASHI et al. 1979a, b) and heterogeneous (separation) procedures.

b) Limitations to Fluoroimmunoassay

Various factors present in biologic fluids can cause apparent enhancement or quenching of the fluorescent signal and include: (a) light scattering; (b) endogenous fluorophores; (c) internal filter effects, where part of the exciting or emitted light is absorbed by compounds present in the reaction mixture; and (d) quantum yield changes due to environmental effects.

3. Chemiluminescence Immunoassay

a) Introduction

The use of chemiluminescent (CL) molecules in immunoassay is of recent application. Luminescence may be defined as the chemical production of light and, as in the case of fluorescence, involves the emission of photons from excited molecules on their return to the ground state. Luminescence can be applied to a wide variety of assays of biologic interest and more information on the characteristics of bioluminescent (BL) and CL reactions, instrumentation, and applications in immunoassay for steroids can be found in reviews by KOHEN et al. (1983) and PAZZAGLI et al. (1984).

b) The Steroid–Chemiluminescent Tracer

Several chemical compounds possess suitable CL characteristics for use as labeling reagents (SCHROEDER and YEAGER 1978; WEEKS et al. 1983). In particular, some aminophthalhydrazides (i.e., luminol, isoluminol, and isoluminol derivatives) participate in simple oxidation reactions (such as the hydrogen peroxide–microperoxidase reaction in the pH range 8.6–13) to produce light with a relatively high quantum efficiency and thus they can be used as universal labeling material for both antigens and haptens.

Following this approach, some isoluminol derivatives have been conjugated to steroid molecules and the resulting steroid–CL tracers have been investigated in terms of both the affinity for the homologous antibody and the CL characteristics. The results of this study have been reported by PAZZAGLI et al. (1983) and can be summarized as follows.

The preparation of the steroid–CL tracer is a simple conjugation reaction, such as a carbodiimide reaction, and it is similar to the methods used in the synthesis of steroid–protein immunogens. The purification step can be performed on silica gel TLC plates and the identity of the final compound can be confirmed by mass spectrometry. An example of the structure of a steroid–CL tracer is shown in Fig. 4.

The CL efficiency of steroid–CL tracers is mainly determined by the isoluminol derivative used and is not significantly affected by the steroid molecule. Some isoluminol derivatives (i.e., aminoethylethylisoluminol AEEI, or aminobutylethylisoluminol ABEI) are more efficient and suitable than others – use of this kind of compound produces CL tracers detectable at the picomolar level (0.1–0.4 fmol per tube).

The affinity of the steroid–CL tracer for the homologous antibody can be affected by the ability of the antibody to recognize both the steroid and the bridge between the steroid and the CL molecule (e.g., carboxymethyloxime or hemisuccinate bridges). Consequently, the affinity can be higher, lower, or similar to the native steroid, and this can affect the sensitivity and the specificity of the assay, as already observed for iodinated steroids (PAZZAGLI et al. 1981 a, c). However, most steroid–CL tracers possess suitable affinity for LIA methods.

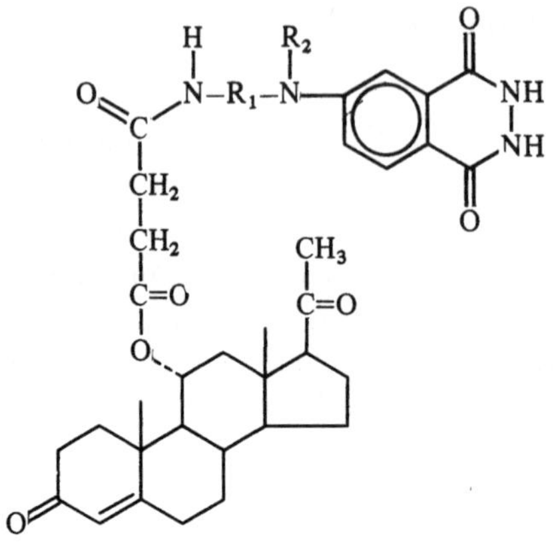

Conjugate	R_1	R_2	M.W. a. u.	FD Mass Spectra			
				M^+	$(M+H)^+$	$(M+H_2)^+$	$(M+N_a)$
P-11α-HS-AEI	$(CH_2)_2$	-H	632	632	633		655
P-11α-HS-AEEI	$(CH_2)_2$	$-CH_2-CH_3$	660	660	661		683
P-11α-HS-ABI	$(CH_2)_4$	-H	660	660	661		683
P-11α-HS-ABEI	$(CH_2)_4$	$-CH_2-CH_3$	688	688	689		711
P-11α-HS-AHEI	$(CH_2)_6$	$-CH_2-CH_3$	716	716	717		739
P-11α-HS-ABMI	$(CH_2)_4$	$-CH_3$	674		675		697

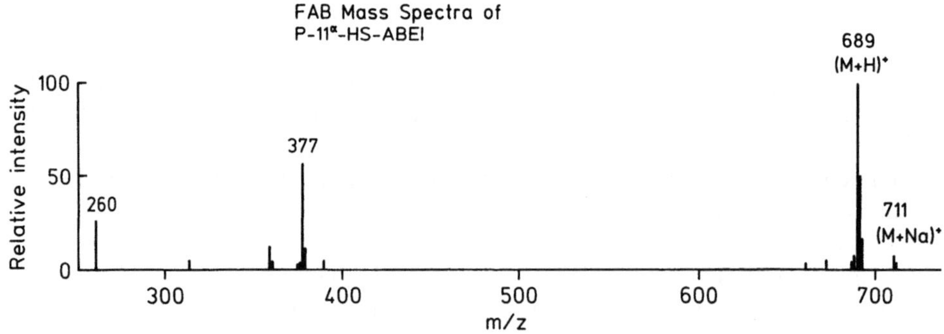

Fig. 4. Example of mass spectra data resulting from a steroid–CL tracer as obtained by the field desorption (FD) or fast atom bombardment (FAB) ionization techniques. (Pazzagli et al. 1983)

c) Homogeneous Luminescence Immunoassay Methods

As we have noted, the steroid–CL tracer emits light on oxidation by the hydrogen peroxide–microperoxidase system at pH values varying between 8.6 and 13.0. pH 8.6 is compatible with monitoring competitive protein-binding reactions in a homogeneous manner when the presence of the specific antibody influences the light yield of the steroid–CL tracer.

This is the case for progesterone (KOHEN et al. 1979), estriol (KOHEN et al. 1980a), and cortisol (KOHEN et al. 1980b), but although homogeneous LIA methods are attractive procedures because they allow full automation of the assay at present they still have some disadvantages, mainly owing to incomplete knowledge of the mechanism of the antibody-enhanced chemiluminescence. In fact, we have already shown that several factors are involved in the enhancement phenomenon, including the chemical structure of the tracer (PAZZAGLI et al. 1982b; PATEL and CAMPBELL 1983) and the immunologic characteristics of the antibody. Consequently, it would appear that different procedures will have to be adopted for each antibody or labeled ligand used, and this is not practical in routine use. Only when well-standardized reagents are available (in particular, monoclonal antibodies), will the practical application of homogeneous LIA in the clinical chemistry laboratory become possible.

d) Heterogeneous Luminescence Immunoassay Methods

In contrast to homogeneous LIA methods based on antibody-enhanced chemiluminescence, heterogeneous LIAs do not utilize the light-enhancing properties of the specific antibody, but introduce a phase separation step into the procedure. Thus, heterogeneous LIAs are closely analogous to conventional RIAs, with the exception of the final stage for detection of chemiluminescence, and hence they are easily learned by technicians already familiar with RIA procedures (KOHEN et al. 1981; LINDSTROM et al. 1982; PAZZAGLI et al. 1981b, d).

e) Limitations to Luminescence Immunoassay Methods

One of the major drawbacks of the use of CL labels is the measurement of the analytic signal which follows the oxidation reaction. As described by TOMMASI et al. (1984), in order to minimize errors due to the CL tracer measurement, the following aspects should be optimized:

1. Choice of the measurement parameter; the CL reaction, because of its kinetics, can be quantified by several parameters which possess different characteristics in terms of counting time, reproducibility, linearity, signal:blank ratio, etc. The most suitable parameter should be selected for the measurement of the CL tracer and this depends on the kind of CL compound used and on the reaction conditions.

2. Evaluation of CL kinetics; the reaction conditions (i.e., the pH of the reaction solution, buffers, the relative concentration of the oxidant and catalyst, and in particular the presence of material of biologic origin such as plasma or urine) can modify the kinetics and consequently the measured light output. Consequently, it seems necessary to have a criterion to evaluate the reaction kinetics.

C. The Adrenal Cortex

I. Physiology

The adrenal cortex secretes five types of steroids, classified according to their biologic action, i.e., glucocorticoids, mineralocorticoids, androgens, progestagens,

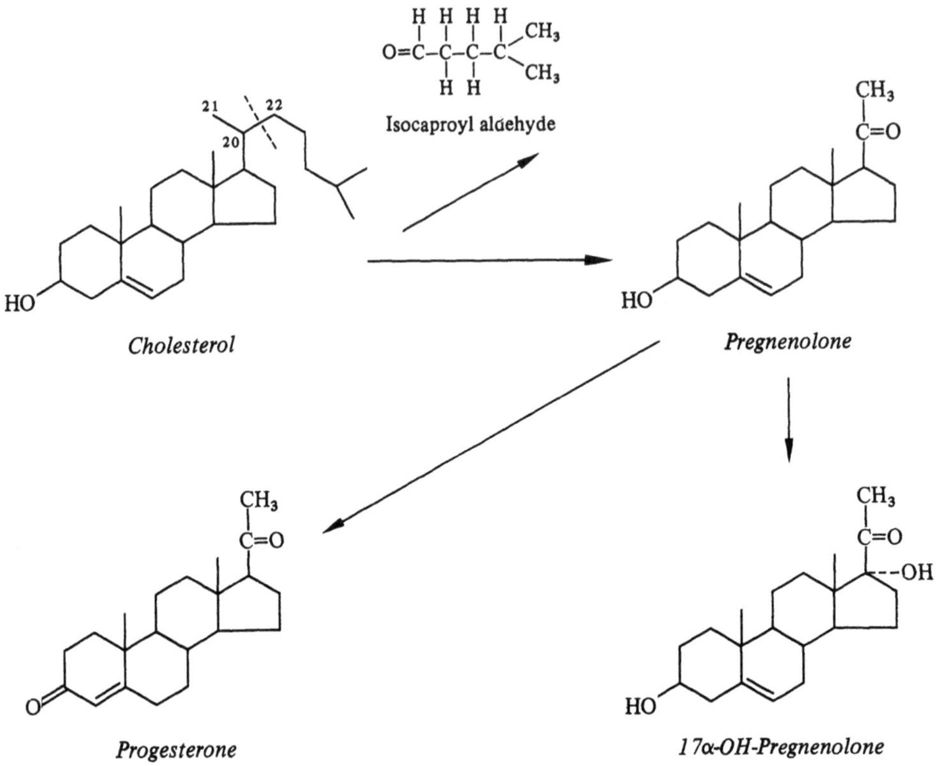

Fig. 5. Biosynthesis of progesterone and 17α-hydroxypregnenolone

and estrogens (Figs. 5–8). However the first two classes are secreted by the adrenal gland solely, while the last three also come from the gonads. Mineralocorticoids, glucocorticoids, and progestagens are C_{21} steroids, the androgens C_{19}, and the estrogens C_{18} steroids.

The glucocorticoids, which include cortisol, cortisone, and corticosterone, have an oxygen substituent at the C-11 position, either as a hydroxyl or an oxo group. They increase the production of glucose from proteins (gluconeogenesis), decrease the peripheral uptake of glucose, and increase the production of glycogen in the liver. In addition, they decrease the protein synthesis of lymphocytes, stimulate the production of red cells and leukocytes from the bone marrow, and maintain normal glomerular filtration (EDWARDS and LANDON 1976). Cortisol is the major glucocorticoid secreted by the gland in human, monkey, horse, and sheep, whereas corticosterone is the major glucocorticoid secreted in rat, rabbit, and other rodents.

The mineralocorticoids, which include aldosterone, 11-deoxycorticosterone (DOC), and 18-hydroxydeoxycorticosterone, are important in the control of electrolyte metabolism. Aldosterone is the most important sodium-retaining steroid in humans. In the rat, however, 18-hydroxy-11-deoxycorticosterone is probably more important. These steroids increase the entry of sodium into the cells of kidney collecting ducts and promote the loss of potassium (EDWARDS and LANDON 1976; NOWACZYNSKI et al. 1977).

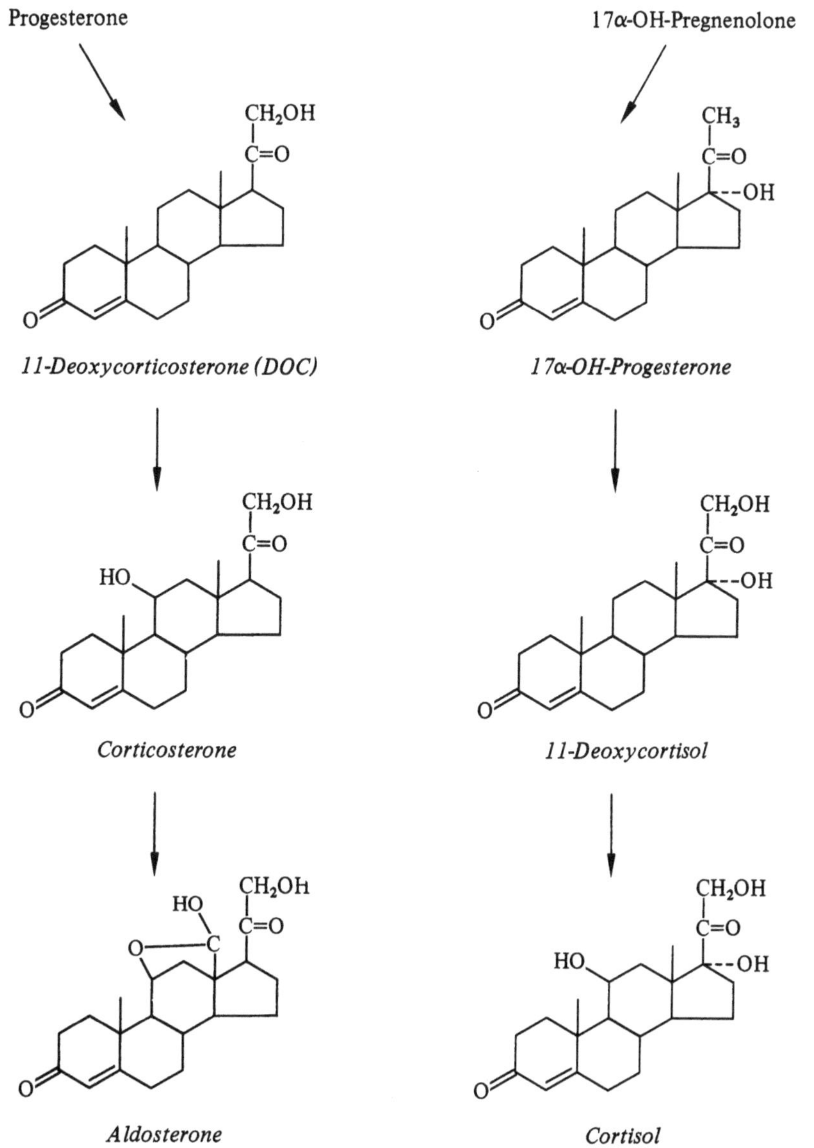

Fig. 6. Biosynthesis of major adrenocortical hormones

The human adrenal cortex secretes about 25 mg cortisol daily as compared with only 2 mg corticosterone and 200 µg aldosterone. The large amount of cortisol provides the major glucocorticoid activity and also contributes to some extent to mineralocorticoid activity. In contrast, the very small amount of aldosterone secreted has negligible glucocorticoid activity, but is the main mineralocorticoid. Corticosterone has both glucocorticoid and mineralocorticoid effects.

There are many species differences in adrenal steroid secretion; therefore the animal model to be used in pharmacologic experiments must be well known as to its physiology and must be well selected before extrapolating the results to hu-

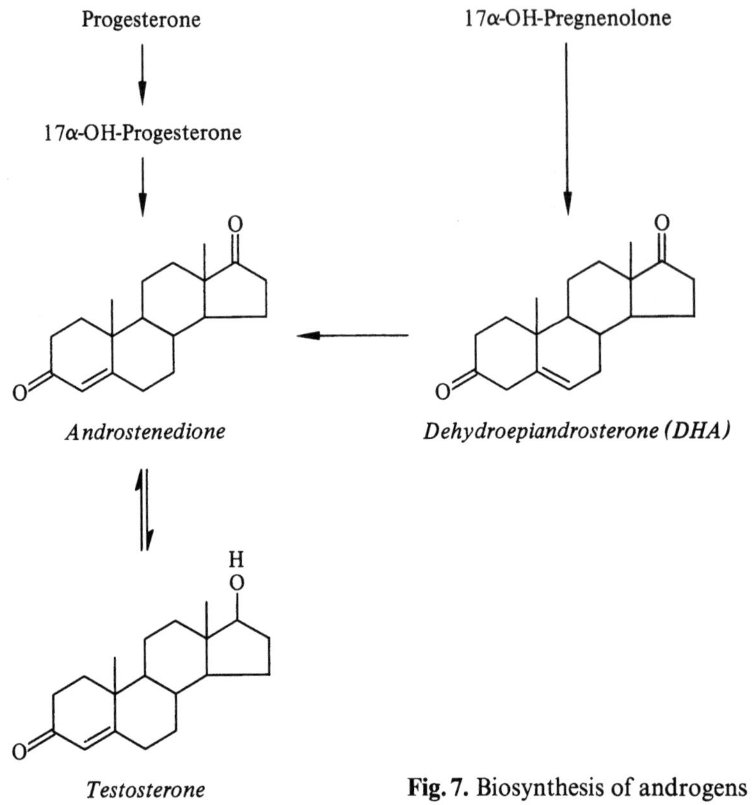

Fig. 7. Biosynthesis of androgens

mans. For instance, the dog is a very good model to study glucocorticoid and mineralocorticoid action.

The secretion of progestogens and estrogens by the human adrenal gland is negligible in terms of a contribution to the biologic action of these hormones. On the contrary, the secretion of adrenal androgens in humans is quantitatively relevant, but their biologic action is very weak. In relation to this, the adrenal androgens are also called weak androgens or preandrogens because they act mainly as precursors of testosterone and 5α-dihydrotestosterone.

II. Control Mechanisms for the Release of Cortisol

There are three major mechanisms that control the secretion of corticotropin-releasing factor (CRF) and, thereby, of adrenocorticotropic hormone (ACTH) and cortisol (EDWARDS and LANDON 1976).

1. Negative Feedback Control

Adrenalectomized animals have high levels of ACTH which fall immediately after administration of corticosteroids. In contrast, it has been known for many years that administration of hydrocortisone or other glucocorticoids results in adrenal

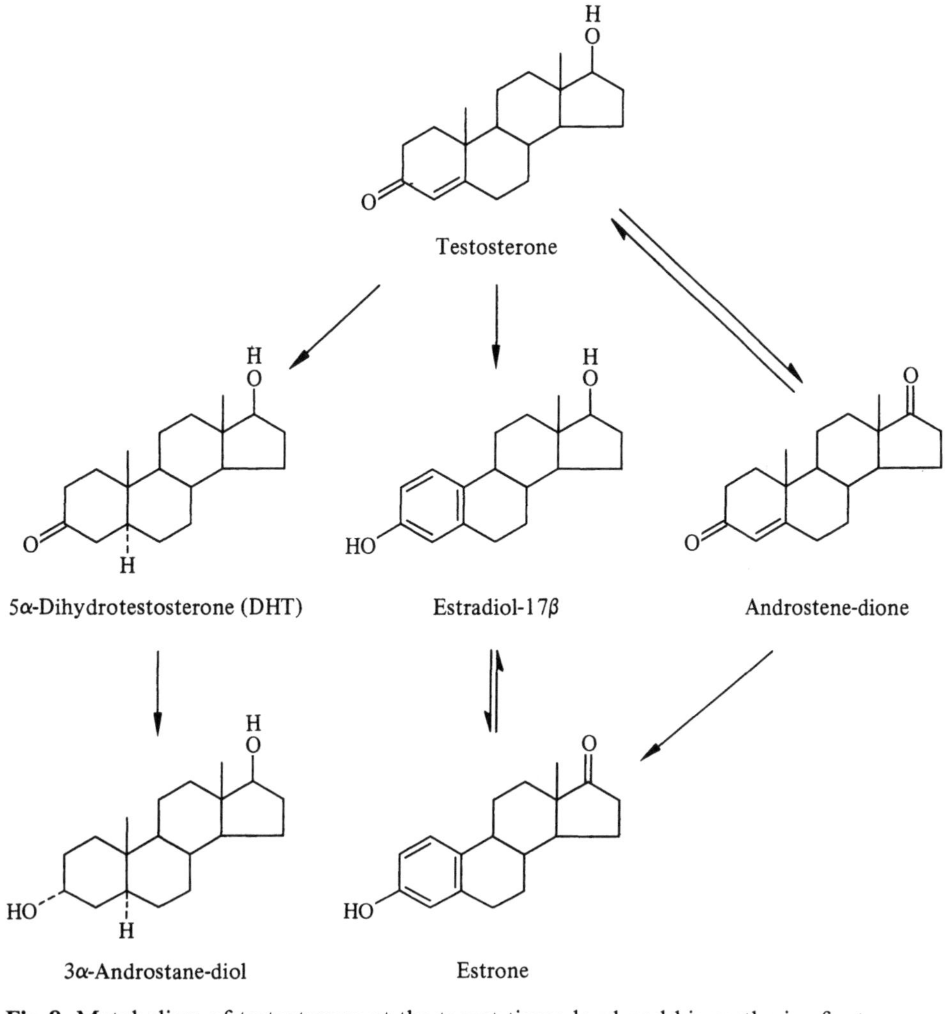

Fig. 8. Metabolism of testosterone at the target tissue level and biosynthesis of estrogens

atrophy. This modulating negative feedback operates via the hypothalamus and pituitary and is termed long feedback control. It is more important than the short feedback control system whereby ACTH itself can alter the secretion of CRF.

2. Nyctohemeral Control

Several animals show a marked rhythm of ACTH secretion during their sleep–wake cycle. The highest levels of ACTH, and thus cortisol or corticosterone, are found in blood samples taken early in the morning and the lowest levels in the evening. This rhythm relates to the sleep–wake cycle rather than the 24-h cycle and it is for this reason that the term nyctohemeral is preferred to circadian or diurnal. Several studies have shown that the secretion of ACTH is not continuous, but episodic during this cycle.

3. Stress Control

This mechanism can override the other two and is brought into action by a wide variety of physiologic, pharmacologic, and psychologic situations, including hypoglycemia, pyrogen administration, apprehension, pain, and fear.

III. Control Mechanisms for the Release of Aldosterone

The major control mechanism for aldosterone release is the renin–angiotensin system, which may be stimulated by many factors, including sodium deprivation or a change from a lying to a standing position. When plasma renin activity is suppressed, as in patients with primary hyperaldosteronism due to an adrenal tumor, ACTH becomes the dominant controlling factor. Aldosterone release may also be stimulated by hyperkalemia. In humans, there are two minor inhibitory mechanisms controlling aldosterone secretion acting directly at the adrenal level, i.e., dopamine and atriopeptins (secreted by the heart) and, in addition, there is a circadian rhythm (EDWARDS and LANDON 1976; WILLIAMS et al. 1972). Consequently, isolated values of plasma aldosterone that are unrelated to posture, sodium intake, or time are of minimal value. In contrast to aldosterone, the major control mechanism for sodium-retaining steroids such as deoxycorticosterone, 18-hydroxy-11-deoxycorticosterone, and corticosterone are regulated by ACTH.

IV. Biosynthesis of Adrenal Steroids

An outline of the biosynthetic pathways for the corticosteroids, estrogens, and androgens is given in Figs. 5–8. Progesterone, 11-deoxycorticosterone, and corticosterone are precursors for aldosterone biosynthesis within the zona glomerulosa. The isolation of 18-hydroxycorticosterone suggested that this might be an intermediate in the synthesis of aldosterone from corticosterone. The role of 11-dehydrocorticosterone in the biosynthesis of aldosterone by the human adrenal is unknown, although it can be converted to aldosterone by the rabbit adrenal.

V. Metabolism of Adrenal Steroids

1. Cortisol

In humans, about 90% of the circulating cortisol is bound to plasma proteins, of which the most important is corticosteroid-binding globulin (CBG), also termed transcortin. There is also weak binding to albumin. Available evidence indicates that only the free fraction is biologically active, with the protein-bound fraction providing a reservoir. The transport proteins are species specific and there are many differences between animals. So the transport proteins are another important factor to be considered in the experimental model.

In humans, when cortisol is injected into the circulation it disappears rapidly with a half-life of 70–110 min and normally only 1% or less of the total amount of cortisol secreted by the adrenal is excreted unchanged in the urine. The cortisol filtered by the renal glomeruli and then excreted reflects the level of circulating non-protein-bound cortisol, which explains the clinical value of urinary free cortisol determinations.

The most important site for corticosteroid breakdown is the liver. Not surprisingly, liver dysfunction alters the rate of cortisol clearance. Enzyme-inducing drugs, such as phenobarbitone, increase the rate of cortisol clearance and, indeed, the production of metabolites such as 6-hydroxycortisol can be used as an index of enzyme induction. Cortisol clearance is also increased in thyrotoxicosis, decreased in hypothyroidism, severe liver disease, malnutrition, and in patients receiving orally active anabolic steroids. Because of the negative feedback control system of ACTH, these changes in the metabolism of cortisol do not normally alter the free cortisol concentration, but are, however, extremely important in patients on cortisol replacement therapy.

2. Aldosterone

The metabolism of aldosterone differs from that of cortisol as the most important route for its excretion is usually via the kidney. In humans, after the injection of radioactive aldosterone, about 75% of the activity is excreted in the first 24 h with nearly 60% in the form of glucuronide-conjugated metabolites, of which the most important is tetrahydroaldosterone. Only about 0.2% of the isotopically labeled aldosterone is excreted in urine unchanged.

VI. Immunochemical Methods for Adrenal Steroids

1. Cortisol

Cortisol is present in human plasma in significantly higher concentrations than other C_{21} steroids and it is therefore possible to assay cortisol directly (i.e., without solvent extraction) and to obtain accurate results (Connolly and Vecsei 1978; Foster and Dunn 1974). The method of choice at present is probably to use antisera raised to a cortisol-3 derivative, probably cortisol-3-(*O*-carboxymethyl)oxime (Fahmy et al. 1975), coupled to a solid-phase support, and to use an ^{125}I-radioligand, because this will reduce running costs. Also, when using an ^{125}I-radioligand and one of the new generation gamma counters with multihead counting, results may be obtained from an immunoassay in less than 2 h. Most of the commercial immunoassay kits now available use methodology of this type, with differences confined to the nature of the solid-phase matrix and the procedure used to deactivate CBG. Current assays using ^{125}I-radioligands appear to have adequate sensitivity, specificity, and precision to measure circulating cortisol concentrations, and sufficient accuracy for clinical purposes, as judged by reference to gas chromatography–mass spectrometry reference procedures (Patterson et al. 1984).

2. Corticosterone

Corticosterone is the major corticosteroid of rats, mice, and birds, but only a minor constituent of normal human plasma, so an assay which is satisfactory for rat plasma (Gomez-Sanchez et al. 1975) obviously cannot be applied to human plasma without extensive reevaluation; even in species such as ducks, whose corticosterone is the predominant C_{21} steroid, a preassay chromatographic step has been required to ensure a reliable determination (EDWARDS and LANDON 1976). A more recent procedure (AL-DUJAILI et al. 1981) allows plasma samples from both rats and humans to be assayed directly, i.e., without extraction or chromatography. The procedure is analogous to that published for the assay of cortisol, using a histamine ^{125}I label. As in the case of the cortisol assay, the use of an antiserum raised to a hapten coupled through the C-3 position gives a highly specific antiserum.

3. 11-Deoxycortisol

A simple direct assay for plasma 11-deoxycortisol (Reichstein's compound S) has been described (PERRY et al. 1982). The assay uses antisera raised to a 3-(*O*-carboxymethyl)oxime conjugate and the homologous ^{125}I-radioligand, deactivation of plasma-binding proteins being achieved by dilution (1:10) with a pH 4 buffer. The determination of 11-deoxycortisol is used with the determination of 17α-hydroxyprogesterone in the localization of the enzyme block in patients with congenital adrenal hyperplasia.

4. Plasma Free Cortisol

Methods in routine use for determining plasma concentrations at this time invariably measure "total" steroid concentrations. Such assays therefore provide information which may not be as clinically useful as that deriving from an assay of the "free," biologically active fraction in plasma.

The principal techniques used for determining plasma "free" steroids are ultrafiltration and equilibrium dialysis, but other procedures using electrophoresis on paper and starch, gel filtration, and indirect methods based on a binding index have also been employed. Regardless of the method of separating the protein-bound and free steroid, these procedures are of two main types: those depending on assay of cortisol in the free fraction, and those in which the distribution between the bound and free fractions is assessed with a radioligand. Purity of the label in the latter procedures is critical, and trace impurities may invalidate the results. Binding of the steroid to membranes may require consideration in studies featuring equilibrium dialysis and ultrafiltration; the need for strict temperature control is a further restriction in ultrafiltration techniques (BAUMANN et al. 1975).

5. Urinary Free Cortisol

The difficulties involved in clinical studies based on time-consuming plasma-sampling regimens have focused attention on urinary free cortisol as an indirect index of circulating free steroid concentrations. Problems of ensuring collection of complete 24-h urine samples, together with the marked dependence of urinary free cortisol levels on fluid intake and normal renal function, limit the usefulness of this technique. Despite these limitations, it has proven to be clinically useful to distinguish between simple exogenous obesity and mild Cushing's syndrome (EDDY et al. 1973).

6. Salivary Cortisol

Determination of salivary cortisol may have an important role to play in future investigations of adrenocortical activity, as both adults and children find it easy to collect saliva by themselves. Salivation into small collection tubes is stress-free and avoids undue perturbation of the hypothalamo–pituitary–adrenal axis. Development of simple nonextraction immunoassays requiring small sample volumes is facilitated by the relatively high cortisol concentrations in saliva and the absence of binding proteins. Cortisol concentration in saliva is independent of flow rate (WALKER et al. 1978) and is in good agreement (UMEDA et al. 1981) with the free fraction in plasma, even though the results are not superimposable.

7. Mineralocorticoids

As this chapter cannot deal with the analysis by RIA of all known mineralocorticoids, we shall report the RIAs of aldosterone, aldosterone metabolites, and the major aldosterone precursors: 18-hydroxycorticosterone, 18-hydroxydeoxycorticosterone, and 11-deoxycorticosterone.

a) Plasma Aldosterone

The techniques are divided into two groups – RIAs with or without a chromatographic separation step following solvent extraction. Results obtained from methods without the chromatographic step are usually well correlated with those obtained from methods involving chromatography (CONNOLLY et al. 1980; JOWETT and SLATER 1977; PRATT et al. 1978; AL-DUJAILI and EDWARDS 1978). However, many laboratories, such as our own, do not accept values as definite if they result from techiques without chromatography. Even though there are some atypical cases with altered steroidogenesis which might cause unexpected interference in the assay and thus misleading results, the great number of clinical samples, on the other hand, together with good practicability and low cost, support the use of simpler techniques. In the discussion for and against methods without chromatography, the specificity of the antibodies plays a key role. The specificity requirements of the various aldosterone assays are stricter than is generally supposed. Numerous kits are available, including several using ^{125}I-labeled aldosterone, but the performance of these procedures should be checked by using reference methods with both extraction of the sample and chromatography, especially when plasma of different species is to be analyzed. SIEKMANN (1979) has described an isotope dilution gas chromatographic–mass spectrometric method which may have some special applications as a reference method.

b) Urinary Aldosterone

The estimation of aldosterone-18-glucuronide (also known as urinary aldosterone) is subject to specificity problems (VECSEI et al. 1972; AL-DUJAILI and EDWARDS 1981). The technique involves the pH 1 hydrolysis of aldosterone-18-glucuronide; this pH can transform other steroids and urinary metabolites into new compounds. The latter can interfere with anti-aldosterone antibodies to different degrees. A practicable multi-microcolumn system for chromatography of aldosterone on Sephadex LH-20 before RIA can be used (KACHEL and MENDELSOHN 1979). Studies with radioactive steroids have shown that, in addition to aldosterone and tetrahydroaldosterone, 18-hydroxycorticosterone, tetrahydro-18-hydroxycorticosterone, and 18-hydroxydeoxycorticosterone also undergo acidic transformation.

Practical experience has confirmed the relevance of interfering factors. In about 5% of 854 samples, the method without chromatography produced clinically misleading values (VECSEI et al. 1982). Many urinary metabolites possibly causing interference are not commercially available, and consequently difficult to evaluate in the assay. Therefore, radioimmunochromatographic analysis seems necessary to reach satisfactory specificity.

A useful approach would be the direct measurement of aldosterone-18-glucuronide or tetrahydroaldosterone glucuronide without hydrolysis by means of specific antibodies (anti-steroid glucuronides) as described by MATTOX and NELSON (1981) and JOWETT and SLATER (1981).

c) Aldosterone Precursors

For the measurement of deoxycorticosterone, 18-hydroxy-11-deoxycorticosterone, and 18-hydroxycorticosterone, the antisera have been obtained by means of 3-carboxymethyloxime conjugates. Owing to the presence of cross-reactions with several other steroids, these methods require a prepurification by paper chromatography of plasma extracts before RIA (VECSEI et al. 1983). The radioactive steroids to be used as internal standard are available from commercial sources. Taking such precautions, these methods can be used in experimental models and in differential clinical diagnosis of salt-wasting disorders of adrenal and renal origin (BIGLIERI and SCHAMBERLAN 1979; VECSEI et al. 1983).

8. Adrenal Androgens

a) Androstenediol

This is the precursor of testosterone in the Δ_5 pathway of steroid biosynthesis. In men and women, about 50% of circulating values originates from direct adrenal secretion. Suitable antisera have been obtained using a carboxymethyloxime (CMO) conjugate in position 16, however, a chromatographic step (paper or celite column) is needed before RIA (FORTI et al. 1978, 1985).

b) Androstenedione

Androstenedione is the direct precursor of testosterone in the pathway of adrenal and gonadal steroidogenesis. In women, it has a mixed origin (adrenal gland and ovary) and its secretion changes with the time of the cycle (maximum concentrations in the late follicular phase). In postmenopausal women, it originates almost entirely from the adrenal secretion. It is a weak androgen, but plays a very important role as the major substrate for aromatization processes, i.e., estrogen formation in extraglandular tissue (central nervous system, skin, adipose tissue, etc.). The aromatization processes in the hypophyseal–hypothalamic system are involved in several biologic activities, such as sexual differentiation and the feedback regulation of gonadotropin secretion. The aromatization in adipose tissue is deeply involved in the pathogenesis of the polycystic ovarian syndrome and of estrogen-related tumors (breast and uterus).

Androstenedione measurement is useful in several clinical and experimental conditions. Several commercial kits are available which require only a simple plasma extraction in the procedure before the assay. Unfortunately, these methods can be affected by interfering substances of various origins which are able to alter the clinical and experimental value of the results. Celite and Sephadex LH-20 columns, and the more sophisticated HPLC procedures can be used after solvent extraction, in order to improve the reliability of the assay.

c) Dehydroepiandrosterone

Dehydroepiandrosterone (DHEA) is the major unconjugated androgen secreted by the adrenal as it originates almost entirely from this gland. In humans, its secretion rate varies between 6 and 10 mg/day. All methods described include a chromatographic step after solvent extraction. Celite column chromatography is the most suitable system as it allows the separation and measurement of several steroids at the same time (ABRAHAM et al. 1977).

d) Dehydroepiandrosterone Sulfate

The adrenal cortex also secretes some conjugated steroids (sulfates and glucuronides); among them the most important is dehydroepiandrosterone sulfate (DHEA-S). This steroid originates solely from adrenal secretion; its concentration in human blood is very high and therefore its measurement is a reliable index of adrenal androgen secretion. DHEA-S can be measured as free DHEA after solvolysis of plasma, but the most practicable approach is the measurement in diluted plasma by means of antisera to DHEA-3-hemisuccinate–BSA and radioactive DHEA-S or DHEA as tracers (CATTANEO et al. 1975; ABRAHAM et al. 1977).

9. Adrenal Progestagens

The C_{21} steroids (pregnenolone, 17α-hydroxypregnenolone, progesterone, and 17α-hydroxyprogesterone) are precursors of aldosterone and cortisol biosynthesis. Their clinical interest is limited to the diagnosis of congenital adrenal hyperplasia in humans. Pregnenolone and 17α-hydroxypregnenolone are increased in 3β-hydrosteroidogenase defect; pregnenolone and progesterone are increased in 17α-hydroxylase defect. All these steroids must be measured after chromatography. However, in relation to the high incidence of the 21-hydroxylase defect in comparison with the other forms of congenital adrenal hyperplasia, the measurement of 17α-hydroxyprogesterone is most important for diagnostic purposes and for assessing the response to the pharmacologic treatment (usually dexamethasone or fluorocortisone) in this disease. The 21-hydroxylase defect in practice represents 98% of all cases of congenital adrenal hyperplasia and therefore in some countries neonatal screening programs have been developed. A practical and rapid method for measuring 17α-hydroxyprogesterone in blood samples absorbed on filter paper disks has been published (SOLYOM 1981).

10. Alternative Immunoassays

Several EIAs have been reported for the measurement of cortisol (COMOGLIO and CELADA 1976; HINDAWI et al. 1980; OGIHARA et al. 1977). A homogeneous FIA for cortisol in serum was described by KOBAYASHI et al. (1979 a) using the fluorescence polarization technique. The nonspecific binding of fluorescent-labeled cortisol was successfully eliminated with sodium dodecylsulfate (KOBAYASHI et al. 1979 b). The sensitivity of the assay, however, was 0.1 ng cortisol per tube.

A direct heterogeneous FIA has been developed for the measurement of cortisol in serum (POURFARZANEH et al. 1980). The method involved cortisol labeled at the 3 position with fluorescein, and antibodies to cortisol coupled to magnetizable cellulose–iron oxide particles. A FIA method for 17α-hydroxyprogesterone has been described by ARAKAWA et al. (1983), which uses dried blood samples on filter paper for the screening of congenital adrenal hyperplasia.

In the case of LIA for cortisol, both homogeneous (KOHEN et al. 1980 b) and heterogeneous assays (PAZZAGLI et al. 1981 b; LINDSTROM et al. 1982) have been described. Finally, LIA methods have been developed for DHEA and its sulfate (ARAKAWA et al. 1981) and for 17α-hydroxyprogesterone (ARAKAWA et al. 1982).

D. The Testis

I. Physiology

The testis can be functionally divided into Leydig cell and seminiferous tubular compartments, which are responsible respectively for the production of testosterone and spermatozoa. The two compartments are regulated by the pituitary gland through gonadotropin secretion. Simultaneously, the testis regulates gonadotropin secretion by a negative feedback system: testosterone regulates LH secretion; inhibin regulates FSH secretion.

The hypothalamus plays an important role in controlling testicular function by secreting gonadotropin-releasing hormone (LHRH) in a pulsatile manner; this serves as the link enabling the central nervous system to influence reproductive function (ISMAIL 1976; FRANCHIMONT 1976). By this mechanism, environmental causes may exert their influence on reproductive processes; seasonal control of reproduction in many mammals presents an excellent example. In considering the details of the hormonal control of the testis, it is simpler to consider each compartment and its function separately, but we should keep in mind that the two processes of spermatogenesis and steroidogenesis are closely linked.

The testis secretes a variety of free and conjugated steroids (sulfates and glucuronides) which are synthesized from cholesterol. However, the principal secretory product in the majority of mammalian species is testosterone, a product of Leydig cells (Fig. 7). Although the testis secretes smaller amounts of androstenedione and dehydroepiandrostenone, they are very weak androgens and masculinization is due principally to testosterone or its target organ metabolite, 5α-dihydrotestosterone (Fig. 8).

Testosterone secretion by Leydig cells is stimulated by luteinizing hormone (LH) and in the majority of mammals a rise in LH secretion is followed by a rise in testosterone. In fact, the secretion of both LH and testosterone is episodic, and hence considerable changes in the levels of these two hormones may be found in a 24-h period. The removal of LH by hypophysectomy or neutralization of its activity by a specific antiserum leads to a cessation of testosterone production. Little is known of the way in which testosterone leaves the Leydig cell, but it is found in high concentration in spermatic venous blood, testicular lymph, and in the fluid within the seminiferous tubules.

In many mammalian species, the testis also secretes estrogens and in some, such as stallion and bear, very large quantities are produced. The circulating estradiol partly derives from the peripheral conversion of testosterone, but in men studies of estradiol levels in spermatic venous blood indicate that approximately 40%–50% of the circulating estradiol is secreted by the testis.

II. Transport of Testosterone in Blood and Its Action at the Target Level

Testosterone secreted into the bloodstream exists mainly in two forms – approximately 99% is bound to plasma proteins and the rest circulates in an unbound form. At least three plasma proteins are involved in testosterone binding, all with different affinities. Thus, at or near physiologic levels, the hormone is largely bound to a low capacity, high affinity globulin, designated sex hormone-binding globulin (SHBG). A smaller fraction is bound to albumin and to transcortin. As in the case of transcortin, there are a lot of differences in SHBGs between animal species. Nevertheless, the binding of testosterone to SHBG is highly specific (the first competitive binding methods for measuring testosterone were in fact assessed using SHBG as binding protein). Thus, only testosterone and other closely related androgens such as androgenically active C_{19} steroids with an intact 17β-hydroxyl group (in particular 5α-dihydrotestosterone) show such strong binding.

In addition, C_{18} estrogenic steroids with a 17β-hydroxyl group, such as 17β-estradiol, are also capable of binding, although with a lower affinity than that of testosterone. Structural alterations in the steroid moiety, particularly at or near position 17, are associated with marked changes in the binding affinity. Oxidation of the 17β-hydroxyl group to a ketone, as in androstenedione or estrone, markedly reduces binding to SHBG.

Only the free testosterone is active at the cellular level. Free testosterone enters the target tissue by a passive diffusion process. Inside the cell, the testosterone is converted to dihydrotestosterone by 5α-reductase (Fig. 8). Then dihydrotestosterone is bound to the high affinity androgen receptor protein in the cytosol. Later on, the hormone–receptor complexes move into the nucleus, where they interact with the acceptor sites on the chromosome and increase transcription of specific structural genes (ISMAIL 1976). In some target structures (such as prostate, hair follicle, prepuce), testosterone is inactive unless reduced to dihydrotestosterone. Therefore, dihydrotestosterone is the most important tissue androgen.

III. Immunochemical Methods for Androgens

1. Testosterone

Testosterone is the most important androgen. In men, it originates entirely from testicular secretion and its secretion rate reaches 7 mg/day. In male animals, its circulating levels are very high, but in females and in the prepubertal period they are generally very low. Many immunoassay methods for testosterone measurement have been published (FURUYAMA et al. 1970; FORTI et al. 1974). The majority are based on the use of antisera to testosterone-3-oxime. The others are based on the use of antisera to testosterone-7α-carboxymethylthioether. Both groups of antisera cross-react significantly with 5α-dihydrotestosterone. However, this cross-reactivity can be disregarded in male human blood as the ratio of testosterone to dihydrotestosterone in this fluid is 10:1. Therefore, with these antisera, plasma testosterone in men can be measured after a simple solvent extraction.

The situation is quite different at the target tissue level. In the human prostate for instance, the ratio of testosterone to dihydrotestosterone is 1:4. Therefore, a chromatographic purification of extracts is needed. The same need exists in female plasma or serum, in children, or prepubertal animals. The preparation of antisera to testosterone-1-carboxymethyloxime with very low cross-reaction with dihydrotestosterone did not solve the problem because, when testosterone is present at low concentrations, nonspecific interference from biologic material can affect the accuracy of the results. Celite or Sephadex LH-20 chromatography can be used to improve the reliability of the assay.

2. 5α-Dihydrotestosterone

This steroid is the active metabolite of testosterone at the level of some androgenic target tissues such as prostate or hair follicle. Even though its concentrations in blood of humans and other species originate almost entirely from peripheral conversion of testosterone (PAZZAGLI et al. 1975), interest in this steroid has increased enormously during the last few years because it seems to be involved in the pathogenesis of prostatic hyperplasia and in the hormonal control of prostatic cancer. Therefore, its measurement in prostatic tissue of man, dog, and rat became widespread, as it is a very good marker of clinical and experimental endocrine manipulation. Antisera to dihydrotestosterone-3-carboxymethyloxime and to dihydrotestosterone-1-carboxymethyloxime have been used. However, as the typical antisera used to measure dihydrotestosterone have considerable cross-reaction with testosterone, the chromatographic separation of dihydrotestosterone from testosterone is a prerequisite for precise determinations of dihydrotestosterone levels by RIA. A nonchromatographic method for dihydrotestosterone has been reported by PURI et al. (1981), based on treating ether extracts of biologic samples with potassium permanganate ($KMnO_4$). Such treatment selectively inactivates the endogenous testosterone whereas dihydrotestosterone remains unaltered.

3. 5α-Androstane-3α-17β-Diol and Its Glucuronide

These steroids are further tissue metabolites of testosterone which can be measured in circulating blood and prostatic tissue after extraction and chromatographic purification. Recently, much interest has been concentrated on the glucuronic acid conjugate (3α-diol glucuronide) as this steroid has been demonstrated to be the most sensitive and specific chemical marker of hirsutism in women. In fact 3α-diol glucuronide originates directly from the intratissue metabolism of dihydrotestosterone which is the active steroid at the hair follicle level. Unfortunately, 3α-diol glucuronide must be measured as free 3α-diol after enzymatic hydrolysis of plasma. The concentrations of the enzyme and the reaction conditions must be carefully controlled to obtain good hydrolysis and hence acceptable accuracy (HORTON et al. 1984).

4. Measurement of Free Testosterone

The importance of measuring free testosterone originates from the hypothesis that the biologic effects of testosterone may be due to the small unbound fraction rather than the total concentration of the circulating hormone. Many methods have been described for the measurement of free testosterone, including equilibrium dialysis, charcoal adsorption, and steady state gel filtration. The determination of free testosterone was not very helpful in the diagnosis of hirsutism and its use is limited to particular experimental conditions.

The measurement of testosterone in saliva has been proposed as an indirect index of the free testosterone fraction in serum since the salivary testosterone correlates well with the latter (BAXENDALE et al. 1981; WALKER et al. 1979; WANG et al. 1981). In spite of the determination of salivary testosterone being easier than the determination of the free fraction in blood, the clinical results were not improved by this technique.

5. Alternative Immunoassay Methods

Heterogeneous EIAs have been reported for the measurement of testosterone (BOSCH et al. 1978b; RAJKOWSKI et al. 1977; TURKES et al. 1979, 1980) whereas YORDE et al. (1976) have described a competitive enzyme-linked immunoassay (CELIA) for the measurement of testosterone involving the use of a soluble enzyme–antibody complex.

EXLEY and EKEKE (1981) described an FIA method for 5α-dihydrotestosterone in which 4-methylumbelliferyl-3-acetic acid was used as a fluorogenic label. 3β-Amino-5α-dihydrotestosterone was conjugated to the terminal carboxyl group of a poly-L-lysine carrier which was labeled with 25 molecules of the fluorogenic compound. The FIA, which involved a double-antibody precipitation technique, achieved the sensitivity and specificity of RIA. Heterogeneous LIA methods have been reported for testosterone (PAZZAGLI et al. 1982a) and testosterone-17β-D-glucuronide (VANNUCCHI et al. 1983).

E. The Ovary

I. Physiology

The ovary is subdivided into a number of specialized compartments and structures, each with its own precisely regulated microenvironment. Although it is often convenient to consider the oogenic and endocrine functions of the ovary separately, the two are strictly interconnected. An ovary devoid of oocytes cannot function normally as an endocrine gland.

In mature animals, the structure and function of the ovary is continually changing during the reproductive cycle. Gonadotropins secreted by the anterior pituitary gland stimulate the growth of graafian follicles (folliculogenesis, ovulation, and the formation of corpora lutea) (Baird 1976; Cooke 1976). The time necessary for follicles and corpora lutea to develop differs from species to species.

In many species, the length of the cycle is largely dependent on the life span of the follicle and corpus luteum, which determine the feedback effects of ovarian hormones on the hypothalamus and the anterior pituitary. Only a mature graafian follicle secretes sufficient estrogen to excite the hypothalamus–pituitary unit to discharge sufficient LH to cause follicular rupture and formation of the corpus luteum. For all species, the length of the luteal phase tends to be fairly constant; since the secretory products of the corpus luteum have an inhibitory effect on the secretion of pituitary gonadotropins, the corpus luteum in turn regulates the degree of follicular development.

The essential feature of the ovary is that its changing anatomic compartments result in an unstable endocrine equilibrium, which is reflected in a succession of estrous or menstrual cycles if pregnancy does not occur. As a consequence, the levels of ovarian hormones are variable throughout the cycle and moreover their secretion is pulsatile. In women, the mature follicle secretes a large amount of androstenedione and estradiol; the corpus luteum secretes a large amount of progesterone and to a lesser extent androstenedione and estradiol.

II. Immunochemical Methods for Ovarian Steroids

1. Progesterone

The method most suited to measure progesterone in blood, on the basis of its accuracy, sensitivity, precision, and convenience, is immunoassay. The literature concerning progesterone assay techniques has been restricted almost entirely to immunoassays and has reflected the continuing importance of these techniques. In a typical progesterone immunoassay, plasma is extracted with a nonpolar solvent, usually hexane or petroleum ether, and an antiserum produced to an 11α-hydroxyprogesterone hemisuccinate–protein immunogen is used with tritium-labeled progesterone (Abraham et al. 1971; Morgan and Cooke 1972). The use of a nonpolar solvent, which selectively extracts nonpolar steroids, acts as a purification step and also enables high and consistent extraction of progesterone (Dighe and Hunter 1974). The need to measure the efficiency of extraction from individual samples can therefore be avoided. For most purposes, however, the accurate measurement of low progesterone concentrations is not necessary as those found in the follicular phase or in male plasma are often at or below the sensitivity limits of clinically useful methods. What is required is the accurate measurement of progesterone in plasma of females, either in the luteal phase in order to detect ovulation, or during pregnancy; nonchromatographic assays are suitable for these purposes.

The use of solvent to extract progesterone from plasma and of liquid scintillation counting adds considerably to the complexity and cost of progesterone RIAs, and tends to restrict progesterone measurement to specialized laboratories which possess the requisite equipment. This is of particular importance because measurement of progesterone has become common practice in the investigation of infertility, and there is now a requirement for assay methods suited to the nonspecialist laboratory. Most of the literature on progesterone assay techniques has therefore been concerned with simplified assays, in which the extraction step is avoided or simplified, or in which an alternative isotope (^{125}I) is used (ALLEN and REDSHAW 1978; DIGHE and HUNTER 1974; KHADEMPOUR et al. 1978; SCARISBRICK and CAMERON 1975).

There are two major obstacles to the successful immunoassay of plasma progesterone without prior solvent extraction. First, the plasma proteins, particularly CBG, to which progesterone binds with high affinity, may decrease assay sensitivity. Second, in the absence of an extraction step, assay specificity is entirely dependent on the specificity of the antiserum. With specific antisera to 11α-hydroxyprogesterone hemisuccinate, the cross-reactions of other steroids, together with their concentrations in plasma, indicate that the direct assay of progesterone in plasma is possible.

Successful direct methods have also been described, and fully validated, for the measurement of progesterone in plasma of nonpregnant women. HAYNES et al. (1980) used an amount of cortisol which reduced to a minimum the binding of progesterone to plasma proteins, but which showed no cross-reaction in the RIA. With this amount of cortisol, standard curves in the presence or absence of plasma were almost identical. Without cortisol, addition of plasma to the standards greatly decreased sensitivity. The method showed only an insignificant positive bias in external quality assessment schemes. In the method described by MCGINLEY and CASEY (1979), the synthetic steroid danazol was used as the blocking agent; this abolished the effect of plasma on progesterone standard curves. The direct method gave values for the progesterone concentration of plasma from women at various stages in their menstrual cycle in good agreement with those from a method using the same antiserum and hexane extraction.

2. Alternative Immunoassay Methods for Progesterone

Immunoassays for progesterone have been described with several types of nonisotopic labels: enzymes (NAKANE 1980; JOYCE et al. 1977, 1981; DRAY et al. 1975; TALLON et al. 1984), a fluorophore (ALLMAN et al. 1981), and chemiluminescent molecules (KOHEN et al. 1979, 1981; PAZZAGLI et al. 1981c, d; DE BOEVER et al. 1984); all these methods have been applied to plasma, but differ in their suitability as routine techniques. In the EIAs described by NAKANE (1980), 11α-hydroxyprogesterone hemisuccinate was labeled with β-galactosidase and used with a homologous antiserum and second antibody separation. The method was applied to bovine plasma; it was precise and gave results in agreement with those by RIA. EIA can therefore give analytic performance as good as RIA. A disadvantage is that the measurement of enzyme activity is tedious and time-consuming.

Allman et al. (1981) have described methods for the FIA of progesterone, using a label prepared by linking fluoresceinamine to progesterone-3-carboxymethyloxime. An antiserum to 11α-hydroxyprogesterone hemisuccinate was used, and separation of the antibody-bound and free antigen was performed by adding ammonium sulfate. Plasma samples were extracted with hexane. The method showed good agreement with RIA. The disadvantages of FIA are that, measurement of the fluorescent label being relatively insensitive, large amounts (500 fmol) of label antigen have to be used; this decreases the sensitivity of the immunoassay and requires the use of larger quantities of antiserum.

For the measurement of progesterone by CL immunoassay, derivatives of isoluminol were linked to 11α-hydroxyprogesterone hemisuccinate and the light emission measured when this labeled progesterone was oxidized by hydrogen peroxide in a reaction catalyzed by microperoxidase. The first assay of this type, described by Kohen et al. (1979), required no physical separation of the antibody-bound and free antigen because binding to antibody enhanced light emission from the label. The method was applied to plasma and appears promising on the basis of a limited comparison with RIA.

The method described by Pazzagli et al. (1981 d), in which charcoal was used to adsorb free antigen, and delayed addition of the label was used to increase sensitivity of the homologous assay, also gave results for the measurement of plasma progesterone in agreement with those by RIA. The advantage of CL for immunoassay is that it allows the measurement of very small quantities of labeled antigen with short reaction times and therefore, as for EIAs, the sensitivity of immunoassay is not limited by the sensitivity of this measurement.

3. Unconjugated Estrogens

RIA is the most widely used method for measuring the concentration of 17β-estradiol in plasma (Abraham and Odell 1970; Abraham et al. 1971). The increasing application of this assay for clinical purposes [detection of ovulation for artificial insemination, in vitro fertilization, gamete intrafallopian transfer (GIFT) etc.] pointed out the necessity of simpler methods to be used routinely. Several direct assays, solid-phase methods, and simple methods after ethyl ether extraction of plasma have been proposed and stressed as reliable (Barnard et al. 1975; Jurjens et al. 1975; Lindberg and Edquist 1974). As a result, the apparent concentration of estradiol in the same plasma sample as measured by RIA can differ considerably from one laboratory to another. To be honest, even today estradiol cannot be measured by RIA in male plasma, prepubertal children, or postmenopausal women without a previous chromatographic step. In ovulatory women, the rapid methods must be carefully controlled using mass spectrometry as reference.

The antisera to the 6-carboxymethyloxime conjugate are still the most widely used because they can easily discriminate between estradiol and estrone (Kuss and GOEBEL 1972). Furthermore, as a consequence of the increasing alteration of the intratissue estrogen metabolism in breast and uterine cancer, estrone and estradiol have been measured in adipose, mammary, and uterine tissue. In fact, estrone is poorly active on cellular proliferative processes while estradiol is very active. At the target levels, estrone can be easily converted into estradiol by the enzyme 17β-hydroxysteroidodehydrogenase (Fig. 8). The activity of this two-direction enzyme can be modified by hormonal or pharmacologic manipulation. For instance, progestagens decrease the rate of conversion of estrone into estradiol, while the androgens have an inverse action.

Unfortunately, estrogen measurements at the tissue level are much more complicated than in plasma in relation to nonspecific interference from different types of tissues. For example, the cytosol and the nuclear fractions must be extracted separately and with different procedures.

4. Conjugated Estrogens

Plasma levels of estrone sulfate (which is in equilibrium with the pool of unconjugated estradiol) have been measured after enzyme or acid hydrolysis of the sulfate moiety. Problems with unacceptably high assay blank values, caused by the use of a sulfatase enzyme preparation from *Helix pomatia*, have been overcome by treating the enzyme preparation with Amberlite XAD-2 or charcoal to ensure specificity (CARLSTRÖM and SKOLDEFORS 1977; ROBERTS et al. 1980). Some assay methods have incorporated a chromatographic step into the assay procedure with either thin layer chromatography (NOEL et al. 1981) or celite column chromatography (FRANZ et al. 1979). Sephadex LH-20 column chromatography has been used to separate estradiol from estrone after hydrolysis of sulfate conjugate (MYKING et al. 1980).

5. Free Estradiol in Blood

Several methods for measuring free estradiol in blood have been described and reviewed by REED and MURRAY (1979). A very accurate technique has been described by HAMMOND et al. (1980) based on incubation of undiluted plasma with estradiol ^{3}H and glucose ^{14}C following ultrafiltration at 37 °C. The percentage of unbound estradiol is calculated by comparing the ratio of estradiol ^{3}H to glucose ^{14}C in the ultrafiltrate with the ratio in serum retained by the membrane. Free estradiol is increased in obese women. The availability of higher free estradiol concentrations at the tissue target levels represents an important risk factor for uterine and mammary cancer in women.

6. Ovarian Hormones in Saliva and Urine

Owing to the variability of hormonal secretion during the cycle and the pulsatility in blood of ovarian hormones, investigations have been made of the use of alternative biologic fluids allowing the determination of ovarian hormones or their main metabolites as well as of a more practicable way of sample collection in comparison with venipuncture, because daily blood sampling during the entire cycle is invasive, expensive, tedious, and interferes with the working activity of women.

Therefore, increasing attention has been focused on saliva and urine as alternative fluids for monitoring the endocrine function of the ovary and detecting the time of ovulation (WALKER et al. 1981; RIAD-FAHMY et al. 1982). Both fluids, however, have advantages and disadvantages.

Saliva seems easy to collect. Women, in fact, can collect samples at home, store them in the refrigerator, and carry them to the laboratory at the end of the cycle. Unfortunately, the levels of estradiol and progesterone are much lower in saliva than in blood. They correlate with the free fraction of these hormones, but they are not superimposable. Furthermore, they reflect the pulsatile secretion of blood hormones and the assay methods are affected by nonspecific interference from different biologic materials. In practice, salivary measurements are limited to the determination of salivary progesterone for the diagnosis of luteal insufficiency. Moreover, even such simple techniques have some as yet unexplained limitations as the differences between the salivary concentration of progesterone in the follicular and luteal phases are very small. In fact, this ratio in saliva is 1:3 whereas in blood it is 1:20 or more. However, salivary progesterone is routinely used in several laboratories with satisfactory clinical results, but extraction of progesterone with petroleum ether or hexane is needed.

For two reasons, the measurement of urinary glucuronide metabolites of ovarian hormones has not been widely used in the past for assessing ovarian function: (a) determination of hormones in urine usually requires a 24-h urine collection; and (b) available methods for urinary metabolites of ovarian hormones based on colorimetry or gas chromatography require previous enzymatic or acid hydrolysis of glucuronides and subsequent extraction of the free steroid. These steps are time-consuming and hydrolysis procedures are not always reliable (MESSERI et al. 1984a). Moreover, the main urinary metabolites of 17β-estradiol and progesterone, estrone-3-glucuronide and pregnanediol-3-glucuronide, represent only a fraction of the daily secretion of such hormones (15%–40%) (BAKER et al. 1978).

Subsequent developments however, based on the pioneer work of KELLIE (1975), have improved the practical and methodological prospecs for the determination of these compounds. The availability of specific monoclonal and polyclonal antibodies to glucuronide steroids has allowed the direct measurement of these metabolites, even in diluted urine (ESHHAR et al. 1981; KOHEN et al. 1980a; BAKER et al. 1978).

RIA methods for pregnanediol-3α-glucuronide (SAMARAJEEWA et al. 1979; STANCZYK et al. 1980), estrone-3α-glucuronide (WRIGHT et al. 1978, 1979; STANCZYK et al. 1980), estriol-16α-glucuronide (LEHTINEN and ADLERCREUTZ 1977; WRIGHT et al. 1979), and estradiol-17-glucuronide (WRIGHT et al. 1979) have already been described. Furthermore, several investigations of ovarian function have demonstrated the adequacy of measuring the concentration of ovarian metabolites in early morning or overnight urine instead of 24-h urine (COLLINS et al. 1979; DENARI et al. 1981). This approach increases the applicability of the tests by facilitating sample collection. In studying ovarian disorders and the response to pharmacologic manipulations (antiestrogens, gonadotropins, GnRH, etc.) we routinely use these methods, with important clinical results. Comparative methods performed in diluted urine with highly sophisticated mass spectrometric techniques gave superimposable results (MONETI et al. 1985).

7. Alternative Immunoassay Methods

Several heterogeneous EIAs have been reported for the measurement of estrogens (VAN WEEMAN and SCHUURS 1975; BOSCH et al. 1978a; NUMAZAWA et al. 1977; EXLEY and ABUKNESHA 1978). SHAH and JOSHI (1982) have reported a rapid, simple, and sensitive enzyme-linked immunosorbent assay for the estimation of estrone-3-glucuronide in diluted urine. The labeled antigen, estrone-3-glucuronide-penicillinase, was prepared by a carbodiimide reaction. SHAH et al. (1984) have also reported a heterogeneous FIA for pregnanediol-3α-glucuronide in diluted urine.

A homogeneous LIA method for estriol-16α-glucuronide (KOHEN et al. 1982) has been developed and the assay achieved sensitivity and specificity similar to that obtained by the conventional RIA. However, an extraction step of the biologic sample was introduced into the procedure to remove nonspecific compounds interfering with the CL reaction. We have developed a homogeneous LIA for total urinary estrogens in diluted hydrolyzed urine samples (MESSERI et al. 1984b); this direct assay was possible because just 0.5 μl of sample was used for the assay and such a quantity did not interfere with the CL reaction.

Heterogeneous LIA methods have been described for estradiol in plasma (KIM et al. 1982; DE BOEVER et al. 1983) and for estriol-16α-glucuronide (BARNARD et al. 1981), estrone-4-glucuronide (LINDNER et al. 1981), and pregnanediol-3-glucuronide (ESHHAR et al. 1981). The solid-phase LIA methods for urinary steroid glucuronides appear likely to find a place in the clinical laboratory. In fact, the use of specific antibodies raised against urinary steroid metabolites, together with solid-phase techniques, can avoid both the hydrolysis and the purification of urine samples which are time-consuming and impractical methodological steps.

Acknowledgments. This work is supported partly by grants from the Region of Tuscany and from the Italian National Research Council special project "Sanitary and Biomedical Technologies."

References

Abraham GE (1969) Solid-phase radioimmunoassay of estradiol-17β. J Clin Endocrinol Metab 29:866–870

Abraham GE (1974) Radioimmunoassay of steroids in biological materials. In: The proceedings of symposium – RIA and related proceedings in medicine, vol II. International Atomic Energy Agency, Vienna, pp 1–29

Abraham GE (1975) Characterisation of anti-steroid sera. In: Cameron EHD, Hillier SG, Griffith K (eds) Steroid immunoassay. Alpha Omega, Cardiff, pp 67–78

Abraham GE, Odell WD (1970) Solid-phase radioimmunoassay of estradiol. In: Peron FG, Caldwell BV (eds) Immunological methods in steroid determination. Appleton-Century Crofts, New York, pp 87–102

Abraham GE, Hopper K, Tulchinsky D, Swerdloff RS, Odell WD (1971) Simultaneous measurement of plasma progesterone, 17-hydroxyprogesterone and estradiol-17β by radioimmunoassay. Analyt Letter 4:325–331

Abraham GE, Manlimos FS, Garza R (1977) Radioimmunoassay of steroids. In: Abraham GE (ed) Handbook of radioimmunoassay. Dekker, New York, pp 591–656

Al-Dujaili EAS, Edwards ERW (1978) The development and application of a direct radioimmunoassay for plasma aldosterone using ^{125}I labelled ligand: comparison of three methods. J Clin Endocrinol Metab 46:105–113

Al-Dujaili EAS, Edwards ERW (1981) Development and application of a simple radioimmunoassay for urinary aldosterone. Clin Chim Acta 116:277–287
Al-Dujaili EAS, Willams BC, Edwards CRW (1981) The development and application of a direct radioimmunoassay for corticosterone. Steroids 37:157–176
Allen RM, Redshaw MR (1978) The use of homologous and heterologous ^{125}I radioligands in the RIA of progesterone. Steroids 32:467–486
Allman BL, Short F, James VHT (1981) Fluoroimmunoassay of progesterone in human serum or plasma. Clin Chem 27:1176–1179
Arakawa H, Maeda M, Tsuji A, Kambegawa A (1981) Chemiluminescence enzyme immunoassay of dehydroepiandrosterone and its sulphate using peroxidase as label. Steroids 38:453–464
Arakawa H, Maeda M, Tsuji A (1982) Chemiluminescence enzyme immunoassay of a 17α-hydroxyprogesterone using glucose oxidase and bis(2,4,6-trichlorophenyl)oxalate-fluorescent dye system. Chem Pharm Bull 30:3036–3039
Arakawa H, Maeda M, Tsuji A, Natuse H, Suzuki E, Kambegawa A (1983) Fluorescence enzyme immunoassay of 17α-hydroxyprogesterone in dried blood samples on filter paper and its application to mass screening for congenital adrenal hyperplasia. Chem Pharm Bull 31:2724–2731
Baird DT (1976) Oestrogens in clinical practice. In: Loraine JA, Bell FT (eds) Hormone assays and their clinical application, 4th edition. Churchill Livingstone, Edinburgh, pp 408–446
Baker TS, Jennison KM, Kellie AE (1978) The direct immunoassay of estrogen glucuronides in human fertile urine. Biochem J 177:729–738
Barnard GJR, Hennam JF, Collins WP (1975) Further studies on radioimmunoassay systems for plasma oestradiol. J Steroid Biochem 6:107–116
Barnard GJR, Collins WP, Kohen F, Lindner HR (1981) The measurement of urinary estriol-16α-glucuronide by a solid-phase chemiluminescence immunoassay. J Steroid Biochem 14:941–948
Baumann G, Rappaport G, Lemarchand-Beraud T, Felber JD (1975) Free cortisol index: a rapid and simple estimation of free cortisol in human plasma. J Clin Endocrinol Metab 40:462–469
Baxendale PM, Reed MJ, James VHT (1981) Inability of human endometrium or myometrium to aromatize androstenedione. J Steroid Biochem 14:305–306
Biglieri EG, Schamberlan M (1979) The significance of elevated levels of plasma 18-hydroxycorticosterone in patients with primary aldosteronism. J Clin Endocrinol Metab 49:87–91
Blake C, Gould BJ (1984) Use of enzymes in immunoassay techniques: a review. Analyst 109:533–547
Bolton AE, Hunter WM (1973) The use of antisera covalently coupled to agarose, cellulose and Sephadex in radioimmunoassay systems for proteins and haptens. Biochim Biophys Acta 329:318–330
Bosch AMG, Dijkhuizen PM, Schuurs AHWM, Van Weeman BK (1978 a) Enzyme-immunoassay for total oestrogens in pregnancy plasma or serum. Clin Chim Acta 89:59–70
Bosch AMG, Stevens WHJM, Van Wimgaarden CJ, Schuurs AHWM (1978 b) Solid-phase enzyme-immunoassay of testosterone. Z Analyt Chem 290:98–106
Cameron EHD, Hillier SG, Griffith K (1975) Steroid immunoassay. Alpha Omega, Cardiff
Carlström K, Skoldefors H (1977) Determination of total estrone in peripheral serum from non-pregnant humans. J Steroid Biochem 8:1127–1128
Carr BR, Mikhail G, Flickinger GL (1971) Column chromatography of steroids on Sephadex LH-20. J Clin Endocrinol Metab 33:358–360
Cattaneo S, Forti G, Fiorelli G, Barbieri U, Serio M (1975) A rapid radioimmunoassay for determination of dehydroepiandosterone sulphate in human plasma. Clin Endocrinol 4:505–512
Collins WP, Barnard GJR, Hennam JF (1975) Factors affecting the choice of separation technique. In: Cameron EHD, Hillier SG, Griffith K (eds) Steroid immunoassay. Alpha Omega, Cardiff, pp 223–228

Collins WP, Collins PO, Kilpatrick MJ, Miaming PA, Pike JM, Tyler JPP (1979) The concentrations of urinary estrone-3-glucuronide LH and pregnanediol-3α-glucuronide as indices of ovarian function. Acta Endocrinol 90:336–348

Comoglio S, Celada F (1976) An immuno-enzymatic assay of cortisol using E. Coli B-galactosidase as label. J Immunol Methods 10:161–170

Connolly TM, Vecsei P (1978) Simple radioimmunoassay of cortisol in diluted samples of human plasma. Clin Chem 24:1468–1472

Connolly TM, Tibor L, Gless KH, Vecsei P (1980) Screening radioimmunoassay for aldosterone in preheated plasma without extraction and chromatography. Clin Chem 26:41–45

Cooke ID (1976) Progesterone and its metabolites. In: Loraine JA, Bell FT (eds) Hormone assays and their clinical application, 4th edition. Churchill Livingstone, Edinburgh, pp 447–518

Corrie JET (1982) Immunoassays for steroid hormones using radioiodinated tracers. Br Vet J 138:439–442

Corrie JET, Hunter WM (1981) ^{125}Iodinated tracers for hapten specific radioimmunoassays. In: Langone JJ, Van Vunakis H (eds) Immunochemical techniques, part B. Methods Enzymol 73:79–112

Dawson EC, Denissen EHC, Van Weeman BK (1978) A simple and efficient method for raising steroid antibodies in rabbits. Steroids 31:357–366

De Boever J, Kohen F, Vanderkerckhove D (1983) A solid-phase chemiluminescence immunoassay for plasma estradiol-17β for gonadotropin therapy compared with two different radioimmunoassays. Clin Chem 29:2068–2072

De Boever J, Kohen F, Vanderkerckhove D, Van Maede G (1984) Solid-phase chemiluminescent immunoassay for progesterone in unextracted serum. Clin Chem 30:1637–1641

Denari H, Farinati Z, Casas PRF, Oliva A (1981) Determination of ovarian function using first morning urine steroid assays. Obstet Gynaecol 58:5–9

Dighe KK, Hunter WM (1974) A solid-phase radioimmunoassay for plasma progesterone. J Biochem 143:219–231

Dray F, Andrieu JM, Renaud F (1975) Enzyme immunoassay of progesterone at the picogram level using β-galactosidase as label. Biochem Biophys Acta 403:131–138

Eddy RL, Jones AL, Gilliland PF (1973) Cushing's syndrome: a prospective study of diagnostic methods. Am J Med 55:621–630

Edwards CRW, Landon J (1976) Corticosteroids. In: Loraine JA, Bell FT (eds) Hormone assays and their clinical application, 4th edition. Churchill Livingstone, Edinburgh, pp 519–579

Erlanger BF (1981) The preparation of antigenic hapten-carrier conjugates: a survey. In: Van Vunakis H, Langone JJ (eds) Immunochemical techniques, part A. Methods Enzymol 70:85–104

Erlanger BF, Borek F, Beiser SM, Liebermann S (1959) Steroid-protein conjugates: preparation and characterization of conjugates of bovine serum albumin with progesterone, deoxycorticosterone and estrone. J Biol Chem 234:1090–1094

Eshhar Z, Kim JB, Barnard G, Collins WP, Gilad S, Lindner HR, Kohen F (1981) Use of monoclonal antibodies to pregnanediol-3α-glucuronide for the development of a solid-phase chemiluminescence immunoassay. Steroids 38:89–109

Exley D (1983) Fluorimmunoassay of steroids. In: Hunter WM, Corrie JET (eds) Immunoassay for clinical chemistry. Churchill Livingstone, Edinburgh, pp 398–400

Exley D, Abuknesha R (1978) A highly sensitive and specific EIA method for oestradiol-17β. FEBS Lett 91:162–165

Exley D, Ekeke GI (1981) Fluorimmunoassay of 5α-dihydrotestosterone. J Steroid Biochem 14:1297–1302

Fahmy D, Read GF, Hillier SG (1975) Some observations on the determination of cortisol in human plasma by radioimmunoassay using antisera against cortisol 3-BSA. Steroids 26:267–280

Fantl VE, Wang DY, Whitehead AS (1981) Production and characterisation of monoclonal antibody to progesterone. J Steroid Biochem 14:405–407

Forti G, Pazzagli M, Calabresi E, Fiorelli G, Serio M (1974) Radioimmunoassay of plasma testosterone. Clin Endoc 3:5–17
Forti G, Calabresi E, Giannotti P, Borrelli D, Gonnelli P, Barbieri U, Serio M (1978) Measurement of 5-androstene-3β,17β-diol in spermatic and peripheral venous blood samples from the same human subjects by a radioimmunoassay method. Horm Res 9:194–203
Forti G, Toscano V, Casilli D, Maroder M, Balducci R, Adamo MV, Santoro S, Grisolia GA, Pampaloni A, Serio M (1985) Spermatic and peripheral venous plasma concentrations of testosterone, 17-hydroxyprogesterone, androstenedione, dehydroepiandrosterone, Δ^5-androstene-3β,17β-diol, dihydrotestosterone, 5α-androstane-3α,17β-diol, 5α-androstane-3β,17β-diol, and estradiol in boys with idiopathic varicocele in different stages of puberty. J Clin Endocrinol Metab 61:322–327
Foster LB, Dunn RT (1974) Single-antibody technique for radioimmunoassay of cortisol in unextracted serum of plasma. Clin Chem 20:365–368
Franchimont P (1976) Releasing hormones in relation to the pituitary gonadal axis. In: Loraine JA, Bell FT (eds) Hormone assays and their clinical application, 4th edn. Churchill Livingstone, Edinburgh, pp 630–656
Franz C, Watzon D, Longcope C (1979) Estrone sulfate and dehydroepiandrosterone sulfate concentrations in normal subjects and men with cirrhosis. Steroids 34:563–573
Furuyama S, Mayes DM, Nugent CA (1970) A radioimmunoassay for plasma testosterone. Steroids 16:415–428
Galfre G, Milstein C (1981) Preparation of monoclonal antibodies: strategies and procedures. Methods Enzymol 73:3–46
Gomez-Sanchez C, Murry BA, Kem DC, Kaplan NM (1975) A direct radioimmunoassay of corticosterone in rat serum. Endocrinology 96:796–798
Hammond GL, Nisker JA, Jones LA, Siiteri PK (1980) Estimation of the percentage of free steroid in undiluted serum by centrifugal ultrafiltration dialysis. J Biol Chem 255:5023–5026
Haynes SP, Corcoran JM, Eastman CJ, Doy FA (1980) Radioimmunoassay of progesterone in unextracted serum. Clin Chem 26:1607–1609
Heftmann E, Hunter IR (1979) High pressure liquid chromatography of steroids. J Chromatogr 165:283–299
Hemmila I, Dakubu S, Mukkala V-M, Siitari H, Lovgren T (1984) Europium as a label in time-resolved immunofluorometric assays. Analyt Biochem 137:335–343
Hindawi RK, Caskell SJ, Read GF, Riad-Fahmy D (1980) A simple, direct, solid-phase enzyme immunoassay for cortisol in plasma. Ann Clin Biochem 17:53–70
Horton R, Endres D, Galmarini M (1984) Ideal conditions for hydrolysis of androstanediol-3α-17β-diol-glucuronide in plasma. J Clin Endocrinol Metab 59:1027
Hunter WM, Nars PW, Rutherford FJ (1975) Preparation and behaviour of ^{125}I-labeled radioligands for phenolic and neutral steroids. In: Cameron EHD, Hillier SG, Griffith K (eds) Steroid immunoassay: fifth Tenovus international workshop. Alpha Omega, Cardiff, pp 141–156
Ismail AAA (1976) Testosterone. In: Loraine JA, Bell FT (eds) Hormone assays and their clinical application, 4th edn. Churchill Livingstone, Edinburgh, pp 580–629
Janoski AH, Shulman FC, Wright EG (1974) Selective 3-(O-carboxymethyl)oxime formation in steroidal 3,20 diones for hapten immunospecificity. Steroids 23:49–64
Jeffcoate SL (1981) Efficiency and effectiveness in the endocrine laboratory. Academic, London
Jowett TP, Slater JDH (1977) Development of radioimmunoassays for the measurement of aldosterone in unprocessed plasma and simple plasma extracts. Clin Chim Acta 80:435–446
Jowett TP, Slater JDH (1981) A radioimmunoassay for the measurement of tetrahydroaldosterone 3-glucosiduronic acid in human plasma. Clin Chim Acta 109:133–144
Joyce BG, Read GF, Riad-Fahmy D (1977) A specific enzymeimmunoassay of progesterone in human plasma. Steroids 29:761–770
Joyce BG, Othick AH, Read GF, Riad-Fahmy D (1981) A sensitive specific, solid-phase enzymeimmunoassay for plasma progesterone. Ann Clin Biochem 18:42–47

Jurjens H, Pratt JJ, Woldring MG (1975) Radioimmunoassay of plasma estradiol without extraction and chromatography. J Clin Endocrinol Metab 40:19–25
Kachel CD, Mendelsohn FA (1979) An automated multicolumn system for chromatography of aldosterone on sephadex LH-20 in water. J Steroid Biochem 10:563–567
Kellie AE (1975) The radioimmunoassay of steroid conjugates. J Steroid Biochem 6:277–281
Khadempour MH, Laing I, Gowenlock AH (1978) A simplified radioimmunoassay procedure for serum progesterone. Clin Chim Acta 82:161–171
Kim JB, Barnard GJ, Collins WP, Kohen F, Lindner HR, Eshhar Z (1982) Measurement of plasma estradiol-17β by solid-phase chemiluminescence immunoassay. Clin Chem 28:1120–1124
Kobayashi Y, Amitani K, Watanabe F, Miyai K (1979a) Fluorescence polarization immunoassay for cortisol. Clin Chem 92:241–247
Kobayashi Y, Miyai K, Tsubota N, Watanabe F (1979b) Direct fluorescence polarization immunoassay of serum cortisol. Clin Chem 24:2139–2144
Kohen F, Pazzagli M, Kim JB, Lindner HR, Boguslaski RC (1979) An assay procedure for plasma progesterone based on antibody enhanced chemiluminescence. FEBS Lett 104:201–205
Kohen F, Kim JB, Barnard G, Lindner HR (1980a) An immunoassay for urinary estriol-16α-glucuronide based on antibody-enhanced chemiluminescence. Steroids 36:405–419
Kohen F, Pazzagli M, Kim JB, Lindner HR (1980b) An immunoassay for plasma cortisol based on chemiluminescence. Steroids 36:421–438
Kohen F, Kim JB, Lindner HR, Collins WP (1981) Development of a solid-phase chemiluminescence immunoassay for plasma progesterone. Steroids 38:73–88
Kohen F, Lichter S, Eshhar Z, Lindner HR (1982) Preparation of monoclonal antibodies able to discriminate between testosterone and 5α-dihydrotestosterone. Steroids 39:453–459
Kohen F, Lindner HR, Gilad S (1983) Development of chemiluminescence monitored immunoassays for steroid hormones. J Steroid Biochem 19:413–418
Kuss E, Goebel R (1972) Determination of estrogens by radioimmunoassay with antibodies to estrogen-6-conjugates. Steroids 19:509–518
Lehtinen T, Adlercreutz H (1977) Solid phase radioimmunoassays of estriol-16α-glucuronide in urine and pregnancy plasma. J Steroid Biochem 8:99–104
Liebermann S, Erlanger BF, Beiser SM, Agate FJ (1959) Steroid protein conjugates: their chemical, immunochemical and endocrinological properties. Recent Prog Horm Res 15:165–196
Lindberg P, Edquist LE (1974) A use of 17β-estradiol-6-(*O*-carboxymethyl)oxine-(^{125}I)tyramine as tracer for a radioimmunoassay of 17β-estradiol. Clin Chim Acta 53:169–174
Lindner HR, Kohen F, Eshhar Z, Kim JB, Barnard G (1981) Novel assay procedure for assessing ovarian function in women. J Steroid Biochem 15:131–136
Lindstrom L, Meurling L, Lovegren T (1982) The measurement of serum cortisol by a solid-phase chemiluminescence immunoassay. J Steroid Biochem 16:577–580
Mattox VR, Nelson AN (1981) Determination of urinary tetrahydro aldosterone glucosiduronic acid by radioimmunoassay. J Steroid Biochem 14:243–249
McGinley R, Casey JH (1979) Analysis of progesterone in unextracted serum: a method using Danazol, a blocker of steroid binding to proteins. Steroids 33:127–138
Messeri G, Cugnetto G, Moneti G, Serio M (1984a) *Helix pomatia* induced conversion of some 3 hydroxysteroids. J Steroid Biochem 20:793–796
Messeri G, Caldini AL, Bolelli GF, Pazzagli M, Tommasi A, Vannucchi PL, Serio M (1984b) Homogeneous luminescence immunoassay for total estrogens in urine. Clin Chem 30:653–657
Moneti G, Agati G, Giovannini MG, Pazzagli M, Salerno R, Messeri G, Serio M (1985) Pregnanediol-3α-glucuronide measured in diluted urine by mass spectrometry with fast atom bombardment/negative-ion ionization. Clin Chem 31:46–49
Morgan CA, Cooke ID (1972) A comparison of the competitive protein-binding assay and radioimmunoassay for plasma progesterone during the normal menstrual cycle. J Endocr 54:445–456

Myking O, Thorsen T, Stoa KF (1980) Conjugated and unconjugated plasma estrogens – estrone, estradiol, estriol – in normal human males. J Steroid Biochem 13:1215–1220

Nakane PK (1980) Future trends and application of immunoassays. In: Keitges PW, Nakamura RM (eds) Diagnostic immunology – current and future trends. College of American Pathologists, Skokie, IL, pp 87–96

Nieschlag E, Kley HK, Usadel K-H (1975) Production of steroid antisera in rabbits. In: Cameron EHD, Hillier SG, Griffith K (eds) Steroid immunoassay. Alpha Omega, Cardiff, pp 87–96

Niswender GD, Midgley AR Jr (1970) Hapten-radioimmunoassay for steroid hormones. In: Peron FG, Caldwell BV (eds) Immunologic methods in steroid determination. Appleton-Century-Crofts, New York, pp 149–174

Noel CT, Reed MJ, Jacobs HS, James VHT (1981) The plasma concentration of estrone sulphate in post-menopausal women: lack of diurnal variation, effect of variectomy, age and weight. J Steroid Biochem 14:1101–1105

Nordblom GD, Counsell RE, England BG (1979) Ligand specificity and bridging phenomena in hapten radioimmunoassays. Ligand Q 2:34–36

Nordblom GD, Webb R, Counsell RE, England BG (1981) A chemical approach to solving bridging phenomena in steroid radioimmunoassays. Steroids 38:161–173

Nowaczynski W, Guthrie GP, Messerli FH, Genest J, Kuchel O, Honda M, Grose J (1977) Effects of ACTH and posture on aldosterone metabolism in essential hypertension. J Steroid Biochem 8:1225–1231

Numazawa M, Haruya A, Kurosaka K, Nambara T (1977) Picogram order enzyme-immunoassay oestradiol. FEBS Lett 79:396–398

Ogihara T, Miyai K, Kumahara K, Ishibashi K, Nishi K (1977) Enzyme-labelled immunoassay of plasma cortisol. J Clin Endocrinol 44:91–95

Patel A, Campbell AK (1983) Homogeneous immunoassay based on chemiluminescence energy transfer. Clin Chem 29:1604–1608

Patterson DG, Patterson MB, Culbreth PH et al. (1984) Determination of steroid hormones in a human-serum reference material by isotope-dilution mass-spectrometry: a candidate definitive method for cortisol. Clin Chem 30:619–626

Pazzagli M, Forti G, Cappellini A, Serio M (1975) Radioimmunoassay of plasma dihydrotestosterone in normal and hypogonadal men. Clin Endocrinol 4:513–520

Pazzagli M, Kim JB, Messeri G, Kohen F, Bolelli GF, Tommasi A, Salerno R, Moneti G, Serio M (1981 a) Luminescence immunoassay (LIA) of cortisol: 1. synthesis and evaluation of the chemiluminescent labels of cortisol. J Steroid Biochem 14:1005–1012

Pazzagli M, Kim JB, Messeri G, Kohen F, Bolelli GF, Tommasi A, Salerno R, Serio M (1981 b) Luminescent immunoassay (LIA) of cortisol: 2. development and validation of the immunoassay monitored by chemiluminescence. J Steroid Biochem 14:1181–1187

Pazzagli M, Kim JB, Messeri G, Martinazzo G, Kohen F, Franceschetti F, Moneti G, Salerno R, Tommasi A, Serio M (1981 c) Evaluation of different progesterone-isoluminol conjugates for chemiluminescence immunoassay. Clin Chim Acta 115:277–286

Pazzagli M, Kim JB, Messeri G, Martinazzo G, Franceschetti F, Tommasi A, Salerno R, Serio M (1981 d) Luminescent immunoassay (LIA) for progesterone in a heterogeneous system. Clin Chim Acta 115:287–296

Pazzagli M, Serio M, Munsun P, Rodbard D (1982 a) A chemiluminescent immunoassay (LIA) for testosterone. In: Radioimmunoassay and related procedures in medicine. IAEA, Vienna, pp 747–755

Pazzagli M, Bolelli GF, Messeri G, Martinazzo G, Tommasi A, Salerno R, Serio M (1982 b) Homogeneous luminescent immunoassay for progesterone: a study on the antibody-enhanced chemiluminescence. In: Serio M, Pazzagli M (eds) Luminescent assays: perspectives in endocrinology and clinical chemistry. Raven, New York, pp 191–200

Pazzagli M, Messeri G, Caldini AL, Moneti G, Martinazzo G, Serio M (1983) Preparation and evaluation of steroid chemiluminescent tracers. J Steroid Biochem 19:407–412

Pazzagli M, Messeri G, Salerno R, Caldini AL, Tommasi A, Magini A, Serio M (1984) Luminescent immunoassay (LIA) methods for steroid hormones. Talanta 10B:901–907

Perry LA, Al-Dujaili EAS, Edwards CRW (1982) A direct radioimmunoassay for 11-deoxy-cortisol. Steroids 39:115–128
Pourfarzaneh M, White GW, Landon J, Smith DS (1980) Cortisol directly determined in serum by fluoroimmunoassay with magnetisable solid-phase. Clin Chem 26:730–733
Pratt JJ, Wiegmann T, Lapphoen RE, Woldring MG (1975) Estimation of plasma testosterone without extraction and chromatography. Clin Chim Acta 59:337–346
Pratt JJ, Boonman R, Woldring MG, Donker AJM (1978) Special problems in the radioimmunoassay of plasma aldosterone without prior extraction and purification. Clin Chim Acta 84:329–337
Puri V, Puri CP, Anaud Kumar TC (1981) Serum levels of dihydrotestosterone in male rhesus monkeys estimated by a nonchromatographic radioimmunoassay method. J Steroid Biochem 14:877–881
Rajkowski KM, Cittanova N, Urios P, Jayle MF (1977) An enzyme-linked immunoassay for testosterone. Steroids 30:129–137
Ratcliffe JG (1983) Requirements for separation methods in immunoassay. In: Hunter WM, Corrie JET (eds) Immunoassays for clinical chemistry. Churchill Livingstone, Edinburgh, pp 135–138
Ratcliffe WA (1983) Direct (non-extraction) serum assays for steroids. In: Hunter WM, Corrie JET (eds) Immunoassays for clinical chemistry. Churchill Livingstone, Edinburgh, pp 401–409
Reed MJ, Murray MAF (1979) The estrogens. In: Gray CH, James VHT (eds) Hormones in blood, 3rd edn. Academic, London, vol 3, pp 263–290
Riad-Fahmy D, Read GF, Walker RF, Griffith K (1982) Steroids in saliva for assessing endocrine function. Endocr Rev 3:367–395
Roberts KD, Rochefort JG, Bleau G, Chapdelaine A (1980) Plasma estrone sulfate levels in post menopausal women. Steroids 35:179–187
Rolleri E, Zannino M, Orlandi S, Malvano R (1976) Direct radioimmunoassay of plasma cortisol. Clin Chim Acta 66:319–322
Samarajeewa P, Cooley G, Kellie AE (1979) The radioimmunoassay of pregnanediol-3α-glucuronide. J Steroid Biochem 11:1165–1171
Scarisbrick JJ, Cameron EHD (1975) Radioimmunoassay of progesterone: comparison of [1,2,6,7-^{3}H4]-progesterone and [^{125}I] progesterone iodohistamine radioligands. J Steroid Biochem 6:51–56
Schroeder HR, Yeager FM (1978) Chemiluminescence yields and detection limits of some iso-luminol derivatives in various oxidation systems. Analyt Chem 50:1114–1120
Shah HP, Joshi UM (1982) A simple, rapid and reliable enzyme-linked immunoabsorbent assay (ELISA) for measuring estrogen-3-glucuronide in urine. J Steroid Biochem 16:283–286
Shah H, Saranko AM, Harkonen M, Adlercreutz H (1984) Direct solid-phase fluoroenzyme immunoassay of 5β-pregnane-3α, 20α-diol-3α-glucuronide in urine. Clin Chem 30:185–187
Siekmann L (1979) Determination of steroid hormones by the use of isotope dilution-mass spectrometry: a definitive method in clinical chemistry. J Steroid Biochem 11:117–123
Smith DS, Al Hakiem HH, Landon J (1981) A review of fluoroimmunoassay and immunofluorometric assay. Ann Clin Biochem 18:253–274
Soini E, Hemmila I (1979) Fluoroimmunoassay: present status and key problems. Clin Chem 25:353–361
Soini E, Kojola H (1983) Time-resolved fluorimeter for lanthanide chelates – a new generation of nonisotopic immunoassays. Clin Chem 29:65–68
Solyom J (1981) Blood spot 17α-hydroxyprogesterone radioimmunoassay in the follow-up of congenital adrenal hyperplasia. Clin Endocrinol 14:547–553
Stanczyk FZ, Miyakawa I, Goebelsmann U (1980) Direct radioimmunoassay of urinary estrogen and pregnanediol glucuronides during the menstrual cycle. Am J Obstet Gynaecol 137:443–450
Tallon OF, Gosling JP, Buckley PM, Dooley MH, Cleere WF, O'Dwyer EM, Fottrell PF (1984) Direct solid-phase enzyme immunoassay of progesterone in saliva. Clin Chem 30:1507–1511

Tommasi A, Pazzagli M, Damiani M, Salerno R, Messeri G, Magini A, Serio M (1984) On-line computer analysis of chemiluminescent reactions with applications to a luminescent immunoassay for free cortisol in urine. Clin Chem 30:1597–1602

Turkes A, Turkes AO, Joyce BG, Riad-Fahmy D (1979) A sensitive solid-phase enzymeimmunoassay for testosterone in plasma and saliva. Steroids 33:347–359

Turkes AO, Turkes A, Joyce BG, Riad-Fahmy D (1980) A sensitive enzymeimmunoassay with a fluorimetric end-point for the determination of testosterone in female plasma and saliva. Steroids 35:89–101

Umeda T, Hiramatsu R, Iwaoko T, Shimada T, Miura F, Sato T (1981) Use of saliva for monitoring unbound free cortisol levels in serum. Clin Chim Acta 110:245–253

Vannucchi PL, Messeri G, Bolelli GF, Pazzagli M, Masala A, Serio M (1983) A solid phase chemiluminescent immunoassay (LIA) for testosterone glucuronide in diluted urines. J Steroid Biochem 18:625–629

Van Weeman BK, Schuurs AHWN (1975) The influence of heterologous combinations of antiserum and enzyme-labeled estrogen on the characteristics of estrogen enzyme-immunoassays. Immunochemistry 12:667–670

Van Weeman BK, Bosch AMG, Dawson EC, Schuurs AHWN (1979) Enzyme-immunoassay of steroids: possibilities and pitfalls. J Steroid Biochem 77:147–151

Vecsei P, Penke B, Joumaah A (1972) Radioimmunoassay of free aldosterone and of its 18-oxo-glucuronide in human urine. Experimentia 28:730–732

Vecsei P, Benraad THJ, Hofman J, Abdelhamid S, Haack D, Lichtwald K (1982) Direct radioimmunoassay for aldosterone and 18-hydroxycorticosterone in unprocessed urine, and their use in screening to distinguish primary aldosteronism from hypertension. Clin Chem 28:453–456

Vecsei P, Abdelhamid S, Bubel P, Haack D, Lewicka S, Lichwald K, Mittelstaedt G (1983) Radioimmunologic methods in aldosterone diagnosis. In: Wikaufmann (ed) Mineralcorticoids and hypertension. Springer, Berlin Heidelberg New York, pp 62–77

Walker RF, Riad-Fahmy D, Read GF (1978) Adrenal status assessed by direct radioimmunoassay of cortisol in whole saliva or parotid saliva. Clin Chem 24:1460–1463

Walker RF, Read GF, Riad-Fahmy D (1979) Salivary progesterone and testosterone concentrations for investigating gonadal function. J Endocrinol 81:1648–1658

Walker S, Mustafa A, Walker RF, Riad-Fahmy D (1981) The role of salivary progesterone in studies of infertile women. Br J Obstet Gynaecol 88:1009–1015

Wang C, Plymate S, Nieschlag E, Paulsen CD (1981) Salivary testosterone in men: further evidence of a direct correlation with free-serum testosterone. J Clin Endocrinol Metab 53:1021–1024

Weeks I, Beheshti I, McCapra F, Campbell AK, Woodhead JS (1983) Acridinium esters as high specific activity labels in immunoassay. Clin Chem 29:1474–1479

White A (1983) Monoclonal antibodies for steroid immunoassay. In: Hunter WM, Corrie JET (eds) Immunoassays for clinical chemistry. Churchill Livingstone, Edinburgh, pp 495–501

Williams GH, Cain JP, Dluly RG, Underwood RH (1972) Studies on the control of plasma aldosterone concentration in normal man. 1. Response to posture, acute and chronic volume depletion and sodium loading. J Clin Invest 51:1731–1742

Wright K, Collins DC, Musey PI, Preedy JRK (1978) Direct radioimmunoassay of specific estrogen glucosiduronates in normal men and non-pregnant women. Steroids 31:407–426

Wright K, Collins DC, Preedy JRK (1979) Urinary excretion of estrone-glucosiduronate, 17β-estradiol 17-glucosiduronate, and estriol 16α-glucosiduronate. Significance of proportionate differences during the menstrual cycle. Steroids 34:445–457

Yorde DE, Sasse EA, Wang TY, Hussa RD, Gavancis JC (1976) Competitive enzyme-linked immunoassay with use of soluble enzyme antibody immune complexes for labelling. Clin Chem 22:1372–1377

Youssefnejadian E, Florensa E, Collins WP, Sommerville IF (1972) Radioimmunoassay of plasma progesterone. J Steroid Biochem 3:893–901

CHAPTER 16

Radioimmunoassay of Thyroid Hormones

L. BARTALENA, S. MARIOTTI, and A. PINCHERA

A. Introduction

The main role of the thyroid gland is to metabolize and to incorporate iodine into a variety of organic compounds, which include the metabolically active thyroid hormones, 3,5,3′,5′-tetraiodothyronine or thyroxine (T_4) and 3,5,3′-triiodothyronine (T_3), and several other precursors and degradation products, as summarized in Table 1 (for review see DEGROOT et al. 1984; INGBAR 1985). T_4 is the most important secretory product of the gland; T_3 is also secreted by the thyroid, although it is mostly (about 80% of the daily production) formed by peripheral outer ring deiodination of T_4. Both T_4 and T_3 are synthesized within a large protein, called thyroglobulin, representing the main constituent of the colloid stored in the follicular lumen. Proteolytic degradation of thyroglobulin releases T_4 and T_3 into the bloodstream, while only very small amounts of thyroglobulin enter the circulation. Minute quantities of diiodotyrosine (DIT) and monoiodotyrosine (MIT) are also secreted by the thyroid, whereas all other iodinated compounds circulating in the serum are almost exclusively produced by peripheral stepwise degradation of T_4 (Fig. 1).

T_4 and T_3 are bound to a group of plasma proteins, including thyroxine-binding globulin (TBG), thyroxine-binding prealbumin (TBPA), and albumin

Table 1. Thyroid hormones and related precursors or degradation products

Substance	Common abbreviation
3,5,3′,5′-Tetraiodothyronine (thyroxine)	T_4
3,5,3′-Triiodothyronine	T_3
3,3′,5′-Triiodothyronine (reverse T_3)	rT_3
3,5-Diiodothyronine	$3,5\text{-}T_2$
3,3′-Diiodothyronine	$3,3'\text{-}T_2$
3′,5′-Diiodothyronine	$3',5'\text{-}T_2$
3′-Monoiodothyronine	$3'\text{-}T_1$
3-Monoiodothyronine	$3\text{-}T_1$
3,5,3′,5′-Tetraiodothyroacetic acid	TETRAC
3,5,3′-Triiodothyroacetic acid	TRIAC
3,5-Diiodotyrosine	DIT
3-Monoiodotyrosine	MIT
Thyroglobulin	Tg

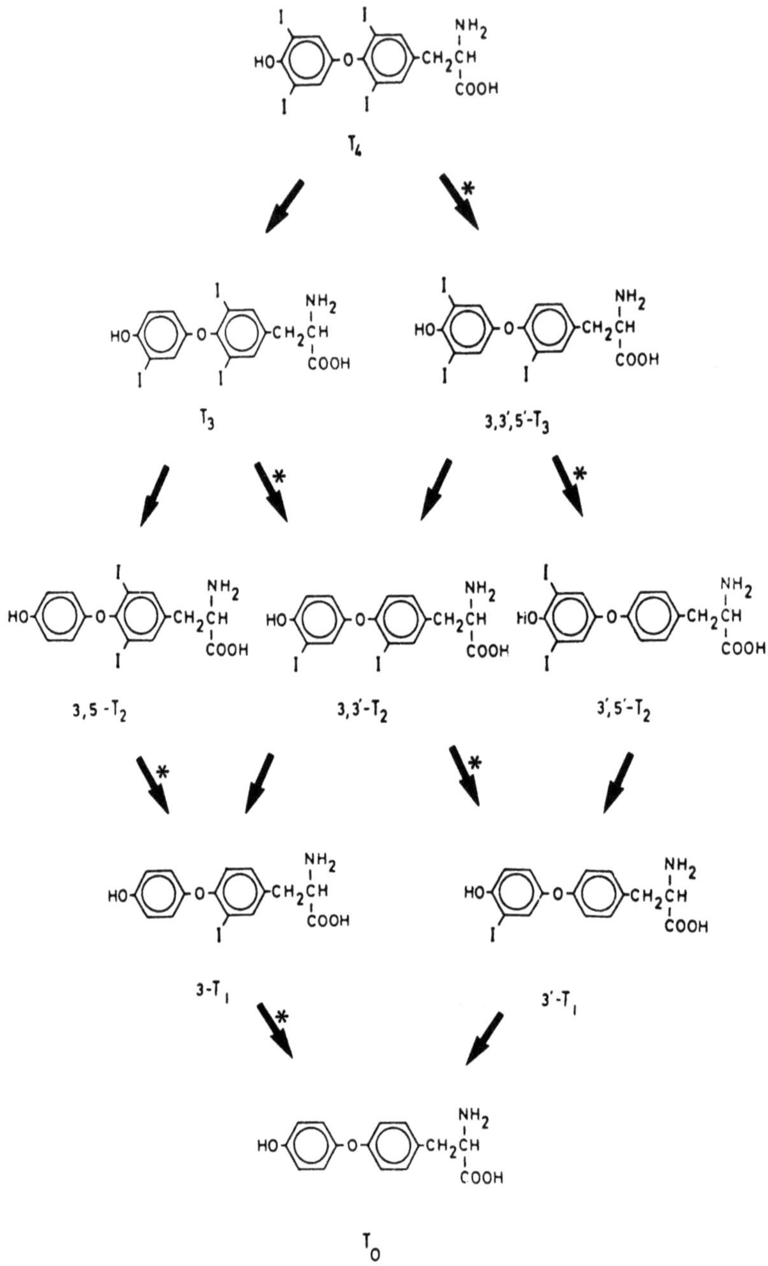

Fig. 1. Structural formulas of thyroid hormones and their metabolic derivatives; pathways of sequential monodeiodination of T_4 are also shown: *plain arrows* → indicate 5-deiodination; *starred arrows* indicate 5'-deiodination

(DEGROOT et al. 1984; DE NAYER and GLINOER 1985; INGBAR 1985; ROBBINS and BARTALENA 1986). Only a very small fraction of thyroid hormones (0.03% for T_4; 0.3% for T_3) is not bound to carrier proteins and circulates free in the blood. It is now widely accepted that free thyroid hormones represent the fraction metabolically active at the tissue level (ROBBINS and RALL 1983; DEGROOT et al. 1984; INGBAR 1985). Because specific tests to assess directly the metabolic action of thyroid hormones at the peripheral level are not available, measurements of circulating total and free T_4 and T_3 are used to assess thyroid status. Increased serum thyroid hormone concentrations are found in hyperthyroidism, whereas hypothyroidism is associated with reduced thyroid hormone levels.

For many years, methods based on iodine content determination have represented the only techniques available for the estimation of total thyroid hormone concentrations in serum. Subsequently, simple, sensitive, and specific radioligand assays for thyroid hormones have replaced these chemical methods. For the purposes of this chapter, iodometric techniques will be only briefly summarized for their historical importance, whereas attention will be focused on radioligand assays.

B. Iodometric Techniques

The rationale for this group of techniques is the fact that iodine is the major component of the thyroid hormone molecule and that iodothyronines account for most naturally occurring serum iodine-containing compounds. The most important assays, which have been used in clinical routine, are the estimation of protein-bound iodine (PBI), butanol-extractable iodine (BEI), and thyroxine–iodine by column (T_4I_c).

I. PBI

This test was the mainstay of thyroid diagnosis for a long period. It estimates the amount of iodine which precipitates with serum proteins (BARKER 1948). Under normal circumstances and in the absence of iodine contamination by exogenous iodinated compounds, PBI reflects the concentration of total T_4, since less than 10% of PBI derives from T_3 and other iodinated substances. Briefly, serum proteins are precipitated and then digested by perchloric acid. The liberated iodine is subsequently measured by a colorimetric technique based on its ability to catalyze the reduction of ceric sulfate by arsenious acid. Normal values for the iodine concentration are 4–8 μg/dl.

II. BEI

In this technique, T_3, T_4, iodotyrosines, and iodide are extracted by acidified butanol, leaving iodoproteins in the residue. The extract is then submitted to reextraction and washing by sodium hydroxide, leaving only T_3 and T_4 in the final residue, whose iodine content is evaluated colorimetrically as in the PBI assay (MAN et al. 1951; BENOTTI and PINTO 1966). This technique was devised to circumvent the interference produced by exogenous iodinated compounds in the PBI test. Normal BEI values for the iodine concentration are slightly lower than PBI values, ranging between 3.5 and 6.5 μg/dl. This is because PBI also measures iodinated proteins. Differences between PBI and BEI (usually referred to as butanol-insoluble iodine, BII) greater than 20% occur in pathologic conditions associated with excess release of thyroglobulin and other iodinated proteins.

III. T_4I_c

In this technique, iodinated compounds are bound to an anion exchange resin column, and then eluted by stepwise acetic acid addition. In this way, iodothyronines can be collected as separate fractions, while iodide remains adsorbed on the column (PILEGGI et al. 1961). Normal T_4I_c values for the iodine concentration are 2.9–6.4 μg/dl.

C. Radioligand Assays of Total T_4 (TT_4) and Total T_3 (TT_3)

All the assays described in this section are based on the same principle, i.e., the competition of a substance to be determined with the same radiolabeled molecule for binding to a specific protein, an antibody in the radioimmunoassays (RIAs), or a high affinity binding protein in the competitive protein-binding analysis (CPBA) methods. The general principles of these techniques, which are grouped under the name of "radioligand assays," are reported in detail elsewhere in this volume (see Chap. 1) and will not be discussed here.

I. TT_4 Assays

1. CPBA

The first radioligand assay for TT_4 determination was the competitive protein-binding analysis (CPBA) method developed by MURPHY and PATTEE (1964). At variance with radioimmunoassay techniques, in CPBA the protein with high affinity for T_4 is not an anti-T_4 antibody, but the major naturally occurring T_4-binding protein, TBG. This procedure requires separation of serum T_4 from endogenous TBG before the assay. This can be obtained by denaturing the protein by enzymatic digestion (CHAN and CUMMINGS 1976), or by extracting T_4 from serum with ethanol (LARSEN 1976) or strong alkaline solution (WONG 1975). The latter procedure was widely used in the past and will be exemplified in the technique detailed here (BRAVERMAN et al. 1971). A 100-μl test sample with a trace amount of ^{125}I-labeled T_4 is added to a small strongly alkaline Sephadex G-25 column. In these conditions, virtually all T_4 is extracted from serum and remains adsorbed to the column, together with ^{125}I-labeled T_4. The column is then washed with 4 ml neutral buffer to lower the pH to 7.3 and to remove any free ^{125}I and serum proteins. Subsequently, 1 ml TBG-containing solution is added to the column. After this step, equilibrium is established between the ligand, the Sephadex column, and TBG. The T_4 bound to TBG is then washed out by adding 4 ml neutral buffer. The ratio of ^{125}I-labeled T_4 bound to Sephadex after elution to that bound initially is directly proportional to the concentration of TT_4 in the serum sample and is established by counting the column before and after elution. The concentration of T_4 in the test sample is determined by running a parallel curve carried out with increasing volumes (20–200 μl) of a standard serum of known TT_4 content.

Normal TT_4 values by this technique range between 4 and 11 μg/dl. The critical steps in this procedure are represented by the extraction of T_4 from serum and by the time of exposure of the TBG–^{125}I-labeled T_4 complex to Sephadex. With careful control of these variables, good precision and reproducibility of determinations can be obtained. However, owing to their higher sensitivity, accuracy, and speed, subsequently developed radioimmunoassays (RIAs) for TT_4 have replaced the CPBA methods.

2. RIA

The prerequisite for TT_4 RIA was the development of specific antibodies to the hormone. Since T_4 per se is poorly immunogenic, it was necessary to immunize animals with T_4 conjugated to carrier proteins, such as bovine serum albumin (Mitsuma et al. 1972), or with thyroglobulin, the thyroid protein naturally rich in iodothyronine residue (Chopra et al. 1971 a). As already mentioned for CPBA, another requirement for TT_4 RIA was the separation of the hormone from endogenous binding proteins. This was initially accomplished by T_4 extraction from sera as in the CPBA methods (O'Connor et al. 1974). A major improvement was represented by the possibility of measuring TT_4 by RIA directly in unextracted serum (Chopra 1972). This can be obtained by the use of substances able to compete with T_4 for binding to TBG, but not to the antiserum, such as 8-anilino-1-naphthalenesulfonic acid (ANS) (Chopra 1972), or salicylic acid (Larsen et al. 1973).

The first TT_4 RIA in unextracted serum was described by Chopra (1972). In this procedure, the following reagents are added in a final volume of 0.5 ml:

300 μl 0.075 *M* barbital buffer with 2% normal rabbit serum (NRS), 150 μg ANS, and about 10000 cpm ^{125}I-labeled T_4

100 μl graded amounts of T_4 (0.05–20 μg/dl) in T_4-free serum (Chopra 1972; Chopra et al. 1972; Larsen 1972) or 25 μl test sera (+75 μl 0.075 *M* barbital buffer with 2% NRS)

100 μl appropriately diluted anti-T_4 serum obtained from rabbits immunized with normal human thyroglobulin.

After incubation for 1 h at room temperature and 5 min in a water bath at 4 °C, 40–50 μl previously titered goat anti-rabbit γ-globulin is added for 20 h at 4 °C. Separation of bound from free ^{125}I-labeled T_4 is achieved by centrifugation at 2000 rpm for 30 min. A standard curve is constructed by plotting the counts obtained with the standards, expressed as a percentage of maximal binding, on the ordinate, and the concentrations of the standards on the abscissa. Knowing the percentage of bound radioactivity in the unknown samples, their serum TT_4 concentration can be calculated from the abscissa.

Several other TT_4 RIAs in unextracted serum have subsequently been developed using different procedures for separation of bound from free T_4. These include dextran-coated charcoal and polyethylene glycol precipitation (MITSUMA et al. 1972; LARSEN et al. 1973; WONG 1975; LARSEN 1976; REFETOFF 1976), which are more rapid than double-antibody precipitation, requiring only a short incubation at room temperature, but are associated with a higher nonspecific binding. Separation of T_4 from the antibody complex has also been obtained by ascending column chromatography, in which the antibody-bound fraction of the hormone moves upward trough the resin, whereas the unbound fraction remains below (WAGNER et al. 1979; MONZANI et al. 1983). Other procedures involve the use of coated tubes in which anti-T_4 antibody is immobilized onto the lower inner wall of the tube (CATT 1969). The unknown samples, standards, controls and ^{125}I-labeled T_4 are incubated within the coated tubes and separation of bound from free is carried out by aspirating or decanting. The major advantage of this procedure is the very short incubation time, since the total time required for the assay is about 60–90 min. Such a short duration of the test is also accomplished by methods employing sheep anti-T_4 serum immunologically bound to a donkey anti-sheep serum covalently linked to a cellulose matrix. This forms a stable, insoluble immune complex. After addition of 100 μl patient's serum or standards and 100 μl tracer, 1 ml double-antibody suspension is added for only 30 min. After brief centrifugation, supernatant is discarded and the radioactivity in the pellet is counted.

Recently, new TT_4 RIAs have been developed, which are almost completely automated, allowing high uniformity of operating conditions. One of these methods is the thyroxine ^{125}I RIA by Becton Dickinson Laboratory Systems using the ARIA II system (REESE and JOHNSON 1978), currently employed in our laboratory. This procedure utilizes a specific anti-T_4 antibody covalently attached to a solid support. After separation of T_4 from endogenous TBG by alkalization, the samples or standards flow over the immobilized antibody, followed by ^{125}I-labeled T_4, which binds to the free T_4-binding sites left on the antibody. The unbound T_4 (including the free labeled T_4) enters a flow cell where it is counted by a γ-detector. An eluting agent (glycine buffer–methanol solution, pH 2.3–2.7) then passes over the immobilized antibody, releasing the bound T_4 which enters the flow cell where it is counted. A built-in computer compares each unknown sample with the previously calculated standard curve, using a weighted regression analysis. Normal values for serum TT_4 by RIA are generally between 4 and 12 μg/dl.

TT_4 RIA has been applied to the measurement of T_4 in small samples of dried blood on filter paper (BUIST et al. 1975; MITCHELL 1976); this technique is presently used in neonatal screening for congenital hypothyroidism (DUSSAULT 1985).

II. TT_3 Assays

1. CPBA

Prior to the development of sensitive and specific RIA techniques for TT_3, CPBA methods were devised. The first of these methods was developed by NAUMANN et l. (1967) and was based on extraction of T_4 and T_3 from serum by methanol–chloroform, followed by separation of T_3 from T_4 by paper chromatography and quantitation of T_3 by a modification of the Murphy–Pattee technique. As pointed out by STERLING et al. (1969), this method yielded falsely high serum TT_3 values, owing to the interference of T_4 methyl ester formed during extraction, which comigrated with T_3 in the chromatogram. A modification of the technique was devised by STERLING et al. (1969), who employed an ion exchange resin column to separate T_3 from T_4. However, even with this modified technique, small but significant artifactual elevations of T_3 concentrations were observed (FISHER and DUSSAULT 1971; LARSEN 1972b). The likelihood of falsely elevated serum TT_3 values by CPBA in the presence of slight contamination by T_4 or its derivatives stems from the much lower (about 50 times) serum T_3 concentration associated with a lower affinity for TBG when compared with T_4. For this reason, even a T_4 contamination as low as 0.5% can substantially affect T_3 determination by CPBA. Furthermore, even during an accurate chromatographic separation, about 0.3%–0.4% of T_4 undergoes deiodination to T_3 (LARSEN 1971). These artifacts explain why serum TT_3 concentrations reported in early studies by CPBA are clearly higher than those subsequently estimated by RIA. Owing to these intrinsic drawbacks, TT_3 CPBA has never attained such wide application as TT_4 CPBA.

2. RIA

A major improvement in the accuracy of serum TT_3 determination was obtained by the development of sensitive and specific RIA methods. As is the case for TT_4, an absolute prerequisite for TT_3 RIA was the production of anti-T_3 antibodies to be used in the test. T_3, as well as T_4, is not antigenic per se. To raise anti-T_3 antisera, the iodothyronine has therefore to be conjugated to some large molecule. To this purpose, different substances have been employed. BROWN et al. (1970) were the first to report successful production of anti-T_3 antibodies by the use of a succinylate polylysine–T_3 conjugate. This antiserum was not, however, highly specific, owing to significant (about 5%) cross-reactivity with T_4. Subsequently, immunization has been carried out by T_3 conjugated to thyroglobulin (CHOPRA et al. 1971b), human serum albumin (GHARIB et al. 1971), bovine serum albumin (LIEBLICH and UTIGER 1972; MITSUMA et al. 1972), or hemocyanin (BURGER et al. 1975). Thyroglobulin alone has also been used for immunization (CHOPRA et al. 1971a), but this procedure produces both anti-T_4 and anti-T_3 antibodies, and the titer of the latter is usually very low.

In any case, the specificity of the antiserum employed should be carefully checked, especially against thyroxine and other naturally occurring iodoamino acids, because of the chemical similarity of these substances. The lack of any cross-reactivity with other compounds is, in fact, really crucial when the substance to be assayed is, like T_3, present in the blood in very small amounts.

Like T_4, T_3 should be stripped from its binding to serum proteins. This can be accomplished by extracting iodothyronines from serum (FANG and REFETOFF 1974) or by inactivating TBG by heating at 56 °C for 1 h (STERLING and MILCH 1974). Alternatively, and more simply, assay mixture can be added with substances competing with T_3 for the binding to serum proteins, but not to the antiserum. To this purpose, nonradioactive T_4 (CHOPRA et al. 1971b), ANS (CHOPRA

1972; CHOPRA et al. 1972), thimerosal (KIRKEGAARD et al. 1974), and tetrachlorothyronine (MITSUMA et al. 1971) have been utilized. This has allowed the direct measurement of T_3 in unextracted serum, as first described by CHOPRA et al. (1971 b, 1972). According to his protocol, the following reagents are sequentially added:

- Standards (10 pg–10 ng T_3) in T_3-free serum (CHOPRA 1972; CHOPRA et al. 1972; LARSEN 1972 b) or unknown samples (200 μl)
- 2 mg/ml ANS (100 μl)
- Anti-T_3 rabbit antiserum diluted so as to bind 30%–50% of the tracer in the absence of nonradioactive T_3 (100 μl)
- ^{125}I-labeled T_3 (about 0.1 ng T_3, 10000 cpm) in 0.075 *M* barbital buffer, pH 8.6, with 1% NRS and 0.1% sodium azide (100 μl)
- The same buffer to adjust to a final mixture volume of 1 ml.

Each assay should include two tubes in which all reagents except nonradioactive T_3 are added (maximal binding or B_0). The tubes are mixed and incubated for 24 h at 4 °C. A second antibody (goat anti-rabbit γ-globulin, 50–100 μl), previously titered, is then added and reincubated. Separation of bound from free is achieved by centrifugation at 2000 rpm for 30 min. Supernatant is discarded and the radioactivity in the pellet is counted. Results are calculated from a standard curve, as described for T_4.

In subsequently developed assays, separation of bound from free T_3 was achieved by the use of dextran-coated charcoal or polyethylene glycol (12%–15%) (CHOPRA 1981), and by ascending column chromatography (WAGNER et al. 1979; MONZANI et al. 1983). Some commercially available kits utilize the coated tubes system, in which anti-T_3 antibody is immobilized onto the lower inner wall of the tube (CATT 1969). Other methods employ sheep anti-T_3 serum immunologically bound to a donkey anti-sheep serum covalently linked to a cellulose matrix. All these methods are essentially the same as those described for T_4. The method currently used in our laboratory is the triiodothyronine ^{125}I RIA by Becton Dickinson Laboratory Systems using the ARIA II system, identical to that described for T_4 RIA. Normal serum TT_3 values measured by RIA methods are usually between 100 and 200 ng/dl.

III. Factors Affecting the Diagnostic Accuracy of Total Thyroid Hormone Determination

Total thyroid hormone measurements by RIA are usually in agreement with the thyroid status, increased or reduced serum concentrations being found in hyperthyroidism and hypothyroidism, respectively. However, inappropriately low or high concentrations are found in a number of physiologic or pathologic conditions associated with changes of thyroid hormone-binding proteins, the presence of circulating anti-iodothyronine autoantibodies, or altered peripheral metabolism of thyroid hormones (Table 2). It should be noted that, in most of these cases, measurement of serum free thyroid hormone concentrations will allow a correct definition of thyroid status, as discussed in Sect. D.III.

Table 2. Conditions affecting the diagnostic accuracy of total thyroid hormone determination

Abnormal TBG concentration
Drugs affecting thyroid hormone binding to TBG and/or TBPA
Abnormal circulating thyroid hormone-binding proteins
Conditions associated with reduced peripheral conversion of T_4 to T_3 and severe nonthyroidal illness
Anti-iodothyronine autoantibodies

1. Abnormal TBG Concentration

Serum TT_4 and TT_3 concentrations almost exclusively reflect circulating hormones bound to carrier proteins. Because TBG is the major thyroid hormone-binding protein, its changes markedly affect total thyroid hormone concentrations. Elevated concentrations of serum TT_4 and, to a lesser extent, of serum TT_3 are, in fact, found when TBG is increased, while the reverse occurs in conditions characterized by low serum TBG concentrations (Table 3).

The most common cause of elevated serum TBG is excess estrogen (ROBBINS and NELSON 1958; DOWLING et al. 1960; HOLLANDER et al. 1963), leading to increased hepatic synthesis of TBG (GLINOER et al. 1977). The inherited X-linked increase in circulating TBG is also due to enhanced synthesis of the protein (REFETOFF et al. 1972b; ROBBINS 1973). Similar changes may result from heroin,

Table 3. Conditions associated with abnormal TBG concentrations

A. Increased TBG concentrations
Inherited, X-linked
Pregnancy and other hyperestrogenic conditions
Drugs:
Estrogens (contraceptive pill)
Heroin
Methadone
Perphenazine
Clofibrate
5-Fluorouracil
Hepatocellular damage
Porphyria cutanea tarda
B. Decreased TBG concentrations
Inherited, X-linked
Drugs:
Androgens
Anabolic steroids
Glucocorticoids
L-Asparaginase
Protein–calorie malnutrition
Nephrotic syndrome

methadone, and perphenazine administration (OLTMAN and FRIEDMAN 1963; WEBSTER et al. 1973) and from hepatocellular dysfunction (VANNOTTI and BERAUD 1959; REFETOFF et al. 1972b).

Besides the X-linked inherited defect or absence of TBG (REFETOFF et al. 1972b; ROBBINS 1973), low serum TBG concentrations due to reduced hepatic synthesis of the protein are found during treatment with androgens (FEDERMAN et al. 1958), anabolic steroids (BARBOSA et al. 1971), large doses of corticosteroids (OPPENHEIMER and WERNER 1966) and L-asparaginase (GARNICK and LARSEN 1979), and in protein–calorie malnutrition (INGENBLEEK et al. 1974). Urine and gastrointestinal losses are responsible for the low serum TBG levels of nephrotic syndrome (ROBBINS et al. 1957) and protein-losing enteropathy (HANSEN et al. 1974), respectively.

2. Drugs Affecting Thyroid Hormone Binding to TBG and/or TBPA

Total thyroid hormone concentrations may also be influenced by drugs competing with the hormone for its binding site on TBG and/or TBPA (Table 4), which in general cause a reduction of serum TT_4 and, to a lesser extent, of serum TT_3.

The most studied drug acting on TBG is the anticonvulsant, diphenylhydantoin (WOLFF et al. 1961). The anti-inflammatory drug, fenclofenac, has a more profound effect on serum TT_4 and TT_3 concentrations, decreasing them to 30%–40% of normal (RATCLIFFE et al. 1980). The salicylates exert their major effect on TBPA, but they interfere with binding to TBG as well, reducing serum TT_4 and TT_3 to 70%–80% of normal (LARSEN 1972a). Heparin mainly affects serum free thyroid hormone concentrations (HERSHMAN et al. 1972) and its effects will be discussed in Sect. D.III.5. Less information is available on other drugs listed in Table 4 (for review see CAVALIERI and PITT-RIVERS 1981).

Table 4. Drugs affecting thyroid hormone binding to TBG and/or TBPA

Drug	TBG	TBPA
Diphenylhydantoin	+	−
Chlorpropamide	+	−
Tolbutamide	+	−
Clofibrate	+	−
Diazepam	+	−
Halofenate	+	−
Fenclofenac	+	−
Sulfobromophthalein	+	−
Iopanoate	+	−
Ipodate	+	−
Heparin	+	−
Salicylate	+	+
Gentisic acid	−	+
γ-Resorcylic acid	−	+
Penicillin	−	+
2,4-Dinitrophenol	−	+

3. Abnormal Serum-Binding Proteins

At least two conditions are presently known to be associated with abnormal thyroid hormone-binding proteins. The first and better described condition is familial dysalbuminemic hyperthyroxinemia (FDH), which is an inherited autosomal dominant syndrome characterized by the presence in the serum of an abnormal albumin which preferentially binds T_4 (HENNEMANN et al. 1979; LEE et al. 1979; BARLOW et al. 1982; RAJATANAVIN and BRAVERMAN 1983). The second condition is a very rare syndrome characterized by an abnormal T_4 binding to TBPA (MOSES et al. 1982). Both these conditions are associated with increased serum TT_4 and normal serum TT_3 concentrations; as discussed in Sect. D.III.2, serum free thyroid hormone concentrations are normal in these conditions.

4. Conditions Associated with Reduced Peripheral Conversion of T_4 to T_3 and Severe Nonthyroidal Illness

A wide variety of physiologic and pathologic conditions are associated with an impaired peripheral conversion of T_4 to T_3 in the absence of any abnormality of the thyroid gland (low T_3 syndrome) (HESCH 1981; ENGLER and BURGER 1984). These include fetal life, acute or chronic caloric deprivation, chronic liver diseases, uncontrolled diabetes, myocardial infarction, and almost all acute or chronic systemic illness, trauma, and surgery. Low T_3 syndrome is also caused by a number of drugs, such as propranolol, large doses of glucocorticoids, propylthiouracil, amiodarone, and other iodinated organic compounds. In all these conditions there is a reduced activity of 5′-monodeiodinase in liver and other tissues, leading to decreased serum T_3, normal or increased serum T_4, and increased serum reverse T_3 (rT_3) concentrations. It has been clearly shown that the latter phenomenon is not due to increased rT_3 production, but to a reduction of its metabolic clearance rate (ENGLER and BURGER 1984).

Beside low serum T_3 concentrations, patients with severe nonthyroidal illness may also show reduced TT_4 levels (WARTOFSKY and BURMAN 1982). This appears to be due to several events, such as the failure of the normal negative feedback control of the pituitary–thyroid axis (WARTOFSKY and BURMAN 1982; WEHMANN et al. 1985), and several abnormalities of thyroid hormone binding to plasma proteins (LUTZ et al. 1972; CHOPRA et al. 1979; OPPENHEIMER et al. 1982; WARTOFSKY and BURMAN 1982).

5. Anti-Iodothyronine Autoantibodies

Although their precise prevalence is still a matter of controversy and varies markedly in different series, anti-thyroid hormone autoantibodies are relatively frequent in patients with thyroid autoimmune disorders, and may be found in other thyroid diseases (SAKATA et al. 1985). These autoantibodies produce different types of interference in TT_4 and TT_3 RIAs, depending on the assay procedure (Table 5). When the assay is based on a single-antibody technique, using either γ-globulin precipitation by PEG or free thyroid hormone absorption by dextran-coated charcoal, autoantibodies have exactly the same behavior as anti-T_4 or anti-T_3 antibodies of the kit. Therefore, an increased binding of the tracer occurs,

Table 5. Interference of anti-iodothyronine autoantibodies in TT_4 and TT_3 RIAs, in relation to the method employed for bound/free (*B*/*F*) separation

B/*F* separation	Interference
Charcoal absorption	Falsely low values
PEG precipitation	Falsely low values
Double-antibody precipitation	Falsely high values (occasionally falsely low)[a]
Solid-phase antibody	Falsely high values (occasionally falsely low)[a]
ARIA II	Falsely low values

[a] Depending on autoantibody affinity and on absolute concentration of endogenous hormone.

and inappropriately low levels of serum TT_4 and TT_3 are determined (SAKATA et al. 1985). Conversely, when a double-antibody technique is employed, provided that the second antibody is strictly specific for the first antibody of the kit and does not react with the human autoantibody, a lower amount of the tracer is precipitated, resulting in falsely elevated values (SAKATA et al. 1985). Finally, in the solid-phase technique, autoantibodies compete with coated anti-thyroid hormone antibodies for binding to tracer hormones. This results in spuriously elevated thyroid hormone values, since low radioactivity is bound to coated antibodies (SAKATA et al. 1985). Occasionally, an apparent lack of interference or falsely low total thyroid hormone concentrations have been reported by double-antibody or solid-phase RIAs (BECK-PECCOZ et al. 1983). This phenomenon has been attributed to the low binding capacity and slow dissociation rate of autoantibodies, as well as to the absolute concentration of the endogenous hormone.

Thus, when serum thyroid hormone concentrations by RIA are discrepant with clinical findings, the possible interference of circulating anti-thyroid hormone autoantibodies should be taken into account and looked for.

D. Free Thyroid Hormone Assays

As mentioned before, thyroid hormones circulate in plasma mostly bound to transport proteins, principally to TBG, but unbound or free thyroid hormones represent the biologically active fraction available to tissues. At variance with total thyroid hormones, circulating free thyroid hormones are unaffected by changes in plasma transport protein concentrations and, therefore, correlate better with the metabolic status (ROBBINS and RALL 1983; DEGROOT et al. 1984; INGBAR 1985). Thus, serum free thyroid hormone determination can be considered one of the most important parameters for the initial assessment of thyroid function.

Several assays have been proposed for serum free thyroid hormone evaluation. As suggested by EKINS (1979), these methods can be divided into indirect and direct techniques, according to the following criteria:

1. Indirect methods. The free thyroid hormone concentration is derived from total thyroid hormone concentration and either: (a) the total or unbound serum-binding protein capacity; or (b) the fraction of serum free hormones, as determined by equilibrium dialysis, ultrafiltration, or other methods.

2. Direct methods. Serum free thyroid hormones are directly measured after being "sequestered" from binding proteins.

Both indirect and direct methods can be further subdivided into dynamic and equilibrium techniques (EKINS 1979). The former include methods in which free thyroid hormone measurements are carried out on the basis of kinetic analysis of a particular phenomenon, whose rate is correlated to free thyroid hormone concentration. In the equilibrium techniques, measurements are carried out at thermodynamic equilibrium by determining either the distribution of total thyroid hormones between bound and free fractions, or the absolute quantity of free thyroid hormones. Finally, all techniques can be further distinguished as absolute or comparative methods. The former measure the hormone concentration per se, while the latter rely on standardization against materials whose free hormone content has been previously established by an absolute method.

I. Indirect Methods for the Estimation of Free Thyroid Hormone

1. Methods Based on Equilibrium Dialysis

The earlier techniques employed for free thyroid hormone determination were based on the indirect method of equilibrium dialysis, as originally proposed by STERLING and HEGEDUS (1962) for T_4. In these assays, the proportion of free/total T_4 or free/total T_3 is estimated by measuring the fraction which is able to diffuse through a dialysis membrane (dialyzable fraction DF). To this purpose, a sample of serum (generally 1 ml) is mixed with a trace amount of radioiodinated T_4 or T_3, which quickly equilibrates with the respective free and bound endogenous hormones. The sample is then dialyzed to equilibrium at 37 °C for 18–24 h against 9 ml phosphate buffer, pH 7.4, and the DF is calculated from the proportion of labeled hormone in the dialysate. Results are expressed as a percentage of the total respective hormones (FT_4 %, FT_3 %). In euthyroid adults, values average 0.02%–0.04% for FT_4 and 0.2%–0.45% for FT_3, in the absence of gross abnormalities of serum-binding protein concentrations. Higher values are found in hyperthyroidism and in TBG deficiency, while lower values are observed in hypothyroidism and in TBG excess. The absolute concentrations of FT_4 and FT_3 are calculated by multiplying DF by the total serum thyroid hormone concentration. By this technique, normal adult values range between 10 and 35 pg/ml for FT_4 and between 2.5 and 6.5 pg/ml for FT_3.

This procedure presents several methodological problems. First, it absolutely requires highly purified radioiodinated hormones, completely devoid of dialyzable radioactive impurities (e.g., radioiodide, labeled iodothyronines, or iodothyronine breakdown products) affecting the tracer distribution between bound and free fractions. To circumvent this problem, radioiodinated T_4 and T_3 can be prepurified by a variety of techniques, including chromatographic separation and/or predialysis, or isolated from the dialysate by magnesium chloride precipitation, anionic exchange adsorption, charcoal adsorption, or trichloroacetic acid precipitation (EKINS 1979).

Furthermore, since free thyroid hormones represent only a minute proportion of total thyroid hormones, small experimental fluctuations may lead to marked errors in the free hormone fraction estimation. This pitfall can be prevented by increasing the ratio between the volume of the dialysate and that of the dialysant (OPPENHEIMER and SURKS 1964).

Various modifications of the dialysis technique have been proposed, including separation by ultrafiltration (SCHUSSLER and PLAGER 1967) or gel filtration (LEE et al. 1964). Independently of the separation method employed, all these techniques for free thyroid hormone determination are cumbersome and technically demanding. They are not, therefore, suitable for use on a routine basis. Thus, before the introduction of methods for direct measurement of FT_4 and FT_3 in serum, in most clinical laboratories free thyroid hormone indices have been used to estimate free hormone concentrations, as described in Sect. D.I.2.

2. Free T_4 and Free T_3 Indices (FT_4I and FT_3I)

FT_4I and FT_3I provide a reliable estimate of the absolute serum free thyroid hormone concentration. They are mathematically derived from determinations of serum total thyroid hormone concentrations and of total or unsaturated serum thyroid hormone-binding capacity. Determination of serum TBG concentrations is generally used to express total thyroid hormone-binding capacity. Serum TBG is commonly assayed by specific radioimmunoassays (GERSHENGORN et al. 1976), whose description is beyond the scope of this chapter. Unsaturated serum thyroid hormone-binding capacity is estimated by special tests called in vitro uptake tests.

a) In Vitro Uptake Tests

These tests are carried out by mixing the patient's serum with a trace amount of radioiodinated T_4 or T_3. The mixture is then incubated with a solid-phase synthetic adsorbent, able to bind thyroid hormones and to compete with endogenous binding proteins. The amount of radioactivity adsorbed onto the solid phase is then determined and expressed as a percentage of total tracer amount. This value inversely correlates with the unsaturated binding sites on TBG. Since an ion exchange resin is often utilized as the synthetic adsorbent (STERLING and TABACHNICK 1961), this test is commonly known as the resin T_3 or T_4 uptake test (RT_3U or RT_4U). Beside resin, other solid-phase matrices, such as red blood cells (used in the original test by HAMOLSKY et al. 1957), coated charcoal, Sephadex, talc, and antibody-coated tubes (BRAVERMAN et al. 1967; CAVALIERI et al. 1969; AZUKIZAWA and PEKARY 1976) can be employed.

Labeled T_3 is used much more commonly than labeled T_4 for the in vitro uptake test because of its weaker binding to serum proteins, which yields a higher counting rate for the adsorbing material, thus decreasing the counting time and the error. By appropriate modifications of the assay conditions, such as the reduction of serum volume, the addition of a greater amount of tracer, or the use of barbital buffer, the in vitro uptake test can also be conveniently performed with ^{125}I-labeled T_4 (REFETOFF et al. 1972a), but this procedure has not gained wide acceptance for routine use. For this purpose, a large number of commercial kits employing ^{125}I-labeled T_3 as a tracer are presently available. Although their principle is basically the same, they differ in several details, including the amount of serum, the time and temperature of incubation, the specific activity and the amount of the tracer, the type and the amount of the adsorbing material. Owing to wide variations of normal T_3 uptake values obtained with the different assays (25%–55%), a convenient procedure is to express them as a fraction of the uptake value observed with a pool of normal sera run in the same assay. This "corrected" value is normally between 0.85 and 1.15 and is utilized to calculate FT_4I and FT_3I (see Sect. D.I.2.b).

RT_3U is inversely related to the amount of unoccupied binding sites of serum TBG and is, therefore, decreased when the unsaturated sites are increased owing to the reduction of endogenous thyroid hormones or to the increase of serum TBG concentrations (see Table 3). Conversely, RT_3U is increased when the amount of unsaturated TBG is decreased owing to an excess of endogenous thyroid hormones or to a decrease of serum TBG levels (see Table 3). An increase of RT_3U may also be the consequence of the interference of pharmacologic agents (see Table 4) competing with thyroid hormone binding to transport proteins, owing to structural similarities.

b) Calculation of FT_4I and FT_3I

The mathematical derivation of FT_4I stems from the mass action equation; at equilibrium, the relationship between TT_4, FT_4, and unsaturated TBG (uTBG) can be expressed as follows

$$K[FT_4] = [TT_4]/[uTBG]. \tag{1}$$

Since RT_3U is inversely proportional to the available binding sites on binding proteins, it can be substituted in Eq. 1 for 1/[uTBG], as follows

$$K[FT_4] = [TT_4]\,RT_3U. \tag{2}$$

FT_4I is defined as equal to a constant K multiplied by FT_4; thus

$$FT_4I = [TT_4]\,RT_3U. \tag{3}$$

Since, as mentioned before, the RT_3U value used for the calculation of FT_4I is a normalized ratio, Eq. 3 can be rearranged as follows

$$FT_4I = [TT_4] \times \text{patient's } RT_3U/\text{mean normal sera } RT_3U. \tag{4}$$

A practical advantage of the FT_4I calculation according to Eq. 4 is that FT_4I is expressed in numerical units in the same range as TT_4, being between 4 and 11 in euthyroid subjects with normal serum TBG concentrations.

A similar procedure can be employed for the calculation of the FT_3I from serum TT_3 and RT_3U values, and in euthyroid subjects generally ranges between 100 and 200. The theoretical assumptions on which the FT_4I calculation was based have been experimentally validated by the observation of an almost perfect correlation between these indices and the absolute FT_4 concentrations directly measured in sera (ROBIN et al. 1971). Similar data on FT_3I are still limited (DEGROOT et al. 1984).

Another method for indirect calculation of circulating FT_4 and FT_3 is the calculation of the ratio between TT_4 or TT_3 and TBG concentrations (GLINOER et al. 1978; BRAVERMAN et al. 1980). These ratios are calculated as: TT_4 (μg/dl)/TBG (μg/ml) × 10 and TT_3 (ng/dl)/TBG (μg/ml). These indices are directly proportional to the free thyroid hormone concentrations and range in euthyroid subjects between 2.2 and 5.8 for T_4 and 3.2 and 10.2 for T_3 (BRAVERMAN et al. 1980).

c) Significance and Limits of FT_4I and FT_3I

The calculation of FT_4I and FT_3I still represents one of the most accurate, yet simple estimations of the thyroid functional status, since these indices are generally elevated in hyperthyroidism and low in hypothyroidism, irrespective of serum TBG concentrations. Some technical limitations must, however, be taken into account. In subjects with extreme TBG deficiency or excess (DEGROOT et al. 1984), RT_3U does not adequately correct the abnormally reduced or increased serum total thyroid hormone levels. This phenomenon is due to the lack of linear correlation between RT_3U values and free thyroid hormone concentrations in these conditions (DEGROOT et al. 1984; INGBAR 1985). This also applies to a greater extent to FT_4I and FT_3I calculated on the basis of TT_4/TBG and TT_3/TBG, respectively.

Other problems may arise from conditions affecting the overall binding affinity of serum-binding proteins for T_4 without influencing the affinity for T_3. One of these conditions is represented by the presence of circulating anti-T_4 autoantibodies, which cause an increase of serum TT_4, while the absolute concentration of FT_4 is unchanged. Provided that a reliable determination of the increased TT_4 is available in this condition, calculation of FT_4I will result in a falsely elevated value, since the affinity of serum proteins for T_3, and hence the RT_3U value, are unchanged. A similar situation is represented by FDH or by the rare condition of increased T_4 binding to TBPA (RAJATANAVIN and BRAVERMAN 1983). In fact, in these syndromes the abnormality of serum thyroid hormone binding results in a greater affinity for T_4 as compared with T_3. Conversely, in patients with severe nonthyroidal illness, one or more still poorly defined serum factors may reduce the binding of T_4 to transport proteins more than that of T_3 (CHOPRA et al. 1979). This interference causes a reduction of serum TT_4, without affecting RT_3U values; as a consequence, FT_4I will underestimate the actual FT_4 concentrations. In these conditions, the use of radioiodinated T_4 instead of T_3 in the in vitro uptake test has been suggested to overcome the interference (WHITE et al. 1980), but the validity of this approach in clinical practice needs further confirmation.

II. Direct Methods for Free Thyroid Hormone Determination

1. Absolute Methods

a) Equilibrium Dialysis Followed by RIA

This method, developed by ELLIS and EKINS (1973), is based on the direct measurement by very sensitive RIAs of the free thyroid hormone concentration in the serum dialysate. The assay is generally carried out as follows: 200 µl serum sample is dialyzed at 37 °C for 18 h against 5 ml 0.01 *M* Hepes buffer, pH 7.4, 0.62% NaCl. The free hormones diffuse through the membrane into the dialysate and their concentration in the dialysant will be readjusted by a net dissociation of bound hormone. This process will continue until a new steady state is achieved with an equal concentration of free thyroid hormones in the dialysate and in the dialysant. After dialysis, the free thyroid hormone concentration is determined in the dialysate by RIA. This technique obviates all the problems due to tracer impurities of the indirect method based on equilibrium dialysis. Furthermore, since it is a direct, absolute, equilibrium assay, it fulfills all the requirements for an "ideal" reference method of free thyroid hormone measurement. However, some methodological drawbacks limit its application on a routine basis. Dialysis is, in fact, a cumbersome procedure needing accurate standardization, mainly because of adsorption of free thyroid hormones on membranes and dialysis cells. Furthermore, the RIA employed must have a high sensitivity and particular care is required in the preparation of absolute thyroid hormone standards used for the calibration curve. By this technique, normal adult values range between 7.5 and 17 pg/ml for FT_4 and between 3.5 and 6.5 pg/ml for FT_3.

b) Column Adsorption Chromatography Followed by RIA

A modification of the equilibrium dialysis method has been subsequently proposed, based on the use of gel chromatography to separate free from bound hormones (ROMELLI et al. 1979). This technique is presently the only direct absolute method for measurement of free thyroid hormones commercially available. In this assay, a 0.5-ml serum sample is submitted to chromatographic adsorption on a Sephadex LH-20 column (150 mg) at 37 °C. During this step, free thyroid hormones are adsorbed on the resin and their concentration is reestablished through dissociation of bound hormones. This adsorption–dissociation process continues until a steady state between the adsorbed hormone and the free hormone in solution is attained. The amount of adsorbed hormone HR is proportional to the free thyroid hormone concentration [H] according to the following equation

$$\mathrm{HR} = \varphi\,[\mathrm{H}].$$

The proportionality constant φ is defined as the "adsorption factor," which corresponds to the product of the adsorption constant K_{ads} of the resin and the number of resin binding sites n_R, as follows

$$\varphi = K_{ads} n_R.$$

Thus, if φ, which can be experimentally determined (ROMELLI et al. 1979), is known, the measurement of HR directly provides the free hormone concentration. It should be noted that, since K_{ads} increases with temperature, φ values are equally influenced by temperature variations, and, therefore, the chromatographic separation must be performed at constant temperature. Practically, after chromatographic adsorption, serum proteins are removed by two washes with 2 ml 0.1 *M* Tris buffer, pH 7.4. Subsequently, T_3 and T_4 adsorbed onto the resin are eluted with 2 ml methanol and the organic solvent is then evaporated to dryness at 37 °C. Dried samples are reconstituted with 4 ml 0.04 *M* phosphate buffer, pH 7.4, and 0.3-ml reconstituted samples are assayed for T_3 and T_4 by RIA using known amounts of thyroid hormones diluted in buffer for the standard curve. Similarly to the technique described in the previous paragraph, this method can be considered as a direct, absolute, equilibrium technique and, hence, as a reference method. Normal values range between 7.1 and 15.4 pg/ml for FT_4 and 2.4 and 6.1 pg/ml for FT_3. Although simpler than the equilibrium dialysis method, this technique is still technically demanding, requiring accurate standardization of the chromatographic separation and of the recovery step. Relative to equilibrium dialysis, the present assay does not require a highly sensitive RIA, since the thyroid hormone concentration in the final sample is much greater than the actual free hormone concentration in the initial sample (ROMELLI et al. 1979).

This method has subsequently been modified to make it suitable for easy routine measurements and it is presently commercially available in this modified version (Liso-phase FT_3 and FT_4 kits, Lepetit-Sclavo). Briefly, the assay is carried out at room temperature, the recovery of the adsorbed hormones is achieved by adding a protein solution (1:4 diluted bovine serum), and the recovered hormone is eluted directly in the RIA tubes. At variance with the original method, in this modified technique, the calibration curve for the RIA is made up with serum standards submitted to the same chromatographic separation as the unknown samples. The actual free thyroid hormone concentrations of the standards are previously determined by an absolute reference method. It is evident that this modified technique is more correctly described as a direct equilibrium comparative method.

2. Comparative Methods

These methods are based on similar principles, involving separation of the serum free thyroid hormone fraction by specific antibody, generally bound to some kind of solid support. The antibodies used in these methods have low affinity for the ligand in order to minimize the antibody binding to free thyroid hormones and, consequently, to limit the dissociation of thyroid hormones from serum binding proteins. Furthermore, all these procedures are comparative, since they utilize standards whose free T_3 and free T_4 concentrations have been predetermined by absolute methods. The assays included in this group can be subdivided as follows (EKINS 1979, 1985): (a) methods based on "labeled hormone antibody uptake"; (b) "two-step/back-titration" methods; and (c) "one-step/analog" methods.

a) Free Thyroid Hormone RIA Based on Labeled Hormone Antibody Uptake

These methods are based on principles similar to those employed for the calculation of FT_4I and FT_3I. In fact, the concentrations of free thyroid hormones are calculated by multiplying the total thyroid hormone concentration in the sample by the hormone fraction bound by exogenous antibody. The most widely known assay of this group was originally developed by ODSTRCHEL (1982) and is presently commercially available, after several modifications, as the Corning Immophase FT_3 and FT_4 kit. This method employs two parallel series of tubes called *A* and *B*. In each tube of the *A* series, 25 µl (for FT_4 determination) or 50 µl (for FT_3) of unknown or standard sera are incubated with 100 µl of the corresponding labeled hormone. In the *B* series tubes, the same amounts of sera and labeled hormone are incubated in the presence of a binding protein blocking agent (thimerosal). After 20 min incubation at room temperature, 0.8 ml anti-T_4 or anti-T_3 antibody coupled to glass particles is added and incubated for 30 min (FT_4) or 60 min (FT_3) at room temperature. The solid phase is then separated by centrifugation and the radioactivity bound to the pellet is counted. In its original version, the ratio of counts recovered in *A* and *B* tubes (*A*/*B*) was plotted against the free thyroid hormone concentrations of the standard sera previously determined by equilibrium dialysis. The *A*/*B* ratio was in fact believed to represent the relative rate of the reaction between the free hormone and the solid-phase antibody, directly related to the free hormone concentration. This method was claimed to be direct; it was also considered a dynamic technique, since it was based on the separation of radiolabeled hormone onto the solid phase before achieving thermodynamic equilibrium. The theoretical basis for this interpretation has been questioned, however, because it relies on several theoretical assumptions and approximations which distort the results, especially in sera with marked abnormalities of TBG concentrations (EKINS 1985). This problem has been subsequently rectified with consequent correction of most abnormal results (ROSS and BENRAAD 1984).

b) Two-Step/Back-Titration Methods

In this approach, serum samples are incubated with a thyroid hormone antibody immobilized on a solid phase to immunoextract free thyroid hormones (step 1). Serum components unbound to solid-phase antibody are washed away and a trace amount of radiolabeled T_4 and T_3 is then added (step 2). Since the labeled hormone will bind to unoccupied sites of the solid-phase antibody, the amount of radioactivity bound to the solid phase will be inversely related to the free thyroid hormone concentration in the test sample.

The "two-step" sequential incubation procedure is needed to avoid the interaction of serum proteins with radiolabeled thyroid hormones. The first method based on this principle was developed by EKINS (1983) employing Sephadex or cellulose-coupled antibody. Presently, the most widely used commercial version of this technique is the Clinical Assay Gammacoat two-step kit (Clinical Assays, Cambridge, Massachusetts, United States), which utilizes antibodies coated on plastic tubes. In this assay, 50-μl serum samples are incubated within tubes coated with anti-T_4 or anti-T_3 antibodies; after addition of 1 ml Tris-buffered saline, the mixture is incubated for 20 min at 37 °C in a water bath. The tubes are then aspirated or decanted and washed with 1 ml incubation buffer. Then 1 ml tracer hormone in Tris-buffered saline is added and incubated for 60 min at room temperature. The tubes are finally aspirated or decanted and their radioactivity determined. Free thyroid hormone concentrations in the unknown samples are determined by the use of standard sera whose FT_4 and FT_3 levels have been previously measured by an absolute method. Normal values range between 6 and 16 pg/ml for FT_4 and 2.1 and 5.8 pg/ml for FT_3.

c) One-Step/Analog Methods

These methods utilize a radioiodinated thyroid hormone analog, which, in theory, should not react with serum proteins, while maintaining its reactivity against antibody to the native hormone. This approach obviates the need of the two-step incubation required by back-titration methods. The most widely studied method employing analogs is the Amersham Amerlex method (Amersham International Ltd., Amersham, Bucks, United Kingdom) (MIDGLEY and WILKINS 1982), but several commercial kits based on the same principle are presently available. The exact nature of the thyroid hormone analog has not been revealed so far, although it is supposed to be a T_4 or T_3–protein conjugate (WITHERSPOON et al. 1984). In the Amerlex assay, 100-μl unknown samples are incubated with 500 μl ^{125}I-labeled analog and 500 μl anti-T_3 or anti-T_4 antibody coupled to latex particles. Incubation is carried out at 37 °C for 1 h (FT_4) or 2 h (FT_3), the solid phase is separated by centrifugation, supernatant is discarded, and pellet radioactivity is counted. A calibration curve is run using standard sera whose free thyroid hormone concentrations have been determined by equilibrium dialysis.

The analog methods, because of their simplicity, are more convenient than the two-step techniques and, therefore, more suitable for routine use. However, the major drawback of these assays is interaction between the analog and endogenous serum proteins, such as albumin (AMINO et al. 1983) or anti-iodothyronine autoantibodies (BECK-PECCOZ et al. 1984). The relevance of these interactions to the estimation of free thyroid hormone concentrations will be detailed in Sect. D.III.3.

III. Factors Affecting the Diagnostic Accuracy of Free Thyroid Hormone Measurements

Several physiologic and pathologic conditions may interfere with free thyroid hormone determination by virtually all the available methods discussed so far. Each assay possesses inherent advantages and disadvantages, depending upon the methodological approach. However, it must be pointed out that the diagnostic

Table 6. Factors affecting the diagnostic accuracy of FT_4 measurement

FT_4 assay	Increased TBG	Decreased TBG	Pregnancy	FDH	Anti-T_4 Ab	NTI	Heparin
ED+RIA	N	N	$\downarrow$ (N)	N	N	N	$\uparrow$
Direct							
Absolute							
Liso-phase	N	N	$\downarrow$ (N)	N	N	N	$\uparrow$
Two-step	N	N	$\downarrow$ (N)	N	N	N	$\uparrow$
Corning Immophase	$\uparrow$	$\downarrow$	$\downarrow$ (N)	N	$\uparrow$	$\downarrow$	
Kinetic							
Mallinckrodt SPAC-ET	N	N	$\downarrow$ (N)	N	N	N	
Analog	N	N	$\downarrow$	$\uparrow$	$\uparrow$	$\downarrow$	$\downarrow$

N = no interference; $\downarrow$ inappropriately high values; $\uparrow$ inappropriately low values; ED+RIA equilibrium dialysis followed by RIA.

accuracy of some methods, such as column adsorption chromatography followed by RIA (Liso-phase, Lepetit-Sclavo) is almost the same as the "gold standard" equilibrium dialysis techniques. Similar results are generally achieved by two-step back-titration methods (e.g., Clinical Assay Gammacoat) and by the Mallinckrodt SPAC-ET kit. Conversely, other techniques, and in particular the analog methods, appear to be grossly affected in several conditions. It should be noted that not all the tests have been fully evaluated in all clinical situations. Furthermore, the majority of the available data concern FT_4 determination, although it is conceivable that a similar interference in FT_3 measurements could be observed. Thus, in this chapter, attention will be focused on FT_4 methods (Table 6). The main factors interfering in FT_4 measurement will be discussed in Sects. D.III.1–5; to this purpose, the results obtained by equilibrium dialysis followed by RIA (direct, absolute methods) will be considered as the reference values.

1. Pregnancy and Other Conditions of Abnormal TBG Concentrations

Most assays give FT_4 values in the normal range in the presence of both TBG excess and deficiency. Exceptions to this rule are methods based on the labeled hormone antibody uptake, which generally provide inappropriately high or low FT_4 values in the presence of TBG excess or deficiency (HELENIUS and LIEWENDAHL 1983; WITHERSPOON et al. 1980). This problem has recently been overcome by the modification of this kind of assay adopted in the Mallinckrodt SPAC-ET FT_4 kit.

Both normal and slightly reduced serum free thyroid hormone concentrations have been reported by different methods (ROMELLI et al. 1979; OBREGON et al. 1981; TUTTLEBEE 1982; HELENIUS and LIEWENDAHL 1983; HOPTON et al. 1983; PACCHIAROTTI et al. 1986). These discrepancies appear to be at least partially due to the different criteria of patient selection. Recently, it has been demonstrated in our laboratory, using two different methods, that a progressive reduction of FT_4 and FT_3 concentrations occurs throughout pregnancy and is associated with small, but significant increases of serum thyrotropin concentrations (PACCHIAROTTI et al. 1986).

2. Familial Dysalbuminemic Hyperthyroxinemia (FDH)

As previously described, this condition, due to abnormal binding of T_4 to albumin, is characterized by elevated serum TT_4 and normal TT_3 concentrations without thyrotoxicosis (RAJATANAVIN and BRAVERMAN 1983). Direct absolute assays, back-titration methods, and labeled hormone antibody uptake tests correctly classify subjects with FDH, giving normal FT_4 values (SPENCER 1985). In contrast, all the currently available methods, based on the use of "analogs," give inappropriately high FT_4 concentrations (RAJATANAVIN et al. 1982; STOCKIGT et al. 1982; DE NAYER et al. 1984). This is due to the previously mentioned interaction of the analog tracer with the abnormal albumin.

3. Anti-Iodothyronine Autoantibodies

As discussed before, anti-thyroid hormone autoantibodies strongly interfere in TT_4 and TT_3 measurement by RIA. In contrast, most methods for free thyroid hormone determination are believed to be substantially unaffected by anti-iodothyronine autoantibodies. However, anti-thyroxine autoantibodies strongly interfere with all FT_4 analog RIAs, giving falsely elevated results, since, similarly to the abnormal albumin of FDH, they are able to bind the tracer analog (Beck-Peccoz et al. 1984, 1985; Spencer 1985). Inappropriately high FT_4 concentrations are also obtained by the Corning Immophase kit (Beck-Peccoz et al. 1984, 1985; Spencer 1985). All other methods are unaffected by anti-T_4 autoantibodies.

4. Severe Nonthyroidal Illness (NTI)

Low serum TT_4 concentrations are frequently observed in patients with severe NTI (Wartofsky and Burman 1982). Several factors appear to be responsible for such phenomena, including dampening of the pituitary–thyroid axis (Wartofsky and Burman 1982; Wehmann et al. 1985), presence of abnormal TBG (Wartofsky and Burman 1982), and presence of a still poorly defined thyroid hormone-binding inhibitor (THBI) (Lutz et al. 1972; Chopra et al. 1979; Oppenheimer et al. 1982). Inappropriately low FT_4 levels are commonly seen in NTI, when analog assays are used (Stockigt et al. 1981; Tuttlebee 1982; Helenius and Liewendahl 1983; Deadman and Evans 1984). This is probably due in part to the decreased albumin levels (Liewendahl et al. 1984) and to increased nonesterified fatty acids (Helenius and Liewendahl 1979). As a consequence, there is a reduced binding of the analog to serum albumin and, therefore, an increased availability of the tracer for the reaction with solid-phase antibody, thus causing falsely low FT_4 values. Inappropriately reduced FT_4 concentrations are also found by the Corning method (Helenius and Liewendahl 1979; Wood et al. 1980; Kaptein et al. 1981; Slag et al. 1981; Tuttlebee 1982; Helenius and Liewendahl 1983). Other available RIAs, including equilibrium dialysis followed by RIA, Liso-phase, and Gammacoat kits, usually result in normal values.

5. Heparin

Heparin interferes with thyroid hormone binding to TBG by altering the steric configuration of the binding site and lowering the affinity of TBG for thyroid hormones (Hollander et al. 1967). This interference results in an acute elevation of serum FT_4 and FT_3 concentrations during intravenous heparin therapy, with reciprocal changes of serum TSH (Hershman et al. 1972), leading to the establishment of a new steady state in which free thyroid hormone concentrations are finally normalized (Refetoff 1979). These changes of free thyroid hormone concentrations are correctly classified by equilibrium dialysis and column adsorption chromatography methods. On the contrary, heparin causes falsely low FT_4 values by the analog methods; this phenomenon probably reflects the presence of a THBI (possibly nonesterified fatty acids) induced by the drug. The consequent interference in the assay would therefore be similar to that described for NTI.

In conclusion, the analysis of factors affecting free thyroid hormone determinations suggests that, in the presence of circulating interfering substances, valid measurements of FT_4 and FT_3 can only be achieved by methods that physically separate the fraction of free thyroid hormone from binding proteins before the assay procedure.

E. Radioimmunoassay of Other Iodothyronines, Iodotyrosines, and Products of Thyroid Hormone Degradation

Several iodinated compounds, either of thyroidal origin or peripherally produced from thyroid hormone degradation, have recently been detected and quantitated in serum by specific RIAs. These substances (see Table 1) are virtually devoid of metabolic activity, but their measurement may be helpful in particular clinical conditions associated with abnormalities of thyroid hormone metabolism. Furthermore, precise measurement of these compounds in serum and in tissues has provided important information on the peripheral metabolism of thyroid hormones. Reverse T_3 (rT_3) is the most widely studied among these substances and its RIA will be discussed in detail; brief mention will also be made of available RIAs of other iodoamino acids.

I. rT_3 RIA

Reverse T_3 is the product of inner ring monodeiodination of T_4. Its measurement in serum was first accomplished by CHOPRA (1974), who developed a specific RIA. Anti-rT_3 antiserum was raised by immunizing rabbits with rT_3 conjugated to human serum albumin. This antiserum proved to be highly specific, since even T_3, with a molecular structure very similar to that of rT_3, showed less than 0.1% cross-reactivity. rT_3 RIA originally implied extraction of sera by ethanol. Subsequently, other methods have been proposed that, like TT_3 RIA, utilize substances inhibiting rT_3 binding to serum proteins (NICOD et al. 1976).

The protocol for rT_3 RIA, as proposed by CHOPRA (1974), is as follows: test sera are extracted by mixing 0.5 ml serum with 1 ml 95% ethanol, followed by centrifugation and collection of supernatants; ethanol extracts of hormone-free serum in the standard tubes or unknown serum samples (300 μl) are added; rT_3 graded amounts (5 pg–3 ng) are then added, together with anti-rT_3 serum appropriately diluted (100 μl) and ^{125}I-labeled rT_3 (10000 cpm, 100 μl) in assay buffer (0.075 *M* barbital buffer, pH 8.6, containing 1% NRS and 0.01% sodium azide). After incubation for 24 h at 4 °C, a second antibody (goat anti-rabbit γ-globulin), previously titered, is pipetted. After incubation, separation of bound from free is achieved by the procedure already described for T_3.

Other RIAs for rT_3 have been subsequently developed and some of them are commercially available. The mean normal serum rT_3 concentrations range from 18 to 60 ng/dl, though different values have been reported by different authors, possibly owing to technical problems, such as the purity of rT_3 standards (CHOPRA et al. 1981). As mentioned previously, increased rT_3 values are found in nonthyroidal illness (WARTOFSKY and BURMAN 1982).

II. RIA of Diiodothyronines

Diiodothyronines (T_2) include 3,5-T_2, 3,3′-T_2, and 3′,5′-T_2. As shown in Fig. 1, these compounds are produced through peripheral deiodination of T_3 and rT_3. In particular, 3,5-T_2 derives from 5′-deiodination of T_3, 3,3′-T_2 is the product either of T_3 5-deiodination or of rT_3 5′-deiodination; 3′,5′-T_2 is produced through inner ring deiodination of rT_3. Early RIAs for T_2s showed marked variations in serum concentrations of these substances, owing to the lack of antisera specificity and to divergent techniques used in the various laboratories (WU et al. 1976; BURMAN et al. 1977, 1978; CHOPRA et al. 1978; MACIEL et al. 1979; PANGARO et al. 1980). More recently, specific and sensitive assays have been developed using unextracted serum, highly specific antibodies, and labeled T_2s of high specific activity (FABER et al. 1979, 1981; ENGLER et al. 1984). By these methods, values were 0.20–0.75 ng/dl for 3,5-T_2, 1–3 ng/dl for 3,3′-T_2, and 1–4 ng/dl for 3′,5′-T_2. Support for the validity of these values also came from turnover studies carried out with labeled T_3 and rT_3 (FABER et al. 1981).

III. RIA of Monoiodothyronines

Monoiodothyronines (T_1) comprise 3′-T_1 and 3-T_1, derived from deiodination of T_2s (see Fig. 1). Serum concentrations of 3′-T_1 by RIA range from 0.6 to 6.7 ng/dl in different reports (CHOPRA 1980; KIRKEGAARD et al. 1981). The major drawback of this assay is the interference produced by the two precursors of 3′-T_1, i.e., 3,3′-T_2 and 3′,5′-T_2. Data on serum 3-T_1 concentrations are very limited. In a recent report using ^{3}H-labeled 3-T_1 as a tracer, values ranging from <0.66 to 7.5 ng/dl were reported (CORCORAN and EASTMAN 1983).

IV. RIA of Tetraiodothyroacetic Acid (TETRAC) and Triiodothyroacetic Acid (TRIAC)

TETRAC and TRIAC are the products of the oxidative deamination (or transamination) and decarboxylation of T_4 and T_3. RIAs have been developed for these substances, giving mean serum concentrations in euthyroid subjects of 1.6–3.0 ng/dl for TRIAC (GAVIN et al. 1980) and <8–60 ng/dl for TETRAC (MAXON et al. 1982).

V. RIA of Iodotyrosines (MIT and DIT)

DIT and MIT derive from intrathyroidal proteolysis of thyroglobulin and are mostly enzymatically deiodinated within the gland (DEGROOT et al. 1984; INGBAR 1985). However, a small fraction of these iodotyrosines escape deiodination and enter the bloodstream (DEGROOT et al. 1984; INGBAR 1985). MIT also derives from peripheral deiodination of DIT. Measurement of MIT and DIT is a relatively new field, whose methodology is still developing. This concept is supported by the marked differences of serum mean DIT values reported using different RIAs (NELSON et al. 1974; MEINHOLD et al. 1981), ranging from 1 to 432 ng/dl in euthyroid subjects. The latter surprisingly high concentrations of DIT are likely due to cross-reactivity of T_4 with DIT antisera. The same problem probably exists for MIT RIA, since the available normal range of serum concentrations of this compound (90–390 ng/dl) (NELSON and LEWIS 1979) appears to be an overestimation.

References

Amino N, Nishi K, Nakatani H, Mizuta K, Ichihara K, Tanizawa O, Miyai K (1983) Effect of albumin concentration on the assay of serum free thyroxine by equilibrium radioimmunoassay with labeled thyroxine analogue (Amerlex Free T_4). Clin Chem 29:321–325

Azukizawa M, Pekary AE, Hershman JM, Parker DC (1976) Plasma thyrotropin, thyroxine, and triiodothyronine relationships in man. J Clin Endocrinol Metab 43:533–542

Barbosa J, Seal US, Doe RP (1971) Effects of anabolic steroids on hormone-binding proteins, serum cortisol and serum nonprotein bound cortisol. J Clin Endocrinol Metab 32:232–240

Barker SB (1948) Determination of protein-bound iodine. J Biol Chem 173:715–724

Barlow JW, Csicmann JM, White EL, Funder JW, Stockigt JR (1982) Familial euthyroid thyroxine excess: characterization of abnormal intermediate affinity thyroxine binding to albumin. J Clin Endocrinol Metab 55:244–250

Beck-Peccoz P, Romelli PB, Faglia G (1983) Circulating antitriiodothyronine autoantibodies in two euthyroid patients. Apparent lack of interference in total T_3 radioimmunoassay based on second antibody or solid phase separation techniques. J Endocrinol Invest 6:333–340

Beck-Peccoz P, Romelli PB, Cattaneo MG, Faglia G, White EL, Barlow JW, Stockigt JR (1984) Evaluation of free thyroxine methods in the presence of iodothyronine-binding autoantibodies. J Clin Endocrinol Metab 58:736–739

Beck-Peccoz P, Romelli PB, Cattaneo MG, Medri G, Persani L, Piscitelli G, Faglia G (1985) Interference in estimation of free thyroid hormones with special reference to antiiodothyronine autoantibodies in serum. Nuc Compact 16:369–372

Benotti J, Pinto S (1966) A simplified method for butanol-extractable iodine and butanol-insoluble iodine. Clin Chem 12:491–496

Braverman LE, Foster AE, Mead LW (1967) The charcoal T_3 ratio. An in vitro test of thyroid function. JAMA 199:169–172

Braverman LE, Vagenakis AG, Foster AE, Ingbar SH (1971) Evaluation of a simplified technique for the specific measurement of serum thyroxine concentration. J Clin Endocrinol Metab 32:497–502

Braverman LE, Abreau CM, Brock P, Kleinmann R, Fournier L, Odstrchel G, Schoemaker HJP (1980) Measurement of serum free thyroxine by RIA in various clinical states. J Nucl Med 21:233–239

Brown BL, Ekins RP, Ellis SM, Reith WS (1970) Specific antibodies to triiodothyronine hormone. Nature 226:359

Buist NRM, Murphey WF, Brandon GR, Foley TP Jr, Penn RL (1975) Neonatal screening for hypothyroidism. Lancet II:872–875

Burger A, Sakoloff C, Staeheli V, Vallotton MB, Ingbar SH (1975) Radioimmunoassay of 3,5,3′-triiodo-L-thyronine with or without a prior extraction step. Acta Endocrinol (Copenh) 80:58–69

Burman KD, Strum D, Dimond RC, Djum Y-Y, Wright FD, Earll JM, Wartofsky L (1977) A radioimmunoassay for 3,3′-L-diiodothyronine (3,3′-T_2). J Clin Endocrinol Metab 45:339–352

Burman KD, Wright FD, Smallridge RC, Green BJ, Georges LP, Wartofsky L (1978) A radioimmunoassay for 3′,5′-diiodothyronine. J Clin Endocrinol Metab 47:1059–1064

Catt KJ (1969) Radioimmunoassay with antibody-coated discs and tubes. Acta Endocrinol (Copenh) [Suppl 142] 63:222–243

Cavalieri RR, Pitt-Rivers R (1981) The effects of drugs on the distribution and metabolism of thyroid hormones. Pharmacol Rev 33:55–80

Cavalieri RR, Castle JM, Searle GM (1969) A simplified method for estimating free-thyroxine fraction in serum. J Nucl Med 10:565–570

Chan KH, Cummings LM (1976) Pepsin: a new extractant for serum thyroxine by enzymatic digestion of thyroxine binding proteins. J Clin Endocrinol Metab 48:189–192

Chopra IJ (1972) A radioimmunoassay for measurement of thyroxine in unextracted serum. J Clin Endocrinol Metab 34:938–947

Chopra IJ (1974) A radioimmunoassay for measurement of 3,3′,5′-triiodothyronine (reverse T_3). J Clin Invest 54:583–592

Chopra IJ (1980) A radioimmunoassay for measurement of 3′-monoiodothyronine. J Clin Endocrinol Metab 51:117–123

Chopra IJ (1981) Triiodothyronines in health and disease. Springer, Berlin Heidelberg New York

Chopra IJ, Nelson JC, Solomon DH, Beall GN (1971 a) Production of antibodies specifically binding triiodothyronine and thyroxine. J Clin Endocrinol Metab 32:299–308

Chopra IJ, Solomon DH, Beall GN (1971 b) Radioimmunoassay for measurement of triiodothyronine in human serum. J Clin Invest 50:2033–2041

Chopra IJ, Ho RS, Lam R (1972) An improved radioimmunoassay of triiodothyronine in serum: its application to clinical and physiological studies. J Lab Clin Med 80:729–739

Chopra IJ, Geola F, Solomon DH, Maciel RMB (1978) 3′,5′-diiodothyronine in health and disease. Studies by a radioimmunoassay. J Clin Endocrinol Metab 47:1198–1207

Chopra IJ, Chua Teco GN, Nguyen AH, Solomon DH (1979) In search of an inhibitor of thyroid hormone binding to serum proteins in nonthyroid illnesses. J Clin Endocrinol Metab 49:63–69

Corcoran JM, Eastman CJ (1983) Radioimmunoassay of 3-L-monoiodothyronine: an application in normal human physiology and thyroid disease. J Clin Endocrinol Metab 57:66–70

Deadman NM, Evans D (1984) "Unbound analogue" radioimmunoassays for free thyroxin (Amerlex Free Thyroxin) and the effect of albumin and nonthyroidal illness. Clin Chem 30:344

DeGroot LJ, Larsen PR, Refetoff S, Stanbury JB (1984) The thyroid and its diseases, 5th edn. Wiley Biomedical, New York

De Nayer P, Glinoer D (1985) Thyroid hormone transport and action. In: Delange F, Fisher DA, Malvaux P (eds) Pediatric thyroidology. Karger, Basel, p 57

De Nayer P, Malvaux P, Beckers C (1984) Familial dysalbuminemic hyperthyroxinemia (FDH): inadequacy of the "analogue" methods for assaying free T_4 levels. Eur J Nucl Med 9:284–285

Dowling JT, Freinkel N, Ingbar SH (1960) Effect of diethylstilbestrol on binding of thyroxine in serum. J Clin Endocrinol Metab 16:1491–1506

Dussault JH (1985) Neonatal screening for congenital hypothyroidism. In: Delange F, Fisher DA, Malvaux P (eds) Pediatric thyroidology. Karger, Basel, p 106

Ekins RP (1979) Methods for measurement of free thyroid hormones. In: Ekins RP, Faglia G, Pennisi F, Pinchera A (eds) Free thyroid hormones. Excerpta, Amsterdam, p 72

Ekins RP (1983) The direct immunoassay of free (non-protein bound) hormones in body fluid. In: Hunter WH, Corrie JET (eds) Immunoassays for clinical chemistry, 2nd edn. Livingstone, Edinburgh, p 319

Ekins RP (1985) Principles of measuring free thyroid hormone concentrations in serum. Nuc Compact 16:305–313

Ellis S, Ekins RP (1973) The direct measurement by radioimmunoassay of the free thyroid hormone concentration in serum. Acta Endocrinol (Copenh) [Suppl 177]:106

Engler D, Burger AG (1984) The deiodination of iodotyrosines and of their derivatives in man. Endocr Rev 5:151–184

Engler D, Merkelbach V, Steiger G, Burger AG (1984) The monodeiodination of triiodothyronine in man: a quantitative evaluation of the pathway by the use of turnover rate techniques. J Clin Endocrinol Metab 58:49–61

Faber J, Kirkegaard C, Lumholtz IB, Siersbaek-Nielsen K, Friis T (1979) Measurements of serum 3′,5′-diiodothyronine and 3,3′-diiodothyronine concentrations in normal subjects and in patients with thyroid and nonthyroid diseases: studies of 3′,5′-diiodothyronine metabolism. J Clin Endocrinol Metab 48:611–617

Faber J, Thomsen HF, Lumholtz IB, Kirkegaard C, Siersbaek-Nielsen K, Friis T (1981) Kinetic studies of thyroxine, 3,5,3′-triiodothyronine, 3,3′,5′-triiodothyronine, 3′,5′-diiodothyronine, 3,3′-diiodothyronine and 3′-monoiodothyronine in patients with liver cirrhosis. J Clin Endocrinol Metab 53:978–984

Fang VS, Refetoff S (1974) Radioimmunoassay for serum triiodothyronine: evaluation of simple techniques to control interference from binding proteins. Clin Chem 20:1150–1154

Federman DD, Robbins J, Rall JE (1958) Effect of methyl testosterone on thyroid function, thyroxine metabolism and thyroxine-binding protein. J Clin Invest 37:1024–1030

Fisher DA, Dussault DH (1971) Contribution of methodologic artifacts to the measurement of T_3 concentration in serum. J Clin Endocrinol Metab 32:675–679

Garnick MB, Larsen PR (1979) Acute deficiency of thyroxine-binding globulin during L-asparaginase therapy. N Engl J Med 301:252–253

Gavin LA, Livermore BM, Cavalieri RR, Hammond ME, Castle JN (1980) Serum concentrations, metabolic clearance and production rates of 3,5,3′-triiodothyroacetic acid in normal and athyreotic man. J Clin Endocrinol Metab 51:529–534

Gershengorn MC, Larsen PR, Robbins J (1976) Radioimmunoassay for serum thyroxine-binding globulin: results in normal subjects and in patients with hepatocellular carcinoma. J Clin Endocrinol Metab 42:907–911

Gharib H, Ryan RJ, Mayberry WE, Hochert T (1971) Radioimmunoassay for triiodothyronine (T_3): I. Affinity and specificity of the antibody for T_3*. J Clin Endocrinol Metab 33:509–516

Glinoer D, Gershengorn MC, Dubois A, Robbins J (1977) Stimulation of thyroxine-binding globulin synthesis by isolated rhesus monkey hepatocytes after in vivo β-estradiol administration. Endocrinology 100:807–813

Glinoer D, Fernandez-Deville M, Ermans AM (1978) Use of direct thyroxine-binding globulin measurement in the evaluation of thyroid function. J Endocrinol Invest 1:329–335

Hamolsky MW, Stein M, Freedberg AS (1957) The thyroid hormone-plasma protein complex in man: II. A new in vitro method for study of "uptake" of labeled hormonal components by human erythrocytes. J Clin Endocrinol Metab 17:33–44

Hansen J (1974) Increased fecal thyroxine losses with protein-losing enteropathy. NY State J Med 74:1993–1995

Helenius T, Liewendahl K (1979) Abnormal thyroid function tests in severe nonthyroidal illness: diagnostic and pathophysiologic aspects. Scand J Clin Lab Invest 39:389–397

Helenius T, Liewendahl K (1983) Improved dialysis method for free thyroxine in serum compared with five commercial radioimmunoassays in nonthyroidal illness and subjects with abnormal concentrations of thyroxine-binding globulin. Clin Chem 29:816–822

Hennemann G, Docter R, Krenning EP, Bos G, Otten M, Visser TJ (1979) Raised total thyroxine and free thyroxine index but normal free thyroxine. A serum abnormality due to inherited increased affinity of iodothyronines for serum binding proteins. Lancet I:639–641

Hershman JM, Jones CM, Bailey AL (1972) Reciprocal changes in serum thyrotropin and free thyroxine produced by heparin. J Clin Endocrinol Metab 34:574–579

Hesch R-D (1981) The "low T_3 syndrome". Academic, London

Hollander CS, Garcia AM, Sturgis SH, Selenkow HA (1963) Effect of an ovulatory suppressant on the serum protein-bound iodine and red-cell uptake of radioactive triiodothyronine. N Engl J Med 269:501–504

Hollander CS, Scott RL, Burgess JA, Rabinowitz D, Merimee TJ, Oppenheimer JH (1967) Free fatty acids: a possible regulator of free thyroid hormone levels in man. J Clin Endocrinol Metab 27:1219–1223

Hopton MR, Ashwell K, Scott IV, Harrop JS (1983) Serum free thyroxine concentration and free thyroid hormone indices in normal pregnancy. Clin Endocrinol (Oxf) 18:431–437

Ingbar SH (1985) The thyroid gland. In: Wilson JD, Foster DW (eds) Textbook of endocrinology, 7th edn. Saunders, Philadelphia, p 682

Ingenbleek Y, De Nayer P, DeVisscher M (1974) Thyroxine-binding globulin in infant protein-calorie malnutrition. J Clin Endocrinol Metab 39:178–180

Kaptein EM, MacIntyre SS, Weiner JM, Spencer CA, Nicoloff JT (1981) Free thyroxine estimates in nonthyroidal illnesses: comparison of eight methods. J Clin Endocrinol Metab 52:1073–1077

Kirkegaard C, Friis T, Siersbaek-Nielsen K (1974) Measurements of serum triiodothyronine by radioimmunoassay. Acta Endocrinol (Copenh) 77:71–81

Kirkegaard C, Faber J, Cohn D, Kolendorf K, Thomsen HF, Lumholtz IB, Siersbaek-Nielsen K, Friis T (1981) Serum 3′-monoiodothyronine levels in normal subjects and in patients with thyroid and non-thyroid disease. Acta Endocrinol (Copenh) 97:454–461

Larsen PR (1971) Technical aspects of the estimation of triiodothyronine in human serum: evidence of conversion of thyroxine to triiodothyronine during assay. Metabolism 20:609–624

Larsen PR (1972a) Salicylates induced increases in free triiodothyronine in serum. Evidence for inhibition of triiodothyronine binding to thyroxine-binding globulin and thyroxine-binding prealbumin. J Clin Invest 51:1125–1134

Larsen PR (1972b) Direct immunoassay of triiodothyronine in human serum. J Clin Invest 51:1939–1949

Larsen PR (1976) Quantitation of triiodothyronine and thyroxine in human serum by radioimmunoassay. In: Antoniades HN (ed) Hormones in human blood. Harvard University Press, Cambridge, p 679

Larsen PR, Dockalova J, Sipula D, Wu FM (1973) Immunoassay of thyroxine in unextracted human serum. J Clin Endocrinol Metab 37:177–182

Lee ND, Henry RJ, Golub OJ (1964) Determination of the free thyroxine content of serum. J Clin Endocrinol Metab 24:486–495

Lee WNP, Golden MP, Van Herle AJ, Lippe BM, Kaplan SA (1979) Inherited abnormal thyroid hormone binding protein causing selective increase of total serum thyroxine. J Clin Endocrinol Metab 49:292–299

Lieblich J, Utiger RD (1972) Triiodothyronine radioimmunoassay. J Clin Invest 51:157–166

Liewendahl K, Tikanoja T, Helenius T, Valimaki M (1984) Discrepancies between serum free triiodothyronine and free thyroxine as measured by equilibrium dialysis and analogue radioimmunoassay in nonthyroidal illness. Clin Chem 30:760–762

Lutz JH, Gregerman SW, Spaulding SW, Harnick RB, Dawkins AT (1972) Thyroxine binding proteins, free thyroxine and thyroxine turnover interrelationships during acute infectious illness in man. J Clin Endocrinol Metab 35:230–249

Maciel RMB, Chopra IJ, Ozawa Y, Geola F, Solomon DH (1979) A radioimmunoassay for measurement of 3,5-diiodothyronine. J Clin Endocrinol Metab 49:399–405

Man EB, Kydd DM, Peters JP (1951) Butanol-extractable iodine of serum. J Clin Invest 30:531–538

Maxon HM, Burman KD, Premachandra BN, Chen I-W, Burger A, Levy P, Georges LP (1982) Familial elevation of total and free thyroxine in healthy, euthyroid subjects without detectable binding protein abnormalities. Acta Endocrinol (Copenh) 100:224–230

Meinhold H, Beckert A, Wenzel KW (1981) Circulating diiodotyrosine: studies of its serum concentration, source, and turnover using radioimmunoassay after immunoextraction. J Clin Endocrinol Metab 53:1171–1178

Midgley JEM, Wilkins TA (1982) An improved method for the estimation of relative binding constants of T_4 and its analogues with serum proteins. Clin Endocrinol (Oxf) 17:523–528

Mitchell ML (1976) Improved thyroxine radioimmunoassay for filter paper discs saturated with dried blood. Clin Chem 22:1912–1919

Mitsuma T, Nihei N, Gershengorn MC, Hollander CS (1971) Serum triiodothyronine measurements in human serum by radioimmunoassay with corroboration by gas-liquid chromatography. J Clin Invest 50:2679–2688
Mitsuma T, Colucci J, Shenkman L, Hollander CS (1972) Rapid simultaneous radioimmunoassay for triiodothyronine and thyroxine in unextracted serum. Biochem Biophys Res Commun 46:2107–2113
Monzani F, Grasso L, Lippi F, Giacomelli T, Pinchera A (1983) Valutazione di una nuova metodica per la determinazione degli ormoni tiroidei. LAB J Res Lab Med 10:11–19
Moses A, Lawlor J, Haddow J, Jackson IMD (1982) Familial euthyroid hyperthyroxinemia resulting from increased thyroxine-binding to thyroxine-binding prealbumin. N Engl J Med 306:966–969
Murphy BEP, Pattee CJ (1964) Determination of thyroxine utilizing the property of protein-binding. J Clin Endocrinol Metab 24:187–196
Naumann JA, Naumann A, Werner SC (1967) Total and free triiodothyronine in human serum. J Clin Invest 46:1346–1355
Nelson JC, Lewis JE (1979) Radioimmunoassay of iodotyrosines. In: Abraham GE (ed) Handbook of radioimmunoassay. Dekker, New York, p 705
Nelson JC, Weiss RM, Lewis JE, Wilcox RB, Palmer FJ (1974) A multiple ligand-binding radioimmunoassay of diiodotyrosine. J Clin Invest 53:416–422
Nicod P, Burger A, Staeheli V, Vallotton MB (1976) A radioimmunoassay for 3,3′,5′-triiodothyronine in unextracted serum. Method and clinical results. J Clin Endocrinol Metab 42:823–829
Obregon MJ, Kurtz A, Ekins RP, Escobar GM (1981) Evaluation of free and total L-thyroxine in serum by a commercial procedure. Clin Chem 27:149–152
O'Connor JF, Wu GY, Gallagher TF, Hellman L (1974) The 24-hour plasma thyroxin profile in normal man. J Clin Endocrinol Metab 39:765–771
Odstrchel G (1982) A kinetic method for the measurement of free thyroxine and its application in the diagnosis of various thyroid disease states. In: Albertini A, Ekins RP (eds) Free hormones in blood. Elsevier, Amsterdam, p 91
Oltman JE, Friedman S (1963) Protein bound iodine in patients receiving perphenazine. JAMA 185:726–727
Oppenheimer JH, Surks MI (1964) Determination of free thyroxine in human serum: a theoretical and experimental analysis. J Clin Endocrinol Metab 24:785–793
Oppenheimer JH, Werner SC (1966) Effect of prednisone on thyroxine-binding proteins. J Clin Endocrinol Metab 26:715–721
Oppenheimer JH, Schwartz HL, Mariash CN, Kaiser FE (1982) Evidence for a factor in sera of patients with nonthyroid disease which inhibits iodothyronine binding to solid matrices, serum proteins, and rat hepatocytes. J Clin Endocrinol Metab 54:757–766
Pacchiarotti A, Martino E, Bartalena L, Buratti L, Mammoli C, Strigini F, Fruzzetti F, Melis GB, Pinchera A (1986) Serum thyrotropin by ultrasensitive immunoradiometric assay and serum free thyroid hormones in pregnancy. J Endocrinol Invest 9:185–188
Pangaro L, Burman KD, Wartofsky L, Cahnmann HJ, Smallridge RL, O'Brian JY, Wright FD, Latham K (1980) Radioimmunoassay for 3,5-diiodothyronine and evidence of dependence on conversion from 3,5,3′-triiodothyronine. J Clin Endocrinol Metab 50:1075–1081
Pileggi VJ, Lee ND, Golub OJ, Henry RJ (1961) Determination of iodine compounds in serum. I. Serum thyroxine in the presence of some iodine contaminants. J Clin Endocrinol Metab 21:1272–1279
Rajatanavin R, Braverman LE (1983) Euthyroid hyperthyroxinemia. J Endocrinol Invest 6:493–505
Rajatanavin R, Fournier L, Decosimo D, Abreau C, Braverman LE (1982) Elevated free thyroxine by thyroxine analogue radioimmunoassays in euthyroid patients with familial dysalbuminemic hyperthyroxinemia. Ann Intern Med 97:865–866
Ratcliffe WA, Hazelton RA, Thomson JA, Ratcliffe JG (1980) The effects of fenclofenac on thyroid function tests in vivo and in vitro. Clin Endocrinol (Oxf) 13:569–575
Reese MG, Johnson LR (1978) Automated measurement of serum thyroxine with the ARIA II as compared with competitive protein binding and radioimmunoassay. Clin Chem 24:342–348

Refetoff S (1976) Principles in competitive binding assay and radioimmunoassay. In: Gottschalk A, Patchen EJ (eds) Diagnostic nuclear medicine. Williams and Wilkins, Baltimore, p 215
Refetoff S (1979) Thyroid function tests. In: DeGroot LJ (ed) Endocrinology, vol 1. Grune and Stratton, New York, p 387
Refetoff S, Hagen SR, Selenkow HA (1972a) Estimation of the T_4-binding capacity of serum TBG and TBPA by a single load-ion-exchange resin method. J Nucl Med 13:2–12
Refetoff S, Robin NI, Alper CA (1972b) Study of four new kindreds with inherited thyroxine-binding globulin abnormalities: possible mutations of a single gene locus. J Clin Invest 51:848–867
Robbins J (1973) Inherited variations in thyroxine transport. Mt Sinai J Med 40:511–519
Robbins J, Bartalena L (1986) Plasma transport of thyroid hormones. In: Hennemann G (ed) Thyroid hormone metabolism. Dekker, New York, p 3
Robbins J, Nelson JH (1958) Thyroxine-binding by serum protein in pregnancy and in the newborn. J Clin Invest 37:153–159
Robbins J, Rall JE (1983) The iodine containing hormones. In: Gray CH, James VHT (eds) Hormones in blood, vol 4, 3rd edn. Academic, London, p 219
Robbins J, Rall JE, Federman ML (1957) Thyroxine-binding by serum and urine proteins in nephrosis. Qualitative aspects. J Clin Invest 36:1333–1342
Robin NI, Hagen SR, Collaco F, Refetoff S, Selenkow HA (1971) Serum tests for measurement of thyroid function. Hormones 2:266–279
Romelli PB, Pennisi F, Vancheri L (1979) Measurement of free thyroid hormones in serum by column adsorption chromatography and radioimmunoassay. J Endocrinol Invest 2:25–40
Ross HA, Benraad TJ (1984) An indirect method for the estimation of free thyroxine by means of monoclonal T_4 antibody-coated tubes. Nuc Compact 15:204–209
Sakata S, Nakamura S, Miura K (1985) Autoantibodies against thyroid hormones or iodothyronines. Implications in diagnosis, thyroid function, treatment and pathogenesis. Ann Intern Med 103:579–589
Schussler GL, Plager JE (1967) Effect of preliminary purification of 131-I-thyroxine in serum. J Clin Endocrinol Metab 27:242–250
Slag MF, Morley JE, Elson MK, Labrosse KR, Crowson TV, Nuttal FQ, Shafer RB (1981) Free thyroxine levels in critically ill patients. A comparison of currently available assays. JAMA 246:2702–2706
Spencer CA (1985) The comparative clinical value of free T_4 estimation using different methodological approaches. Nuc Compact 16:321–327
Sterling K, Hegedus A (1962) Measurement of free thyroxine concentration in human serum. J Clin Invest 41:1031–1040
Sterling K, Milch PD (1974) Thermal inactivation of thyroxine binding globulin for direct radioimmunoassay of triiodothyronine in serum. J Clin Endocrinol Metab 38:866–875
Sterling K, Tabachnick M (1961) Resin uptake of I-131 triiodothyronine as a test of thyroid function. J Clin Endocrinol Metab 21:456–464
Sterling K, Bellabarba D, Newman ES, Brenner M (1969) Determination of triiodothyronine concentration in human serum. J Clin Invest 48:1150–1158
Stockigt JR, Degaris M, Csicmann J, Barlow JW, White EL, Hurley DM (1981) Limitations of a new free thyroxine assay (Amerlex free T_4). Clin Endocrinol (Oxf) 15:317–318
Stockigt JR, Degaris M, Barlow JM (1982) "Unbound analogue" methods for free T_4: a note of caution. N Engl J Med 307:126
Tuttlebee JW (1982) Further experience with free thyroxine assays with particular reference to pregnancy. Ann Clin Biochem 13:374–378
Vannotti A, Beraud T (1959) Functional relationships between the liver, the thyroid binding protein of serum and the thyroid. J Clin Endocrinol Metab 18:466–477
Wagner D, Alspector B, Feingers J, Pick A (1979) Direct radioimmunoassay with capillary chromatography tubes. Clin Chem 25:1337–1339

Wartofsky L, Burman KD (1982) Alterations in thyroid function in patients with systemic illness: the "euthyroid sick syndrome". Endocr Rev 3:164–217
Webster JB, Coupal JJ, Cushman P Jr (1973) Increased serum thyroxine levels in euthyroid narcotic addicts. J Clin Endocrinol Metab 37:928–934
Wehmann RE, Gregerman RI, Burns WM, Saral L, Santos GW (1985) Suppression of thyrotropin in the low-thyroxine state of severe nonthyroidal illness. N Engl J Med 312:546–552
White EL, Barlow JW, Burke CW, Funder JW, Stockigt JR (1980) A T_4-loaded resin uptake system to detect familial euthyroid thyroxine excess. In: Stockigt JR, Nagataki S (eds) Thyroid research VIII. Australian Academy of Science, Canberra, p 513
Witherspoon LR, Shuler SE, Garcia MM, Zollinger LA (1980) An assessment of methods for the estimation of free thyroxine. J Nucl Med 21:529–539
Witherspoon LL, Shuler SE, Gilbert SS (1984) Evaluation of an immunoextraction procedure for the estimation of free thyroxine concentration. J Nucl Med 25:188–196
Wolff J, Standaert ME, Rall JE (1961) Thyroxine displacement from serum proteins and depression of serum protein-bound iodine by certain drugs. J Clin Invest 40:1373–1377
Wong WH (1975) Determination of serum thyroxine after dissociation from thyroxine binding globulin in alkaline solution and adsorption on dextran coated charcoal. Clin Chem 21:216–220
Wood DG, Cyrus J, Samols G (1980) Low T_4 and low FT_4I in seriously ill patients: concise communication. J Nucl Med 21:432–435
Wu SY, Chopra IJ, Nakamura Y, Solomon DH, Bennett LR (1976) A radioimmunoassay for measurement of 3,3′-L-diiodothyronine (3,3′-T_2). J Clin Endocrinol Metab 43:682–685

CHAPTER 17

Radioimmunoassay of Catecholamines

E. KNOLL and H. WISSER

A. Introduction

Even nowadays, the determination of catecholamines in biologic samples, especially in plasma and cerebrospinal fluid, is beset with great difficulties. For a long time, fluorometry (particularly the trihydroxyindole method) has been the most widely used method for the determination of catecholamines (WEIL-MALHERBE 1968; RENZINI et al. 1970; WISSER 1970; MIURA et al. 1977). The sensitivity of this method is certainly sufficient for the analysis of urine and brain tissue, but not for the determination of unconjugated epinephrine and dopamine in plasma. Gas chromatography, using either flame ionization detection (LOVELADY and FOSTER 1975) or electron capture detection (IMAI et al. 1973; WONG et al. 1973; LHUGUENOT and MAUME 1974) has also been introduced to determine catecholamines, but it has not found widespread application. Gas chromatography combined with mass spectrometry (referred to as mass fragmentography) shows high specificity, but requires expensive equipment (KAROUM et al. 1972; KOSLOW et al. 1972; LHUGUENOT and MAUME 1974; EHRHARDT and SCHWARTZ 1978). Moreover, its sensitivity is not as high as that of the radioenzymatic methods.

The double-isotope derivative technique (ENGELMAN and PORTNOY 1970) was a very interesting innovation, but proved to be extremely laborious and too insensitive. A modification of this technique, the radioenzymatic single-isotope derivative method, represented an essential improvement in practicability and sensitivity. The separation of the enzymatically methylated catecholamines was carried out by thin layer chromatography (PASSON and PEULER 1973; DA PRADA and ZUERCHER 1976; PEULER and JOHNSON 1977; BOSAK et al. 1980) or by high pressure liquid chromatography (ENDERT 1979; RATGE et al. 1983). An improved double-isotope technique for enzymatic assay of catecholamines permitted high precision, sensitivity, and plasma sample capacity (BROWN and JENNER 1981). During the last 6 years, liquid chromatography with electrochemical detection has increasingly been used for the estimation of the catecholamines in biologic specimens (HJEMDAHL et al. 1979; GOLDSTEIN et al. 1981; CAUSON et al. 1983; HAMMOND and JOHNSTON 1984). At present, this method cannot reach the sensitivity of the radioenzymatic procedure. However, after instrumental difficulties have been overcome, it may become the method of choice, because it is less laborious and needs no radioactive material.

It might seem surprising that radioimmunoassay, which has found widespread application in clinical chemistry and endocrinology during the last 10–15 years, has scarcely been used for the determination of catecholamines. A radioimmunologic approach, using specific antibodies of high avidity, seems on theoretical grounds to meet all requirements for the measurement of catecholamines in small quantities of plasma. However, no suitable antibodies have yet been produced. This chapter surveys experiments on the production of antibodies to the catecholamines and some of their metabolites.

B. Radioimmunoassay of Catecholamines and Their Metabolites

I. Antibodies to Catecholamines

There is only a small number of papers dealing with the production of antibodies to catecholamines. However, the number of scientists who have struggled with this problem in vain, without publishing any results, may well be much higher.

The first report on the development of antibodies against catecholamines originates from SPECTOR et al. (1973). They synthesized an immunogen of dopamine by binding β-(3,4-dimethoxyphenyl)ethylamine (I) to the γ-carboxyl group of the L-glutamic acid of a copolymer of lysine and glutamic acid and, by subsequent treatment with BBr_3, removing the two *O*-methyl groups of I. After immunization of rabbits, the authors demonstrated the presence of antibodies by their neutralizing capacity on some of the biologic activities of catecholamines. However, when unlabeled norepinephrine or epinephrine was incubated with the antibody, it became apparent that the determinant group which the antibody recognized was the catechol nucleus as it bound the latter as well as dopamine. No details were given on the titer of the antiserum, but it can be assumed that it was very low, owing to the use of ^{14}C-labeled dopamine with a very low specific activity. After this initial success in the production of antibodies to catecholamines, no improvements were published by these authors.

VERHOFSTAD et al. (1980) reported on the production of antibodies to epinephrine and norepinephrine and their applicability in immunohistochemistry. Rabbits and sheep were injected with immunogens, which were simply synthesized by coupling epinephrine or norepinephrine to bovine serum albumin, using formaldehyde as a coupling reagent. The authors demonstrated that, after appropriate fixation, immunohistochemical staining of epinephrine- or norepinephrine-containing cells could be effected by their antisera. No detailed characterization of the antisera was given with respect to a radioimmunologic application.

Some years later, GEFFARD et al. (1984a–c) were able to raise antibodies against dopamine and *p*-tyramine in rabbits by dopamine (or *p*-tyramine) conjugated to albumin either via formaldehyde or via glutaraldehyde in a similar manner to that described by VERHOFSTAD et al. (1980). They also used the two antibodies for immunohistochemical studies; it was possible to visualize dopamine specifically in glutaraldehyde-fixed rat brains. In addition, GEFFARD et al. investigated the specificity of the antibodies by equilibrium dialysis competition experiments, using an immunoreactive tritiated derivative synthesized by coupling dopamine or *p*-tyramine to *N*-α-acetyl-L-lysine *N*-methylamide with glutaraldehyde. Hence, these radiolabeled ligands mimicked the antigenic determinant of conjugated immunogens. The anti-dopamine antibodies recognized dopamine–glutaraldehyde, but not *p*-tyramine–glutaraldehyde; the opposite occurred for the anti-*p*-tyramine antibodies. The cross-reactivity ratios between the glutaraldehyde derivatives of dopamine and norepinephrine on the one hand, and the corresponding derivatives of *p*-tyramine and 1-(*p*-hydroxyphenyl)-2-aminoethanol (octopamine) on the other hand, were of the same order. This can be explained by the same structural difference between dopamine and norepinephrine and between *p*-tyramine and octopamine: There is a hydroxyl group on the side chain of norepinephrine and octopamine.

The antibody recognition site was markedly concentrated on the ring structure and much less on the linear part of the molecule. Thus, this simple approach of GEFFARD et al. is unsuited to obtaining antibodies which can differentiate between epinephrine and norepinephrine.

The only genuine description of the development of antibodies to catecholamines which could be used for a radioimmunoassay was given by MIWA et al. (1976, 1977, 1978a, b) and by YOSHIOKA et al. (1978). In a series of papers, they reported on the synthesis of antigens of catecholamines as well as on the production and characterization of antibodies. Preparation of the various antigens is schematically represented in Fig. 1.

The amino group of the side chain was tentatively protected by a maleyl group to avoid possible cyclization. The coupling of the *N*-maleyl-haptens to bovine serum albumin was carried out by a Mannich reaction. The authors demonstrated that the site of attachment to catecholamines in the conjugates was the 6 position (*ortho* to the aminoethyl side chain) and not the 5 position (*ortho* to the hydroxyl group) which is predominant in aminomethylation by the Mannich reaction of phenols (MIWA et al. 1978a). The maleyl group was removed before immunization.

Antibodies to catecholamines were prepared by immunization of rabbits with these conjugates. Separation of bound and free antigen was carried out by use of a microfiltration method. The antiserum to L-epinephrine showed a B/T value of 0.5 at antiserum dilutions of 1:500–1:1000. Addition of ascorbic acid (10 mmol/l) as antioxidant in the incubation mixture proved to be necessary.

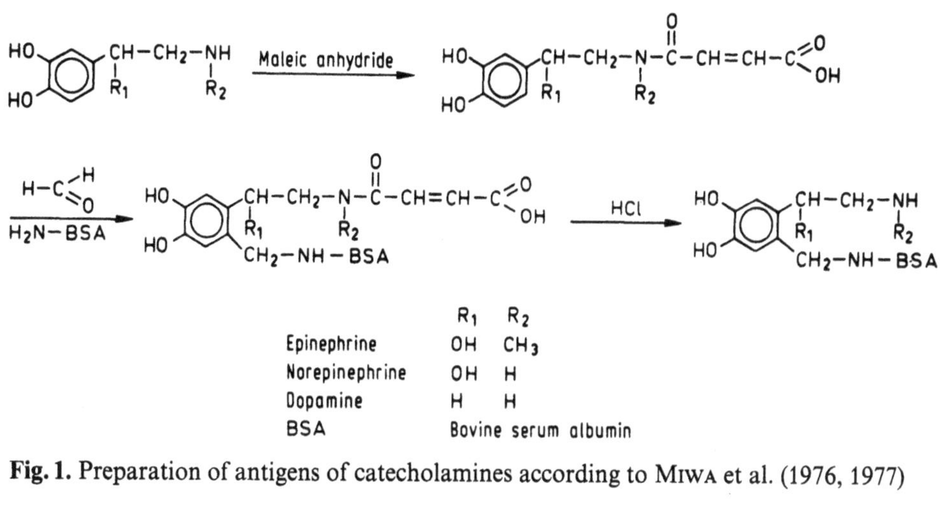

Fig. 1. Preparation of antigens of catecholamines according to MIWA et al. (1976, 1977)

The antiserum to epinephrine showed a very high specificity. Only metanephrine and synephrine, having the same side chain as epinephrine, had slight cross-reactivity (1.25%). Norepinephrine and dopamine showed a cross-reactivity of <0.1%. The high specificity of the antisera might be due to the structure of the antigens in which all functional groups are left intact. The coupling to the carrier protein was carried out by means of an additional methylene bridge. The standard curve of epinephrine, obtained by drawing a logit-log plot, showed linearity in the range 0.1–10 pmol (20–2000 pg) epinephrine. As can be concluded from the sensitivity of this standard curve and from epinephrine concentrations in plasma measured radioenzymatically, about 1 ml plasma would be needed for a radioimmunologic epinephrine determination. Thus, a radioimmunoassay of epinephrine, using this antiserum, would be considerably less sensitive than the radioenzymatic single-isotope derivative methods, which allow the simultaneous determination of the three catecholamines in only 50 µl plasma (DA PRADA and ZUERCHER 1976; BOSAK et al. 1980; ENDERT 1979; RATGE et al. 1983). YOSHIOKA et al. (1978) developed an antiserum to epinephrine with higher sensitivity (0.05 pmol = 10 pg) and specificity, but they have not so far published the use of this method in a clinical context, i.e., the measurement of the catecholamines in biologic specimens.

The development of antibodies to the catecholamines has also been intensively investigated in the laboratories of the present authors. Various alternatives in the synthesis of immunogens were tested:

1. By analogy with the preparation of antibodies to 3,4-dimethoxyphenylethylamine (WISSER et al. 1978) two antigens of dopamine were synthesized (KNOLL et al. 1979). The course of synthesis is outlined in Fig. 2. In view of the susceptibility to oxidation of the catechol structure, dopamine was reacted with acetone to form an isopropylidene derivative. Antigen A was synthesized by coupling 3,4-dimethoxyphenylethylamine to the carrier protein by means of succinic anhydride, and subsequent removal of the isopropylidene group by hydrolysis with diluted hydrochloric acid. Synthesis of antigen B was started by introducing the nitro group in ring position 6 of 3,4-dimethoxyphenylethyl-

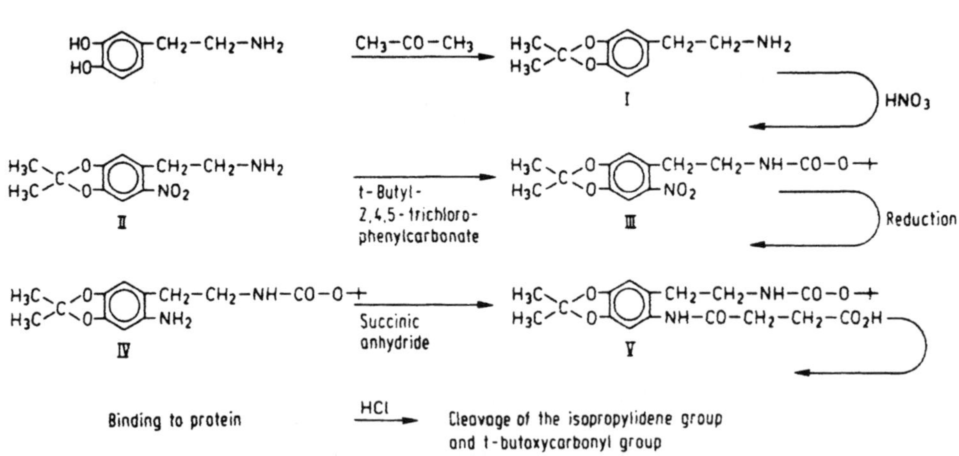

Fig. 2. Preparation of an antigen of dopamine according to KNOLL et al. (1979)

amine. Then the amino group of the aliphatic side chain was protected with a *t*-butoxycarbonyl group (*t*-BOC), followed by reduction of the nitro group to the amino group by hydrogen and Raney nickel. Finally, the amino group was coupled to the carrier protein and the protecting groups were removed by hydrochloric acid.

2. An antigen of dopamine was prepared according to the method of MIWA et al. (1976).
3. To circumvent the susceptibility to oxidation of the catecholamines, antigens were converted to structurally similar compounds lacking the catechol structure. The aim of these experiments was to produce antibodies to these compounds and to develop a radioimmunoassay using the cross-reactivity of the antibodies to the catecholamines. Three groups of compounds were used for this investigation: (a) mono-4-hydroxyphenylethylamines; (b) compounds whose 3-hydroxyl groups were replaced by substituents of similar size, e.g., 3-fluorotyramine; and (c) compounds in which both phenolic hydroxyl groups were replaced by substituents of similar size, e.g., 3,4-dichlorophenylethylamine.

Preparation of antigens of the compounds of group 3(b) and 3(c) is exemplified by 3,4-dichlorophenylethylamine, as shown in Fig. 3. The hapten was synthesized by Tolkachev's method for the synthesis of β-(3,4-dimethoxy-5-bromophenyl)ethylamine (TOLKACHEV et al. 1964). 3,4-Dichlorobenzaldehyde was reacted with nitromethane to result in 3,4-dichloro-ω-nitrostyrene which was reduced with Zn/HCl. 3-Fluorotyramine was similarly prepared, starting with 3-fluoro-4-methoxybenzaldehyde. Antigens of both haptens were synthesized by coupling the amino group of the side chain to the carrier protein with succinic anhydride. 3-Fluorotyramine was also conjugated through the ring by the Mannich reaction.

With the exception of the antigens of the monohydroxyphenylethylamines (described in Sect. B.II.2), no antibody to catecholamines could be detected after attempted immunization with the prepared antigens. Different immunization

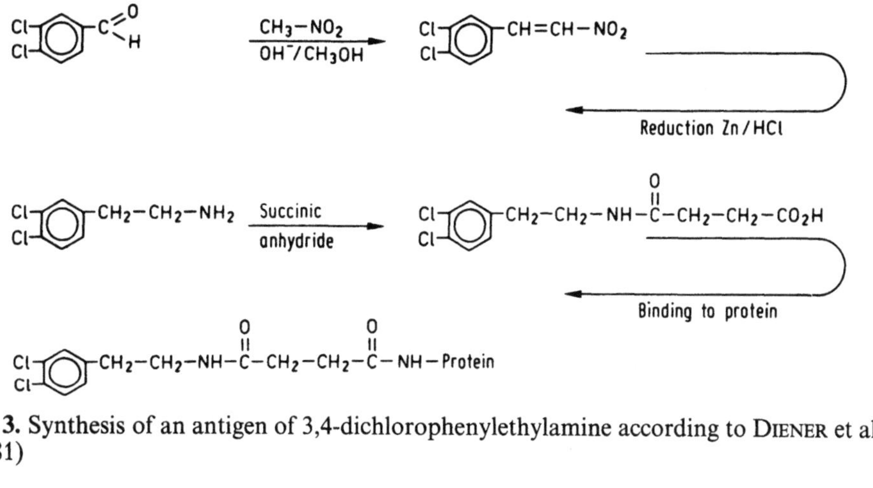

Fig. 3. Synthesis of an antigen of 3,4-dichlorophenylethylamine according to DIENER et al. (1981)

schedules and animal species (rabbit, sheep) were used, and tritiated dopamine was used as tracer. After immunization with the antigen prepared according to MIWA et al. (1976), detectable antibodies were still absent. Various techniques for the separation of bound from free hapten were used, such as dextran-coated charcoal, ammonium sulfate precipitation, the protein A method (YING and GUILLEMIN 1979), and the microfiltration method of MIWA et al. (1978 b).

Variation in the pH value of the incubation medium (pH 4.8–7.4) and the incubation time (2–120 h) did not improve the results. The antisera were pretreated with charcoal to remove any endogenous catecholamines that may have been occupying binding sites, but there was still no significant binding of catecholamines after this treatment. One possible explanation for the failure to obtain antibodies specific for catecholamines is that the immunogen might become oxidatively degraded in the tissue after injection. Oxidation of the antigen in the course of synthesis could be excluded by the introduction of the protective isopropylidene group, by working under a nitrogen atmosphere, and by the presence of an antioxidant.

II. Antibodies to Metabolites of Catecholamines

In contrast to the failure to raise antibodies to the catecholamines themselves, a series of authors have described methods for the production of antibodies to metabolites of the catecholamines. The reason for this difference might be the lesser susceptibility to oxidation of these compounds in which one or both phenolic hydroxyl groups are methylated.

1. 3,4-Dimethoxyphenylethylamine

VAN VUNAKIS et al. (1969) and RICEBERG and VAN VUNAKIS (1975), as well as KNOLL and WISSER (1973), first described the production of antibodies to 3,4-dimethoxyphenylethylamine. Owing to its stability, this compound proved to be suitable for investigations of the interdependence of the structure of the antigen and the specificity of the antibody. Antigens were prepared by succinylation of 3,4-dimethoxyphenylethylamine and subsequent coupling to poly-L-lysine (RICEBERG and VAN VUNAKIS 1975), or bovine γ-globulin (KNOLL and WISSER 1973).

Although ^{125}I-labeled tracer with a very high specific activity was used, the antiserum produced by RICEBERG and VAN VUNAKIS (1975) had a titer of only 1:50 and the detection level for 3,4-dimethoxyphenylethylamine was 100 pg. Using tritiated 3,4-dimethoxyphenylethylamine as tracer, KNOLL and WISSER (1973) obtained an antiserum with a titer of 1:16400 which allowed the detection of as little as 15 pg 3,4-dimethoxyphenylethylamine. As could be expected from the structure of the antigens, both antisera had a high specificity for the dimethoxyphenyl nucleus, whereas structural changes in the ethylamine side chain were not so well recognized. Therefore, we synthesized an additional antigen of 3,4-dimethoxyphenylethylamine by introducing an amino group to the benzene ring and by linking this group with the carrier protein as demonstrated in Fig. 2 (WISSER et al. 1978). Table 1 contains a comparison of the specificities of various antisera to 3,4-dimethoxyphenylethylamine.

Table 1. Specificity of various antisera to 3,4-dimethoxyphenylethylamine

Compound	Cross-reactivity (%)		
	Hapten coupled via the side chain (RICEBERG and VAN VUNAKIS 1975)	Hapten coupled via the side chain (KNOLL and WISSER 1973)	Hapten coupled through the ring (KNOLL and WISSER 1976)
3,4-Dimethoxyphenylethylamine	100	100	100
3,4-Dimethoxybenzyl alcohol		47	0.1
3,4-Dimethoxycinnamic acid		52	0.1
3,4-Dimethoxyphenylacetic acid	0.8	15	0.1
3,4-Dimethoxybenzoic acid	0.9	7.5	0.1
3,4-Dimethoxybenzylamine	36		
3,4-Dimethoxyaniline	90		
3,4,5-Trimethoxyphenylethylamine (mescaline)	0.9	< 0.1	< 0.8
3,5-Dimethoxyphenylethylamine	< 0.1	< 0.1	< 0.1
4-Methoxyphenylethylamine	< 0.1	< 0.1	< 0.1
4-Hydroxyphenylethylamine (tyramine)	< 0.1	< 0.1	< 0.1
3,4-Dihydroxyphenylethylamine (dopamine)		< 0.1	< 0.1
Epinephrine		< 0.1	< 0.1
Norepinephrine		< 0.1	< 0.1
Metanephrine	< 0.1	< 0.1	< 0.1
Normetanephrine	< 0.1	< 0.1	< 0.1
3-Methoxy-4-hydroxybenzoic acid (vanillic acid)	< 0.1	< 0.1	< 0.1
3-Methoxy-4-hydroxyphenylacetic acid (homovanillic acid)		< 0.1	< 0.1
3-Methoxy-4-hydroxymandelic acid (vanillylmandelic acid)	< 0.1	< 0.1	< 0.1

The higher specificity of the antiserum produced from the hapten bound through the ring, is very apparent. This antiserum recognizes structural changes not only of ring substituents, but also of the aminoethyl side chain. Thus, the four compounds with the 3,4-dimethoxyphenyl structure: 3,4-dimethoxybenzyl alcohol, 3,4-dimethoxycinnamic acid, 3,4-dimethoxyphenylacetic acid, and 3,4-dimethoxybenzoic acid show cross-reactivities of only 0.1%.

A radioimmunoassay for the determination of 3,4-dimethoxyphenylethylamine in urine was developed using this antiserum (KNOLL and WISSER 1976). In a comparative study, the 3,4-dimethoxyphenylethylamine excretion of healthy volunteers, of patients with schizophrenic psychoses, and of psychiatric patients without schizophrenia was measured (KNOLL et al. 1978). No significant differences were observed between these three groups. The previous claim by FRIEDHOFF and VAN WINKLE (1962), using insensitive and nonspecific methods, that 3,4-dimethoxyphenylethylamine is an abnormal metabolite in schizophrenics, was thereby clearly refuted.

2. 3-*O*-Methylcatecholamines

a) 3-*O*-Methyldopamine (3-Methoxytyramine)

FARAJ et al. (1977) produced an antiserum to 3-*O*-Methyldopamine. The synthesis of the antigen is shown in Fig. 4. The carboxyl group of *p*-aminohippuric acid was coupled to hemocyanin with the aid of carbodiimide. After diazotizing the amino group of the *p*-aminohippuric acid–hemocyanin conjugate, the latter was introduced in position 5 of 3-*O*-methyldopamine by an azo coupling reaction. After immunization of rabbits, an antiserum to 3-*O*-methyldopamine was obtained with a titer of 1:1000. A concentration of 5 µg/l (0.5 ng/0.1 ml) of 3-*O*-methyldopamine could be detected with this antiserum. Cross-reactivities of 3,4-dimethoxyphenylethylamine and 3,4-dimethoxy-*N*-methylphenylethylamine were 30% and 15%, respectively. All other tested analogs of 3-*O*-methyldopamine had cross-reactivities of less than 0.3%. This antiserum was used to develop a radioimmunoassay for the quantitation of 3-*O*-methyldopamine in urine and plasma. The values for 3-*O*-methyldopamine (3-methoxytyramine) in urine (determined by radioimmunoassay) were similar to those obtained with other methods. The sensitivity of the radioimmunologic method (detection limit 5 µg/l) allowed direct analysis in urine without prior extraction, but the low levels of 3-*O*-methyldopamine in plasma necessitated a preliminary extraction with ethyl acetate: 2 ml plasma was usually used for the extraction. FARAJ et al. (1978) also applied this antiserum for the radioimmunologic determination of DOPA and dopamine in urine and plasma. This enzyme radioimmunoassay was based on the

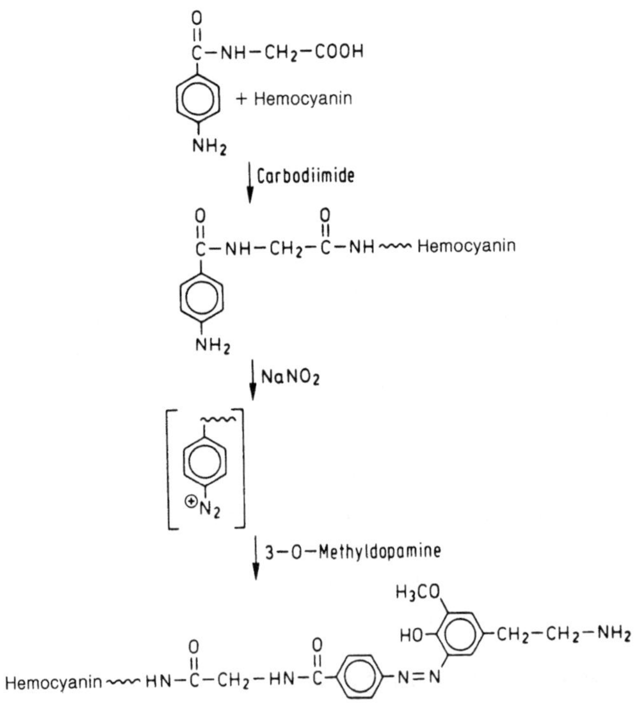

Fig. 4. Synthesis of an antigen of 3-*O*-methyldopamine according to FARAJ et al. (1977)

incubation of urine or plasma in the presence of catechol-*O*-methyltransferase (EC 2.1.1.6), aromatic L-amino acid decarboxylase (EC 4.1.1.28), and *S*-adenosylmethionine to convert dopa and dopamine into 3-*O*-methyldopamine.

Their results for the excretion of dopamine in 24-h urine were in agreement with those of other workers. However, the determination of endogenous plasma levels of dopamine (10 ml blood, deproteinization and alumina separation) resulted in values which were approximately four times higher than those obtained by other authors (WEISE and KOPIN 1976; BOSAK et al. 1980) using radioenzymatic methods. The sensitivity of the assay was 0.5 ng dopamine per 1.1 ml incubation mixture. The results of FARAJ et al. (1978) indicated that the average plasma level of dopamine was 211 ± 52 ng/l. This is astonishing in that less than 2 ml plasma, equivalent to approximately 400 pg dopamine, was used for the assay. Taking into consideration the detection limit of 500 pg, it must be seriously questioned whether dopamine plasma levels of healthy normal volunteers could be measured with this assay. In a recent paper (FARAJ et al. 1981), this method was applied to the determination of the excretion of DOPA, dopamine, and 3-*O*-methyldopamine in patients with melanoma. However, fluorometric methods also proved adequate for this investigation with regard to both specificity and sensitivity.

b) 3-*O*-Methylnorepinephrine (Normetanephrine)

PESKAR et al. (1972) described the production of antibodies to 3-*O*-methylnorepinephrine (normetanephrine). In a one-step reaction, they linked the amino group of normetanephrine to the amino groups of the carrier macromolecules poly ($Tyr^{40}Glu^{60}$), using glutaraldehyde. KNOLL et al. (1979) and DIENER et al. (1981) also developed antibodies to normetanephrine using a similar technique. They used the succinyl group to couple normetanephrine to bovine γ-globulin. A comparison between the specificities of two antisera to normetanephrine can be seen in Table 2. Specificity, especially of antiserum I, was predictable from the antigen structure; i.e., a high specificity for ring substituents. *N*-Methylnormetanephrine (metanephrine), on the other hand, had a higher affinity to the antibody than nor-

Table 2. Specificity of two antisera to normetanephrine. Antiserum I according to PESKAR et al. (1972); antiserum II according to KNOLL et al. (1979)

Compound	Cross-reactivity (%)	
	Antiserum I	Antiserum II
Normetanephrine	100	100
Metanephrine	120	153
3-Methoxytyramine	65.5	0.6
3-Methoxy-4-hydroxy-phenylglycol	15	
Norepinephrine	< 7	< 0.1
Epinephrine		< 0.1
Dopamine		< 0.1
Vanillylmandelic acid	< 7	< 0.1
Homovanillic acid		< 0.1

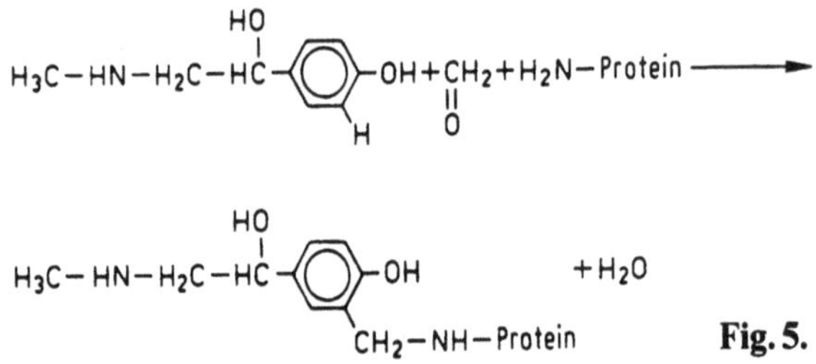

Fig. 5. Mannich reaction with synephrine

metanephrine. This is an expression of the greater structural similarity of normetanephrine to the immunogen of normetanephrine. Additionally, antiserum II had a high specificity for the side chain, i.e., less cross-reactivity to 3-methoxytyramine (0.6%), which has the same ring substituents as normetanephrine, but lacks the hydroxyl group of the side chain.

c) 3-*O*-Methylepinephrine (Metanephrine)

GROTA and BROWN (1976) were the first to produce antibodies to 3-*O*-methylepinephrine (metanephrine). For the preparation of the antigen they used the corresponding monohydroxyphenylethylamine, synephrine. This compound was conjugated to bovine serum albumin by the formaldehyde condensation Mannich reaction (Fig. 5).

This principle was first applied by these authors to couple the indoles, serotonin and melatonin, to a carrier protein (GROTA and BROWN 1974). Synephrine, containing a phenolic hydroxyl group, was attached to the carrier by means of a methylene bridge *ortho* to the phenolic hydroxyl substituent. After immunization of rabbits with this antigen, the authors were able to detect antibodies whose specificity and sensitivity were investigated utilizing ^{3}H-labeled metanephrine as tracer. An antibody dilution of 1:1000 bound approximately 34% of the tracer. The standard curve for metanephrine indicated a minimum sensitivity of 0.1–0.5 pmol (20–100 pg). Subsequently, several workers reported the production of antibodies by the technique of GROTA and BROWN (1976). LAM et al. (1977) developed a radioimmunoassay for the measurement of free and conjugated metanephrine in urine with ^{3}H-labeled metanephrine as tracer. The antiserum was produced by immunization of rabbits with a synephrine–bovine serum albumin conjugate (GROTA and BROWN 1976). Using this antiserum, LAM et al. (1977) developed a specific radioimmunoassay which allowed urine samples to be measured directly. The detection limit of the assay was 40 pg metanephrine.

Human amniotic fluid was also investigated for free metanephrine. After 30 weeks' gestation, there was a positive correlation of free metanephrine concentrations in amniotic fluid with gestational age (ARTAL et al. 1979). RAUM and SWERDLOFF (1977, 1981 a) utilized a synephrine antiserum which had been prepared in the same way, but involving an ^{125}I ligand. Specificity and sensitivity of their urinary metanephrine radioimmunoassay were comparable to the assay of LAM et al. (1977). Therefore, it is peculiar that the values of the metanephrine excretion of normal volunteers reported by RAUM and SWERDLOFF (1981 a) are three times higher than those reported by LAM et al. (1977). Aside from this discrepancy, both methods had the disadvantage that only metanephrine, not normetanephrine, could be measured. RAUM and SWERDLOFF (1981 b) and RAUM (1984) described a radioimmunoassay for epinephrine and norepinephrine in tissues and plasma. Tracer and anti-metanephrine antiserum were identical with those of the method of RAUM and SWERDLOFF (1981 a). However, by analogy to the DOPA and dopamine assay of FARAJ et al. (1978, 1981), epinephrine and norepinephrine had to be converted to metanephrine with the enzymes catechol-*O*-methyltransferase and phenylethanolamine-*N*-methyltransferase prior to the radioimmunoassay. The specified plasma volume for one measurement of the two catecholamines is incomprehensible. The authors speak of "less than 500 µl of normal human plasma" for their radioimmunoassay, but for the alumina extraction of the deproteinized plasma (30 µl 60% $HClO_4$ per milliliter plasma) they need 1–3 ml perchloric acid-extracted plasma. In comparison, only 50 µl plasma is sufficient for a simultaneous measurement of epinephrine, norepinephrine, and dopamine with the radioenzymatic assay. The standard curves of epinephrine and norepinephrine had fairly gentle slopes which might contribute to the high coefficients of variation both within and between assays (up to 36.4% for norepinephrine and 33.4% for epinephrine at 80% $B/B0$). The statement of the authors that 100–200 samples could easily be processed per day by the radioimmunoassay seems to be too optimistic. In our opinion, taking all the data into consideration, the radioimmunoassay is no alternative to the radioenzymatic assay which represents the present state of the art in the measurement of catecholamines.

KNOLL et al. (1979) and DIENER et al. (1981) also produced antibodies to metanephrine following the procedure of GROTA and BROWN (1976). The aim of this study was to develop a radioimmunoassay for epinephrine utilizing the cross-reaction of the metanephrine antiserum to epinephrine. An antiserum with a titer of 1 : 13 000 was obtained using ^{3}H-labeled metanephrine as tracer. Table 3 represents a comparison of the specificities of various antisera to metanephrine.

Table 3. Specificity of three antisera to metanephrine

Compound	Cross-reactivity (%)			
	Antiserum I GROTA and BROWN 1976)	Antiserum II (KNOLL et al. 1979; DIENER et al. 1981)		Antiserum III (RAUM and SWERDLOFF 1981b)
	nephrine ^{3}H	Tracer: metanephrine ^{3}H	Tracer: metanephrine ^{3}H	Tracer: 3-iodosynephrine ^{125}I
Metanephrine	800	78.6	250	100
Synephrine	100	100	360	20
Epinephrine	60	2.0	100	12.8
Normetanephrine	0.08	< 0.2	< 0.2	4.8×10^{-3}
Norepinephrine	0.01	< 0.2	< 0.2	9.7×10^{-4}
3-Methoxytyramine		< 0.2	< 0.2	10^{-5}
p-Tyramine	0.08		< 0.2	10^{-5}
Dopamine		< 0.2	< 0.2	10^{-5}
Octopamine			< 0.2	10^{-5}
Vanillylmandelic acid		< 0.2	< 0.2	10^{-5}

Antiserum I and III show very similar specificities. Both antisera have a higher affinity to metanephrine than to synephrine. The reason for this is that metanephrine resembles the structure of the immunogen more closely than that of synephrine. Epinephrine, having a 3-hydroxy group instead of a 3-methoxy group, shows 60% cross-reactivity (compared with synephrine) with both antisera. Antiserum II, with ^{3}H-labeled metanephrine as tracer, shows different behavior. Metanephrine has a cross-reactivity of 78.6% (synephrine taken as 100%) and in epinephrine cross-reactivity is reduced to 2%. If additional alterations are made to the side chain, as in normetanephrine and 3-methoxytyramine, the cross-reactivity decreases strongly in all antisera. Antiserum II was also investigated, using ^{3}H-labeled epinephrine as tracer, to test the possibility of a radioimmunoassay for epinephrine. However, the sensitivity of the antiserum was not sufficient for a promising application.

GROTA and BROWN (1976) and later on DIENER et al. (1981) applied the formaldehyde condensation reaction to tyramine and octopamine, the monohydroxy derivatives of dopamine and norepinephrine. Only the antiserum raised against the tyramine conjugates had a titer which was high enough for further characterization. Both antisera showed a specificity which was similar to that of the synephrine conjugates outlined in Table 3.

3. 3-Methoxy-4-hydroxyphenylethyleneglycol

KEETON et al. (1978, 1981) produced antibodies to 3-methoxy-4-hydroxyphenylethyleneglycol, the major terminal metabolite of norepinephrine in the central nervous system of humans and rats. In order to obtain a derivative of 3-methoxy-4-hydroxyphenylethyleneglycol that possessed a free carboxyl group, 6-bromohexanoic acid was reacted with 3-methoxy-4-hydroxyphenylethyleneglycol to form 3-methoxy-4-(5-carboxypentoxy)phenylethyleneglycol (Fig. 6).

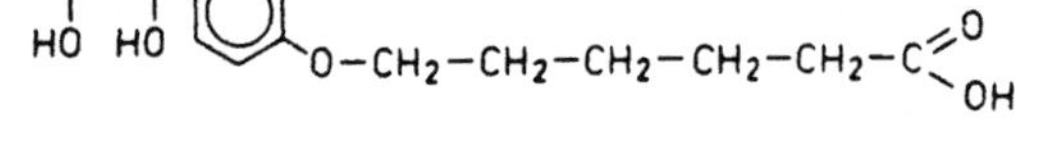

Fig. 6. Structural formula of 3-methoxy-4-(5-carboxypentoxy) phenylethyleneglycol

After immunization of rabbits with the thyroglobulin conjugate of the 3-methoxy-4-hydroxyphenylethyleneglycol derivative, antibodies to 3-methoxy-4-hydroxyphenylethyleneglycol could be raised and characterized using ^{3}H-labeled 3-methoxy-4-hydroxyphenylethyleneglycol as tracer. The specificity of the antiserum was very high. In spite of the low final dilution of the antiserum (1:180), a radioimmunoassay for 3-methoxy-4-hydroxyphenylethyleneglycol in brain tissue was developed. The sensitivity of the assay was sufficient to determine 3-methoxy-4-hydroxyphenylethyleneglycol in discrete brain regions. Comparison of the values for the total 3-methoxy-4-hydroxyphenylethyleneglycol concentration of different brain regions of rats determined by radioimmunoassay and by gas chromatography–mass spectrometry showed good agreement of the two methods.

C. Summary

This chapter describes experiments to produce antibodies to the catecholamines and some catecholamine metabolites with the intention of developing radioimmunologic methods. First, attempts to produce antibodies to the catecholamines themselves are described. In the course of synthesis of immunogens of catecholamines it is necessary to exercise special care, e.g., the introduction of protecting groups, owing to the great susceptibility of the catechol structure to oxidation. Despite many efforts, only one working group (MIWA et al. 1976, 1977, 1978 a, b) has reported the successful production of antibodies to catecholamines, but they did not develop a radioimmunoassay. In contrast, antibodies to some metabolites of the catecholamines, such as 3,4-dimethoxyphenylethylamine, the 3-*O*-methylated catecholamines (normetanephrine, metanephrine, and 3-methoxytyramine), as well as 3-methoxy-4-hydroxyphenylethyleneglycol, could be produced whose avidity and specificity were high enough to permit the development of sensitive radioimmunoassays.

References

Artal R, Hobel CJ, Lam R, Oddie TH, Fisher DA (1979) Free metanephrine in human amniotic fluid as an index of fetal sympathetic nervous system maturation. Am J Obstet Gynecol 133:452–454

Bosak J, Knoll E, Ratge D, Wisser H (1980) Single-isotope enzymatic derivative method for measuring catecholamines. J Clin Chem Clin Biochem 18:413–421

Brown MJ, Jenner DA (1981) Novel double-isotope technique for enzymatic assay of catecholamines, permitting high precision, sensitivity and plasma sample capacity. Clin Sci 61:591–598

Causon RC, Brown MJ, Boulos PM, Perret D (1983) Analytical differences in measurement of plasma catecholamines. Clin Chem 29:735–737

Da Prada M, Zuercher G (1976) Simultaneous radioenzymatic determination of plasma and tissue adrenaline, noradrenaline and dopamine within the femtomole range. Life Sci 19:1161–1174

Diener U, Knoll E, Wisser H (1981) Preparation of antibodies to catecholamines and metabolites – syntheses of various immunogens and characterization of the resulting antibodies. Clin Chim Acta 109:1–11

Ehrhardt JD, Schwartz J (1978) A gas chromatography-mass spectrometry assay of human plasma catecholamines. Clin Chim Acta 88:71–79

Endert E (1979) Determination of noradrenaline and adrenaline in plasma by a radioenzymatic assay using high pressure liquid chromatography for the separation of the radiochemical products. Clin Chim Acta 96:233–239

Engelman K, Portnoy B (1970) A sensitive double-isotope derivative assay for norepinephrine and epinephrine. Circ Res 26:53–57

Faraj BA, Camp VM, Pruitt AW, Isaacs JW, Ali FM (1977) The measurement of 3-*O*-methyldopamine in urine and plasma by a rapid and specific radioimmunoassay. J Nucl Med 18:1025–1031

Faraj BA, Walker WR, Camp VM, Ali FM, Cobbs WB (1978) Development of an enzyme-radioimmunoassay for the measurement of dopamine in human plasma and urine. J Nucl Med 19:1217–1224

Faraj BA, Lawson DH, Nixon DW, Murray DR, Camp VM, Ali FM, Black M, Stacciarini W, Tarcan Y (1981) Melanoma detection by enzyme-radioimmunoassay of L-dopa, dopamine, and 3-*O*-methyldopamine in urine. Clin Chem 27:108–112

Friedhoff AJ, van Winkle E (1962) Isolation and characterization of a compound from the urine of schizophrenics. Nature 194:897–898
Geffard M, Buijs RM, Seguela P, Pool CW, Le Moal M (1984a) First demonstration of highly specific and sensitive antibodies against dopamine. Brain Res 294:161–165
Geffard M, Seguela P, Heinrich-Rock AM (1984b) Antisera against catecholamines: specificity studies and physicochemical data for antidopamine and anti-*p*-tyramine antibodies. Mol Immunol 21:515–522
Geffard M, Kah O, Onteniente B, Seguela P, Le Moal M, Delaage M (1984c) Antibodies to dopamine: radioimmunological study of specificity in relation to immunocytochemistry. J Neurochem 42:1593–1599
Goldstein DS, Feuerstein G, Izzo JL, Kopin IJ, Keiser HR (1981) Validity and reliability of liquid chromatography with electrochemical detection for measuring plasma levels of norepinephrine and epinephrine in man. Life Sci 28:467–475
Grota LJ, Brown GM (1974) Antibodies to indolealkylamines: serotonin and melatonin. Can J Biochem 52:196–202
Grota LJ, Brown GM (1976) Antibodies to catecholamines. Endocrinology 98:615–622
Hammond VA, Johnston DG (1984) A semi-automated assay for plasma catecholamines using high-performance liquid chromatography with electrochemical detection. Clin Chim Acta 137:87–93
Hjemdahl P, Daleskog M, Kahan T (1979) Determination of plasma catecholamines by high performance liquid chromatography with electrochemical detection: comparison with a radioenzymatic method. Life Sci 25:131–138
Imai I, Wang MT, Yoshiue S, Tamura Z (1973) Determination of catecholamines in the plasma of patients with essential hypertension and of normal persons. Clin Chim Acta 43:145–149
Karoum R, Cattabeni F, Costa E, Ruthven CRJ, Sandler M (1972) Gas chromatographic assay of picomole concentrations of biogenic amines. Anal Biochem 47:550–561
Keeton TK, Krutzsch H, Lovenberg W (1978) A specific radioimmunoassay for 3-methoxy-4-hydroxyphenylethyleneglycol (MOPEG). In: Usdin E, Kopin IJ, Barchas J (eds) Catecholamines: basic and clinical frontiers, vol I. Pergamon, New York, p 871
Keeton TK, Krutzsch H, Lovenberg W (1981) Specific and sensitive radioimmunoassay for 3-methoxy-4-hydroxyphenylethyleneglycol (MOPEG). Science 211:586–588
Knoll E, Wisser H (1973) Gewinnung und Charakterisierung von Antikörpern gegen 3,4-Dimethoxyphenyläthylamin. Clin Chim Acta 48:183–192
Knoll E, Wisser H (1976) Radioimmunologische Bestimmung von 3,4-Dimethoxyphenyläthylamin im Urin. Clin Chim Acta 68:327–332
Knoll E, Wisser H, Emrich HM (1978) 3,4-Dimethoxyphenylethylamine excretion of normals and schizophrenics, behaviour during total fasting. Clin Chim Acta 89:493–502
Knoll E, Wisser H, Diener U, Herrmann R (1979) Investigations of the production of antibodies towards catecholamines and metabolites. In: Albertini A, Da Prada M, Peskar BA (eds) Radioimmunoassay of drugs and hormones in cardiovascular medicine. Elsevier, Amsterdam, p 217
Koslow SH, Cattabeni F, Costa E (1972) Norepinephrine and dopamine: assay by mass fragmentography in the picomole range. Science 176:177–180
Lam RW, Artal R, Fisher DA (1977) Radioimmunoassay for free and conjugated urinary metanephrine. Clin Chem 23:1264–1267
Lhuguenot JC, Maume BF (1974) Improvements in quantitative gas phase analysis of catecholamines in the picomole range by electron-capture detection and mass fragmentography of their pentafluorobenzylimine-trimethylsilyl derivatives. Chromatogr Sci 12:411–418
Lovelady HG, Foster LL (1975) Quantitative determination of epinephrine and norepinephrine in the picogram range by flame ionization gasliquid chromatography. J Chromatogr 103:43–52
Miura Y, Campese V, DeQuattro V, Meijer D (1977) Plasma catecholamines via an improved fluorimetric assay: comparison with an enzymatic method. J Lab Clin Med 89:421–427

Miwa A, Yoshioka M, Shirahata A, Nakagawa Y, Tamura Z (1976) Preparation of specific antibody to each one of catecholamines and L-Dopa. Chem Pharm Bull 24:1422–1424

Miwa A, Yoshioka M, Shirahata A, Tamura Z (1977) Preparation of specific antibodies to catecholamines and L-3,4-dihydroxyphenylalanine. I. Preparation of the conjugates. Chem Pharm Bull 25:1904–1910

Miwa A, Yoshioka M, Tamura Z (1978 a) Preparation of specific antibodies to catecholamines and L-3,4-dihydroxyphenylalanine. II. The site of attachment on catechol moiety in the conjugates. Chem Pharm Bull 26:2903–2905

Miwa A, Yoshioka M, Tamura Z (1978 b) Preparation of specific antibodies to catecholamines and L-3,4-dihydroxyphenylalanine. III. Preparation of antibody to epinephrine for radioimmunoassay. Chem Pharm Bull 26:3347–3352

Passon PG, Peuler JD (1973) A simplified radiometric assay for plasma norepinephrine and epinephrine. Anal Biochem 51:618–631

Peskar BA, Peskar BM, Levine M (1972) Specificities of antibodies to normetanephrine. Eur J Biochem 26:191–195

Peuler JD, Johnson GA (1977) Simultaneous single isotope radioenzymatic assay of plasma norepinephrine, epinephrine and dopamine. Life Sci 21:1161–1174

Ratge D, Baumgard G, Knoll E, Wisser H (1983) Plasma free and conjugated catecholamines in diagnosis and localisation of pheochromocytoma. Clin Chim Acta 132:229–243

Raum WJ (1984) Methods of plasma catecholamine measurement including radioimmunoassay. Am J Physiol 247:E4–E12

Raum WJ, Swerdloff RS (1977) Catecholamine 125-J radioimmunoassay. IRCS Med Sci 5:413–413

Raum WJ, Swerdloff RS (1981 a) Urinary metanephrine radioimmunoassay: comparison with the colorimetric assay. Clin Chem 27:43–47

Raum WJ, Swerdloff RS (1981 b) A radioimmunoassay for epinephrine and norepinephrine in tissues and plasma. Life Sci 28:2819–2827

Renzini V, Brunori CA, Valori C (1970) A sensitive and specific fluorimetric method for the determination of noradrenalin and adrenalin in human plasma. Clin Chim Acta 30:587–594

Riceberg LJ, van Vunakis H (1975) Estimation of 3,4-dimethoxyphenylethylamine and related compounds in urine extracts by radioimmunoassay. Biochem Pharmacol 24:259–265

Spector S, Dalton C, Felix AM (1973) Development of antibodies against catecholamines. In: Usdin E, Snyder SH (eds) Frontiers in ctecholamine-research. Pergamon, New York, p 345

Tolkachev ON, Chernowa VP, Kuznetsova EV, Fang-Ling P, Preobrazhenskii NA (1964) Synthesis of 5-bromosubstituted-β-phenylethylamines (in Russian). Zh Obshch Khim 34:545–548

Verhofstad AAJ, Steinbusch HWJ, Penke P, Varga J, Joosten HWJ (1980) Use of antibodies to norepinephrine and epinephrine in immunohistochemistry. Adv Biochem Psychopharmacol 25:185–193

Van Vunakis H, Bradvica H, Benda P, Levine L (1969) Production and specificity of antibodies directed toward 3,4,5-trimethoxyphenylethylamine, 3,4-dimethoxyphenylethylamine, and 2,5-dimethoxy-4-methylamphetamine. Biochem Pharmacol 18:393–404

Weil-Malherbe H (1968) The estimation of total (free + conjugated) catecholamines and some catecholamine metabolites in human urine. In: Glick D (ed) Methods of biochemical analysis, vol 16. Interscience, New York, p 293

Weise VK, Kopin IJ (1976) Assay of catecholamines in human plasma: studies of a single isotope radioenzymatic procedure. Life Sci 19:1673–1685

Wisser H (1970) Bestimmung der freien Katecholamine im Harn. Z Klin Chem Klin Biochem 8:637–648

Wisser H, Herrmann R, Knoll E (1978) Methodical investigation of the production of antibodies towards 3,4-dimethoxyphenylethylamine. Clin Chim Acta 86:179–185

Wong KP, Ruthven CRJ, Sandler M (1973) Gas chromatographic measurement of urinary catecholamines by an electron capture detection procedure. Clin Chim Acta 47:215–222
Ying SY, Guillemin R (1979) Dried staphylococcus aureus as a rapid immunological separating agent in radioimmunoassays. J Clin Endocrinol Metab 48:360–362
Yoshioka M, Miwa A, Tamura Z (1978) Radioimmunoassay of catecholamines. In: Usdin E, Kopin IJ, Barchas J (eds) Catecholamines: basic and clinical frontiers, vol I. Pergamon, New York, p 868

CHAPTER 18

Radioimmunoassay of Prostaglandins and Other Cyclooxygenase Products of Arachidonate Metabolism

B. A. PESKAR, TH. SIMMET, and B. M. PESKAR

A. Introduction

Since the first description of radioimmunoassays for prostaglandins (PG) of the E (LEVINE and VAN VUNAKIS 1970; JAFFE et al. 1971), B (LEVINE et al. 1971), A (JAFFE et al. 1971), and F series (LEVINE and VAN VUNAKIS 1970; CALDWELL et al. 1971; LEVINE et al. 1971), immunologic methods have been extensively used for quantitative determination of PG and related compounds. The first critical comparative evaluation of radioimmunoassays for the determination of PG in biologic material was reported by SAMUELSSON et al. (1973). Seven laboratories had analyzed four human plasma samples containing different amounts of $PGF_{2\alpha}$. Large variations in the results were observed for all four samples. No obvious explanation could be given for the large differences between the radioimmunoassay results obtained by some of the laboratories. Furthermore, a considerable discrepancy was observed between the measured basal plasma levels of $PGF_{2\alpha}$ and those expected on the basis of production rate. The lowest basal value reported in this study was obtained by gas chromatography–mass spectrometry (GC–MS). From these results, it had been concluded that radioimmunoassay results are less reliable than data obtained by GC–MS. It should be pointed out, however, that a similar study comparing results obtained in different laboratories by GC–MS has not been published. Since then, considerable efforts have been made to improve the validity of PG radioimmunoassay results by clearly defining validation criteria (see Sects. D and E).

B. Biosynthesis and Metabolism of Prostaglandins and Thromboxanes

The cyclooxygenase products of arachidonic acid metabolism comprise a family of closely related compounds. These are illustrated in Fig. 1. The primary products of arachidonic acid metabolism formed via the enzyme fatty acid cyclooxygenase are the PG endoperoxides PGG_2 and PGH_2. These can be converted enzymatically or nonenzymatically to PGE_2, PGD_2, and $PGF_{2\alpha}$. In various organs and cells, the PG endoperoxides are metabolized to thromboxane A_2 (TXA_2) (HAMBERG et al. 1975) or PGI_2 (MONCADA et al. 1976). The chemically labile TXA_2 and PGI_2 are rapidly converted to the stable, biologically inactive degradation products TXB_2 and 6-keto-$PGF_{1\alpha}$, respectively (Fig. 1). The enzyme TX-synthetase metabolizes PG endoperoxides not only to TXA_2, but also to 12-L-hy-

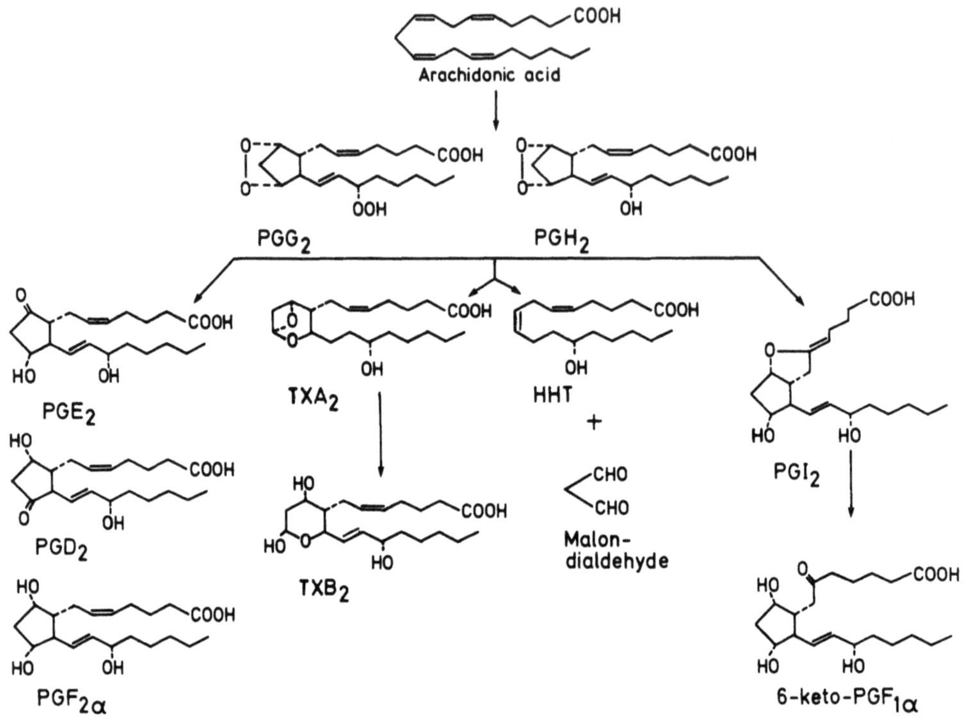

Fig. 1. Metabolism of arachidonic acid via the cyclooxygenase pathway

droxy-5,8,10-heptadecatrienoic acid (HHT) and malondialdehyde (Hammarström 1982). From the fact that both compounds lack groups with predictable immunodominance one can assume that the production of specific antibodies is either impossible or at least extremely difficult (Granström and Kindahl 1978). In fact, while radioimmunoassays for all other stable products of arachidonic acid metabolism have been developed, immunoassays for HHT or malondialdehyde have never been described. The number of closely related compounds occurring in biologic material is further increased by the fact that, besides arachidonic acid, other polyunsaturated fatty acids such as dihomo-γ-linolenic acid and 5,8,11,14,17-eicosapentaenoic acid are also substrates for the enzyme fatty acid cyclooxygenase, resulting in the formation of PG of the 1 series, like PGE_1 and $PGF_{1\alpha}$ and PG of the 3 series, like PGI_3.

PGE_2 and $PGF_{2\alpha}$ are inactivated (Fig. 2) by the enzymes 15-hydroxy-PG-dehydrogenase and PG-Δ^{13}-reductase (Samuelsson et al. 1975). These enzymes are widely distributed in the bodies of animals and humans (Änggård 1971; Samuelsson et al. 1975) with an especially high activity in lung tissue. The pulmonary enzymes are mainly responsible for the short half-life of E- and F-type PG in the circulation, with more than 90% inactivation during one single passage through the lung (Ferreira and Vane 1967). The biologically inactive 15-keto-13,14-dihydro metabolites of PG have a longer half-life (about 8–10 min) in the circulation than the primary PG (Samuelsson et al. 1975; Bothwell et al. 1982). Consequently, their plasma levels reflect PG biosynthesis more accurately than

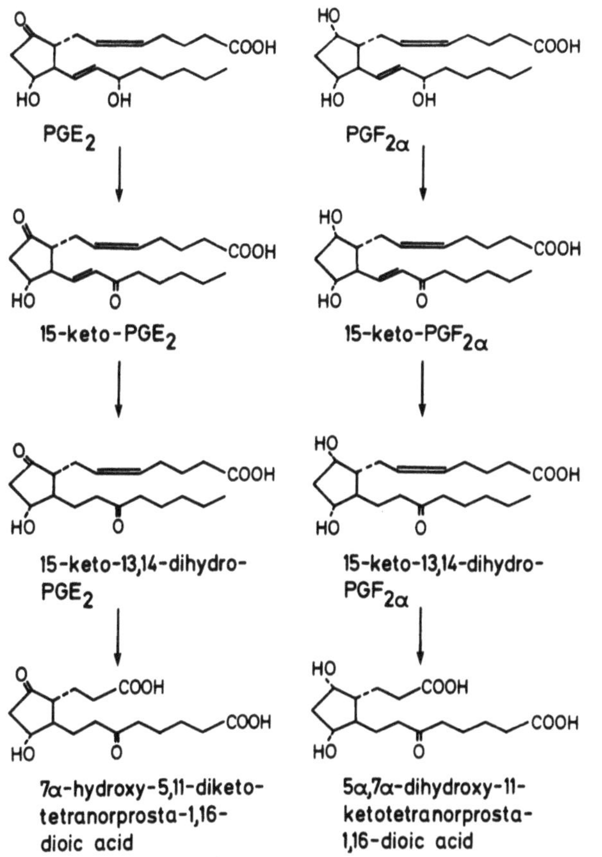

Fig. 2. Products of enzymatic inactivation of PGE_2 and $PGF_{2\alpha}$

levels of primary PG. Determination of 15-keto-13,14-dihydro metabolites of PG in blood has been considered a useful method for the quantification of arachidonic acid metabolism in vivo (SAMUELSSON et al. 1975). The 15-keto-13,14-dihydro-PG are further metabolized by β- and ω-oxidation (SAMUELSSON et al. 1975). These compounds have a much longer half-life (of the order of several hours) in the circulation than the 15-keto-13,14-dihydro metabolites of PG (GRANSTRÖM et al. 1982; GRANSTRÖM and KINDAHL 1982) and are finally excreted in the urine (SAMUELSSON et al. 1975). Determination of the urinary metabolites in plasma is particularly advantageous if blood samples cannot be collected in short intervals. Quantitative determination of 15-keto-13,14-dihydro metabolites of PG and their main urinary metabolites in plasma has the additional advantage that these compounds, contrary to primary PG and TX, are not formed as artifacts during blood sampling.

Exogenous TXB_2, PGI_2, and 6-keto-$PGF_{1\alpha}$ have also been shown to be metabolized via the 15-hydroxy-PG-dehydrogenase–PG-Δ^{13}-reductase pathway (ROBERTS et al. 1978, 1981; ROSENKRANZ et al. 1980; BRASH et al. 1983a). However, with exogenous substrates, ketodihydro metabolites comprise only a minor portion of the total metabolites formed. The metabolism of endogenous PGI_2 and TXA_2 might, however, differ from that of exogenous substrates and deserves further investigation. Major urinary metabolites of exogenous TXB_2 and PGI_2 are 2,3-dinor-TXB_2 and 2,3-dinor-6-keto-$PGF_{1\alpha}$. MS assays for these compounds have been used as an index of in vivo production of TXB_2 and PGI_2, respectively (ROBERTS et al. 1981; OATES et al. 1981; BRASH et al. 1983a). The dinor metabolites differ from the parent compounds only in the length of the carboxyl side chain. Development of highly specific radioimmunoassays for these metabolites seems, therefore, to be very difficult. Only recently, PATRONO et al. (1985) have developed an assay for urinary 2,3-dinor-TXB_2. However, the relative cross-reaction of unmetabolized TXB_2 was 51%.

From Figs. 1 and 2, it becomes evident that assay specificity is a major requirement for any method used for quantitative PG and TX determination. The methods employed should recognize even slight differences in the ring structure and the side chain configuration of the prostanoid molecule. Furthermore, concentrations of prostanoids in tissues and body fluids are generally very low. Additional important requirements for assay methods used to measure these compounds in biologic material are, therefore, high sensitivity with detection limits in the low picogram range as well as high precision and accuracy.

C. Development of Radioimmunoassays for Prostaglandins and Thromboxanes

I. Preparation of Immunogens

Since PG and related compounds are small molecules that are not immunogenic by themselves, they have to be coupled to macromolecular carriers in order to elicit an immune response. Although PG have several reactive groups, which could be used for coupling procedures, PG immunogens have been synthesized exclusively by chemical reactions with the single carboxyl group. The advantage of this method is that the ring structure and the 15-hydroxyl group remain unaffected and thus retain their immunodominance. Three methods have been mainly used to couple the carboxyl group of prostanoids to amino groups of a protein carrier, usually bovine serum albumin. A simple and very effective method is the use of water-soluble carbodiimides (GOODFRIEND et al. 1964). In fact, most authors seem to have used these coupling reagents, particularly 1-ethyl-3-(3-dimethylaminopropyl)carbodiimide, for the synthesis of prostanoid–protein conjugates. The only disadvantage of this method is dehydration, which easily occurs in the presence of carbodiimides. Such dehydration seems to be responsible for the production of anti-PGA_2 and anti-PGB_2 antibodies after immunization with PGE_2–protein conjugates prepared by the use of carbodiimides (LEVINE et al. 1971; JOBKE et al. 1973; GRANSTRÖM and KINDAHL 1978). Similarly, dehydration of 15-keto-13,14-dihydro-PGE_2 was observed when 1-ethyl-3-(3-dimethylaminopropyl)carbodiimide was the coupling reagent for synthesis of the immunogen (PESKAR et al. 1974). Another method for the preparation of prostanoid–protein conjugates is the use of *N,N'*-carbonyldiimidazole for the coupling reaction (AXEN 1974). This method has been successfully applied for the preparation of immunogens for the production of antibodies against TXB_2 (GRANSTRÖM et al.

1976b; ANHUT et al. 1977) and PGD_2 (ANHUT et al. 1978c). *N,N'*-Carbonyldiimidazole is a milder coupling reagent than the carbodiimides and dehydration of PGE_2 to PGA_2 was not observed (AXEN 1974). A further method for coupling prostanoids to proteins, which is, however, only rarely used, is the mixed anhydride method (ERLANGER 1973). This method requires an incubation step at alkaline pH and is, therefore, not suitable for PG and related compounds which are not stable under such conditions (LANGE and PESKAR 1984).

As to macromolecular carriers, bovine serum albumin is the most widely used backbone molecule. However, some authors have observed a more specific immune response with thyroglobulin (RAZ et al. 1975). LEVINE and VAN VUNAKIS (1970) have used poly-L-lysine as a synthetic polypeptide carrier. This molecule has the theoretical advantage of containing many free amino groups which could be bound covalently to the carboxyl group of prostanoids. However, no superior immune response seems to have been obtained with this type of conjugate.

Only a few authors have calculated the molar ratio of hapten to carrier of the immunogenic prostanoid conjugates. Successful immunizations have been described for conjugates with molar ratios as low as 1.5:1 (JAFFE et al. 1971) and as high as 40:1 (DRAY et al. 1972). For prostanoids, the optimal number of hapten molecules coupled to one carrier molecule has never been determined. The work of LANDSTEINER (1945) has demonstrated that conjugation of too many or too few hapten molecules to one carrier molecule may result in a suboptimal antihapten immune response.

II. Immunization

For immunization, usually rabbits, rarely guinea pigs or goats, are used. Inoculation with the immunogen is routinely done by subcutaneous injection, in either the back or the footpads. The latter method is painful for the animals, and, therefore, many scientists prefer injection in the back. Immunization by other routes such as intravenous or intramuscular injection is only rarely performed (GRANSTRÖM and KINDAHL 1978). For maximal immune response, the hapten–carrier conjugate is dissolved in distilled water and emulsified with complete Freund's adjuvant. Injections are repeated at intervals which should be long enough (several weeks to months) to avoid the induction of tolerance. Blood should be taken 1–2 weeks after booster injections. During blood collection, it is advisable to inhibit generation of endogenous eicosanoids, which could occupy antibody-binding sites. Therefore, the blood is collected into a mixture of a cyclooxygenase inhibitor such as aspirin or indomethacin with an anticoagulant such as ETDA. Then the antiplasma is separated from blood cells by immediate centrifugation at 4 °C.

III. Labeled Ligands

Tritium-labeled PG, PG metabolites, and TXB_2 of sufficiently high specific activity for use in sensitive radioimmunoassays are commercially available. Other labeled ligands can be prepared from the corresponding labeled precursor by chemical or biosynthetic methods. A convenient method for preparing tritium-labeled methylesters of eicosanoids is methylation with [^{3}H]methyl iodide (MOONEN et al. 1985). Furthermore, tritium-labeled amino acid conjugates of PG and TX have been employed in radioimmunoassays (SAUTEBIN et al. 1985; PATRONO et al. 1985). Tritiated urinary metabolites of PG can be extracted from the urine of various species after intravenous injection of tritiated PG (GRANSTRÖM and KINDAHL 1978). In order to obtain radioactive tracers with very high specific activities, biosynthesis of endogenous PG should be suppressed by the administration of a cyclooxygenase inhibitor such as aspirin or indomethacin. Stable tritiated degradation products of 15-keto-13,14-dihydro-PGE_2 and PGD_2 can be prepared by in vitro incubation of the parent compounds at alkaline pH (BOTHWELL et al. 1982; LANGE and PESKAR 1984). Labeled ligands with even higher specific activity can be prepared by iodination with ^{125}I. For this purpose, the prostanoid has to be coupled covalently to histamine or tyramine. These conjugates can be iodinated, e.g., by the chloramine-T method (GREENWOOD et al. 1963). This approach has been used by MACLOUF et al. (1976) and SORS et al. (1977) for the preparation of labeled ligands with very high specific activity for radioimmunoassays of PGE_1, PGE_2, $PGF_{2\alpha}$, and 15-keto-13,14-dihydro-$PGF_{2\alpha}$. Similarly, iodinated ligands for radioimmunoassays of TXB_2 (SORS et al. 1978; KOH et al. 1978) and 15-keto-13,14-dihydro-TXB_2 (PESKAR and HOLLAND 1979) have been synthesized. OHKI et al. (1974) prepared iodinated ligands for radioimmunoassays of $PGF_{2\alpha}$ and its main urinary metabolite using conjugates of these compounds with tyrosine methylester amide. Some authors (LEVINE and VAN VUNAKIS 1970; ANHUT et al. 1978b) have used polyvalent iodinated tracers. In these experiments, the advantage of the high specific activity of iodinated tracers is balanced by the fact that several hapten molecules are linked to one iodinated carrier molecule, e.g., bovine serum albumin. As a consequence, radioimmunoassays using such labeled ligands are relatively insensitive (LEVINE and VAN VUNAKIS 1970; VAN VUNAKIS et al. 1972; ANHUT et al. 1978b). Nevertheless, such assays have been successfully used to determine haptens like prostanoids in biologic material.

Chemical identity of the radioactive ligand and the compound to be measured is not essential for a valid radioimmunoassay (YALOW 1973a; GRANSTRÖM and KINDAHL 1978). Thus, heterologous tracers have often been used in prostanoid radioimmunoassays (GRANSTRÖM and KINDAHL 1978), especially when homologous tracers were not available and the use of a heterologous tracer did not seriously limit the sensitivity and specificity of the assay.

More recent developments in the field of prostanoid immunoassays are the use of chemiluminescence (WEERASEKERA et al. 1983; LANGE et al. 1985) and enzyme tracers (HAYASHI et al. 1981, 1983; YANO et al. 1981; YAMAMOTO et al. 1983; SAWADA et al. 1985; PRADELLES et al. 1985). These assays have the advantage of avoiding the use of radioactive material. Furthermore, such assays are often considerably more sensitive than conventional radioimmunoassays (LANGE et al. 1985). An excellent review on chemiluminescence and bioluminescence immunoassays has recently been published (KOHEN et al. 1985).

IV. Separation of Antibody-Bound and Free Fractions of Ligands

Numerous methods for separation of the antibody-bound and free fractions of ligands have been described for various radioimmunoassays (YALOW 1973a, b). In the PG field, the method most frequently used is adsorption of the free ligand fraction by charcoal, either coated with dextran (HERBERT et al. 1965) or uncoated (PALMIERI et al. 1971). These methods, although technically simple, may easily lead to erroneous results. Especially for the analysis of protein-rich samples, the charcoal method may be unsuitable. A more generally applicable method for the separation of antibody-bound and free fractions of ligands is the double-antibody method (MORGAN and LAZAROW 1962; HALES and RANDLE 1963). This method has the advantage that it can be adjusted to work satisfactorily even in the presence of higher concentrations of proteins, although the protein content should not differ between the samples analyzed. The method is, however, influenced by nonspecific factors such as the presence or absence of EDTA or various cations. Furthermore, one has to consider that the low dilutions of the second antiserum, usually necessary for complete precipitation, might contain significant amounts of PG and TXB_2 which interfere specifically with the radioimmunoassays. In order to avoid this problem, one should use dialyzed samples of the second antiserum or antiplasma that was obtained in the presence of a cyclooxygenase inhibitor. A further method to separate free and antibody-bound fractions of the ligand, often used for the radioimmunologic analysis of cyclooxygenase products of arachidonate metabolism in plasma or serum, is precipitation of the immunoglobulin fraction of the antiserum by polyethyleneglycol (DESBUQUOIS and AURBACH 1971; VAN ORDEN and FARLEY 1973; GRANSTRÖM et al. 1976b; BOTHWELL et al. 1982). An additional method is the use of nitrocellulose membranes, which adsorb proteins such as the antibody molecules (GERSHMAN et al. 1972). Furthermore, gel filtration separation of antibody-bound and free antigen has been described for a radioimmunoassay of $PGF_{2\alpha}$ (DEHENY et al. 1981). Solid-phase immunoassays have only rarely been developed for PG and related compounds (DIGHE et al. 1975; FITZPATRICK and WYNALDA 1976). LEVINE et al. (1979) have used immobilized ligands and protein A ^{125}I for the immunoassay of various PG and TXB_2. In this procedure, the iodinated protein A binds specifically to the Fc region of rabbit IgG without inhibiting the ability of the rabbit antibody to bind antigen. Antibody-coated polystyrene tubes have been used in chemiluminescence immunoassays for 15-keto-13,14-dihydro-PGE_2 and a stable degradation product of PGD_2 (LANGE et al. 1985).

D. Validation of Radioimmunoassays for Cyclooxygenase Products of Arachidonic Acid Metabolism

A minimum requirement for a valid radioimmunoassay is parallelism between standards and unknowns determined at various dilutions (YALOW 1973a, b). However, although necessary, fulfillment of just this requirement is not a sufficient validation of the radioimmunoassay (YALOW 1973a, b). Parallelism of the dilution curves demonstrates that the antibody used cannot discriminate between the immunoreactivity of standards and unknowns. This does not completely exclude the possibility that the immunoreactive material in biologic samples is heterogeneous (CIABATTONI et al. 1979). Further validation criteria for radioimmunoassays of low molecular weight compounds like cyclic nucleotides or PG have been defined by LANDS et al. (1976). Thus, particularly in the early stages of the development of such radioimmunoassays, the validity of data obtained should be tested by the use of independent assay methods like bioassay or combined GC–MS. A similar method, which is, however, less reliable, is the use of several antisera with different specificities (CIABATTONI et al. 1979). A valid radioimmunoassay procedure should give identical results for biologic samples. The accuracy of a radioimmunoassay has to be tested by the addition of various amounts of exogenous standard compound to the biologic sample, which should produce the expected increment in inhibition of binding of label to the antibody. The endogenous level of the compound measured in the biologic material should then be represented by the intercept of the y-axis and the regression line. An important further requirement for the validity of radioimmunoassay results is that the immunoreactivity found in biologic material cochromatographs with the corresponding standard compound, preferably in several solvent systems. For PG and TX, such experiments have usually been performed using thin layer chromatography (ANHUT et al. 1978a; SALMON 1978; CIABATTONI et al. 1979; PATRONO et al. 1982). More recently, high pressure liquid chromatography has been used to separate the compound to be measured from related compounds, which could interfere in the radioimmunoassay (LEVINE and ALAM 1979; ALAM et al. 1979). Such a procedure might be particularly important if small amounts of a PG are measured in the presence of much larger amounts of closely related compounds. A typical example is the effect of a TX-synthetase inhibitor on the generation of 6-keto-$PGF_{1\alpha}$ by whole blood during clotting (PEDERSEN et al. 1983; CERLETTI et al. 1984). An additional validation, which is only rarely used in the PG field, is the decrease in immunoreactivity upon addition of an enzyme capable of destroying the compound to be measured (LANDS et al. 1976). Such a procedure has been used for $PGF_{2\alpha}$ found in serum and metabolized by 15-hydroxy-PG-dehydrogenase (KIRTON et al. 1972). The most important prerequisite for a valid radioimmunoassay is, however, identical physicochemical composition such as pH, temperature, and ionic strength of buffer in standard and unknown samples (YALOW 1973a, b).

E. Factors that Affect the Validity of Prostaglandin and Thromboxane Radioimmunoassay Results

I. Extraction and Purification Procedures

Only rarely are the concentrations of PG and TX in biologic material sufficiently high to be measured correctly at various dilutions by radioimmunoassay. Examples are the determination of TXB_2 in serum after clotting (PATRONO et al. 1980) or the levels of the main metabolite of $PGF_{2\alpha}$, $5\alpha,7\alpha$-dihydroxy-11-ketotetranorprosta-1,16-dioic acid, in urine (OHKI et al. 1975, 1976; GRANSTRÖM and

KINDAHL 1976). On the other hand, for determination of the primary PGE_2 and $PGF_{2\alpha}$, human urine samples have to be extracted at acidic pH before radioimmunologic analysis (SCHERER et al. 1978; CIABATTONI et al. 1979). Extractions are performed in order to concentrate the compounds to be measured and to remove nonspecifically interfering substances present in urine. The same reasons are valid for the extraction of PG from gastric juice (PESKAR et al. 1974). For extraction, several organic solvents have been used. While chloroform and diethyl ether are suitable for the extraction of the less polar PG of the A and E series, ethyl acetate is more suitable for the extraction of PG of the F series (GRANSTRÖM and KINDAHL 1978). For the determination of recovery rates, tritium-labeled PGs are added and carried through the procedure. An aliquot of each sample of extract has then to be used for quantification of radioactivity (JAFFE et al. 1973). It is important to note, however, that extraction of PG and related compounds into organic solvents not only removes material which might interfere in radioimmunoassays, but also introduces new sources of error. Thus, residues of organic solvents have been shown to affect the antigen–antibody binding reaction nonspecifically. It has been suggested that addition of excess rabbit immunoglobulin could prevent this undesirable effect (LEYENDECKER et al. 1972). However, usually the solvent residue effect is not completely abolished by addition of immunoglobulin. Further problems arise if the extract is not washed and therefore contains traces of acid. The acid is concentrated during evaporation of the organic solvent and can affect both the stability of PGs such as those of the E series and the antigen–antibody binding reaction. The rapid dehydration of PGE during extraction from human plasma had resulted in the assumption that significant concentrations of PGA occur in the circulation (for review see GRANSTRÖM and KINDAHL 1978). Finally, one has to remember that extraction into organic solvents at acidic pH is not specific for PG. In this context, arachidonic acid may present a particular problem. While in biologic material like plasma arachidonic acid is bound to proteins such as albumin, and, therefore, its interference in radioimmunoassays for PG is low or absent, free arachidonic acid after extraction can interfere to a much greater extent. This interference can be so marked that radioimmunoassay of extracted prostanoids may be impossible and a further separation step to eliminate arachidonic acid has to be introduced. In view of this complication, radioimmunoassay of prostanoids in unextracted plasma has been favored by a number of authors (PATRONO 1973; GRANSTRÖM and KINDAHL 1978; PESKAR et al. 1979; MORRIS et al. 1981; STRICKLAND et al. 1982). Alternatively, extraction followed by group separation of PG and related compounds by column chromatography, e.g., on silicic acid (JAFFE et al. 1973) or by celite partition chromatography (MELDRUM and ABRAHAM 1976) have been used. Other purification and separation methods employ ion exchange resins (FRETLAND 1974), Sephadex LH-20 (CHRISTENSEN and LEYSSAC 1976), Sephadex G-25 (THOMAS et al. 1978), acid-washed florisil (BANSCHBACH and LOVE 1979), or high pressure liquid chromatography (LEVINE and ALAM 1979; ALAM et al. 1979).

II. The Blank Problem

If an extraction step precedes radioimmunologic prostanoid analysis, the preparation of correct procedure blank samples is especially important. These procedure blanks should contain all the possibly interfering material originating in the extraction and chromatography procedure, such as solvent residues. Furthermore, procedure blank samples should be used for the generation of standard curves in order to account for nonspecifically interfering material, the effects of which might add to the inhibitory effect of standard amounts of prostanoids on the antigen–antibody reaction. A more difficult problem in the PG field is the correct preparation of sample blanks. For example, in many cases it will be difficult to obtain completely prostanoid-free plasma samples. Attempts have been made to use charcoal-adsorbed plasma for such purposes. If this type of blank is used, one should remember that charcoal adsorption removes not only prostanoids from the plasma, but also other low molecular weight compounds, including fatty acids like arachidonic acid. Therefore, such a procedure can introduce more errors into the radioimmunoassay than it removes. Similar problems with biologic blanks occur with other material such as gastric juice or various tissue incubates. Better biologic blanks would be plasma or other biologic material from patients treated chronically with high doses of nonsteroidal anti-inflammatory drugs.

III. Problems Associated with the Determination of Tissue and Plasma Concentrations of Prostanoids

Many groups have determined tissue concentrations of PG and related compounds (for references see GRANSTRÖM 1978). Since PG are not stored in the tissues, but are newly synthesized upon stimulation, it is not advisable to determine tissue concentrations. These concentrations mostly reflect PG and TX formed during preparation and homogenization of the tissue samples. Such values, therefore, mainly represent artifacts and no valid conclusions can be drawn from these results. GRANSTRÖM (1978) has suggested determining the synthetic capacity of a tissue rather than tissue concentrations of prostanoids. In fact, determination of the synthetic capacity for PG has given valid results, e.g., for human gastrointestinal tissue (PESKAR et al. 1980b). Similarly, PG-inactivating enzyme activity in a tissue can be determined rather than tissue concentrations of PG metabolites (PESKAR and PESKAR 1976).

A special problem is the artifactual generation of TXB_2 and PG during blood sampling. In order to prevent the synthesis of these compounds by platelets in vitro, a cyclooxygenase inhibitor such as aspirin or indomethacin is drawn into a syringe together with an anticoagulant. For the latter purpose, sodium–EDTA has been found to be superior to heparin (MORRIS et al. 1981; SINZINGER et al. 1985). However, even with these precautions, plasma levels of TXB_2 are generally highly variable and their determination cannot be recommended. Similarly, artifactual generation of 6-keto-$PGF_{1\alpha}$ might occur during blood sampling (RITTER et al. 1983). Contrary to TXB_2 and the primary PGs, 15-keto-13,14-dihydro metabolites of PGE_2 and $PGF_{2\alpha}$ and their urinary metabolites are not formed by blood cells. Their determination is, therefore, not influenced by artifacts during blood sampling. On the other hand, determination of primary PG in plasma may be useful, if the venous drainage of an organ is analyzed (GRANSTRÖM 1978).

Storage conditions are only rarely reported for biologic samples obtained for PG and TXB_2 determination. The most reliable radioimmunoassay results seem to be obtained if plasma is analyzed immediately after sampling. If plasma has to be stored, freezing at −70 °C (SINZINGER et al. 1985) or −80 °C (VIERHAPPER et al. 1985) is recommended. With storage at −20 °C time-dependent increases in PG-like immunoreactivity have been observed if the storage time exceeds several weeks (JUBIZ and FRAILEY 1974; METZ et al. 1979; SINZINGER et al. 1985; VIERHAPPER et al. 1985). Thus, prolonged storage or repeated freezing and thawing (WAITZMANN and LAW 1975; SINZINGER et al. 1985) should be avoided.

F. Radioimmunoassay for Various Prostaglandins and Thromboxanes

I. Prostaglandin $F_{2\alpha}$

Since the first description of radioimmunoassays for PG of the F series by LEVINE and VAN VUNAKIS (1970) and CALDWELL et al. (1971) a great number of radioimmunoassays for this particular prostanoid has been described. Since $PGF_{2\alpha}$ is generally stable in biologic material, radioimmunoassays for this compound present less problems than radioimmunoassays for PGD_2 or PGE_2. In most assays, $PGF_{1\alpha}$ is the only related PG interfering significantly with the binding of $PGF_{2\alpha}$ to the antibodies. However, owing to the generally low amounts of the precursor fatty acid dihomo-γ-linolenic acid in biologic material, concentrations of $PGF_{1\alpha}$ can be assumed to be very low. Serologic differentiation of $PGF_{2\alpha}$ and $PGF_{1\alpha}$ can be achieved by the simultaneous use of two antisera of different specificity (LEVINE et al. 1971; RAFFEL et al. 1976). The antibodies are generally specific for the hydroxyl configuration at C-9 in the cyclopentane ring of the PG molecule. Thus, by the use of both anti-$PGF_{2\alpha}$ and anti-$PGF_{2\beta}$ antibodies it could be demonstrated that the product of reduction of PGE_2 by chicken heart 9-keto-reductase is $PGF_{2\alpha}$ (LEVINE et al. 1975). Several nonradioactive labels have been synthesized for immunoassays for $PGF_{2\alpha}$. Thus, YANO et al. (1981) have described a sensitive and specific enzyme immunoassay for $PGF_{2\alpha}$, in which the hapten was labeled with β-galactosidase. The detection limit and specificity of the enzyme immunoassay were comparable to that of a radioimmunoassay for $PGF_{2\alpha}$. DRAY et al. (1972) developed a method for the immunochemical detection of $PGF_{2\alpha}$ with PG-coated bacteriophage T_4. These authors demonstrated that this nonradioactive method compared favorably with radioimmunoassay.

II. 15-Keto-13,14-dihydro-prostaglandin $F_{2\alpha}$

Since the first description of a radioimmunoassay for the $PGF_{2\alpha}$ metabolite 15-keto-13,14-dihydro-$PGF_{2\alpha}$ by GRANSTRÖM and SAMUELSSON (1972), a great number of immunoassays for this compound has been developed (LEVINE and GUTIERREZ-CERNOSEK 1973; STYLOS et al. 1973; CORNETTE et al. 1974; LIEBIG et al. 1974; FAIRCLOUGH and PAYNE 1975; MITCHELL et al. 1976; SORS et al. 1977;

HANING et al. 1977; YOUSSEFNEJADIAN et al. 1978). The 15-keto group and the reduced Δ^{13}double bond are generally immunodominant, resulting in antibodies which discriminate quite well between the metabolite and unmetabolized $PGF_{2\alpha}$. Owing to the stability of 15-keto-13,14-dihydro-$PGF_{2\alpha}$ in biologic material, radioimmunoassays for this compound are generally precise and accurate. Such assays have often been used to determine the level of the circulating $PGF_{2\alpha}$ metabolite as an indicator of endogenous $PGF_{2\alpha}$ biosynthesis (GRANSTRÖM 1979). Interestingly, the major metabolites of exogenous PGD_2 formed during infusion were identified as 15-keto-13,14-dihydro-$PGF_{2\alpha}$ and 15-keto-$PGF_{2\alpha}$ (BARROW et al. 1984). Thus, quantitative determination of these metabolites may prove useful not only in studies on the biosynthesis and metabolism of $PGF_{2\alpha}$, but also of PGD_2.

III. Prostaglandin E_2

The production of specific antibodies against E series PGs seems to be very difficult (LEVINE et al. 1971; YU and BURK 1972; ZUSMAN et al. 1972; JOBKE et al. 1973; BAUMINGER et al. 1973; RITZI and STYLOS 1974; RAZ et al. 1975). PGE_1 and PGE_2 are easily dehydrated at the 11-hydroxyl position, resulting in the formation of A and B series PG. This reaction can occur during the preparation of the immunogen as well as by reaction of the injected conjugate with plasma enzymes (HORTON et al. 1971; POLET and LEVINE 1975) in the immunized animals. It has been suggested (LEVINE et al. 1971) that the production of anti-PGE_2 antibodies may be so difficult that PGE_2 should rather be measured after quantitative conversion to the stable PGB_2. Alternatively, LINDGREN et al. (1974) measured PGE_2 by a method based on $NaBH_4$ reduction of samples, followed by radioimmunoassay of $PGF_{2\alpha}$ and $PGF_{2\beta}$. Several laboratories, however, have demonstrated that the production of rather specific anti-PGE_2 antibodies in rabbits is possible (DRAY et al. 1975; CHRISTENSEN and LEYSSAC 1976), the reason for their success being unknown. The most elegant strategy for producing specific polyclonal anti-PGE_2 antibodies is the use of a stable hapten mimic instead of the unstable PGE_2 for the preparation of an immunogenic conjugate (FITZPATRICK and BUNDY 1978). A more recent successful approach is the production of specific monoclonal anti-PGE_2 antibodies (BRUNE et al. 1985; DAVID et al. 1985; TANAKA et al. 1985).

IV. 15-Keto-13,14-dihydro-prostaglandin E_2

Although the 15-keto-13,14-dihydro metabolites of PGE_2 and $PGF_{2\alpha}$ in plasma reflect PG biosynthesis more accurately than levels of the primary PG (SAMUELSSON et al. 1975; LEVINE 1977), there are only a few reports on the development and use of radioimmunoassays for 15-keto-13,14-dihydro-PGE_2 (B. A. PESKAR et al. 1974; LEVINE 1977). The reliability of these assays is considerably lower than the reliability of radioimmunoassays of the corresponding $PGF_{2\alpha}$ metabolite (MITCHELL et al. 1977; GRANSTRÖM 1979; METZ et al. 1979). The low assay accuracy seems to be a consequence of the chemical instability of 15-keto-13,14-dihydro-PGE_2, which is rapidly dehydrated to 15-keto-13,14-dihydro-PGA_2. This compound can bind to albumin and further, e.g., at alkaline pH, rearrange to the stable bicyclic degradation end product 11-deoxy-15-keto-13,14-dihydro-11β-16ξ-cyclo-PGE_2 (GRANSTRÖM 1979; FITZPATRICK et al. 1980; GRANSTRÖM et al. 1980). Several radioimmunoassays for this nonenzymatic degradation product of 15-keto-13,14-dihydro-PGE_2 have been described (GRANSTRÖM and KINDAHL 1979; PESKAR et al. 1980a; BOTHWELL et al. 1982; MITCHELL et al. 1982; DEMERS et al. 1983). Contrary to direct radioimmunoassays for 15-keto-13,14-dihydro-PGE_2, the radioimmunoassays for 11-deoxy-15-keto-13,14-dihydro-11β,16ξ-cyclo-PGE_2, which require complete conversion of 15-keto-13,14-dihydro-PGE_2 to the degradation end product before quantitative determination, have been shown to be generally highly specific, sensitive, and precise. Furthermore, a high accuracy of such assays has been repeatedly demonstrated (BOTHWELL et al. 1982; STARCZEWSKI et al. 1984). Finally, it should be pointed out that the bicyclic degradation product of the PGE_2 metabolite is a racemic mixture of two C-16 epimers. It has been demonstrated that the two epimers differ slightly in their activity to displace a racemic tritiated ligand from antibody (BOTHWELL et al. 1982). The difference is, however, not important for practical purposes, since the two epimers are formed in a highly reproducible ratio at pH 10–11 (BOTHWELL et al. 1982). The only compound cross-reacting significantly in this radioimmunoassay is the corresponding metabolite of PGE_1. This is no limitation of the assay, since basal levels of 15-keto-13,14-dihydro-PGE_1 can be assumed to be about as low as the normal levels of PGE_1 (BOTHWELL et al. 1982). On the other hand, owing to the extensive cross-reaction, the assay permits studies on the pharmacokinetic disposition of exogenous PGE_1 as well as of PGE_2 (BOTHWELL et al. 1982).

V. Prostaglandin E_1

Since, contrary to arachidonic acid, dihomo-γ-linolenic acid, the precursor of the 1 series of PG, occurs only in trace amounts in mammalian tissues, endogenous levels of PGE_1 should be much lower than levels of PGE_2. However, PGE_1 has interesting biologic effects, which differ from those of PGE_2 (KARIM 1976). Highly variable results have been obtained when plasma or serum levels of PGE_1 were determined radioimmunologically and compared with levels of PGE_2 (RITZI and STYLOS 1974; RAZ et al. 1975; DRAY et al. 1975; VAN ORDEN et al. 1977). Generally, antibodies directed against PGE_1 also recognize PGE_2 to a significant extent. It is, therefore, difficult to measure PGE_1 specifically in the presence of PGE_2. It has been possible to differentiate PGE_1 and PGE_2 to a certain extent by the simultaneous use of anti-PGE_1 and anti-PGE_2 antibodies (RAZ et al. 1975; DIEKMANN et al. 1977). VAN ORDEN et al. (1977) using an anti-PGE_1 antiserum which recognizes mainly PGE_1 with 28% relative cross-reaction of PGE_2, tried to increase the specificity for PGE_1 by adding absorbing amounts of PGE_2. This approach was successful to a limited extent only. It requires selection of absorbing amounts of PGE_2 which produce the greatest shift of the PGE_2 standard curve to the right with the least perturbation of the PGE_1 standard curve. The study revealed that the antiserum did not contain two distinct populations of antibodies directed against PGE_1 and PGE_2, respectively, but rather contained antibody molecules reacting with equal capacity for both haptens and differing only in their avidity for PGE_1 and PGE_2. MACLOUF et al. (1975) raised antibodies with high affinity for PGE_1 in a sheep, while rabbits immunized with the same immunogen and the same schedule of immunization produced antibodies with much lower affinity and poor specificity. Consequently, analysis of human plasma samples by radioimmunoassay for PGE_1 resulted in considerably lower levels with the sheep antiserum than with the rabbit antiserum.

VI. Prostaglandin D_2

Only few radioimmunoassays for PGD_2 have been described so far (ANHUT et al. 1978c; LEVINE et al. 1979; NARUMIYA et al. 1982). The first two of these assays have been found to lack the sensitivity and accuracy generally achieved for radioimmunoassays of other cyclooxygenase products of arachidonic acid metabolism. The reason for the low assay reliability may be the fact that PGD_2 is unstable under various conditions (GRANSTRÖM and KINDAHL 1978; LEVINE et al. 1979; FITZPATRICK and WYNALDA 1983; KIKAWA et al. 1984). It has been demonstrated that alkaline treatment of biologic samples containing PGD_2 by the method used for the conversion of 15-keto-13,14-dihydro-PGE_2 to its stable bicyclic degradation product (BOTHWELL et al. 1982) results in the formation of one or more stable degradation products of PGD_2 (LANGE and PESKAR 1984). The degradation products of PGD_2 formed under alkaline conditions have, however, not been formally identified so far. Only recently, a new radioimmunoassay for PGD_2 has been developed using an antiserum against the stabilized 11-methoxime derivative of PGD_2 (MACLOUF et al. 1986). The procedure requires the immediate derivatization of the PGD_2 contained in biologic samples into 11-methoxime-PGD_2, thus avoiding the instability problems inherent in the parent compound.

Interest in the physiologic and pathophysiologic role of PGD_2 has considerably increased, since it has been demonstrated that PGD_2 is the predominant arachidonic acid metabolite in mast cells (LEWIS et al. 1982). Patients with mastocytosis have been shown to excrete large amounts of a PGD_2 metabolite, 9α-hydroxy-11,15-dioxo-2,3,18,19-tetranorprost-5-ene-1,20-dioic acid (ROBERTS et al. 1980). This metabolite has been measured by a stable isotope dilution mass spectrometric assay. Unfortunately, a radioimmunoassay for this PGD_2 metabolite is not available so far.

VII. Indirect Determination of Prostaglandin Endoperoxides and Thromboxane A_2

The PG endoperoxides PGG_2 and PGH_2, as well as TXA_2, are highly unstable under physiologic conditions. Antibodies which recognize and bind PG endoperoxides have been raised in rabbits immunized with a bovine serum albumin conjugate of the stable hapten mimic 9,11-azo-15-hydroxy-prosta-5,13-dienoic acid. These antibodies were successfully used to study the role of PG endoperoxides in platelet aggregation (FITZPATRICK et al. 1978). However, owing to instability of PG endoperoxides, such antibodies cannot be used for quantitative determination of these compounds by radioimmunoassay procedures. PG endoperoxides are instead determined by bioassay methods. For indirect radioimmunologic determination of PG endoperoxides, it has been proposed (GREEN et al. 1976; GRANSTRÖM and KINDAHL 1978) to convert them to $PGF_{2\alpha}$ by treatment with stannous chloride or triphenylphosphine (HAMBERG et al. 1976). The difference in $PGF_{2\alpha}$ content in treated and untreated samples is then a measure of the amount of PG endoperoxides present. This method for PG endoperoxide determination has, however, only rarely been used.

For the radioimmunologic determination of TXA_2, GRANSTRÖM et al. (1976a) have developed an elegant indirect method. They converted TXA_2 to mono-*O*-methyl-TXB_2 by addition of methanol (HAMBERG et al. 1975). This treatment results in the formation of two epimers of mono-*O*-methyl-TXB_2, the methoxy group being attached to C-3 of the propyl side chain in TXB_2. GRANSTRÖM et al. (1976a) have used the major, less polar, epimer for the development of a radioimmunoassay. The antibodies obtained after immunization of a rabbit were highly specific for the major epimer of mono-*O*-methyl-TXB_2 with little cross-reaction by the minor epimer and TXB_2 itself. Cross-reaction of various PGs, including PGD_1 and PGD_2, was less than 0.1%. The negligible cross-reaction of PG of the D series as compared with the cross-reaction in radioimmunoassays for TXB_2 (GRANSTRÖM et al. 1976b; ANHUT et al. 1977) can be explained by the fact that the ring structure in the TXB_2 derivative is stable and an open form cannot occur. Using this radioimmunoassay, kinetic studies on the formation of TXA_2 by aggregating platelets were performed. It was demonstrated (as expected) that the kinetics of TXA_2 formation by stimulated platelets differ considerably from the time course of TXB_2 formation. Furthermore, the half-life of TXA_2 under various conditions could be determined exactly by the radioimmunoassay for mono-*O*-methyl-TXB_2 (GRANSTRÖM et al. 1976a).

VIII. Thromboxane B_2 and 11-Dehydro-thromboxane B_2

Contrary to its mono-*O*-methyl derivative, TXB_2 can exist in an open and a closed ring form. Usually, PGD_2 is the compound interfering most strongly in radioimmunoassays for TXB_2 (GRANSTRÖM et al. 1976b; ANHUT et al. 1977; FITZPATRICK et al. 1977). It has been suggested that this cross-reaction is due to the fact that the hemiacetal structure of TXB_2 is in equilibrium with the corresponding acyclic aldehyde–alcohol, the chemical structure of which resembles to a certain extent that of PGD_2 (GRANSTRÖM et al. 1976b).

Even if a radioimmunoassay for TXB_2 is correctly performed, results may give misleading information about TX production if a protein-rich environment is used. An example has been described by FITZPATRICK and GORMAN (1977) who studied the production of immunoreactive TXB_2 from PGH_2 in platelet-rich plasma. These authors demonstrated that, under certain experimental conditions, the intermediate TXA_2 binds covalently to proteins and is not rapidly hydrolyzed to TXB_2.

Correct determinations of TXB_2 in plasma samples are extremely difficult. As discussed in Sect. F, rapid and extensive artifactual generation of TXB_2 may occur during blood sampling, resulting in falsely high plasma levels. Nevertheless, a few radioimmunoassays for the determination of TXB_2 in plasma have been published (VIINIKKA and YLIKORKALA 1980; MCCANN et al. 1981; KUZUYA et al.

1985). GRANSTRÖM et al. (1985) have investigated the metabolism of TXB_2 in different species in order to find degradation products that might serve as better parameters of TX production in vivo. Such a metabolite was identified as 11-dehydro-TXB_2. The dehydrogenase catalyzing the formation of this product was not detected in blood cells. The authors developed a radioimmunoassay for this compound and measured formation of 11-dehydro-TXB_2 after induction of platelet aggregation in rabbits. It was concluded that measurement of 11-dehydro-TXB_2 in blood and urine reflected TX synthesis far better than measurement of TXB_2. This conclusion is further supported by a recent study in humans by the same group of investigators (WESTLUND et al. 1986).

IX. 15-Keto-13,14-dihydro-thromboxane B_2

Exogenous TXB_2 remains in the circulation as such for a considerable length of time after intravenous injection. Contrary to PGE_2 and $PGF_{2\alpha}$, it is not converted into a 15-keto-13,14-dihydro derivative in the cynomolgus monkey (KINDAHL 1977; ROBERTS et al. 1978) and humans (ROBERTS et al. 1981). However, the metabolism of endogenous TX may differ in this respect. Thus, GC–MS analysis has shown that 15-keto-13,14-dihydro-TXB_2 is by far the major cyclooxygenase product of arachidonic acid metabolism released from isolated perfused anaphylactic guinea pig lungs (DAWSON et al. 1976; BOOT et al. 1978; ROBINSON et al. 1984). Using a radioimmunoassay for 15-keto-13,14-dihydro-TXB_2, ANHUT et al. (1978b) obtained results on the concentrations of the TX metabolite in anaphylactic guinea pig lung perfusates, which were essentially in agreement with the GC–MS data. The radioimmunoassay was performed with either a polyvalent (ANHUT et al. 1978b) or a monovalent (PESKAR and HOLLAND 1979) iodinated tracer. With the monovalent ligand, considerable concentrations of 15-keto-13,14-dihydro-TXB_2 could be detected in the circulation of anesthetized ovalbumin-sensitized guinea pigs after, but not before, antigenic challenge (PESKAR and HOLLAND 1979). Interestingly, the TX metabolite has recently also been detected in human lung (DAHLÉN et al. 1983). However, nothing is known about its possible occurrence in the human circulation.

X. 6-Keto-prostaglandin $F_{1\alpha}$

6-Keto-$PGF_{1\alpha}$ is the stable degradation product of the biologically active, but unstable PGI_2 (JOHNSON et al. 1976). Antibodies recognizing and binding PGI_2 have been raised by immunization of rabbits with the stable 5,6-dihydro-PGI_2 (FITZPATRICK and GORMAN 1978; LEVINE et al. 1979). These antibodies have been used to inhibit biologic effects of PGI_2 such as inhibition of platelet aggregation (FITZPATRICK and GORMAN 1978) or to trap PGI_2 in the circulation (PACE-ASCIAK et al. 1980). However, owing to the instability of PGI_2, such antibodies cannot be used for its quantitative determination by radioimmunoassay procedures.

The first radioimmunoassay for 6-keto-$PGF_{1\alpha}$ was described by SALMON (1978). The antibodies were raised in rabbits by immunization with a 6-keto-$PGF_{1\alpha}$–bovine serum albumin conjugate prepared by the carbodiimide method (GOODFRIEND et al. 1964). The antibodies exhibited marked cross-reaction with several PGs and their metabolites. This cross-reaction necessitated a purification step prior to immunoassay. The thin layer chromatographic method used clearly separated 6-keto-$PGF_{1\alpha}$ from related cross-reacting compounds, but the low recovery reduced the assay sensitivity considerably. Since then, a great number of radioimmunoassays for 6-keto-$PGF_{1\alpha}$ has been developed in various laboratories (DIGHE et al. 1978; DRAY et al. 1978; BULT et al. 1978; MITCHELL 1978; PESKAR et al. 1979; LEVINE et al. 1979; DEMERS and DERCK 1980; MORRIS et al. 1981; YLIKORKALA and VIINIKKA 1981). Although most of these radioimmunoassays are rather specific for the homologous hapten, some of them have measured surprisingly high basal plasma levels of 6-keto-$PGF_{1\alpha}$. In view of the calculated production rate (OATES et al. 1981) and the low levels of circulating 6-keto-$PGF_{1\alpha}$ measured by GC–MS (BLAIR et al. 1982; RITTER et al. 1983), a reevaluation of several of the radioimmunoassay data seems necessary.

6-Keto-$PGF_{1\alpha}$ exists in an open form and a hemiketal form (JOHNSON et al. 1976). It is not known if one form prevails under radioimmunoassay conditions or if antibodies are produced predominantly against one of the two forms. SALMON (1978) has stressed the importance of using standard incubation conditions for all samples analyzed, ensuring that a consistent equilibrium between the two forms is maintained. HENSBY et al. (1979) have explained the fact that radioimmunoassay gave lower values for circulating 6-keto-$PGF_{1\alpha}$ than their GC–MS method by the possibility that immunoassay measures only one of the isomers of 6-keto-$PGF_{1\alpha}$. In order to avoid problems arising from the two forms of 6-keto-$PGF_{1\alpha}$ and to develop a radioimmunoassay for this compound with increased specificity, OLIW (1980) utilized an antiserum against 6-methoxime-$PGF_{1\alpha}$. This assay requires quantitative conversion of PGI_2 and 6-keto-$PGF_{1\alpha}$ to the methoxime derivative before assay. Owing to the three-dimensional structure of the nitrogen ligands, the derivative may exist as two isomers. Nevertheless, the assay was specific and sensitive, and radioimmunoassay results obtained with these antibodies were in excellent agreement with GC–MS data.

If an extraction step is used before immunoassay, correct quantitative determination of 6-keto-$PGF_{1\alpha}$ is additionally complicated by the fact that this compound is not completely stable under such conditions. Up to five radioactive peaks, all binding anti-6-keto-$PGF_{1\alpha}$ antiserum to a highly variable degree, were observed when 6-keto-$PGF_{1\alpha}$ ^3H was subjected to thin layer chromatography after extraction (MITCHELL et al. 1981).

XI. Metabolites of Prostaglandin I_2 and 6-Keto-prostaglandin $F_{1\alpha}$

Only a few radioimmunoassays for 6,15-diketo-$PGF_{1\alpha}$ and 6,15-diketo-13,14-dihydro-$PGF_{1\alpha}$ have been described (PESKAR et al. 1980c; MACHLEIDT et al. 1981; MYATT et al. 1981). It was found that brief incubation of rat vascular tissue induced the release of significant amounts of both 6,15-diketo-13,14-dihydro-$PGF_{1\alpha}$ and 6,15-diketo-$PGF_{1\alpha}$ (PESKAR et al. 1980c). After intravenous administration of PGI_2 to healthy volunteers, 6,15-diketo-$PGF_{1\alpha}$ remained undetectable in the plasma throughout the infusion. On the other hand, levels of immunoreactive 6,15-diketo-13,14-dihydro-$PGF_{1\alpha}$ showed a similar, though delayed, increase as the nonenzymatic PGI_2 breakdown product 6-keto-$PGF_{1\alpha}$. The metabolite, however, disappeared with an appreciably longer half-life (PATRONO et al. 1981). Qualitatively similar results were obtained by MYATT et al. (1981) when they infused either PGI_2 or 6-keto-$PGF_{1\alpha}$ into healthy volunteers. The results of these authors differed from those of PATRONO et al. (1981) by the much longer half-life of 6-keto-$PGF_{1\alpha}$, and the 15-keto-13,14-dihydro metabolite in plasma, and by the fact that a basal circulating level of both compounds was detected. Similar basal levels were determined by MACHLEIDT et al. (1981) in the anesthetized cat. After administration of PGI_2, a significant increase in the plasma levels of 6-keto-$PGF_{1\alpha}$ and 6,15-diketo-13,14-dihydro-$PGF_{1\alpha}$ was observed in this species (MACHLEIDT et al. 1981) as in humans (PATRONO et al. 1981; MYATT et al. 1981). From these studies, it could be concluded that not only chemical instability, but also enzymatic degradation is a major determinant of the fate of PGI_2 in the circulation. This result is remarkable since, in contrast to PGE_2 and $PGF_{2\alpha}$, pulmonary enzymatic PGI_2 inactivation seems to be of only minor importance (ARMSTRONG et al. 1978). However, it has been demonstrated that a substantial part of PGI_2 is metabolized via the 15-hydroxy-PG-dehydrogenase/Δ^{13}-reductase pathway in other organs in vitro (WONG et al. 1978, 1979) and in vivo (SUN and TAYLOR 1978). In this context it is of great interest that EDLUND et al. (1982) found release of not only immunoreactive 6-keto-$PGF_{1\alpha}$, but also of 6,15-diketo-13,14-dihydro-$PGF_{1\alpha}$ from the heart of patients during cardioplegia.

In some organs, e.g., rabbit liver, 6-keto-PGE_1 can be generated from either PGI_2 or 6-keto-$PGF_{1\alpha}$ (WONG et al. 1981). 6-Keto-PGE_1 has biologic effects on platelets, blood pressure, and vascular resistance similar to PGI_2 (WONG et al. 1981). A radioimmunoassay for this interesting compound has been developed (CHANG and TAI 1982). In human urine, however, no PGI_2 metabolites with the 6-keto-PGE_1 structure could be detected after infusion of tritiated PGI_2 (BRASH et al. 1983b).

XII. Main Urinary Metabolites of Prostaglandin $F_{2\alpha}$ and Prostaglandin E_2

The main urinary metabolites of $PGF_{2\alpha}$ and PGE_2 are 1,16-dioic acids (see Fig. 2). Normal coupling procedures, e.g., by the carbodiimide method (GOODFRIEND et al. 1964), result in heterogeneous immunogens, since conjugation can be expected to occur randomly at either carboxyl group (GRANSTRÖM and KINDAHL 1978). For the main urinary metabolite of $PGF_{2\alpha}$, 5α,7α-dihydroxy-11-ketotetranorprosta-1,16-dioic acid, the problem has been solved by formation of a δ-lactone between the carboxyl group in the upper side chain and the 5α-hydroxyl group at acidic pH. Coupling then occurs specifically at the ω end of the molecule and, in fact, a specific radioimmunoassay for this compound has been developed (GRANSTRÖM and KINDAHL 1976). The antisera produced recognized the tetranor side chain, but did not distinguish the ω end of the molecule. As pointed out by the authors, the assay is, however, not specific for the lactone form of 5α,7α-dihydroxy-11-ketotetranorprosta-1,16-dioic acid. It can be expected that an equilibrium between the lactone form and the open form is reestablished as soon as the compound is dissolved in water or after injection of the immunogenic conjugate into an animal. Thus, antibodies against both forms are most probably produced. Since an equilibrium between both forms also occurs in the material to be analyzed, this antibody heterogeneity may be an advantage rather than a drawback of the immunoassay. OHKI et al. (1974–1976) and CORNETTE et al. (1975) took no special precautions as to the site of coupling during the preparation of their immunogens. However, OHKI et al. (1976) compared the levels of the main urinary metabolite of $PGF_{2\alpha}$ in untreated human urine and after incubation at pH 10 to hydrolyze the δ-lactone. They did not observe significant differences and concluded that the main urinary metabolite of $PGF_{2\alpha}$ exists predominantly as the 1,16-dioic acid in the urine.

GRANSTRÖM et al. (1982) and GRANSTRÖM and KINDAHL (1982) have pointed out that, after intravenous administration of $PGF_{2\alpha}$ to various species, including humans, 15-keto-13,14-dihydro-$PGF_{2\alpha}$ was the dominant circulating metabolite during the first few minutes only. Later, the metabolic pattern in the circulation resembled more and more the urinary pattern of products. From these studies it was concluded that measurement of the later metabolites may give more reliable results on PG biosynthesis, since these compounds are detectable in the circulation for longer periods. This is particularly important if blood samples cannot be obtained frequently and thus short bursts of PG release may be overlooked (GRANSTRÖM et al. 1982).

INAGAWA et al. (1983) have described a rather specific radioimmunoassay and enzyme immunoassay for the determination of the main urinary metabolite of PGE_1 and PGE_2, 7α-hydroxy-5,11-diketotetranorprosta-1,16-dioic acid, in humans. Similar to the measurement of 15-keto-13,14-dihydro-PGE_2 (BOTHWELL et al. 1982), the method involves formation of a bicyclic degradation product by alkali treatment before assay. The radioimmunoassay was shown to be highly specific and precise, and an excellent agreement of data obtained with radioimmunoassay and enzyme immunoassay was observed.

G. Comparison of Radioimmunoassay with Other Methods for Quantitative Determination of Cyclooxygenase Products of Arachidonic Acid Metabolism

Three methods have been proven useful for the determination of cyclooxygenase products of arachidonate metabolism in biologic material, bioassay, combined GC–MS, and radioimmunoassay. These methods differ with respect to sensitivity, specificity, precision, and speed. MS is the only method that permits formal identification of a compound. In addition to its high specificity, this method is generally characterized by high precision. It has to be mentioned, however, that variable results have been obtained with combined GC–MS. An example is the level of 6-keto-$PGF_{1\alpha}$ in human blood. Circulating concentrations in humans have been reported as 189 ± 9 pg/ml (HENSBY et al. 1979) and <0.50–2.49 pg/ml (BLAIR et al. 1982) by the same laboratory. The discrepancy has been explained by variable and poor recovery of 6-keto-$PGF_{1\alpha}$ and an interfering peak derived from the internal standard in one of the channels used for quantitation. However, these problems could not fully account for the large discrepancy in the results obtained (BLAIR et al. 1982). Furthermore, even interference by nonprostanoid compounds is not completely excluded by the use of MS profiling. Thus, CATTABENI et al. (1979) found that mass fragmentographic analysis of PGE_2 was severely hampered by the presence of interfering compounds in the biologic material to be analyzed. The interfering compounds were assumed to be phthalic acid esters used as plasticizers. The recent development of GC–MS profiling with negative ion chemical ionization detection now permits the determination of several cyclooxygenase-derived products of arachidonic acid metabolism at the same time and in the same sample, with detection limits comparable to those of radioimmunoassays (PACE-ASCIAK and MICALLEF 1984; ROBINSON et al. 1984; SCHWEER et al. 1985). Major disadvantages of GC–MS methods for the determination of PG and related compounds are the high costs and the long time required for analysis.

Bioassay is considered to have only limited specificity and precision. Furthermore, less samples can be analyzed per day than by radioimmunoassay. An advantage of bioassay is the detection of biologically active compounds, including those which are highly unstable. In fact, TXA_2 (PIPER and VANE 1969; HAMBERG et al. 1975) and PGI_2 (MONCADA et al. 1976) were discovered by the use of bioassay methods. Furthermore, bioassay permits continuous monitoring of PG and TXA_2 release from organs and tissues and the results are obtained immediately. In order to increase specificity, bioassay of prostanoids in various smooth muscle preparations is performed in the presence of receptor antagonists such as mepyramine, phenoxybenzamine, propranolol, and atropine which block responses to other mediators like histamine, adrenaline, and acetylcholine (MONCADA et al. 1978). The bioassay superfusion technique (GADDUM 1953; VANE 1964) has been modified (FERREIRA and DE SOUZA COSTA 1976) by a laminar superfusion technique, in which baseline stability of the assay organs is increased by bathing the tissues in mineral oil. Furthermore, this modification resulted in a significant increase in sensitivity of the bioassay into the picogram range. Various biologic systems have been used for the detection and quantification of the different PG and TXA_2. Thus, PGE_2 can be measured by its contractile effect on the rat stomach strip and $PGF_{2\alpha}$ is specifically active on the rat colon. PGI_2 is determined by relaxation of the bovine coronary artery or by its inhibitory action on platelet aggregation (MONCADA et al. 1978). The biologic activity of rabbit aorta-contracting substance has been identified as a mixture of mainly TXA_2 with smaller amounts of PG endoperoxides (SAMUELSSON 1976).

Radioimmunoassay is generally considered to be highly sensitive and sufficiently specific to permit reliable determination of picogram amounts of cyclooxygenase products of arachidonate metabolism in biologic material. It should be pointed out, however, that radioimmunoassay does not formally identify a compound, but, in principle, determines inhibition of binding of a label to antibodies. Such inhibition of binding can be caused by specific and/or nonspecific factors. The strict observance of various validation criteria, as discussed in Sects. D and E, is thus necessary for the correct radioimmunologic determination of PG and TX. Advantages of radioimmunoassay are that large numbers of samples can be analyzed per day and that, in contrast to bioassay, biologically inactive compounds such as PG metabolites can be measured. Furthermore, owing to the high sensitivity of the method, small amounts of tissues or biologic fluids are usually sufficient for analysis, a fact that is especially important in clinical research. In this field, sample size and frequency of collection may be limited.

H. Radioimmunoassay of Cyclooxygenase Products of Arachidonic Acid Metabolism in Basic and Clinical Pharmacology

The radioimmunoassay technique has been used for quantitation of cyclooxygenase products of arachidonic acid metabolism in an immense number of investigations in basic as well as clinical pharmacology. In the following paragraphs, several examples are given.

In basic pharmacology, isolated perfused organs such as the heart, lung, or spleen of various species have frequently been employed as models to study generation of eicosanoids and their pharmacologic modification. In general, the perfusates from isolated organs can readily be assayed for prostanoids, since interference by blood contents with the radioimmunoassay is minimized. Similar advantages are offered by incubation of tissue pieces (e.g., from the gastrointestinal tract, lung, or brain) in buffer solutions, even though this method is hampered by the fact that the mechanical stimulation of cutting or chopping the tissue induces an artifactual release of prostanoids. Nevertheless, the method seems suitable to estimate the total synthetic capacity of a tissue for various prostanoids. In many instances, such a method is to be preferred instead of attempting to measure the tissue content of prostanoids (see Sect. E.III). However, the synthetic capacity does not necessarily correlate with the actual rate of biosynthesis in vivo. For example, although surgical specimens of atherosclerotic vessels have a lower capacity to synthesize 6-keto-$PGF_{1\alpha}$ than healthy vascular tissues in vitro (D'ANGELO et al. 1978; SINZINGER et al. 1979), endogenous PGI_2 biosynthesis, monitored as urinary excretion of a PGI_2 metabolite, was found to be enhanced in patients with severe atherosclerosis (FITZGERALD et al. 1984). Tissue culture is another commonly employed experimental setup, offering the opportunity to investigate prostanoid release from cell populations and homogeneous cell lines under well-defined conditions. However, in contrast to organ perfusates or tissue incubation buffers, the radioimmunologic determination of prostanoids in cell culture supernatants is complicated by proteins or fatty acids usually present in the media. In addition, cell culture media may contain significant amounts of exogenous prostanoids derived from sera with which the media are supplemented (SMETHURST and WILLIAMS 1977). Therefore, it is absolutely necessary to establish correct blank and binding values, i.e., in the presence of blank media (see Sects. B and E.II).

In clinical pharmacology, there is a considerable need for analytic procedures for prostanoid determination. Unfortunately, in this field which actually requires the most rigid validation of radioimmunoassay, this is often insufficiently performed or even neglected. In numerous studies, plasma levels of prostanoids such as 6-keto-$PGF_{1\alpha}$ and TXB_2 have been determined. However, these data should be interpreted with great caution since physiologic concentrations of these compounds are extremely low and inappropriate sample handling rapidly stimulates synthesis of TX (FITZGERALD et al. 1983 a; see Sect. E.III). In studies on PGI_2 and TXA_2 production in the human coronary circulation, samples are usually obtained after insertion of catheters. Although it has been claimed that blood sampling through catheters is reliable with respect to TX concentrations (HIRSH et al. 1982), others have reported PGI_2 biosynthesis to be stimulated by cardiac catheterization and angiography (ROY et al. 1985).

Owing to the inherent problems of plasma determination of prostanoids, attempts have been made to establish more specific indexes of tissue-related prostanoid release. Thus, TXB_2 formation in clotting blood (PATRONO et al. 1980; PATRIGNANI et al. 1982) or in stimulated platelet-rich plasma (FITZGERALD et al. 1983 b) represent elegant ex vivo methods of studying the effect of drugs on platelet TX production. In order to investigate human gastric mucosal PG generation and metabolism, levels of primary PG and the 15-keto-13,14-dihydro metabolites in gastric juice have been determined (BENNETT et al. 1973; B. M. PESKAR et al. 1974; TONNESEN et al. 1974; PESKAR et al. 1980 b). By such methods it has been possible to demonstrate that the PG synthesizing and metabolizing enzyme systems are targets for drugs acting on the gastric mucosa such as carbenoxolone or aspirin (CHILD et al. 1976; RASK-MADSEN et al. 1983). Another example is the attempt to correlate cerebral vasospasm induced by subarachnoid hemorrhage with intracranial prostanoid production. In these investigations, radioimmunoassays have been used for the determination of PG in cerebrospinal fluid (LA TORRE et al. 1974; WALKER et al. 1983).

Under physiologic conditions, urinary 6-keto-$PGF_{1\alpha}$ and TXB_2 as well as PGE_2 and $PGF_{2\alpha}$ have been said to reflect predominantly renal biosynthesis (FRÖLICH et al. 1975; PATRONO et al. 1982, 1983; FITZGERALD et al. 1983 a; CIABATTONI et al. 1984; PATRONO and CIABATTONI 1985). On the other hand, measurements of urinary metabolites of primary prostanoids represent the only noninvasive approach to quantitation of systemic prostanoid biosynthesis. Thus, urinary levels of 2,3-dinor-6-keto-$PGF_{1\alpha}$ and 2,3 dinor-TXB_2 have been used to quantify the effects of nonsteroidal anti-inflammatory drugs on systemic PGI_2 and TXA_2 synthesis in humans (FITZGERALD et al. 1983 b). Furthermore, by parallel radioimmunologic determination of primary prostanoids as well as the metabolite 2,3 dinor-TXB_2 in urine, the functional significance of renal PGI_2 and TXA_2 production in patients with systemic lupus erythematosus has been demonstrated (PATRONO et al. 1985).

J. Concluding Remarks

Radioimmunoassay has been the most widely used method for quantitative determination of PG and related compounds in biologic material. The method is technically simple, permits the analysis of a great number of samples per day, is generally sensitive and precise, and can be adapted to the measurement of eicosanoids in various experimental and clinical materials. However, highly variable results may be obtained if validation criteria are not considered. Future developments in this rapidly growing field will include better validation of immunoassay methods, attempts to improve the reproducibility of immunoassay results, e.g., by the use of monoclonal antibodies, and the replacement of radioactive by non-radioactive tracers such as chemiluminescence and enzyme labels.

References

Alam I, Ohuchi K, Levine L (1979) Determination of cyclooxygenase products and prostaglandin metabolites using high-pressure liquid chromatography and radioimmunoassay. Anal Biochem 93:339–345

Änggård E (1971) Studies on the analysis and metabolism of the prostaglandins. Ann NY Acad Sci 180:200–215

Anhut H, Bernauer W, Peskar BA (1977) Radioimmunological determination of thromboxane release in cardiac anaphylaxis. Eur J Pharmacol 44:85–88

Anhut H, Bernauer W, Peskar BA (1978a) Pharmacological modification of thromboxane and prostaglandin release in cardiac anaphylaxis. Prostaglandins 15:889–900

Anhut H, Peskar BA, Bernauer W (1978b) Release of 15-keto-13,14-dihydro-thromboxane B_2 and prostaglandin D_2 during anaphylaxis as measured by radioimmunoassay. Naunyn-Schmiedebergs Arch Pharmacol 305:247–252

Anhut H, Peskar BA, Wachter W, Gräbling B, Peskar BM (1978c) Radioimmunological determination of prostaglandin D_2 synthesis in human thrombocytes. Experientia 34:1494–1496

Armstrong JM, Lattimer N, Moncada S, Vane JR (1978) Comparison of the vasodepressor effects of prostacyclin and 6-oxo-prostaglandin $F_{1\alpha}$ with those of prostaglandin E_2 in rats and rabbits. Br J Pharmacol 62:125–130

Axen U (1974) N,N′-carbonyldiimidazole as coupling reagent for the preparation of bovine serum albumin conjugates. Prostaglandins 5:45–47

Banschbach MW, Love PK (1979) Use of mini-columns packed with acid-washed florisil for the rapid separation of A, E, and F series prostaglandins. Prostaglandins 17:193–200

Barrow SE, Heavey DJ, Ennis M, Chappel CG, Blair IA, Dollery CT (1984) Measurement of prostaglandin D_2 and identification of metabolites in human plasma during intravenous infusion. Prostaglandins 28:743–754

Bauminger S, Zor U, Lindner HR (1973) Radioimmunological assay of prostaglandin synthetase activity. Prostaglandins 4:313–323

Bennett A, Stamford IF, Unger WG (1973) Prostaglandin E_2 and gastric acid secretion in man. J Physiol (Lond) 229:349–360

Blair IA, Barrow SE, Waddell KA, Lewis PJ, Dollery CT (1982) Prostacyclin is not a circulating hormone in man. Prostaglandins 23:579–589

Boot JR, Cockerill AF, Dawson W, Mallen DNB, Osborne DJ (1978) Modification of prostaglandin and thromboxane release by immunological sensitisation and successive immunological challenges from guinea-pig lung. Int Arch Allergy Appl Immunol 57:159–164

Bothwell W, Verburg M, Wynalda M, Daniels EG, Fitzpatrick FA (1982) A radioimmunoassay for the unstable pulmonary metabolites of prostaglandin E_1 and E_2: an indirect index of their in vivo disposition and pharmacokinetics. J Pharmacol Exp Ther 220:229–235

Brash AR, Jackson EK, Saggese CA, Lawson JA, Oates JA, FitzGerald GA (1983a) Metabolic disposition of prostacyclin in humans. J Pharmacol Exp Ther 226:78–87

Brash AR, Jackson EK, Lawson JA, Branch RA, Oates JA, FitzGerald GA (1983b) Quantitative aspects of prostacyclin metabolism in humans. Adv Prostaglandin Thromboxane Leukotriene Res 11:119–122

Brune K, Reinke M, Lanz R, Peskar BA (1985) Monoclonal antibodies against E- and F-type prostaglandins. High specificity and sensitivity in conventional radioimmunoassays. FEBS Lett 186:46–50

Bult H, Beetens J, Vercruysse P, Herman AG (1978) Blood levels of 6-keto-PGF$_{1\alpha}$, the stable metabolite of prostacyclin during endotoxin-induced hypotension. Arch Int Pharmacodyn 236:285–286
Caldwell BV, Burstein S, Brock WA, Speroff L (1971) Radioimmunoassay of the F prostaglandins. J Clin Endocrinol 33:171–175
Cattabeni F, Borghi C, Folco GC, Nicosia S, Spagnuolo C (1979) RIA of PGF$_{2\alpha}$ and PGE$_2$ in biological samples of different origin: comparison with the mass fragmentographic technique. In: Albertini A, DaPrada M, Peskar BA (eds) Radioimmunoassay of drugs and hormones in cardiovascular medicine. Elsevier, Amsterdam, pp 281–289
Cerletti C, Chiabrando C, Fanelli R, de Gaetano G (1984) Is immunoreactive 6-keto-PGF$_{1\alpha}$ measured in serum after treatment with thromboxane-synthetase inhibitors an index of prostacyclin synthesis? Lancet I:967–968
Chang DGB, Tai H-H (1982) A radioimmunoassay for 6-ketoprostaglandin E$_1$. Prostaglandins Leukotrienes Med 8:11–19
Child C, Jubiz W, Moore JG (1976) Effects of aspirin on gastric prostaglandin E (PGE) and acid output in normal subjects. Gut 17:54–57
Christensen P, Leyssac PP (1976) A specific radioimmunoassay for PGE$_2$ using an antibody with high specificity and a Sephadex LH-20 microcolumn for the separation of prostaglandins. Prostaglandins 11:399–420
Ciabattoni G, Pugliese F, Cinotti GA, Patrono C (1979) Methodologic problems in the radioimmunoassay of prostaglandin E$_2$ and F$_{2\alpha}$ in human urine. In: Albertini A, DaPrada M, Peskar BA (eds) Radioimmunoassay of drugs and hormones in cardiovascular medicine. Elsevier, Amsterdam, pp 265–280
Ciabattoni G, Cinotti GA, Pierucci A, Simonetti BM, Manzi M, Pugliese F, Barsotti P, Pecci G, Taggi F, Patrono C (1984) Effects of sulindac and ibuprofen in patients with chronic glomerular disease. Evidence for the dependence of renal function on prostacyclin. N Engl J Med 310:279–283
Cornette J, Harrison K, Kirton K (1974) Measurement of prostaglandin F$_{2\alpha}$ metabolites by radioimmunoassay. Prostaglandins 5:155–164
Cornette J, Kirton K, Schneider W, Sun F, Johnson R, Nidy E (1975) Preparation and quantitation of urinary metabolites of PGF$_{2\alpha}$ by radioimmunoassay. Prostaglandins 9:323–338
Dahlén S-E, Hansson G, Hedqvist P, Björck T, Granström E, Dahlén B (1983) Allergen challenge of lung tissue from asthmatics elicits bronchial contraction that correlates with the release of leukotrienes C$_4$, D$_4$ and E$_4$. Proc Natl Acad Sci USA 80:1712–1716
D'Angelo V, Villa S, Mysliwiec M, Donati MB, de Gaetano G (1978) Defective fibrinolytic and prostacyclin-like activity in human atheromatous plaques. Thromb Haemost 39:535–536
David F, Sommé G, Provost-Wisner A, Roth C, Astoin M, Dray F, Thèze J (1985) Characterization of monoclonal antibodies against prostaglandin E$_2$: fine specificity and neutralization of biological effects. Mol Immunol 22:339–346
Dawson W, Boot JR, Cockerill AF, Mallen DNB, Osborne DJ (1976) Release of novel prostaglandins and thromboxanes after immunological challenge of guinea pig lung. Nature 262:699–702
Deheny TP, Murdoch WS, Boyle L, Walters WAW, Boura ALA (1981) Gel filtration separation of bound and free antigens in radioimmunoassay of prostaglandin F$_{2\alpha}$. Prostaglandins 21:1003–1006
Demers LM, Derck DD (1980) A radioimmunoassay for 6-keto-prostaglandin F$_{1\alpha}$. Adv Prostaglandin Thromboxane Res 6:193–199
Demers LM, Brennecke SP, Mountford LA, Brunt JD, Turnbull AC (1983) Development and validation of a radioimmunoassay for prostaglandin E$_2$ metabolite levels in plasma. J Clin Endocrinol Metab 57:101–106
Desbuquois B, Aurbach GD (1971) Use of polyethylene glycol to separate free and antibody-bound peptide hormones in radioimmunoassays. J Clin Endocrinol 33:732–738
Diekmann JM, Jobke A, Peskar BA, Hertting G (1977) Angiotensin II-induced contractions of rabbit splenic capsular strips and release of prostaglandins. Use of radioimmu-

noassays for prostaglandins E_1 and E_2. Naunyn Schmiedebergs Arch Pharmacol 297:177–183
Dighe KK, Emslie HA, Henderson LK, Simon L, Rutherford F (1975) The development of antisera to prostaglandins B_2 and $F_{2\alpha}$ and their analysis using solid-phase and double antibody radioimmunoassay methods. Br J Pharmacol 55:503–514
Dighe KK, Jones RL, Poyser NL (1978) Development of a radioimmunoassay for measuring 6-oxo-prostaglandin $F_{1\alpha}$. Br J Pharmacol 63:406P
Dray F, Maron E, Tillson SA, Sela M (1972) Immunochemical detection of prostaglandins with prostaglandin-coated bacteriophage T_4 and by radioimmunoassay. Anal Biochem 50:399–408
Dray F, Charbonnel B, Maclouf J (1975) Radioimmunoassay of prostaglandins F_α, E_1, and E_2 in human plasma. Eur J Clin Invest 5:311–318
Dray F, Pradelles P, Maclouf J, Sors H, Bringuier A (1978) Radioimmunoassay of 6-keto-$PGF_{1\alpha}$ using an iodinated tracer. Prostaglandins 15:715
Edlund A, Bomfim W, Kaijser L, Olin C, Patrono C, Pinca E, Wennmalm Å (1982) Cardiac formation of prostacyclin during cardioplegia in man. Prostaglandins 24:5–19
Erlanger BF (1973) Principles and methods for the preparation of drug protein conjugates for immunological studies. Pharmacol Rev 25:271–280
Fairclough RJ, Payne E (1975) Radioimmunoassay of 13,14-dihydro-15-keto-prostaglandin F in bovine peripheral plasma. Prostaglandins 10:266–272
Ferreira SH, de Souza Costa F (1976) A laminar flow superfusion technique with much increased sensitivity for the detection of smooth muscle-stimulating substances. Eur J Pharmacol 39:379–381
Ferreira SH, Vane JR (1967) Prostaglandins: their disappearance from and release into the circulation. Nature 216:868–873
FitzGerald GA, Pedersen AK, Patrono C (1983a) Analysis of prostacyclin and thromboxane biosynthesis in cardiovascular disease. Circulation 67:1174–1177
FitzGerald GA, Oates JA, Hawiger J, Maas RL, Roberts LJ II, Lawson JA, Brash AR (1983b) Endogenous biosynthesis of prostacyclin and thromboxane and platelet function during chronic aspirin administration in man. J Clin Invest 71:676–688
FitzGerald GA, Smith B, Pedersen AK, Brash AR (1984) Increased prostacyclin biosynthesis in patients with severe atheroslecrosis and platelet activation. N Engl J Med 310:1065–1068
Fitzpatrick FA, Bundy GL (1978) Hapten mimic elicits antibodies recognizing prostaglandin E_2. Proc Natl Acad Sci USA 75:2689–2693
Fitzpatrick FA, Gorman RR (1977) Platelet rich plasma transforms exogenous prostaglandin endoperoxide H_2 into thromboxane A_2. Prostaglandins 14:881–889
Fitzpatrick FA, Gorman RR (1978) An antiserum against 9-deoxy-6,9-epoxy-$PGF_{1\alpha}$ recognizes and binds PGI_2 (prostacyclin). Prostaglandins 15:725–735
Fitzpatrick FA, Wynalda MA (1976) A rapid, solid phase radioimmunoassay for prostaglandin $F_{2\alpha}$ and its main circulating metabolite. Anal Biochem 73:198–208
Fitzpatrick FA, Wynalda MA (1983) Albumin-catalyzed metabolism of prostaglandin D_2. J Biol Chem 258:11713–11718
Fitzpatrick FA, Gorman RR, McGuire JC, Kelly RC, Wynalda MA, Sun FF (1977) A radioimmunoassay for thromboxane B_2. Anal Biochem 82:1–7
Fitzpatrick FA, Gorman RR, Bundy GL (1978) An antiserum against 9,11-azo-15-hydroxy-prosta-5,13-dienoic acid recognises and binds prostaglandin endoperoxides. Nature 273:302–304
Fitzpatrick FA, Aguirre R, Pike JE, Lincoln FH (1980) The stability of 13,14-dihydro-15-keto-PGE_2. Prostaglandins 19:917–931
Fretland DJ (1974) Use of ion-exchange resins for removing prostaglandins from human urine prior to radioimmunoassay. Prostaglandins 6:421–425
Frölich JC, Wilson TW, Sweetman BJ, Smigel M, Nies AS, Carr K, Watson JT, Oates JA (1975) Urinary prostaglandins. Identification and origin. J Clin Invest 55:763–770
Gaddum JH (1953) The technique of superfusion. Br J Pharmacol 8:321–326
Gershman H, Powers E, Levine L, van Vunakis H (1972) Radioimmunoassay of prostaglandins, angiotensin, digoxin, morphine and adenosine-3′,5′-cyclic-monophosphate with nitrocellulose membranes. Prostaglandins 1:407–423

Goodfriend TL, Levine L, Fasman GD (1964) Antibodies to bradykinin and angiotensin: a use of carbodiimides in immunology. Science 144:1344–1346
Granström E (1978) Radioimmunoassay of prostaglandins. Prostaglandins 15:3–17
Granström E (1979) Sources of error in prostaglandin and thromboxane radioimmunoassay. In: Albertini A, DaPrada M, Peskar BA (eds) Radioimmunoassay of drugs and hormones in cardiovascular medicine. Elsevier, Amsterdam, pp 229–238
Granström E, Kindahl H (1976) Radioimmunoassay for urinary metabolites of prostaglandin $F_{2\alpha}$. Prostaglandins 12:759–783
Granström E, Kindahl H (1978) Radioimmunoassay of prostaglandins and thromboxanes. Adv Prostaglandin Thromboxane Res 5:119–210
Granström E, Kindahl H (1979) Radioimmunologic determination of 15-keto-13,14-dihydro-PGE_2: a method for its stable degradation product, 11-deoxy-13,14-dihydro-15-keto-11β-16ε-cycloprostaglandin E_2. Abstract Fourth International Prostaglandin Conference, Washington D.C., 1979, p 42
Granström E, Kindahl H (1982) Species differences in circulating prostaglandin metabolites. Relevance for the assay of prostaglandin release. Biochim Biophys Acta 713:555–569
Granström E, Samuelsson B (1972) Development and mass spectrometric evaluation of a radioimmunoassay for 9α,11α-dihydroxy-15-ketoprost-5-enoic acid. FEBS Lett 26:211–214
Granström E, Kindahl H, Samuelsson B (1976a) A method for measuring the unstable thromboxane A_2: radioimmunoassay of the derived mono-*O*-methyl-thromboxane B_2. Prostaglandins 12:929–941
Granström E, Kindahl H, Samuelsson B (1976b) Radioimmunoassay for thromboxane B_2. Anal Lett 9:611–627
Granström E, Hamberg M, Hansson G, Kindahl H (1980) Chemical instability of 15-keto-13,14-dihydro-PGE_2: the reason for low assay reliability. Prostaglandins 19:933–957
Granström E, Kindahl H, Swahn M-L (1982) Profiles of prostaglandin metabolites in the human circulation. Identification of late-appearing, long-lived products. Biochim Biophys Acta 713:46–60
Granström E, Westlund P, Kumlin M, Nordenström A (1985) Monitoring of thromboxane production in vivo: metabolic and analytical aspects. Adv Prostaglandin Thromboxane Leukotriene Res 15:67–70
Green K, Hamberg M, Samuelsson B (1976) Quantitative analysis of prostaglandins and thromboxanes by mass spectrometric methods. Adv Prostaglandin Thromboxane Res 1:47–58
Greenwood FC, Hunter WM, Glover JS (1963) The preparation of ^{131}I-labelled human growth hormone of high specific radioactivity. Biochem J 89:114–123
Hales CM, Randle PJ (1963) Immunoassay of insulin with insulin-antibody precipitate. Biochem J 88:137–146
Hamberg M, Svensson J, Samuelsson B (1975) Thromboxanes: a new group of biologically active compounds derived from prostaglandin endoperoxides. Proc Natl Acad Sci USA 72:2994–2998
Hamberg M, Svensson J, Samuelsson B (1976) Novel transformations of prostaglandin endoperoxides: formation of thromboxanes. Adv Prostaglandin Thromboxane Res 1:19–27
Hammarström S (1982) Biosynthesis and biological actions of prostaglandins and thromboxanes. Arch Biochem Biophys 214:431–445
Haning RV Jr, Kieliszek FX, Alberino SP, Speroff L (1977) A radioimmunoassay for 13,14-dihydro-15-keto prostaglandin $F_{2\alpha}$ with chromatography and internal recovery standard. Prostaglandins 13:455–477
Hayashi Y, Yano T, Yamamoto S (1981) Enzyme immunoassay of prostaglandin $F_{2\alpha}$. Biochim Biophys Acta 663:661–668
Hayashi Y, Ueda N, Yokota K, Kawamura S, Ogushi F, Yamamoto Y, Yamamoto S, Nakamura K, Yamashita K, Miyazaki H, Kato K, Terao S (1983) Enzyme immunoassay of thromboxane B_2. Biochim Biophys Acta 750:322–329

Hensby CN, FitzGerald GA, Friedman LA, Lewis PJ, Dollery CT (1979) Measurement of 6-oxo-PGF$_{1\alpha}$ in human plasma using gas chromatography-mass spectrometry. Prostaglandins 18:731–736
Herbert V, Lau K-S, Gottlieb CW, Bleicher SJ (1965) Coated charcoal immunoassay of insulin. J Clin Endocrinol 25:1375–1384
Hirsh PD, Firth BG, Campbell WB, Willerson JT, Hillis LD (1982) Influence of blood sampling site and technique on thromboxane concentrations in patients with ischemic heart disease. Am Heart J 104:234–237
Horton E, Jones R, Thompson C, Poyser N (1971) Release of prostaglandins. Ann NY Acad Sci 180:351–362
Inagawa T, Imaki K, Masuda H, Morikawa Y, Hirata F, Tsuboshima M (1983) Simplified immunoassays of prostaglandin E main metabolite in human urine. Adv Prostaglandin Thromboxane Leukotriene Res 11:191–196
Jaffe BM, Smith JW, Newton WT, Parker CW (1971) Radioimmunoassay for prostaglandins. Science 171:494–496
Jaffe BM, Behrman HR, Parker CW (1973) Radioimmunoassay measurement of prostaglandins E, A and F in human plasma. J Clin Invest 52:398–405
Jobke A, Peskar BA, Peskar BM (1973) On the specificity of antisera against prostaglandin A_2 and E_2. FEBS Lett 37:192–196
Johnson RA, Morton DR, Kinner JH, Gorman RR, McGuire JC, Sun FF, Whittaker N, Bunting S, Salmon J, Moncada S, Vane JR (1976) The chemical structure of prostaglandin X (prostacyclin). Prostaglandins 12:915–928
Jubiz W, Frailey J (1974) Prostaglandin E generation during storage of plasma samples. Prostaglandins 7:339–344
Karim SMM (ed) (1976) Prostaglandins: physiological, pharmacological and pathological aspects. MTP Press, Lancaster
Kikawa Y, Narumiya S, Fukushima M, Wakatsuka H, Hayaishi O (1984) 9-Deoxy-Δ^9, Δ^{12}-13,14-dihydroprostaglandin D_2, a metabolite of prostaglandin D_2 formed in human plasma. Proc Natl Acad Sci USA 81:1317–1321
Kindahl H (1977) Metabolism of thromboxane B_2 in the cynomolgus monkey. Prostaglandins 13:619–629
Kirton KT, Cornette JC, Barr KL (1972) Characterization of antibody to prostaglandin $F_{2\alpha}$. Biochem Biophys Res Commun 47:903–909
Koh H, Inoue A, Numano F, Maezawa H (1978) A radioimmunoassay for thromboxane B_2 with thromboxane B_2-^{125}I-tyramide. Proc Jpn Acad (Ser B) 54:553–558
Kohen F, Pazzagli M, Serio M, De Boevers J, Vandekerckhove D (1985) Chemiluminescence and bioluminescence immunoassay. In: Collins WP (ed) Alternative immunoassays. Wiley, Chichester, pp 103–121
Kuzuya T, Matsuda H, Hoshida S, Yamagishi M, Tada M (1985) Determination of plasma thromboxane B_2 by radioimmunoassay: methodological problems and accuracy. Adv Prostaglandin Thromboxane Leukotriene Res 15:99–101
Lands WEM, Hammarström S, Parker CW (1976) Limitations of prostaglandin assays. J Invest Dermatol 67:658–660
Landsteiner K (1945) The specificity of serological reactions. Harvard University Press, Cambridge, USA
Lange K, Peskar BA (1984) Radioimmunological determination of prostaglandin D_2 after conversion to a stable degradation product. Agents Actions 15:87–89
Lange K, Simmet T, Peskar BM, Peskar BA (1985) Determination of 15-keto-13,14-dihydro-prostaglandin E_2 and prostaglandin D_2 in human colonic tissue using a chemiluminescence enzyme immunoassay with catalase as labelling enzyme. Adv Prostaglandin Thromboxane Leukotriene Res 15:35–38
La Torre E, Patrono C, Fortuna A, Grossi-Belloni D (1974) Role of prostaglandin F_2 in human cerebral vasospasm. J Neurosurg 41:293–299
Levine L (1977) Levels of 13,14-dihydro-15-keto-PGE$_2$ in some biological fluids as measured by radioimmunoassay. Prostaglandins 14:1125–1139
Levine L, Alam I (1979) Arachidonic acid metabolism by cells in culture: analyses of culture fluids for cyclooxygenase products by radioimmunoassay before and after separation by high pressure liquid chromatography. Prostaglandins Med 3:295–304

Levine L, Gutierrez-Cernosek RM (1973) Levels of 13,14-dihydro-15-keto-PGF$_{2\alpha}$ in biological fluids as measured by radioimmunoassay. Prostaglandins 3:785–804
Levine L, van Vunakis H (1970) Antigenic activity of prostaglandins. Biochem Biophys Res Commun 41:1171–1177
Levine L, Gutierrez-Cernosek RM, van Vunakis H (1971) Specificities of prostaglandins B$_1$, F$_{1\alpha}$, and F$_{2\alpha}$ antigen-antibody reactions. J Biol Chem 246:6782–6785
Levine L, Wu K-Y, Pong S-S (1975) Stereospecificity of enzymatic reduction of prostaglandin E$_2$ to F$_{2\alpha}$. Prostaglandins 9:531–544
Levine L, Alam I, Langone JJ (1979) The use of immobilized ligands and (^{125}I) protein A for immunoassays of thromboxane B$_2$, prostaglandin D$_2$, 13,14-dihydro-prostaglandin E$_2$, 5,6-dihydroprostaglandin I$_2$, 6-keto-prostaglandin F$_{1\alpha}$, 15-hydroxy-9α,11α-(epoxymethano)prosta-5,13-dienoic acid and 15-hydroxy-11α,9α-(epoxymethano)prosta-5,13-dienoic acid. Prostaglandins Med 2:177–189
Lewis RA, Soter NA, Diamond PT, Austen KF, Oates JA, Roberts JL II (1982) Prostaglandin D$_2$ generation after activation of rat and human mast cells with Anti-IgE$_1$. J Immunol 129:1627–1631
Leyendecker G, Wardlaw S, Nocke W (1972) Gamma globulin protection of radioimmunoassay and competitive protein binding saturation analysis of steroids. J Clin Endocrinol 34:430–433
Liebig R, Bernauer W, Peskar BA (1974) Release of prostaglandins, a prostaglandin metabolite, slow-reacting substance and histamine from anaphylactic lungs, and its modification by catecholamines. Naunyn Schmiedebergs Arch Pharmacol 284:279–293
Lindgren JÅ, Kindahl H, Hammarström S (1974) Radioimmunoassay of prostaglandins E$_2$ and F$_{2\alpha}$ in cell culture media utilizing antisera against prostaglandins F$_{2\beta}$ and F$_{2\alpha}$. FEBS Lett 48:22–25
Machleidt C, Förstermann U, Anhut H, Hertting G (1981) Formation and elimination of prostacyclin metabolites in the cat in vivo as determined by radioimmunoassay of unextracted plasma. Eur J Pharmacol 74:19–26
Maclouf J, Andrieu JM, Dray F (1975) Validity of PGE$_1$ radioimmunoassay by using PGE$_1$ antisera with differential binding parameters. FEBS Lett 56:273–278
Maclouf J, Pradel M, Pradelles P, Dray F (1976) ^{125}I derivatives of prostaglandins. A novel approach in prostaglandin analysis by radioimmunoassay. Biochim Biophys Acta 431:139–146
Maclouf J, Corvazier E, Wang Z (1986) Development of a radioimmunoassay for prostaglandin D$_2$ using an antiserum against 11-methoxime prostaglandin D$_2$. Prostaglandins 31:123–132
McCann DS, Tokarsky J, Sorkin RP (1981) Radioimmunoassay for plasma thromboxane B$_2$. Clin Chem 27:1417–1420
Meldrum DR, Abraham GE (1976) Separation of A and E prostaglandins by celite partition chromatography for radioimmunoassay. Clin Biochem 9:42–45
Metz SA, Rice MG, Robertson RP (1979) Applications and limitations of measurement of 15-keto-13,14-dihydro prostaglandin E$_2$ in human blood by radioimmunoassay. Prostaglandins 17:839–861
Mitchell MD (1978) A sensitive radioimmunoassay for 6-keto-prostaglandin F$_{1\alpha}$: preliminary observations on circulating concentrations. Prostaglandins Med 1:13–21
Mitchell MD, Flint APF, Turnbull AC (1976) Plasma concentrations of 13,14-dihydro-15-keto-prostaglandin F during pregnancy in sheep. Prostaglandins 11:319–329
Mitchell MD, Sors H, Flint APF (1977) Instability of 13,14-dihydro-15-keto-prostaglandin E$_2$. Lancet I:558
Mitchell MD, Brunt JD, Webb R (1981) Instability of 6-keto-prostaglandin F$_{1\alpha}$ when subjected to normal extraction procedures. Prostaglandins Med 6:437–440
Mitchell MD, Ebenhack K, Kraemer DL, Cox K, Cutrer S, Strickland DM (1982) A sensitive radioimmunoassay for 11-deoxy-13,14-dihydro-15-keto-11,16-cyclo-prostaglandin E$_2$: application as an index of prostaglandin E$_2$ biosynthesis during human pregnancy and parturition. Prostaglandins Leukotrienes Med 9:549–557
Moncada S, Gryglewski R, Bunting S, Vane JR (1976) An enzyme isolated from arteries transforms prostaglandin endoperoxides to an unstable substance that inhibits platelet aggregation. Nature 263:663–665

Moncada S, Ferreira SH, Vane JR (1978) Bioassay of prostaglandins and biologically active substances derived from arachidonic acid. Adv Prostaglandin Thromboxane Res 5:211–236

Moonen P, Klok G, Keirse MJNC (1985) An easy method for preparing radioactive methyl esters of eicosanoids suitable as ligands in radioimmunoassays. Prostaglandins 29:443–448

Morgan CR, Lazarow A (1962) Immunoassay of insulin using a two-antibody system. Proc Soc Exp Biol Med 110:29–32

Morris HG, Sherman NA, Shepperdson FT (1981) Variables associated with radioimmunoassay of prostaglandins in plasma. Prostaglandins 21:771–788

Myatt L, Jogee M, Lewis PJ, Elder MG (1981) Metabolism of prostacyclin and 6-oxo-$PGF_{1\alpha}$ in man. In: Lewis PJ, O'Grady J (eds) Clinical pharmacology of prostacyclin. Raven, New York, pp 25–35

Narumiya S, Ogorochi T, Nakao K, Hayaishi O (1982) Prostaglandin D_2 in rat brain, spinal cord and pituitary: basal level and regional distribution. Life Sci 31:2093–2103

Oates JA, Falardeau P, FitzGerald GA, Branch RA, Brash AR (1981) Quantification of urinary prostacyclin metabolites in man: estimates of the rate of secretion of prostacyclin into the general circulation. In: Lewis PJ, O'Grady J (eds) Clinical pharmacology of prostacyclin. Raven, New York, pp 21–23

Ohki S, Hanyu T, Imaki K, Nakazawa N, Hirata F (1974) Radioimmunoassays of prostaglandin $F_{2\alpha}$ and prostaglandin $F_{2\alpha}$-main urinary metabolite with prostaglandin-^{125}I-tyrosine methylester amide. Prostaglandins 6:137–148

Ohki S, Imaki K, Hirata F, Hanyu T, Nakazawa N (1975) Radioimmunoassay of main urinary metabolite of prostaglandin $F_{2\alpha}$. Prostaglandins 10:549–555

Ohki S, Nishigaki Y, Imaki K, Kurono M, Hirata F, Hanyu T, Nakazawa N (1976) The levels of main urinary metabolite of prostaglandin $F_{1\alpha}$ and $F_{2\alpha}$ in human subjects measured by radioimmunoassay. Prostaglandins 12:181–186

Oliw E (1980) A radioimmunoassay for 6-keto-prostaglandin $F_{1\alpha}$ utilizing an antiserum against 6-methoxime-prostaglandin $F_{1\alpha}$. Prostaglandins 19:271–284

Pace-Asciak CR, Micallef S (1984) Gas chromatographic-mass spectrometric profiling with negative-ion chemical ionization detection of prostaglandins and their 15-keto- and 15-keto-13,14-dihydro catabolites in rat blood. J Chromatogr 310:233–242

Pace-Asciak CR, Carrara MC, Levine L (1980) Antibodies to 5,6-dihydroprostaglandin I_2 trap endogenously produced prostaglandin I_2 in the rat circulation. Biochim Biophys Acta 620:186–192

Palmieri GMA, Yalow RS, Berson SA (1971) Adsorbent techniques for the separation of antibody-bound from free peptide hormones in radioimmunoassay. Horm Metab Res 3:301–305

Patrignani P, Filabozzi P, Patrono C (1982) Selective cumulative inhibition of platelet thromboxane production by low-dose aspirin in healthy subjects. J Clin Invest 69:1366–1372

Patrono C (1973) Radioimmunoassay of prostaglandin $F_{2\alpha}$ in unextracted human plasma. J Nucl Biol Med 17:25–29

Patrono C, Ciabattoni G (1985) Measurement of arachidonate metabolites in urine by radioimmunoassay. Adv Prostaglandin Thromboxane Leukotriene Res 15:71–73

Patrono C, Ciabattoni G, Pinca E, Pugliese F, Castrucci G, De Salvo A, Satta MA, Peskar BA (1980) Low dose aspirin and inhibition of thromboxane B_2 production in healthy subjects. Thromb Res 17:317–327

Patrono C, Ciabattoni G, Peskar BM, Pugliese F, Peskar BA (1981) Is plasma 6-keto-prostaglandin $F_{1\alpha}$ a reliable index of circulating prostacyclin? Clin Res 29:276A

Patrono C, Pugliese F, Ciabattoni G, Patrignani P, Maseri A, Chierchia S, Peskar BA, Cinotti GA, Simonetti BM, Pierucci A (1982) Evidence for a direct stimulatory effect of prostacyclin on renin release in man. J Clin Invest 69:231–239

Patrono C, Ciabattoni G, Patrignani P, Filabozzi P, Pinca E, Satta MA, Van Dorne D, Cinotti GA, Pugliese F, Pierucci A, Simonetti BM (1983) Evidence for a renal origin of urinary thromboxane B_2 in health and disease. Adv Prostaglandin Thromboxane Leukotriene Res 11:493–498

Patrono C, Ciabattoni G, Remuzzi G, Gotti E, Bombardieri S, Di Munno O, Tartarelli G, Cinotti GA, Simonetti BM, Pierucci A (1985) The functional significance of renal prostacyclin and thromboxane A_2 production in patients with systemic lupus erythematosus. J Clin Invest 76:1011–1018

Pedersen AK, Watson ML, FitzGerald GA (1983) Inhibition of thromboxane biosynthesis in serum: limitations of the measurement of immunoreactive 6-keto-$PGF_{1\alpha}$. Thromb Res 33:99–103

Peskar BA, Holland A (1979) Plasma levels of immunoreactive 15-keto-13,14-dihydro-thromboxane B_2 in guinea pigs during anaphylaxis and after histamine injection. Agents and Actions, Suppl. 6. Prostaglandins Inflammation, pp 51–56

Peskar BM, Peskar BA (1976) On the metabolism of prostaglandins by human gastric fundus mucosa. Biochim Biophys Acta 424:430–438

Peskar BA, Holland A, Peskar BM (1974) Antisera against 13,14-dihydro-15-keto-prostaglandin E_2. FEBS Lett 43:45–48

Peskar BM, Holland A, Peskar BA (1974) Quantitative determination of prostaglandins in human gastric juice by radioimmunoassay. Clin Chim Acta 55:21–27

Peskar BA, Steffens CH, Peskar BM (1979) Radioimmunoassay of 6-keto-prostaglandin $F_{1\alpha}$ in biological material. In: Albertini A, DaPrada M, Peskar BA (eds) Radioimmunoassay of drugs and hormones in cardiovascular medicine. Elsevier, Amsterdam, pp 239–250

Peskar BM, Günter B, Steffens CH, Kröner EE, Peskar BA (1980a) Antibodies against dehydration products of 15-keto-13,14-dihydro-prostaglandin E_2. FEBS Lett 115:123–126

Peskar BM, Seyberth HW, Peskar BA (1980b) Synthesis and metabolism of endogenous prostaglandins by human gastric mucosa. Adv Prostaglandin Thromboxane Res 8:1511–1514

Peskar BM, Weiler H, Schmidberger P, Peskar BA (1980c) On the occurrence of prostacyclin metabolites in plasma and vascular tissue as determined radioimmunologically. FEBS Lett 121:25–28

Piper PJ, Vane JR (1969) Release of additional factors in anaphylaxis and its antagonism by anti-inflammatory drugs. Nature 223:29–35

Polet H, Levine L (1975) Metabolism of prostaglandins E, A and C in serum. J Biol Chem 250:351–357

Pradelles P, Grassi J, Maclouf J (1985) Enzyme immunoassay of eicosanoids using acetylcholine esterase as label: an alternative to radioimmunoassay. Anal Chem 57:1170–1173

Raffel G, Clarenbach P, Peskar BA, Hertting G (1976) Synthesis and release of prostaglandins by rat brain synaptosomal fractions. J Neurochem 26:493–498

Rask-Madsen J, Bukhave K, Madsen PER, Bekker C (1983) Effect of carbenoxolone on gastric prostaglandin E_2 levels in patients with peptic ulcer disease following vagal and pentagastrin stimulation. Eur J Clin Invest 13:351–356

Raz A, Schwartzman M, Kenig-Wakshal R, Perl E (1975) The specificity of antisera to conjugates of prostaglandins E with bovine serum albumin and thyroglobulin. Eur J Biochem 53:145–150

Ritter JM, Blair IA, Barrow SE, Dollery CT (1983) Release of prostacyclin in vivo and its role in man. Lancet I:317–319

Ritzi EM, Stylos WA (1974) The simultaneous use of two prostaglandin E radioimmunoassays employing two antisera of differing specificity. I. Determination of prostaglandin content in sera and culture medium of simian virus 40 transformed cells. Prostaglandins 8:55–66

Roberts LJ II, Sweetman BJ, Oates JA (1978) Metabolism of thromboxane B_2 in the monkey. J Biol Chem 253:5305–5318

Roberts LJ II, Sweetman BJ, Lewis RA, Austen KF, Oates JA (1980) Increased production of prostaglandin D_2 in patients with systemic mastocytosis. N Engl J Med 303:1400–1404

Roberts LJ II, Sweetman BJ, Oates JA (1981) Metabolism of thromboxane B_2 in man. J Biol Chem 256:8384–8393

Robinson C, Hoult JRS, Waddell KA, Blair IA, Dollery CT (1984) Total profiling by GC/NICIMS of the major cyclo-oxygenase products from antigen and leukotriene-challenged guinea-pig lung. Biochem Pharmacol 33:395–400

Rosenkranz B, Fischer C, Weimer KE, Frölich JC (1980) Metabolism of prostacyclin and 6-keto-prostaglandin $F_{1\alpha}$ in man. J Biol Chem 255:10194–10198

Roy L, Knapp HR, Robertson RM, FitzGerald GA (1985) Endogenous biosynthesis of prostacyclin during cardiac catheterization and angiography in man. Circulation 71:434–440

Salmon JA (1978) A radioimmunoassay for 6-keto-prostaglandin $F_{1\alpha}$. Prostaglandins 15:383–397

Samuelsson B (1976) Introduction: new trends in prostaglandin research. Adv Prostaglandin Thromboxane Res 1:1–6

Samuelsson B, Axen U, Behrman H, Granström E, Gréen K, Jaffe BM, Kirton K, Levine L, Skarnes RC, Speroff L, Wolfe LS (1973) Round-table discussion on analytical methods. In: Bergström S, Bernhard S (eds) Adv Biosci 9:121–123

Samuelsson B, Granström E, Gréen K, Hamberg M, Hammarström S (1975) Prostaglandins. Annu Rev Biochem 44:669–695

Sautebin L, Kindahl H, Kumlin M, Granström E (1985) Use of tritium labeled amino acid conjugates of prostaglandins and thromboxanes as labeled ligands in prostanoid radioimmunoassay. Prostaglandins 30:435–456

Sawada M, Inagawa T, Frölich JC (1985) Enzyme-immunoassay of thromboxane B_2 at the picogram level. Prostaglandins 29:1039–1048

Scherer B, Schnermann J, Sofroniev M, Weber PC (1978) Prostaglandin (PG) analysis in urine of humans and rats by different radioimmunoassays: effect on PG-excretion by PG-synthetase inhibitors, laparotomy and furosemide. Prostaglandins 15:255–266

Schweer H, Kammer J, Seyberth HW (1985) Simultaneous determination of prostanoids in plasma by gas chromatography-negative-ion chemical ionization mass spectrometry. J Chromatogr 338:273–280

Sinzinger H, Feigl W, Silberbauer K (1979) Prostacyclin generation in atherosclerotic arteries. Lancet II:469

Sinzinger H, Reiter S, Peskar BA (1985) Removal, preparation, and storage of human plasma for radioimmunological detection of prostaglandins. In: Schrör K (ed) Prostaglandins and other eicosanoids in the cardiovascular system. Proc 2nd Int Symposium Nürnberg-Fürth 1984, Karger, Basel, pp 62–67

Smethurst M, Williams DC (1977) Levels of prostaglandin E and prostaglandin F in samples of commercial serum used for tissue culture. Prostaglandins 13:719–722

Sors H, Maclouf J, Pradelles P, Dray F (1977) The use of iodinated tracers for a sensitive radioimmunoassay of 13,14-dihydro-15-ketoprostaglandin F_{α}. Biochim Biophys Acta 486:553–564

Sors H, Pradelles P, Dray F, Rigaud M, Maclouf J, Bernard P (1978) Analytical methods for thromboxane B_2 measurement and validation of radioimmunoassay by gas liquid chromatography-mass spectrometry. Prostaglandins 16:277–290

Starczewski M, Voigtmann R, Peskar BA, Peskar BM (1984) Plasma levels of 15-keto-13,14-dihydro-prostaglandin E_2 in patients with bronchogenic carcinoma. Prostaglandins Leukotrienes Med 13:249–258

Strickland DM, Brennecke SP, Mitchell MD (1982) Measurement of 13,14-dihydro-15-keto-prostaglandin $F_{2\alpha}$ and 6-keto-prostaglandin $F_{1\alpha}$ in plasma by radioimmunoassay without prior extraction or chromatography. Prostaglandins Leukotrienes Med 9:491–493

Stylos WA, Burstein S, Rosenfeld J, Ritzi EM, Watson DJ (1973) A radioimmunoassay for the initial metabolites of the F prostaglandins. Prostaglandins 4:553–565

Sun FF, Taylor BM (1978) Metabolism of prostacyclin in rat. Biochemistry 17:4096–4101

Tanaka T, Ito S, Hiroshima O, Hayashi H, Hayaishi O (1985) Rat monoclonal antibody specific for prostaglandin E structure. Biochim Biophys Acta 836:125–133

Thomas CMG, van den Berg RJ, de Koning Gans HJ, Lequin RM (1978) Radioimmunoassays for prostaglandins. I. Technical validation of prostaglandin $F_{2\alpha}$ measurements in human plasma using Sephadex G-25 gelfiltration. Prostaglandins 15:839–847

Tonnesen MG, Jubiz W, Moore JG, Frailey J (1974) Circadian variation of prostaglandin E (PGE) production in human gastric juice. Am J Dig Dis 19:644–648

Vane JR (1964) The use of isolated organs for detecting active substances in the circulating blood. Br J Pharmacol 23:360–373

Van Orden DE, Farley DB (1973) Prostaglandin $F_{2\alpha}$ radioimmunoassay utilizing polyethylene glycol separation technique. Prostaglandins 4:215–233

Van Orden DE, Farley DB, Clancey CJ (1977) Radioimmunoassay of PGE and an approach to the specific measurement of PGE_1. Prostaglandin 13:437–453

Van Vunakis H, Wasserman E, Levine L (1972) Specificities of antibodies to morphine. J Pharmacol Exp Ther 180:514–521

Vierhapper H, Jörg J, Waldhäusl W (1985) Generation of prostaglandin E_2-like radioimmunoreactive material in human plasma during storage at −20 °C but not at −80 °C. Prostaglandins Leukotrienes Med 18:113–117

Viinikka L, Ylikorkala O (1980) Measurement of thromboxane B_2 in human plasma or serum by radioimmunoassay. Prostaglandins 20:759–766

Waitzmann MB, Law ML (1975) Changes in prostaglandin concentration in blood subjected to repetitive freezing and thawing. Prostaglandins 10:949–957

Walker V, Pickard JD, Smythe P, Eastwood S, Perry S (1983) Effects of subarachnoid haemorrhage on intracranial prostaglandins. J Neurol Neurosurg Psychiatry 46:119–125

Weerasekera DA, Koullapis EN, Kim JB, Barnard GJ, Collins WP, Kohen F, Lindner HR (1983) Chemiluminescence immunoassay of thromboxane B_2. Adv Prostaglandin Thromboxane Leukotriene Res 11:185–190

Westlund P, Granström E, Kumlin M, Nordenström A (1986) Identification of 11-dehydro-TXB_2 as a suitable parameter for monitoring thromboxane production in the human. Prostaglandins 31:929–960

Wong PY-K, Sun FF, McGiff JC (1978) Metabolism of prostacyclin in blood vessels. J Biol Chem 253:5555–5557

Wong PY-K, McGiff JC, Cagen L, Malik KU, Sun FF (1979) Metabolism of prostacyclin in the rabbit kidney. J Biol Chem 254:12–14

Wong PY-K, Lee WH, Quilley CP, McGiff JC (1981) Metabolism of prostacyclin: formation of an active metabolite in the liver. Fed Proc 40:2001–2004

Yalow RS (1973a) Radioimmunoassay. Practices and pitfalls. Circ Res [Suppl I] 32/33:I/116–I/128

Yalow RS (1973b) Radioimmunoassay methodology: Application to problems of heterogeneity of peptide hormones. Pharmacol Rev 25:161–178

Yamamoto S, Hayashi Y, Yano T, Ueda N, Kawamura S, Ogushi F, Nakamura K (1983) Enzyme immunoassay of thromboxane B_2 and 6-keto-prostaglandin $F_{1\alpha}$ as a substitute for radioimmunoassay. Adv Prostaglandin Thromboxane Leukotriene Res 11:181–184

Yano T, Hayashi Y, Yamamoto S (1981) β-Galactosidase-linked immunoassay of prostaglandin $F_{2\alpha}$. J Biochem 90:773–777

Ylikorkala O, Viinikka L (1981) Measurement of 6-keto-prostaglandin $F_{1\alpha}$ in human plasma with radioimmunoassay: effect of prostacyclin infusion. Prostaglandins Med 6:427–436

Youssefnejadian E, Brodovcky H, Johnson M, Craft I (1978) Radioimmunoassay of 13,14-dihydro-15-keto-prostaglandin $F_{2\alpha}$. Prostaglandins 15:239–253

Yu S, Burk G (1972) Antigenic activity of prostaglandins: specificities of prostaglandin E_1, A_1, and $F_{2\alpha}$ antigen-antibody reactions. Prostaglandins 2:11–22

Zusman RM, Caldwell BV, Speroff L, Behrman HR (1972) Radioimmunoassay of the A prostaglandins. Prostaglandins 2:41–53

CHAPTER 19

Radioimmunoassay of Leukotrienes and Other Lipoxygenase Products of Arachidonate

H. J. ZWEERINK, G. A. LIMJUCO, and E. C. HAYES

A. Introduction

Arachidonic acid is the precursor of a number of important mediators in the regulation of the microenvironment. It is released by the action of phospholipases on membrane phospholipids and then modified by enzymes of the cyclooxygenase pathway to produce prostaglandins, thromboxanes, and prostacyclin, or by enzymes of the lipoxygenase pathway to produce leukotrienes and hydroxyeicosatetraenoic acids. Prostaglandins, identified in 1963 by BERGSTROM et al. have been shown to possess a wide spectrum of biologic activities, most notably as mediators in inflammation.

Leukotrienes, identified only 10 years ago (BORGEAT et al. 1976; BORGEAT and SAMUELSSON 1979a, b; MURPHY et al. 1979) were initially isolated in very small quantities from biologic fluids (for example, from extracts of polymorphonuclear cells that were stimulated with the ionophore A23187) and the unavailability of materials made the definition of their precise biologic function very difficult. More recently, stereospecific chemical synthesis (COREY et al. 1980a, b, 1981; ROKACH et al. 1980; GUINDON et al. 1982) has provided substantial quantities of many of the leukotrienes and rapid progress has been made in our understanding of their diverse biologic activities. Chemical synthesis has also provided leukotrienes for the preparation of leukotriene–protein conjugates for the production of leukotriene-specific antisera, and radiolabeled leukotrienes that are essential for the development of sensitive and convenient radioimmunoassays. These radioimmunoassays have been used in conjunction with reverse-phase high pressure liquid chromatography (RP-HPLC) and biologic assays to measure leukotrienes in vitro in complex reaction mixtures or ex vivo in clinical samples drawn from humans and experimental animals.

The purpose of this chapter is to review radioimmunoassays for lipoxygenase products that have been reported in the literature, to discuss their applications, and to assess problems that may be encountered in their use. Other important areas in the field of leukotriene research such as the biosynthesis and metabolism of lipoxygenase products and their mode of action will be touched on only to the extent that they constitute relevant background for the discussion on radioimmunoassays. A number of excellent reviews provide more detailed information on these topics (LEWIS and AUSTEN 1981, 1984; SAMUELSSON 1983; PIPER 1983, 1984; MALMSTEN 1984; STENSON and PARKER 1984; FORD-HUTCHINSON 1985; BORGEAT et al. 1985).

B. Biosynthesis and Metabolism of Lipoxygenase Products

Figure 1 shows that some of the leukotrienes are end products in the lipoxygenase pathway whereas others are substrates for the synthesis of other leukotrienes. In contrast to the wide distribution of cyclooxygenase, the enzyme that initiates the conversion of arachidonic acid to prostaglandins, lipoxygenase activity has been found in a limited number of cell types (neutrophils, monocytes, macrophages, eosinophils, and granulocytes). Furthermore, cell types differ in enzyme specificity in regard to the addition of the hydroperoxy group to arachidonic acid. Lipoxygenase in platelets adds this group at the C-12 position (HAMBERG and SAMUELSSON 1974) to form 12-hydroperoxy-6,8,11,14-eicosatetraenoic acid (12-HPETE); the major lipoxygenase in neutrophils generates 5-HPETE (BORGEAT et al. 1976); and a third form of the enzyme generates 15-HPETE (FORD-HUTCHINSON 1985). Figure 1 shows that 5-HPETE is the precursor of a number of prod-

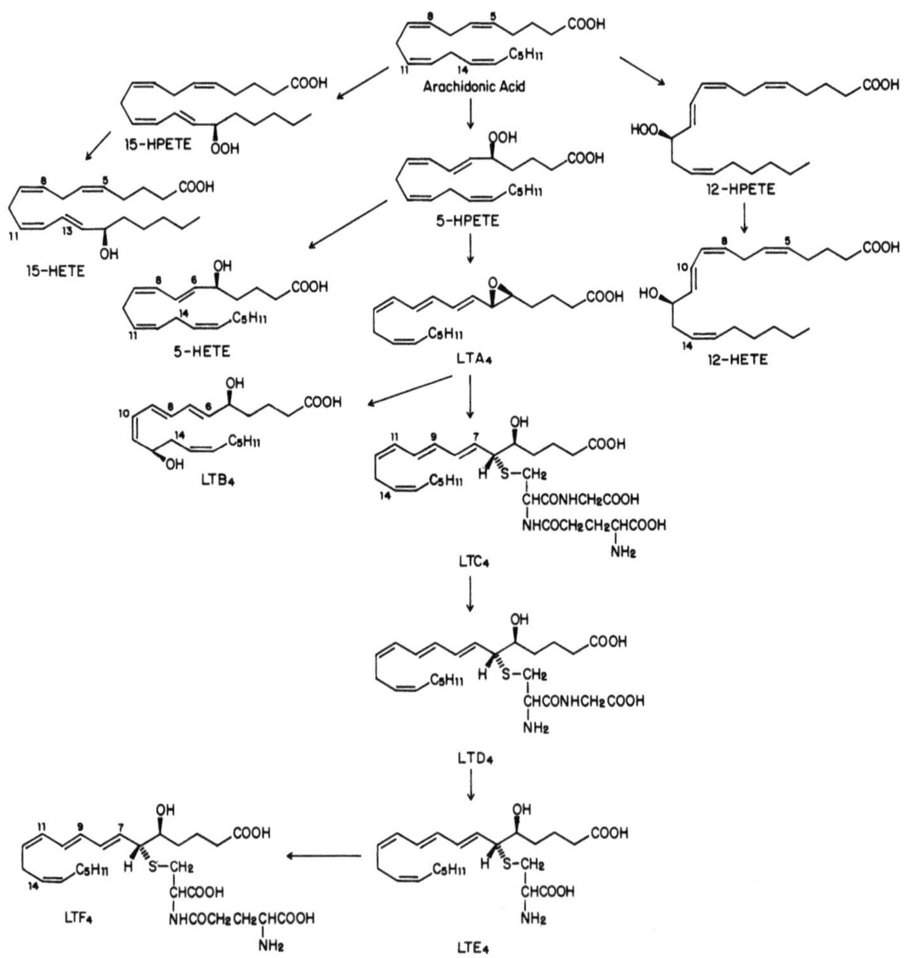

Fig. 1. Biosynthesis of lipoxygenase products

Table 1. Catabolism of the biologically active leukotrienes

Leukotriene	Catabolite
$LTC_4(LTD_4, LTE_4)$ ⟶	Corresponding diastereoisomeric sulfoxides of peptidoleukotriene 6-*trans*-LTB_4 6-*trans*-12-*epi*-LTB_4
LTB_4 ⟶	20-*OH*-LTB_4 ⟶ 20-*COOH*-LTB_4

ucts with a wide spectrum of biologic activities. It is converted either into a stable product, 5-HETE, or an unstable epoxide intermediate leukotriene A_4 (LTA_4) which is metabolized to generate LTB_4 or LTC_4. The formation of LTC_4 involves the addition of glutathione via the sulfur linkage at the C-6 position and requires the enzyme glutathione-*S*-transferase. Formation of the next two compounds in the pathway (LTD_4 and LTE_4) requires the sequential removal of glutamic acid and glycine, respectively. LTC_4, LTD_4, and LTE_4 are also known as sulfidopeptidoleukotrienes or slow reacting substance of anaphylaxis (SRS-A). More recently, it has been shown in vitro that LTE_4 can be the substrate for γ-glutamyltransferase and that this results in the incorporation of a glutamic acid residue and the formation of LTF_4 (ANDERSON et al. 1982; DENIS et al. 1982).

In vitro studies have shown that the half-life of lipoxygenase products in cells or tissue is short. Part of this is due to the fact that they are intermediates in the lipoxygenase pathway and to catabolic inactivation where the sulfidopeptide leukotrienes LTC_4, LTD_4, and LTE_4 are converted to their *S*-diastereoisomeric sulfoxides and to biologically inactive 6-*trans*-LTB_4 and 6-*trans*-12-*epi*-LTB_4 (Table 1). This conversion is mediated by hypochlorous acid that is released from activated neutrophils (LEE et al. 1982, 1983b; WELLER et al. 1983; LEWIS and AUSTEN 1984). Synthetic leukotriene-derived sulfones have strong biologic activities (DENIS et al. 1982; JONES et al. 1982), but their metabolic relation to the corresponding sulfoxides has not been established. LTB_4 is metabolized in neutrophils by ω-oxidation to 20-hydroxy- and 20-carboxy-LTB_4 (HANSSON et al. 1981) which have some of the biologic activities that are associated with LTB_4 (FORD-HUTCHINSON 1983).

C. Biologic Activities of Lipoxygenase Products

The biologic activities of the lipoxygenase products have been summarized and generalized in Table 2. These activities had to be generalized because variations exist between compounds within each broad class, the tissue-specific response varies among different animal species and various tissues within the same animal species may also respond differently. For details readers are referred to a number of reviews (MALMSTEN 1984; HAMMARSTROM 1983; PIPER 1984; STENSON and PARKER 1984; PETERS 1985).

Table 2. Major biologic activities of lipoxygenase products

1. HETEs and HPETEs
   a) Stimulate chemotaxis, chemokinesis, and aggregation of neutrophils
   b) Increase neutrophil degranulation
   c) Stimulate chemotaxis and chemokinesis in eosinophils
   d) Inhibit synthesis of other lipoxygenase and cyclooxygenase products
   e) Stimulate mucus secretion
   f) Increase Ca^{2+} uptake
   g) Stimulate the release of mediators from basophils
2. LTB_4 and ω-oxidation products
   a) Stimulate chemotaxis, chemokinesis, aggregation, and adherence of neutrophils
   b) Stimulate contraction of human bronchus and guinea pig parenchyma
   c) Stimulate plasma exudation and vascular permeability
   d) Stimulate degranulation of neutrophils
   e) Stimulate eosinophil chemotaxis
   f) Stimulate Ca^{2+} flux
3. Sulfidopeptidoleukotrienes (LTC_4, LTD_4, LTE_4, LTF_4, and their sulfones)
   a) Stimulate smooth muscle contraction
   b) Stimulate bronchoconstriction
   c) Stimulate vaso- and arteriolar constriction
   d) Stimulate mucus secretion
   e) Stimulate increase in vascular permeability and induce edema
   f) Induce wheal-and-flare reaction in skin

D. Biologic Assays to Measure Lipoxygenase Products

The most commonly used biologic assay is based on one of the properties of the lipoxygenase products, namely the stimulation of the contraction of smooth muscle by SRS-A (BROCKLEHURST 1962; DRAZEN et al. 1980). In the guinea pig ileal smooth muscle assay, strips of tissue are suspended in an oxygenated bath while attached to a transducer to determine the force of muscle contraction. Results are standardized in comparison with the contraction induced by known concentrations of histamine and leukotrienes (PARKER et al. 1982).

To establish that contraction is leukotriene mediated, the experimental sample is assayed in the presence of compounds that inhibit the action of histamine and acetylcholine (e.g., pyrilamine maleate and atropine sulfate). The specificity of the SRS-A reaction can be established further with FPL-55712, an SRS-A antagonist (AUGSTEIN et al. 1973; DRAZEN et al. 1980; PONG and DEHAVEN 1983) and with indomethacin, an inhibitor of prostaglandin synthesis (VANE 1971). All leukotrienes are active in the assay, but the relative potency of LTD_4, LTC_4, LTE_4, and LTF_4 differs (1:7:170:280, respectively on a molar basis; LORD et al. 1985). Because specific antagonists for each one of these compounds that are required to establish specificity in mixed samples are not available, it is necessary to fractionate prior to analysis, for example by HPLC (see Sects. E and H). Partial purification may also be necessary to eliminate substances that interfere with the assay or that will degrade the leukotrienes. Although the guinea pig ileum assay is very sensitive (detection thresholds for LTD_4, LTC_4, and LTE_4 are 0.1, 0.2, and 2 pmol, respectively; PARKER et al. 1982) it is quite laborious and lacks specificity.

Biologic assays for LTB_4 utilize its ability to induce smooth muscle contraction, neutrophil aggregation, and vascular permeability. LTB_4 is measured in the guinea pig ileum strip assays under conditions that antagonize SRS-A (FORD-HUTCHINSON et al. 1983). Neutrophil aggregation utilizes peritoneal cells from sodium caseinate-stimulated rats and the aggregation on exposure to LTB_4 is evaluated by nephelometry (FORD-HUTCHINSON et al. 1983). An in vivo assay for LTB_4 has been described (FORD-HUTCHINSON 1983) that measures local extravasation of radiolabeled human serum albumin after the intradermal injection of LTB_4 and PGE_2. These assays are specific for LTB_4 and some of its metabolic products, with sensitivities of 1, 20, and 40 pmol for the neutrophil aggregation, smooth muscle contraction, and vascular permeability assays, respectively.

E. Physicochemical Separation and Identification of Lipoxygenase Products

High pressure liquid chromatography (HPLC) is very effective in separating and identifying lipoxygenase products. Conjugated double bonds absorb in the ultraviolet region and this allows for detection and quantitation within the picomole range (1–10 ng). Minimal sample preparation is required (see Sects. G and H) and the lipoxygenase products need not be modified chemically. Although straight-phase silica gel HPLC will separate the less polar compounds (e.g., HETES and LTB_4), reverse-phase HPLC (RP-HPLC), with an octadecylsilyl stationary phase on a silica support, is used to separate more polar leukotrienes. Chromatography conditions are dictated to some degree by the lipoxygenase products that are present in the sample (PETERS et al. 1983). Details are discussed in two reviews (HAMMARSTROM 1983; MURPHY 1985).

More recently, chromatography–mass spectrometry, electron impact and negative ion chemical ionization techniques, and fast atom bombardment mass spectrometry have been employed. Although the usefulness of latter technique for quantitative analysis is limited by its requirements for sample sizes of several nanomoles, the spectral data generated are extremely useful in determining structures, and in monitoring the metabolism and degradation of leukotrienes (MURPHY 1985; TAYLOR et al. 1983).

F. Radioimmunoassays for Lipoxygenase Products

I. Introduction

The eicosatetraenoic acid moieties of the various lipoxygenase products differ in the position of the double bonds: at positions 6, 8, 11, and 14 in 5-HETE, positions 6, 8, 10, and 14 in LTB_4, and positions 7, 9, 11, and 14 in LTA_4, LTC_4, LTD_4, LTE_4, and LTF_4 (see Fig. 1). In addition *cis–trans* arrangements differ and side chains range from simple hydroxy groups for the HETEs and LTB_4 to more complex glutathione for LTC_4. These differences suggest sufficient antigenic variability to obtain immune sera that are specifically reactive with individual lipoxygenase products, and thus are useful reagents for specific radioimmunoassays.

Because of their low molecular weight, lipoxygenase products must be coupled to high molecular weight carrier molecules such as bovine serum albumin (BSA) or keyhole limpet hemocyanin (KLH) to be immunogenic. Such coupling is possible because all lipoxygenase products contain at least one carboxyl group for coupling and some (LTC_4, LTD_4, LTE_4, or LTF_4) contain additional carboxyl groups and an amino group. These haptens can be coupled directly to carrier proteins, but the use of interspacing conjugating reagents with a few carbon atoms is generally preferred. This approach will often lead to the generation of immunoglobulins that are specific for the hapten rather than antigenic regions that are shared between the hapten and the carrier molecule (PINCKARD 1978). The site on the hapten that is used for coupling and the spacer arm will determine the hapten's orientation in relation to the carrier protein, and its subsequent recognition by antibody-producing cells. This may have a profound effect on the antibody responses against the various antigenic determinants on the molecule. Immune sera have been prepared against five different lipoxygenase products: LTC_4, LTD_4, LTB_4, 12-HETE, and 15-HETE. The preparation of each serum and its use will be discussed separately.

II. Radioimmunoassays to Measure LTC_4

1. Preparation of Immunogens

A number of laboratories have published LTC_4-specific radioimmunoassays (AEHRINGHAUS et al. 1982; LINDGREN et al. 1983; HAYES et al. 1983; WYNALDA et al. 1984) and each used a different method of coupling the antigen to carrier protein. AEHRINGHAUS et al. (1982) conjugated LTC_4 to BSA via the free amino group, using glutaraldehyde as the coupling reagent. The advantages of this method are convenience and efficiency, and the fact that glutaraldehyde provides a 6-carbon spacer; the disadvantage is the formation of multimolecular heterogeneous aggregates. LINDGRÉN et al. (1983) blocked the free amino group on the LTC_4 molecule with acetic anhydride and coupled this acetylated hapten to polyamino-bovine serum albumin (PABSA) with the conjugating reagent 1-ethyl-3(3-dimethylaminopropyl)carbodiimide. (PABSA was prepared by derivatizing BSA with triethylenetetramine.) Any of the three carboxylic acid groups on the acetylated LTC_4 could have been reacted to form amide bonds. HAYES et al. (1983) utilized LTC_4 coupled via its free amino group to KLH with the bifunctional coupling agent 6-*N*-maleimidohexanoic acid chloride to yield a homogeneous conjugate (Fig. 2a; YOUNG et al. 1982). First, the 6-*N*-maleimidohexanoic acid amide of LTC_4 was prepared and this was reacted with KLH that had been thiolated with *S*-acetylmercaptosuccinic anhydride. As will be discussed in Sect. F.II.4, the sera raised against these conjugates exhibited some degree of cross-reactivity with LTD_4 and LTE_4. In an attempt to circumvent this, WYNALDA et al. (1984) used a structural analog of LTC_4, 7-*cis*-9,10,11,12,14,15-hexahydroleukotriene C_4, and coupled this via the free amino group to thiolated KLH through 6-*N*-maleimidohexanoic acid. The hapten is structurally identical to LTC_4 except that it

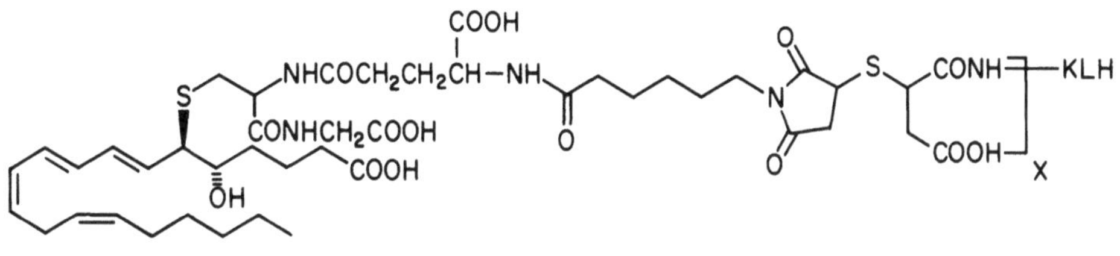

LTC$_4$ conjugate

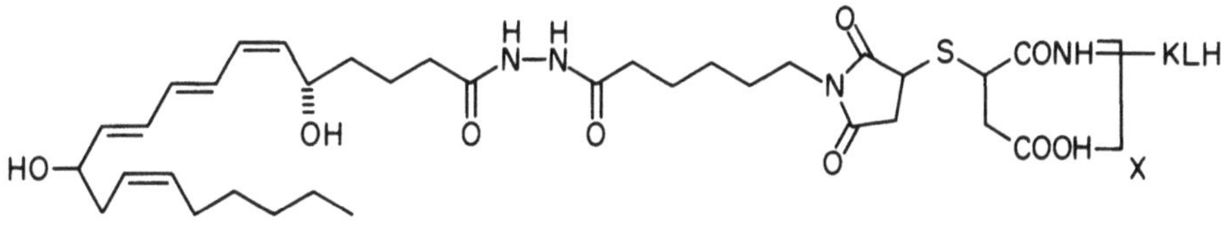

LTB$_4$ conjugate

Fig. 2 a, b. Structures of LTC$_4$–KLH and LTB$_4$–KLH immunogens. **a** LTC$_4$–(6-*N*-maleimidohexanoic acid)–KLH (HAYES et al. 1983), **b** LTB$_4$–(6-*N*-maleimidohexanoic acid)–KLH (ROKACH et al. 1984)

differs in the number of double bonds and their geometry. It was felt that the hexahydroleukotriene C$_4$ would be less susceptible to metabolic degradation and therefore would increase the specificity of the antiserum.

2. Immunizations

The immunization protocols that were used by the various investigators were very similar. The immunogen was injected with complete Freund's adjuvant (CFA) into rabbits at multiple sites followed by boosting immunizations in incomplete Freund's adjuvant (IFA) or in CFA at various times thereafter. Animals were bled 7–9 days after the boosts. Multiple injections were required to induce acceptable antibody titers (Table 3). It has been our experience and that of others (HERBERT 1978) that, if there is sufficient time between booster immunizations, one obtains sera with both high antibody titers and high affinities. For example, when rabbits that had been immunized as described in Table 3 were rested for 26 weeks after the second booster immunization, a third boost with LTC$_4$–KLH in IFA increased the antibody titer from 1280 to 10240.

3. Details of Assay Conditions

Radiolabeled LTC$_4$ ^{3}H with high specific activity is available commercially and sensitive liquid phase radioimmunoassays have been developed. Generally LTC$_4$ ^{3}H is mixed first with varying antibody concentrations to establish binding curves. Next, an antibody concentration is chosen that results in 50%–80% binding of LTC$_4$ ^{3}H and the reactions are then carried out in the presence of different

Table 3. LTC_4-specific antibody titers in rabbits after immunization with LTC_4–KLH (HAYES et al. 1983)

Week	Immunization	Antibody titer[a]	
		Rabbit 1	Rabbit 2
0	200 µg LTC_4–KLH[b]	ND[c]	ND
3	200 µg LTC_4–KLH[d]	ND	ND
5		20	28
14		32	8
23		22	8
27	200 µg LTC_4–KLH[d]	ND	ND
29		1280	1280
32		320	640
35		240	320
44		240	320

[a] Determined by double-antibody RIA and expressed as reciprocal of the dilution of sera in the assay to precipitate 50% of the LTC_4 ^3H.
[b] Antigen emulsified in 1:1 PBS–CFA.
[c] Titer not determined.
[d] Antigen emulsified in 1:1 PBS–IFA.

amounts of the experimental sample. The antibody concentration which is higher than what is used generally (35–50%) was chosen because some of the LTC_4 ^3H preparations, especially those obtained during the early stages of the work, were impure and unstable. Standard inhibition curves are established by adding known concentrations of LTC_4 to the mixture of LTC_4 ^3H and antibody. Inhibition of the formation of complexes of LTC_4 ^3H and antibody is in relation to the LTC_4 concentrations as determined from standard curves. A typical protocol described by HAYES et al. (1983) is the following: 25 µl LTC_4 ^3H (approximately 20000 dpm in 5×10^{-13} mol LTC_4) was incubated for 1 h at 22 °C with 25 µl appropriately diluted immune rabbit serum raised against LTC_4–KLH (in this particular system, a dilution of 1:1000–1:2000 resulted in binding of 50% of LTC_4 ^3H), 100 µl competing ligand diluted in phosphate-buffered saline (PBS, pH 7.2), and 100 µl normal rabbit serum (diluted 1:25 in PBS). LTC_4–antibody complexes were precipitated for 18 h at 4 °C with commercially obtained goat anti-rabbit immunoglobulin (800 µl 1:8 dilution in PBS; this dilution should be determined experimentally to assure complete precipitation). Next, the precipitates were pelleted by centrifugation at 12000 *g* for 3 min and washed and pelleted twice in PBS. The pellet was dissolved in 500 µl sodium dodecylsulfate (0.1%) and radioactivity determined in a liquid scintillation counting solution.

HAYES et al. (1983) also investigated the use of dextran-coated charcoal for the separation of free LTC_4 from LTC_4 bound to antibody. Components of the reaction mixture (LTC_4 ^3H, antibody, and competing ligands) were identical to those already described, except that 25 µl 20% horse serum was used instead of the normal rabbit serum. This mixture was incubated for 18 h at 4 °C after which 1 ml

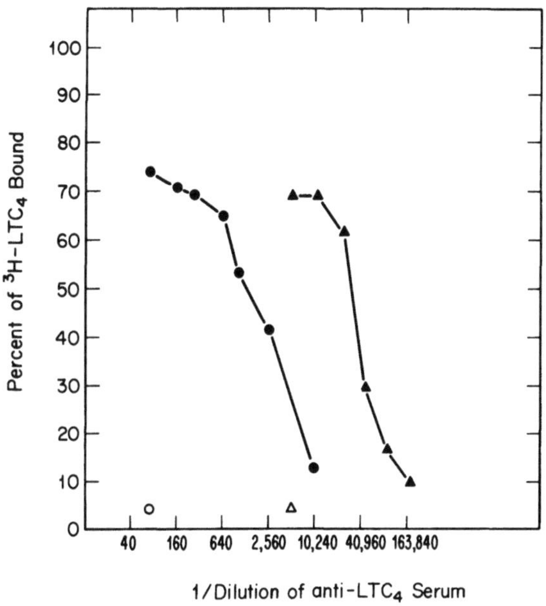

Fig. 3. Separation of free LTC_4 ^3H from antibody-bound LTC_4. Titration of rabbit anti-LTC_4 serum, comparing charcoal and goat anti-rabbit immunoglobulin for the separation of bound and unbound ligand. LTC_4 ^3H was incubated with dilutions of preimmune or immune serum. The dilutions indicated in the figure are final dilutions in the assay itself. *Full circles* immune serum, charcoal assay; *full triangles* immune serum, double-antibody assay; *open circle* preimmune serum, charcoal assay; *open triangle* preimmune serum, double-antibody assay. (HAYES et al. 1983)

of a suspension (1.5%) of dextran-coated charcoal was added. The preparation of the charcoal suspension has been described by HAYES et al. (1983) and by HUMES (1982). After 5 min at 22 °C, charcoal was pelleted by centrifugation at 10000 *g* for 1 min. Radioactivity in the supernatant represents ^3H-labeled LTC_4–antibody complexes and radioactivity in the pellet represents LTC_4 ^3H not bound to antibody. A comparison of the two methods to separate free LTC_4 ^3H from LTC_4 ^3H bound to antibody shows that the double-antibody method is to be preferred since LTC_4 ^3H binding was observed at much higher serum dilutions (Fig. 3). This probably reflects the tendency of charcoal to "strip" antigen from antigen–antibody complexes (HUNTER 1978).

To determine LTC_4 concentrations in experimental samples, a standard inhibition curve with nonradioactive LTC_4 was prepared and it was observed that 0.70×10^{-12} mol unlabeled LTC_4 reduced binding of LTC_4 ^3H to antibody by 50% (Fig. 4). Similar inhibition curves were obtained for other leukotrienes to determine the extent of their reactivity with this serum. For example, LTD_4 and LTE_4 reduced the binding of LTC_4 ^3H with the antibody to 50% at concentrations of 1.25×10^{-12} and 10.05×10^{-12} mol, respectively (Fig. 4). Cross-reactivity was defined as $(0.70/1.25) \times 100\% = 56\%$ for LTD_4 and $(0.70/10.05) \times 100\% = 6.9\%$ for LTE_4. (Values of 43% and 6% were obtained with serum from another rabbit that had been immunized with the same LTC_4–KLH conjugate.) A number of other compounds were tested for their ability to inhibit ^3H-labeled LTC_4–antibody complexes. None, including LTB_4, HETES, arachidonic acid, prostaglandins, or glutathione, showed significant inhibition (Table 4).

If the use of radiolabeled LTC_4 is not feasible, a solid-phase immunoassay can be developed which requires the binding of the antigen to surfaces, generally polystyrene or polyvinylchloride, although other supports such as nitrocellulose may be considered (ENGVALL and PERLMAN 1971; HAWKES et al. 1982). High molecu-

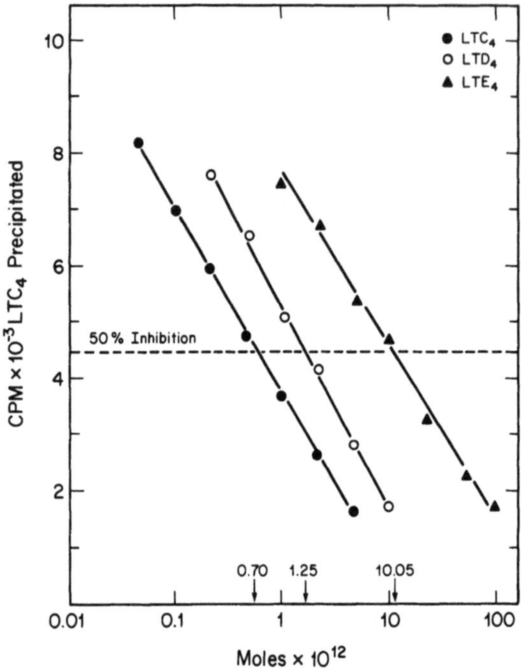

Fig. 4. Inhibition of the binding of LTC_4 ^3H to rabbit anti-LTC_4 serum by LTC_4 (*full circles*); LTD_4 (*open circles*); and LTE_4 (*full trangles*). The concentration of each leukotriene that inhibits binding by 50% is indicated by an *arrow* on the *x*-axis. (HAYES et al. 1983)

lar weight proteins and carbohydrates generally bind readily to such surfaces. However, low molecular weight compounds such as leukotrienes may require special treatment of the surface, or they need to be coupled covalently to high molecular weight polymers that themselves bind well. LEVINE et al. (1980) and MORGAN and LEVINE (1982) described a method for the covalent coupling of 12-HETE to beads and MILLER et al. (1985) described the use of LTC_4–BSA conjugates for noncovalent adsorption. Conjugates used for attachment must differ from those

Table 4. Properties of LTC_4-specific antisera

	AEHRINGHAUS et al. (1982)	HAYES et al. (1983)	LINDGREN et al. (1983)	WYNALDA et al. (1984)
Sensitivity (pmol)	0.25	0.10	0.05	0.15
Serum dilution for 50% precipitation	1:75	1:10000	1:1000	1:100
Specific immunoglobulin per milliliter serum (μg)	NR[a]	1200	NR	NR
$K_a (M^{-1})$	NR	2.1×10^9	NR	NR
Reactivity (%)[b] with:				
LTC_4	100	100	100	100
11-*trans*-LTC_4	NR	NR	100	0.6
LTD_4	16	43	0.08	1.6
11-*trans*-LTD_4	NR	39	NR	0.4
LTE_4	NR	8.3	0.07	0.66
LTB_4	NR	<0.2	<0.01	<0.1
HETEs	NR	<0.025	<0.01	<0.05

[a] NR not reported. [b] Determined as described in the text.

used for immunization both in regard to the carrier protein and the spacer molecule since leukotriene-specific sera will contain antibodies that react with these molecules. Therefore, YOUNG et al. (1982) prepared LTC_4–(1,5-difluoro-2,4-dinitrobenzene)–BSA as an antigen for solid-phase radioimmunoassays. (The KLH–maleimidohexanoic acid–LTC_4 conjugate described in Sect. F.II.1 was used for immunization.) The reaction between immunoglobulin and immobilized LTC_4 is determined with probes such as protein A or anti-rabbit immunoglobulin that are radiolabeled. Alternatively, an enzyme-linked immunosorbent assay (ELISA) can be employed (ENGVAL and PERLMAN 1971) with an enzyme that is coupled to the probe. Binding of the probe is detected with a chromogenic or fluorogenic substrate.

In a liquid immunoassay, both the radioactive and competing ligands are in solution and therefore recognized equally by the LTC_4-specific immunoglobulin whereas in a solid-phase assay, antibody and competing ligand are in solution and the primary antigen is linked to the surface. If the antibody-binding sites on the latter are sequestered owing to its surface attachment, then binding of the antibody may be impaired significantly. Attachment to a solid surface may also lead to closely spaced antibody-binding sites and this may result in very strong bivalent binding of the antibody and the competition by soluble antigen in the experimental sample may then be less efficient. In view of these considerations, it is of interest that MILLER et al. (1985), who developed LTC_4- and LTB_4-specific solid-phase ELISA, found that the LTB_4-specific radioimmunoassay was significantly more sensitive than the LTB_4-specific ELISA, whereas both assays were equally sensitive for LTC_4.

4. Comparison of LTC_4-Specific Antisera

Table 4 compares some of the properties of LTC_4-specific sera that have been published by various investigators. These sera were used under experimental conditions unique for each laboratory and therefore some of the parameters (serum dilution for 50% precipitation and sensitivity of the assay) do not necessarily reflect properties of the immunoglobulin preparations. However, others (affinity, antibody concentration, and cross-reactivity with other lipoxygenase products) do reflect properties of the LTC_4-specific immunoglobulin. All sera appear to be very similar in their ability to measure LTC_4. The most potent serum seems that developed by HAYES et al. (1983), but this serum reacts significantly with LTD_4 and LTE_4. Serum described by LINDGREN et al. (1983) exhibits the greatest degree of specificity in that very little reactivity was observed with LTD_4 or LTE_4. None of these sera exhibit any reactivity with compounds such as prostaglandins, arachidonic acid, or glutathione.

III. Radioimmunoassays to Measure LTD_4

The approaches toward the generation of LTD_4-specific immunoglobulin and experimental details for the radioimmunoassay are very similar to those described for LTC_4. Therefore, this section will only highlight those points that are unique to the LTD_4-specific assays.

Table 5. Properties of LTD_4-specific sera

	Levine et al. (1981)	Beaubien et al. (1984)[a]	Beaubien et al. (1984)[b]
Sensitivity (pmol)	0.5	1×10^{-3}	9×10^{-3}
Serum dilution for 50% precipitation	1:40	1:5000[c]	1:5000[c]
Specific immunoglobulin per milliliter serum (μg)	0.32	NR[d]	NR
K_a (M^{-1})	2.8×10^9	1.1×10^{10}	1×10^9
Reactivity (%)[e] with:			
LTD_4	100	100	100
11-*trans*-LTD_4	300	NR	NR
LTC_4	200	7	8
11-*trans*-LTC_4	170	NR	NR
LTE_4	92	17	48
11-*trans*-LTB_4	161	NR	NR
LTB_4	< 0.1	< 0.1	< 0.1

[a] Immunized with LTD_4–thyroglobulin and LTD_4–KLH.
[b] Immunized with LTD_4–KLH only.
[c] Approximate values.
[d] NR not reported.
[e] Determined as described in the text.

Levine et al. (1981, 1982) conjugated LTD_4 to BSA via the carboxyl group on the eicosatetraenoic acid using the C-1 monoacid of the dimethylester of *N*-trifluoroacetyl-LTD_4 as the starting material. Other reactive groups on the molecule were protected to obtain a specific LTD_4–BSA conjugate without a spacer arm. LTD_4–thyroglobulin and LTD_4–KLH immunogens were prepared by Beaubien et al. (1984) using glutaraldehyde as a cross-linking reagent. The latter authors used two immunization protocols: rabbits were immunized twice with LTD_4–thyroglobulin, after which they received multiple booster immunizations with LTD_4–KLH; or they were immunized and boosted with LTD_4–KLH only.

The radioimmunoassays employed by both groups differed in that Levine et al. (1981) used LTC_4 ^3H as the radiolabeled ligand and goat anti-rabbit IgG for the precipitation of the antigen–antibody complexes, whereas Beaubien et al. (1984) used LTD_4 ^3H and dextran-coated charcoal. The results in Table 5 show that rabbit serum obtained after combined immunization with LTD_4–thyroglobulin and LTD_4–KLH is very effective in detecting LTD_4 (10^{-3} pmol); and that the sera show various degrees of reactivity with LTC_4 and LTE_4, but they do not react significantly with LTB_4. Interestingly, the serum described by Levine et al. (1981), although raised against LTD_4–BSA, is more reactive with LTC_4 than with LTD_4.

IV. Radioimmunoassays to Measure LTB_4

Three groups have reported on LTB_4-specific radioimmunoassays. Salmon et al. (1982) prepared an immunogen by first synthesizing a mixed anhydride of the carboxyl group on LTB_4 and the isobutyl ester of carbonic acid. This material was

Table 6. Properties of LTB_4-specific sera

	LEWIS et al. (1982)	ROKACH et al. (1984)	SALMON et al. (1982)
Sensitivity (pmol)	0.3	0.1	0.01
Serum dilution for 50% precipitation	>1:25	1:2000	1:4000
Specific immunoglobulin per milliliter serum (μg)	0.37	38	NR[a]
K_a (M^{-1})	3.2×10^9	5.8×10^8	NR
Reactivity (%)[b] with:			
LTB_4	100	100	100
5-*S*-12*S*-6-*trans*-8-*cis*-LTB_4	~ 30	6	NR
5-*S*-12-*epi*-6-*trans*-8-*cis*-LTB_4	< 3	0	NR
LTC_4	< 5	0.02	0.03
LTD_4	NR	0.04	0.03
LTE_4	NR	0.15	NR
5-HETE	0	0.2	0.03
5,12-di-HETE	NR	< 7	0.14

[a] NR not reported. [b] Determined as described in the text.

reacted with BSA to generate amide bonds (without spacer molecules) with free amino groups on the carrier protein. YOUNG et al. (1983) converted the carboxyl group on LTB_4 to the ω-lactone by treatment with dicyclohexylcarbodiimide which in turn was reacted with hydrazine to generate LTB_4–hydrazide. This was coupled to thiolated KLH with 6-*N*-maleimidohexanoic acid chloride, thus providing a spacer between LTB_4 and KLH (see Fig. 2b). In contrast, LEWIS et al. (1982) used the 12-oxy group on the LTB_4 molecule for coupling to BSA. LTB_4 was activated by reacting the 5-benzoate of the LTB_4 methyl ester with *p*-nitrophenylchloroformate, and the resulting 12-*p*-nitrophenoxycarbonyl derivative was reacted with BSA to generate a conjugate with a single spacer carbon atom.

LTB_4 ^3H can be obtained commercially, but initially it was generated by incubating polymorphonuclear cells and arachidonic acid ^3H in the presence of the calcium ionophore A23187. It was then purified after methanol–ether extraction by chromatography with silicic acid and HPLC (SALMON et al. 1982). Conditions for the radioimmunoassays reported in the literature are similar to those described for LTC_4 and experimental results are summarized in Table 6. The assay is quite specific in that very little reactivity is observed with LTC_4, LTD_4, and LTE_4; and it is sensitive to 0.01 pmol.

An ELISA was described by MILLER et al. (1985) that used LTB_4 coupled to BSA without a spacer arm and serum from rabbits immunized with LTB_4–KLH (YOUNG et al. 1983; the same serum was used in radioimmunoassays by ROKACH et al. 1984). It was as specific as the radioimmunoassay, but less sensitive (a detection limit of 8 pmol/ml).

V. Radioimmunoassays to Measure 12-HETE

Rabbit serum was raised against 12-HETE after immunization with a conjugate of 12-HETE and human serum albumin (HSA). The conjugate was prepared in a two-step procedure as described originally by BAUMINGER et al. (1973) for the generation of conjugates of prostaglandins and proteins. First 12-HETE, *N*-hydroxysuccinimide, and *N,N'*-dicyclohexylcarbodiimide were reacted to generate the hydroxysuccinimide ester with the carboxyl group on 12-HETE. This active ester was purified and conjugated via the free amino acid groups on HSA. The conjugate was injected with CFA into rabbits, followed by booster immunizations 3 and 6 months later with the antigen in CFA. Two radioimmunoassays with this serum were reported. Initially unlabeled 12-L-HETE was covalently coupled to a solid support and antibody binding was measured with ^{125}I-labeled protein A (LEVINE et al. 1980). Under these conditions, approximately 0.4 pmol 12-L-HETE could be detected. With the exception of the methyl ester of 12-HETE (approximately 20% cross-reactive) very little, if any, reactivity was found with a number of compounds, including 5-HETE, arachidonic acid, and prostaglandins. ESKRA et al. (1980) and, more recently, MORGAN and LEVINE (1982) have reported the use of 12-L-HETE ^{3}H in a liquid radioimmunoassay with the sensitivity of the assay being approximately 1 pmol.

VI. Radioimmunoassays to Measure 15-HETE

15-HETE was coupled without a spacer arm to BSA using 1-ethyl-3(3-dimethylaminopropyl)carbodiimide as the coupling agent (GRANSTRÖM et al. 1976). It was injected into rabbits in the presence of CFA with five booster immunizations at monthly intervals (BRYANT and HWANG 1983). The resulting antiserum bound 40% of 15-HETE ^{3}H when diluted 1 : 600 and deteced 15-HETE at approximately 0.1 pmol. No reactivity was found with other monohydroxyeicosatetraenoic acids, prostaglandins, or arachidonic acid, but there was significant reactivity with various 15-hydroxy di- and mono-HETEs and related 15-hydroxyeicosanoids.

G. Sample Preparation for Radioimmunoassays

Experimental samples may contain substances that interfere with the radioimmunoassays (for example, salts or enzymes that metabolize leukotrienes). The presence of such substances can be evaluated by determining the recovery of exogenously added lipoxygenase products from the experimental sample. Preferably, the recovery of both radiolabeled (for example by RP-HPLC) and unlabeled material (by RIA) should be determined. Removal of interfering substances is accomplished most conveniently by extraction with organic solvents (PARKER et al. 1982; ROUZER et al. 1982; METZ et al. 1982; PETERS et al. 1983). Proteins are precipitated or inactivated whereas the lipoxygenase products remain in solution. Solvents are evaporated prior to the assay and the sample can be concentrated severalfold by redissolving it in a small volume of buffer. Alternatively, leukotrienes in aqueous samples can be adsorbed to small octadecylsilyl–silica columns (Sep-Pak C_{18}) and eluted with organic solvents (MORRIS et al. 1983; CRETICOS et al. 1984). If necessary, further purification can be accomplished by chromatography on Sephadex LH-20 or Amberlite XAD-7 columns (PARKER et al. 1982). The efficacy of extraction and purification should be evaluated by determining the recovery of exogenously added lipoxygenase products.

H. Combination of Radioimmunoassays with Physicochemical Separation Methods

As has been discussed, many of the radioimmunoassays for lipoxygenase products, although sensitive and convenient, lack absolute specificity. Therefore, samples that contain or are suspected of containing multiple lipoxygenase products have been analyzed by combining radioimmunoassays with RP-HPLC. As described in Sect. G, lipoxygenase products in aqueous samples are extracted with organic solvents and concentrated prior to RP-HPLC (ROUZER et al. 1982; METZ et al. 1982; PETERS et al. 1983; MORRIS et al. 1983). Alternatively, aqueous samples are diluted in water, applied to the Sep-Pak C_{18} column, and eluted with organic solvents (G. A. LIMJUCO, D. L. LOMBARDO, H. J. ZWEERINK, and E. C. HAYES 1984, unpublished work). The presence of lipoxygenase products in each HPLC fraction is then determined by RIA. Because of the sensitivity of the RIA, it is often possible to assay fractions after dilution which will adjust the pH and lower the concentrations of organic solvents below inhibitory levels.

An analysis was carried out to determine leukotriene levels in synovial fluids and the results are shown in Figs. 5 and 6. A Bondapak C_{18} analytic RP-HPLC column was calibrated with a number of leukotrienes and PGB_2 as an internal standard (Fig. 5). Synovial fluid was diluted with water, concentrated on a Sep-Pak C_{18} column, eluted with methanol, and fractionated on an RP-HPLC column under the conditions described in Fig. 5. The same synovial fluid sample was also fractionated after the addition of 100 ng LTC_4 and LTB_4. After each HPLC

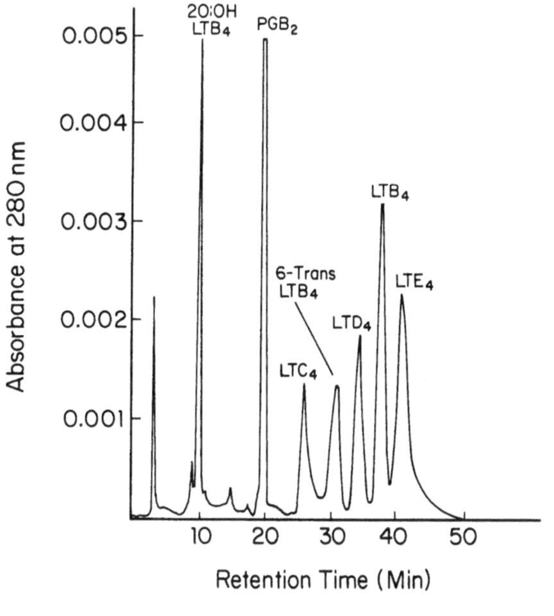

Fig. 5. Separation of leukotrienes and PGB_2 by RP-HPLC. A mixture containing 200 ng PGB_2 and 50 ng each of LTC_4, LTB_4, LTD_4, LTE_4, 20-*OH*-LTB_4, and 6-*trans*-LTB_4 was resolved on a Bondapak C_{18} column using methanol, water, acetic acid, and ammonium hydroxide (67/33/0.08/0.04, v/v, pH 5.45) at a flow rate of 0.8 ml/min

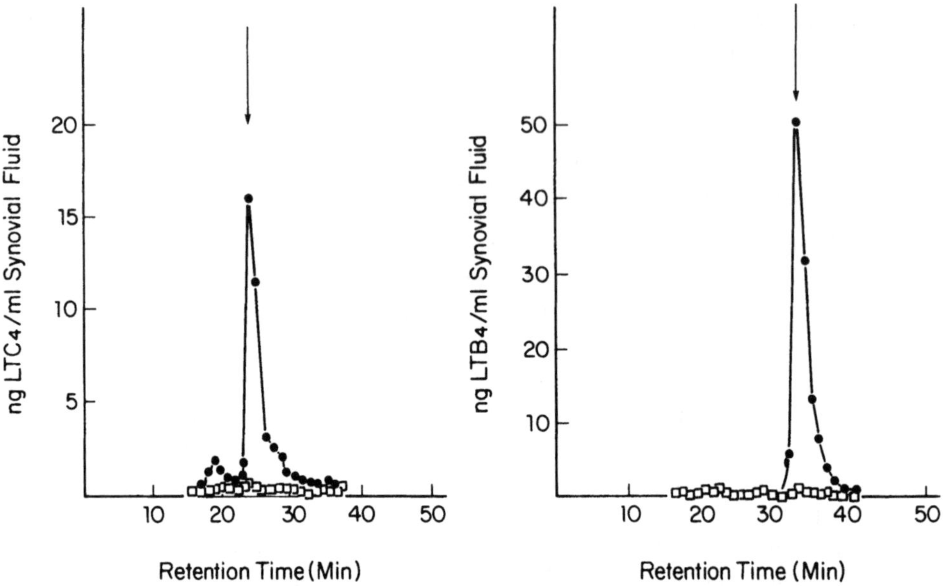

Fig. 6 a, b. LTC_4 and LTB_4 in synovial fluid samples. **a** LTC_4-specific RIA of the RP-HPLC fractions (see Fig. 5) of synovial fluid (*open squares*) and synovial fluid with 100 ng LTC_4 (*full circles*) added to it prior to analysis. *Arrow* indicates the position where LTC_4 ^3H eluted in a separate run, **b** LTB_4-specific RIA of the RP-HPLC fractions of synovial fluid (*open squares*) and synovial fluid with 100 ng LTB_4 (*full circles*) added to it prior to analysis. *Arrow* indicates the position where LTB_4 ^3H eluted in a separate run

run, all fractions were assayed by LTC_4- and LTB_4-specific radioimmunoassays. DAVIDSON et al. (1983) used a combination of HPLC and biological assays to measure LTB_4 in synovial fluid samples from patients with active rheumatoid arthritis. Most of these samples contained LTB_4 at very low levels (less than 0.25 ng/ml).

Figure 6 a and b show that HPLC fractions of synovial fluid (see Fig. 5) without the addition of LTC_4 or LTB_4 contained very little immunoreactive LTC_4 or LTB_4. After the addition of 100 ng LTC_4, immunoreactive material eluted as a single peak with a recovery of 35% with the appropriate retention time for LTC_4. The recovery of immunoreactive LTB_4 was greater than 90% and this material was present in a single peak with the appropriate retention time for LTB_4. A similar approach was taken by BEAUBIEN et al. (1984) to measure LTC_4, LTD_4, and LTE_4 in plasma of human volunteers.

J. Conclusions

This chapter demonstrates the utility of radioimmunoassays for rapid, convenient, and sensitive (in the range of much less than 1 pmol) measurements of lipoxygenase products. These compounds share an eicosatetraenoic acid moiety, but differ in the position of the double bonds and their *cis–trans* arrangements; and in the nature of side chains. Each one of these differences appears to contribute to the antigenic uniqueness of the molecules. For example, LTB_4 differs from the sulfidopeptido eicosatetraenoic acids (LTC_4, LTD_4, and LTE_4), in the configuration of the eicosatetraenoic acid, and the absence of the glutathione side chain. The failure of LTB_4 to react with LTC_4- and LTD_4-specific immunoglobulin as well as the unreactivity of LTC_4- and LTD_4-specific immunoglobulin with LTB_4 illustrates the importance of the eicosatetraenoic acid configuration in conferring antigen specificity. The sulfidopeptido side chain also contributed to antigen specificity, but less so than the rest of the molecule. This is suggested by the fact that several of the LTC_4-specific sera show incomplete, but often significant reactivity with LTD_4 and LTE_4 and that the same is true for the reactivity of LTD_4-specific sera with LTC_4 and LTE_4. However, the role of the sulfidopeptido chain is difficult to evaluate because glutamic acid and glycine may have been removed from the glutathione moiety after immunization. Furthermore, it is not obvious how the position of the hapten relative to the carrier protein in the immunogen affects the specificity of the antibody response in the case of the compounds within the SRS-A group.

Immune sera prepared by various investigators against LTC_4 and LTD_4 differ in the degree of specificity (see Table 4). Radioimmunoassays employing sera that exhibit significant reactivity with other lipoxygenase products need to be carried out in conjunction with fractionation methods such as RP-HPLC, unless it is established that the experimental sample does not contain any of the cross-reactive compounds. In the case of LTC_4-specific sera, those reported by LINDGREN et al. (1983) and WYNALDA et al. (1984) appear to be quite specific. This suggests that they can be used without fractionation of the sample prior to the radioimmunoassays. However, it should be kept in mind that the apparent lack of cross-reactivity of these sera needs to be verified rigorously. Inhibition studies were done in well-defined buffers and not in biologic samples that may contain substances that affect the recognition between antigen and antibody.

Acknowledgments. We acknowledge Dr. D.R. Robinson of the Massachusetts General Hospital for providing the synovial fluid samples for the studies shown in Figs. 5 and 6; and Ms. D. L. Lombardo for technical assistance. We also thank Mrs. Eileen Miller for her help in the preparation of this manuscript. We acknowledge Williams and Wilkins Inc. and the American Association of Immunologists for their permission to incorporate in Tables 3 and 4 and Figs. 2a, 3, and 4 data previously published in HAYES et al. (1983); and we acknowledge Churchill Livingston Inc. for their permission to incorporate in Table 6 and Fig. 2b data previously published in ROKACH et al. (1984).

References

Aehringhaus U, Wolbling RH, Konig W, Patrono C, Peskar BM, Peskar BA (1982) Release of leukotriene C_4 from human polymorphonuclear leukocytes as determined by radioimmunoassays. FEBS Lett 146:111–114

Anderson ME, Allison RD, Meister A (1982) Interconversion of leukotrienes catalyzed by purified γ-glutamyl transpeptidase; concomitant formation of leukotriene D_4 and γ-glutamyl amino acids. Proc Natl Acad Sci USA 79:1088–1091

Augstein J, Farmer JB, Lee TB, Sheard P, Tattersall ML (1973) Selective inhibition of slow reacting substance of anaphylaxis. Nature New Biol 245:215–218

Bauminger S, Zor U, Lindner HR (1973) Radioimmunological assay of prostaglandin synthetase activity. Prostaglandins 4:313–324

Beaubien BC, Tippins JR, Morris HR (1984) Leukotriene biosynthesis and metabolism detected by the combined use of HPLC and radioimmunoassay. Biochem Biophys Res Commun 125:97–104

Bergstrom S, Ryhage R, Samuelsson B, Sjovall J (1963) Prostaglandins and related factors. The structures of prostaglandins E_1, F_{1alpha}, and F_{1beta}. J Biol Chem 238:3555–3564

Borgeat P, Samuelsson B (1979a) Transformation of arachidonic acid by rabbit polymorphonuclear leukocytes. Formation of novel dihydroxyeicosatetraenoic acid. J Biol Chem 254:2643–2646

Borgeat P, Samuelsson B (1979b) Metabolism of arachidonic acid in polymorphonuclear leukocytes: structural analysis of novel hydroxylated products. Biol Chem 254:7865–7869

Borgeat P, Hamberg H, Samuelsson B, (1976) Transformation of arachidonic acid and homo-γ-linolenic acid by rabbit polymorphonuclear leukocytes: monohydroxy acids from novel lipoxygenase. J Biol Chem 251:7816–7820

Borgeat P, Nadeau M, Salari H, Poubelle P, Fruteau de LaClos B (1985) Leukotrienes: biosynthesis, metabolism and analysis. In: Pavletti R, Kritchevsky D (eds) Advances in lipid research, vol 21. Academic, New York, pp 47–77

Brocklehurst WE (1962) Slow reacting substance and related compounds. Prog Allergy 6:539–558

Bryant RW, Hwang DH (1983) Development of a radioimmunoassay for 15-HETE and its application to 15-HETE production by reticulocytes. Prostaglandins 26:375–386

Corey EJ, Albright JO, Barton AE, Hashimoto SI (1980a) Chemical and enzymic synthesis of 5-HPETE, a key biological precursor of slow reacting substance of anaphylaxis (SRS), and 5-HETE. J Am Chem Soc 102:1435–1436

Corey EJ, Clark DA, Goto G, Marfat A, Mioskowski C, Samuelsson B, Hammerstrom S (1980b) Stereospecific total synthesis of a "slow reacting substance" of anaphylaxis, leukotriene C-1. J Am Chem Soc 102:1436–1439

Corey EJ, Marfat A, Munro J, Kim KS, Hopkins PB, Brion F (1981) A stereocontrolled and effective synthesis of leukotriene B. Tetrahedron Lett 22:1077–1083

Creticos PS, Peters SP, Adkinson NF, Naclerio RM, Hayes EC, Norman PS, Lichtenstein LM (1984) Peptide leukotriene release after antigen challenge in patients sensitive to ragweed. N Engl J Med 310:1626–1630

Davidson EM, Rae SA, Smith JH (1983) Leukotriene B_4, a mediator of inflammation present in synovial fluid in rheumatoid arthritis. Annals Rheumatic Diseases 42:677–679

Denis D, Charleson S, Rackham A, Jones TR, Ford-Hutchinson AW, Lord A, Cirino M, Girard Y, Larue M, Rokach J (1982) Synthesis and biological activities of leukotriene F_4 sulfone. Prostaglandins 24:801–814

Drazen JM, Austen KF, Lewis RA, Clark DA, Goto G, Marfat A, Corey EJ (1980) Comparative airway and vascular activities of leukotrienes C-1 and D *in vivo* and *in vitro*. Proc Natl Acad Sci USA 77:4354–4358

Engvall E, Perlman P (1971) Enzyme-linked immunosorbent assay (ELISA). Quantitative assay of immunoglobulin G. Immunochemistry 8:871–874

Eskra JD, Levine L, Carty TJ (1980) Preparation of (^{3}H)-12-L-hydroxyeicosatetraenoic acid and its use in radioimmunoassays. Prostaglandins Med 5:201–207

Ford-Hutchinson AW (1983) Biological properties of leukotriene and leukotriene sulfones. In: Piper PJ (ed) Leukotrienes and other lipoxygenase products. Wiley, Chichester, pp 152–160

Ford-Hutchinson AW (1985) Leukotrienes: their formation and role as inflammatory mediators. Fed Proc 44:25–29

Ford-Hutchinson AW, Rackham A, Zamboni R, Rokach J, Roy S (1983) Comparative biological activities of synthetic leukotriene B_4 and its ω-oxidation products. Prostaglandins 25:29–37

Granström E, Kindahl H, Samuelsson B (1976) Radioimmunoassay for thromboxane B_2. Anal Lett 9:611–627

Guindon Y, Zamboni R, Lau CK, Rokach J (1982) Stereospecific synthesis of leukotriene B_4 (LTB_4). Tetrahedron Lett 23:739–742

Hamberg M, Samuelsson B (1974) Prostaglandin endoperoxides: novel transformation of arachidonic acid in human platelets. Proc Natl Acad Sci USA 71:3400–3404

Hammarstrom S (1983) Leukotrienes. In: Snell EE, Boyer PD, Meister A, Richardson CC (eds) Annual review of biochemistry, vol 52. Annual Reviews, Palo Alto, pp 355–377

Hansson G, Lindgren JA, Dahlen S-E, Hedqvist P, Samuelsson B (1981) Identification and biological activity of novel ω-oxidized metabolites of leukotriene B_4 from human leukocytes. FEBS Lett 130:107–112

Hawkes R, Niday E, Gordon J (1982) A dot-immunobinding assay for monoclonal and other antibodies. Anal Biochem 119:142–147

Hayes EC, Lombardo DL, Girard Y, Maycock AL, Rokach J, Rosenthal AS, Young RN, Egan RW, Zweerink HJ (1983) Measuring leukotrienes of slow reacting substance of anaphylaxis: development of a specific radioimmunoassay. J Immunol 131:429–433

Herbert W (1978) Mineral-oil adjuvants and the immunization of laboratory animals. In: Weir DM (ed) Cellular immunology, 3rd edn. Blackwell, Oxford, (Handbook of experimental immunology, vol 2, pp A3.1–A3.15)

Humes J (1982) Prostaglandins. In: Adams DO, Edelson PJ, Karen HS (eds) Methods for studying mononuclear phagocytes. Academic, New York, pp 641–654

Hunter WM (1978) Radioimmunoassay. In: Weir DM (ed) Immunochemistry, 3rd edn. Blackwell, Oxford (Handbook of experimental immunology, vol 1, pp 14.1–14.40)

Jones T, Masson P, Hamel R, Brunet G, Holme G, Girard Y, Larue M, Rokach J (1982) Biological activity of leukotriene sulfones on respiratory tissues. Prostaglandins 24:279–291

Lee CW, Lewis RA, Corey EJ, Barton A, Oh H, Tauber AJ, Austen KF (1982) Oxidative inactivation of leukotriene C_4 by stimulated human polymorphonuclear leukocytes. Proc Natl Acad Sci USA 79:4166–4170

Lee CW, Lewis RA, Corey EJ, Austen KF (1983a) Conversion of leukotriene D_4 to leukotriene E_4 by a dipeptidase released from the specific granule of human polymorphonuclear leukocytes. Immunology 48:27–35

Lee CW, Lewis RA, Tauber AI, Mehrotra MM, Corey EJ, Austen KF (1983b) The myeloperoxidase-dependent metabolism of leukotriene C_4, D_4, and E_4 to 6-*trans*-leukotriene B_4 diastereoisomers and the subclass-specific *S*-diastereoisomeric sulfoxides. J Biol Chem 258:15004–15010

Levine L, Alam I, Gjika H, Carty TJ, Goetzl EJ (1980) The development of a radioimmunoassay for 12-L-hydroxyeicosatetraenoic acid. Prostaglandins 20:923–934

Levine L, Morgan RA, Lewis RA, Austen KF, Clark DA, Marfat A, Corey EJ (1981) Radioimmunoassay of the leukotrienes of slow reacting substance of anaphylaxis. Proc Natl Acad Sci USA 78:7692–7996

Levine L, Lewis RA, Austen KF, Corey EJ (1982) Radioimmunoassay of the 6-sulfidopeptide leukotrienes and serologic specificity of the antileukotriene D_4 plasma. Meths Enzymol 86:252–258

Lewis RA, Austen KF (1981) Mediation of local homeostasis and inflammation by leukotrienes and other mast cell dependent compounds. Nature 293:103–108

Lewis RA, Austen KF (1984) The biologically active leukotrienes. Biosynthesis, metabolism, receptors, functions and pharmacology. J Clin Invest 73:889–897

Lewis RA, Mencia-Huerta J-M, Soberman RJ, Hoover D, Marfat A, Corey EJ, Austen KF (1982) Radioimmunoassay for leukotriene B_4. Proc Natl Acad Sci USA 79:7904–7908

Lindgren JA, Hammerstrom S, Goetzl EJ (1983) A sensitive and specific radioimmunoassay for leukotriene C_4. FEBS Lett 152:83–88

Lord A, Charleson S, Letts LG (1985) Leukotriene F_4 and the release of arachidonic acid metabolites from perfused guinea pig lungs in vitro. Prostaglandins 29:651–660

Malmsten CL (1984) Leukotrienes: mediators of inflammation and immediate hypersensitivity reactions. CRC Crit Rev Immunol 4:307–333

Metz SA, Hall ME, Harper TW, Murphy RC (1982) Rapid extraction of leukotrienes from biologic fluids and quantitation by high-performance liquid chromatography. J Chromatogr 233:193–201

Miller DK, Sadowski S, DeSousa D, Maycock AL, Lombardo DL, Young RN, Hayes EC (1985) Development of enzyme-linked immunosorbent assays for measurement of leukotrienes and prostaglandins. J Immunol Methods 81:169–185

Morgan RA, Levine L (1982) Radioimmunoassay and immunochromatography of 12-L-hydroxyeicosatetraenoic acid. Meths Enzymol 86:246–252

Morris HR, Clinton PM, Tailor GW, Piper PJ, Tippins JR, Barrett K (1983) Extractions of leukotrienes from biological fluids: studies with ^{3}H-radiolabelled and unlabelled leukotrienes. In: Piper PT (ed) Leukotrienes and other lipoxygenase products. Wiley, New York, pp 64–69

Murphy RC (1985) Measurement of 5-lipoxygenase products in the lung. Prog Biochem Pharmacol 20:84–100

Murphy RC, Hammerstrom S, Samuelsson B (1979) Leukotriene C: a slow reacting substance from murine mastocytoma cells. Proc Natl Acad Sci USA 76:4275–4279

Parker CW, Huber MM, Falkenhein SF (1982) Pharmacological characterization of slow reacting substances. Meths Enzymol 86:655–667

Peters SP (1985) The cyclooxygenase and lipoxygenase pathways and inflammatory mediators. In: Kaplan AP (ed) Allergy. Livingstone, New York, pp 111–130

Peters SP, Schulman ES, Liu MC, Hayes EC, Lichtenstein LM (1983) Separation of major prostaglandins, leukotrienes and mono HETE's by high performance liquid chromatography. J Immunol Methods 64:335–343

Pinckard RN (1978) Equilibrium dialysis and preparation of hapten conjugates. In: Weir DM (ed) Immunochemistry, 3rd edn. Blackwell, Oxford (Handbook of experimental immunology, vol 1, pp 17.1–17.23)

Piper PJ (1983) Leukotrienes and other lipoxygenase products. Research Studies Press, Wiley, New York

Piper PJ (1984) Formation and action of leukotrienes. Physiol Rev 64:744–761

Pong S-S, DeHaven RN (1983) Characterization of a leukotriene D_4 receptor in guinea pig lung. Proc Natl Acad Sci USA 80:7415–7419

Rokach J, Girard Y, Guindon Y, Atkinson JG, Larue M, Young RN, Masson P, Holme G (1980) The synthesis of a leukotriene with SRS-like activity. Tetrahedron Lett 21:1485–1488

Rokach J, Hayes EC, Girard Y, Lombardo DL, Maycock AL, Rosenthal AS, Young RN, Zamboni R, Zweerink HJ (1984) The development of sensitive and specific radioimmunoassays for leukotrienes. Prostaglandins Leukotrienes Med 13:21–25

Rouzer CA, Scott WA, Griffith OW, Hamill AL, Cohn ZA (1982) Arachidonic acid metabolism in glutathione-deficient macrophages. Proc Natl Acad Sci USA 79:1621–1625

Salmon JA, Simmons PA, Palmer RMJ (1982) A radioimmunoassay for leukotriene B_4. Prostaglandins 24:225–235

Samuelsson B (1983) Leukotrienes: mediators of immediate hypersensitivity reactions and inflammation. Science 220:568–575

Stenson WF, Parker CW (1984) Leukotrienes. In: Stollerman GH (ed) Advances in internal medicine, vol 30. Year Book Medical Publishers, Chicago, pp 175–199

Taylor GW, Morris HR, Beaubien B, Clinton PM (1983) Fast atom bombardment mass spectrometry of eicosanoids. In: Piper PJ (ed) Leukotrienes and other lipoxygenase products. Wiley, New York, pp 277–282

Vane JR (1971) Inhibition of prostaglandin synthesis as a mechanism of action for aspirin-like drugs. Nature [New Biol] 231:232–235

Weller PF, Lee CW, Foster DW, Corey EJ, Austen KF, Lewis RA (1983) Generation and metabolism of 5-lipoxygenase pathway leukotrienes by human eosinophils: predominant production of leukotriene C_4. Proc Natl Acad Sci USA 80:7626–7630

Wynalda MA, Brashler JR, Bach MK, Morton DR, Fitzpatrick FA (1984) Determination of leukotriene C_4 by radioimmunoassay with a specific antiserum generated from a synthetic hapten mimic. Anal Chem 56:1862–1865

Young RN, Kakushima M, Rokach J (1982) Studies on the preparation of conjugates of leukotriene C_4 with proteins for development of an immunoassay for SRS-A. Prostaglandins 23:603–613

Young RN, Zamboni R, Rokach J (1983) Studies on the conjugation of leukotriene B_4 with proteins for development of a radioimmunoassay for leukotriene B_4. Prostaglandins 26:605–613

CHAPTER 20

Radioimmunoassay of Cyclic Nucleotides

C.W. PARKER

A. Introduction

Adenosine 3′,5′-cyclic monophosphate (cAMP) mediates a remarkable array of physiologic responses, fulfilling the role of a second messenger in virtually every cell type and tissue through its ability to stimulate protein kinases (ROBISON et al. 1968). Cyclic 3′,5′-guanosine monophosphate (cGMP), like cAMP, is very widely distributed, although its physiologic function remains uncertain despite clear-cut changes in a few tissues in response to physiologic stimuli. Both of these cyclic nucleotides are present in mammalian tissues at extremely low concentrations (cAMP 0.2–1.5 μmol/kg; cGMP 0.008–0.06 μmol/kg) and their measurement has therefore been very difficult (STEINER et al. 1970). In the mid-1960s, a number of enzymatic procedures were developed for measuring cyclic nucleotides, all of which required large amounts of tissue and extensive chromatographic purification before analysis, making it impossible to conduct measurements in large numbers of tissue samples. In the late 1960s, competitive binding assays were developed for the measurement of cAMP, involving either the use of anti-2′-*O*-succinyl cAMP antibodies (STEINER et al. 1969, 1972) or naturally occurring cAMP-binding proteins (GILMAN 1970). The radioimmunoassay was both highly specific for cAMP and sensitive to 1–2 pmol, permitting rapid measurements of large numbers of samples in as little as 10 mg tissue (STEINER et al. 1969, 1972).

The immunoassay was considerably improved in sensitivity by Cailla, Delaage, and their colleagues (CAILLA et al. 1973) by reacting the cAMP in tissue samples with succinic anhydride. The resulting 2′-*O*-succinyl cAMP more closely resembles the original immunogen than unmodified cAMP and was bound with a higher affinity by the antibody to cAMP, accounting for the increase in sensitivity. HARPER and BROOKER (1975) modified the original succinylation procedure of Cailla and Delaage to a more reproducible and equally sensitive acetylation procedure which permits the routine measurement of cGMP and cAMP in extremely small amounts without purification or concentration of the sample. A similar acetylation procedure has also been developed for cGMP, also permitting its measurement in small amounts of tissue despite its presence at very low concentrations. This chapter will review the radioimmunoassays for cAMP and cGMP, covering such aspects as the preparation of the immunogen, method of immunization, preparation of radioactive markers, possible methods for tissue extraction, application of the assays to enzyme measurements, routine immunoassay procedures, validation of the assay, and possible problems in interpretation. For previous reviews of cyclic nucleotide immunoassays see STEINER et al.

(1970) and Brooker et al. (1979). For general reviews of immunoassay procedures see O'Dell and Daughaday (1971), Parker (1976), and Gilman (1970).

B. Preparation of the Immunogen

The cyclic nucleotides are rendered immunogenic by conjugating their succinylated derivatives to a high molecular weight protein carrier. The cyclic nucleotides can be succinylated at the 2′-*O* position with succinic anhydride in anhydrous pyridine (Fig. 1; Steiner et al. 1969). Substitution at this position leaves the portions of the molecule most important for immunologic discrimination unaltered (Parker 1976). The method of succinylation is a modification of the method of Falbriard et al. (1967). In the preparation of succinyl cAMP (SCAMP), *N,N′*-dicyclohexylcarboxyamidine (0.76 mmol) (4′-morpholine or triethylamine may be used instead) is dissolved in 7.5 ml hot anhydrous pyridine and 0.7 mmol cAMP (free acid) is added slowly over the following 30–60 min. After cooling, 10 mmol succinic anhydride is added and the suspension is stirred at room tem-

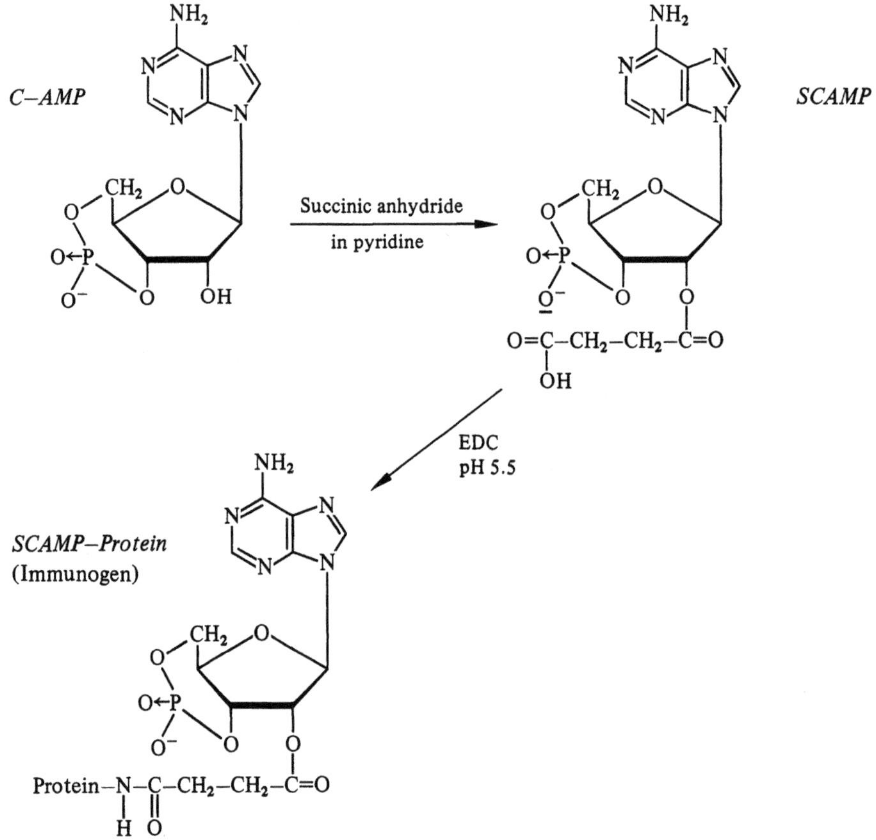

Fig. 1. Synthesis of 2′-*O*-succinyl cyclic nucleotide and of cyclic nucleotide immunogen. SCAMP succinyl cAMP; EDC 1-ethyl-3-(3-dimethylaminopropyl)carbodiimide. (Steiner et al. 1970)

perature for 18 h. Unreacted succinic anhydride is hydrolyzed by the addition of 3.75 ml water and allowing the reaction mixture to stand for an additional 2 h at 4 °C. At this stage, thin layer chromatography (TLC) of the reaction mixture on cellulose, developing with butanol–acetic acid–water 12:3:5 (v/v) demonstrates a new spot absorbing under short-wavelength ultraviolet light with an R_f of 0.42 (unmodified cAMP has an R_f of 0.3). The pyridine is removed by repeated rotatory evaporation under vacuum at 40 °C or below. The residue is dissolved in 3 ml water and, after adjusting the pH to 4.5, the product is purified by chromatography on a 1.5 × 44 cm Dowex 50 (H^+form) column using distilled water at a flow rate of 30 ml/h for elution; 8-ml fractions are collected and read at 258 nm. A single peak containing the SCAMP is obtained at or near fractions 30–45. (Cyclic AMP elutes later at or near fractions 50–65.) The fractions containing SCAMP are pooled, lyophilized, and evaluated by TLC. If necessary, the SCAMP can be further purified by precipitation as the barium salt. The yield of SCAMP ranges from 45% to 60%.

Succinyl cGMP (SCGMP) is prepared similarly (STEINER et al. 1969, 1972) but using trioctylammonium hydroxide for solubilization instead of *N,N'*-dicyclohexylcarboxyamidine. Since the cGMP is incompletely solubilized, the yield of SCGMP is lower than for SCAMP. The SCGMP is usually purified by TLC on cellulose using the solvent system described. The SCGMP has the same R_f as the cAMP derivative in this system. SCAMP and SCGMP and their tyrosine methyl esters (TME) are now available commercially from the Sigma Chemical Company or Boehringer Mannheim at reasonable cost.

In preparing the immunogen, the succinylated cyclic nucleotides are coupled to protein through their free carboxylate groups using a water-soluble carbodiimide for activation (Fig. 1; STEINER et al. 1970, 1969; PARKER 1976). Either SCAMP or SCGMP (10 mg) and solid protein (20 mg) – usually either human serum albumin (HSA) or keyhole limpet hemocyanin (KLH) – are dissolved in 2 ml water. The pH is adjusted to 5.5, then 10 mg 1-ethyl-3-(3-dimethylaminopropyl)carbodiimide-HCl (EDC) is added, and the pH is readjusted to 5.5 (avoiding extreme swings of pH). The reaction mixture is gently stirred in the dark for 18–24 h at room temperature and then dialyzed against phosphate-buffered saline (PBS, 0.15 *M* NaCl–0.01 *M* sodium phosphate, pH 7.4) at 4 °C with frequent changes of the dialyzing solution. The conjugate is analyzed by UV spectroscopy reading from 250 to 280 nm. The dialyzed conjugate exhibits absorption maxima at or near 258 nm (nucleotide) and 280 nm (protein). SCAMP has an extinction coefficient of 15000 at 258 nm, whereas cGMP has a value of 13100 at 255 nm. Typical conjugates have approximately 5 or 6 cyclic nucleotide residues per HSA molecule (based on a molecular weight of 67000 for the protein) or 60–70 residues per KLH molecule (based on a molecular weight of approximately 10^6 for the hemocyanin monomer). The dialyzed cyclic nucleotide–protein conjugate is preferably stored in aliquots in a lyophilized form since slow hydrolysis of the ester bond linking the nucleotide to the succinyl group may occur even when solutions are stored frozen. An aliquot of conjugate is reconstituted to isotonicity with distilled water or PBS (depending on the final dialysis solution) on the day of use. Use of diacid phosphate (pH ~5.75) for buffering the PBS is desirable if conjugates are stored in solution for extended periods.

C. Immunization

Rabbits, goats, and rats have all provided polyclonal antisera of good quality (STEINER et al. 1970; BROOKER et al. 1979) and monoclonal antibodies have been produced in mice (O'HARA et al. 1982). For the immunization of rabbits 0.25 mg immunogen is emulsified in complete Freund's adjuvant and injected into each footpad (STEINER et al. 1970; BROOKER et al. 1979). Booster injections with 0.2–0.4 mg cyclic nucleotide–protein conjugate are given into two footpads at 4- to 6-week intervals and the animals are bled 10–14 days after each boost. Booster injections may also be given into the back if inflammation of the foodpads becomes a problem. Antisera of good titer are usually present after a single booster injection, but later sera may be better. Blood is preferably obtained by arterial ear puncture. Cardiac puncture may also be used, but is associated with a significant mortality rate.

D. Preparation of the Radioindicator Molecule

Radioindicator molecules of very high specific activities are prepared by conjugating the succinylated cyclic nucleotides to tyrosine and iodinating with ^{125}I (Fig. 2; STEINER et al. 1969) which is similar to the radioimmunoassay for digitoxin described earlier by OLIVER et al. (1968). In the original method, the aglycone of digitoxin was succinylated, attached to TME, and iodinated. This marker could be used to measure serum levels of digitoxin despite their very low levels, thus providing the first practical radioimmunoassay for a drug present at low concentrations in body fluids. Radioiodinated haptens have subsequently been used in measurements of cAMP, cGMP, testosterone, and a variety of other ligands, indicating the broad applicability of this approach (PARKER 1976). Because of the presence of radioiodine in these markers, specific activities in excess of 150 Ci/mmol can be easily obtained.

While it might be assumed that the introduction of an iodinated benzene ring not present in the original immunogen would interfere with antibody binding, this has not been a problem, apparently because the attachment to protein and to tyrosine is made at the same position on the hapten. The conjugation of succinyl cyclic nucleotide to tyrosine involves a mixed carboxylic–carbonic acid anhydride reaction using ethylchloroformate for activation of the succinyl carboxyl group of SCAMP or SCGMP (STEINER et al. 1972). If the cAMP derivative is in the form of the barium salt, it is dissolved in dimethylformamide (DMF) and chromatographed on Dowex 50 (H^+ form) to remove the barium. Succinylated cyclic nucleotide (5 μmol) is dissolved (SCAMP) or suspended (SCGMP) in 100 μl DMF containing 15 μmol trioctylamine at 0 °C and 5 μmol ethylchloroformate in 25 μl DMF is added. After 15 min at 0 °C, 10 μmol TME hydrochloride and 10 μmol triethylamine are added in 100 μl DMF. The reaction mixture is maintained for 3 h at room temperature with continuous stirring. The tyrosine derivatives of the succinylated cyclic nucleotides migrate with R_f values of 0.57 in the cellulose thin layer system already described and can be obtained essentially pure by preparative TLC. The presence of tyrosine in the product can be verified by nitrosonaphthol staining (HAIS and MACEK 1963). The yield is 40%–55%.

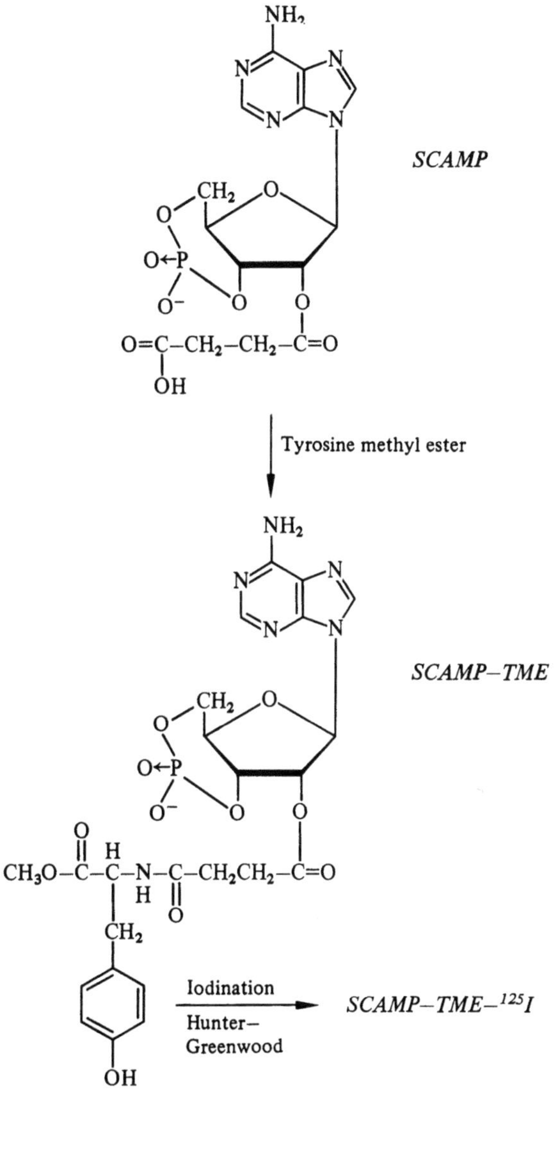

Fig. 2. Synthesis and iodination of cAMP ligand (SCAMP–TME). The coupling agent is a mixed carboxylic acid–carbonic acid anhydride. (STEINER et al. 1970)

For later immunoassay use, the TME derivatives can be distributed in aliquots containing 2 μg per tube in 50 μl 5 m*M* sodium acetate, pH 4.75 (BROOKER et al. 1979). The cyclic nucleotide concentrations in these solutions can be verified by UV absorption spectrometry using molar absorption coefficients of 14100 at 258 nm for cAMP and 12950 at 254 nm for cGMP.

For iodination, 1 mCi carrier-free sodium ^{125}I at a level of 100–500 mCi/ml without reducing agent is placed in a pointed minireaction vial; 20 μl 0.5 *M* potassium phosphate, pH 7, and 20 μl TME of the appropriate cyclic nucleotide (SCAMP–TME and SCGMP–TME) containing approximately 800 ng are added followed by 5 μl freshly dissolved chloramine-T, 1 mg/ml, in 0.05 *M* potassium phosphate, pH 7.2 (BROOKER et al. 1979). After 45 s at room temperature, excess sodium metabisulfite is added in a volume of 50 μl. The iodination is performed in a well-ventilated fume hood with all the usual precautions used in working with radioactive iodine. The radioactive cAMP–TME conjugate is readily purified by column chromatography on Sephadex G-10 (STEINER et al. 1969, 1972) or TLC on cellulose or polyethylimine- (PEI)-impregnated cellulose plates (SCHMIDT and BAER 1984). Probably the most convenient procedure is TLC on PEI cellulose. Sheets of PEI cellulose are cut to 20 × 10 cm squares and washed ahead of time with water. The iodination mixture is applied and the sheets are developed with 1 *M* LiCl. The radioactive spots are located by radioautography, cut out with scissors, and extracted twice with 1 ml 0.1 *M* NaCl. SCHMIDT and BAER (1984) have shown that TLC on PEI separates both iodinated SCAMP–TME and SCGMP–TME, each into two fractions, presumably corresponding to the mono- and diiodinated products. The slower band is apparently the diiodo product and provides somewhat higher immunoassay sensitivity than the faster monoiodo product, because of its greater specific activity. The iodinated cyclic nucleotide derivatives can also be purified by chromatography on Sephadex G-10 previously primed with PBS containing 1 ml 3% HSA and washed with PBS (STEINER et al. 1972). When the column is eluted with PBS, three distinct peaks of radioactivity are obtained. The iodinated TME elutes last after the main iodine peak on this column because of adherence of the tyrosine moiety to the column (Fig. 3).

After dilution with 0.02 *M* sodium acetate, pH 5.0, the iodinated cyclic nucleotides are stored in small aliquots at −20 °C. Iodinated preparations of SCAMP and SCGMP can be stored for as long as 2 months with no detectable loss of immunoreactivity, provided they are kept at −20 °C and not subjected to repeated freezing and thawing (BROOKER et al. 1979).

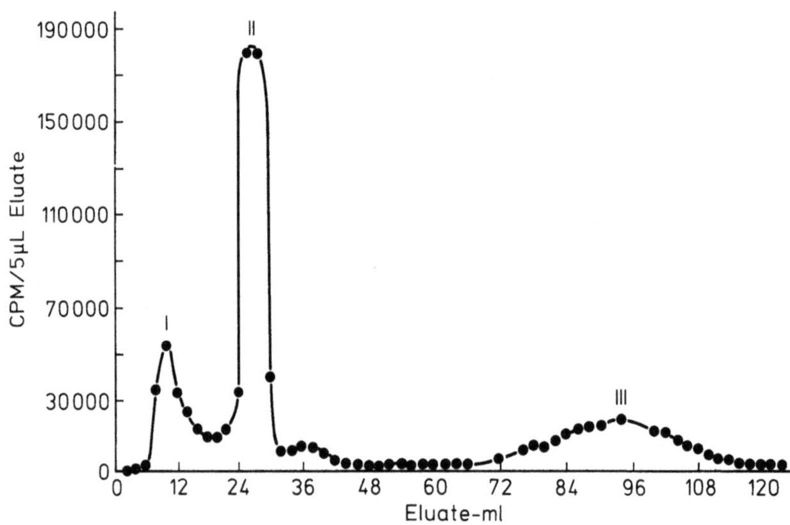

Fig. 3. Elution pattern of ^{125}I-SCAMP–TME from a Sephadex G-10 column (0.9 × 16 cm) using 0.15 *M* PBS as the eluent. Fraction I ^{125}I-TME; fraction II ^{125}I; fraction III ^{125}I-SCAMP–TME. (STEINER et al. 1969)

E. Evaluation of Antisera and Routine Immunoassay Procedures

Antisera are screened using the ^{125}I-labeled TME derivatives as radioindicator molecules. Standard curves are constructed with various nucleotides and nucleosides present in tissue samples as well as with the cyclic nucleotides themselves (Figs. 4 and 5; STEINER et al. 1970; BROOKER et al. 1979). Comparisons are made both with acetylated and unacetylated standards (Fig. 6). In preparing standard solutions, the extinction coefficients already given can be used to determine the precise concentrations of concentrated standards. Concentrated standards may be stored frozen at −70 °C for as long as 1 year (BROOKER et al. 1979). Diluted solutions of the standards can be stored for a few days at 4 °C. The assay is performed in 0.05 *M* acetate buffer which is used for all dilutions of standards or samples as well. Acetate buffer at pH 6.2 was utilized initially, but more recently pH 5.0 buffer, which provides a greater neutralization capacity following acetylation, has been used. The acetylated cyclic nucleotide standards are added in 50 µl, delivering 0, 3, 10, 30, 100, 300, and 1000 fmol per tube (BROOKER et al. 1979). For other nucleotides, much higher concentrations may and should be used. Evaluation of cross-reactivity of the acetylated nucleotides is of particular importance. With careful screening, antisera which are 10^6-fold more sensitive to cAMP than to other nucleotides have been identified (STEINER et al. 1970). Screening of the anti-cGMP antibodies must be done with even greater care since cAMP cross-reacts quite significantly with cGMP with most antisera. Moreover,

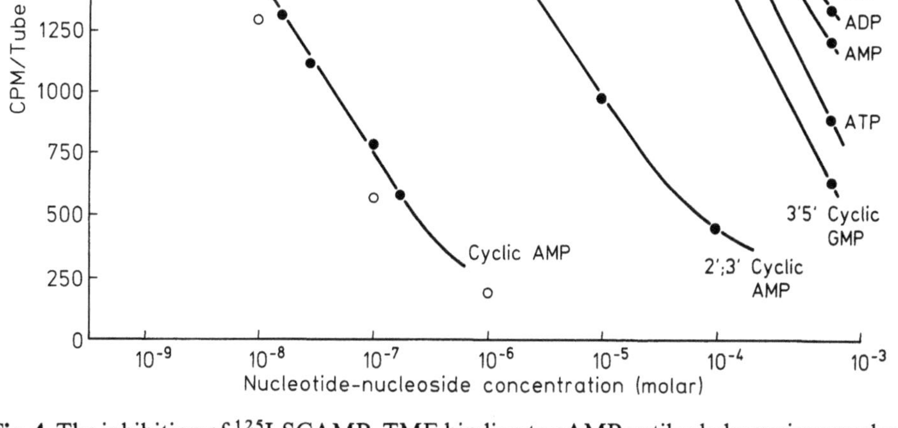

Fig. 4. The inhibition of ^{125}I-SCAMP–TME binding to cAMP antibody by various nucleotides. *Open circles* 2-deoxyribose 3′,5′-cAMP. (STEINER et al. 1970)

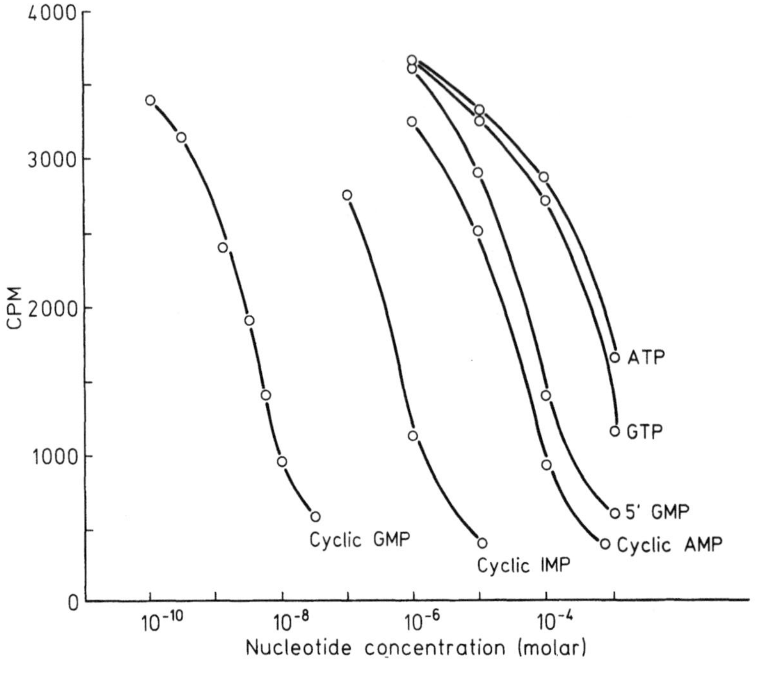

Fig. 5. The inhibition of ^{125}I-SCGMP–TME binding to cGMP antibody by various nucleotides. (Steiner et al. 1970)

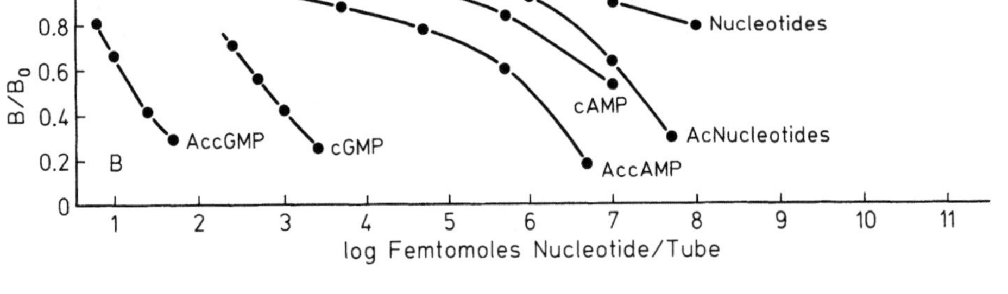

Fig. 6. Sensitivity and specificity of anti-cAMP and anti-cGMP sera. The effects of various nucleotides on the binding of SCAMP–^{125}I-TME or SCGMP–^{125}I-TME to their respective antibodies are shown. The amount of nonradioactive nucleotide per assay tube (in fmol) is plotted on the logarithmic abscissa. Displacement is denoted by B/B_0, the amount of radioactivity bound in the presence of competing nucleotide B divided by the amount bound in the absence of nonreactive nucleotide B_0. (Brooker et al. 1979)

concentrations of cAMP in tissues exceed those of cGMP by 10- to 100-fold, so greater sensitivity is needed in the cGMP assay.

Assays are performed in nonsiliconized disposable glass culture tubes (10 × 75 mm). Tubes contain (in order of addition): 50–100 µl cyclic nucleotide standard, unknown solution, or buffer; 100 µl diluted antibody in buffer containing 5 mg/ml bovine serum albumin (usually the dilution of antibody is in the range 1:1000 to 1:10000, sufficient to bind 30%–50% of the radiolabeled ligand); 100 µl ^{125}I-labeled ligand (approximately 10000 cpm) in buffer containing 5 mg/ml bovine serum albumin; and sufficient buffer to bring the final volume to 500 µl (BROOKER et al. 1979). The bovine albumin present with the radioactive marker and antiserum provide a final concentration of 2 mg/ml albumin in the final immunoassay mixture. After mixing, the incubation is allowed to proceed 3–24 h at 4 °C. Depending on the method of separating bound from free radioactive ligand, other additions may be made early in the assay.

I. Double-Antibody Procedure

In the double-antibody procedure, the second antibody (directed against the IgG of the animal species used for preparing the anti-cyclic nucleotide antibody) is added, either at the beginning of the assay or at some later time in the first 6 h and then the incubation is continued for another 18 h at 4 °C (PARKER 1976). An amount of second antibody sufficient to precipitate all of the first antibody must be used. The following day, the precipitate is sedimented by centrifugation, washed twice with PBS containing 2 mg/ml bovine serum albumin, and counted.

II. Ammonium Sulfate Precipitation

Carrier IgG, 200 µg per tube, preferably from the species of animal used in preparing the anti-cyclic nucleotide antibody, is included in the incubation mixture at the beginning of the assay. The antibody–ligand complex is precipitated by the addition of 2.5 ml cold 60% ammonium sulfate, mixing each tube individually on a vibrator as the ammonium sulfate solution is added. After incubation at 4 °C for 30 min, the precipitated protein is isolated by centrifugation. If desired, an additional wash with cold 50% ammonium sulfate may be included.

III. Charcoal

After 3 h or more at 4 °C, 1 ml of a cold suspension of charcoal (2 mg/ml in 100 m*M* potassium phosphate, pH 6.3, containing 2 mg/ml bovine albumin) is added (BROOKER et al. 1979). After another 20 min at 4 °C, the tubes are centrifuged at 1500 *g* at 4 °C for 10 min and the supernate or pellet is counted. When the charcoal immunoassay procedure is used, charcoal–albumin–buffer mixtures need to be made up every few days and centrifuged to remove fines on the day of the assay. With screening it is possible to identify antisera which permit measurements of cGMP in the presence of considerably higher concentrations of cAMP.

F. Tissue Extraction

Radioimmunoassays for cyclic nucleotides are generally performed on crude tissue extracts, although in working with a new tissue it is important to show that similar values are obtained with highly purified samples. The most frequent extraction method involves using trichloracetic acid (TCA) or perchloric acid (PCA) as a protein and tissue denaturant (STEINER et al. 1970; BROOKER et al. 1979). However, the acid that is added must be removed or neutralized before acetylation or succinylation and if it is neutralized the salt content of the sample will be increased. To extract a tissue with TCA, the samples ar either chilled to 4 °C in buffer or frozen and concentrated TCA is added to a final concentration of 5% or 10%. The tissue is homogenized with a Dounce or Polytron homogenizer or sonicated, and precipitated protein is removed by centrifugation. The TCA is removed from the supernate by extraction at least three times with five volumes diethyl ether saturated with water. Residual ether is removed by heating to 70 °C for 2 min and the aqueous phase is either sampled directly or the residue is dried under a stream of nitrogen and immediately or later eluted with buffer. PCA can also be used for tissue extraction, but the methods for its removal are generally less satisfactory than the ether extraction procedure for TCA. Use of diluted HCl (0.1–0.5 *M*, depending on the volume and concentration of buffer present with the tissue) may also be advantageous for processing tissue samples since hydrochloric acid is also a protein denaturant as well as being volatile. Still another method of tissue extraction which is suitable for some tissues involves taking 20–200 mg frozen tissue and boiling it for 10 min in the buffer used in the immunoassay. However, the main virtue of this procedure is its simplicity and it is primarily of value for samples that are not going to be acetylated or succinylated. A problem has been observed with cAMP measurements in human plasma where a protein (probably albumin) interferes with detectability of the cyclic nucleotide (EASTMAN and AURBACH 1982). A relatively efficient procedure for circumventing this problem is to prepare ultrafiltrates of plasma samples using a Gamme-Flo system and then to perform cAMP measurements in the filtrates.

G. Derivatization of Tissue Samples

To improve the sensitivity of the assay, the cyclic nucleotides present in tissue samples may be succinylated (CAILLA et al. 1973) or acetylated (HARPER and BROOKER 1975; BROOKER et al. 1979) to increase their reactivity with antibody. Since the two procedures produce comparable increases in assay sensitivity and acetylation occurs more readily and is more reproducible, acetylation is probably preferable. Acetylation of tissue samples improves assay sensitivity approximately 50- 100-fold. Samples for acetylation include 500 μg tissue sample, 10 μl undiluted triethylamine, and 5 μl undiluted acetic anhydride. The triethylamine is added with an Eppendorf pipette positioned with its tip just above the surface of the fluid. The acetic anhydride is most reproducibly added as a small bead at the bottom of the tube (BROOKER et al. 1979). The contents of the tube are mixed on a vibrator immediately after adding the acetic anhydride to avoid its loss by hydrolysis before the reaction with cyclic nucleotide is completed. Over-acetylation with reaction at the N-6 position of cGMP must be avoided since cGMP derivatized at this position shows considerably reduced immunologic reactivity (STEINER et al. 1970). At very high ratios of acetic anhydride to cGMP, an additional over-acetylation product is formed. Acetylated tissue samples must be adjusted without delay back to the pH at which the assay will be performed to maximize assay sensitivity and to minimize hydrolysis of the acetate ester bond in the acetylated cyclic nucleotide on standing. Even at an acidic pH, samples are stable for a maximum of 6 h, although frozen samples may retain full immunoreactivity for up to 3 days (BROOKER et al. 1979). When 0.05 *M* acetate, pH 5.0, is used as the immunoassay buffer and acetylated tissue samples are small, the adjustment in pH is not normally a problem (BROOKER et al. 1979).

Individual samples of acetylated cAMP and cGMP can be separated by high pressure liquid chromatography or TLC to verify that the immunoreactivity is present in these fractions and to compare different acetylation procedures. After acetylation, the immunoassay curve is linear between 5 and 100 fmol of either acetylated cAMP or cGMP. The increase in immunoassay sensitivity for cAMP or cGMP is usually as great with acetylation as with succinylation, presumably because the charge on the carboxylate group of the succinylated cyclic nucleotide is not present after conjugation to protein in the immunogen.

H. Pitfalls in Immunoassay Measurements

There are a number of important potential pitfalls in the performance of radioimmunoassays for cyclic nucleotides, some of which are problems for immunoassays in general and some of which are related to cyclic nucleotide assays per se (STEINER et al. 1970; BROOKER et al. 1979; PARKER 1976):

1. Rapid processing of samples is necessary to be sure that the levels of cyclic nucleotide present when the experiment is completed do not change as the tissue is being extracted (PARKER 1976). Depending on the tissue, significant losses of cAMP can occur within a period of time as short as 1 s. Nor can it be assumed that extraction methods that are satisfactory in one tissue are necessarily adequate for another. Rapid freezing in liquid nitrogen, boiling, or homogenization in TCA while still frozen may be necessary to avoid breakdown of cyclic nucleotides.

2. Considerable experience would indicate that crude tissue samples containing cyclic nucleotides for later measurements should be stored at −70 °C or below. Otherwise significant losses may occur.

3. Standards for acetylated tissue samples should be acetylated on the day of the assay because of slow deacetylation during storage, even at an acidic pH.

4. Triethylamine must be added before acetic anhydride during acetylation to ensure that the pH is high enough for the reaction to occur. Once the acetic anhydride is added, immediate mixing is necessary.

5. Once the acetylation is completed, failure to acidify the tissue samples rapidly is likely to result in accelerated hydrolysis of the newly introduced ester linkage.

6. Large aliquots of tissue samples should generally be avoided in the immunoassay because of possible alterations in ionic strength or pH which may interfere with the antigen–antibody reaction. High concentrations of bovine serum albumin seem to reduce interference from large amounts of salt or buffer in the immunoassay, but problems may nonetheless arise.

7. As in all immunoassays, the standards and blanks should be as comparable to the unknown tissue samples as possible. Therefore, any further dilutions of tissue samples should be performed in mock acetylation mixtures containing the amounts of triethylamine and acetic anhydride which are used in the standard acetylation procedure so that these components remain constant in the radioimmunoassay.

8. The reactivity of anti-cAMP antibodies may be affected by the divalent cation content of the medium (STEINER et al. 1970; VOLKER et al. 1985), probably because of complexing of the phosphate moiety of the cyclic nucleotide with cations such as magnesium and calcium ions or nonspecific effects in the immunoassay system. Antisera differ in their sensitivity to these effects, but certainly when measurements are performed on tissue samples containing chelating agents for divalent cations, consideration has to be given to possible nonspecific alterations in cyclic nucleotide measurements.

9. The amount of measured cyclic nucleotide should be directly proportional to the amount of tissue extract used in the assay.

10. Regardless of the procedure that is used to separate free and antibody-bound radioindicator molecules, the total processing time and temperature should be maintained as constant as possible between samples and standards.

11. Every immunoassay that is performed must be performed with a standard curve to ensure that the sensitivity and specificity of the assay are as expected. The standard curve with marker should detect unexpected problems in labeling of the radioactive marker or loss of its immunoreactivity due to deterioration. It will also detect problems associated with improper storage or dilution of the antibody.

12. Where a later bleeding from an immunized animal is used, new bleedings have to be assessed for possible changes in the titer, sensitivity, or specificity of the antibody (PARKER 1976). Antisera suitable for immunoassay purposes are available from a number of commercial sources. However, even when the previous experience has been favorable, the individual laboratory must recheck the immunoassay system from time to time.

J. Validation of the Immunoassay

Apart from the standard curve per se and a detailed analysis of the immunoassay system with a series of dilutions of possible cross-reacting nucleotides and nucleosides, a number of procedures can be used to help validate immunoassay results:

1. The addition of known amounts of cyclic nucleotide to tissue samples at the time the tissue is extracted (or even in some instances as the tissue is in its final stages of incubation) should give the expected increase in cyclic nucleotide con-

centration in the immunoassay. This helps verify that no interfering substance is present and that the added cyclic nucleotide is being recovered.

2. If ^{3}H-labeled cAMP or cGMP of high enough specific activity is available, the radioactive cyclic nucleotide may be added as an internal standard to permit the recovery of radioactivity as well as of immunoreactivity to be measured. Samples with or without added radioactive standard can be processed by high pressure liquid chromatography or some other fractionation system to determine if the immunoreactivity is recovered in the expected place and in the amounts expected from the column. Of course, where chromatographic procedures are being used for partial purification of a cyclic nucleotide, parallel column elutions without cyclic nucleotide in the sample should also be evaluated for possible reactivity in the immunoassay. This will help identify problems due to salt or buffer in the column fractions altering the ionic strength or pH in the immunoassay and interfering with the antigen–antibody reaction.

3. It is also frequently possible to add known amounts of a partially purified cyclic nucleotide phosphodiesterase to samples and show that the immunoreactivity is destroyed (STEINER et al. 1969). This procedure is more satisfactory in some types of tissue samples than others, perhaps because of variable amounts of divalent cations which may affect the activity of the enzyme.

4. In the initial radioimmunoassay measurements of a cyclic nucleotide in a laboratory, it may be helpful to study a tissue which has already been well characterized in terms of its cyclic nucleotide responses. Pharmacologic stimuli affecting cyclic nucleotide concentrations, such as PGE_2 or β-adrenergic catecholamines for cAMP, or azide or nitroprusside for cGMP, may be used to verify that the expected large cyclic nucleotide response can be readily detected in the immunoassay. Phosphodiesterase inhibitors may also produce substantial increases in cyclic nucleotides, although caution is needed because methylxanthines may interfere nonspecifically in the immunoassay.

5. In some instances, blank tissue samples may also be obtained where low levels of immunoreactivity are expected. For example, plasma that has been properly treated with charcoal should have a low cAMP content (EASTMAN and AURBACH 1982).

K. Special Applications of the Immunoassays

The radioimmunoassay for cAMP can be applied to the measurement of adenylate and guanylate cyclases (STEINER et al. 1970; BROOKER et al. 1979; VOLKER et al. 1985). Tissues containing these enzyme activities are incubated in the usual way with ATP or GTP (typically 1 m*M* ATP and 0.6 m*M* Mg^{2+} for adenylate cyclase or 1 m*M* GTP and 10 m*M* Mn^{2+} for guanylate cyclase). Theophylline or methylisobutylxanthine is present in both instances to inhibit cyclic nucleotide phosphodiesterase. In some tissues, it may be desirable to have an ATP-generating system present and 12 m*M* creatine phosphate and 1 mg/ml creatine phosphokinase may be used in combination for this purpose. The use of a radioimmunoassay to measure adenylate cyclase activity avoids expensive ^{32}P-containing assays and the use of column chromatography and barium zinc precipitations, or in fact any purification procedure to separate the cAMP before it is measured. In performing the cAMP measurements, controls are needed for possible effects of ADP, AMP, or adenosine hydrolyzed from ATP in the immunoassay (VOLKER et al. 1985). In addition, artifactual formation of cAMP from ATP–Mg^{2+} complexes under the alkaline conditions used for succinylation or acetylation is possible. EDTA can be used to complex the Mg^{2+}, avoiding this complication, but then consideration has to be given to the possible effect of removal of Mg^{2+} (or other divalent cations) on the reactivity of anti-cyclic nucleotide antibodies. Cross-reactivity from unhydrolyzed ATP in the immunoassay also has to be considered, which may necessitate the absorption of the ATP or oxidized ATP on alumina. In some instances, this has not been necessary owing to exceptionally low cross-reactivity between ATP and cAMP with the particular anti-cAMP antibodies available (STEINER et al. 1970). Another procedure that can be used is periodic acid oxidation of samples prior to succinylation which prevents the formation of cAMP, even in the presence of $MgCl_2$ (VOLKER et al. 1985). Once the cyclization problem is controlled, the radioimmunoassay can be as sensitive or more sensitive than methods in which α-^{32}P ATP is used as a substrate.

References

Brooker G, Harper JF, Terasaki WL, Moylan RD (1979) Radioimmunoassay of cyclic AMP and cyclic GMP. Adv Cyclic Nucleotide Res 10:2

Cailla HL, Racine-Weisbuch MS, Delaage MA (1973) Adenosine 3′,5′-cyclic monophosphate assay at 10^{-15} mole level. Anal Biochem 56:394–407

Eastman ST, Aurbach GD (1982) Determination of plasma cyclic AMP with automatic radioimmunoassay system (Gamma-Flo). J Cyclic Nucleotide Res 5:297–307

Falbriard JG, Posternak TH, Sutherland EW (1967) Preparation of derivatives of adenosine 3′,5′-phosphate. Biochim Biophys Acta 148:99–105

Gilman AG (1970) A protein binding assay for adenosine 3′,5′-cyclic monophosphate. Proc Natl Acad Sci USA 67:305–312

Hais IM, Macek K (1963) Paper chromatography. A comprehensive treatise. Academic, New York

Harper JF, Brooker G (1975) Femtomole sensitive radioimmunoassay for cyclic AMP and cyclic GMP after 2′O acetylation by acetic anhydride in aqueous solution. J Cyclic Nucleotide Res 1:207–218

O'Dell WD, Daughaday WH (1971) Principles of competitive protein-binding assays. Lippincott, Philadelphia

O'Hara J, Sugi M, Fujimoto M, Watanabe T (1982) Microinjection of macromolecules into normal murine lymphocytes by means of cell fusion. II. Enhancement and suppression of mitogenic responses by microinjection of monoclonal anti-cyclic AMP into B lymphocytes. J Immunol 129:1227–1232

Oliver GC Jr, Parker BM, Basfield DL, Parker CW (1968) The measurement of digitoxin in human serum by radioimmunoassay. J Clin Invest 47:1035–1042

Parker CW (1976) Radioimmunoassay of biologically active compounds. Prentice-Hall, New York

Robison GA, Butcher RW, Sutherland EW (1968) Cyclic AMP. Annu Rev Biochem 37:149–174

Schmidt K, Baer HP (1984) Purification of radioiodinated succinyl cyclic nucleotide tyrosine methyl esters by anion-exchange thin-layer chromatography. Anal Biochem 141:499–502

Steiner AL, Kipnis DM, Utiger R, Parker C (1969) Radioimmunoassay for the measurement of adenosine 3′,5′-cyclic phosphate. Pro Natl Acad Sci USA 64:367–373

Steiner AL, Parker CW, Kipnis DM (1970) The measurement of cyclic nucleotides by radioimmunoassay. In: Greengard P, Costa E (eds) Advances in biochemical psychopharmacology, vol 3. Raven, New York, p 89
Steiner AL, Parker CW, Kipnis DM (1972) Radioimmunoassay for cyclic nucleotides. J Biol Chem 247:1106–1113
Volker TT, Viratelle OM, Delaage MA, Labouesse J (1985) Radioimmunoassay of cyclic AMP can provide a highly sensitive assay for adenylate cyclase, even at very high ATP concentrations. Anal Biochem 144:347–355

CHAPTER 21

Radioimmunoassay of Platelet Proteins

D. S. PEPPER

A. Introduction

The platelets, formed by fragmentation from the megakaryocytes in the bone marrow, circulate in the blood and play a significant role in normal haemostasis as well as in pathological processes such as thrombosis, atherosclerosis, inflammation and immune rejection. Within the platelet are secretory storage vesicles of both dense and α granules containing respectively lower and higher molecular weight species. The contents of the α granules are largely proteins and include not only the familiar plasma proteins such as albumin and fibrinogen, but also a large proportion of unique proteins that have been referred to as platelet-specific granule proteins. Since normal plasma contains only low levels of these latter proteins, their assay measurement in plasma ex vivo constitutes a valuable method of measuring platelet activation and secretion as it occurs in vivo.

It has been known for a long time (for review see WALZ 1984) that both human and animal platelets contain a platelet-specific protein, platelet factor 4 or PF4, that could be detected by its ability to neutralise heparin (NIEWIAROWSKI et al. 1968). Subsequently, during purification of this molecule (MOORE et al. 1975a) another physically similar molecule termed β-thromboglobulin or β-TG was isolated, but it lacked anti-heparin activity (MOORE et al. 1975b). These molecules have been purified to homogeneity by a number of groups and their primary amino acid sequences are known (BEGG et al. 1978). Both proteins are basic, heat stable, acid soluble, have low contents of aromatic amino acids, no carbohydrate and are of rather low molecular weight. β-TG has a polypeptide subunit of 81 residues giving a subunit molecular weight of 8851, but under normal conditions these associate noncovalently to form a reversible tetramer structure of molecular weight 36000 (BOCK et al. 1982).

By contrast PF4 has 70 residues giving a polypeptide molecular weight of 7780 which also associates to form a tetramer of molecular weight 30000. However, unlike β-TG, PF4 is not readily soluble under physiological conditions, but is stored and secreted as a complex of 8 molecules of PF4 tetramer ionically complexed to a chondroitin-4-SO_4 proteoglycan carrier of molecular weight 110000 (BARBER et al. 1972) giving a total molecular weight of 350000 under physiological conditions. In the presence of highly sulphated molecules like heparin and heparan sulphate, PF4 can reversibly dissociate from its carrier and bind more strongly to the presumed target molecules (LUSCOMBE et al. 1981). Both PF4 and β-TG contain conserved amino acid residues giving a homology of 50%–60% and having similar secondary structures comprising a double loop in the form of a butterfly with both the NH_2 and COOH termini free (LAWLER 1981). This structure is stabilised by a pair of closely opposed disulphide bonds. At the COOH terminus of PF4 there exists a sequence of pairs of basic lysine residues alternating with pairs of hydrophobic residues whilst in β-TG this sequence is broken up by the insertion of acidic and hydrophilic residues. These structures are the basis of the heparin-binding site in PF4 and β-TG and explain the difference in magnitude of binding to heparin (PF4 > > β-TG). Fortunately for the purposes of radioimmunoassay, both molecules contain a single tyrosine residue which is amenable to radioiodination by any of the well-known methods.

A third protein thrombospondin (TSP) which is quite unrelated to the two already mentioned was first thought to be a thrombin-sensitive membrane glycoprotein (BAENZIGER et al. 1971), but was subsequently shown to be an α granule glycoprotein of molecular weight 450000 (LAWLER et al. 1978) composed of three disulphide-bonded polypeptide subunits of molecular weight 150000. This glycoprotein has calcium-binding sites, which induce conformational changes and is a substrate for thrombin. It binds to the platelet surface and can, under appropriate conditions, act as a "lectin", agglutinating cells in the presence of fibrinogen and may be involved in platelet–platelet or platelet–subendothelium interactions (PHILLIPS et al. 1980; MARGOSSIAN et al. 1981; JAFFE et al. 1982).

All three platelet proteins, β-TG, PF4 and TSP, have been purified, radiolabelled and made the basis of RIAs in several laboratories. Significant amounts of data have been accumulated on their concentrations and behaviour in bodily fluids (Table 1). More recently, the platelet-derived growth factor (PDGF), which is responsible for much of the cell growth stimulatory action of serum, has been purified in a number of laboratories and a radioimmunoassay (HELDIN et al. 1981) as well as a ligand-binding assay (BOWENPOPE and ROSS 1985) have been described. Unfortunately the amounts of PDGF in plasma and serum are very

Table 1. Platelet protein content of various bodily fluids in *Homo sapiens*[a]

Fluid	β-TG (ng/ml)	PF4 (ng/ml)	TSP (ng/ml)	PDGF (ng/ml)
	Mean SD	Mean SD	Mean SD	Mean SD
Plasma	37.2±10.9	14.7±10.1	105±31	< 0.2
Serum	18100±3650	14800±3900	17500±5500	17.5±3.1
Whole blood	10900±1860	10890±2470	18000±3400	
Urine	0.14± 0.09	0.43±0.2	<2–25	
Amniotic fluid[b]	48.9 ±35	5.6±6.1		
Synovial fluid[b]	56.5 ±19.8	8.8±6.2		
Cerebrospinal fluid[b]	1.08± 0.47	< 0.5		
Saliva	0.3 ± 0.2		136±24	
Colostrum			65000±10700	
Milk			3975±780	
Intraplatelet content (ng per/10^9 platelets)	55900±1200	63700±4350	89000±28300	42±9

[a] Data were taken from various publications and are therefore prone to standardisation errors.
[b] Some samples were necessarily pathological specimens.

small and the presence of interfering substances in plasma rules out the use of PDGF as a marker of platelet activation. However, the available data have been included in Table 1 for completeness (BOWENPOPE et al. 1984).

Despite the apparent close homology between PF4 and β-TG, the polyclonal animal sera raised in a variety of species have never shown any significant cross-reactivity (<1%) so presumably the surface epitopes are very different despite the primary sequence homology, which must reside in the core of the molecule (LAWLER 1981).

More recently (KAPLAN and NIEWIAROWSKI 1985), precursor forms of β-TG have been found which contain increasing numbers of NH_2 terminal residue additions, 4 in the case of "low affinity PF4" (LAPF4) and 13 in the case of "platelet basic protein" (PBP). A further possible candidate in this group is "connective tissue activating protein" (CTAP). Fortunately, none of these precursor forms is distinguishable by RIA with polyclonal antisera used in existing assay though DAVIS et al. (1985) have prepared an antiserum to a synthetic NH_2 terminus with the 4 residue additions of LAPF4. In practice, endogenous proteases in the body will degrade both PBP and LAPF4 to the stable end product β-TG. At present the only satisfactory way of quantitating these various homologues is by isoelectric focusing prior to RIA (the respective isoelectric points of PBP, LAPF4 and β-TG are ~9.0, 8.0, and 7.0).

For the purposes of laboratory investigation and pharmaceutical testing, it would be highly desirable if the common laboratory animals (mice, rats, guinea pigs, rabbits) had assayable levels of platelet-specific proteins. Whilst it is true that all vertebrate species tested do indeed express a heparin-neutralising activity in platelet-rich blood serum and this can likely be associated with the animal analogues of human PF4, it is unfortunate that no species cross-reactivity of any useful degree exists except in the higher primates (DAWES et al. 1983). Thus if such a test is needed it is necessary for most laboratory workers to start out with the purification of the animal protein and to prepare new tracers and antibodies. The smaller the animal the more difficult this becomes and so far only rabbit PF4 (GINSBERG et al. 1979) and porcine PF4 radioimmunoassays (RUCINSKI et al. 1983) have been described in the literature.

The situation with β-TG is even less clear, for no obvious biological activity has been associated with it, although a report exists of inhibition of PGI_2 synthesis in cultured cells (HOPE et al. 1979) and more recently of a partly homologous mRNA transcript in γ-interferon-stimulated monocytes (LUSTER et al. 1985) both pointing to a possible role in inflammation. Since no satisfactory bioassay exists for β-TG and since, with the exception of higher primates, the human RIA does not cross-react, it has not been easy to purify the small animal analogues of human β-TG.

However, based on physicochemical properties alone, MUGGLI et al. (1981) have been able to isolate from rabbit platelets, by means of gel filtration and heparin affinity chromatography, a protein or mixture which has the appropriate molecular weight, solubility, isoelectric point, etc. of an anticipated β-TG analogue, there is as yet little published data on how sensitive their RIA might be for platelet activation in the rabbit. A possibly homologous bovine β-TG has been isolated in impure form by CIAGLOWSKI et al. (1981).

In the case of TSP, however, the situation is quite different. DAWES et al. (1983) have shown that in higher primates no cross-reacting material can be detected, yet in bovine, ovine, caprine, porcine and canine species a considerable degree of cross-reaction is evident with samples of animal serum. In the case of bovine and foetal bovine serum an independent report confirms this (MCPHERSON et al. 1981). The cross-reactivity with canine platelet TSP, although not as complete as bovine TSP, is nevertheless sufficient to be useful, e.g. in a dog model of coronary thrombosis (KORDENAT et al. 1983). Unfortunately, TSP is also secreted by several other cells (WIGHT et al. 1985) including endothelial cells, at least in culture, so the possible contribution of these to plasma TSP levels is presently unknown.

The usefulness of any one of these three markers is dependent on a number of factors which cannot be predicted from simple in vitro measurements. Thus simple measurements of the serum:plasma ratio show a decreasing series PF4 > β-TG > TSP (actual values are respectively 1000 : 1, 500 : 1, 170 : 1), but the measured equilibrium levels in normal subjects are TSP > β-TG > PF4 (actual values are respectively 100, 30, 10 ng/ml) and finally the half-lives of the markers are respectively 9 h, 1.5 h and <10 min for TSP, β-TG and PF4. Another complicating matter is the catabolic route of each marker, which being different are also differently influenced in pathological conditions. It has been proposed (DAWES et al. 1978) that β-TG is cleared via the kidneys and thence tubular resorption, whilst PF4 is cleared via endothelial cell heparan sulphate (and secondarily via liver and/or kidneys) whilst TSP is cleared via the liver (LANE et al. 1984). It has been clearly shown that deteriorating kidney function results in abnormal elevation of β-TG (>60 ng/ml) (ANDRASSY et al. 1980) even in the absence of any platelet activation; thus, false positives can be generated simply by poor kidney function. To some extent this can be corrected for by calibrating the elevated β-TG against the observed creatinine levels.

Over the previous decade, the development of platelet-specific protein RIAs has made a significant impact on research and clinical studies of platelet function. However, the assays so developed have also provided additional useful information on the fundamental nature of the markers themselves. Progress in this area has been reviewed by DEBOER et al. (1982, 1983) and KAPLAN and OWEN (1981, 1982).

B. Sample Collection and Processing

The development of sensitive tests for in vivo platelet activation carries with it the need for an equally rigorous sample collection and processing system which will reliably prevent spurious activation of blood samples in vitro (LEVINE et al. 1981 a). Experience over the last decade has shown that it is this area, rather than RIA methodology per se, which has caused the majority of technical problems.

There is no absolute way of knowing what the "true" normal levels of β-TG and PF4 are in human plasma. Rather, an empirical, iterative approach is used whereby successive modifications in anticoagulants and procedures are tested against previous "best" systems until no further lowering in observed levels is seen. In our laboratories, for β-TG, this is in the range 20–30 ng/ml. However,

Table 2. Composition of different anticoagulants useful in platelet RIAs[a]

Name	Anti-coagulant	Antiplatelet drugs	Other additives	References
ETP (Edinburgh cocktail)	78 mM EDTA	10 mM theophylline, 0.33 µg/ml PGE_1		LUDLAM and CASH (1976)
ET (Amersham β-TG kit)	78 mM EDTA	10 mM theophylline		
Thrombotect (Abbott)	2.5% w/v EDTA	0.025% 2-chloro-adenosine	7% w/v procaine	
CAPE	113 mM citrate	150 mM aspirin, 5 µM PGE_1	550 mM ethanol	FILES et al. (1981)
CTAD (Stago)	110 mM citrate	3.7 mM adenosine, 15 mM theophylline, 0.2 mM dipyridamole		CONTANT et al. (1983)

[a] All anticoagulants are used in the ratio of 1 volume anticoagulant to 9 volumes whole blood except CAPE which is 1 volume to 4 volumes. Blood samples collected into any one of these anticoagulants are suitable for use in all three platelet protein RIAs, i.e. for β-TG, PF4 and TSP assays.

this approach has one serious disadvantage, for when comparing lowest numerical values from different laboratories, differences due to assays and standards will not readily be estimated, thus comparison of such empirical iteration is best limited to studies within a particular laboratory.

Early studies by LUDLAM and CASH (1976) established the need for cooling to 0 °C during both sample collection and centrifuging and the need to use EDTA as anticoagulant together with a pair of complementary platelet-inhibiting drugs such as theophylline and prostaglandin E_1, now widely known as ETP or the "Edinburgh cocktail". In our hands, this combination has not been bettered, but a number of alternatives exist (Table 2) which have other useful features. Thus, Abbott Laboratories have developed an anticoagulant which can be stored and used at room temperature. This anticoagulant (Thrombotect, Abbott catalogue 720301) contains EDTA as anticoagulant, 2-chloroadenosine as an anti-platelet inhibitor and the local anaesthetic procaine as a membrane-stabilising agent. In principle, this allows the blood samples to be processed without refrigerated centrifuges, but in practice it is difficult to keep blood below the 30 °C limit specified, especially using nonrefrigerated centrifuges in summer. In an attempt to improve room temperature processing, citrate has been used in place of EDTA as anticoagulant and platelet inhibitors of theophylline, adenosine and dipyridamole have been proposed by CONTANT et al. (1983) and this will be available commercially from Stago. The disadvantages of the "Edinburgh cocktail" ETP include the need to store unused anticoagulant at −40 °C before use because of the instability of PGE_1 in aqueous solution, and its nonavailability predispensed from commercial sources. However, the Amersham β-TG RIA kit (IM-88) does include a stable version of ETP from which the prostaglandin E_1 has been omitted. One disadvantage of the Abbott Thrombotect anticoagulant is that it is packaged in Vacutainers which cause unnecessary trauma to the blood sample and may result in enhanced in vitro activation.

Studies by LUDLAM and CASH (1976), FILES et al. (1981), and FABRIS et al. (1985) show that if the plasma samples obtained from suitable anticoagulants are diluted (to reduce their density) and further centrifuged at 50000 g for 1 h, a significant reduction in β-TG levels can be obtained. This probably reflects the removal of small and/or light, empty microvesicles of platelet-derived membrane structures with platelet protein either entrapped within or bound to their surface. Unfortunately, this technique is rather time-consuming and requires an ultracentrifuge, not normally available in the majority of hospital or coagulation laboratories.

An observation by RASI (1979) showed that even in low speed spinning ($\sim$1000 g) a proportion of light (i.e., empty or lipid-bound) platelet vesicles will accumulate in the meniscus layer of plasma, often associated with a visible pellicle of turbid, white, lipid material, probably chylomicrons. It is desirable, in the interests of assay reproducibility, to avoid this layer when pipetting off aliquots and to sample once only from the middle third of the plasma layer, avoiding disturbance to both the buffy coat and meniscus layers.

It sometimes happens that heparin is included in the sample, either as an in vitro experimental addition or from a previously heparinised patient. Since PF4 is not itself very soluble under physiological conditions, it tends to bind to negatively charged surfaces such as red cell and platelet membrane. At low (i.e. normal) levels of plasma PF4, a significant amount of PF4 will cosediment with the cell and if heparin is included in the sample prior to separation a higher (approximately twofold) level of PF4 may be found in the plasma owing to its preferential removal by heparin from the cell surface. This should not be confused with the platelet activation, or the heparin-induced release which occurs in vivo (see Sect. F).

Another source of artifactual platelet activation which should not be overlooked is the indwelling catheter. Studies have shown that 30–45 min after insertion of short, venous catheters, reliable, normal values of β-TG and PF4 can no longer be obtained. If at all possible, catheters should be avoided, but if this is not possible, e.g. in cardiac catheterisation, some form of antiplatelet drug should be present within the lumen either by slow perfusion between sampling or by the use of a double-lumen catheter, whereby anticoagulant is mixed with blood at the catheter tip (MANT et al. 1984).

The basic sampling procedure operated in the author's laboratory is as follows: 5 ml venous blood is taken with minimal trauma into a polypropylene syringe with a 19-gauge needle. The volume collected can be in the range 5–50 ml, but the entire sampling time should not exceed 60 s. With the needle still in place on the end of the syringe, 2.7 ml blood is dispensed into the bottom of a siliconised 12 × 75 mm glass tube, held upright in crushed moist ice with a premarked calibration to show the correct fill volume. The tubes already contain 0.3 ml ETP anticoagulant (see Table 2) and have been marked with the patient's name, etc. in waterproof pen. A glass tube is superior to a plastic one since the rate of cooling is much faster. The needle is kept on so that blood never contacts the upper (warm) half of the tube nor the rubber stopper area. After dispensing the blood sample (or samples) the tubes are capped and pushed down further into the ice mixture for 60 s to cool the entire tube and then gently shaken by tapping to mix the ETP anticoagulant and blood, yet without allowing any blood to contact the stopper or the upper half of the tube where it may warm up during subsequent processing. The tubes are then (within 2 h) transferred to a precooled swing-out centrifuge fitted with rubber adaptors to carry the entire set of experimental samples (e.g. 24) at one time, and the samples are spun at 1000 *g* for 45 min at 0 °C. It is always worthwhile to check the temperature calibration of the centrifuge at the end of a run, e.g. with a thermocouple inserted into one spare sample, or a small crystal ($\sim 5\ mm^3$) of wet ice, placed on top of the cap of one tube. At the end of the run, the ice crystals should still be present and wet. Experience has shown that cooling is more important than anticoagulation and should take precedence in any sequence of handling events.

The spun samples are carefully transferred without any shaking to a rack where the stoppers are gently removed and a disposable pipette capable of sampling 0.7 ml is inserted gently through the meniscus region until the tip is in the central third volume of plasma, when the sample is aspirated slowly and smoothly without mixing either the meniscus region or the buffy coat. The sample is then dispensed into 2-ml plastic tubes (prelabelled with patient's name) either as a single sample of 0.7 ml or as two aliquots of ~0.35 ml. It is not permissible to resample a second time from the same plasma volume on the red cells as mixing of meniscus lipid and/or buffy coat are more likely. The same sample of plasma can be used for both β-TG and PF4 assays. The samples after spinning and separation are remarkably stable as regards the two antigens and can be stored at −40 °C for several years without noticeable loss.

It often happens that β-TG and PF4 assays are required at the same time as other RIAs (e.g. fibrinopeptide A) or clotting assays (e.g. activated partial thromboplastin time). It is thus desirable that if a large volume syringe is filled (e.g. 50 ml) the first sample to be dispensed should be the β-TG/PF4 sample followed by progressively less labile samples into their appropriate anticoagulants. It is often possible to perform other assays on the sample collected into ETP, as the EDTA, etc. do not interfere in the assay. For those assays where interference is expected, it may be possible to use one of the citrate-based anticoagulants (see Table 2) e.g. CTAD (Stago), or more simply, a separate citrated sample tube.

Many alternative membrane-stabilising drugs apart from procaine have been described which can prevent α granule release (Prowse et al. 1982) and the antimalarials are particularly active, the best being hydroxychloroquine. Being stable for prolonged periods in aqueous solution, these represent a useful basis when combined with citrate for "nonrefrigerated" sample collection systems. However, it should be stressed that they have not yet given normal ranges as low as ETP at 0 °C and further work is still required.

A totally different approach to the problem of obtaining artifact-free samples is to use a body fluid other than blood which contains equilibrium levels of marker protein, but is devoid of platelets. In this respect β-TG is better than PF4 or TSP because it circulates freely in a low molecular weight form and should pass readily through semipermeable membranes into various bodily fluids (see Table 1). Significant data have only been obtained so far with urine (Deboer et al. 1981; Van Oost et al. 1983) and to a lesser extent with saliva (Cooke et al. 1981) when some correlation with deep venous thrombosis has been claimed. A problem with both these approaches is knowing how to allow for various secretion rates of the respective fluids. Despite this, they represent a valuable alternative to blood sampling, especially in situations such as mass screening where single samples must be obtained with the least risk of false positives. Unfortunately, this reliability is gained at the expense of sensitivity (i.e. more false negatives will occur) since there appears to be a threshold level of β-TG in plasma at about 80 ng/ml (significantly elevated) at which normal urine values can still be obtained (J. DAWES, J. ANDERTON and C. V. RUCKLEY 1981, unpublished work).

C. Assay of β-Thromboglobulin

The method of BOLTON et al. (1976a) is the basis of many "in-house" assays and is performed as follows. Purified β-TG is obtained from fresh platelet concentrates by the method of MOORE et al. (1975b) and 10 μg is iodinated to a specific activity of 50–80 μCi/μg by the chloramine-T method. No problems are normally encountered with this reagent, which can be stored frozen in liquid nitrogen (−196 °C) for several months after adding 30% w/v sucrose.

Specific antibody to β-TG can be "in-house", or a suitable sheep antibody (SA 1140) can be obtained from the Scottish Antibody Production Unit, Law Hospital, Carluke, Lanarkshire, Scotland. A solid-phase second antibody, e.g. commercially available donkey anti-sheep, is coupled to cyanogen bromide-activated Sepharose 4B in the ratio of, say, 5 ml whole serum to 35 ml hydrated activated gel.

After determining the dilution of primary antiserum needed to bind 50% of the tracer, appropriate dilutions of samples (20–50 μl) were mixed with 50 μl tracer (diluted to contain about 1 ng/ml β-TG and giving 10000 counts in <100 s) and 50 μl diluted antibody and incubated together at room temperature for ~18 h at 20 °C in disposable polystyrene 12×75 mm tubes. All dilutions were made in an assay buffer composed of 0.05 *M* phosphate, pH 7.4 and 2% v/v normal horse serum. After overnight incubation, the bound and free tracer were separated with the solid-phase antibody as follows. A slurry of solid phase (~20% v/v beads in dilution buffer) was constantly agitated with a magnetic stirrer bar and 100-μl volumes of slurry were dispensed via a disposable pipette whose tip had been cut off to increase the available lumen diameter and avoid clogging by beads. The assay tubes with solid phase were shaken on a horizontal rotary shaker at 1000 rpm for 1 h and then bound and free tracer were separated by injecting 1.8 ml 10% w/v sucrose as an underlayer below the samples and beads were simultaneously washed and separated by sedimentation through the sucrose layer at unit gravity, which normally is complete in 20 min (HUNTER 1977). Supernatant fluid was aspirated via a vacuum line leaving only the bound tracer attached to the solid-phase pellet in ~0.3 ml. Both the sucrose addition (underlayering) and supernatant removal (vacuum aspiration) were conveniently performed with a pair of eight-way manifolds constructed to the same spacing as the test tube rack. Nonspecific binding in this assay is routinely less than 2% and the assay calibration curve (Fig. 1) is sensitive down to <0.2 ng tube (≡3 ng/ml for a 50-μl sample).

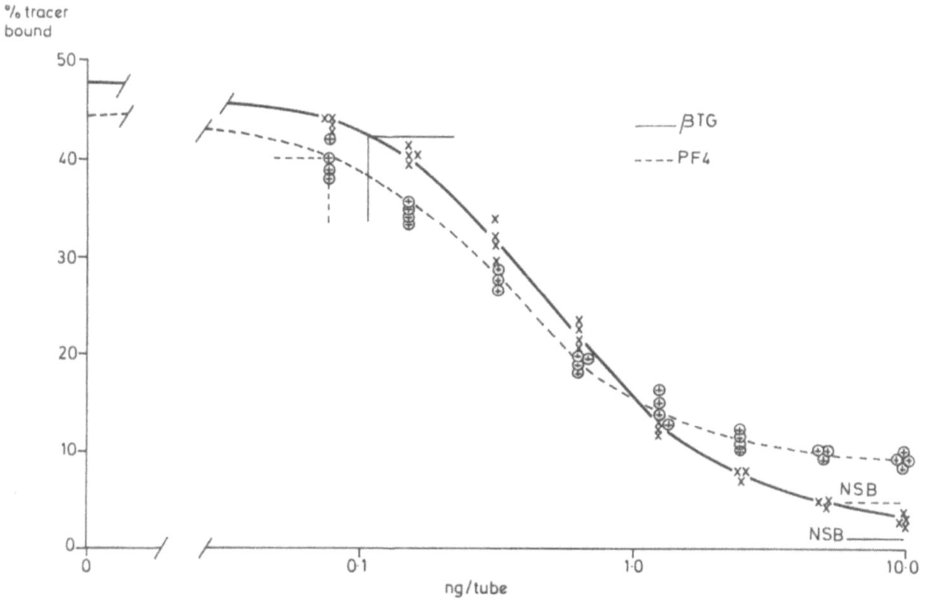

Fig. 1. Standard assay curves for the radioimmunoassay of β-TG (*full curve*) and PF4 (*broken curve*) using simultaneous addition of sample and tracer with 18 h first-stage incubation followed by 45 min second antibody separation. See Sects. C and D for experimental details. Lower sensitivity limits are shown by *full* and *broken right angles* for β-TG and PF4, respectively, and nonspecific binding (NSB) by *full* and *broken horizontal lines*, respectively

Two alternative forms of the assay have also been described (BOLTON et al. 1976a; BROWN et al. 1980) a more sensitive version (down to ~0.05 ng/ml) using delayed addition of tracer and taking 48 h to complete, and a rapid assay completed in 90 min, in which primary antibody is coupled to solid-phase Sepharose 4B and the bound tracer is separated rapidly before any drift of percentage binding in the sample batch can be detected. This constraint limits both the sample numbers and the precision.

For those laboratories which do not have prior experience of developing and running their own RIAs, the commercial kit for β-TG assay from Amersham (β-TG kit IM-88) is a satisfactory alternative. The kit contains anticoagulant, sample tubes, antibody at the appropriate dilution, tracer, ammonium sulphate separation reagent and a complete set of standards to construct a calibration curve from 10 to 225 ng/ml. The kit will perform 48 assays in about 90 min which, allowing for 5 standards in duplicate corresponds to 19 unknown samples in duplicate. Where unusual samples are present (e.g. serum or other artificially elevated plasma samples) it is sometimes necessary to predilute the sample to obtain values within the normal range (i.e. <225 ng/ml). It is essential that this be done with a reagent that contains the same amount of protein as the sample, but is devoid of cross-reacting β-TG antigen. Such a suitable diluent is 20%–100% v/v normal horse serum; under no circumstances must the sample be diluted in buffered salt solutions devoid of protein; even washed platelet pellets must be resuspended in horse serum prior to assay to avoid incomplete precipitation of bound tracer at the final ammonium sulphate separation step. This rapid assay kit can also be used with the same reagents as a slower (6 h), but more sensitive (>0.6 ng/ml) assay for the detection of β-TG in urine. The assay is simply operated in the delayed addition of tracer mode and full details can be obtained from Amersham on their information sheet for β-TG urine assays or in the publication of VAN OOST et al. (1983).

Even for those laboratories which do operate their own assays the commercial kits represent an alternative source of quality controlled tracer and assigned standards which are invaluable for "troubleshooting" when an "in-house" assay goes inoperative for whatever reason. An interlaboratory study on assay performance with the β-TG and/or PF4 assay as performed in different laboratories has been published (KERRY and CURTIS 1985).

D. Assay of Platelet Factor 4

The method of BOLTON et al. (1976b) as modified by DAWES et al. (1978) is the basis of several "in-house" assays for PF4. The modification of the original method is necessary when PF4 is prepared from outdated rather than fresh platelet concentrates, as proteolysis of the protein can produce a spectrum of tracer molecules that do not bind to heparin or antibody. Other published RIA methods include those of LEVINE and KRENTZ (1977), LEVINE et al. (1981b), and HANDIN et al. (1978).

The assay buffer and tubes are similar to those described for β-TG, as are the solid-phase second antibody and fluid-dispensing manifolds for the final separation, except that 1 *M* NaCl is added to the assay buffer to improve the solubility of PF4. Tracer is prepared from 10 μg PF4 dissolved in 10 μl 1 *M* NaCl and iodinated by the chloramine-T method. Labelled protein is separated from free iodide by gel filtration on a column of Sephadex G-50 swollen and washed in 0.05 *M* phosphate buffer, pH 7.4, containing 2% v/v normal horse serum and 1 *M* NaCl. A specific activity of 30–50 μCi/μg is so obtained. When the tracer is prepared from PF4 that is purified from fresh platelet concentrates, this is normally sufficient, but when the PF4 is purified from outdated platelet concentrates, an alternative immunopurification or affinity purification step is desirable.

Affinity purification on heparin–Sepharose (Pharmacia) is possible, the intact PF4 eluting as the last peak, between 1.0 and 1.5 *M* NaCl. However, some of these molecules may still be antigenically deficient and a better method is immunopurification on a column of solid-phase low affinity polyclonal antibody to PF4. In this method, 5 ml whole antiserum is coupled to 35 ml CNBr-activated Sepharose 4B gel and 2 ml of this gel is packed into a disposable polypropylene chromatographic column (Amicon-Wright) and eluted with a sequence of: 0.1 *M* acetate buffer, pH 4.7; 0.1 *M* borate buffer, pH 9.0; and finally 3 *M* KSCN. The precycled column is then washed with starting buffer (phosphate–horse serum–1 *M* NaCl) and a suitable quantity of PF4 tracer applied. After washing out unbound tracer with starting buffer, the immunologically bound PF4 is eluted with

3 *M* KSCN, pooled and diluted in starting buffer. Tracers were stable at 4 °C for up to 4 weeks and after adding 30% w/v sucrose, could be frozen as aliquots in liquid nitrogen and remained stable for several months, as judged by the maximum binding remaining above 90%.

The appropriate dilution of anti-PF4 serum is chosen so as to give approximately 50% binding and the assay is performed with 20- to 50-μl volumes of samples and standards in duplicate to which are added 50-μl volumes of tracer (~0.5 ng) containing sufficient radiolabel to give 10000 counts in less than 100 s. The antiserum samples and tracer are all diluted in assay buffer consisting of 0.05 *M* phosphate, pH 7.4+1 *M* NaCl+2% v/v normal horse serum and incubated overnight for 18 h at 20 °C without agitation.

Separation is obtained exactly as described in the β-TG assay, i.e. a slurry of excess second antibody solid phase is added, shaken for 45 min and sedimented through a sucrose underlayer prior to aspiration of unbound counts and counting of the pellet. This assay normally has a nonspecific binding (NSB) of <5% and a sensitivity of <0.2 ng per tube (≡3 ng/ml for a 50-μl sample; see Fig. 1).

For those laboratories not wishing to set up their own assays, a commercially available kit is obtainable from Abbott and also separately, the anticoagulant Thrombotect. This kit contains all the reagents for the assay of 100 samples and standards in the range 2.5–100 ng/ml in 2 h; thus, after allowing for 5 standards in duplicate, this means a maximum of 45 unknown samples in duplicate. As with the β-TG kit it is essential to carry out any predilutions of "off-scale" samples with the kit diluent provided or 20%–100% v/v horse serum to prevent inadequate precipitation of antigen–tracer antibody complexes in the final separation step. The presence of normal (<50 units/ml) concentrations of heparin does not interfere with this assay, probably because the ammonium sulphate concentration used is of sufficiently high ionic strength to dissociate the noncovalent heparin–PF4 complex. But with "in-house" assays where polyethylene glycol is used as a separation step (Woodhams and Kernoff 1983) PEG should not be used for PF4 assay in the presence of heparin. The Abbott kit's lowest standard for PF4 is 10 ng/ml and values below 2.5 ng/ml are unreliable, thus the practice of extrapolating the standard curve below 2 ng/ml by means of the zero standard is not recommended; instead, values should be reported as <2.5 ng/ml. It is not known if these same reagents can be used in a delayed addition of tracer mode to enhance sensitivity, but in any case, such an assay is not needed for plasma and in the case of urine, the measurement of PF4 is not useful.

A fast assay of PF4 has also been described in the literature (Brown et al. 1980). Again, as with the β-TG kit, the PF4 kit represents a source of quality controlled tracer PF4 and preassigned standards which are useful not only for those troubleshooting their own assays, but also for nonimmunological binding studies where phenomena such as ^{125}I-labelled PF4 binding to heparin columns or cell surfaces are being studied. The presence of heparin in the commercial tracer preparation should be made known to the experimenter concerned in these applications as it may interfere with binding.

Suitable supplies of high affinity sheep antibody to human PF4 are available as item SA1145 from the Scottish Antibody Production Unit, Law Hospital, Carluke, Lanarkshire, Scotland. Another source of goat anti-human PF4 is Atlantic

Antibodies, Bar Harbor, Maine, United States, but the author has not evaluated this source for use as either an immunopurification or immunoassay reagent.

E. Assay of Other Platelet Proteins

Only two other platelet proteins apart from β-TG and PF4 have had RIAs described for them in the literature, namely thrombospondin (TSP) and platelet-derived growth factor (PDGF), but since so little is published about the latter assay (HELDIN et al. 1981) only the former will be described in any detail.

The method used in this laboratory is that of DAWES et al. (1983). Two other assays have been described in the literature by MOSHER et al. (1982) and SAGLIO and SLAYTER (1982). Purified TSP is obtained most easily by a two-step sequential chromatographic process on heparin agarose followed by HPLC on the Pharmacia ion exchanger Mono Q (CLEZARDIN et al. 1984) using the thrombin-released products of fresh washed human platelets. Recently, large quantities (>1 mg/ml) of a cross-reacting TSP-like antigen have been found in bovine and caprine colostrum (DAWES et al. 1985) and this may be a valuable alternative source. A 20-μg sample of the purified TSP is iodinated by the chloramine-T method, separated from free iodide on Sephadex G-50 in 0.05 *M* NaH_2PO_4, pH 7.4+2% v/v horse serum+1% v/v Tween 20 and after adding 30% w/v sucrose can be stored as frozen aliquots in liquid nitrogen. Specific activities of 20–30 μCi/μg are routinely obtained and 85%–90% of counts can be bound to excess antibody. Suitable specific antibodies to TSP can easily be raised in rabbits (LAWLER et al. 1978) and do not appear to present any problems as far as RIA is concerned, though none of the antibodies appears to be precipitating. No commercial source of antibody to TSP is known.

The assay diluent is 0.05 *M* PO_4 pH 7.4+2% v/v horse serum + 1% v/v Tween 20. To each assay tube is dispensed 50 μl of a suitable dilution of antiserum, 50 μl of tracer dilution (equivalent to 0.5 ng TSP protein tracer and giving 10000 counts in <100 s) and 50 μl standard or sample plus 100 μl assay diluent to give a final volume of 250 μl. The assay tubes are incubated overnight at 20 °C without agitation, and then separated as in the β-TG and PF4 assays by the addition of 100 μl of a slurry of solid-phase donkey anti-rabbit IgG second antibody on Sepharose 4B. After shaking for 45 min, the bound and free tracer were separated by sedimentation at unit gravity through an underlayer of 10% w/v sucrose as described by HUNTER (1977). This assay has a nonspecific binding of 1.0%–1.5% and a sensitivity of 0.5 ng/tube (equivalent to 10 ng/ml with a 50-μl sample), the working range of the assay was 10–250 ng/ml. By using delayed addition of tracer, the assay sensitivity can be increased threefold, though time of assay increases from 24 to 48 h. Purified TSP tends to precipitate easily on freeze–thawing so pure antigen standards and tracer should be stored at 4 °C or snap-frozen in 30% w/v sucrose solution in liquid nitrogen with subsequent rapid thawing and no repeated freeze–thaw cycles. In our hands, frozen plasma samples have stable TSP antigen levels over several years' storage at −20 °C, but some instability has been reported by SAGLIO and SLAYTER (1982).

F. Applications of Platelet-Specific Protein Radioimmunoassays

In a review of this length it is not possible to detail specific studies, instead general points can be highlighted and recent publications which mention pharmacological intervention together with β-TG, PF4 and TSP studies are summarised in Table 3 together with the relevant references and pathology. Several reviews should be consulted for a more detailed discussion of the past literature (KAPLAN and OWEN 1981, 1982; DEBOER et al. 1982, 1983; MESSMORE et al. 1984; WALZ 1984; BOWENPOPE et al. 1984).

Because of the ephemeral nature of coagulation markers, it is essential that single assays of platelet markers be treated with the greatest caution, bearing in mind that artifactual in vitro activation is always a possibility and that a single assay may miss an important event because of its short duration and the half-life of the marker substance. As a general rule two or three samples should be taken to determine a baseline value for any individual before and after an event. When this is done with normal persons, it is found that the markers are remarkably stable for that individual, but different individuals show wide ranges (typically 10–50 ng/ml for β-TG). So far as is known, these markers show no sex-, race- or age-related differences except that several reports note an increase in "normal" persons over 60 years of age. However, this is difficult to separate from occult pathology in the older person. Platelet count will obviously be a variable influencing marker level, so this should be performed at the same time as an RIA. If the platelet count is in the normal range ($200–400 \times 10^9$/l) then no correction is necessary, but if platelet counts well below or above this range are found, correction to the norm is necessary. There is one report of β-TG/PF4 varying during the menstrual cycle (MOTOMIYA and YAMAZAKI 1981).

Where the underlying pathological process is of a chronic nature (diabetes, hypertension, etc.) or where the stimulation applied is chronic (cardiac valve prostheses) then estimation of markers and their interpretation is generally quite reliable, repeat assays giving good consistency. On the other hand, where the event is acute (myocardial infarction, transient ischaemic attacks) then it is very difficult to know how to interpret a single abnormal value, unless this is consistently raised on subsequent sampling or the event is repeated during sampling. Another category of importance in studies of platelet markers is where the event may be precipitated by the experimenter or subject at will (migraine, asthma, emotional stress) or where a drug is to be administered. The latter is a particularly useful type of experiment since the usual pre/post-baseline, crossover, placebo and washout all give useful information when repetitive sampling can be carried out.

Frequently, the experimenter will have available both β-TG and PF4 assays (as well as other coagulation RIAs such as fibrinopeptide A) and the more assays that are performed together as a group, the more useful the study becomes for each assay, especially when sampling is repeated at regular intervals. In this way, it is possible to gain some insight into what is a normal, abnormal or artifactual value. For instance, if both the β-TG and PF4 values are raised and the ratio is close to 1 : 1 in mass concentration terms, then the possibility of an in vitro artifact should be considered. On the other hand if both are elevated, but the mass ratio

Table 3. Recent clinical reports on the levels of β-TG, PF4 and TSP and their response to treatment by a variety of drugs[a]

Condition	Markers	Drug(s)	References[b]
Angina	β-TG/PF4		SOBEL et al. (1981)
Arterial disease	β-TG	Aspirin	KAPLAN et al. (1978)
Atherosclerosis	β-TG/PF4	Thromboxane antagonist	RIESS et al. (1984)
Cardiac catheterisation	β-TG/PF4	Antiplatelet	MANT et al. (1984)
Cardiomyopathy	β-TG		LONGO et al. (1984)
Charcoal haemo-perfusion	β-TG/PF4	PGI_2	CORDOPATRI et al. (1982)
Coronary artery disease	β-TG	Aspirin, Sulphin-pyrazone, β-blocker	STEELE (1980)
	β-TG	Coumarin + β-blockers	MUHLHAUSER et al. (1981)
	PF4	Propranolol	LEVINE et al. (1984)
	β-TG/PF4	Aspirin + sulphinpyrazone	ALMONDHIRY et al. (1985)
Coronary bypass surgery	β-TG/PF4	PGI_2	WALKER et al. (1981)
	β-TG/PF4	Aspirin + dipyridamole	SALTER et al. (1982)
	β-TG/PF4	PGI_2	AREN et al. (1983)
	β-TG/PF4	PGI_2	DITTER et al. (1983)
Crohn's disease	PF4		SIMI et al. (1982)
Deep vein thrombosis (DVT)	β-TG		DEBOER et al. (1981)
	β-TG		WOJCIECHOWSKI et al. (1983)
	β-TG	Stanozolol	DOUGLAS et al. (1985)
Diabetes	β-TG	Dipyridamole	SCHERNTHANER et al. (1979)
	β-TG/PF4	PGE_1	MUHLHAUSER et al. (1980)
	β-TG/PF4	Ticlopidine	BAELE et al. (1982)
	β-TG		VAN OOST et al. (1983)
	β-TG		JANKA et al. (1983)
	β-TG		PORTA et al. (1983)
	β-TG		BORSEY et al. (1984)
	β-TG/PF4	Insulin	HOLAN et al. 1984)
	β-TG/PF4/TSP		LANE et al. (1984)
	β-TG/PF4		REYNDERS et al. (1984)
	β-TG/PF4		ROSOVE et al. (1984)
Eclampsia	β-TG/PF4		AROCHA-PINANGO et al. (1985)
Grafted artery	β-TG/PF4	Sulphinpyrazone	CADE et al. (1982)
Hip replacement	β-TG/PF4/TSP		LANE et al. (1984)
Hyperlipidaemia	β-TG/PF4		STRANO and DAVI (1982)
Hypertension	β-TG/PF4	PGI_2	BUGIARDINI et al. (1982)
	β-TG		KJELDSEN et al. (1983)
	β-TG	Sulphinpyrazone	DAVI et al. (1984)
	β-TG	Prazosin	IKEDA et al. (1985)
Ischaemic heart disease	β-TG	Dipyridamole	SANO et al. (1980)
	β-TG/PF4	Unsaturated lipids	HAY et al. (1982)
Leukaemia and polycythaemia rubra vera	β-TG/PF4/TSP		LANE et al. (1984)
	β-TG/PF4	Ticlopidine	NAJEAN and POIRIER (1984)
	β-TG/PF4		VIERO et al. (1984)
Liver disease	β-TG/PF4/TSP		LANE et al. (1984)
Menstrual cycle irregularities	β-TG/PF4	Oestrogen	MOTOMIYA and YAMAZAKI (1981)

Table 3 (continued)

Condition	Markers	Drug(s)	References[b]
Migraine	β-TG	Aspirin	D'ANDREA et al. (1983)
	β-TG		MANNOTTI et al. (1983)
	β-TG		VIOLI et al. (1984)
Mitral valve replacement	β-TG	Suloctidil	TURPIE et al. (1984)
Moya-moya disease	β-TG	Aspirin, dipyridamole, ticlopidine, bencyclane	TAOMOTO et al. (1983)
	β-TG/PF4	Aspirin	ALBALA and Levine (1984)
Myocardial infarction	β-TG		VAN HULSTEIJN et al. (1984)
Peripheral vascular disease	β-TG/PF4	PGI_2	BLATTLER et al. (1981)
	β-TG/PF4	PGI_2	SINZINGER et al. (1983)
	β-TG/PF4		BLATTLER et al. (1984)
Prosthetic heart valves	β-TG	Aspirin + dipyridamole	CHESEBRO et al. (1981)
	β-TG	Sulphinpyrazone	LUDLAM et al. (1981)
Psoriasis	β-TG		BERRETINI et al. (1985)
Raynaud's disease	β-TG	Prostacyclin analogue	BELCH et al. (1985)
Renal dialysis	β-TG/PF4	Indobufen, heparin	BUCCIANTI et al. (1982)
	β-TG/PF4	Indobufen, heparin	POGLIANI et al. (1982)
	PF4	Heparin	FLICKER et al. (1982)
Renal disease	PF4	Sulphinpyrazone	BERN and GREEN (1982)
	β-TG/PF4/TSP		LANE et al. (1984)
Scleroderma	β-TG	Dipyridamole	KAHALEH et al. (1982)
Sickle cell disease	β-TG/PF4	Ticlopidine	SEMPLE et al. (1984)
Smoking	β-TG	Oral contraceptives	DUNCAN et al. (1981)
	β-TG/PF4		DAVIS et al. (1983)
	β-TG		LONGENECKER et al. (1984)
	β-TG/PF4		SCHMIDT and RASMUSSEN (1984)
Stress	PF4	Catecholamines, β-blockers	LEVINE et al. (1979)
	PF4	Catecholamines, β-blockers	LEVINE et al. (1982)
	PF4	Epinephrine	SUAREZ et al. (1982)
	β-TG	Catecholamines	FITCHETT et al. (1983)
	PF4	Catecholamines, β-blockers	LEVINE et al. (1985)
Stroke	β-TG	Antiplatelet	TAOMOTO et al. (1983)
Thrombocytosis	β-TG		FABRIS et al. (1984)
Transient ischaemic attacks	β-TG	Aspirin + dipyridamole	AUSHRI et al. (1983)
	β-TG/PF4	Aspirin + dipyridamole	FABRIS et al. (1983)
	β-TG	Antiplatelet	TAOMOTO et al. (1983)
Normal	PF4	Histamine	BRINDLEY et al. (1983)
	PF4	Histamine	GOETZL et al. (1983)
	TSP	Trifluoperazine	GARTNER and WALZ (1983)
	β-TG/PF4	Membrane stabilisers	PROWSE et al. (1982)
	β-TG	Unsaturated lipids	VALLES et al. (1982)

[a] The majority of these studies are in vivo.
[b] References given usually report a response (lowering) of the platelet RIA marker to the drug. There are many other reports of no response to drugs.

is 2:1, then a genuine in vivo activation can be anticipated. If β-TG values are elevated, but PF4 is normal, kidney failure should be considered as a contributory cause, whilst elevated PF4 levels in the presence of normal β-TG values are normally seen following heparin infusion. A study of individual β-TG:PF4 ratios in obstetric patients has been described (AROCHA-PINANGO et al. 1985) and in myeloproliferative diseases (VIERO et al. 1984), and for single samples from one individual it is clear that the β-TG:PF4 ratio has little value, though this may improve when linear studies are carried out in an individual.

One of the more unexpected findings which arose from the availability of RIAs for platelet proteins was the discovery of the "heparin effect" on PF4 levels in vivo (DAWES et al. 1982). When heparin (or other highly sulphated polymers) are infused to humans in the total dose range 500–10000 units, a rapid increase in the circulating plasma level of PF4 is seen without any change in the levels of β-TG or TSP, i.e. it is not due to direct release from the platelets and no such increase is seen when heparin is added to whole blood in vitro. This PF4 effect was shown to have two important features:

1. It came from a strictly limited reservoir, which was readily exhausted by doses of heparin above 5000 units.
2. The half-life refill time of this pool was about 24 h, i.e. no subsequent response was obtained if heparin was reinfused a short time (e.g. 90 min) after the first dose.

From other experiments with tissue-cultured endothelial cells (BUSCH et al. 1980) it was shown that endothelial cells are the most likely source of this reservoir, and the PF4 is probably derived from the normal platelet background activation in blood and then cleared rapidly to the endothelial cell surface where heparan sulphate acts as a receptor, and heparin is capable of displacing the PF4 back into the circulation by virtue of its higher affinity complex with PF4 (BARBER et al. 1972; LUSCOMBE et al. 1981). An unexpected conclusion of this work is that although the biological activity of heparin disappears from the plasma with a half-life of about 90 min, the "immobilised anticoagulant" in the form of heparan sulphate newly exposed on the endothelium disappears (or is neutralised) with a half-life of about 24 h, i.e. although the drug cannot be detected in the blood, it may well have a persistent anticoagulant (or antithrombotic) effect.

As can be seen from Table 3, a remarkably wide application of RIAs for platelet proteins has been seen in diverse clinical studies over the last decade. This is largely due to the commercial availability of simple kits for β-TG and PF4. Ideally, both assays should be performed whenever possible as they tend to validate each other. Where this is not possible, backup samples of plasma can be kept for assay of the second marker when results of the first marker are considered interesting. Finally, if only one marker is to be studied, the author prefers β-TG because in most situations it appears to be the most sensitive to changes in the haemostatic/thrombotic mechanisms.

The consensus of opinion of clinical studies over the last decade is that although with appropriate control groups, significant differences between patient and control groups are readily found, the degree of overlap between individuals often prevents the appropriate clinical management of a particular patient on the basis of a single RIA measurement of a plasma platelet marker level.

Despite this, there are particular circumstances where effective management is possible, particularly where, for example, the effect of a particular drug is being investigated, then the dose and time period to achieve a response are readily obtained via objective measurements. Interestingly these studies sometimes show that drugs (e.g. antiplatelet drugs) do produce a normalisation of the platelet activation marker and yet the clinical symptoms do not subside. Evidently in these cases the platelet activation is either caused by some other disease or is independent of it. More rarely, the disease responds to therapy, but yet the markers remain elevated long after objective clinical improvement. In these cases perhaps, the markers are evidence of an underlying chronic stimulation which is a necessary, but not sufficient cause without another precipitating event. More commonly, both the clinical symptoms and the platelet markers either respond or do not respond to treatment together.

Complicating factors abound which are themselves capable of modulating β-TG and PF4 levels and thus blurring the correlation between marker levels and the presence of disease. These include emotional stress (LEVINE et al. 1982, 1985), septicaemia (VAN HULSTEIN et al. 1982) and hospitalisation itself (SMITH et al. 1978). Smoking (DUNCAN et al. 1981; DAVIS et al. 1983; LONGENECKER et al. 1984) and oral contraceptives have also been implicated (DUNCAN et al. 1981) but the evidence is still conflicting.

A number of workers (O'BRIEN et al. 1984) have used the intraplatelet content of β-TG and/or PF4 as an indicator of activation of the release reaction on the grounds that a circulating, depleted platelet will have a half-life similar to other platelets ($\sim$ 6 days) and this persistence might overcome the artifacts of plasma sampling. However, the technical problems of platelet isolation and counting as well as the small decrease measured against a large background tend to cancel out any theoretical advantage of this approach.

G. Discussion

Over the previous decade, the radioimmunoassay of platelet proteins has moved out of the research laboratory into the clinic where the availability of good reliable and simple home-made or commercial assays has played an important role. During the next few years we can anticipate an ever widening clinical application of these tests, but it is important to bear in mind that their intelligent deployment requires a clear understanding of their strengths and weaknesses by both the RIA service laboratory and the clinician.

The assays themselves, though technically excellent, will be severely compromised if good sample collection technique is not taught and maintained. In the design of studies, as many other assays as possible should be performed to back up or validate the RIA markers, at least until the persons concerned have confidence that they can be dispensed with. Such markers might include platelet count, both β-TG *and* PF4 assays, complete coagulation screens and whole blood haematology, fibrin degradation products, fibrinopeptide A RIA, etc. Finally, interpretation of results is still as much an art as a science, requiring experience on the part of those involved, preferably under local conditions and with the particular patient–disease–drug combination. As a first step in this direction, the normal range of the local population should be determined by the investigating team, e.g. both in blood donors and in the hospital population devoid of haemostatic/thrombotic abnormality. Interlaboratory training schemes for both technical and clinical staff are a convenient way to acquire initial skills, and such a scheme has been operated by the European Economic Community under the auspices of the ECAT programme whereby various blood tests are being subjected to trials of their predictive value in large patient subgroups suffering either angina or deep venous thrombosis. As with all RIAs, international standardisation is an important goal, and the recent establishment of satisfactory technical material (KERRY and CURTIS 1985) opens the way for a future ICTH/WHO international standard for both β-TG and PF4.

In planning studies with the various platelet marker RIAs, an understanding of their various unique properties is desirable and these are conveniently summarised as follows: in higher primates β-TG and PF4 can be assayed with the same reagents as used for humans, but TSP cannot; however, in contrast TSP can be assayed in several domestic animals (see Table 2 in DAWES et al. 1983 for details). Heparin administered intravenously rapidly raises PF4, but is without effect on β-TG and TSP. Both urine and saliva samples may be useful for β-TG assays, but are probably not useful for PF4 and TSP assays. Kidney function which is impaired will raise β-TG levels artifactually by virtue of reduced clearance whereas this has little effect on PF4 and TSP. In contrast, poor liver function may result in raised TSP levels for similar reasons. There appear to be no significant sex, race or age differences for any of the three markers except a tendency to rise above the age of 60 years. Platelet counts within the normal range do not require any correction to the observed values of markers. All three proteins appear to be stored in the same granule compartment and released simultaneously from the platelet. The half-lives of the markers decrease in the series TSP $>>$ β-TG $>>$ PF4.

Thus, in conclusion, the radioimmunoassay of platelet-specific proteins has proven to be an excellent way of monitoring platelet activation in vivo. In contrast to earlier methods such as aggregometry, which has been the major tool used in the evaluation of antiplatelet drugs, the RIAs are capable of working with samples which have been subjected to physiological conditions such as haematocrit, oxygen tension, shear rate and ionised calcium concentration. Also, in contrast to aggregometry, no choice of agonist is necessary. Thus, for the first time it has been possible to monitor the effects of therapeutic intervention with drugs upon the platelet release reaction in vivo. It seems reasonable to equate the release reaction in vivo with activation in vivo, though the stimuli necessarily remain unknown. Nevertheless, the fact that a significant number of the compounds mentioned in Table 3 are indeed capable of reducing platelet activation in vivo and that this effect can be measured objectively is a major step forward in our understanding of platelet pharmacology. Two important goals remain to be achieved however, the establishment of nonhuman animal models for the evaluation of newer compounds in vivo and the longer-term goal of proving in the clinical setting the relevance or otherwise of platelet activation per se to the clinical outcome of a particular disease. In this respect, the availability of accurate, reliable and specific radioimmunoassays has a central role.

References

Albala MM, Levine PH (1984) Platelet factor 4 and beta-thromboglobulin in moya-moya disease. Am J Pediatr Haematol Oncol 6:96–99

Almondhiry H, Pierce WS, Pennock JL (1985) Platelet release in coronary heart disease – effect of antiplatelet drugs and coronary artery bypass graft. J Lab Clin Med 105:397–402

Andrassy K, Deppermann D, Ritz E, Koderisch J, Seelig H (1980) Different effects of renal failure on beta-thromboglobulin and high affinity platelet factor 4 (HA-PF4) concentrations. Thromb Res 18:469–475

Aren C, Fedderson K, Tegernilsson AC, Radegran K (1983) Effects of prostacyclin infusion of platelet release during extracorporal circulation. Int J Microcirc 2:268–269

Arocha-Pinango CL, Ojeda A, Lopez G, Garcia L, Linares J (1985) Beta-thromboglobulin (beta-TG) and platelet factor 4 (PF4) in obstetrical cases. Acta Obstet Sci 2:115–120

Aushri Z, Berginer V, Nathan I, Dvilansky A (1983) Plasma thromboglobulin and platelet aggregation index in transient ischemic attack – effect of aspirin and dipyridamole therapy. Int J Clin Pharmacol Ther Toxicol 3:339–342

Baele G, Rottiers R, Rubens R, Priem H (1982) Effect of ticlopidine on plasma beta-thromboglobulin and platelet factor 4 levels in diabetes mellitus – preliminary finds. Haemostasis 12:138

Baenziger NL, Brodie GN, Majerus PW (1971) A thrombin sensitive protein of human platelet membranes. Proc Natl Acad Sci USA 68:240–243

Barber AJ, Kaserglanzmann R, Jakabova M, Luscher E (1972) Characterization of a chondroitin 4-sulphate proteoglycan carrier for heparin neutralizing activity (platelet factor 4) released from human blood platelet. Biochim Biophys Acta 286:312–329

Begg GS, Pepper DS, Chesterman CN, Morgan FJ (1978) Complete covalent structure of human β-thromboglobulin. Biochemistry 17:1739–1744

Belch JJF, Greer I, McLaren M, Saniabadi AR, Miller S, Sturrock RD, Forbes CD (1985) The effects of intravenous ZK36-374, a stable prostacyclin analogue, on normal volunteers: a potential treatment for Raynaud's syndrome. Scot Med J 30:124

Bern MM, Green J (1982) Effect of sulfinpyrazone upon antithrombin III and platelet factor 4 in chronic renal failure. Thromb Res 27:457–465

Berrettini M, Parise P, Costantini V, Grasselli S, Nenci GG (1985) Platelet activation in psoriasis. Thromb Haemost 53:195–197

Blattler W, Furrer K, Schriber K, Kofmehl R, Massini C (1981) Platelet proteins during and after prostacyclin therapy for lower limb ischaemia: suggestion of a rebound platelet activation. Vasa 10:261–263

Blattler W, Diener R, Haeberli A, Straub PW (1984) Plasma levels of beta-thromboglobulin and platelet factor 4 in peripheral vascular disease. Vasa S12:75–80

Bock PE, Luscombe M, Marshall SE, Pepper DS, Holbrook JJ (1982) The effect of concentrations and ionic strength on the aggregation of β-thromboglobulin. Thromb Res 25:169–172

Bolton AE, Ludlam CA, Moore S, Pepper DS, Cash JD (1976a) Three approaches to the radioimmunoassay of human β-thromboglobulin. Br J Haematol 33:233–238

Bolton AE, Ludlam CA, Pepper DS, Moore S, Cash JD (1976b) A radioimmunoassay for platelet factor 4. Thromb Res 8:51–58

Borsey DQ, Prowse CV, Gray RS, Dawes J, James K, Elton RA, Clarke BF (1984) Platelet and coagulation factors in proliferative diabetic retinopathy. J Clin Pathol 37:659–664

Bowenpope DF, Ross R (1985) Methods for studying the platelet derived growth factor receptor. Methods Enzymol 109:69–100

Bowenpope DF, Malpass TW, Foster DM, Ross R (1984) Platelet derived growth factor in vivo – levels, activity and rate of clearance. Blood 64:458–469

Brindley LL, Sweet JM, Goetzl EJ (1983) Stimulation of histamine release from human basophils by human platelet factor 4. J Clin Invest 72:1218–1223

Brown TR, Ho TTS, Walz DA (1980) Improved radioimmunoassay of platelet factor 4 and beta-thromboglobulin in plasma. Clin Chim Acta 101:225–233

Buccianti G, Pogliani E, Miradoli R, Colombi MA, Valenti G, Lorenz M, Polli EE (1982) Reduction of plasma levels of betathromboglobulin and platelet factor 4 during hemodialysis – a possible role for a short acting inhibitor of platelet aggregation. Clin Nephrol 18:204–208

Bugiardini R, Chierchia S, Roskovec A, Wild S, Lenzi S, Maseri A (1982) Response of platelet release factors and aggregation to catheter induced activation and prostacyclin infusion in man. Clin Sci 63:P73

Busch C, Dawes J, Pepper DS, Wasteson A (1980) The binding of PF4 to cultured human umbilical vein endothelial cells. Thromb Res 19:129–137

Cade JF, Doyle DJ, Chesterman CN, Morgan FJ, Rennie GC (1982) Platelet function in coronary artery disease: effects of coronary surgery and sulfinpyrazone. Circulation 66:29–32

Chesebro JH, Fuster V, Pumphrey CW, McGoon DC, Pluth JR, Puga FJ, Orszulak TA, Piehler JM, Schaff HV, Danielson GK (1981) Combined warfarin platelet inhibitor antithrombotic therapy in prosthetic heart valve replacement. Circulation 64:76

Ciaglowski RE, Snow JW, Walz DA (1981) Bovine platelet antiheparin protein – platelet factor 4. Ann NY Acad Sci 370:668–679

Clezardin P, McGregor JL, Manach M, Robert F, Dechavanne M, Clemetson KJ (1984) Isolation of thrombospondin released from thrombin stimulated human platelets by fast protein liquid chromatography on an anion exchange Mono-Q chain. J Chromatogr 296:249–256

Contant G, Gouault-Heilmann M, Martinoli JL (1983) Heparin inactivation during blood storage: its prevention by blood collection in citric acid, theophylline, adenosine, dipyridamole-CTAD mixture. Thromb Res 31:365–374

Cooke ED, Bolton AE, Levack B (1981) Thromboglobulin levels in postoperative deep vein thrombosis. Lancet 2:1427

Cordopatri F, Boncinelli S, Marsili M, Lorenzi P, Fabbri LP, Paci P, Salvadori M, Morfini M, Cinotti S, Casparini P (1982) Effects of charcoal hemoperfusion with prostacyclin on the coagulation-fibrinolysis system and platelets of patients with fulminant hepatic failure – preliminary observations. Int J Artif Organ 5:243–247

D'Andrea G, Cananzi D, Toldo M, Cortelazzo S, Ferro-Milone F (1983) Platelet behaviour in classic migraine: responsiveness to small doses of aspirin. Thromb Haemost 49:153

Davi G, Novo S, Gullotti A, Avellone G, Sofia MA, Strano A (1984) Sulfinpyrazone and platelet function in hypertensive patients. Thromb Haemost 51:413

Davis JW, Hartman CR, Shelton L, Ruttinger HA (1983) Effect of tobacco smoking on release of platelet specific proteins. Clin Res 31:A768

Davis LE, Castor CW, Tinney FJ, Anderson B (1985) Preparation and characterisation of antibodies with specificity for the amino-terminal tetrapeptide sequence of the platelet derived connective tissue activating peptide-III. Biochem Int 10:395–404

Dawes J, Smith RC, Pepper DS (1978) The release distribution and clearance of human β-thromboglobulin and platelet factor 4. Thromb Res 12:851–861

Dawes J, Pumphrey CW, McLaren KM, Prowse CV, Pepper DS (1982) The in vivo release of human platelet factor 4 by heparin. Thromb Res 27:65–76

Dawes J, Clemetson KJ, Gogstad GO, McGregor J, Clezardin P, Prowse CV, Pepper DS (1983) A radioimmunoassay for thrombospondin used in a comparative study of thrombospondin, β-thromboglobulin and platelet factor 4 in healthy volunteers. Thromb Res 29:569–581

Dawes J, Miller WR, Brown K (1985) Thrombospondin in milk and other breast secretions; a new marker for breast cyst classification. Thromb Haemost 54:64

Deboer AC, Han P, Turpie AGG, Butt R, Zielinski A, Genton E (1981) Plasma and urine β-thromboglobulin concentration in patients with deep vein thrombosis. Blood 58:693–698

Deboer AC, Genton E, Turpie AGG (1982) Chemistry, measurement and clinical significance of platelet specific proteins. CRC Crit Rev Lab Sci 18:183–211

Deboer AC, Han P, Turpie AGG, Butt R, Gent M, Genton E (1983) Platelet tests and antiplatelet drugs in coronary artery disease. Circulation 67:500–504

Ditter H, Heinrich D, Matthias FR, Sellmann-Richter R, Wagner WL, Hehrlein FW (1983) Effects of prostacyclin during cardiopulmonary bypass in men on plasma levels of β-thromboglobulin, platelet factor 4, thromboxane B_2, 6-keto-prostaglandin $F_{1\alpha}$ and heparin. Thromb Res 32:393–408

Douglas JT, Blamey SL, Lowe GDO, Carter DC, Forbes CD (1985) Plasma beta-thromboglobulin, fibrinopeptide A and Bβ15-42 antigen in relation to postoperative DVT, malignancy and stanozolol treatment. Thromb Haemost 53:235–238

Duncan A, Depratti VJ, George RR (1981) Elevated β-thromboglobulin levels associated with smoking and oral contraceptive agents in normal healthy women. Thromb Res 21:425–430

Fabris F, Randi ML, Crociani ME, Manzoni S, Tonin P, Dezanche L, Cella G, Girolami A (1983) Platelet specific proteins in patients with transient ischemic attacks – effects of anti-platelet drugs. Ric Clin Lab 13:437–442

Fabris F, Randi ML, Casonato A, Zanon RDB, Bovicini P, Girolami A (1984) Clinical significance of beta-thromboglobulin in patients with high platelet count. Acta Haematol 71:32–38

Fabris F, Luzzatto G, Girolami A (1985) The value of β-TG/PF4 ratio in patients with high platelet count. Thromb Haemost 53:288

Files JC, Malpass TW, Yee EK, Ritchie JL, Harker LA (1981) Studies of human platelet alpha-granule release in vivo. Blood 58:607–618

Fitchett D, Toth E, Gilmore N, Ehrman M (1983) Platelet release of beta-thromboglobulin within the coronary circulation during cold pressor stress. Am J Cardiol 52:727–730

Flicker W, Milthorpe BK, Schindhelm K, Odell RA, McPherson J, Farrell PC (1982) Platelet factor release following heparin administration and during extracorporeal circulation. Trans Am Soc Artif Intern Organs 28:431–436

Gartner TK, Walz DA (1983) TFP inhibits the expression of the endogenous platelet lectin and secretion of alpha-granules. Thromb Res 29:63–74

Ginsberg MH, Hoskins R, Sigrist P, Painter RG (1979) Purification of a heparin neutralizing protein from rabbit platelets and its homology with human platelet factor 4. J Biol Chem 254:2365–2371

Goetzl EJ, Brindley LL, Sweet JM, Handin RI (1983) Unique mechanism of stimulation of human basophils by platelet factor 4. Clin Res 31:A457

Handin RI, McDonough M, Lesch M (1978) Elevation of platelet factor four in acute myocardial infarction: measurement by radioimmunoassay. J Lab Clin Med 91:340–349

Hay CRM, Durber AP, Saynor R (1982) Effect of fish oil on platelet kinetics in patients with ischaemic heart disease. Lancet I:1269–1272

Heldin CH, Westerman B, Wasteson A (1981) Demonstration of an antibody against platelet derived growth factor. Exp Cell Res 136:255–261

Holan J, Kubisz P, Cronberg S, Parizek M (1984) Platelet function studies in diabetics with beta-thromboglobulin (beta-TG) platelet factor 4 (PF4) and thromboxane-B2 (TXB2) RIA assay. Eur J Nucl Med 9:A75

Hope W, Martin TJ, Chesterman CN, Morgan FJ (1979) Human β-thromboglobulin inhibits PGI_2 production and binds to a specific site in bovine aortic endothelial cells. Nature 282:210–212

Hunter WM (1977) Simplified solid-phase radioimmunoassay without centrifugation. Acta Endocrinol 85:463

Ikeda T, Nonaka Y, Goto A, Ishii M (1985) Effects of praxosin on platelet aggregation and plasma beta-thromboglobulin in essential hypertension. Clin Pharmacol 37:601–605

Jaffe EA, Leung LLK, Nachman RL, Levin RI, Mosher DF (1982) Thrombospondin is the endogenous lectin of human platelets. Nature 295:246–248

Janka HU, Standl E, Schramm W, Mehnert H (1983) Platelet enzyme activities in diabetes mellitus in relation to endothelial damage. Diabetes 32:47–51

Kahaleh MB, Osborn I, Leroy EC (1982) Elevated levels of circulating platelet aggregates and beta-thromboglobulin in scleroderma. Ann Intern Med 96:610–613

Kaplan KL, Niewiarowski S (1985) Nomenclature of secreted platelet proteins – report of the working party on secreted platelet proteins of the subcommittee on platelets. Thromb Haemost 53:282–284

Kaplan KL, Owen J (1981) Plasma levels of β-thromboglobulin and PF4 as indexes of platelet activation in vivo. Blood 57:199–202
Kaplan KL, Owen J (1982) Radioimmunoassay of β-thromboglobulin. Methods Enzymol 84:93–102
Kaplan KL, Nossel HL, Drillings M, Lesznik G (1978) Radioimmunoassay of platelet factor 4 and β-thromboglobulin: development and application to studies of platelet release in relation to fibrinopeptide A generation. Br J Haematol 39:129–146
Kerry PJ, Curtis AD (1985) Standardisation of β-thromboglobulin (β-TG) and platelet factor 4 (PF4): a collaborative study to establish international standards for β-TG and PF4. Thromb Haemost 53:51–55
Kjeldsen SE, Gjesdal K, Eide I, Aakesson I, Amundsen R, Foss OP, Leren P (1983) Increased beta-thromboglobulin in essential hypertension – interactions between arterial plasma adrenaline, platelet function and blood lipids. Acta Med Scand 213:369–373
Kordenat K, Dawes J, Pepper DS (1983) Canine thrombospondin (TSP) levels measured in plasma by radioimmunoassay before and during acute coronary thrombosis. Thromb Haemost 49:250
Lane DA, Ireland H, Wolff S, Ranasinghe E, Dawes J (1984) Detection of enhanced in vivo platelet α-granule release in different patient groups – comparison of β-thromboglobulin, platelet factor 4 and thrombospondin assays. Thromb Haemost 52:183–187
Lawler JW (1981) Prediction of the secondary structure of platelet factor 4 and β-thromboglobulin from their amino acid sequences. Thromb Res 21:121–127
Lawler JW, Slayter HS, Coligan JE (1978) Isolation and characterization of a high molecular weight glycoprotein from human blood platelets. J Biol Chem 253:8609–8616
Levine SP, Krentz LS (1977) Development of a radioimmunoassay for human platelet factor 4. Thromb Res 11:673–686
Levine SP, Lindenfeld J, Crawford M, Horwitz LD (1979) Effects of catecholamines on plasma platelet factor 4 in normal human subjects. Clin Res 27:A462
Levine SP, Suarez AJ, Sorenson RR, Knieriem LK, Raymond NM (1981 a) The importance of blood collection methods for assessment of platelet activation. Thromb Res 24:433–443
Levine PH, Fisher M, Fullerton AL, Duffy CP, Hoogasian JJ (1981 b) Human platelet factor 4 – preparation from outdated platelet concentrates and application in cerebral vascular disease. Am J Haematol 10:375–385
Levine SP, Knieriem LK, Harris MA, George JN (1982) Emotional stress, catecholamines (CATS) and platelet activation. Circulation 66:52
Levine SP, Suarez AJ, Sorenson RR, Raymond NM, Knieriem LK (1984) Platelet factor 4 release during exercise in patients with coronary artery disease. Am J Haematol 17:117–127
Levine SP, Towell BL, Suarez AM, Knieriem LK, Harris MM, George JN (1985) Platelet activation and secretion associated with emotional stress. Circulation 71:1129–1134
Longenecker GL, Bowen RJ, Swift IA, Beyers BJ (1984) Increased beta-thromboglobulin in the serum of smokers occurs because of increased platelet counts. Res Commun Sub Abuse 5:147–152
Longo G, Cecchi F, Grossi A, Matucci M, Rafanelli D, Vannucchi AM, Casprini P, Morfini M, Dolara A (1984) Coagulation and platelet function in hypertrophic cardiomyopathy. Thromb Haemost 51:299
Ludlam CA, Cash JD (1976) Studies on the liberation of β-thromboglobulin from human platelets in vitro. Br J Haematol 33:239–247
Ludlam CA, Allan N, Blandford RB, Dowdle R, Bentley NJ, Bloom AL (1981) Platelet and coagulation function in patients with abnormal cardiac valves treated with sulphinpyrazone. Thromb Haemost 46:743–746
Luscombe M, Marshall SE, Pepper DS, Holbrook JJ (1981) The transfer of platelet factor 4 from its proteoglycan carrier to natural and synthetic polymers. Biochim Biophys Acta 678:137–142
Luster AD, Unkeless JC, Ravetch JV (1985) γ-Interferon transcriptionally regulates an early response gene containing homology to platelet proteins. Nature 315:672–676

Manotti C, Quintavalla R, Manzoni GC (1983) Platelet activation in migraine. Thromb Haemost 50:758

Mant MJ, Kappagoda CT, Taylor RF, Quinlan JE (1984) Platelet activation caused by cardiac catheter blood collection and its prevention. Thromb Res 33:177–187

Margossian SS, Lawler JW, Slayter HS (1981) Physical characterisation of platelet thrombospondin. J Biol Chem 256:7495–7500

McPherson J, Sage H, Bornstein P (1981) Isolation and characterization of a glycoprotein secreted by aortic endothelial cells in culture. Apparent identity with platelet thrombospondin. J Biol Chem 256:11330–11336

Messmore HL, Walenga JM, Fareed J (1984) Molecular markers of platelet activation. Semin Thromb Hemost 10:264 – 269

Moore S, Pepper DS, Cash JD (1975a) Platelet antiheparin activity: the isolation and characterisation of platelet factor 4 released from thrombin-aggregated washed human platelets and its dissociation into subunits and the isolation of membrane bound antiheparin activity. Biochim Biophys Acta 379:370–384

Moore S, Pepper DS, Cash JD (1975b) The isolation and characterisation of a platelet specific β-globulin (β-thromboglobulin) and the detection of anti-urokinase and antiplasmin released from thrombin aggregated washed human platelets. Biochim Biophys Acta 379:360–369

Mosher DF, Doyle MJ, Jaffe EA (1982) Synthesis and secretion of thrombospondin by cultured human endothelial cells. J Cell Biol 93:343–348

Motomiya T, Yamazaki H (1981) Plasma β-thromboglobulin and platelet factor 4 during the normal menstrual cycle. Acta Haematol Jpn 44:193–195

Muggli R, Glatthaar B, Mittelholzer E, Tschopp TB (1981) Purification, characterization and radioimmunoassay of a new antiheparin protein secreted by rabbit platelets. Thromb Haemost 46:20

Muhlhauser I, Schernthaner G, Silberbauer K, Sinzinger H, Kaliman J (1980) Platelet proteins (Beta-TG and PF4) in atherosclerosis and related diseases. Artery 8:73–79

Muhlhauser I, Schernthaner G, Silberbauer K, Sinzinger H, Kaindl F (1981) Plasma concentrations of platelet specific proteins in coronary artery disease. Cardiology 68:129–138

Najean Y, Poirier O (1984) Beta-thromboglobulin and platelet factor 4 in polycythemia patients treated by ticlopidine. Acta Haematol 72:83–89

Niewiarowski S, Poplawski A, Lipinski B, Farbiszewski R (1968) The release of platelet clotting factors during aggregation and viscous metamorphosis. Exp Biol Med 3:121–128

O'Brien JR, Etherington MD, Pashley M (1984) Intra-platelet platelet factor 4 (IP.PF4) and the heparin mobilisable pool of PF4 in health and atherosclerosis. Thromb Haemost 51:354–357

Phillips DR, Jennings LK, Prasanna HR (1980) Ca^{2+} mediated association of glycoprotein G (thrombin-sensitive protein, thrombospondin) with human platelets. J Biol Chem 255:11629–11632

Pogliani EM, Colombi M, Cristoforetti G, Valenti G, Miradoli R, Buccianti G (1982) Betathromboglobulin and platelet factor 4 plasma levels during haemodialysis: effect of indobufen. Pharmatherapeutica 3:127–132

Porta M, Peters AM, Cousins SA, Cagliero E, Fitzpatrick ML, Kohner EM (1983) A study of platelet relevant parameters in patients with diabetic micro-angiopathy. Diabetalogia 25:21–25

Prowse C, Pepper DS, Dawes J (1982) Prevention of the platelet alpha-granule release reaction by membrane active drugs. Thromb Res 25:219–227

Rasi V (1979) Beta-thromboglobulin in plasma – false high values caused by platelet enrichment of the top layer of plasma during centrifugation. Thromb Res 15:543–552

Reynders J, Rodzynek JJ, Leautaud P, Delcourt A (1984) Comparison between platelet function tests in diabetic patients with special reference to beta-thromboglobulin. Acta Clin Belg 39:314

Riess H, Hiller E, Reinhardt B, Brauning C (1984) Effects of BM 13.177, a new antiplatelet drug in patients with atherosclerotic disease. Thromb Res 35:371–378

Rosove MH, Frank HJL, Haewig SSL (1984) Plasma beta-thromboglobulin, platelet factor 4, fibrinopeptide A and other hemostatic functions during improved, short-term glycemic control in diabetes mellitus. Diabetes Care 7:174–179
Rucinski B, Poggi A, James P, Holt JC, Niewiarowski S (1983) Purification of 2 heparin binding proteins from porcine platelets and their homology with human secreted platelet proteins. Blood 61:1072–1080
Saglio SD, Slayter HS (1982) Use of a radioimmunoassay to quantify thrombospondin. Blood 59:162–166
Salter MCP, Crow MJ, Learoyd P, Rajah SM (1982) Platelet function and survival changes before and after coronary artery revascularisation: assessment of the influence of dipyridamole and acetylsalicylic acid. Haemostasis 12:191
Sano T, Motomiya T, Yamazaki H (1980) Platelet release reaction in vivo in patients with ischaemic heart disease after isometric exercise and its prevention with dipyridamole. Thromb Haemost 42:1589–1597
Schernthaner G, Muhlhauser I, Silberbauer K (1979) β-Thromboglobulin lowered by dipyridamole in diabetes. Lancet II:748–749
Schmidt KG, Rasmussen JW (1984) Acute platelet activation induced by smoking: in vivo and ex vivo studies in humans. Thromb Haemost 51:279–282
Semple MJ, Al-Hasani SF, Kioy P, Savidge GF (1984) A double-blind trial of ticlopidine in sickle cell disease. Thromb Haemost 51:303–306
Simi M, Leardi S, Castelli M, Tebano MT, Vitale M, Prantera C, Speranza V (1982) Plasmatic levels of platelet factor 4 (PF4) in Crohn's disease. World J Surg 6:662
Sinzinger H, Horsch AK, Silberbauer K (1983) The behaviour of various platelet function tests during long-term prostacyclin infusion in patients with peripheral vascular disease. Thromb Haemost 50:885–887
Smith RC, Duncanson J, Ruckley CV, Webber RG, Allan NC, Dawes J, Bolton AE, Hunter WM, Pepper DS, Cash JD (1978) β-Thromboglobulin and deep vein thrombosis. Thromb Haemost 39:338–345
Sobel M, Salzman EW, Davies GC, Handin RI, Sweeney J, Ploetz J, Kurland G (1981) Circulating platelet products in unstable angina-pectoris. Circulation 63:300–306
Steele PP (1980) Platelet suppressants inhibit release of beta-thromboglobulin during isometric exercise in coronary disease. Clin Res 28:A214
Strano A, Davi G (1982) Platelet release reaction, platelet sensitivity to prostacyclin and thromboxane production by platelets in hyperlipidemia. Haemostasis 12:87
Suarez AJ, Levine SP, George JN (1982) Emotional stress, catecholamines and platelet activation. Clin Res 30:A331
Taomoto K, Asada M, Kanazawa Y, Matsumoto S (1983) Usefulness of the measurement of plasma beta-thromboglobulin (beta-TG) in cerebrovascular disease. Stroke 14:518–524
Turpie AGG, Gent M, DeBoer AC, Giroux M, Kinch D, Butt R, Gunstensen J, Saerens E, Genton E (1984) Effect of suloctidil on platelet function in patients with shortened platelet survival time. Thromb Res 35:397–406
Valles J, Aznar J, Santos MT (1982) Relationship between platelet fatty acid pattern and plasma β thromboglobulin. Thromb Res 27:737–742
Van Hulsteijn H, Fibbe W, Bertina R, Briet E (1982) Plasma fibrinopeptide A and beta-thromboglobulin in major bacterial infections. Thromb Haemost 48:247–249
Van Hulsteijn H, Kolff J, Briet E, Vanderlaarse A, Bertina R (1984) Fibrinopeptide A and beta thromboglobulin in patients with angina pectoris and acute myocardial infarction. Am Heart J 107:39–45
Van Oost BA, Veldhuyzen B, Timmermans APM, Sixma JJ (1983) Increased urinary β-thromboglobulin excretion in diabetes assayed with a modified RIA kit-technique. Thromb Haemost 49:18–20
Viero P, Cortelazzo S, Barbui T (1984) Does the β-TG/PF4 ratio have any value in myeloproliferative diseases? Thromb Haemost 51:411
Violi F, Juliano L, Alessandri C, Frattaroli S, Bonavita MS (1984) Beta-thromboglobulin plasma levels in common migraine. Thromb Haemost 51:295

Walker ID, Davidson JF, Faichney A, Wheatley DJ, Davidson KG (1981) A double blind study of prostacyclin in cardiopulmonary bypass surgery. Br J Haematol 49:415–423
Walz DA (1984) Platelet released proteins as molecular markers for the activation process. Semin Thromb Hemost 10:270–279
Wight TN, Raugi GJ, Mumby SM, Bornstein P (1985) Light microscopic immunolocation of thrombospondin in human tissues. J Histochem Cytochem 33:295–302
Wojciechowski J, Olausson M, Korsanbengtsen K (1983) Fibrinopeptide A, beta-thromboglobulin and fibrin degradation products as screening test for the diagnosis of deep vein thrombosis. Haemostasis 13:254–261
Woodhams BJ, Kernoff PBA (1983) The application of polyethylene glycol to radioimmunoassays used in haemostasis. Thromb Res 29:333–341

CHAPTER 22

A Competitive Binding Assay for Heparin, Heparan Sulphates, and Other Sulphated Polymers

J. DAWES and D. S. PEPPER

A. Introduction

Heparin and heparan sulphate are sulphated polysaccharides present in connective tissue and on cell surfaces, either as the free carbohydrate (glycosaminoglycan) chain, or covalently bonded to protein as proteoglycans. They consist of alternating uronic acid and glucosamine residues, and are sulphated both at the O and N positions. Heparan sulphate contains less iduronic acid and has a lower sulphate content than heparin, but they should in fact be considered as a single class of glycosaminoglycan, including a continuous spectrum of molecules of differing uronic acid composition and degree of sulphation. Their structures are reviewed by LINDAHL and HOOK (1978), HASCALL and HASCALL (1981) and HOOK et al. (1984).

These glycosaminoglycans fulfil a range of biological functions. They are intimately involved in maintenance of the electronegativity of cell surfaces and contribute to tissue viscoelasticity (WIGHT 1980). They act as receptors for many biological macromolecules and may have a role in the control of cell growth. However, their most obvious pharmacological function is as anticoagulants. Heparin has been used therapeutically as an anticoagulant for many years, and it is becoming increasingly clear that heparan sulphate (THOMAS et al. 1981) and another glycosaminoglycan, dermatan sulphate (TEIEN et al. 1976a), may exert anticoagulant effects under normal physiological conditions.

However, the selection of appropriate assays for heparin and related carbohydrates is controversial. Most of the sensitive techniques available are based either on overall anticoagulant activity (LEE and WHITE 1913; THOMSON 1970; ANDERSSON et al. 1976) or on more specific properties such as the inactivation of thrombin (EIKA et al. 1972) or factor Xa (YIN et al. 1973; TEIEN et al. 1976b).

Unfortunately, such assays are affected by factors other than the concentration of heparin present; they are subject to interference from clotting factors and inhibitors of coagulation (YIN et al. 1973; GODAL 1975) and there is considerable variation in the detection of different preparations of heparin by the different assays (TEIEN and LIE 1975). Moreover, it is not possible to predict the in vivo effects of a heparin preparation with confidence on the basis of in vitro assays, particularly for the low molecular weight heparins and synthetic and semisynthetic heparinoids which are being increasingly used clinically. Although these compounds have antithrombotic activity in vivo, the apparent discrepancies between their properties ex vivo and in vitro are in some cases very marked (LANE et al. 1977; FISCHER et al. 1982; THOMAS and MERTON 1982; TORNGREN et al. 1983).

The interpretation of these observations can be greatly facilitated by measurements of the actual mass concentrations of glycosaminoglycan, without any assumptions as to its biological activity. These could then be related to the results of the biological assays. However, the classical chemical methods such as the uronic acid–carbazole reaction (BITTER and MUIR 1962) or the metachromatic interaction of heparin with azure A (JAQUES et al. 1947) and toluidine blue (JAQUES and WOLLIN 1967) are relatively insensitive and often only qualitative when applied to impure substrates. Sensitivity has been greatly improved by the application of ^{103}Ru-labelled ruthenium red as a detection system on cellulose acetate electrophoretograms (CARLSON 1982), but the method is too cumbersome for routine use.

WU and COHEN (1984) have reported a different type of assay based on the ability of glycosaminoglycans to form precipitates when incubated with serum and zinc acetate. Heparin, heparan sulphate and chondroitin sulphates competed with radiolabelled heparan sulphate for binding in a dose-dependent manner, but concentrations less than 10 μg/ml could not be detected.

We have developed a sensitive competitive binding assay which can be applied to therapeutic heparins and heparinoids, and to endogenous heparin-like glycosaminoglycans (DAWES and PEPPER 1982; MACGREGOR et al. 1985; DAWES et al. 1985a, b). The technique is based on the same principle as radioimmunoassays, but, as it has not yet proved possible to raise antibodies against unmodified heparin, the binding agent is not an antibody but Polybrene, poly (*N,N,N',N'*-tetramethyl-*N*-trimethylenehexamethylenediammonium bromide), a highly positively charged synthetic polymer. The methodology employed and the results it has yielded to date are described in this chapter.

B. Radiolabelling of Glycosaminoglycans

Glycosaminoglycans are not amenable to radiolabelling with isotopes of iodine by the standard techniques, as they contain neither hydroxylated aromatic rings nor significant numbers of free amino groups. However, ^{125}I remains the isotope of choice for assays; it is a γ-emitter of useful half-life (80 days) and can be detected easily, accurately and with high sensitivity, without any of the interfering factors experienced in β-counting. We have therefore developed techniques for derivatising glycosaminoglycans so that the overall chemical and biological properties of the molecule are unaltered, but new groups are incorporated which can be radioiodinated.

I. Derivatisation

Most naturally occurring glycosaminoglycans contain a few unsubstituted amino groups, which will react with an excess of unlabelled *N*-succinimidyl-3-(4-hydroxyphenyl)propionate (SHPP). Heparin can be successfully derivatised by this method (DAWES and PEPPER 1979). Sodium heparin (10 mg; Sigma catalogue H3125) was reacted with 10 mg SHPP (Pierce 27710) in 5 ml 0.05 *M* sodium borate (pH 9.2) at 4 °C for 20 h. Unreacted ester and its hydrolysis products were removed by gel filtration in water on Sephadex G-25 (Pharmacia), and the material eluting in the void volume was chromatographed on protamine–agarose (PEPPER and DAWES 1977) to isolate the biologically active heparin derivatives. Heparin was detected in the fractions by means of the azure A metachromatic assay (JAQUES et al. 1947), and the incorporated hydroxyphenyl group was identified by its absorption at 280 nm. The derivative is finally dialysed in 3000 molecular weight cut-off tubing (Spectrapor 3) against water and freeze-dried, and has an indefinite (> 5 years) shelf life.

The same chemistry has recently been used to prepare an SHPP conjugate of the therapeutic heparinoid Org 10172, a mixture of glycosaminoglycans from animal intestinal mucosa. The reagent concentration was increased to 25 mg/ml and the reaction time to 72 h, and the product was chromatographed by fast protein liquid chromatography (Pharmacia) using an anion exchange Mono Q column and a 0–2 *M* NaCl gradient rather than protamine–agarose; alternatively DEAE–Sepharose (Pharmacia) can be used. ^{125}I-labelled heparin was added as a marker, and material eluting in front of the heparin peak was pooled, dialysed and freeze-dried.

However, other sulphated polysaccharides do not contain any amino groups and require a different approach, which is illustrated by the derivatisation of pentosan polysulphate (SP54) described by MACGREGOR et al. (1984). SP54 was converted from the sodium to the tetrabutylammonium salt, freeze-dried and dissolved in dimethylsulphoxide (DMSO) at a concentration of 40 mg/ml. Then, 20 mg was reacted with 14 mg carbonyldiimidazole (Aldrich catalogue 11,553-3) at 56 °C for 45 min, after which 36 mg tyrosylhydrazide (Aldrich, catalogue 13,804-5) was added and allowed to react overnight at 20 °C. Excess sodium chloride (1 *M* in water) was then added to convert the derivative to the Na^+ salt, which was gel filtered on Sephadex LH-20 (Pharmacia) in distilled water, and freeze-dried. Tyrosyl incorporation was determined using a molar extinction coefficient $\varepsilon_{1\,cm}^{280}$ of 4.55×10^3, and was typically 0.5 mol for 1 mol SP54.

II. Iodination

These derivatives are readily iodinated by the chloramine-T method of GREENWOOD et al. (1963). Iodination of 10 μg with 18500 kBq carrier-free $Na^{125}I$ (Amersham) yielded products with specific activities in the range 20–2000 kBq/

Table 1. Properties of radioiodinated glycosaminoglycans

Glycosaminoglycan	Source	Derivative	Specific activity (kBq/μg)
Unfractionated heparin	Sigma	SHPP	1000–2000
SP54 (pentosan polysulphate)	Benechemie, Munich	Tyrosylhydrazide	30–80
Org 10172 (mainly heparan sulphate)	Organon, Oss	SHPP	200–250
HS I (heparan sulphate)	E. A. Johnson, NIBSAC	SHPP	300–400

μg. Heparins from different commercial sources apparently varied widely in their content of unsubstituted amino groups and that supplied by Sigma was selected for use in the assay as it routinely yielded radiolabelled tracer of specific activity 1000–2000 kBq/μg. The properties of the tracers currently in use in this laboratory are detailed in Table 1.

C. Synthesis of Binding Reagent

This assay was originally developed using protamine linked to Sepharose as the binding reagent (DAWES and PEPPER 1982). However, the protamine was unstable on storage and susceptible to proteolysis during the assay, and has been replaced by Polybrene. Polybrene (Aldrich catalogue 10,768-9) is reacted with Epoxy-Sepharose (Pharmacia) at a loading of 12 μg Polybrene per milliliter slurry at pH 12 for 16 h at 37 °C, a process which is subject to United Kingdom Patent Application 8516570.

D. Assay for Therapeutic Heparins and Heparinoids

I. Method

The assay mixture contains:

50 μl Polybrene-Sepharose suspension
50 μl ^{125}I-labelled glycosaminoglycan (10 ng/ml)
50 μl standard glycosaminoglycan or sample
50 μl assay buffer (0.05 *M* phosphate/1% v/v Tween 20/10% v/v normal human serum)
200 μl DMSO

Tubes are shaken for 2 h at room temperature, and Sepharose-bound material is separated from that remaining in the liquid phase either by centrifugation and washing or by sedimentation through assay buffer containing 10% sucrose, and removal of the supernatant by aspiration (HUNTER 1977).

The concentration of Polybrene–Sepharose which bound 35%–50% of the tracer under these assay conditions in the absence of unlabelled glycosaminoglycan was used. At a Polybrene loading of 12 μg/ml this was usually a dilution of 1:10–1:50, which gave good reproducibility. At higher dilutions reproducibility was poor, presumably because of the difficulty of dispensing low numbers of Sepharose beads with accuracy, and it was also compromised by omission of serum from the buffer.

The technique described was developed from that of DAWES and PEPPER (1982) specifically to measure the concentration of administered heparin or heparinoid in biological fluids, without interference from endogenous glycosaminoglycans. This was achieved by carrying out the assay in the presence of 50% DMSO. Under these conditions no endogenous glycosaminoglycan was detected in the plasma and serum of ten healthy volunteers, and urine could be reliably assayed at a dilution of 1:5. Plasma proteins did not interfere in the assay, which was unaffected by pretreatment of plasma and serum samples with pronase.

Moreover, platelet factor 4 (PF4) and antithrombin III, which have high affinities for heparin, did not affect the assay at concentrations present in plasma. It was insensitive to pH over the range 6.0–8.5, but was affected by ionic strengths above 0.4; samples containing high salt concentrations must therefore be dialysed or diluted.

II. Sensitivity

The sensitivity of this type of assay is limited by the affinity of the ligand for the binding reagent. DMSO reduces the affinity of binding, but it is still possible to measure heparin at concentrations as low as 60 ng/ml. The affinities of different glycosaminoglycans for Polybrene are determined both by their degree of sulphation and their molecular weight, though charge density appears to be a more important determining factor than size. Thus larger, more highly sulphated molecules can be assayed with higher sensitivity than smaller molecules of lower charge density.

As all gycosaminoglycan preparations encompass a spectrum of different carbohydrate chains, this necessarily implies some discrimination within the sample, but in most cases this is not of practical importance. The different affinities of different glycosaminoglycans can, however, be exploited to increase the sensitivity of some assays by using a tracer of lower affinity than the glycosaminoglycan in the sample. Thus, in the assay of SP54 the sensitivity can be improved from 150 ng/ml to 40 ng/ml by substituting ^{125}I-labelled heparin – a less highly sulphated molecule – for ^{125}I-labelled SP54. This effect is illustrated in the standard curves shown in Fig. 1, where the use of ^{125}I-labelled heparin gives a sensitive assay for SP54, but a relatively insensitive test for a low molecular weight heparin

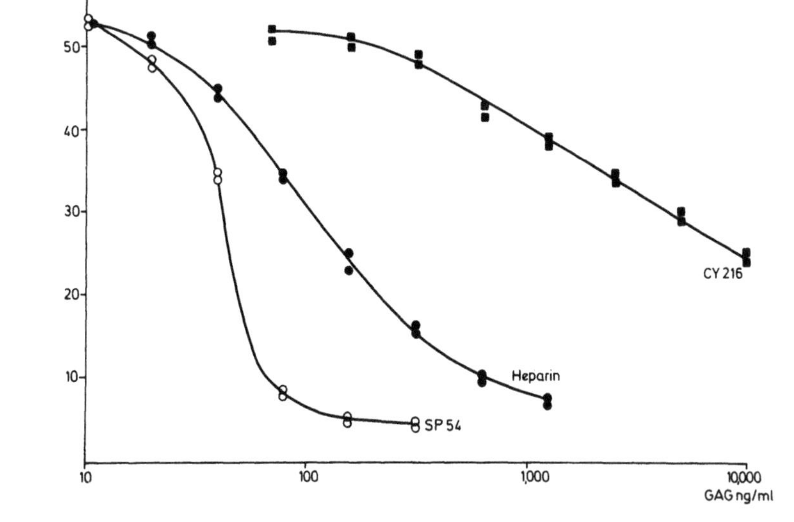

Fig. 1. Competitive binding assay: dilution curves of pentosan polysulphate (SP54), heparin and low molecular weight heparin (CY 216) using ^{125}I-labelled heparin as tracer

Table 2. Assay characteristics for different heparins and heparinoids

Analyte	Source	Mean molecular weight	SO_4 (% w/w)	Tracer	Sensitivity (ng/ml)	Range (ng/ml)
Heparin	Sigma	12000	31	Heparin ^{125}I	60	60–600
SP54 (pentosan polysulphate	Benechemie, Munich	6000	48	SP54 ^{125}I	150	150–1500
				Heparin ^{125}I	40	40–150
PK 10169	Pharmuka, Gennevilliers	5000	30	Heparin ^{125}I	300	300–30000
CY 216	Choay, Paris	4500	29	Heparin ^{125}I	300	300–50000
CY 222	Choay, Paris	2500	29	Heparin ^{125}I	1000	1000–100000

of the same degree of sulphation but smaller size than the tracer. The assay characteristics and properties of different heparins and heparinoids are given in Table 2.

III. Specificity

It is probably clear already to readers of this chapter that the competitive binding assay described is not specific for any one glycosaminoglycan. However, it has been designed to give maximum specificity under defined conditions, reflecting not absolute specificity, but relative sensitivity to different glycosaminoglycans. This is illustrated in Fig. 2. All sulphated glycosaminoglycans (hyaluronic acid is

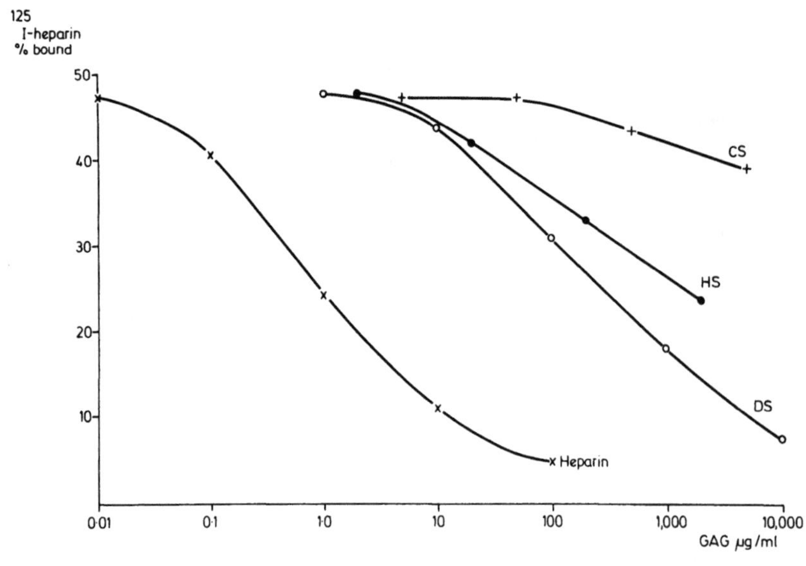

Fig. 2. Competitive binding assay with different glycosaminoglycans (GAGs) in the presence of 50% DMSO, and using ^{125}I-labelled heparin as tracer. *DS* dermatan sulphate (E. A. Johnson, NIBSAC); *HS* heparan sulphate (Opocrin); *CS* chondroitin 4-sulphate (Sigma)

not sulphated) are detectable by the binding assay, but in the presence of 50% DMSO the concentrations of chondroitin, dermatan and heparan sulphates present in normal plasma (CALATRONI et al. 1969) are not detectable and as a result only heparin and highly sulphated heparinoids are measured.

The specificity is thus artificially imposed by the assay conditions rather than intrinsic. It is no less real for that, but care must be taken when using the assay under different circumstances to bear this constraint in mind. Thus, urine samples can contain enough endogenous glycosaminoglycans to interfere with the measurement of heparin, and to avoid this problem they must either be pretreated with chondroitinase ABC or assayed at a dilution of 1 : 5. Other types of sample, such as tissue homogenates, may present similar difficulties.

IV. Studies in Human Volunteers

This assay has been applied to studies of the pharmacokinetics of heparin, low molecular weight heparins and pentosan polysulphate in healthy human subjects. After intravenous bolus injections covering a wide dose range, both heparin and SP54 were cleared with apparent first-order kinetics at all doses. However the half-life, which did not differ significantly between volunteers, was strongly dose dependent (Fig. 3); it ranged from 45 min after administration of 5000 units hep-

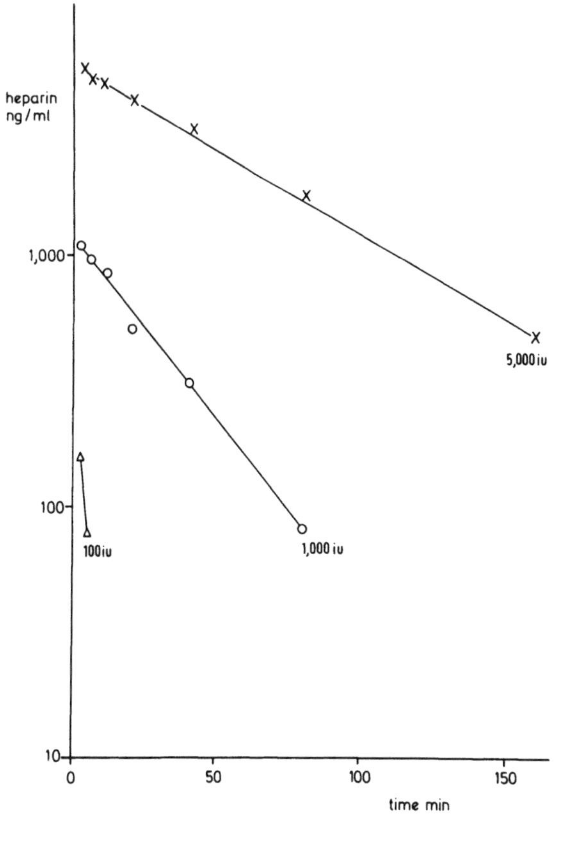

Fig. 3. Clearance of heparin following intravenous bolus injection in healthy volunteers

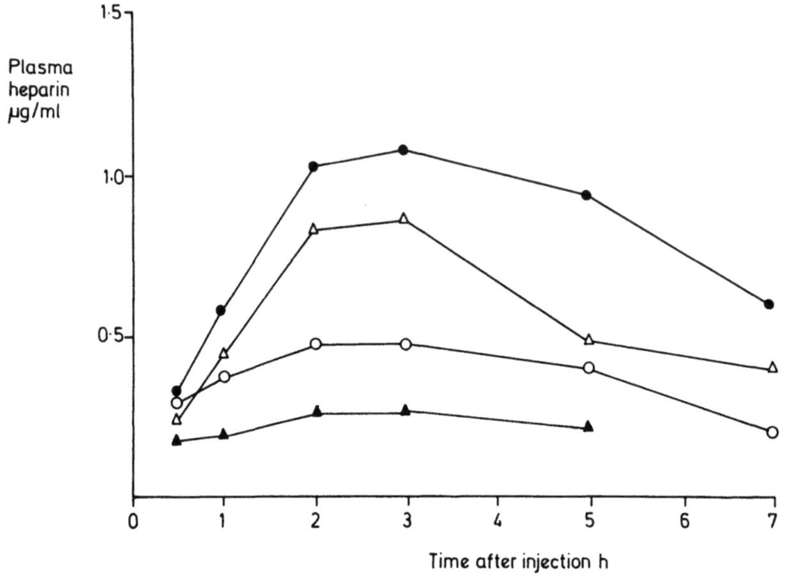

Fig. 4. Plasma levels of heparin following subcutaneous injection of 75 mg in four healthy volunteers

arin to 3 min after 100 units, and from 55 min after administration of 100 mg SP54 to 7 min after 1 mg. It is noteworthy that half-lives can be measured after such low doses with this technique. No other method except the injection of radio-labelled heparin offers this potential.

The response of healthy volunteers to subcutaneous injections of unfractionated heparin varied widely (Fig. 4; DAWES et al. 1985a, b), although the peak plasma level always occurred 2–3 h after injection. The intravenous studies had shown that this was not due to differences in clearance rate; rather, it reflects individual variation in the rate and extent of absorption of heparin from the subcutaneous injection site. This characteristic was reproducible, but could not be correlated with any other obvious physical characteristic including age, sex, weight or skinfold thickness. Subcutaneous injection is a preferred route for therapeutic and prophylactic use of heparin, and the variation in absorption efficiency is so great that it is clearly a major contributor to the overall variation which makes patient response to heparin so unpredictable. The significance of absorption could only be unequivocally demonstrated using an assay such as the one described here, where concentration is measured independently of biological activity and variable contributions from plasma proteins can be ruled out.

Measurement of plasma levels after subcutaneous injection of low molecular weight heparins (CY 216, CY 222; see Tables 2 and 3) and pentosan polysulphate revealed that the efficiency of absorption increased markedly as the molecular weight was reduced (Fig. 5). As there was no increase in clearance except for the smallest heparin fraction, CY 222, the bioavailability of these products was considerably greater than that of unfractionated heparin. This provides at least a partial explanation of the greater response to low molecular weight heparin than to

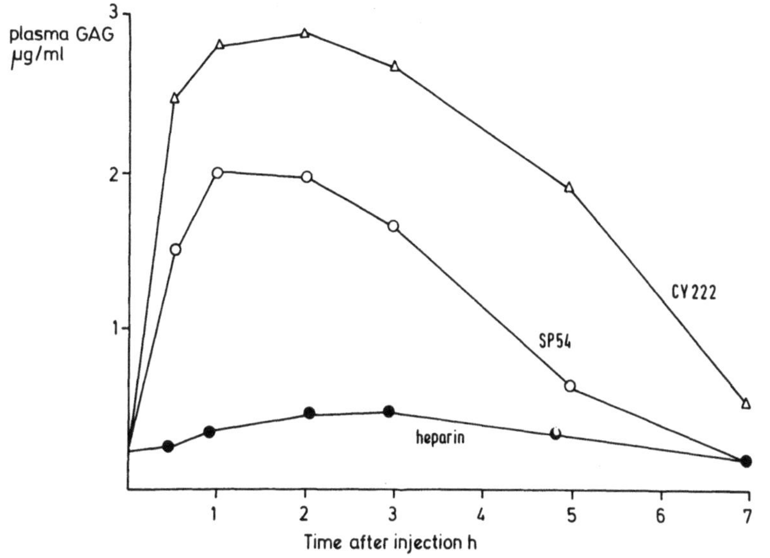

Fig. 5. Plasma levels of therapeutic glycosaminoglycans in a healthy volunteer after subcutaneous injection of 75 mg: effect of molecular weight

unfractionated heparin reported by several groups (KAKKAR et al. 1982; VERSTRAETE 1984; BRATT et al. 1985). Moreover, the individual variation decreased at lower molecular weights, making it easier to predict dosage (Table 3).

Comparison of the plasma concentration of therapeutic heparins and heparinoids with some of their biological activities has yielded some interesting data. Concentration of unfractionated heparin correlated well with two clotting assays, the activated partial thromboplastin time (APTT) and the anti-Xa assay of YIN et al. (1973) (DAWES et al. 1985b). The APTT also correlated reasonably well with levels of low molecular weight heparins and SP54 (DAWES et al. 1985a; MACGREGOR et al. 1985), but there were major discrepancies between concentration and anti-Xa clotting activity, which was higher than expected (MACGREGOR et al.

Table 3. Dependence of bioavailability on molecular weight of therapeutic glycosaminoglycans administered subcutaneously[a]

Drug	Mean molecular weight	Bioavailability ($\mu g\,h^{-1}\,ml^{-1}$) Subject			
		1	2	3	4
Heparin	12000	2.66	3.35	0.99	5.13
SP54	6000	8.70	7.13	6.72	12.03
CY 216	4500	14.43	15.87	25.55	22.18
CY 222	2500	14.41	11.01	11.19	15.75

[a] Recipients were healthy volunteers of both sexes aged 29–42 years; bioavailability was measured in plasma.

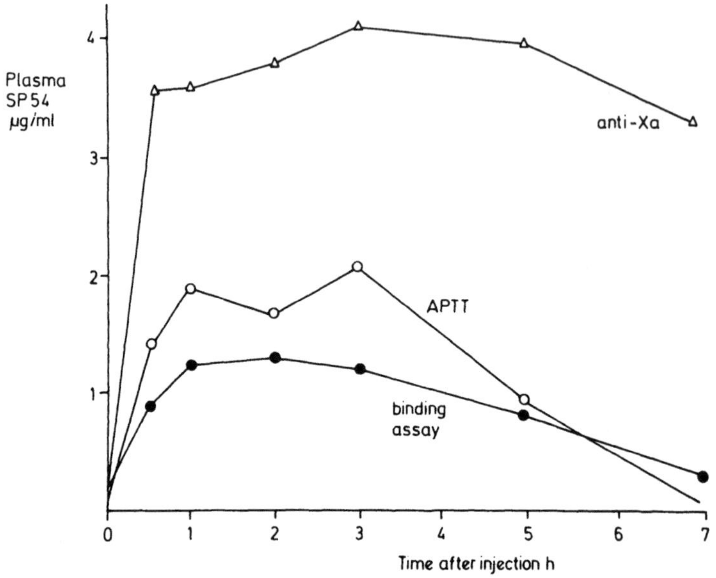

Fig. 6. Levels of SP54 in a healthy volunteer after subcutaneous injection of 75 mg measured by different assay methods: persistence of anti-Xa activity

1985) and persisted after the clearance of all detectable administered glycosaminoglycan. This effect is illustrated in Fig. 6. Similar results were obtained after both intravenous and subcutaneous injection of the low molecular weight heparin PK 10169 (see Table 2), but in this case the anti-Xa activity was measured by a chromogenic substrate assay (DAWES et al. 1986). Amidolytic anti-IIa activity was cleared slightly faster and absorbed less well than total PK 10169; it is probably contributed by a higher molecular weight fraction of the material.

However, the discrepant anti-Xa activity is too great to be accounted for by a fraction of the administered heparin, and may represent a secondary activity, released by the therapeutic agent and persisting independently of it. Possible candidates include heparan sulphate released from the vessel wall (THOMAS 1984), or one of the lipases (OLIVECRONA et al. 1977).

E. Assay for Endogenous Heparan Sulphate

I. Method

This assay closely resembles that for heparin and highly sulphated heparinoids, except that DMSO is omitted; the final assay volume is therefore 200 µl rather than 400 µl. There are two additional aspects of this variant of the competitive binding assay that warrant attention, preparation of the sample and selection of the tracer and standard.

All biological samples contain proteins which interfere in this assay and must be removed. Precipitation of the protein invariably resulted in major losses of en-

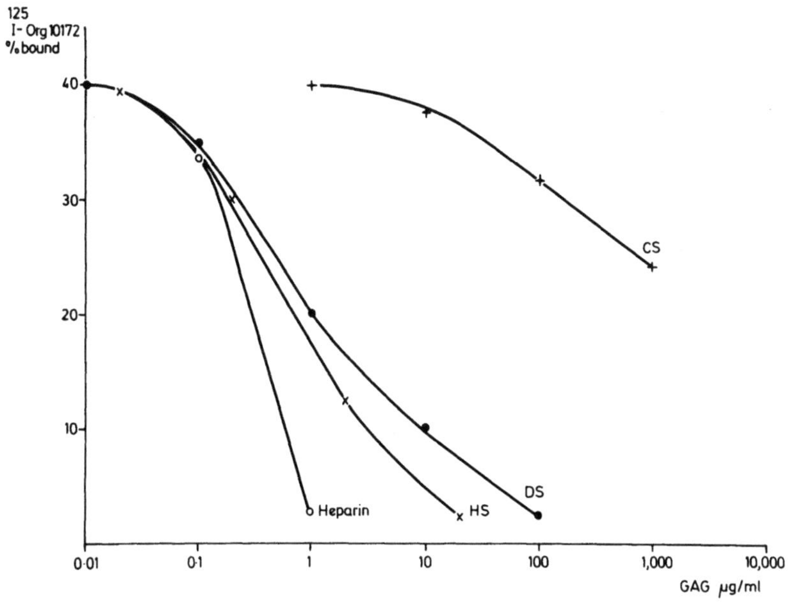

Fig. 7. Competitive binding assay with different glycosaminoglycans (GAGs) using ^{125}I-labelled Org 10172 as tracer. *DS* dermatan sulphate (E. A. Johnson, NIBSAC); *HS* heparan sulphate (Opocrin); *CS* chondroitin 4-sulphate (Sigma) and 6-sulphate (Sigma)

dogenous glycosaminoglycan, which coprecipitated either because it was covalently bound to protein as proteoglycan or because strong ionic interaction with sample protein survived the denaturing procedures. The technique currently used in this laboratory to eliminate interfering proteins is digestion with pronase E (*Streptomyces griseus;* Sigma catalogue P5147) at a final concentration of 2 mg/ml for 16 h at 37 °C. Inactivation of residual pronase is unnecessary, since the Polybrene solid phase is not a protein. Recent data (VANNUCCHI et al. 1985) suggest that phosphatidylcholine may also be a source of nonspecific interference, but we have no evidence as yet on this point.

Two radiolabelled tracer conjugate preparations have been tested in this system, heparan sulphate I, a preparation from bovine lung with a low sulphate content, kindly donated by Dr. E. A. Johnson (National Institutes for Biological Standards and Control, London), and the therapeutic heparinoid Org 10172, which is a mixture of glycosaminoglycans, but contains mainly heparan sulphate. The two are equally suitable as tracers in this assay. The selection of a standard, however, is more difficult and to some extent arbitrary. The slope of the dilution curve obtained with this version of the assay is sensitive to the degree of sulphation and composition of the glycosaminoglycan tested (Fig. 7). Very little is known about the endogenous heparan sulphate in plasma or other biological fluids, and until more information is available the standard has therefore been selected as that preparation which gives a dilution curve parallel with the sample. Org 10172 and Opocrin (Alfaricerche, Bologna; a gift of Dr. R. Mastacchi) both fulfil this criterion, and either can be used.

II. Sensitivity

Because the standard is arbitrary, it is invalid to express the results of this assay on a weight basis, and nanogram equivalents of the stated standard have therefore been adopted as the units. They probably approximate quite closely to nanograms, as the standards adopted are likely to resemble endogenous heparan sulphate, and the maximum sensitivity obtained is 150 ng equiv./ml.

III. Specificity

This version of the assay is less specific than that containing DMSO, for not only is the discriminating effect of DMSO omitted, but in addition other glycosaminoglycans compete more effectively with the less highly sulphated heparan sulphate than they do with heparin (see Fig. 7). Thus, chondroitin 4-sulphate can be detected at a concentration of 10 μg/ml, whereas with the DMSO assay and a heparin tracer it is not measurable below 100 μg/ml. Dermatan sulphate is detected with similar sensitivity to heparan sulphate, but its concentration is usually very low (CALATRONI et al. 1969). Despite theoretical drawbacks, treatment with chondroitinase ABC (Sigma) which degrades both chondroitin and dermatan sulphates has shown that in plasma and most other biological fluids other than urine, heparan sulphate is usually the only glycosaminoglycan measured. This is a very useful method of verifying the specificity of the assay with a particular sample, and is necessary for measuring heparan sulphate in urine where the concentrations of other glycosaminoglycans are high. The sample is incubated with chondroitinase ABC (0.2 units/ml) for 16 h at 37 °C before treatment with pronase. The difference between the measured level after treatment with chondroitinase ABC and the level in an untreated sample reflects the contribution of chondroitin and dermatan sulphates to the measured heparan sulphate level in the untreated sample. In normal plasma there was no difference between heparan sulphate concentrations before and after digestion with chondroitinase ABC.

IV. Results of Rat and Human Studies

Normal human plasma contains 220–560 ng equiv./ml heparan sulphate, with a median value of 330 ng equiv./ml, but the source of this material is not yet clear, as most cells, including the vascular endothelium, synthesise a range of glycosaminoglycans. When whole blood was fractionated by density gradient centrifugation, however, heparan sulphate could only be detected in the plasma and the basophils (Table 4); small amounts in the other leukocyte fractions can be attributed to basophil contamination. Some or all of the plasma heparan sulphate may therefore originate in the basophils, which have been shown to release it during clotting (serum contains 1700–3200 ng equiv./ml) and on storage. Release from stored cells is accentuated by the presence of free Ca^{2+} at ionic concentrations greater than 85 μM, and by reduction of the ratio of plasma to cells; in the latter instance increased release occurs on storage even at low free Ca^{2+} concentrations (Fig. 8) and there may be a plasma component which inhibits basophil release under normal physiological conditions.

Table 4. Concentrations of heparan sulphate in blood fractions following density gradient centrifugation

Fraction	Heparan sulphate concentration[a]	
	(μg equiv./ml)	(μg equiv./10^6 cells)
Whole blood	12.0	
Plasma	0.5	
Erythrocytes		< 0.0002
Platelets		< 0.006
Lymphocytes		6.96
Neutrophils + eosinophils		< 0.30
Basophils		243.3

[a] Purity was >85% in each case; cells were lysed in 1% Triton X-100.

Nevertheless one of the most interesting areas for application of this assay is in the study of allergic reactions. When rats primed with *Nippostrongylus brasiliensis* were challenged intravenously with worm antigen, glycosaminoglycan was released in parallel with the marker for mucosal mast cells, rat mast cell protease II (KING et al. 1985). Levels are currently being studied in human atopic individuals, and after adverse reactions to blood products and anaesthetics.

The assay is also being used in the identification and purification of naturally occurring glycosaminoglycans. It is invaluable for monitoring fractionation, and can be used to obtain some information about the glycosaminoglycan without further processing. Human follicular fluid, for example, contains 25 μg equiv./ml, and the slope obtained on serial dilution is parallel with that of the heparan sulphate standard.

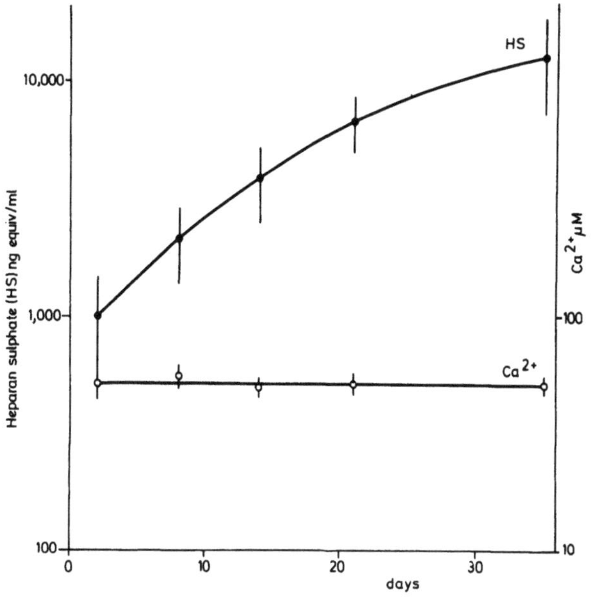

Fig. 8. Release of heparan sulphate in stored red cell concentrates at constant low ionised calcium concentrations

Table 5. Comparison of relative competitive binding of various synthetic and semisynthetic polymers

Polymeric acid

Substance	Molecular weight	SO_4 (% w/w)	ID_{50}[a] (µg/tube)	ID_{50} (pmol/tube)
Pentosan sulphate (SP54, Benechemie)	6000	48.0	0.0035	1.000
Dextran sulphate	40000	53.7	0.0037	0.093
Oversulphated chondroitin sulphate (SSHA, Luitpold)	9000	40.0	0.0054	0.594
Polyanetholesulphonate (Liquoid, Roche)	30000	43.0	0.012	0.400
Dextran sulphate	500000	52.0	0.040	0.080
Polyvinylsulphate	100000	64.0	0.042	0.420
ι-Carrageenan	45000	38.0	0.800	40.0
Polystyrenesulphonate	100000	48.0	11.2	112

Polymeric base

Substance	Molecular weight	ID_{50}[a] (µg/tube)	ID_{50} (pmol/tube)
Polybrene	10000	0.026	2.6
Platelet factor 4 (PF4)	28000	0.028	1.0
Protamine	5000	0.028	5.6
DEAE–dextran	500000	0.160	0.32
Polyethyleneimine	100000	0.56	5.6
Polydiallyldimethylammonium bromide	100000?	1.32	13.2
Poly-L-lysine (4 discrete batches)	26000–144000	1.00–1.90	13.2–38.0
Poly-D-lysine	194000	1.20	6.0
Poly-L-arginine (2 discrete batches)	60000 and 100000	1.33 and 1.20	22.2 and 12.0
Polyethyloxazoline (50% dealkylated)	100000	0.17	1.7
Poly(2,5-tetramethylene-1-amino-1,3,4-triazole)	10000	>10	>1000

[a] Actual minimum detectable sensitivity is 5- to 10-fold better than ID_{50} values.

It is not, therefore, heparin, which gives a much steeper slope, or chondroitin sulphate, which is less steep (see Fig. 7). The absence of heparin is confirmed by finding that little or no material is measurable in the DMSO-containing assay. Further characterisation requires enzymic and chemical degradation and electrophoretic analysis, with the continuing use of the competitive binding assay to measure undegraded material.

F. Other Applications

The competitive binding assay also has wider applications outside heparan (for instance dextran sulphates and carrageenans are readily assayed; Table 5) and outside pharmacology, for instance in tissue culture, agriculture and foodstuffs. In addition, the assay may be used in the reverse manner to rank the relative abilities (ID_{50}) of any soluble polybasic compound to inhibit the binding of tracer to the solid phase (Table 5). In this manner we were able to show that Polybrene had a higher affinity for heparin than protamine without the need to first synthesise a Polybrene solid phase.

G. Summary

A sensitive competitive binding assay based on the same principles as radioimmunoassay has been developed using novel techniques for iodinating glycosaminoglycans and for synthesising the binding reagent, Polybrene–Sepharose. It has been adapted to the measurement either of heparin and other highly sulphated therapeutic heparinoids, or of endogenous heparan sulphate. The assay is ideal for studying pharmacokinetics, as, unlike biological assays, it is not affected by variations in other plasma components.

Heparin, low molecular weight heparin, and pentosan polysulphate were all cleared with apparent first-order kinetics and dose-dependent half-lives. The bioavailability of these materials after subcutaneous injection was inversely related to the molecular weight, and the wide variation in absorption demonstrated between individuals after subcutaneous heparin administration was also reduced at lower molecular weights. Comparison of data obtained using this assay with the results of biological assays indicates that anti-Xa activity persists after clearance of the administered glycosaminoglycan. This may be a secondary effect, mediated by material released by the glycosaminoglycan and cleared independently of it.

Endogenous heparan sulphate has been demonstrated in plasma, serum and follicular fluid. The assay has been used to demonstrate release from basophils in vitro and from basophils and/or mast cells as a result of immunological stimulus in vivo. Moreover, it is proving invaluable in monitoring the purification and characterisation of endogenous glycosaminoglycans which will fulfil the functions described here, and in both areas further application is expected to yield much information where knowledge is currently sparse.

Acknowledgments. We are very grateful to those who generously donated glycosaminoglycans: Dr. E.A. Johnson, National Institutes for Biological Standards and Control, London; the late Dr. E. Halse, Benechemie, Munich; Dr. A.T. Irvine, Sanofi UK, for Choay, Paris; Dr. H. Moelker, Organon, Oss; and Dr. R. Mastacchi, Alfaricerche, Bologna.

References

Andersson LO, Barrowcliffe TW, Holmer E, Johnson EA, Sims GEC (1976) Anticoagulant properties of heparin fractionated by affinity chromatography on matrix-bound antithrombin III and by gel filtration. Thromb Res 9:575–583

Bitter T, Muir HM (1962) A modified uronic acid carbazole reaction. Anal Biochem 4:330–334

Bratt G, Tornebohm E, Lockner D, Bergstrom K (1985) A human pharmacological study comparing conventional heparin and a low molecular weight heparin fragment. Thromb Haemost 53:208–211

Calatroni A, Donnelly PV, Di Ferrante N (1969) The glycosaminoglycans of human plasma. J Clin Invest 48:332–343

Carlson SS (1982) ^{103}Ruthenium red, a reagent for detecting glycosaminoglycans at the nanogram level. Anal Biochem 122:364–367

Dawes J, Pepper DS (1979) Catabolism of low-dose heparin in man. Thromb Res 14:845–860

Dawes J, Pepper DS (1982) A sensitive competitive binding assay for exogenous and endogenous heparins. Thromb Res 27:387–396

Dawes J, Prowse CV, Pepper DS (1985a) Heparin, LMW heparin and SP54 concentrations after subcutaneous injection. Thromb Haemost 54:94

Dawes J, Prowse CV, Pepper DS (1985b) The measurement of heparin and other therapeutic sulphated polysaccharides in plasma, serum and urine. Thromb Haemost 54:630–634

Dawes J, Bara L, Billaud E, Samama M (1986) Relationship between biological activity and concentration of a LMW heparin (PK 10169) and unfractionated heparin after intravenous and subcutaneous administration. Haemostasis 16:116–122

Eika C, Godal HC, Kierulf P (1972) Detection of small amounts of heparin by the thrombin clotting time. Lancet II:376

Fischer A-M, Barrowcliffe TW, Thomas DP (1982) A comparison of pentosan polysulphate (SP54) and heparin I: mechanism of action on blood coagulation. Thromb Haemost 47:104–108

Godal HC (1975) Heparin assay methods for control of in vivo heparin effect. Thrombos Diathes Haemorrh 33:77–80

Greenwood FC, Hunter WM, Glover JS (1963) The preparation of ^{131}I-labelled human growth hormone of high specific radioactivity. Biochem J 89:114–123

Hascall VC, Hascall CK (1981) Proteoglycans. In: Hay E (ed) Cell biology of the extracellular matrix. Plenum, New York, p 39

Hook M, Kjellen L, Johansson S, Robinson R (1984) Cell surface glycosaminoglycans. Annu Rev Biochem 53:847–869

Hunter WM (1977) Simplified solid-phase radioimmunoassay without centrifugation. Acta Endocrinol 85, Suppl 212:463

Jaques LB, Wollin A (1967) Modified method for the colorimetric determination of heparin. Can J Physiol Pharmacol 45:787–794

Jaques LB, Bruce-Mitford M, Ricker AG (1947) Metachromatic activity of heparin. Rev Can Biol 6:740–754

Kakkar VV, Djazaeri B, Fok J, Scully M, Westwick J (1982) Low molecular weight heparin and the prevention of postoperative deep vein thrombosis. In: Witt I (ed) Heparin. New biochemical and medical aspects. de Gruyter, Berlin, p 175

King SJ, Reilly K, Dawes J, Miller HRP (1985) The presence in blood of both glycosaminoglycan and mucosal mast cell protease following systemic anaphylaxis in the rat. Int Arch Allergy Appl Immunol 76:286–288

Lane DA, Michalski R, van Ross ME, Kakkar VV (1977) Comparison of heparin and a semisynthetic heparin analogue, A73025. Br J Haematol 37:239–245
Lee RI, White PK (1913) A clinical study of the coagulation time of blood. Am J Med Sci 145:495–503
Lindahl V, Hook M (1978) Glycosaminoglycans and their binding to biological macromolecules. Annu Rev Biochem 47:385–417
MacGregor IR, Dawes J, Paton L, Pepper DS, Prowse CV, Smith M (1984) Metabolism of sodium pentosan polysulphate in man – catabolism of iodinated derivatives. Thromb Haemost 51:321–325
MacGregor IR, Dawes J, Pepper DS, Prowse CV, Stocks J (1985) Metabolism of sodium pentosan polysulphate in man measured by a new competitive binding assay for sulphated polysaccharides – comparison with effects upon anticoagulant activity, lipolysis and platelet α-granule proteins. Thromb Haemost 53:411–414
Olivecrona T, Bengtsson G, Marklund S-E, Lindahl U, Hook M (1977) Heparin-lipoprotein lipase interactions. Fed Proc 36:60–65
Pepper DS, Dawes J (1977) Preparation and applications of ^{125}I-labelled heparin. Thromb Haemost 38:15
Teien AN, Lie M (1975) Heparin assay in plasma: a comparison of five clotting methods. Thromb Res 7:777–780
Teien AN, Abildgaard U, Hook M (1976a) The anticoagulant effect of heparan sulfate and dermatan sulfate. Thromb Res 8:859–867
Teien AN, Lie M, Abildgaard U (1976b) Assay of heparin in plasma using a chromogenic substrate for activated factor X. Thromb Res 8:413–416
Thomas DP (1984) Heparin, low molecular weight heparin and heparin analogues. Br J Haematol 58:385–390
Thomas DP, Merton RE (1982) A low molecular weight heparin compared with unfractionated heparin. Thromb Res 28:343–350
Thomas DP, Johnson EA, Barrowcliffe TW (1981) The antithrombotic action of heparan sulphate. In: Tesi M, Dormandy J (eds) Vascular occlusion: epidemiological, pathophysiological and therapeutic aspects. Academic, London, p 377
Thomson JM (1970) A practical guide to blood coagulation and haemostasis. Churchill, London
Torngren S, Noren I, Forsskahl B, Sipila H (1983) Effects of heparin and a semisynthetic heparin analogue on platelet aggregation, lipoprotein lipase and other laboratory tests in surgical patients. Thromb Res 30:527–534
Vannucchi S, Ruggiero M, Chiarugi V (1985) Complexing of heparin with phosphatidylcholine. A possible supramolecular assembly of plasma heparin. Biochem J 227:57–65
Varadi DP, Cifonelli JA, Dorman A (1967) The acid mucopolysaccharides in normal urine. Biochim Biophys Acta 141:103–117
Verstraete M (1984) Unresolved questions with heparin. Triangle 23:49–55
Wight T (1980) Vessel proteoglycans and thrombogenesis. Prog Haemost Thromb 5:1–39
Wu V-Y, Cohen MP (1984) A competitive binding assay for measurement of heparan sulfate in tissue digests. Anal Biochem 139:218–222
Yin ET, Wessler S, Butler JV (1973) Plasma heparin: a unique, practical, submicrogram-sensitive assay. J Lab Clin Med 81:298–310

CHAPTER 23

Radioimmunoassay of the Somatomedins/Insulin-like Growth Factors

L. E. UNDERWOOD and M. G. MURPHY

A. Introduction and Nomenclature of the Somatomedins

The somatomedins, also referred to as insulin-like growth factors, are a family of peptides whose serum concentrations are regulated principally by growth hormone and nutrient status. Evidence is now emerging that most or all of the biologic effects of the somatomedins can be attributed to two peptides, somatomedin-C/insulin-like growth factor I (Sm-C/IGF-I) and IGF-II (VAN WYK 1984). Sm-C/IGF-I has a molecular weight of 7649 and a pI of 8.0–8.7. It contains 70 amino acid residues in a single chain with 3 disulfide bridges. It is highly growth hormone dependent, and has potent growth-promoting activity in many in vitro systems (ZAPF et al. 1984). IGF-II is also a single-chain peptide, which has a more nearly neutral pI and a molecular weight of 7471. IGF-II is less growth hormone dependent and appears to have less growth-promoting activity than Sm-C/IGF-I. Both Sm-C/IGF-I and IGF-II possess nearly 50% homology with proinsulin in regions of the molecule that correspond to the A and B chains of insulin. As in humans, two forms of somatomedin have been isolated from rat plasma. One of these, multiplication-stimulating activity (MSA) is the rat homolog of human IGF-II, differing from human IGF-II by only five amino acid residues (MARQUARDT et al. 1981). Although not certain, it is expected that other species will be found to have two somatomedins, similar to those in humans and rats. The term somatomedin-A has been applied to a substance that now appears to be a deamidated form of Sm-C/IGF-I, and the peptide referred to as somatomedin-B proved to be a fragment of a plasma-spreading factor contaminated with epidermal growth factor (EGF) (HELDIN et al. 1981; BARNES et al. 1984).

In plasma, the somatomedin peptides of molecular weight 7500 are bound to larger carrier proteins, and to date, no free somatomedins have been identified (SMITH 1984). The major class of plasma-binding protein is growth hormone dependent, has a molecular weight of approximately 150000, and can be dissociated with acid into at least two subunits. The other distinct somatomedin-binding protein is not growth hormone dependent, and has a molecular weight of approximately 40000 (DROP et al. 1984). These binding proteins account for the fact that the concentrations of somatomedins in the circulation are constant throughout the day, and are relatively high, being 100–180 ng/ml for Sm-C/IGF-I and even higher for IGF-II. The effects of these binding proteins on measurement of somatomedins in RIAs and membrane-binding assays will be discussed in Sect. B.II.

While it is clear that the somatomedins (particularly Sm-C/IGF-I) in plasma are responsive to growth hormone, nutritional status, and a variety of other modulators, it is less clear what biologic function the somatomedins in plasma might serve. The reason for this is that there is considerable circumstantial evidence that the somatomedins might act by paracrine and/or autocrine mechanisms, as well as (or instead of) the traditional endocrine mechanisms (UNDERWOOD et al. 1986). The evidence for paracrine/autocrine mechanisms of action include the observations that: (a) no somatomedin-rich organ reservoir has been found; (b) somatomedins appear to be produced by many cells in many tissues; (c) the somatomedins exert biologic effects on diverse types of cells; and (d) after growth hormone injection, Sm-C/IGF-I concentrations in tissues of hypophysectomized rats rise prior to the rise in blood concentration.

I. Methods for Measuring Somatomedins Before the Development of Radioimmunoassays

1. Bioassays

The biologic assays that have been used for measurement of somatomedins in serum reflect the diversity of actions of these peptides and the focus of early investigations into these biologic effects. Daughaday and colleagues (see VAN WYK 1984), exploring the effects of somatomedins on cartilage growth in vitro, showed that serum from hypopituitary animals had little stimulatory effect on uptake of sulfate by cartilage. Likewise, growth hormone added to cartilage had no significant stimulatory effect. However, after hypopituitary animals were treated with growth hormone, their serum stimulated sulfate uptake in vitro. The "sulfation factor," and subsequently the "thymidine factor" assays were developed from these somatomedin effects and became the principal means for study of somatomedin. While such assays are quite sensitive, they are not specific for somatomedins and they measure the net effect of somatomedins, somatomedin-like stimulators, and inhibitors of somatomedin action.

Because of their interest in the serum insulin-like activity that cannot be removed by addition of antibodies to insulin, Froesch and colleagues (see ZAPF et al. 1984) developed a bioassay dependent on the oxidation of radiolabeled glucose to CO_2 by fat cells. This assay was used to purify the nonsuppressible insulin-like activity (NSILA) of serum, and led to the characterization of IGF-I and IGF-II.

Researchers interested in cell culture have monitored purification of serum factors that stimulate cell growth by measuring the incorporation of thymidine into cultured chick embryo fibroblasts (PIERSON and TEMIN 1972). This line of investigation led eventually to the purification of MSA and the characterization of IGF-II in the rat. In addition to their lack of specificity, these bioassays are time-consuming, expensive to perform, and somewhat unpredictable when used to assay serum samples.

2. Radioreceptor Assays and Plasma Protein Binding Assays

The first radioreceptor assay for Sm-C/IGF-I was based on the observation that partially purified Sm-C/IGF-I competed with radiolabeled insulin for the insulin receptor (HINTZ et al. 1972). This assay proved quite useful for monitoring the purification of Sm-C/IGF-I, but is not specific for individual somatomedins. Subsequently, the placental membrane receptor assay for Sm-C/IGF-I was developed, utilizing radiolabeled Sm-C/IGF-I (MARSHALL et al. 1974), and cell membranes prepared from human placenta. In this assay, IGF-II cross-reacts by about 30%. It is possible to show differences in the concentrations of Sm-C/IGF-I among sera from hypopituitary, normal, and acromegalic individuals with the placental receptor assay (D'ERCOLE et al. 1977). With the addition of 10 ng Sm-C/IGF-I per milliliter, 50% displacement of tracer occurs (VAN WYK et al. 1980). A cell membrane assay for somatomedin-A (a deamidated form of Sm-C/IGF-I) has also been described (TAKANO et al. 1976), as have membrane receptor assays for IGF-II (ZAPF et al. 1978) and MSA (RECHLER et al. 1978). As in the Sm-C/IGF-I placental receptor assay, all these assays have a high degree of cross-reactivity with other somatomedins.

Competitive protein binding assays for the somatomedins are based on the observation that purified, iodinated somatomedin peptides have the capacity to bind to higher molecular weight carrier proteins in serum. These assays use crude serum fractions prepared by stripping endogenous somatomedins away from the binding proteins by gel chromatography in acid (ZAPF et al. 1977; SCHALCH et al. 1978). In general, the specificity of these assays is no better than that of cell membrane receptor assays. Furthermore, it is essential that the serum sample to be quantitated have all the binding protein removed before assay, a procedure that may result in variable loss of the active peptide.

B. Radioimmunoassay for Sm-C/IGF-I

I. Antisera Production, Sensitivity, and Specificity

The first radioimmunoassay (RIA) for Sm-C/IGF-I was described by FURLANETTO et al. (1977). The antibody used in this assay was produced in rabbits by conjugating partially purified Sm-C/IGF-I with ovalbumin and administering the primary injection intradermally. The antibody produced has a K_a of 4.6×10^{10} l/mol, an effective titer of 1:10000, exhibits approximately 2% cross-reactivity with IGF-II, and has virtually no cross-reactivity with a variety of other peptides (Fig. 1). The Sm-C/IGF-I activity in many nonhuman species can also be detected with this assay. Because of interference by serum binding proteins, nonequilibrium assay conditions are needed to obtain parallel dose–response curves between purified Sm-C/IGF-I and serum samples. Specifically, when sample and antibody are permitted to incubate for a period of time before addition of tracer, parallel dose–response curves can be obtained. Subsequently, similar RIAs for Sm-C/IGF-I have been developed by ZAPF et al. (1978) and BALA and BHAUMICK (1979), and RIAs directed at specific portions of the Sm-C/IGF-I molecule have been developed by HINTZ et al. (1980a, b) using synthetic peptide fragments.

II. Influence of Binding Proteins

The plasma proteins that bind Sm-C/IGF-I and IGF-II have varied effects on the quantitation of these peptides by RIA. On the one hand they are responsible for

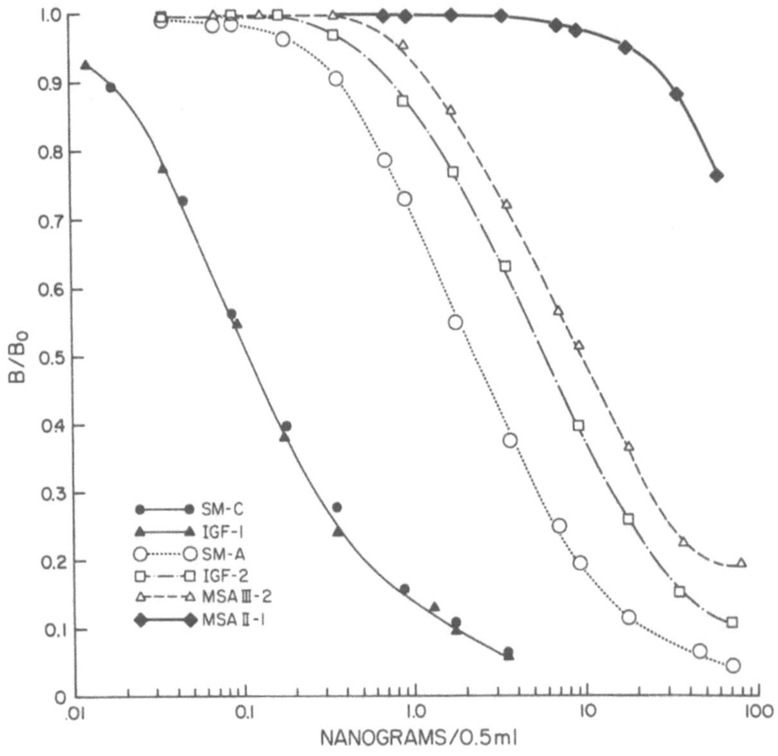

Fig. 1. Curves of competition in the Sm-C/IGF-I radioimmunoassay. Cross-reactivity of IGF-II is approximately 1.5%. (VAN WYK et al. 1980)

the relatively high concentrations in serum and for the lack of rapid fluctuations. It has been proposed that the binding proteins might play an important physiologic role, in that they might produce a serum reservoir for Sm-C/IGF-I that constantly provides needed quantities of this peptide to tissues. On the other hand, the binding proteins interfere to variable degrees with the quantitation of the peptide in serum, amniotic fluid, and conditioned media from cell cultures. Because our antibody is of sufficiently high affinity, it is possible to obtain parallel dose–response curves between serum and pure Sm-C/IGF-I, and thereby estimate the quantity of peptide in serum without separating or destroying the binding proteins. With some antibodies this is not possible.

We have adapted an assay system that utilizes polystyrene tubes and protamine-containing buffer (0.2% protamine sulfate). Compared with assay systems using glass tubes and protamine-free buffers, this assay produces low nonspecific binding of trace (approximately 1%) and parallelism of the serum and pure Sm-C/IGF-I dose–response curves. However, only 30%–40% of the Sm-C/IGF-I present in the serum sample is measured (CHATELAIN et al. 1983).

A series of studies in our laboratory have shed light not only on the importance of assay conditions, but also on the nature of the Sm-C/IGF-I–binding protein interaction. Although our assay measures only 30%–40% of the Sm-C/IGF-I

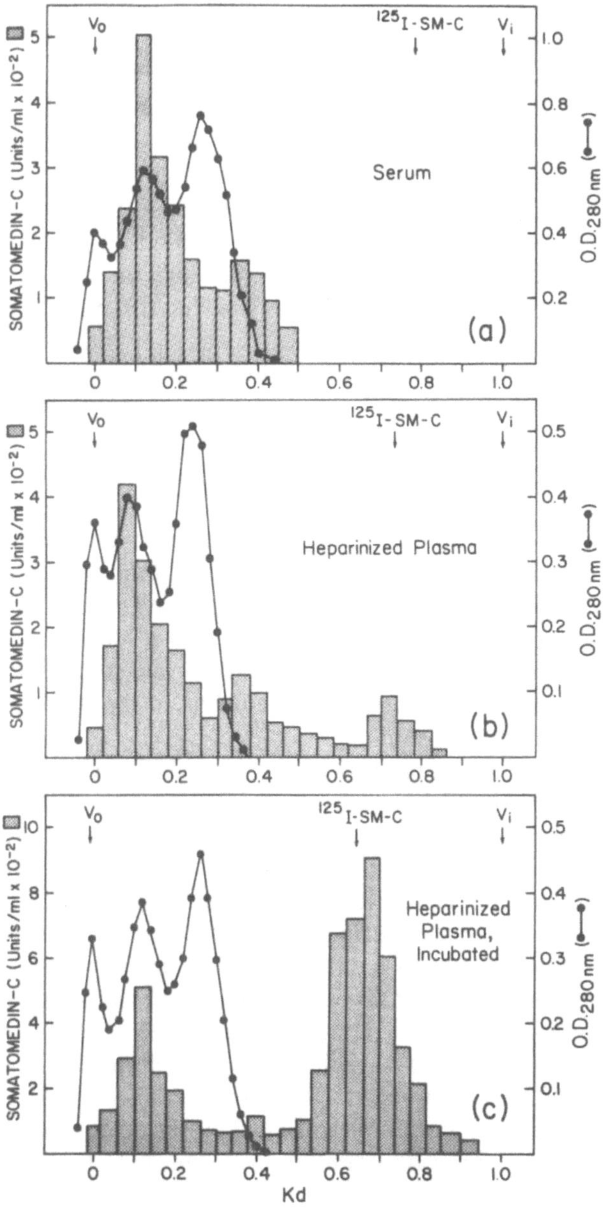

Fig. 2 a–c. Gel chromatography in Sephadex G-150 of **a** serum, **b** heparinized plasma, and **c** heparinized plasma that had been incubated at 37 °C for 24 h. The elution buffer contained 0.5 units heparin per milliliter. The *full lines* are the protein profiles obtained by absorption at 280 nm. The *bars* are Sm-C/IGF-I activity. Note that Sm-C/IGF-I can be separated from the binding protein following incubation. (CLEMMONS et al. 1983)

in serum, this amount can be increased 2- to 3-fold if the sample is allowed to incubate for several hours at 37 °C or is exposed transiently to acid (CHATELAIN et al. 1983). The increase observed after incubation at neutral pH is dependent on the action of endogenous serum proteases, since it can be inhibited by a variety of chelating agents and protease inhibitors. However, as judged by gel filtration, the Sm-C/IGF-I of serum does not change in size following incubation at neutral pH. It was postulated, therefore, that the limited proteolysis had either changed the conformation of the binding proteins, permitting access of Sm-C/IGF-I to antibody, or decreased the affinity of Sm-C/IGF-I for the binding proteins so that the peptide could react more readily with the antibody. CLEMMONS et al. (1983) then showed that endogenous plasma proteases disturb the binding proteins so that their affinity for Sm-C/IGF-I is reduced, but this reduction is not sufficient to lead to spontaneous dissociation of the two components. This was done by showing that if serum that had been incubated at 37 °C was subjected to gel chromatography in a heparin-containing buffer, more than 80% of the low molecular weight immunoreactive material could be dissociated from the binding protein and eluted in the region corresponding to free Sm-C/IGF-I. Without incubation, however, heparin alone had no such effect (Fig. 2). To confirm that this effect on the elution profile was dependent on proteolytic enzymes, it was shown that the effect of incubation of serum could be inhibited by the thiolprotease inhibitor, antipain.

1. Methods of Extracting Serum

Because of the interference of binding proteins in the RIA, many investigators have developed methods for extracting the low molecular weight, immunoreactive Sm-C/IGF-I from serum before submitting the samples to assay. Foremost among these is gel chromatography in 1 *M* acetic acid (ZAPF et al. 1978). While this is an effective means for separating binding proteins from low molecular weight Sm-C/IGF-I, a small amount of the active peptide is lost in the binding protein fraction and the method is time-consuming.

Extraction of samples with acid–ethanol has also been reported (DAUGHADAY et al. 1980). This technique is simple and less labor intensive than column chromatography, however it is performed at a pH (1.3) that is far below the pH that is optimal for disruption of the binding protein–Sm-C/IGF-I bond (pH 3.6–3.8). Furthermore, extraction is incomplete because of losses of the low molecular weight immunoreactive peptide in the bulky precipitate produced by acid–ethanol. Finally, BALA and BHAUMICK (1979) reported that much of the effect of extraction procedures could be accomplished by simply incubating serum samples at acid pH. We have made similar observations (CHATELAIN et al. 1983, 1986) on human serum, and have used this technique in heterologous RIAs for Sm-C/IGF-I in sheep (UNDERWOOD et al. 1982a) and in rabbits (D'ERCOLE et al. 1984a). A similar technique has been used to prepare samples of chicken serum for assay (HUYBRECHTS et al. 1985), and in our laboratory for monkey serum (L. E. UNDERWOOD and J. J. VAN WYK, unpublished work).

2. What is the Best Method for Measuring Sm-C/IGF-I in Serum?

From the issues already discussed, it should be clear that there is uncertainty about the best method for measuring the Sm-C/IGF-I in serum. This is partly because it is not known what form of the peptide best reflects the physiologically active, clinically relevant form. Is it the total or free Sm-C/IGF-I that is measured after acid–ethanol extraction or acid gel chromatography? Is it binding protein complexed peptides? Or is it neither of these? While the answers to these questions are not known, measurements made after extraction should provide a more precise estimate of the total Sm-C/IGF-I in the sample and can be compared more readily with a pure peptide standard. On the other hand, it is conceivable that the "available" Sm-C/IGF-I measured on blood samples collected and stored so as to eliminate in vitro proteolytic activity might have more clinical significance than that measured after the effects of proteolysis have been maximized. Furthermore, measurements on unextracted serum avoid the errors introduced by extraction procedures.

We have elected to measure the Sm-C/IGF-I immunoreactivity in unextracted plasma (COPELAND et al. 1980). Because endogenous proteolytic enzymes disrupt the bond between binding proteins and Sm-C/IGF-I, we collect blood samples in EDTA, which has been shown to inhibit the proteolytic process (CHATELAIN et al. 1983). The assay is then carried out in polystyrene tubes and protamine-containing buffer, with delayed addition of ^{125}I-labeled Sm-C/IGF-I tracer. We have compared results obtained with this system with those obtained after extraction. When compared with acid gel chromatography they show a correlation coefficient r of 0.92 ($P<0.001$; 31 samples). The correlation between unextracted sera and acid–ethanol-extracted sera is 0.89 ($P<0.001$; 33 samples). Similar correlations are observed between unextracted sera and sera incubated at neutral pH or at acid pH (pH 3.6–3.8) (CHATELAIN et al. 1986).

III. Clinical Utility of Measurements of Sm-C/IGF-I in Plasma

The Sm-C/IGF-I radioimmunoassay has been of incalculable value in understanding how this peptide is regulated in plasma. Concentrations are relatively low (30%–40% of adult values) in human fetuses, newborns, and infants (see CLEMMONS and VAN WYK 1984). As the child ages, the concentration of Sm-C/IGF-I rises, reaching maximal values during puberty (Fig. 3). These peak values, which correspond to midpubertal development and occur at the time of maximum pubertal growth, are 2–4 units/ml (1 unit Sm-C/IGF-I is the activity in 1 ml of a pool of a plasma standard). With advancing age, concentrations decline, so that in the sixth and seventh decades, values have often declined to approximately one-half the young adult mean. Values fluctuate minimally throughout the day (Fig. 4).

Measurement of Sm-C/IGF-I provides insight into growth hormone status, being low in hypopituitary individuals (Fig. 3) and high in patients with active acromegaly. Low values may result from a variety of factors other than growth hormone deficiency. These include hypothyroidism, chronic illness, and poor nutritional intake. Therefore, a low Sm-C/IGF-I value in a short child is not proof of growth hormone deficiency, and it is still necessary to assess growth hormone secretion directly. However, a normal value in a short child makes growth hormone deficiency unlikely, and suggests that other diagnoses should be pursued. We know of two exceptions to this rule: some children who have had removal of a craniopharyngioma are growth hormone deficient, but have normal growth associated with severe hyperphagia. Such children may have Sm-C/IGF-I values within the normal range. Also, adults with large pituitary tumors, growth hormone deficiency, and massive elevations in serum prolactin concentrations may

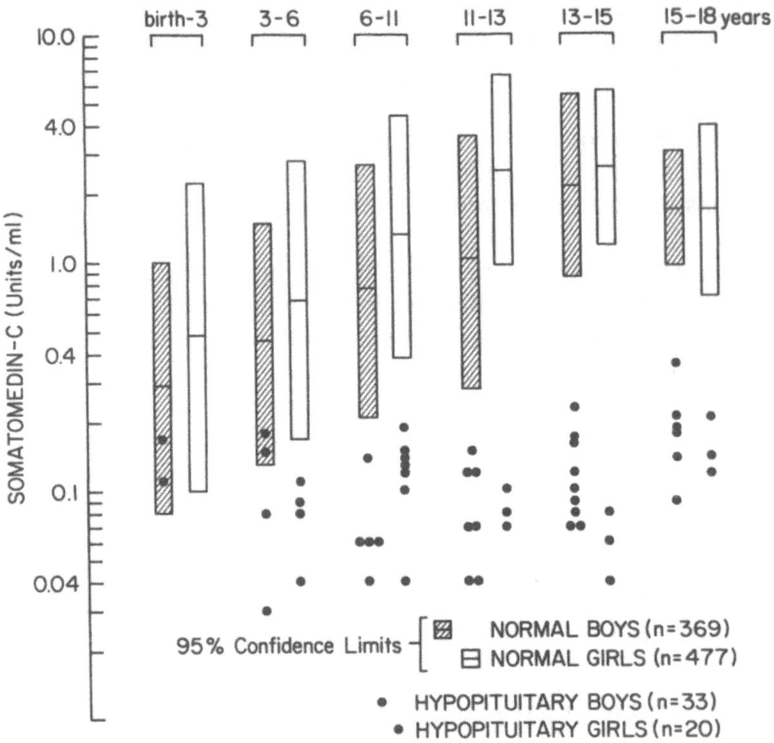

Fig. 3. Plasma Sm-C/IGF-I concentrations in normal children and children with hypopituitarism. (UNDERWOOD and VAN WYK 1985)

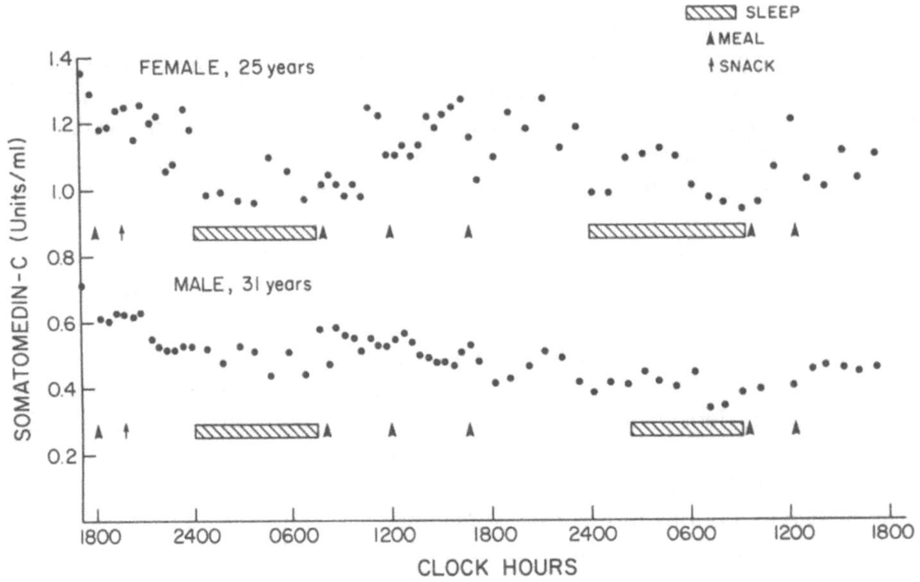

Fig. 4. Fluctuation in plasma Sm-C/IGF-I concentrations in two normal adults over periods of 48 h. (UNDERWOOD et al. 1982 b)

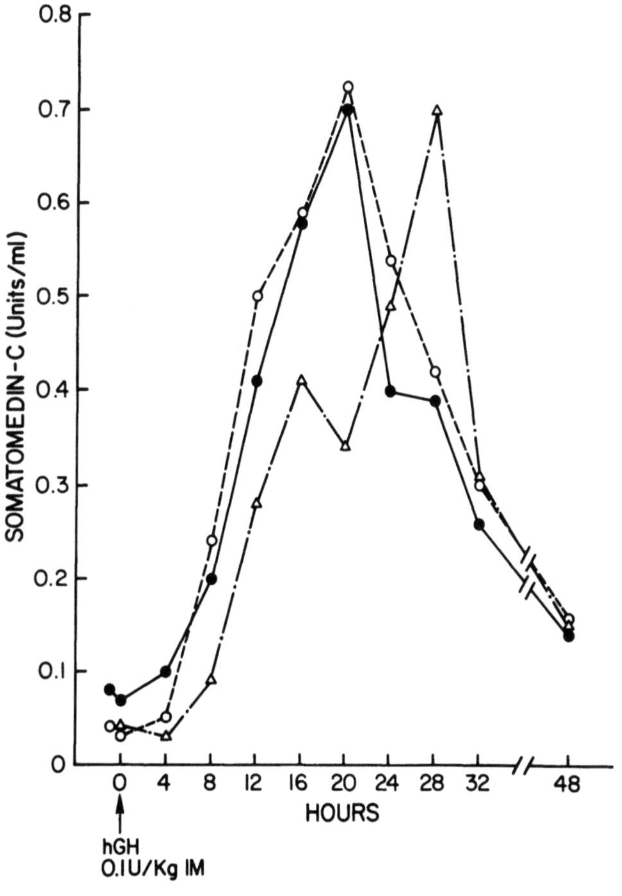

Fig. 5. Rise in plasma Sm-C/IGF-I in hypopituitary children following the injection of growth hormone. (COPELAND et al. 1980)

have normal Sm-C/IGF-I values. We believe that this is due to the weak somatotrophic properties of prolactin.

When growth hormone is given to hypopituitary or normal individuals, a rise in plasma Sm-C/IGF-I is observed in 4–6 h, with peak values being reached between 24 and 28 h (Fig. 5). Values usually return to basal levels within 72 h of the injection of growth hormone.

Sm-C/IGF-I measurements are quite useful in the diagnosis of active acromegaly, being several times higher than normal in most cases (Fig. 6; CLEMMONS et al. 1979). Furthermore, we have observed that the Sm-C/IGF-I concentration is a better indicator of disease activity than growth hormone itself, and is a better predictor of the success of therapy than growth hormone (CLEMMONS et al. 1980; WASS et al. 1982).

Besides growth hormones, the other principal regulator of Sm-C/IGF-I is the dietary intake. It has long been known that starvation decreases the serum concentrations of bioactive somatomedin (PHILLIPS and VASSILOPOULOU-SELLIN 1979). In a series of studies using our RIA, we have shown that short-term fasting for several days reduces Sm-C/IGF-I markedly (CLEMMONS et al. 1981); the reduction in Sm-C/IGF-I is due to insufficiency of both energy and protein (ISLEY et al. 1983, 1984); the quality of protein is a determinant of the Sm-C/IGF-I (CLEM-

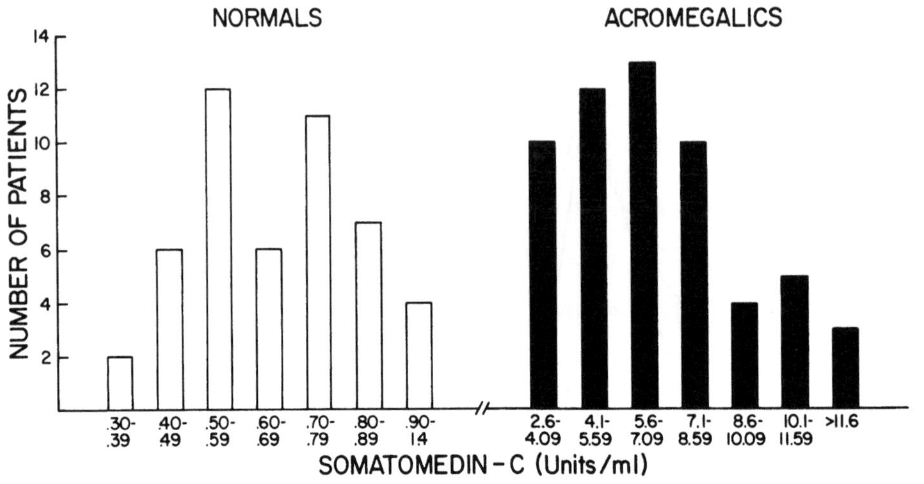

Fig. 6. Plasma Sm-C/IGF-I in normal adults and 57 patients with active acromegaly. (CLEMMONS et al. 1979)

MONS et al. 1985a); and the concentrations of this peptide are a good, early indicator of the adequacy of refeeding in chronically malnourished patients (CLEMMONS et al. 1985b). We believe, therefore, that this test may prove to be useful clinically as an indicator of change in nutritional status following dietary manipulation.

IV. Heterologous Radioimmunoassays for Sm-C/IGF-I

Sm-C/IGF-I appears to be highly conserved in nature. Therefore, it can be measured in the serum of many of the species in which studies have been attempted. However, the relationship between the low molecular weight, immunoreactive Sm-C/IGF-I peptide and its binding protein varies with each species. In the rat (D'ERCOLE et al. 1984b), mouse (D'ERCOLE and UNDERWOOD 1980), and baboon (COPELAND et al. 1983), this relationship appears to be much like the human. It is possible in these species to measure serum concentrations without performing extraction procedures. On the other hand, the binding of the small immunoreactive Sm-C/IGF-I by the binding protein appears to be quite tight in the sheep, where prolonged incubation of the serum sample at pH 3.6–3.8 is needed before assay is possible (UNDERWOOD et al. 1982a). Once this bond is disrupted, however, the cross-reactivity of sheep Sm-C/IGF-I with the human antibody is good, since the values measured are approximately three times greater than those found normally in humans. In rabbits and chickens extraction is also required (D'ERCOLE et al. 1984a; HUYBRECHTS et al. 1985). In the latter, however, the quantities of Sm-C/IGF-I measured after extraction are lower than in humans.

Before the Sm-C/IGF-I concentrations in the serum of a given species are quantitated, several steps are imperative. First, it must be shown that the curve of competition produced by graded doses of serum of the species under study is parallel to purified Sm-C/IGF-I. If parallel, it must be shown that the activity being measured is growth hormone dependent, i.e., low in hypopituitary animals

Table 1. Extractable Sm-C/IGF-I (units/g) in rats (D'ERCOLE et al. 1984b)

Tissue	Normal animals	Hypophysectomized animals
Serum[a]	28.7±0.98[b]	0.74±0.12
Liver	1.91±0.23	0.23±0.08
Lung	2.04±0.86	0.57±0.13
Kidney	2.59±0.88	0.77±0.29
Heart	0.92±0.33	0.48±0.14
Skeletal muscle	0.42±0.05	<0.08
Brain	0.26±0.09	0.28±0.04
Testes	1.88±0.42	0.52±0.32

[a] Serum values are in units/ml.
[b] Mean±SD; differences are significant ($P<0.005$) in liver, lung, kidney, and testes.

and increased by growth hormone. Finally, it must be shown that pure Sm-C/IGF-I added to serum of the species under study can be measured fully in the RIA. If not fully measurable, it must be assumed that the binding proteins are reacting with and covering up the active peptide. If the serum under study does not produce a parallel dose–response curve, extraction of the sample is mandatory, and after extraction, growth hormone dependency must be established.

V. Measurement of Sm-C/IGF-I in Tissues and Media Extracts

The quantities of Sm-C/IGF-I in tissues are small and the amount of binding protein that is present appears to be relatively high. Therefore, it is necessary to carry out extraction of tissue samples, not only to gain access to the Sm-C/IGF-I, but also to destroy binding protein. We have done this successfully in rats, showing that a variety of tissues have measurable amounts of the peptide, and that the concentration of Sm-C/IGF-I is reduced in hypophysectomized animals (Table 1). Following the injection of growth hormone, tissue levels rise before an increase in serum is observed (D'ERCOLE 1984b). We also have recently completed studies on Sm-C/IGF-I extracts of human fetal tissues (D'ERCOLE et al. 1986).

C. Radioimmunoassay for IGF-II

In this chapter, most of the focus has been on Sm-C/IGF-I because it is the peptide with which the authors have the most familiarity, and on which more studies have been done. In general, the information gained on the RIA for Sm-C/IGF-I applies to IGF-II. Both peptides have the same (or similar) serum binding proteins, and both are regulated by growth hormone and dietary status. For the most part, the concentrations of IGF-II seem to be less susceptible to the factors known to modulate Sm-C/IGF-I. The first RIA for IGF-II was described by ZAPF et al. (1981). The antibody used in this assay binds 35%–45% of trace amounts of IGF-II at a final dilution of 1:2000 and Sm-C/IGF-I cross-reacts by approximately

10%. Half-maximal displacement of IGF-II occurs at 1000 pg per tube. All serum samples are filtered in Sephadex G-50 at acidic pH before assay. The mean concentration of IGF-II in serum of normal adults is 641 ± 189 ng/ml.

Like Sm-C/IGF-I, serum IGF-II concentrations are relatively low in fetal and cord blood, but unlike Sm-C/IGF-I, there is no clear increment in IGF-II during puberty. The finding that IGF-II values in fetal blood were not high was surprising, because MOSES et al. (1980) had reported that the concentration of MSA (the rat homolog of human IGF-II) in fetal serum was more than 20 times that of maternal serum.

IGF-II appears to be less growth hormone dependent than Sm-C/IGF-I because IGF-II concentrations are nearly 40% of normal in hypopituitary patients, and values are not elevated consistently in acromegaly (ZAPF et al. 1981). Similarly, the decline of IGF-II with prolonged fasting is less pronounced than that observed with Sm-C/IGF-I (MERIMEE et al. 1982).

Acknowledgments. Research summarized in this chapter was supported by USPHS Research Grants HD 08299 and AM 01022 and training grant AM 07129. The Clinical Research Unit of the University of North Carolina is supported by grant RR 00046 from the General Clinical Research Centers Program of the Division of Research Resources, National Institutes of Health. The authors are indebted to Eyvonne Bruton and Karen Koerber for technical assistance and to Christine Silva for assistance in preparation of the manuscript.

References

Bala RM, Bhaumick B (1979) Radioimmunoassay of a basic somatomedin: comparison of various techniques and somatomedin levels in various sera. J Clin Endocrinol Metab 49:770–777

Barnes DW, Foley TP Jr, Shaffer MC, Silnutzer JE (1984) Human serum spreading factor: relationship to somatomedin B. J Clin Endocrinol Metab 59:1019–1021

Chatelain PG, Van Wyk JJ, Copeland KC, Blethen SL, Underwood LE (1983) Effect of in vitro action of serum proteases or exposure to acid on measurable immunoreactive somatomedin-C in serum. J Clin Endocrinol Metab 56:376–383

Chatelain PG, Underwood LE, Murphy MG, Van Wyk JJ (1986) Measurement of circulating human immunoreactive somatomedin-C: implications of various assay methods (to be published)

Clemmons DR, Van Wyk JJ (1984) Factors controlling blood concentrations of somatomedin-C. Clin Endocrinol Metab 13:113–143

Clemmons DR, Van Wyk JJ, Ridgway EC, Kliman B, Kjellberg RN, Underwood LE (1979) Evaluation of acromegaly by radioimmunoassay of somatomedin-C. N Engl J Med 301:1138–1142

Clemmons DR, Underwood LE, Ridgway EC, Kliman B, Kjellberg RN, Van Wyk JJ (1980) Estradiol treatment of acromegaly: reduction of immunoreactive somatomedin-C and improvement in metabolic status. Am J Med 69:571–575

Clemmons DR, Klibanski A, Underwood LE, McArthur JW, Ridgway EC, Beitins IZ, Van Wyk JJ (1981) Reduction of plasma immunoreactive somatomedin-C during fasting in humans. J Clin Endocrinol Metab 53:1247–1250

Clemmons DR, Underwood LE, Chatelain PG, Van Wyk JJ (1983) Liberation of immunoreactive somatomedin-C from its binding proteins by proteolytic enzymes and heparin. J Clin Endocrinol Metab 56:384–389

Clemmons DR, Seek MM, Underwood LE (1985 a) Supplementary essential amino acids augment the somatomedin-C/insulin-like growth factor I response to refeeding after fasting. Metabolism 34:391–395

Clemmons DR, Underwood LE, Dickerson RN, Brown RO, Hak LJ, MacPhee RD, Heizer WD (1985b) Use of plasma somatomedin-C/insulin-like growth factor I measurements to monitor the response to nutritional repletion in malnourished patients. Am J Clin Nutr 41:191–198

Copeland KC, Underwood LE, Van Wyk JJ (1980) Induction of immunoreactive somatomedin-C in human serum by growth hormone: dose response relationships and effect on chromatographic profiles. J Clin Endocrinol Metab 50:690–697

Copeland KC, Johnson DM, Underwood LE, Van Wyk JJ (1983) Radioimmunoassays of somatomedin-C in the baboon (*Papio cynocephalus*): a comparison of multiple techniques of measurement. Am J Primatol 51:161–169

Daughaday WH, Mariz IK, Blethen SL (1980) Inhibition of access of bound somatomedin to membrane receptors and immunobinding sites: a comparison of radioreceptor and radioimmunoassay of somatomedin in native and acid-ethanol extracted serum. J Clin Endocrinol Metab 51:781–788

D'Ercole AJ, Underwood LE (1980) Ontogeny of somatomedin during development in the mouse: serum concentrations, molecular forms, binding proteins, and tissue receptors. Dev Biol 79:33–45

D'Ercole AJ, Underwood LE, Van Wyk JJ (1977) Serum somatomedin-C in hypopituitarism and in other disorders of growth. J Pediatr 90:375–381

D'Ercole AJ, Bose CL, Underwood LE, Lawson EE (1984a) Serum somatomedin-C concentrations in a rabbit model of diabetic pregnancy. Diabetes 33:590–595

D'Ercole AJ, Stiles AD, Underwood LE (1984b) Tissue concentrations of somatomedin-C: further evidence for multiple sites of synthesis and paracrine/autocrine mechanisms of action. Proc Natl Acad Sci USA 81:935–939

D'Ercole AJ, Hill DJ, Strain AJ, Underwood LE (1986) Tissue and plasma somatomedin-C/insulin-like growth factor I (Sm-C/IGF-I) concentrations in the human fetus during the first half of gestation. Pediatr Res 20:253–255

Drop SLS, Kortleve DJ, Guyda HJ (1984) Isolation of a somatomedin-binding protein from preterm amniotic fluid: development of a radioimmunoassay. J Clin Endocrinol Metab 59:899–907

Furlanetto RW, Underwood LE, Van Wyk JJ, D'Ercole AJ (1977) Estimation of somatomedin-C levels in normals and patients with pituitary disease by radioimmunoassay. J Clin Invest 60:648–657

Heldin C-H, Wasteson A, Fryklund L, Westermark B (1981) Somatomedin B: mitogenic activity derived from contaminant epidermal growth factor. Science 213:1122–1123

Hintz RL, Clemmons DR, Underwood LE, Van Wyk JJ (1972) Competitive binding of somatomedin to the insulin receptors of adipocytes, chondrocytes and liver membranes. Proc Natl Acad Sci USA 69:2351–2353

Hintz RL, Liu F, Marshall LB, Chang D (1980a) Interaction of somatomedin-C with an antibody directed against the synthetic C-peptide region of insulin-like growth factor I. J Clin Endocrinol Metab 50:405–407

Hintz RL, Liu F, Rinderknecht E (1980b) Somatomedin-C shares the carboxyterminal antigenic determinants with insulin-like growth factor I. J Clin Endocrinol Metab 51:672–673

Huybrechts LM, King DB, Lauterio TJ, Marsh J, Scanes CG (1985) Plasma concentrations of somatomedin-C in hypophysectomized, dwarf and intact growing domestic fowl as determined by heterologous radioimmunoassay. J Endocrinol 104:233–239

Isley WL, Underwood LE, Clemmons DR (1983) Dietary components that regulate serum somatomedin-C concentrations in humans. J Clin Invest 71:175–182

Isley WL, Underwood LE, Clemmons DR (1984) Changes in plasma somatomedin-C in response to ingestion of diets with variable protein and energy content. JPEN 8:407–411

Marquardt H, Todaro GJ, Henderson LE, Oroszlan S (1981) Purification and primary structure of a polypeptide with multiplication-stimulating activity from rat liver cell cultures: homology with human insulin-like growth factor II. J Biol Chem 256:6859–6865

Marshall RN, Underwood LE, Voina SJ, Foushee DB, Van Wyk JJ (1974) Characterization of the insulin and somatomedin-C receptors in human placental cell membranes. J Clin Endocrinol Metab 39:283–292

Merimee TJ, Zapf J, Froesch ER (1982) Insulin-like growth factors in the fed and fasted states. J Clin Endocrinol Metab 55:999–1002

Moses AC, Nissley SP, Short PA, Rechler MM, White RM, Knight AB, Higa OZ (1980) Increased levels of multiplication-stimulating activity, an insulin-like growth factor, in fetal rat serum. Proc Natl Acad Sci USA 77:3649–3653

Phillips LS, Vassilopoulou-Sellin R (1979) Nutritional regulation of somatomedin. Am J Clin Nutr 32:1082–1096

Pierson RW Jr, Temin HM (1972) The partial purification from calf serum of a fraction with multiplication-stimulating activity for chicken fibroblasts in cell culture and with non-suppressible insulin-like activity. J Cell Physiol 79:319–329

Rechler MM, Fryklund L, Nissley SP, Hall K, Podskalny JM, Skottner A, Moses AC (1978) Purified human somatomedin A and rat multiplication stimulating activity: mitogens for cultured fibroblasts that crossreact with the same growth peptide receptors. Eur J Biochem 82:5–12

Schalch DS, Heinrich UE, Koch JG, Johnson CJ, Schlueter RJ (1978) Nonsuppressible insulin-like activity (NSILA) I. Development of a new sensitive competitive protein-binding assay for determination of serum levels. J Clin Endocrinol Metab 46:664–671

Smith GL (1984) Somatomedin carrier proteins. Mol Cell Endocrinol 34:83–89

Takano K, Hall K, Ritzen M, Iselius L, Sievertsson H (1976) Somatomedin-A in human serum, determined by radioreceptor assay. Acta Endocrinol (Copenh) 82:449–459

Underwood LE, Van Wyk JJ (1985) Normal and aberrant growth. In: Wilson J, Foster DW (eds) Williams textbook of endocrinology. Saunders, Philadelphia, pp 155–205

Underwood LE, D'Ercole AJ, Copeland KC, Van Wyk JJ, Hurley T, Handwerger S (1982a) Development of a heterologous radioimmunoassay for somatomedin-C in sheep blood. J Endocrinol 93:31–39

Underwood LE, Clemmons DR, D'Ercole AJ, Minuto F, Chatelain P, Copeland KC, Van Wyk JJ (1982b) Bulletin of the Swiss Academy of Medicine. Proceedings of the first combined meeting of the European Society for Pediatric Endocrinology and the Lawson Wilkins Pediatric Endocrine Society, pp 183–198

Underwood LE, D'Ercole AJ, Clemmons DR, Van Wyk JJ (1986) Paracrine functions of somatomedins. Clin Endocrinol Metab 15:59–77

Van Wyk JJ (1984) The somatomedins: biological actions and physiologic control mechanisms. In: Li CH (ed) Hormonal proteins and peptides vol XII. Academic, New York, pp 82–125

Van Wyk JJ, Svoboda ME, Underwood LE (1980) Evidence from radioligand assays that somatomedin-C and insulin-like growth factor I are similar to each other and different from other somatomedins. J Clin Endocrinol Metab 50:206–208

Wass JAH, Clemmons DR, Underwood LE, Barrow I, Besser GM, Van Wyk JJ (1982) Changes in circulating somatomedin-C levels in bromocriptine-treated acromegaly. Clin Endocrinol 17:369–377

Zapf J, Kaufmann U, Eigenmann EJ, Froesch ER (1977) Determination of nonsuppressible insulin-like activity in human serum by a sensitive protein-binding assay. Clin Chem 23:677–682

Zapf J, Schoenle E, Froesch ER (1978) Insulin-like growth factors I and II: some biological actions and receptor binding characteristics of two purified constituents of nonsuppressible insulin-like activity of human serum. Eur J Biochem 87:285–296

Zapf J, Walter H, Froesch ER (1981) Radioimmunological determination of insulin like growth factors I and II in normal subjects and in patients with growth disorders and extrapancreatic tumor hypoglycemia. J Clin Invest 68:1321–1330

Zapf J, Schmid CH, Froesch ER (1984) Biological and immunological properties of insulin-like growth factors (IGF) I and II. Clin Endocrinol Metab 13:3–30

CHAPTER 24

Radioimmunoassay of Drugs and Neurotransmitters

S. SPECTOR

A. Introduction

Pharmacologists pose a number of questions for themselves regarding drugs. They are interested in the role of absorption, distribution, metabolism, excretion, and localization of the compound. In order to answer these questions, it is necessary that methods be available, but these methods must also have the attributes of high sensitivity, specificity, and reproducibility. An added attribute of an assay would be if the procedure had a simplicity that neither required extraction from biologic fluids nor column chromatography nor high pressure liquid chromatography instrumentation currently available. The radioimmunoassay techniques as developed by BERSON and YALOW (1959) possess most of the attributes enumerated above. However, it needs to be emphasized that the specificity of the assay is not absolute.

The principles of radioimmunoassay are predicated on the development of specific antibodies and the competition of labeled and unlabeled antigens for the specific antibodies. Drugs are generally low molecular weight molecules which are incapable of stimulating the generation of antibodies. However, the work of LANDSTEINER (1945) showed that low molecular weight compounds could be made antigenic by coupling them to a protein by covalent bonds; the antibodies formed by such a conjugated protein can be directed against the low molecular weight substance. There are various functional groups on the carrier, i.e., protein or polyamino acid, which lend themselves to the formation of covalent linkages. The most readily available groups are the ε-amino group of lysine, the α-amino groups, the phenolic portion of tyrosine, the carboxyl groups of dicarboxylic amino acids, the hydroxyl group of serine, sulfhydryl groups of cysteine, and the imidazole ring of histidine. The ring structures of tyrosine, tryptophan, and histidine lend themselves to substitutions by diazotization. The development of extremely sensitive, specific assays for neurotransmitters have stimulated investigations on the role of the neurotransmitter in both the peripheral and central nervous systems in various pathophysiologic conditions.

Radioimmunoassay offers advantages of sensitivity, specificity, and ease of operation in the determination of blood and tissue levels of endogenous levels of the neurotransmitter. The sensitivity of the radioimmunoassay is a function of the avidity of the antiserum used and the specific activity of the labeled drug. Since the assay depends on a competition between labeled and unlabeled drug for the limited available binding sites on the antibody, the detection of minute amounts of the unlabeled drug is markedly enhanced when the specific activity of the labeled material is high. The sensitivity of the assay can also be enhanced by initially incubating the unlabeled drug with the antibody, and then adding labeled drug.

Radioimmunoassay measures the inhibition of binding of labeled ligand by the antibody. Anything that competes for the available sites on the antibody or affects the binding of the ligand will influence the results. Thus, one has to be especially cognizant of possible nonspecific inhibition that could give spurious results. Antibody–ligand interactions are usually performed at 4°–40 °C and pH 7–8. The assay can be affected by factors that can influence the binding of the ligand to the antibody. Among the nonimmunologic factors that can influence the antibody–ligand interaction are acid or alkaline pH, high osmolality, and high concentrations of non-antibody protein. All these factors can lead to discrepancies in the radioimmunoassay and must be avoided.

B. Methods for Coupling Neurotransmitters to Carrier Proteins

I. Acetylcholine

The neurotransmitter of the cholinergic nervous system has been measured by a number of analytic methods and each has its advantages. There has always been an interest in the cholinergic nervous system in pharmacology and neurochemistry. However, there is increasing interest in acetylcholine today as there is evidence that, in patients with senile dementia of the Alzheimer's type, cortical cholinergic degeneration of the nucleus basalis of Meynert occurs and that there may be a cholinergic basis for the disease. Thus, it would be most advantageous to have available a rapid, sensitive, and specific assay for the determination of acetylcholine (SPECTOR et al. 1978).

The acetylcholine molecule lacks any reactive group that would lend itself to conjugation to a protein carrier. Therefore, an immunogenic compound was synthesized according to the procedure described in Fig. 1. As each new compound (A, B, C, and D) was synthesized, it was characterized by infrared and nuclear magnetic resonance spectroscopy as well as by elemental analysis.

The *N*-benzyloxycarbonyl-protected derivative of 6-aminocaproic acid (A) was prepared. A solution of 6-aminohexanoic acid 0.1 mol in 40 ml 2.5 *M* NaOH was cooled to 0 °C and 0.11 mol benzyloxycarbonylchloride in 25 ml ether together with 50 ml 2.5 *M* NaOH was added in several portions with vigorous stirring over a 1-h period. The reaction mixture was extracted with ether, cooled, and acidified with 6 *M* HCl. Recrystallization from light ether–petroleum yielded 19.4 g product (melting point 54°–55 °C).

The 2-trimethylaminoethyl ester (C) was prepared in two stages. A solution of 0.01 mol A in 60 ml ethyl acetate was treated with 0.02 mol triethylamine and 0.025 mol β-dimethylaminoethylchloride, refluxed for 16 h, filtered, and the solvent evaporated in vacuo, leaving a viscous oil (B). A solution of 4.46 mol B in 10 ml ethyl acetate was treated with 4.46 mmol methyl-*p*-toluenesulfonate. After standing at 25 °C for 3 h, the product precipitated and was recrystallized from ethanol–ether to give a white crystalline product, C (melting point 105°–106.5 °C).

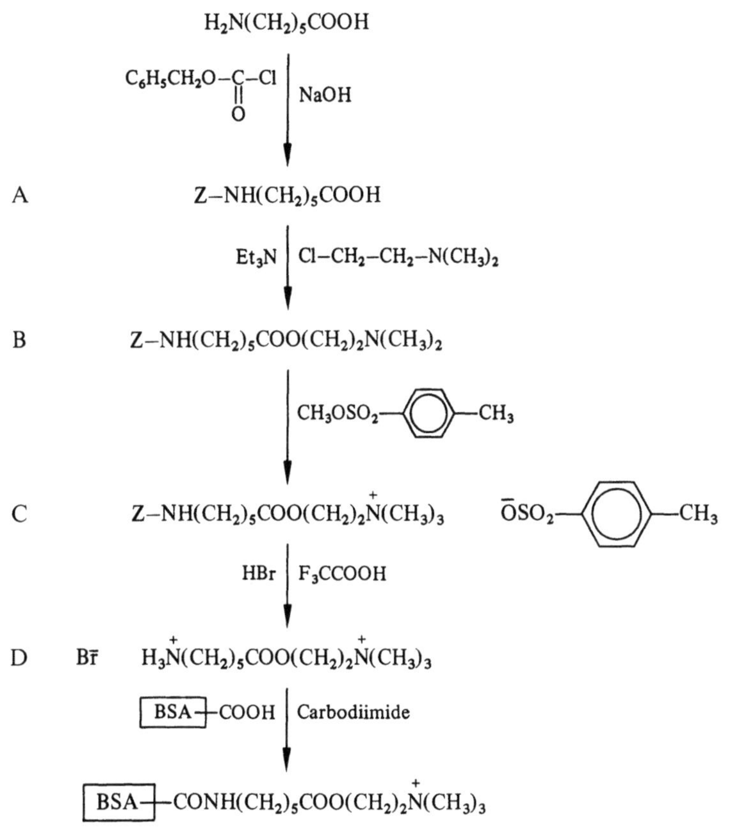

Fig. 1. Synthesis of an acetylcholine immunogen

The benzyloxycarbonyl group was removed from compound C by dissolving it in 25 ml trifluoroacetic acid and passing hydrogen bromide gas through the solution, which was then evaporated to dryness. The residue was crystallized from ethanol–ether and recrystallized from dimethylformamide–ether to give a white, crystalline product, D (melting point 139.5°–143 °C). Compound D was coupled to bovine serum albumin (BSA) by dissolving 0.2 mmol D and 25 mg BSA in 3 ml H_2O, adjusting to pH 5.5, and adding 0.2 mmol 1-ethyl-3-(3-dimethylaminopropyl)carbodiimide. The reaction proceeded at 25 °C for 2 days, was then dialyzed against water for 3.5 days, and freeze-dried to afford 20.7 mg product.

Quantitative determination of the extent of coupling of the hapten to BSA was achieved by acid hydrolysis (6 *M* HCl at 110 °C for 24 h) of an aliquot of the hapten–albumin complex. The 6-aminohexanoic acid was resolved from the constituent amino acids derived from BSA by means of a Beckman Model 121C amino acid analyzer. Quantitation of 6-aminohexanoic acid relative to the amino acids present in BSA revealed that 9.9 residues of 6-aminohexanoic acid (and therefore of hapten) were associated with each BSA equivalent.

The BSA–hapten complex was dissolved in phosphate-buffered saline (PBS), pH 7.4, and emulsified with an equal volume of complete Freund's adjuvant. New Zealand white rabbits were immunized with 1 mg of the conjugated protein injected into the footpad followed by 1 mg intramuscularly every 2 weeks for 1 month and then once a month. An antibody dilution curve was prepared in 5 °C normal rabbit serum in the presence of 2×10^{-4} *M* eserine. At a dilution of 1:1000, the antibody bound 40% of a 10 pmol tracer of acetylcholine ^{3}H (250 mCi/mmol, Amersham-Searle).

The procedure for the synthesis of immunogen is somewhat difficult. KAWASHIMA et al. (1980) have utilized a simpler procedure to prepare their immunogen. Glutaric anhydride 4 g was added to a mixture of 1 g choline chloride and 10 ml dry pyridine. This insoluble mixture was solubilized with the addition of 10 ml *N,N*-dimethylformamide. On stirring the solution, a white precipitate formed. The mixture was stirred overnight at room temperature. The precipitate was collected by filtration and washed with ethyl acetate. The precipitate was crystallized from acetone–methanol.

The choline hemiglutarate was conjugated to BSA by adding 20 mg choline hemiglutarate and 100 mg 1-ethyl-3-(3-dimethylaminopropyl)carbodiimide to a solution of 100 mg BSA dissolved in 10 ml 0.2 *M* phosphate buffer, pH 6.0. The reaction was allowed to proceed for 2 h at room temperature with constant stirring. It was then kept overnight at 4 °C with constant stirring. The reaction mixture was dialyzed against water at 4 °C for 2 days, changing the water three times a day. The immunogen was lyophilized and stored at −20 °C.

II. Serotonin

The biogenic amine serotonin has been implicated as having a physiologic role and has been suggested as being involved in many human diseases (COPPEN 1973). Serotonin was conjugated to a carrier protein by the following procedure: to 50 mg DL-*p*-aminophenylalanine, dissolved in 5 ml distilled water, 50 mg BSA and 50 mg 1-ethyl-3-(3-dimethylaminopropyl)carbodiimide were added (PESKAR and SPECTOR 1973a). The mixture was incubated at room temperature overnight and then dialyzed for 5 days against distilled water with three changes (2 l each) per day. Then the pH was adjusted to 1.5 with 1 *M* HCl, and the following procedure was performed at 0°–4 °C. $NaNO_2$ 100 mg, dissolved in 1 ml distilled water, was added slowly dropwise, followed by 50 mg ammonium sulfamate in 1 ml distilled water. The diazotized protein solution was then added dropwise to 100 mg serotonin creatinine sulfate, dissolved in 10 ml 0.1 *M* borate buffer, pH 9.0, with constant stirring. The pH was maintained above 8.0 by the addition of borate buffer. A dark red color developed almost immediately. The preparation was stirred overnight in the dark at 4 °C and then dialyzed exhaustively against distilled water. For determination of the amount of serotonin coupled to the protein carrier, 1 µCi serotonin ^{3}H was added together with unlabeled serotonin, and the above procedure was followed. By measuring the radioactivity of the dialyzates and the antigen solution, it was calculated that 5.7% of the serotonin had been coupled, corresponding to about 20 mol serotonin to 1 mol BSA (molecular weight 70000).

GROTA and BROWN (1974) used to formaldehyde condensation reaction suggested by RANADIVE and SEHEN (1967) to couple serotonin to BSA. In this procedure the indolealkylamine is dissolved in water and added to a solution containing BSA. To this is added 3 *M* sodium acetate and 7.5% formaldehyde. The reaction is run at room temperature. It is then dialyzed against water for 3 days with continuous changes of water. The extent of hapten bound to protein is determined by spectrophotometric analysis at 280 and 300 nm.

III. Melatonin

Interest has been shown in the pineal gland because of its influence on the endocrine system. The hormone of the pineal which is considered most likely to be responsible for these physiologic effects is the indolealkylamine melatonin (5-methoxy-*N*-acetyltryptamine). The melatonin immunogen was prepared by dissolving 1 mmol (137 mg) *p*-aminobenzoic acid (PABA) in 5 ml 1.0 *M* HCl at room temperature and cooling the solution to 0°–4 °C in an ice bath. Then, 69 mg (1 mmol) sodium nitrite was dissolved in 1 ml water and this was added dropwise to the PABA. The reaction was allowed to proceed for 20 min in the cold at which time a 0.5 *M* solution of ammonium sulfamate was added dropwise to stop the reaction. The reaction was considered to have reached completion when there were no more bubbles being evolved and when the test with potassium iodide–starch paper was negative. At this point, 232 mg (1 mmol) melatonin was dissolved in 1 ml 0.1 *M* HCl and this solution was cooled. The diazotized PABA was added dropwise to the solution of melatonin and stirred in an ice bath in the cold room for 2 h. A precipitate formed and the material was then transferred to a centrifuge tube. The reaction beaker was washed once with 10 ml 0.1 *M* HCl and the washing added to the reaction mixture which was centrifuged for 10 min at 12000 *g*. The precipitate was washed another 5–6 times using 10 ml water for each wash, and the centrifugation was repeated after each wash. The washed precipitate (hapten) was then dried under vacuum. The hapten was chromatographed by thin layer chromatography using a Redi-Plate, silica gel GF, 250 μm, in a solvent system of methanol–chloroform–ammonium hydroxide (20:80:5). Chromatography gave one spot at the origin and showed that no free melatonin ($R_f = 0.8$) was present.

The diazotized melatonin hapten was conjugated to BSA by the mixed anhydride procedure (Erlanger et al. 1957, 1959). First, 100 mg (0.26 mmol) hapten was dissolved in 2 ml dry dioxane. The dioxane had been passed over basic alumina in order to eliminate dioxane peroxides. Then, 0.36 ml (0.26 mmol) triethylamine was added to the hapten solution which was then cooled to 8°–10 °C and 8 ml dry dioxane was slowly added to the BSA solution, maintaining the pH at about 9.0 with 0.1 *M* NaOH. The mixed anhydride was then added dropwise to the BSA solution, maintaining the pH at 9.0 and the temperature at 8°–10 °C; the reaction mixture was kept under constant stirring for 30 min and then stored overnight at 4 °C. The solution was dialyzed for 2–3 days against 80–100 volumes dioxane–water (1:1), changing the solution twice a day, and then against water for 1 day, changing the solution twice. The immunogen was then lyophilized and stored desiccated at 4 °C.

C. Methods for Coupling Drugs to Carrier Proteins

The term drug has been extended in its definition so that it is not limited to those chemicals which are used for medical purposes. However, irrespective of the use for which the chemical is taken, one is interested in the rate of absorption, distribution, metabolism, excretion, and localization since many of the drugs of therapeutic interest are low molecular weight substances. Generally, molecules with a molecular weight below 1000 are nonimmunogenic, unless they are covalently linked to a higher molecular weight carrier substance such as a protein, polypeptide, or polysaccharide, to cite three illustrations. There exist a number of procedures to prepare drug–carrier conjugates which can then be used to produce antibodies. One can use the diazo reaction to substitute the amino acids tyrosine, tryptophan, lysine, and histidine residues of the protein. ERLANGER (1973) points out that there are many ε-amino groups of lysine in BSA which are available for amide linkages and uses them as a site to conjugate haptenic groups. The review by BUTLER (1975) offers general principles regarding the development of antibodies to small molecules, immunoassay methods, and specific applications of the assay.

I. Barbiturates

Although barbiturates are being replaced by safer sedative hypnotic drugs, they are still widely used. The barbiturate–protein conjugate was formed by converting the barbiturate, 5-allyl-5-(1-carboxyisopropyl)barbituric acid to 5-allyl-5-(1-*p*-nitrophenyloxycarbonylisopropyl)barbituric acid by reacting the free base with *p*-nitrophenol in *N,N*-dimethylformamide at 4 °C. This forms an active ester of barbituric acid. The active ester of barbituric acid was then coupled to bovine gamma-globulin in a glycerine–water solution (1 : 1 by volume) in the presence of dicyclohexylcarbodiimide. Thus, there is a nucleophilic attack on the ester by a pair of electrons of the NH_2 of the protein to displace *p*-nitrophenol and replace it with an amide linkage (Fig. 2).

II. Curare

The neuromuscular blocking agent D-tubocurarine (TC), which is used extensively as an adjunct in the practice of clinical anesthesia, was coupled to a protein (BSA) to generate an immunogen by the following procedure. PABA (137 mg) was dissolved in 10 ml 1 *M* HCl at 4 °C. Sodium nitrite (100 mg in 0.5 ml distilled water) was added dropwise over 10 min and allowed to stand for 30 min. In order to remove the nitrous acid, ammonium sulfamate (600 mg/ml) was added dropwise until no more nitrogen bubbles were given off. TC 50 mg was dissolved in 5 ml distilled water and the diazotized PABA was added dropwise over a 2-min period. The pH was then adjusted to 7.0 with 5 *M* NaOH and the mixture was allowed to stand at 0 °C for 2 h. The diazotized TC (0.5 ml) was coupled to 10 mg BSA in 10 ml PBS, in the presence of 10 mg 1-ethyl-3-(3-dimethylaminopropyl)-carbodiimide, stirred overnight at 4 °C in the dark, and dialyzed for 3 days against distilled water with three or four changes per day.

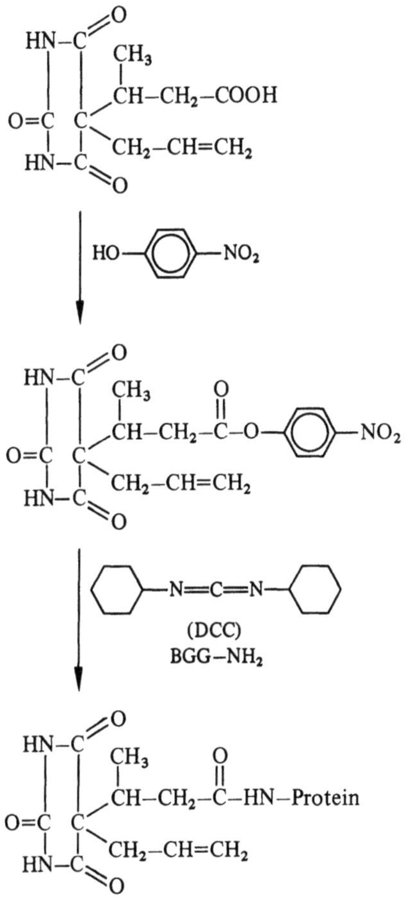

Fig. 2. Formation of a barbiturate–protein conjugate

The question whether quarternary compounds traverse the blood–brain barrier was examined by the radioimmunoassay for TC, which can measure as little as 5 ng/ml in serum and 1 ng/ml in CSF (HOROWITZ and SPECTOR 1973). It can be shown that the intravenous administration of 0.3 mg/kg does pass into the CSF of humans (MATTEO et al. 1977). Since the concentrations are in the nanogram range, it requires an extremely sensitive analytic method (Fig. 3).

III. Clonidine

Clonidine is a clinically important antihypertensive drug which is believed to lower mean blood pressure by an action on the central nervous system (KOBINGER and WALLAND 1967; STRUYKER BOUDIER et al. 1975; STRUYKER BOUDIER and VAN ROSSUM 1972; REID et al. 1977). It is a highly potent drug, being effective in microgram doses and has other central actions such as sedation (LAVERTY and TAYLOR 1969; DELBARRE and SCHMITT 1971; STROMBOM 1975; DOLLERY et al. 1976), hypothermia (LAVERTY and TAYLOR 1969; TSOUCARIS-KUPFER and SCHMITT 1972), inhibition of conditioned avoidance response (LAVERTY and TAYLOR 1969), and analgesia (PAALZOW 1974; SCHMITT et al. 1974). Clonidine was coupled to BSA

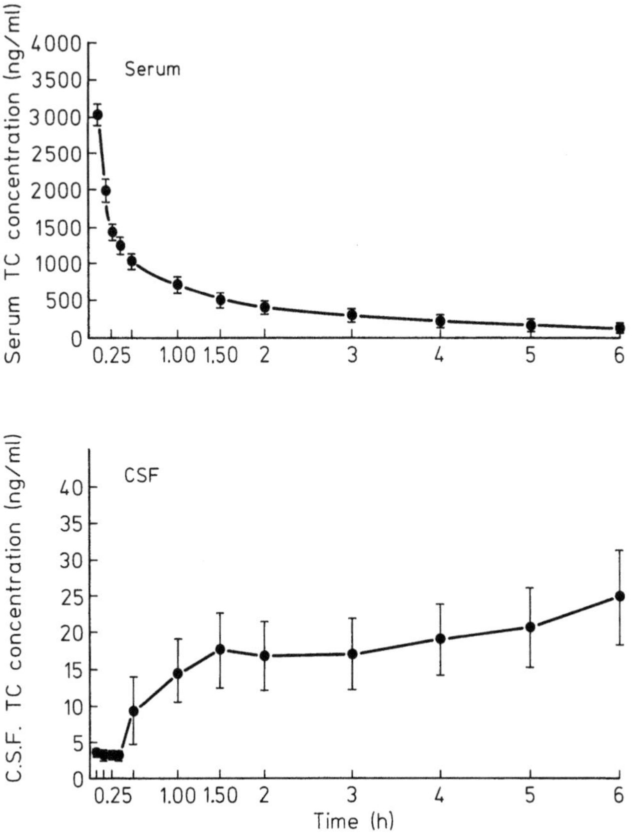

Fig. 3. Passage of D-tubocurarine into the CSF after a single, intravenous injection. Patients (n = 9) received a single injection of dTc, 0.3 mg/kg, i.v. Upper graph represents concentration of dTc in the serum (mean ± SE). Lower graph shows the concentration of dTc in the CSF (mean ± SE). (MATTEO et al. 1977)

to prepare an immunogen by reacting 4-carboxybenzenediazonium cgloride with 4-hydroxyclonidine to form 4-[6-[2,4-dichloro-3-(4,5-dihydro-1-*H*-imidazol-2-yl)-amino]hydroxyphenylazol]benzoic acid and then an amide bond between the carboxyl group of the benzoic acid derivative and the amino group of the protein (JARROT and SPECTOR 1978).

PABA 0.2 mmol was dissolved in 3 ml 1 *M* HCl and the solution cooled to 2°–4 °C. An equal volume of ice-cold 0.75 *M* sodium nitrite solution was added, with constant stirring. After 1 min, 3 ml 0.5 mmol ammonium sulfamate was added. The pH was adjusted to 10 with 5 *M* sodium hydroxide. The resultant solution was stirred in an ice bath for 1 h and then added dropwise to a solution of 4-hydroxyclonidine (0.2 mmol in 5 ml water). After stirring for 2 h at room temperature, the solution was added dropwise to a solution of BSA (10 mg in 5 ml distilled water) and then 10 ng 1-ethyl-3-(3-dimethylaminopropyl)carbodiimide was added. This was stirred for 18 h at room temperature and then dialyzed against 2 l PBS, pH 7.4 at 4 °C with four changes over a 48-h period. It was then dialyzed against distilled water and then lyophilized.

IV. Phenothiazines

When dealing with drugs that influence the central nervous system, one would like to be able to correlate behavioral effects and plasma concentrations and to try to ascertain the amount of drug responsible for the pharmacologic effects in the brain. With the phenothiazines one also has to consider the many metabolites in relation to their clinical effectiveness.

KAWASHIMA et al. (1975) coupled chlorpromazine to BSA by two methods. The first involved initially conjugating the diazotized PABA to chlorpromazine and forming an amide linkage between the amine groups of the protein and the carboxyl groups of benzoic acid. The procedure for diazocoupling was as follows: 2.4 mmol PABA was dissolved in 3 ml water and 0.7 ml concentrated hydrochloric acid. Then, 1 ml water containing 2.5 mmol sodium nitrite was added dropwise and stirred for 30 min. To remove the excess nitrous acid, 0.15 ml 1 *M* ammonium sulfamate was added slowly with constant stirring for 10 min, 2.5 mmol chlorpromazine hydrochloride dissolved in water was added dropwise to the diazonium salt of PABA with continuous stirring, and 1 ml water containing 2 mmol sodium acetate was then added. The pH of the solution was 1.5. The reaction was allowed to continue overnight at 4 °C with constant stirring. The pH was then adjusted to 5.6 with 0.5 *M* NaOH and extracted with 30 ml methylene chloride. The organic layer was washed with water and evaporated to dryness under vacuum. The diazo coupled chlorpromazine was coupled to BSA by the mixed anhydride technique. Triethylamine 0.05 mmol and isobutylchloroformate 0.05 mmol were added to 0.05 mmol diazo coupled chlorpromazine which was dissolved in dioxane at 4 °C. The reaction was allowed to continue for 20 min. With occasional stirring, 8 ml dioxane was added to 50 mg BSA in 10 ml water, maintaining the pH around 9 with 0.1 *M* NaOH. The mixed anhydride solution was then added dropwise to the BSA solution with constant stirring for 30 min at 4 °C, maintaining the pH at 9. The solution was stirred at 4 °C overnight, then dialyzed for 3 days against dioxane–water (1:1) and for 1 day against water.

HUBBARD et al. (1978) published another procedure for conjugating chlorpromazine to a protein. They initially prepared 7-(3-methoxycarbonylpropionyl) chlorpromazine by preparing a solution of 13.5 mmol 3-methoxycarbonylpropionyl chloride in methylene dichloride and adding it to a solution of 27 mmol aluminum chloride dissolved in methylene dichloride. The mixture was stirred at room temperature for 30 min. Chlorpromazine 10 mmol dissolved in methylene dichloride was added and the mixture stirred in the dark for 72 h. The solvent was then decanted and discarded. To the solid residue, ice water was added, followed by 100 ml 1 *M* NaOH The aqueous suspension was extracted with ether. The ether extracts were washed with water and then dried over anhydrous magnesium sulfate. Then, 5 ml 1 *M* NaOH was added to 100 ml water containing 2.3 ml 1 *M* HCl and the 7-(3-methoxycarbonylpropionyl)chlorpromazine. A fine oily precipitate was kept in suspension by rapid stirring and heated to a gentle reflux. The solution was then allowed to cool to room temperature. It was washed with 5×100 ml ether at pH 11.5. The pH was then adjusted to 6.0. The compound was coupled to BSA by carbodiimide.

The phenothiazine compounds have many metabolites, including aromatic hydroxylation at position 7 of the phenothiazine ring, *S*-oxidation, *N*-dealkylation, and *N*-oxidation. KLEINMAN et al. (1980) have reported that the 7-hydroxy metabolite is as effective as chlorpromazine. An RIA for 7-hydroxytrifluoroperazine was developed by ARAVAGIRI et al. (1985). They prepared 7-hydroxy-10-

([3-(4-(2-carboxyethyl)-1-piperazinyl)]-2-trifluoromethyl)-1-*OH*-phenothiazine by adding 1.2 mmol 1-(2-methoxycarbonylethyl)piperazine and 50 mg NaCl to a solution of 1.1 mmol 10-(3-chloropropyl)-7-tetrahydropyranyloxy-2-trifluoromethyl-1-*OH*- phenothiazine in 20 ml methylethylketone. The mixture was heated under reflux overnight. The organic solvent was evaporated and the residue taken up in dichloromethane. They then dried the organic phase over $MgSO_4$ and evaporated the solvent. It was then dissolved in 2 *M* HCl and the pH was adjusted to 7.5–8.0 by addition of saturated NaH CO_3 and then extracted with dichloromethane. The extract was dried. The hapten was then coupled to BSA by the previously described mixed anhydride method.

V. Benzodiazepines

The antianxiety drugs are some of the most frequently prescribed. There are many compounds in this broad category, but the most widely used drugs in this class are the benzodiazepines (diazepam and congeners). The first RIA to measure diazepam in plasma was developed by PESKAR and SPECTOR (1973b). DIXON and CREWS (1978) modified the procedure for the synthesis of the hapten diazepam. To a mixture of 15 mmol concentrated HCl 6 mmol *p*-aminoacetanilide was added and this was followed by the dropwise addition of 2 mmol sodium nitrite solution. This is done in an ice bath. The diazonium salt solution is then added dropwise, again at 4 °C, to a solution of 1.67 mmol 7-chloro-1,3-dihydro-5(4-hydroxyphenyl)1-methyl-2-*H*-1,4-benzodiazepine-2-one in tetrahydrofuran, with constant stirring. The mixture is kept at 4 °C for 2–3 days. The precipitate is collected and recrystallized from dichloromethane–methanol. It is then dissolved in methanol–HCl (5:1) and heated for 10 min on a steam bath. Following which it remains at room temperature for 2 h and then the methanol is evaporated and the solution made basic with ammonium hydroxide and extracted with dichloromethane. The organic phase is dried and evaporated. The hapten is coupled to the protein by dissolving 0.05 mmol hapten in dimethylformamide and 0.5 ml water. To this is added 1 *M* HCl and 0.05 ml 1 *M* sodium nitrite solution to diazotize the *p*-amino group. This is done in an ice bath. The mixture is stirred for 10 min and 0.05 ml 1 *M* ammonium sulfamate solution is added to remove the excess nitrous acid. The diazonium salt of the hapten is then added dropwise to a solution of 200 mg BSA in 0.16 *M* borate buffer, pH 9. This is then stirred overnight at 4 °C, followed by sequential dialysis against 2 × 2 l 0.05 *M* sodium bicarbonate and 2 × 3 l water to remove any uncoupled hapten.

Another benzodiazepine, clonazepam, was coupled to BSA by DIXON and CREWS (1977) to develop a RIA. They suspended 6 g clonazepam in 400 ml dry tetrahydrofuran and added it to 1.2 g 50% suspension of sodium hydride 27 nmol in mineral oil. Succinic anhydride 33 nmol was added with constant stirring. After 1 h, the tetrahydrofuran was evaporated and water was added to the residue; it was then acidified with acetic acid and extracted with 300 ml dichloromethane. The acidification and extraction were done rapidly. The extract was dried over anhydrous sodium sulfate. Addition of hexane generates the desired hapten, 5-(2-chlorophenyl)1,3-dihydro-3-hemisuccinyloxy-7-nitro-2-*H*-1,4-benzodiazepine-2-one. Employing the mixed anhydride method of ERLANGER et al. (1959), the hapten is coupled to BSA. The hapten is dissolved in 2 ml dioxane, 0.16 ml trimethylamine–dioxane (1 : 10) is added, and the solution is then cooled to 4 °C in an ice bath. Then, 0.14 ml isobutylchloroformate–dioxane (1 : 10) is added and the mixture kept at 5°–10 °C for 20 min. The mixed anhydride is then added with rapid stirring to a solution of 70 mg BSA in 10 ml water and 8 ml dioxane at pH 8. During the coupling the pH must be maintained at 8–9 with addition of 0.1 *M* NaOH. The reaction mixture is stirred for 2 h in the cold and then dialyzed against 2 l 0.05 *M* borate buffer, pH 9, followed by dialysis against water (5 × 2 l).

VI. Butyrophenones

Haloperidol is a potent neuroleptic drug of the butyrophenone class. Its primary mode of action is by blocking the postsynaptic dopamine receptor. Since haloperidol is widely used for the treatment of psychiatric disorders it would be advantageous to have a relatively simple method for the measurement of blood concentrations. WURZBURGER et al. (1980) initially converted haloperidol to haloperidol hemisuccinate. Haloperidol 0.3 mmol was dissolved in 4 ml chloroform. The solution was stirred and 0.6 mmol succinic chloride was added. The reaction continued for 30 min and was stopped by the addition of 2.0 ml water. Then, 3.0 ml 0.1 *M* sodium phosphate buffer, pH 6.8, was added and the pH of the reaction mixture was adjusted to 6.8. The haloperidol hemisuccinate was extracted into 15.0 ml chloroform, leaving the unreacted reagents in the aqueous phase. The dried chloroform residue was dissolved in 4.0 ml dry dioxane to which was added 0.3 mmol triethylamine. The reaction mixture was then cooled to 8 °C and 0.3 mmol isobutylchloroformate was added. The reaction continued for 30 min and then the mixture was added dropwise to a solution of 3.0 ml 0.1 *M* sodium borate buffer and 2.0 ml dioxane, pH 9.0, containing 30 mg BSA. The pH was kept at 9.0 with 0.1 *M* NaOH. The reaction continued for 40 min at 8 °C and then overnight at 4 °C. The reaction mixture was dialyzed against 100 volumes 50% dioxane for 24 h. Dialysis was continued against 100 volumes water for an additional 36 h, changing the water every 12 h. Finally, the reaction mixture was dialyzed against PBS, pH 7.2, for 8 h.

The other widely used butyrophenone is pimozide. MICHIELS et al. (1975) have developed a RIA for this neuroleptic. They converted pimozide to the more reactive form of 3-[1-(4-bis-(4-fluorophenyl)butyl)-4-piperidinyl]-2,3-dihydro-2-oxo-1-*H*-benzimidazole-1-acetic acid, by preparing a mixture of equimolar amounts of chloroacetic acid and sodium amide (0.05 mmol), dissolved in toluene, and adding it slowly to pimozide. The mixture was allowed to react overnight under reflux. After cooling, the reaction mixture was acidified to pH 5 with dilute acetic acid and extracted with chloroform. The hapten was then coupled to BSA by first dissolving it in 5 ml boiling dioxane and adding it dropwise to a solution of 400 mg 1-ethyl-3-(3-dimethylaminopropyl)carbodiimide hydrochloride dissolved in 5 ml water. To this was added dropwise a solution of 200 mg BSA in 20 ml 0.05 *M* phosphate buffer, pH 6.0. The solution was then stirred overnight at 4 °C. The resultant mixture was dialyzed against 0.005 *M* phosphate buffer, pH 7.4, for 12 h at 4 °C.

VII. Atropine

Atropine is an ester of tropic acid and tropine and is a widely used drug. Atropine antagonizes the action of acetylcholine on muscarinic receptors of smooth and cardiac muscle and exocrine glands. This results in increased heart rate and depressed salivation. These properties have resulted in the frequent use of the drug as a preoperative medication. It also reduces gastric secretions and motility and, when applied to the eye, causes mydriasis and cycloplegia.

Since atropine is widely used clinically, a radioimmunoassay was developed to measure plasma levels associated with therapeutic dosages (WURZBURGER et al. 1977). The procedure involved the diazotization of PABA which is then coupled to atropine and the diazotized atropine is coupled to BSA with a water-soluble carbodiimide reagent. The procedure for the preparation of the immunogen is similar to that mentioned previously for clonidine and as a consequence will not be repeated.

VIII. Opiates

1. Fentanyl

The synthetic narcotic fentanyl is currently used widely as it is more potent than morphine. It acts at the same receptor as morphine. HENDERSON et al. (1975) have developed a radioimmunoassay to detect the small amounts in tissue and to do pharmacokinetic studies as it has a rapid onset and short duration of action. They initially prepared carboxyfentanyl, *N*-phenyl-*N*-4-(β-phenethyl)piperidine succinamic acid. First, 5 g 1-(β-phenethyl)-4-piperidone dissolved in 50 ml toluene was mixed with 3.1 ml freshly distilled aniline. This was refluxed under nitrogen and after 15 min a catalytic amount of *p*-toluenesulfonic acid was added. The mixture was refluxed for 20 h and then reduced in volume under a stream of nitrogen. The resulting imine, 1-(β-phenethyl)-4-piperidylidene aniline was dissolved in 50 ml isopropanol and refluxed with 4 equiv. sodium borohydride. After 10 h the mixture was cooled, the alcohol removed by rotary evaporation, and 30 ml 7% $NaHCO_3$ was added. The mixture was extracted with benzene and purified. Then, 1 g was dissolved in 50 ml benzene and reacted with 0.36 g succinic anhydride by refluxing for 40 h under an atmosphere of nitrogen. The product, carboxyfentanyl, was crystallized from a benzene–hexane (3:1) system. The carboxyfentanyl (50 mg) was coupled to 100 mg bovine gamma-globulin with 200 mg 1-ethyl-3-(3-dimethylaminopropyl)carbodiimide in 32 ml PBS (SPECTOR and PARKER 1970; SPECTOR 1971).

2. Morphine

The morphine molecule has a number of sites that are amenable to the generation of an immunogen. The phenolic OH at C-3 can be used, as can the alcoholic OH at C-6. One can also prepare an immunogen through the nitrogen atom. 3-*O*-Carboxymethylmorphine was formed by dissolving 0.02 mol morphine in 20 ml KOH. The solution was evaporated at 60° and 25 mm Hg, 100 ml benzene–azeotropically distilled water was added, and the mixture was distilled until the benzene was evaporated off. To the solids 120 ml absolute ethanol plus 0.025 *M* chloroacetic acid sodium salt was added and the mixture was refluxed for 3 h. It was then cooled to room temperature and the solids were filtered off. The solids were washed with 100 ml distilled water. The pH was adjusted to 7 with 1 *M* NaOH, the solution was evaporated to dryness and 100 ml absolute ethanol was added to the solids. After refluxing for 3 h the mixture was filtered while the solution was still hot.

The carboxymethylmorphine (8 mg) was dissolved in 2 ml distilled water containing 10 mg BSA. The pH was adjusted to 5.5 and 8 mg 1-ethyl-3-(3-dimethylaminopropyl)carbodiimide was added. The mixture was incubated at room temperature overnight and then dialyzed for 7 days against distilled water with 4–5 changes per day.

The *N*-carboxymethylmorphine immunogen was prepared by adding 0.03 mol normorphine to 2.5 g sodium bicarbonate and 60 ml dimethylformamide at room temperature, following which 0.03 mol ethylbromoacetate was added. The mixture was refluxed for 4 h. The residue was suspended in 60 ml water and then extracted with chloroform (1 : 1). The extract was dried and recrystallized from acetone–ether. The dried material was heated and refluxed for 2 h in 100 ml 2 *M* HCl. The acid was evaporated, the residue dissolved in 80 ml hot water, and some Norite A was added. The hot suspension was filtered and the filtrate was allowed to stand at room temperature for 72 h.

The *N*-carboxymethylmorphine (10 mg) was dissolved in 2 ml 0.01 *M* HCl which was then heated at 65 °C for 10 min. To the solution, 10 mg BSA was added, and the pH adjusted to 5.5, immediately followed by the addition of 10 mg 1-ethyl-3-(3-dimethylaminopropyl)carbodiimide. The solution was left for 24 h at room temperature and then dialyzed against distilled water for 2 days with 4–5 changes per day.

3. Naloxone

Naloxone is a potent and specific antagonist of narcotic analgesics. However, owing to its short duration of action, its utility as a narcotic antagonist is limited. As to the basis for this short duration of action, the explanation is lacking. A sensitive assay is required to analyze the very low tissue and fluid drug concentrations which occur in vivo.

A sensitive and specific radioimmunoassay for naloxone was developed by BERKOWITZ et al. (1975). The naloxone immunogen was linked to BSA with ethyl-*N*-carbamylcyanomethylacetimidate. Ethyl-*N*-carbamylcyanomethylacetimidate 30 mg was dissolved in 3 ml water. To this solution was added 30 mg BSA and 2 ml 0.5 *M* sodium bicarbonate. The solution was stirred for 48 h at 4 °C and dialyzed extensively against PBS, pH 7.4. The protein derivative solution was placed in an ice bath and diazotized by the addition of 3 ml 4% sodium nitrite and 3 ml 2 *M* HCl. After 30 s, 3 ml 2% ammonium sulfamate solution was added and the solution was mixed for another 30 s. Then, 20 mg naloxone hydrochloride was dissolved in 2 ml 0.5 *M* sodium acetate buffer, pH 4.8, and added to the mixture. The solution was stirred for 10 min, maintaining the pH at 4.8 with 1 *M*

NaOH. The pH was then adjusted to 7.4 with 0.5 *M* NaOH and kept at 4 °C for several hours. Insoluble residues were removed by centrifugation and it was then dialyzed against water for 48 h.

The antibodies generated by conjugating the hapten to a carrier can be and have been used for localization purposes by immunocytochemical and immunofluorescent techniques. They have also been utilized as surrogate receptors to look for endogenous substances. I would also like to point out their potential as specific antagonists of drugs, toxins, and hormones. Antibodies have been shown to be effective in blocking the pharmacologic and toxic effects of many drugs (BERKOWITZ et al. 1975; DECARLO and ADLER 1973; FLYNN et al. 1977; GUNNE et al. 1975; SMITH et al. 1976). BUTLER et al. (1977) demonstrated that anti-digoxin antibodies prevented death induced by a dose of digoxin in dogs. Also, that it restored sinus rhythm in most of the dogs 2–3 h after a toxic dose of digoxin, and by 6 h all dogs receiving the digoxin antibodies were clinically improved. Thus, we see that antibodies have proven to be powerful diagnostic agents with a great deal of specificity and sensitivity. They also have great potential as therapeutic agents.

References

Aravagiri N, Hawkes EM, Midha KK (1985) Radioimmunoassay for the 7-hydroxy metabolite of trifluoperazine and its application to a kinetic study in human volunteers. J Pharmaceut Sci 74:1196–1202

Berkowitz BA, Ngai SH, Hempstead J, Spector S (1975) Disposition of naloxone: use of a new radioimmunoassay. J Pharmacol Exp Ther 195:499–504

Berson SA, Yalow RS (1959) Quantitative aspects of the reaction between insulin and insulin binding antibody. J Clin Invest 38:1996

Butler V (1975) Drug immunoassays. J Immunol Methods 7:1–24

Butler VP Jr, Smith TW, Schmidt DH, Haber E (1977) Immunological reversal of the effects of digoxin. Fed Proc 36:2235–2241

Coppen A (1973) Role of serotonin in affective disorders. In: Barchas J, Usdin E (eds) Serotonin and behavior. Academic, New York, pp 523–527

DeCarlo I Jr, Adler FI (1973) Neutralization of morphine activity by antibody. Res Commun Chem Pathol Pharmacol 5:775–788

Delbarre B, Schmitt H (1971) Sedative effects of α-sympathomimetic drugs and their antagonism by adrenergic and cholinergic blocking drugs. Eur J Pharmacol 13:356–363

Dixon R, Crews T (1977) An ^{125}I radioimmunoassay for the determination of the anticonvulsant agent clonazepam directly in plasma. Res Commun Chem Pathol Pharmacol 18:477–486

Dixon R, Crews T (1978) Diazepam determination in micro samples of blood, plasma and saliva by radioimmunoassay. J Anal Toxicol 2:210–213

Dollery CT, Davies DS, Draffan GH, Dargie HJ, Dean CR, Reid JL, Clare RA, Murray S (1976) Clinical pharmacology and pharmacokinetics of clonidine. Clin Pharmacol Ther 19:11–17

Erlanger BF (1973) Principles and methods for the preparation of drug protein conjugates for immunological studies. Pharmacol Rev 25:271–280

Erlanger BF, Borek F, Beiser S, Lieberman S (1957) Steroid protein conjugates I. Preparation and characterization of bovine serum albumin with testosterone and cortisone. J Biol Chem 228:713

Erlanger BF, Borek F, Beiser S, Lieberman S (1959) Steroid protein conjugates II. Preparation and characterization of conjugates of bovine serum albumin with progesterone, deoxycorticosterone and estrone. J Biol Chem 234:1090

Flynn EJ, Cerreta KV, Spector S (1977) Pharmacologic response to pentobarbital in actively immunized mice. Eur J Pharmacol 42:21–29
Grota LJ, Brown GM (1974) Antibodies to indolealkylamines: serotonin and melatonin. Can J Biochem 52:196–202
Gunne IM, Jonsson J, Paalzow L, Anggard E (1975) Antibody bound nalorphine released on challenge with morphine. Nature 255:418–419
Henderson GL, Frincke J, Leung CY, Torten M, Benjamini E (1975) Antibodies to fentanyl. J Pharmacol Exp Ther 192:489–496
Horowitz P, Spector S (1973) Determination of serum d-tubocurarine concentration by radioimmunoassay. J Pharmacol Exp Ther 184:94–100
Hubbard JW, Midha KK, McGilvery IJ, Cooper GK (1978) Radioimmunoassay for psychotropic drugs. J Pharmaceut Sci 67:1563–1571
Jarrott B, Spector S (1978) Disposition of clonidine in rats as determined by radioimmunoassay. J Pharmacol Exp Ther 207:195–202
Kawashima K, Dixon R, Spector S (1975) Development of radioimmunoassay for chlorpromazine. Eur J Pharmacol 32:195–202
Kawashima K, Ishikawa H, Mochizuki M (1980) Radioimmunoassay for acetylcholine in the rat brain. J Pharmacol Methods 2:115–123
Kleinman JE, Biglow LB, Rogel A, Weinberger DR, Nasrallah HA, Wyatt RJ, Gillis CJ (1980) In: Usdin E, Eckert H, Forrest IS (eds) Phenothiazine and structurally related drugs: basic and clinical studies. Elsevier/North-Holland, New York, pp 275–278
Kobinger W, Walland A (1967) Investigation into the mechanism of the hypotensive effect of 2-(2,6-dichlorophenylamino)-2-imidazoline HCl. Eur J Pharmacol 2:155–162
Landsteiner K (1945) The specificity of serological reactions. Harvard University Press, Cambridge, MA
Laverty R, Taylor KM (1969) Behavioral and biochemical effects of 2(2,6-dichloropheny-(amino)2-imidazodine HCl (ST 155) on the central nervous system. Br J Pharmacol 35:253–264
Matteo RS, Pua EK, Khambatta HJ, Spector S (1977) Cerebrospinal fluid levels of d-tubocurarine in man. Anesthesiology 46:396–399
Michiels LJM, Heykants JJ, Knaeps AG, Jansen PAJ (1975) Radioimmunoassay of the neuroleptic drug pimozide. Life Sci 16:937–944
Paalzow L (1974) Analgesia produced by clonidine in mice and rats. J Pharm Pharmacol 26:361–363
Peskar B, Spector S (1973a) Serotonin: radioimmunoassay. Science 179:1340–1341
Peskar B, Spector S (1973b) Quantitative determination of diazepam in blood by radioimmunoassay. J Pharmacol Exp Ther 186:167–172
Ranadive NS, Sehon AH (1967) Antibodies to serotonin. Can J Biochem 45:1701–1710
Reid JL, Wing LMH, Malthias CJ, Frankel HL, Neill E (1977) The central hypotensive effect of clonidine. Clin Pharmacol Ther 21:375–381
Schmitt H, Le Donarec JC, Petillot N (1974) Antinociceptive effects of some α-sympathomimetic agents. Neuropharmacology 13:389–394
Smith TW, Haber E, Yeatman L, Butler VP Jr (1976) Reversal of advanced digoxin intoxication with Fab fragments of digoxin specific antibodies. N Engl J Med 294:797–800
Spector S (1971) Quantitative determination morphine in serum by radioimmunoassay. J Pharmacol Exp Ther 178:253–258
Spector S, Parker CW (1970) Morphine: radioimmunoassay. Science 168:1347–1348
Spector S, Felix AM, Semenuk G, Finberg JPM (1978) Development of a specific radioimmunoassay for acetylcholine. J Neurochem 30:685
Strombom U (1975) Effects of low doses of catecholamine receptor agonists on exploration in mice. J Neural Transm 37:229–235
Struyker Boudier H, Van Rossum J (1972) Clonidine induced cardiovascular effects after stereotoxic application in the hypothalamus of rats. J Pharm Pharmacol 24:410–411
Struyker Boudier H, De Boer J, Smeets G, Lien EJ, Van Rossum J (1975) Structure activity relationships for central and peripheral alpha adrenergic activities of imidazoline derivitives. Life Sci 17:355–386

Tsoucaris-Kupfer D, Schmitt H (1972) Hypothermic effect of α-sympathomimetic agents and their antagonism by adrenergic and cholinergic blocking drugs. Neuropharmacology 11:625–635

Wurzburger RJ, Miller RL, Boxenbaum HG, Spector S (1977) Radioimmunoassay of atropine in plasma. J Pharmacol Exp Ther 203:435–441

Wurzburger RJ, Miller RL, Marcum EA, Colburn WA, Spector S (1980) A new radioimmunoassay for haloperidol; direct measurement of serum and striatal concentrations. J Pharmacol Exp Ther 217:757–763

Subject Index

Handbook of Experimental Pharmacology

Continuation of "Handbuch der experimentellen Pharmakologie"

Volume 44
Heme and Hemoproteins

Volume 45: Part 1
Drug Addiction I

Part 2
Drug Addiction II

Volume 46
Fibrinolytics and Antifibronolytics

Volume 47
Kinetics of Drug Action

Volume 48
Arthropod Venoms

Volume 49
Ergot Alkaloids and Related Compounds

Volume 50: Part 1
Inflammation

Part 2
Anti-Inflammatory Drugs

Volume 51
Uric Acid

Volume 52
Snake Venoms

Volume 53
Pharmacology of Ganglionic Transmission

Volume 54: Part 1
Adrenergic Activators and Inhibitors I

Part 2
Adrenergic Activators and Inhibitors II

Volume 55
Psychotropic Agents

Part 1
Antipsychotics and Antidepressants

Part 2
Anxiolytics, Geronto-psychopharmacological Agents and Psychomotor Stimulants

Part 3
Alcohol and Psychotomimetics, Psychotropic Effects of Central Acting Drugs

Volume 56, Part 1 + 2
Cardiac Glycosides

Volume 57
Tissue Growth Factors

Volume 58
Cyclic Nucleotides

Part 1
Biochemistry

Part 2
Physiology and Pharmacology

Volume 59
Mediators and Drugs in Gastrointestinal Motility

Part 1
Morphological Basis and Neurophysiological Control

Part 2
Endogenous and Exogenous Agents

Springer

Handbook of Experimental Pharmacology

Continuation of "Handbuch der experimentellen Pharmakologie"

Springer-Verlag
Berlin Heidelberg New York
London Paris Tokyo

Springer

MIX
Papier aus verantwortungsvollen Quellen
Paper from responsible sources
FSC® C105338

If you have any concerns about our products,
you can contact us on
ProductSafety@springernature.com

In case Publisher is established outside the EU,
the EU authorized representative is:
Springer Nature Customer Service Center GmbH
Europaplatz 3, 69115 Heidelberg, Germany

Printed by Libri Plureos GmbH
in Hamburg, Germany